The Fire Ants

The Fire Ants

Walter R. Tschinkel

The Belknap Press of Harvard University Press
Cambridge, Massachusetts, and London, England
2006

Printed in the United States of America

Library of Congress Cataloging-in-Publication Data
ISBN: 0-674-02207-6
Cataloging-in-Publication data
available from the Library of Congress

Contents

Foreword by Edward O. Wilson vii

I. Origin and Spread, Present and Future Range 1

Prelude: And What Do You Do for a Living? 3
1. A Quick Tour of Fire Ant Biology 5
2. The Species of Fire Ants and Their Biogeography 13
3. An Atlas of Fire Ant Anatomy 22
4. Getting There 23
Interlude: Beachhead Mobile 34
5. La Conquista: Spreading Out 37
Interlude: Another Immigrant Moves West 75
6. Predicting Future Range Limits 83

II. Basic Needs and the Monogyne Colony Cycle 93

An Important Note: Monogyne and Polygyne Social Forms 95
7. Shelter 96
Interlude: There's Nothing Like Getting Plastered 103
8. Space 106
9. Food 121
Interlude: Mundane Methods 130
10. Mating and Colony Founding 136
Interlude: Spring among the Fire Ants 147
11. The Claustral Period 150
Interlude: Sharon's House of Beauty 165
12. The Incipient Phase and Brood Raiding 168
13. Dependent Colony Founding 186
14. Colony Growth 193
15. Relative Growth and Sociogenesis 205
Interlude: The Porter Wedge Micrometer 227
16. Colony Reproduction and the Seasonal Cycle 229

III. Family Life 247

Interlude: Deby Discovers Ants 249
17. Nestmate and Brood Recognition 251
Interlude: Ant ID Systems 269
18. Division of Labor 272
Interlude: Moving Up in a Harvester Ant Colony 292
19. Adaptive Demography 295
Interlude: Driving to Work with Odontomachus 304
20. The Organization of Foraging 308
Interlude: Who's in Charge Here? 328

21. Food Sharing within the Colony 330
Interlude: The Fire Ant on Trial 358
22. Venom and Its Uses 364
Interlude: You Call That Pain!? 384
23. Social Control of the Queen's Egg-laying Rate 386
Interlude: Catching Queens 396
24. Necrophoric Behavior 398

IV. Polygyny 403

25. Discovery of Polygyny 405
Interlude: I Want This to Be Accurate 412
26. The Suppression of Independent Colony Founding in Polygyne Colonies 415
27. The Nature and Fate of Polygyne Alates 442
Interlude: A Useful Tool 454
28. Polygyne Mating, Adoption, Execution 456
29. Biological Consequences of Polygyny 495

V. Populations and Ecology 501

30. Hybridization between *Solenopsis invicta* and *S. richteri* 503
31. Populations of Monogyne Fire Ants 512
Interlude: Gang Wars 541
32. Territorial Behavior and Monogyne Population Regulation 544
33. Ecological Niche 550
34. *Solenopsis invicta* and Ant Community Ecology 568
Interlude: Membership in a Prestigious Organization 595
35. *Solenopsis invicta* and Other Communities 597
36. Fire Ants and Vertebrates 612
Interlude: A Microsafari in Antland 628
37. Biological Control 629
Interlude: The Heartbreak of Parasitoids 667
Some Final Words 671

References 673
Acknowledgments 707
Index 709

Color plates follow page 280

Foreword by Edward O. Wilson

◂●●

During the decade this much-anticipated book was in development, entomologists knew that it would be the definitive work on fire ants, destined to remain a stimulus for additional research and necessary reference work for many years. Walter Tschinkel was the only person who could have written *The Fire Ants*. He has lived among the ferocious little pests for a large part of his distinguished career, coming at them through almost every available avenue of research. In a phrase, he has inhabited their biology for a long time. And, on a personal note, he and his students have surely surpassed many times over the record for number of fire ant sings suffered, one that I first set during my own earlier and comparatively meager research.

Tschinkel has chronicled the story of a species well worth the telling, but he has done much more. He has delivered a masterpiece. *The Fire Ants* is an example of how future biology will be written: a holistic account of a species or group of related species across all levels of biological organization, from the molecules that regulate the behavior of the organisms (in social insects, also colonies), to their life cycles, their role in the environment, their classification, evolution, and biogeography—and, in this case especially, their importance to humanity.

Tschinkel earlier introduced the concept of the sociogram, the description and measure of caste and behavior in fine detail to produce a systematic record treated as important for its own sake. His approach differs from those who search species for the theoretical principles the species may or may not illustrate. His kind of research is grueling at times, but it discovers new phenomena more readily, allowing us not just to test theories but to break outside the established boxes of theoretical thinking and forge new directions of theoretical reasoning. In organismic and evolutionary biology at least, such scientific natural history almost always precedes theoretical advance. Pursued species by species, it produces the enduring foundation of the discipline.

The research of biologists who have focused on fire ants is here synthesized in its entirety for the first time. Tschinkel makes clear that this research has moved the fire ant, *Solenopsis invicta*, to equal rank with the honeybee *(Apis mellifera)* as the best understood of all social insect species. *Solenopsis invicta* is now positioned to yield still more information, and often of a kind not immediately accessible in species addressed to lesser degree. We are reminded of Karl von Frischís loving characterization of the honeybee as a magic well: the more you draw from it, the more there is to draw.

Solenopsis invicta rose to prominence in science not just for its intrinsic interest but for its importance as a major pest. Its entomological stardom illustrates an important trend in global ecology: most countries are awash in nonnative species that have been inadvertently introduced by human activity. The flood is intensifying, even in the face of strict quarantine procedures, because travel and commerce are accelerating all over the world, and borders remain far from alien-tight. Only on the order of one in a hundred alien species becomes noticeably destructive to health or the environment, and probably fewer than one in a thousand become major pests of *Solenopsis invicta* magnitude. Yet the flood is such that, unless stronger measures than heretofore are taken to curb future invasions and to eradicate potential new pest species that do establish a beachhead, there will be many more sagas like that reported in *The Fire Ants*. Biologists will love them for the secrets they yield, but the rest of the population, not to mention the environment, will suffer.

I

Origin and Spread, Present and Future Range

◂•• *Prelude: And What Do You Do for a Living?*

Tallahassee, at the southern end of the red hills country, is the seat of the government of Florida, the government of Leon County, and the government of Tallahassee (of course). It is also home to two universities, a regional medical center, a lot of lawyers, doctors, and state employees, and quite a few professors. Many years ago, I noticed a pattern that was repeated with only small variations whenever I met new people. What do you do for a living?, someone will ask, drink in hand. And when I answer, the reaction is almost always the short laugh, before politeness takes over, or at least the puzzled smile, as if to say, What a strange way to make a living. Really strange.

And you?, I ask. What do you do? He says, I'm a lawyer. Litigator. Which, in plain language, means he has just spent the whole day arguing with another guy who also makes his living by arguing. Another time it might be a young woman who sells insurance or someone who has a business selling Christmas stuff.

Sure I do something uncommon for a living. How could I deny it when there are only a couple hundred of us myrmecologists in the world. There are probably a million lawyers in the USA, untold insurance agents, and armies of retail clerks. All these professions seem "normal" to most people. But when most people think you do something "funny" for a living, you start to reflect more deeply on "funny."

The truth is that we all do funny things for a living. Some of us put mystic marks on pieces of paper, then move the papers to the other side of the desk. Some of us argue for a living. Others take green pieces of paper and press buttons on a machine that goes beep beep. No one I know makes a living doing anything that would even remotely qualify as "basic." Nobody hunts for a living in my neck of the woods, at least not legally. Nobody even *remembers* how to gather. Farming comes closer, but what is a farmer going to do with 500 hectares of corn? Surely not eat it. Nobody subsistence-farms around here either. No, from a remote vantage, we all do funny things for a living. We are all specialists in some tiny little fraction of the total activity that goes into making a human society work, and this specialization invites a comparison with the ants I study. Each of us, in fact, is much more specialized than a worker ant, who, in the full course of her life, will carry out most of the activities it takes to run an ant society. It seems unlikely that I will ever rivet steel girders, that my wife will butcher a cow, that my daughter will smelt iron, or that anyone I know will make tires for trucks.

An ant alone can survive and can live out her normal life span. Perhaps she pines for her nestmates, but she survives. With very rare exceptions, a human alone, truly alone, without the products, services, and contrivances produced by other humans, is sunk, finished, and doomed. The price of our extreme specialization is an extreme interdependence. A consequence of that mutual dependence is that we all do funny things for a living.

1

A Quick Tour of Fire Ant Biology

I love fire ants. Most of my neighbors in the American South find this love completely incomprehensible, because most of them hate fire ants with a passion equally strong. My passion has fed and been fed by 35 years of poking into fire ants' secrets, of trying to understand the rhythms of their everyday lives as well as the drama of their several rites of passage. I have written this book both for professional biologists and for people still open-minded enough to be intrigued, charmed, or fascinated by the many results of biological research on fire ants. By "fire ants," I mean mostly *Solenopsis invicta* Buren, an alien invader from South America that has become a pest in its new homeland in the USA. This species is one of about 20 species of "fire ants" native to the Americas (see Chapter 2). At least three of these have become world travelers through human commerce, but *S. invicta* has received most of the attention. The notoriety of pesthood has certainly created a large fire-ant folklore and scores of amusing factoids, but it has also stimulated a lot of research on fire-ant biology. My purpose in this book is to knit together these diverse research findings into a coherent story of fire-ant social biology and ecology. There is so much to tell that it is hard to know where to start. When journalists want to know "the truth" about fire ants, I often take them on a field trip to see the ant in its habitat. We poke a couple of nests and then have rambling conversations about fire-ant biology while the ants panic at our feet. Let's start this book with a similar, although imaginary, tour. I'll expound on some topics as we come to them, and others will be covered in greater detail in later chapters. So just sweep that junk off the car seat onto the floor, hop in, and let's go.

We are now at Southwood Plantation, a corporate cattle farm just east of Tallahassee, Florida, on this warm April day. For 30 years, I have had permission to do fire-ant research on this property and have seen four managers come and go. The rolling pastures are dotted here and there with isolated live oaks, wider than tall, and are ideal habitat for fire ants. The disturbance caused by cattle grazing has kept it that way (Plate 1).

We check in with Stella in the office, head back to the car, unlock the gate to the pasture, and drive in, bumping slowly over the conservation terracing to park the car in the shade of a huge live oak. Two cows eye us curiously but don't rise from their bramble beds under the oak.

This particular pasture covers about 14 hectares. I guess it's pretty typical of what I think of as good fire-ant habitat—open and grassy with a small proportion of other herbaceous plants mixed in (Chapter 33). The pines and the oaks on the perimeter try mightily to get plant succession moving. You can see the scattering of pine seedlings, and here and there some sapling pines too, especially at the east end of the pasture, where a haze of young pines almost obscures the grass. Young live oaks up to waist height dot the edges, and the fences are entangled with rows of cherry trees, each a memento of a bird that perched briefly on the fence to defecate a cherry stone many years ago. In the middle of the pasture, young persimmon trees struggle against fungi and caterpillars. A possum must have made a deposit at that spot several years before. Scattered patches of blackberry and dewberry grow like slowly expanding bacterial colonies.

Left alone, this pasture would face slow extinction as these arboreal invaders turned it back into forest, killing the grass with shade and carpeting the ground with fallen leaves and branches, gradually turning to soil. Perhaps a century later, a forest of spruce pine, beech, magnolia, hickories, and oaks and about two dozen other broadleaf tree species would resonate with the songs of birds, katydids, and cicadas, and the pasture would have vanished even from human memory. Even the last of the early invaders, the loblolly pines, water oaks, and persimmons would be in their dotage, with few years left to live and no replacements in the shade below.

But it is not left alone. Cows and people assure that the footholds gained by pines, oaks, and persimmons are temporary. Every few years, the farm workers "bush hog" the pasture, severing all the trees and brambles at ground level and leaving coarse wood chips and splintered stumps. Many of the pines die, but the oaks and persimmons sprout into vigorous bonsai shrubs, the bushy sprouts fighting among themselves over which one will be the leader. The decision takes several years. Then the mower comes back.

The land manager's intention is to keep the habitat appropriate for cows by holding back plant succession. In the process, he creates a paradise for fire ants (Chapter 33). If you cast your eye over this pasture, you will see that it is dotted with fire ant mounds from fence to fence. Ecologically speaking, humans and fire ants have similar needs, but only one of the two species can actually alter the native forest to suit its needs. Providing for the needs of fire ants is an unintended consequence of providing for the needs of cows. Throughout the southeastern USA, providing for the real or perceived habitat-needs of humans and cows has created fire-ant habitat on a staggering scale. Only the few remnants of forest, or the patches of secondary forest, are unsuitable still.

Most human ecological activities favor weeds, creatures of disturbance, and the fire ant is an animal weed. We are the fire ant's best friend.

Let's take a closer look at a fire ant mound. Here is a nice, large mound, about the size of a very large, overturned steel mixing bowl. The bare, red-soil dome pokes up like a bald head through the fringe of grass on its lower slopes. The mound was built by the workers out of the soil excavated from subterranean chambers as far as a meter and a half below (Chapter 7). It is not just a waste depot but encloses a maze of sinuous tunnels. On a warm, sunny, spring morning like this, the mound warms faster than the surrounding soil, and a large part of the colony moves into the mound, bringing their brood with them to speed its development (Chapter 7). I have brought a shovel and this large tray so that we can observe the ants without being stung. In case you're wondering, the sides of the tray have been painted with Fluon, a paint of Teflon emulsion so slippery that ants can't climb it. Even at $245 per gallon, Fluon has become a tool no myrmecologist can afford to be without.

So let's cut into the mound with this shovel and toss the dirt into the tray, where we spread it out in a thin layer. I hate to mention it, but I'm not sure it was wise to use your bare hands to spread the dirt, but since you just did, you have discovered that fire ants are energetic stingers, the trait that accounts for their unpopularity. Those hunched-over workers you see on your arm and in the webbing between your fingers are injecting a toxin into your skin (Chapter 22). Yes, I know it burns; that's why they're called fire ants. Tomorrow you will have little white pustules that you will not be able to resist popping (it's only human).

You can see pandemonium in the tray; panicked ants rush wildly about, seeking cover and rescuing brood. We've caught about 20 to 30 thousand of them, only a fraction of the 150,000 or so that call this mound home. Note the different classes of fire ants scattered in the tray. The queen has a fair chance of being among them, but it will be a few minutes before we can spot her. The great majority of the ants are adult workers. They are a ruddy brown and from 3 to 6 mm in length. Being insects, they have three pairs of legs, one pair on each thoracic segment, but unlike most adult insects, they lack wings (Chapter 3). If we were to dissect one, we would find that they also lack ovaries and other reproductive structures. Fire-ant workers are sterile, as are, more or less, the workers of all ants. The function of worker ants is summarized in their names—they provide the labor for all of colony life. They build the nest, rear the next generation of ants, causing colonies to grow and reproduce (Chapters 14, 16), forage for food (Chapter 20), process and distribute food (Chapter 21), defend the nest (Chapter 22), care for the queen, build the nest (Chapter 7), and provide sanitation services.

You will notice that the workers come in a large range of sizes, and I can tell you, from having measured many, that the largest are 15 to 20 times the weight of the smallest. This tremendous size variation is called polymorphism (from Greek *poly* and *morphos*, many forms), and is the first foundation of the division of labor (Chapter 18) in fire ant colonies. The tasks carried out by large, medium, and small workers differ, although they frequently overlap. The second foundation for division of labor is not visible, so I will tell you that it is worker age. As an adult worker ages, she moves gradually from taking care of brood in the central brood chambers to more general duties (food processing and distribution, nest maintenance, sanitation, etc.) in more peripheral chambers of the nest. This outward movement continues until, during the final weeks of her life, the worker leaves the nest to forage, eventually to die. Only a minority of the workers is engaged in foraging at any one time, but this minority brings back the insect and sugary food for the rest of the colony, including other workers, larvae, and the queen (Chapters 20, 21).

These foragers search for food in a territory to which this colony lays exclusive claim and which we are standing in the middle of. An invisible boundary separates this colony from its neighbors, and intruding workers from other colonies may be attacked (and vice versa). Workers recognize nonnestmates by the odor of their bodies, an odor derived from the waxes that waterproof their cuticles as well as from the food they eat and the soil they dwell in (Chapter 17). As a consequence, within this territory, only workers of this colony can be found. Larger colonies claim and hold larger territories, and larger territories probably mean more food, so territory size and resources are inextricably linked (Chapters 8, 14). Our colony is mature and has a territory of about 60 m^2. Like those of most colonies, the nest mound is fairly close to the center of the territory. The spacing thus created distributes mounds evenly on the landscape, a condition ecologists refer to as overdispersed or uniformly dispersed. In turn, suitable habitat is covered with a continuous mosaic of fire ant territories, without gaps, like a platted subdivision (but with more fluid boundaries).

Because workers fan out over the surrounding territory from one central location where 150,000 to 200,000 ants are waiting for food, you might expect to see a constant stream of workers coming and going from the mound, but if you squat down near that undisturbed mound over there, you will see very few fire ants on the ground or in the grass. The reason is that nearly all traffic to and from the foraging territory takes place within an amazing system of underground tunnels that branch and rebranch to service all parts of the territory (Chapter 20). Recruits wait within these tunnels, ready to respond to the signals and chemical trails laid by scouts upon the discovery of food. Then they surge forth to help secure the food against invaders and return it to the mound along these subterranean superhighways. Such defense and mass retrieval of food is a central feature of fire-ant life (Chapter 20).

The tray has more to show us. In addition to the workers, you can see several thousand "white things." Myrmecologists refer to these collectively as brood. Brood is immature ants and can be in one of three stages: egg, larva, or pupa. The same is true of all insects that, like ants, undergo complete metamorphosis. We can't see eggs under these conditions, but the larvae are the glistening, creamy grubs. They are legless and helpless and must be fed by adult workers (Chapter 21). As in all insects, all growth takes place in the larval stage. As the larva grows, it molts (that is, sheds its cuticle, or skin) four times, passing through four larval stages, called instars. The last larval molt begins the process of metamorphosis, transforming and molting to the pupal stage. In the pupa, most adult structures are visible, so the pupa looks like a trussed-up, white adult. In our tray, the pupae are being piled up like seeds in the shade of dirt clods. Looking closely, we can see that pupae are of three types. The most abundant are the pupae of workers; their size range is similar to that of the adult workers that carry them to safety. The other two kinds of pupae are much larger, and the rudiments of future wings clasp the sides of their bodies. These are the male and female sexuals or alates (from the Latin, *ala*, wing; having wings), which will soon attempt to mate and found new colonies. As adults, both will be capable of flight and sexual reproduction.

The destiny of a larva depends first on whether it hatched from a fertilized or an unfertilized egg. Sex is determined by fertilization in the ants, wasps, and bees, a system known as haplodiploid sex determination. Each gamete (egg or sperm) has a single set of chromosomes and is said to be haploid. When an egg is fertilized, the chromosome sets of the egg and sperm are combined, so the resulting fertilized egg has two sets of chromosomes and is said to be diploid. Diploid eggs become female ants. In most animals, unfertilized eggs do not develop, but in the ants, wasps, and bees, they develop into males, a phenomenon known as arrhenotokous parthenogenesis (a nice phrase to drop at cocktail parties; from the Greek *arrheno,* male, and *tokia,* production of offspring, fecundity). Males only occur in the sexual form. The diploid, female eggs, on the other hand, can develop into either sterile workers or reproductive females.

Which of these developmental paths a female follows depends on her feeding and chemical environment (Chapter 16). Most larvae become adult workers, but especially in the spring, some grow much larger and become sexual larvae. Because this is April, you can see quite a few sexual larvae being dragged to shelter in the tray below us. You can't tell the sex of a larva until it molts into the pupal stage, at which time the differences between male and female pupae are clear: females are reddish and have large heads, males are black and have smaller heads. The number of alates of each sex that a colony produces is very important to its reproductive success, or fitness (Chapters 16, 31).

In about a month, this nest will be teeming with mature adult sexuals, poised to try to found new colonies, one of the really dramatic rites of passage of fire-ant life, full of danger, death, and (for a rare few) triumph. When the late-spring weather fronts bring heavy rains followed by warm days, the thousands of fire-ant nests in this pasture, and untold thousands in the surrounding countryside, will each launch hundreds or thousands of male and female alates into the air, to mate in the sky in a free-for-all no human has ever seen (Chapter 10). Each female mates with a single male and stores a lifetime supply of sperm in a little sac in her body. Then she descends to the ground, breaks off her wings, and finds or digs a suitable chamber in which to begin the colony-founding process. From reserves stored in her body, she rears one or two dozen tiny workers, who immediately take over all brood-rearing duties, foraging, and queen care (Chapter 11). They and the queen also engage in complex competitive interactions with other newly founded colonies in the neighborhood, a confusing mixture of nest invasion, brood theft, fighting, and execution. Ultimately, a single queen heads each surviving colony.

Wait, you say! If each colony has a single queen who has mated with a single male, and the workers are sterile females, then all the workers must be the daughters of the queen, and the entire colony is one huge, yea verily, colossal family. Yes, that is quite correct, and with this insight you have just pointed out the most basic truth about sociality in insects (and most animals, for that matter). Sociality has evolved most frequently on the foundation of family relatedness. Because the male contributes only one set of chromosomes in haplodiploid sex determination, and every daughter contains this entire set while getting only half of the mother's chromosomes, daughters share many more genes with one another than do daughters produced by "ordinary" diploid sex determination (Chapter 16). In the meantime, look around this pasture at the approximately 3000 fire-ant mounds. Each of them shelters a single family of ants.

Surviving colonies wedge themselves into a territory and grow by converting the food they collect into new workers. The territory grows more or less in parallel; additional area is wrested from the grip of weaker neighbors by force or simply claimed from unoccupied terrain. When colonies reach about 10% of their maximum size, they begin to produce sexuals, at least during the spring and summer. Once a colony begins to reproduce in this way, its size declines in the spring because it dedicates its growth capacity to producing sexuals, allowing the worker population to decline because workers die at a steady rate. The great majority of eggs develop into workers who simply replace those that have lived out their lives and passed on to ant heaven. Only workers in excess over this replacement number cause colony growth (Chapter 14). After midsummer, colonies switch back to worker production, causing the colony to grow larger than it was at the beginning of the year. Fully grown colonies merely return to their midwinter maximum size. This seasonal cycle is an integral part of colony reproduction—the part of the year when colonies produce few sexuals is spent amassing reserves for the part when they do. Many colony characteristics, especially the size and variability in size of workers, change gradually as the colony grows by over five orders of magnitude. Changing worker size has many profound effects on the physiology, behavior, and labor distribution of colonies (Chapter 15).

Populations of colonies, like the one we see around us in this pasture, have characteristics that emerge from the nature of the colonies of which they are composed. Because queens eventually run out of sperm, colonies die, to be replaced by newly founded ones (Chapters 15, 31). A population of colonies can therefore be described by the size distribution, age distribution, average longevity, and replacement rate of its members. Of special interest is the total biomass of ants supported and the area from which the support is drawn. These determine the ecological impact of the ant.

Now enough time has elapsed for us to look for the queen. Chances are, if she is in the tray, she is sheltering under this litter or these clods. So let's gently turn these over, and . . . well, dang, there she is! You see she's surrounded by a knot of excited workers, eager to protect her against harm and attracted by her body smell. Let's suck her into the aspirator here and dump her into this little dish with just a few workers. Notice that she is quite capable of moving under her own power, unlike the famous termite queens. The queen's abdomen is so swollen with her immense ovaries that large expanses of the intersegmental membrane are visible between the plates of her gaster (her "abdomen"). With a microscope, you can see through these stretched membranes to the ovaries slowly pulsing inside. This queen is the only egg layer in the colony, a rather unromantic egg-laying machine, popping out an egg, in the case of the one before us, about every 30 seconds, without pause. She is also a chemical factory whose pheromonal products control the fates of larvae and the behavior of workers. Because the queen lays all the eggs and the workers rear all the brood, queen fertility (egg-laying rate) must be coordinated with the available labor force, and the manner in which the coordination is achieved is surprising (Chapter 23).

The region around Tallahassee, indeed most of the southeastern USA, is occupied by colonies with single queens—simple families like this one—but some areas, especially in Texas, harbor enclaves in which colonies have multiple queens, all laying eggs simultaneously. The tolerance of multiple queens by workers results from a difference in a single gene. Discovery of this gene is one of the most interesting recent discoveries of social insect biology (Chapter 28). The tolerance has a cascade of profound effects on the biology of the fire ant, converting populations from a mosaic of staunchly territorial colonies, aggressive toward their neighbors, to a freely mixing, mutually tolerant population of nests without colony boundaries. The former reproduce by staging mating flights, and the latter bud new nests like yeasts. The ecological effects of this switch are also profound—densities soar, mating flights are curtailed, colony growth rates increase, and the population spreads as a solid front, on foot, like marauding armies. Within this new social form, queens vie for reproductive advantage by waging chemical war against one another, using their pheromones to stunt each other's reproduction (Chapter 26). These are not cozy, cooperative families but a collection of unrelated, competing families cohabiting in a common nest. These silent struggles are dramatic in their own way, although we only dimly understand the rules and the prize.

With this, our brief tour must come to an end. It must suffice as an introduction and guide for the chapters that follow. Topics cannot all be arranged in a linear, logical order, and this overview is intended to provide the necessary background when they cannot. So let's hop back into the car and bump our way back to the road to Tallahassee, where a glass of tea awaits us. Iced, of course. And sweet. This is, after all, the South.

2

The Species of Fire Ants and Their Biogeography

The genus *Solenopsis* (from Greek, *solen*, a pipe or channel, and *opsis*, appearance, sight), to which the fire ants belong, currently contains about 185 described species. Taxonomists, the scientists concerned with classification and naming, place the genus within the largest and most diverse of the 21 ant subfamilies, the Myrmicinae, and have long considered the classification of *Solenopsis* species something of a nightmare. Creighton commented that Emery had considered the genus to be the *crux myrmecologorum* and that, at the end of three years of study, he still found the cross a heavy burden. The primary problem is that, throughout the genus, worker ants, especially the smaller ones, are rather similar; few morphological characters (that is, physical features) are useful for classification. As a result, the generic limits of *Solenopsis*, and the affinities of its species and subgeneric groupings, have been fluid and uncertain. Although he remarked that the genus had always been "extraordinarily intractable in the matter of subdivision," Creighton divided the genus into five subgenera and several related genera (Creighton, 1930), but later taxonomists, working on all or part of the genus, collapsed these back into a single genus (Wilson, 1952; Snelling, 1963; Ettershank, 1966; Buren, 1972; Trager, 1991). Currently, in recognition of the taxonomic disarray of this genus, most authorities recognize no formal subgeneric divisions. Therefore, in his world catalog of ants, Bolton (1995) listed no subgenera within *Solenopsis*, but consensus seems to have settled on three "natural" groupings within the genus. The first includes the 20 or so New World species known as "fire ants," most of which have polymorphic (highly size-variable) workers. These correspond more or less to Creighton's subgenus *S. Solenopsis*. The second comprises the cosmopolitan "small *Solenopsis*," commonly known as thief ants because many nest near other ants and plunder their brood. Some authorities previously placed the thief ants in the subgenus *Diplorhoptrum*, and many myrmecologists still refer to them informally in this manner (perhaps because old habits die hard). The third includes the social parasites, previously a separate genus, *Labauchena*.

Our current taxonomic understanding of the fire ant species is based on the thorough revision by James Trager (1991) and the more recent phylogenetic (i.e., evolutionary) analysis by James Pitts (2002). Both Pitts and Trager recognized that the simplification of body morphology

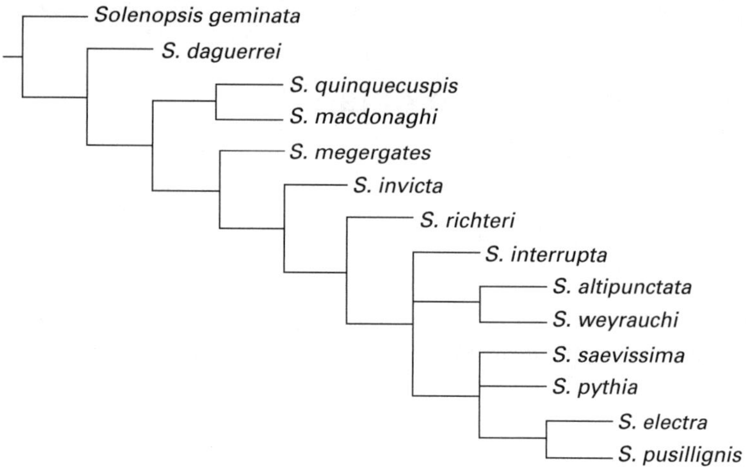

Figure 2.1. A consensus tree diagram showing the most likely phylogenetic (evolutionary) relationships among the species of fire ants in the *saevissima* species group of the genus *Solenopsis* (and *S. geminata*; a species from outside the group is always included to serve as an "attachment point," showing where the small tree under investigation joins the larger "tree of life"). Reprinted with permission from Pitts (2002).

characteristic of a worker caste, as well as their small body size, greatly reduced the number of characters useful for distinguishing species. Speciation (evolutionary diversification) may also have been too rapid for workers to diverge morphologically. Minor (small) workers of most fire-ant species look more or less alike, and have few characters we can use to distinguish among them, but the major workers that make up the larger end of the worker size distribution present less of a problem. Among these, variations in head shape and coloration have proved a sufficient basis for revising the fire-ant species. Trager placed all of the fire-ant species, as well as two close relatives (*S. substituta* and *S. tridens*), in an informal grouping he called the *geminata* species group, separate from the thief ants and the social parasites within *Solenopsis*. He believed this species group, together with the social parasites, to be monophyletic, that is, to have descended from a single, common ancestor. Within the *geminata* species group, he recognized four "species complexes" (an informal category) that reflected the degree of similarity among the species: the *saevissima* complex (which includes *S. invicta* and *S. richteri*), the *virulens* complex, the *tridens* complex, and the *geminata* complex (which includes all the native North American fire-ant species). Within some of these complexes, he further distinguished subcomplexes.

James Pitts's (2002) revision of the *saevissima* complex built on Trager's work by including characters of males, females, and larvae in addition to those of workers, in the expectation that these additional, complex characters would facilitate separation and identification. Unfortunately, none of the new characters revealed themselves as panaceas for fire ant classification, because none differed greatly among the species within the *saevissima* group. Pitts then subjected a large number of characters from all life stages and castes, separately and together, to phylogenetic analysis designed to reveal the probable evolutionary relationships among the species. One of the resulting "consensus" tree diagrams is shown in Figure 2.1. The closer relationship

Table 2.1. ▸ Two classifications of the genus *Solenopsis,* the New World fire ants.

Older classification, by Trager (1991)	***Newer classification, by Pitts (2002)***	***Range***
S. virulens complex	*S. tenuis* species group	
S. virulens	*S. virulens*	Amazonian Brazil
S. tridens complex	*S. tridens* species group	
S. substituta	*S. substituta*	east central Brazil
S. tridens	*S. tridens*	northeastern Brazil
S. geminata complex	*S. geminata* species group	
S. geminata subcomplex		
S. geminata	*S. geminata*	southern USA, Mexico, Central America, Caribbean
S. xyloni subcomplex		
S. amblychila	*S. amblychila*	southwestern USA, northern Mexico
S. aurea	*S. aurea*	southwestern USA, northern Mexico
S. xyloni	*S. xyloni*	southern USA, northern Mexico
S. gayi subcomplex		
S. bruesi	*S. bruesi*	Peru
S. gayi	*S. gayi*	Chile, southern Peru
S. saevissima complex	*S. saevissima* species group	
S. saevissima subcomplex		
S. interrupta	*S. interrupta*	normally lowland species in western Argentina and Bolivia
S. invicta	*S. invicta*	lowland species, western Amazonia, south through Mato Grosso, eastern Bolivia, Paraguay, and southeastern Brazil to Santa Fe Province
S. macdonaghi	*S. macdonaghi*	Uruguay, Entre Rios Province and adjacent parts of bordering provinces in Argentina
S. megergates	*S. megergates*	southeastern Brazil
S. pythia	*S. pythia*	Mato Grosso do Sul to southeastern Brazil and Misiones, Argentina
S. quinquecuspis	*S. quinquecuspis*	Buenos Aires and La Pampa provinces, Argentina, Uruguay, north to Santa Catarina, Brazil
S. richteri	*S. richteri*	southeastern Brazil to central eastern Argentina
S. saevissima	*S. saevissima*	Orinoco drainage, Guianas, Amazonia and along rivers in bordering regions, also southeastern Brazil
S. weyrauchi	*S. weyrauchi*	Peruvian and Bolivian Andes, 2000–3500 m elevation
S. electra subcomplex		
S. electra	*S. electra*	western Argentina and Paraguay, north to Bolivia in Andean foothills
S. pusillignis	*S. pusillignis*	southern half of Mato Grosso, Mato Grosso do Sul, Brazil

Table 2.1. ▶ (*Continued*)

Older classification, by Trager (1991)	*Newer classification, by Pitts (2002)*	*Range*
Not assigned to a subcomplex		
S. daguerrei	*S. daguerrei*	south from Mato Grosso do Sul, Brazil, to Buenos Aires Province, Argentina
S. hostilis	*S. hostilis*	Jacarepaguá, Rio de Janeiro, Brazil, southwest to Rolândia, Paraná, Brazil

between *S. invicta* and *S. richteri* than between either of these and *S. geminata* may explain why the first two have hybridized in the USA, but neither has hybridized with *S. geminata*.

Pitts retained Trager's "complexes" but renamed them "species groups." He did not further divide these categories as had Trager. In addition, he improved Trager's key for identifying fire ants by including the male, queen, and larval characters, making the identification of fire ants somewhat easier. Major workers remain the best basis for species identification. Trager's and Pitts's classifications are summarized and compared in Table 2.1. Figures 2.2 and 2.3 show the natural distributions of the species in North and South America. All but four of the 17 South American species occur east of the Andes, but only two occur primarily in the tropical rain forest. Most species are found in the tropical monsoon or warm-temperate regions. North America is home to four native species, two introduced species, and two hybrids, all of which occur from the warmer regions of the southern USA south through Central America and the West Indies.

All in all, species of fire ants are best represented in warm regions of the New World where the rainfall is not extreme. In addition, *S. geminata* has become a pantropical tramp species found in far-flung places such as India and Hawaii and has even been introduced in some cities in Brazil, the cradle of *Solenopsis* species (MacKay et al., 1994; Fowler et al., 1995a). It appeared in the Galapagos Islands as early as the 1890s and is currently causing concern for the fragile ecosystems of these islands (Williams and Whelan, 1991). *Solenopsis invicta* is, of course, on the way to becoming an accomplished traveler as well, having made it to the USA, Puerto Rico, the Lesser Antilles, and most recently, via the USA to Australia, New Zealand, Hong Kong, and southern China. No doubt its travels are not yet over. And who knows? Other *Solenopsis* species may be waiting in the wings to become successful tramps.

All taxonomies are subject to revision as a result of subsequent studies. Pitts improved Trager's keys, but effective keys for field identification remain a pressing need for fire-ant biologists, because reliable progress in biological control and ecological research depends on them. A number of species of *Solenopsis* almost surely remain unrecognized because they are insufficiently distinct in body form. Chemical characters such as the composition of the cuticular hydrocarbons and venom alkaloids have proven useful. Without doubt, future studies will increasingly rely on several kinds of genetic data to determine species boundaries and relationships among species. Such genetic methods have confirmed the genetic distinctness of several of Trager's species in their native ranges, for example, *S. invicta* and *S. richteri* (Ross and Trager, 1990; Ross and Shoemaker, 2005). Such distinctness not only supports the species boundaries erected on the basis of morphological differences but also indicates that the species are reproductively isolated from one another. Genetic methods have also suggested that genes have recently flowed from *S. richteri* and *S. invicta* into *S. quinquecuspis* but not the other way (Ross and Shoemaker, 2005), and they detected a population in Uruguay that appeared to be *S. invicta* but showed little or no signs of gene exchange with the surrounding *S. invicta* population. Such cryptic species suggest that fire ants are evolving rapidly, spinning off new species whose morphological differences lag well behind their genetic distinctness. No doubt, future studies will detect more such cryptic species. The days of basing classification on a few morphological characters are, as they should be, behind us. Future progress will depend on analysis of many types of characters: morphological, genetic, biogeographic, behavioral, chemical, and ecological.

We can now bid most of these fire-ant species a fond adieu, because this book is, first and foremost, about the introduced *S. invicta* and, to a lesser extent, the introduced *S. richteri* and the native *S. geminata.* This departure does not represent a major case of neglect—nearly nothing is known about the biology of other fire-ant species anyway.

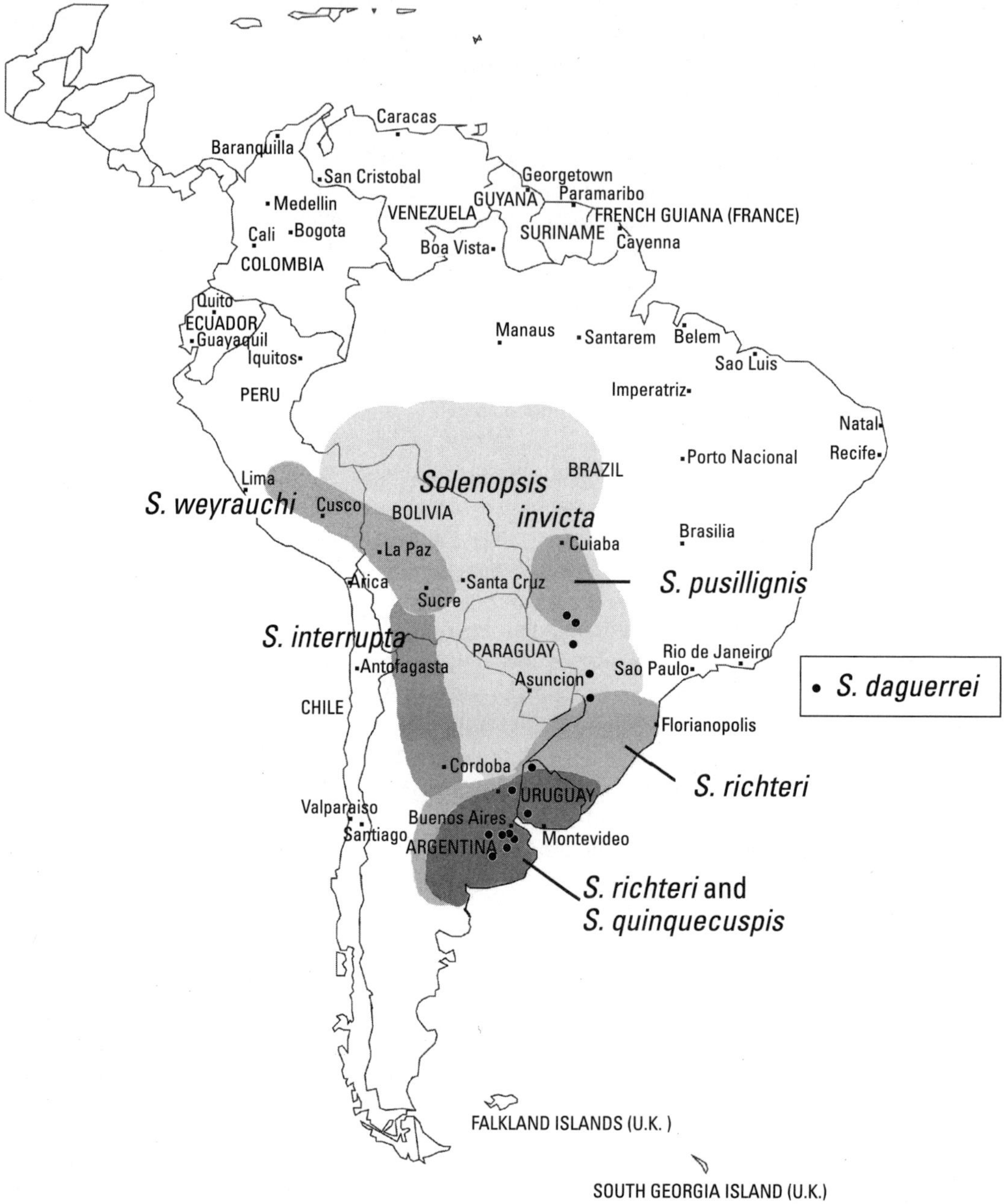

Figure 2.2. The geographic ranges of South American species of fire ants (genus *Solenopsis*) of the *saevissima* species group. Adapted from Pitts (2002).

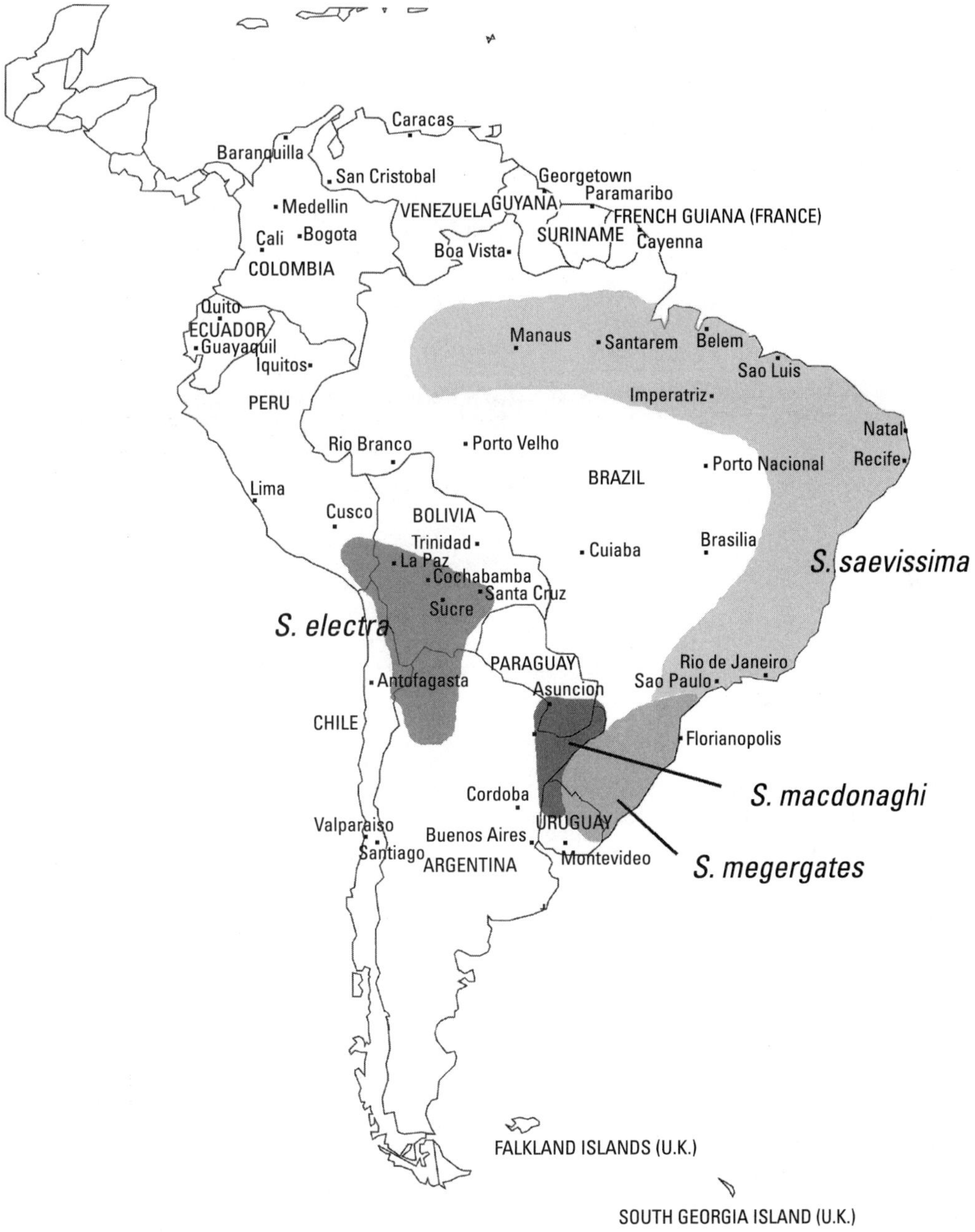
Caracas
Baranquilla
San Cristobal
Georgetown
Paramaribo
Medellin
VENEZUELA
GUYANA
FRENCH GUIANA (FRANCE)
Cali
Bogota
SURINAME
Cayenna
Boa Vista
COLOMBIA
Quito
ECUADOR
Guayaquil
Iquitos
Manaus
Santarem
Belem
Sao Luis
Imperatriz
PERU
Natal
Rio Branco
Porto Velho
Porto Nacional
Recife
BRAZIL
Lima
Cusco
BOLIVIA
Trinidad
Brasilia
Cuiaba
La Paz
Cochabamba
Santa Cruz
S. saevissima
Sucre
S. electra
PARAGUAY
Rio de Janeiro
Antofagasta
Asuncion
Sao Paulo
CHILE
Florianopolis
S. macdonaghi
Cordoba
Valparaiso
URUGUAY
Buenos Aires
Santiago
ARGENTINA
Montevideo
S. megergates
FALKLAND ISLANDS (U.K.)
SOUTH GEORGIA ISLAND (U.K.)

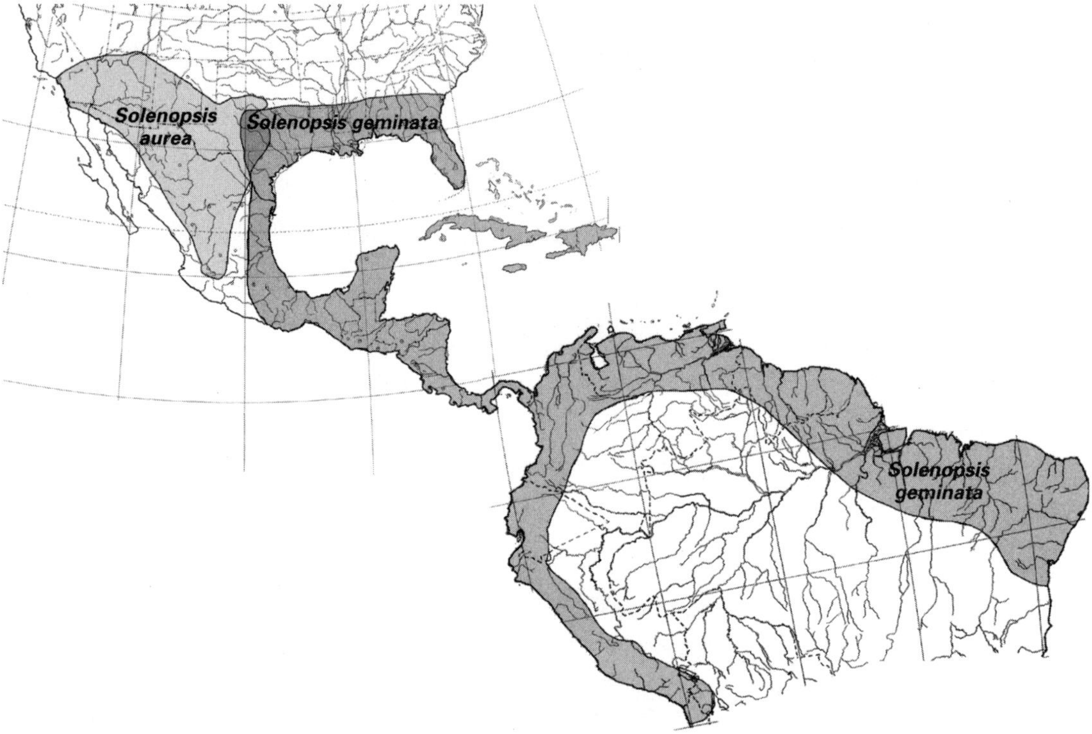

Figure 2.3. The ranges of the North American species (*geminata* species group) of fire ants. The historic range of *S. xyloni* from which it has been extirpated by *S. invicta* is shown in light shading, its surviving range in medium. The presumed natural range for *S. geminata* includes the Caribbean and northern South America. The world distribution of this tramp species is not shown.

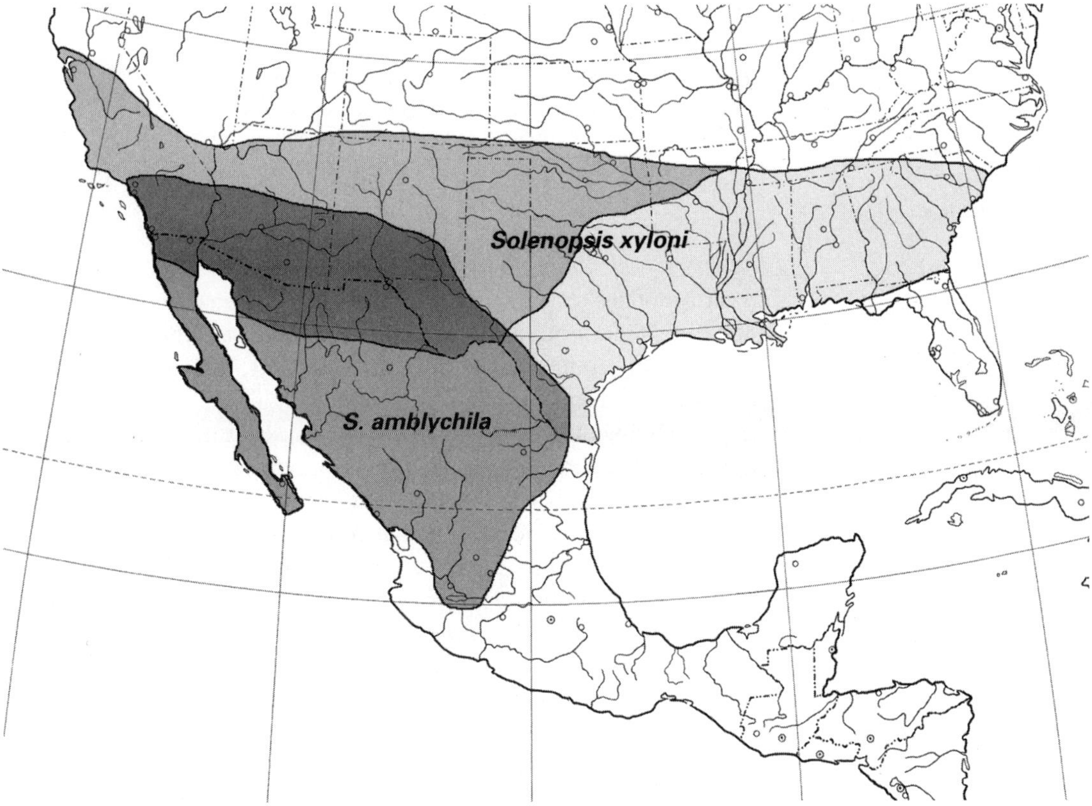
Solenopsis xyloni
S. amblychila

3

An Atlas of Fire Ant Anatomy

As arthropods, insects have a segmented body, a legacy of their annelid ancestry. The various arthropods differ from one another in how these segments are combined to form functional body regions. Insects differ from other arthropods in having 20 body segments that form three distinct body regions—a six-segmented head with three pairs of mouth appendages, a three-segmented thorax with three pairs of walking legs (and two pairs of wings in winged insects), and an 11-segmented abdomen lacking appendages (though limb bases remain associated with the genital openings). In ants, these patterns have been modified further. The first abdominal segment has become integrated into the thorax, so the middle body region is not equivalent to the usual insect thorax and is called the mesosoma. The second abdominal segment, and in some ant subfamilies the third as well, has been constricted into a narrow "waist" or petiole. The remaining abdominal segments form an articulated "gaster" that is not the equivalent of the usual insect abdomen, consisting as it does of abdominal segments 3–10 (11 has been lost) or 4–10, depending on the particular ant subfamily. This posterior body region is called the metasoma. The gaster is not equivalent in all ants because it may or may not include abdominal segment 3.

All ant colonies include a variety of forms and life stages. All colony members begin as eggs and develop through four larval stages and a pupal stage, but the end product of this development can be a sterile, wingless worker of any of a range of sizes; a winged adult female; or a winged adult male. Plates 2 through 7 show images of all of these forms and stages for *Solenopsis invicta.*

4

Getting There

The rapid increase in human commerce and human movement in the last half of the 20th century has brought the problem of invasive exotics to center stage. The challenge is urgent because many of these exotics contribute to severe environmental degradation and loss of habitat and biodiversity in their newfound homes. Interest is growing in learning how these invaders travel, what they need to become established, how they interact with native fauna and flora, and what makes them invasive, that is, what makes people sit up and take notice. The fire ant is one of the early members of this horde of immigrants and one that has been notoriously successful. When the fire ants arrived in the USA, consciousness of the potential destruction wreaked by exotic organisms was not high; in fact, large numbers of them were being introduced deliberately because they were pretty or useful (remember kudzu and water hyacinths?). Fire ants were probably introduced several times, but no one saw them enter the United States—they traveled without passport or visa, probably as stowaways on a ship. No one took note of their embarkation or disembarkation or of their efforts to settle into their new neighborhoods. No one tracked their early fortunes or took note of their struggles to survive. Can we nevertheless reconstruct how and when the fire ant arrived in the United States? Interestingly, we can.

Even though fire ants slipped into the country unnoticed, a little detective work reveals several important facts: (1) *S. invicta* arrived in the 1930s; (2) the original home of this ant is northern Argentina or southern Brazil; (3) it traveled on some form of ocean shipping; (4) several other species of ants have made the same trip; (5) the original immigration included several mated queens; (6) the original immigrants included at least one mature colony; (7) immigration probably occurred in several waves, several times; and (8) the makings of both social forms were introduced as part of these original inocula. I will present evidence for each of these in turn.

Arrival

In 1930, Creighton identified an exotic ant in the Mobile, Alabama, area as *S. saevissima richteri* (Creighton, 1930) and recognized it as a "chronic nuisance" and a threat to The American Way of Life, because "their unsightly nests disfigure lawns." E. O. Wilson was growing up in Alabama during this era, and much of what we know about the early history of fire ants in the USA is due to this fortuitous circumstance. When the student Wilson began his fire ant studies in 1949 (Wilson and Eads, 1949; Wilson, 1951, 1953b; Wilson and Brown, 1958), the fire-ant population was undergoing explosive growth, the situation had become complex, and the taxonomy had become muddled. As a result, Wilson's studies require reinterpretation. Hindsight reveals that the difficulties arose as a result of the appearance of a second species that was initially regarded as a color phase or variant, of local origin, of *S. saevissima richteri*. Wilson later concluded that the lighter color phase was probably a second introduction. Creighton eventually came to believe that two separate species were involved, but he never published this opinion (Buren, 1972). It was not until 1972 that Bill Buren formally recognized that two different species of *Solenopsis* had been introduced into North America. Building on and revising Creighton's and Wilson's work, he determined that the "dark phase" *S. saevissima richteri* was *S. richteri* and that the light phase was an undescribed species (Buren, 1972). With a touch of ironic humor for which the world can be forever grateful, he named this new ant *Solenopsis invicta*, the unvanquished, because by that time this fire ant had won two major wars waged against it by humans. Despite a recent challenge maintaining that *S. invicta* should be called *S. wagneri* (Anonymous, 2001), the name still stands today. *Solenopsis richteri* can be distinguished from *S. invicta* by three rather distinct characters: its minor workers are about 15% larger, their blackish-brown color contrasts with the reddish-brown of *S. invicta*, and they have a conspicuous yellowish stripe on the dorsal side of the gaster that is lacking in minor workers of *S. invicta*. We can be reasonably confident that the "dark phase" in the earlier literature is *S. richteri* and the "red" or "light phase" is *S. invicta*, and I will use these species names hereafter to avoid perpetuating confusion.

Against this taxonomic background, we can now try to sort out the arrival dates of these two species. *Solenopsis richteri* is generally agreed to have arrived about 1918. Creighton first reported it from Mobile, Alabama, in 1930 and noted that the amateur myrmecologist H. P. Löding, who had studied the Mobile region intensively, believed that it had been in the USA since 1918 or 1919. A USDA report (U.S. Department of Agriculture, 1958) suggests, on the basis of an interview with an old salt, that the ant arrived even earlier in dunnage mats used in the transport of coffee from South America. The sailor claims to have seen fire-ant mounds on the Mobile bayfront in 1900, but of course, no specimens are available to confirm this observation. From its rate of spread of about 2 km per year in 1935, backward extrapolation from its 1935 range does indeed suggest an arrival date of 1915 to 1920.

Establishing the date of arrival of *S. invicta* is made more difficult by the taxonomic confusion described above. Creighton collected

Solenopsis species throughout the southeastern USA in the early 1930s for his revision of the genus (Creighton, 1930) and continued actively collecting in the Mobile area until 1933. He found no *S. invicta.* If it was present in 1933, it must have been very rare. The first specimens of *S. invicta* were collected near Mobile in 1945 by Bill Buren, but the first reliable observation of *S. invicta* in this country was actually made earlier, in 1942, by E. O. Wilson, who found fire-ant mounds containing a light, or red, form rather than the familiar dark workers of *S. richteri.* Wilson noted that this red form was abundant in the Mobile Harbor area. His observation brackets *S. invicta*'s arrival between about 1933 and 1942.

Can we corroborate these estimates? By 1949 (Wilson and Eads, 1949), the rapid centrifugal expansion of *S. invicta* had, like advancing Roman legions, eliminated most of the *S. richteri* in its path. After about 1945, most of the range increase of fire ants was due to *S. invicta.* If we extrapolate the range expansion of fire ants between 1945 and 1953 backwards, we conclude that *S. invicta* began its population expansion in 1937 or so. If we allow about four years for the initial immigrants to establish an incipient population, the most likely arrival date is 1933, the year Creighton stopped collecting *Solenopsis* in the Mobile area. Perhaps he should have continued collecting for one more year.

Home

Home is Argentina and/or Brazil. Creighton (1930), by comparing specimens, became the first to recognize the South American origin of the exotic *Solenopsis* that he and Löding collected during the late 1920s and early 1930s. Once Buren cleared up the taxonomic confusion, real progress could be made toward determining the ranges of the two invaders in South America and just where within those ranges the North American populations had originated. Several collecting expeditions showed that the range of *S. richteri* was northern Argentina, Uruguay, and extreme southern Brazil and that *S. invicta* occupied an elongate range centered on the Paraguay River (Figure 2.2) (Allen et al., 1974; Buren et al., 1974). The two ranges overlapped only in a small area. Buren's task of pinpointing the exact source of *S. invicta* was greatly aided by Father Kempf, a priest and amateur myrmecologist who provided him with *Solenopsis* specimens from all over central South America. The morphology of *S. invicta* from the USA matched only a single one of these diverse samples from South America, a sample that came from Chapada in the state of Mato Grosso in southern Brazil. Later samples from the nearby city of Cuiaba matched *S. invicta* from the USA very well. Several subsequent collecting expeditions strengthened the case for Cuiaba as the source of the initial migrants to the USA. Buren chose as the holotype (the individual on which the species description is based) a major worker collected in the city of Cuiaba, Mato Grosso, Brazil, on 16 February 1971, by W. Whitcomb and R. Williams. At the same time, he designated as paratypes (additional individuals supporting the description) other workers from the same colony, as well as workers from other colonies collected nearby and from Daphne, near Mobile, Alabama. He deposited the holotype in the U.S. National Museum in Washington, D.C., and various paratypes were sent to museums at Harvard University and the Universities of Florida and Georgia.

The city of Cuiaba is near the headwaters of two major rivers. The Paraguay River flows southward through Brazil, Paraguay, and Argentina to empty into the Rio de la Plata, the great commercial river of Argentina and Uruguay, and the Guaporé River flows northward to join the Amazon. Only a low divide separates their sources. The range of *S. invicta* stretches over 3000 km north and south along these rivers (Figure 2.2). The core of the range is the headwaters of the Paraguay River, which arises in one of the largest wetlands in the world, the fabled Pantanal, an expansive mosaic of flooded savannas and wetlands. The region is home to the richest variety of vertebrates in the Americas and to the fire ant as well. During the dry season, the area is dotted with small lakes, but every year during the rainy season, the waters of the Paraguay River flood an area approximately as large as the state of Missouri. During these floods, the wildlife seeks shelter on wooded mounds while fish swim freely where cattle once roamed. These watery origins have left their mark on *S. invicta*, whose colonies are capable of floating on the water's surface for weeks, until land is once more available. Still, *S. invicta*'s nautical nature has limits—the regularly flooded parts of the Pantanal do not support large populations of fire ants. Rather, the ants occupy mostly the margins and adjacent regions.

Can the origin of the North American *S. invicta* be further narrowed within this 3000-km-long range? Although this question may seem of only academic interest, it probably has great practical importance for the development of biological control agents against *S. invicta*. Many natural enemies are adapted to local genetic races of their host species and are much less successful at attacking hosts with which they have not coevolved. The search for the North American population's exact South American street address and house number therefore continue, primarily by means of attempts to match the genetic patterns of North American and candidate South American populations. Genetic variation across the entire North American range of *S. invicta* is small, because this population is descended from a few individuals and bears their gene ratios. In South America, on the other hand, the ants have been evolving and adapting to local conditions for millions of years, and large differences in the gene ratios of local populations have arisen. If we could match the gene patterns in the North American fire ants to one of these local variants in South America, we would have strong evidence that we had found the source of North America's *S. invicta*.

In practice, common metabolic enzymes are used as surrogates for the genes that code for them. After gel electrophoresis, specific biochemical reactions make even very small quantities of the enzymes visible as colored bands. Variants of enzymes (allozymes) show up as bands of different mobility. Only enzymes that occur in two or more variants (alleles) can differ geographically in their ratios, from 0 to 1.0. When an enzyme has more than two alleles, more complex variation is possible. Therefore, at each locality, the alleles occur in signature proportions, yielding a combination like a local dialect. Of special value are variants that are present in some localities and entirely absent in others—if the imported fire-ant population contains such an allele, it cannot have come from a region in which this allele is absent.

In this way, Ross and Trager (1990) analyzed the allozymes from over 200 colonies of fire ants collected from 48 sites along the drainage of the Rio Paraná from near its mouth to the Paraguayan border. Five of these allozyme loci were polymorphic enough to allow estimation of geographic differences in allele frequencies. *Solenopsis richteri* occurred only in the southernmost part of this range, where it overlapped slightly with the *S. invicta* range to the north. To increase the sample size for each geographic area, Ross and Trager lumped the *S. invicta* sites into five "populations" lined up south to north along the river drainage (Figure 4.1) and compared their allele frequencies. The allele frequency of all five loci changed in a regular, rather gradual way in progressing from south to north, that is, as a geographic cline. Along this cline, four of the alleles increased in frequency, and one decreased. For our purposes, the most interesting finding is that the allele frequency of the North American population most closely resembled the northernmost sample of Argentine *S. invicta* at all five loci, suggesting that the original home of the progenitors of the introduced population lies just south of the Paraguay border. This conclusion is strengthened by the presence of the *Agp-1*^{41} allele, the only *Agp* allele in the USA, only in the northernmost Argentine population of *S. invicta*. These specimens were also judged by Trager (1991), without prior knowledge of the genetic patterns, to resemble the North American *S. invicta* most closely on the basis of morphological characters. In this conclusion, Trager disagreed somewhat with his mentor Bill Buren, who placed the source population further north, in the Brazilian province of Mato Grosso. Between Mato Grosso and northern Argentina lies the country of Paraguay, where the bureaucratic, political, and cultural aspects of travel and collecting are so difficult that no collections have yet been made there. Trager recognized this information vacuum and proposed that the source of North American *S. invicta* can be pinpointed no more precisely than the region that includes northernmost Argentina, all of Paraguay, and the Brazilian states of Mato Grosso and Mato Grosso do Sul. Ken Ross and his coworkers have recently undertaken a more complete genetic analysis

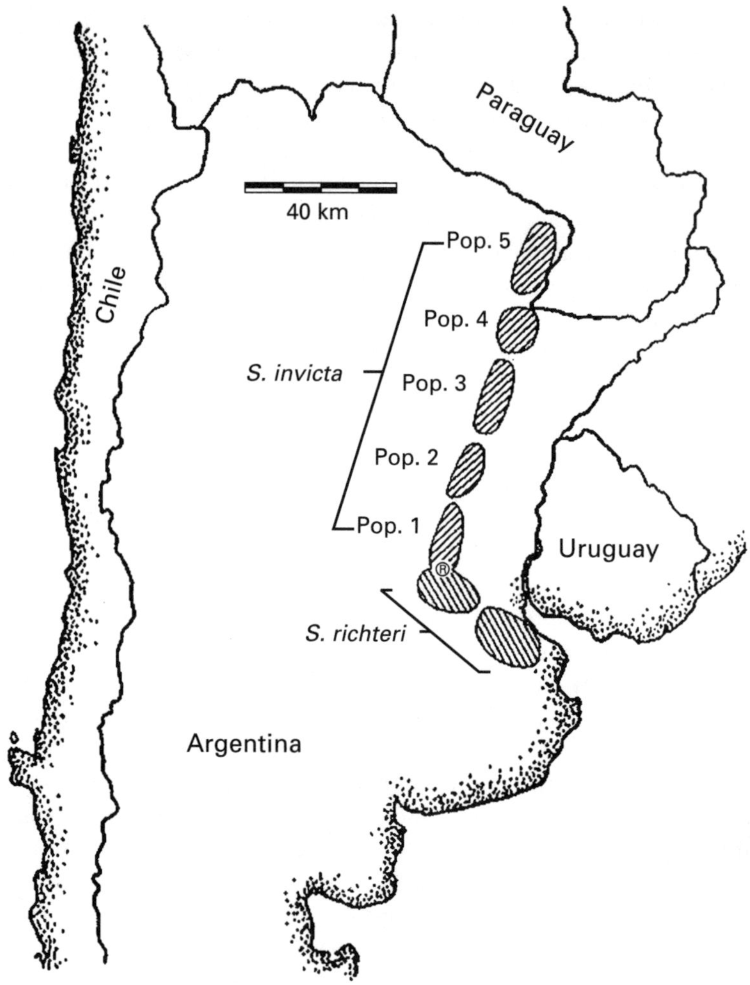

Figure 4.1. Locations in South America of the regionally pooled collections of *Solenopsis invicta* and *S. richteri* used by Ross and Trager (1991) to narrow down the geographic origin of North American populations. Reprinted from Ross and Trager (1991), with permission from the Society for the Study of Evolution.

of all the fire ant species in central South America. The genes for polygyny (allowing more than one functional queen per colony) occur only in the southern half of *S. invicta*'s range, so at least some of the original pioneers must have come from this region. Additional genetic and morphological data strengthen the case for northeastern Argentina as the most likely source area, agreeing with the earlier genetic studies. Continued work in this region will clarify the picture in a few years.

Mode of Transport

Both *S. richteri* and *S. invicta* traveled from South America by cargo ship. This point seems hardly worth discussing, because both species made their first North American appearance in the harbor area of Mobile. Neither air transport nor road travel between Argentina or Brazil and the USA existed in the 1930s. Why these two species arrived and other *Solenopsis* did not is more mysterious.

In what cargo might the ants have stowed away? Lennartz attempted to associate cargoes arriving in Mobile from Montevideo, Buenos Aires, and other nearby ports with fire ants, by researching harbor records. Unfortunately, these often left much to be desired. In any case, she found no associations consistent enough to make us confident of how the ant got on board ships (Lennartz, 1973)—quebracho, ipecac, cattle, coffee—none seemed likely products. Perhaps soil was sometimes used as ballast. Taken from dockside, it would very probably have contained fire ants. Or perhaps they stowed away in animal feed, hollow logs, or potted plants. Judging from the products on which fire ants hitch rides to California, the potential list is long.

Low Resistance Transport

The runs from Argentine and Uruguayan ports to Mobile seem to have been a sort of underground railway for exotic ants between South and North America. At least five species from the Paraguay River drainage have successfully made the trip (Naves, 1985): most conspicuously, *S. invicta*, and the earlier migrant *S. richteri*, but also *Brachymyrmex patagonicus*, a highly mobile and opportunistic, though less conspicuous, species. Although the exact source is not known, the Argentine ant, *Linepithema humile*, also hails from that general part of the world and probably set sail via the ports on the Rio de la Plata. Finally, so did the rather spectacular but slowly spreading *Pheidole obscurithorax*, which shared its native home with *S. richteri*. I should point out that the detection of exotic ants, and the determination of their ports of entry, is a very casual and occasional business. Evidence of a new exotic usually takes the form of the sudden appearance of a strange species in a long historical collection series from a locality and the matching of that species with one from another country. Regular and thorough collections of ants at various geographic localities are the tiniest of exceptions, not the rule, because at any time, only a small handful of myrmecologists are collecting and identifying ants. Only when an ant is very conspicuous, or comes into conflict with humans, as was the case with fire ants, is it likely to be detected early. Because most ants fulfill neither of these conditions, decades may pass before exotics are recognized. For example, the fungus-gardening ant *Cyphomyrmex rimosus*, huddled under cowpies or loose bark, tended its garden of dead insect parts for years before Mark Deyrup (pers. comm.) recognized it as a naturalized citizen of the USA. The Mobile area was exceptional in another important way, however—it was the focus of attention of at least three myrmecologists, including the young E. O. Wilson, who lived and collected in this area from the early 1940s through the 1950s.

However it occurred and whatever the conditions, a low-resistance conduit for ants existed (and perhaps still exists) from the Paraguay River basin to Mobile. This conduit must have included at least three elements. First, the likelihood of embarkation, probably sequestered in some type of cargo, had to be high; second, the probability of surviving the trip had to be high; and third, the situation at disembarkation in Mobile had to favor immigrants. Establishment in Mobile would have depended on favorable physical and biotic conditions. Climate and latitude are similar and would not have posed barriers. On the other hand, an alien ant does not enter a biotic vacuum but invades an existing community of organisms. The most important of these would be the ants already established there, whether native or earlier alien invaders.

Perhaps we should ask not, Why Mobile? but rather Why not New Orleans? or Pensacola? During the first half of the 20th century, New Orleans received more cargo from South America than did either Pensacola or Mobile. During the same period, New Orleans was swept by another exotic species, the Argentine ant, which became so abundant that it was colloquially known as the "New Orleans ant." This virulent invader wiped out most of the native ants in the New Orleans area (Newell, 1908) and may have prevented *S. invicta* and *S. richteri* from becoming established there (Wilson, 1951; Buren et al., 1974). If so, then why not Pensacola, which was free of Argentine ants? Why did both fire-ant species first appear in Mobile and only in Mobile? Wilson (1951) speculates that *S. richteri* "preconditioned" the native ant community to ease the entry of *S. invicta.* In view of the superior competitiveness of *S. invicta,* it would hardly seem to have needed the help. We will never really understand why Mobile. Whatever charm the city holds for ants must simply be accepted, much like the southern hospitality on which Mobile prides itself.

Mature Colony

The original immigrants included at least one mature colony (and probably more). How can you possibly know this? the reader may ask. The answer rests with two small beetles that live in fire-ant nests in South America and that came with them to the USA. The staphylinid *Myrmecosaurus ferrugineus* and the scarab *Martinezia dutertrei* spend most of their lives in the nests of fire ants (Wojcik, 1975, 1980; Frank, 1977; Wojcik et al., 1977). Although both species briefly leave their host nests to disperse (I have seen *M. ferrugineus* following fire-ant odor trails), both need the ants for successful completion of their life cycles (perhaps not entirely true of *M. dutertrei*). The beetles are therefore highly likely to have been transported together with the ants as guests of the family. Their establishment in the USA from such a small inoculum may have been aided by their use of the *S. richteri* already present in Mobile as an alternate host. If both species accept *S. richteri* as a host in South America as they do in the USA, these myrmecophiles could have come with either or both ant species.

The transport of mature colonies suggests that fire ants came in a product that had contact with the soil or that was transported with soil. Fire ants readily make their homes in piles of litter, sacks of granular goods, under loose bark or in rot cavities of logs piled on the ground, in potted plants, or in any number of other goods that they might find on the ground in temporary piles awaiting shipment. From there, they would be loaded unnoticed onto ships, surviving the trip by foraging on shipboard insects or food scraps or by cannibalizing their own brood. Arriving in Mobile, they would have been unloaded with the rest of the cargo to be piled on the ground, soon excavating their first nest in the soil of their new home. Even if dumped overboard in the harbor, they might have drifted ashore, adapted as they are to high water.

Multiple Introductions

The original arrivals included several mature colonies or combinations of mature colonies and colony-founding queens. Again, the footprints of genes tell the story. In all populations, the total number of alleles for a gene locus fixes the total amount of genetic information in that population. Each individual, of course, carries no more than two alleles, so when a new population of an organism is founded by a very small number of individuals, as is the case with most accidentally introduced organisms, the genetic result is known as a population bottleneck (the brand of bottle is not specified, but we can assume that it had a long, narrow neck, like Lone Star or Coors beer): they carry with them only a fraction of the species' total genetic variation. For example, if a single queen founds a new population, she can carry no more than three alleles for each locus (her own two plus one in the sperm stored in her spermatheca). All the remaining versions of this gene are left behind. If two individuals form the initial population, they can carry *up to* six different alleles for a locus. And so on. When a locus has few alleles, and they occur at different frequencies, the most common ones are the most likely to be carried to the new land, and the rare ones are likely to remain behind.

In most Hymenoptera, including ants, sex is determined by the number of alleles present at the (usually single) sex-determining gene locus in the embryo. Unfertilized, haploid eggs have only a single allele, so they develop into males. Diploid embryos (formed by fertilization of haploid eggs) develop into females as long as the two alleles are different. Normally, they are, because the population includes a very large number of alleles for this gene locus, so the chances that a male will mate with a female that carries the same allele is very, very small. The vast majority of fertilized eggs therefore develop into females. Rarely, just by chance, an ant queen will mate with a male that carries the same allele at the sex-determining locus—a matched mating, but not a match made in heaven. In spite of being diploid, the developing embryo has only a single version of the gene and develops into a male, albeit a diploid male. The smaller the number of different sex-determining alleles in the population, the more likely is matched mating and diploid-male production. The frequency of diploid-male production is thus a measure of the number of different sex alleles (and therefore, queens) that passed through the immigration bottleneck.

Ross and his coworkers estimated the frequency of matched matings by noting what proportion of newly mated monogyne queens produced male larvae or pupae among their broods. From the result, they estimated the number of queens that founded the original population (Ross and Fletcher, 1985b; Ross et al., 1993). In the USA, 17% of these queens were match-mated and produced diploid males. These colonies never survive long because the diploid males sap the founding queen's reserves but perform no work and are therefore useless for further colony growth.

When the colony is polygyne (i.e., contains multiple fertile queens), the situation is different. In such a colony, a match-mated queen is readily adopted and her diploid male offspring reared to adulthood. Isolating polygyne queens with a few workers showed that, in Argentina, only about 2.3% of polygyne queens are match-mated, whereas in the USA, the figure for polygyne queens is 17%, just as for monogyne queens. Clearly, the number of sex alleles has been greatly reduced through a population bottleneck. If we assume existence of only a single sex-determining locus, we can calculate from these data that the Argentine population contains 86 sex alleles but the North American population only 10 to 13, an 80% loss of genetic diversity. These 10 to 13 alleles, at three per queen, tell us that at least three or four mated queens (and probably more) arrived to found the North American population.

The 76 allozyme-coding genes in the USA underwent similar loss of allelic diversity—from 120 alleles in Argentina to 96 in Georgia—and rare alleles are less than half as common in the USA. Rarity decreases the chance of being among the chosen few. Although allozyme-coding alleles confirm a genetic bottleneck, they do not allow easy estimation of the size of the introduced fire ant population.

Several Immigration Events

Several immigration events probably took place over a period of years. When at least five different species of ants (and who knows how many other organisms) entered the USA at Mobile, how can we doubt the repeated transport of immigrants for many years? Are we to believe instead that all five species came on the same boat? No, we already know that *S. richteri* arrived about 1918 to 1920 and that *L. humile* made shore in New Orleans in the late 19th century. Most probably, some products that were frequently contaminated with one or more species of ants from the Paraguay River drainage were being regularly transported, with ants, from South America to Mobile. The genetic diversity present in the North American populations of introduced fire ants was therefore probably not established all at once but was initially lower than it is today and was increased by a succession of later arrivals.

Social Form

Social form in *S. invicta* is determined by the two alleles at a single locus, an amazing discovery described in detail in a later chapter. The ants in monogyne colonies are homozygous for one of the alleles (that is, carry two copies of the "normal" allele), and all polygyne queens are heterozygous (carry both alleles). Because the second allele is found only in polygyne colonies, at least one of the first immigrant colonies must have been polygyne. In the USA, the two populations are genetically similar and yield similar bottleneck estimates of three to five mated queens in the initial inoculum.

◂•• *Beachhead Mobile*

For much of the 20th century, Mobile, Alabama was one of those southern port cities noticed primarily by people who live in the region. Although it had its share of gracious southern society and English, Irish, French, Spanish, African, and Creole heritage, it lacked the uncorked vitality of New Orleans, with its more overt and creative blend of cultures and races. Oh, perhaps Mobilians could feel superior to their counterparts in smaller and less energized Pensacola to the east, but outside the gulf coast region, the differences wouldn't have seemed very important.

During the 20th century, and, who knows, perhaps during previous centuries too, certain enemies have sailed unchallenged past the forts defending the entrance to Mobile Harbor to make their beachheads on the shores of Mobile Bay. These waves of invaders were not people but exotic animals and plants, carried by ships in products unknown, from their homes in South America. Nobody stopped them because nobody was aware of their presence or thought it important.

The fire ant *Solenopsis invicta* was one of these fortunate immigrants, coming ashore between the two World Wars. Mobile represented escape from the burdens of its own history, a place of rebirth, as it did and perhaps does for other immigrant species. If *S. invicta* constructed monuments, it would surely put a formicid version of the Statue of Liberty on the shores of Mobile Bay. A mere decade or two after establishing itself at the water's edge in Mobile Harbor, *S. invicta* went from a struggling minor member of a cut-throat ant community in South America to a major dominant creature in an immense new land. It is therefore fitting that we make a pilgrimage to the beginnings of fire-ant history in North America, to the moment when humans first consciously encountered *S. invicta* in the USA and our shared history began.

Can we pinpoint this moment in time and space? Amazingly, we can, because the story of *S. invicta* in the USA is intertwined with the youth of a man named Edward O. Wilson, who has been generous in sharing his memories of Mobile with me and in checking the facts in this essay. The successes of both the ant and the youth are rooted in Mobile—the first went on to become a major pain in the collective North American neck, the second a giant in myrmecology and biology.

The young Ed Wilson, 13 years of age in 1942, was already acutely interested in ants and many other animals. He scarcely noticed the stratified society in which he lived, the Azalea Trail upper-crust families with their circular drives and festivals; the solid middle class of his own neighborhood; the flood of rural poor come to seek wartime jobs in the shipyards, the military bases, and the docks; and the blacks, like a separate nation, forbidden and shut out, in the southern end of town. Like many a young man whose destiny is biology, Ed had eyes primarily for nature, wherever he could find it. On his single-gear Schwinn bicycle, he often pedaled the 20 miles to the high ground of Spanish Fort, where his grandfather had finally surrendered to the Yankees. Sometimes it was the woodland south of Brookley Air Force Base to catch snakes. At other times he would head south on Washington Avenue, then cross through the L&N rail yards to Bay Street and beyond to the waterfront, where sandy little trails led down to the bay through a tangled scrub arched over by groves of live oaks, their huge branches bearing gardens of resurrection fern. A palisade of reeds guarded the water's edge, beyond which the bay water rose and fell with the tides. On hot days, a smell of oil hung over the area like a cloak, and half-buried bricks and concrete blocks provided high spots on which swift-running skinks paused momentarily to survey their tiny domains.

If he raised his eyes above the hollow grasses inside which nested the bizarre ant *Colobopsis,* whose soldiers plugged the nest entrance with their flattened heads, Ed could see the huge ships slide past, to and from the docks to the north, where war-driven commerce pulsed around the clock. Behind him in the rail yard, switching engines throbbed and hollow boxcars banged and thundered with each new coupling. These man-made wonders did not distract him for long from the sulfurs, skippers, and swallowtails that flitted through the vegetation or from the many ants, both native and recently arrived, that waged their tiny wars underfoot. Little pockets like this were probably as congenial to alien ants as to boys. Linked by water to South America and by rail to the rest of the USA, what better place could a subtropical ant find to make landfall?

But *Solenopsis invicta* met Ed Wilson much closer to home. The Wilson house at 550 Charleston St., was only nine blocks from L&N rail yards and the docks, in a neighborhood of old Victorian houses in genteel decline. The house next door at 552 Charleston St. had been "tore down" in 1938, leaving a weedy, fecund vacant lot surrounded by a crumbling fence that marked out limits but allowed secret entry. Stifling hot in summer, with bare spots between clumps of grass and a reclining fig tree in one corner, it was a place alive with butterflies, beetles, and spiders. It was a young boy's private nature reserve. In 1942, Ed undertook a boyish version of a complete survey of the ants of this vacant lot. Waves of *Linepithema humile,* the Argentine ant, streamed for cover when he lifted broken bricks or pulled apart a broken fence. Under an empty whiskey bottle, he discovered a colony of *Pheidole floridanus,* its brood scattered like a handful of seeds. Carefully replacing the bottle, he rechecked it every few days until, one day, he found the queen. And like anyone who has had the experience, he was amazed by the trap-jawed *Odontomachus insularis,* whose crowbar jaws snap shut with an audible click, with a force that sometimes hurls the ant backward through the air.

But of all of these, one ant glows in his memory—in the very center of the vacant lot rose a towering, unmistakable mound that, at a mere touch, exploded into a frenzy of furious ants. This moment in May of 1942 at 552 Charleston St. was the first conscious encounter between a human and *S. invicta,* the first awareness that this ant was something unfamiliar, something alien (Wilson, 1994). It would take a few more years for humans to realize that the South would never again be the same.

More than 60 years have passed since that moment. The ant and the South have had their civil war, which ended only after southerners finally recognized that the alien was unbeatable. Hating the fire ant has become a major industry among southerners, who spend millions of dollars on endless, futile guerrilla actions against the ant. Little by little, the ant has insinuated itself into southern lore. In my opinion, and in recognition of this fact, southerners should erect a monument to the fire ant, much as they did to the boll weevil. If they ever do so, it would celebrate not so much the fire ant itself as the way that it has drawn attention to their own cherished attributes—the will to endure, the ability to see humor in adversity, the taste for tall stories and eccentricity, and the magnet pull that dangerous, destructive, and wild things exert on all true southerners. It will be a monument to yet another defeat that merely stiffened southern defiance another notch.

5

La Conquista: Spreading Out

Each having made a beachhead in Mobile, *Solenopsis invicta* and *S. richteri* began to spread, and their increasing contact with people unleashed the Fire Ant Wars in all their political, social, and technological color and tumult. Although this saga has little to do with fire-ant social biology, it is a fine story, and telling it once more will do no harm. It is full of selective perception and wishful thinking, arrogance and misjudgment, myth and outrageous exaggeration, unwavering faith in technology, unintended consequences, and chips falling where they may—but also clear-sightedness, stubborn skepticism, and dogged opposition. Not least, had the Fire Ant Wars never been fought, this book could never have been written, because we would know little to nothing about fire ants. The USA harbors dozens of exotic ant species, some quite common, and we know little about the rest of them, but *S. invicta* drew notice with its mound and burned itself into our consciousness with its sting. Controversy and politics spawned the first research into the biology of the fire ant, and for many years, controversy fueled it. Although southerners have largely accepted (if not become fond of) the new immigrant, the ant still has a high profile, and the research continues. People from all walks of life are *interested* in fire ants, and along the way, we have discovered that this ant is a wonderful experimental animal.

Mine is not the first account of the Fire Ant Wars. I have taken the facts largely from documents and previous histories, colored by my own experience as a Fire Ant Warrior. From the beginning, people were deeply divided on the issues. The accounts of fire-ant history given by employees of the U.S. Department of Agriculture (USDA; Lofgren, 1986b) give an impression very different from those provided by conservationists (Hinkel, 1982) or scientists (Brown, 1961; Davidson and Stone, 1989). The tone of news accounts at the time was typically hysterical ("red menace advancing," "entire Mississippi counties scarred," "killing people and animals, destroying crops," "cobra-like venom," and perhaps the worst indictment of all, because it showed deep disrespect for authority, "fire ant spreading, postman attacked"). A few were fairly even-handed (Shapley, 1971a, b). The wars were a clash of the subcultures and ideologies of the era. At one extreme were the technologists with their unshakable belief in the ability and the right of humans to manipulate nature for their own benefit, to the exclusion of all else. At the other were (broadly defined) environmentalists and ecologists. To them, nature itself mattered; excessive tampering was unwise, unethical, and subject to unpredictable consequences. The Fire Ant Wars were born in politics, but it was science that finally ended them. They were not the only entomological wars—similar battles raged over DDT and several other persistent pesticides. They demonstrated the futility of chemical control as well as its high environmental costs and eventually turned science, government, and society away from "control" to "management" (i.e., prophylactic measures such as quarantine and "integrated pest management"). The Fire Ant Wars can be seen as a case study of how the effective application of scientific method can accrue information that slowly overcomes the inertia of ignorance. And of course, it is a great story.

Early Spread and Skirmishes

In the decade or so after its arrival, *S. richteri* gradually spread outward from Mobile Harbor. By 1928, it occupied the northwestern part of what is greater Mobile today. By 1932, it had spread 8 km northwest from the harbor to Whistler and 25 km southwest and southeast to St. Elmo and Fairhope on opposite shores of Mobile Bay (Murphree, 1947). By 1937, it occupied much of Mobile and Baldwin counties and had spread across the state line into Mississippi (Figure 5.1). It had become abundant enough and annoying enough that four county, state, and federal agencies joined to do battle against it. The opening skirmish (one cannot dignify it as a battle) of the Fire Ant Wars had begun. On about 800 hectares (ha), agents injected nest mounds with "grade A" calcium cyanide dust and claimed to have eliminated more than three-quarters of the colonies (Eden and Arant, 1949; Green, 1952). During this same period but farther east, Travis (1939a, b, 1940) waged a similar war against the native *S. geminata* on the quail-hunting plantations of southern Georgia and northern Florida, believing that the ants killed nestling quail. Killing quail was strictly reserved for wealthy humans with shotguns, so he applied cyanide mound drenches as well as poisoned baits containing thallium salts. His instructions suggest that "cyanide should not be taken internally" (good advice). His success was not high, but one wonders what else these deadly poisons eliminated. In 1936, in what amounted to a kind of "preview of coming attractions," myrmecologist M. R. Smith fingered the southern fire ant, *S. xyloni*, as an important pest causing crop damage, stinging people, eating okra and eggplant, gnawing on dahlia stems, stealing seeds from gardens, encouraging mealybugs, killing chickens, and gnawing holes in dirty clothes (Smith, 1936). He had only limited success in eliminating it with poisoned baits. The opening skirmish was abandoned during the early 1940s as American attention turned to a much larger war among humans. Compared to those to come, this first effort at fire-ant control

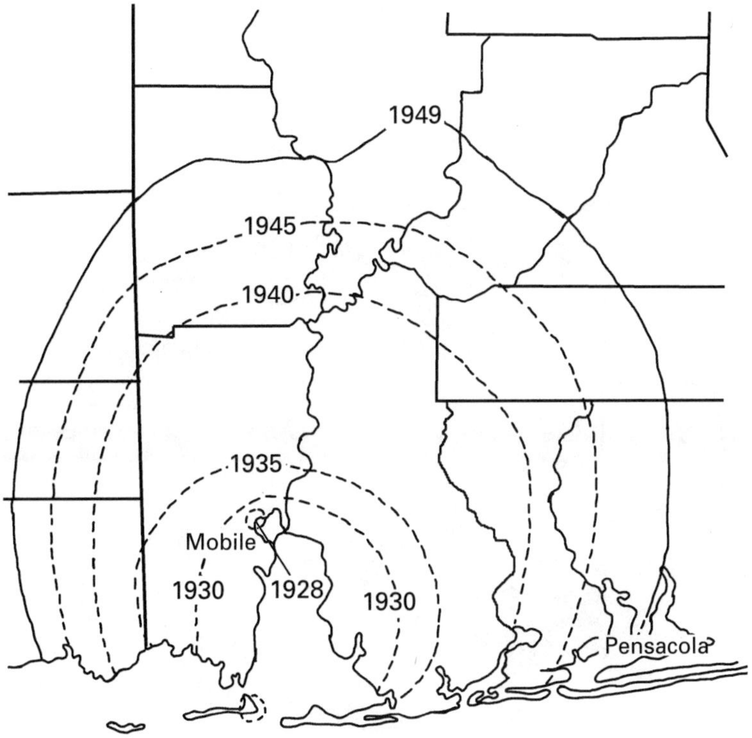

Figure 5.1. Estimated ranges of exotic fire ants between 1928 and 1949. Most of the range increase after about 1940 was probably by the more rapidly spreading *Solenopsis invicta*. Adapted from Wilson (1951).

was but a minor action, a patrol firing scattered shots over the heads of a hostile force, very few of which hit their targets.

This was probably also the era when *S. richteri* established three separate northern populations, one each in the Artesia and Meridian areas of Mississippi and one in the Selma area in Alabama. All of these northern populations were separated from the main Mobile population by 150 to 300 km and were almost certainly the result of transport by shipping of some type. They also all centered on towns. Either fire ants found it easier to hitch rides to towns, or the environment there was more conducive to their establishment, or fire ants were less likely to be spotted in rural areas. In 1949, these fire-ant populations were only 20 to 40 km across, and densities were high only at their centers, suggesting that they had only existed for a few years. Wilson estimated that the Artesia population arose about 1935 and the Meridian one about 1940. The ants did not appear in Selma, Alabama, until 1944.

The Arrival of *S. invicta* and the First Fire Ant War

The arrival of *S. invicta* about 1933 was far worse news for *S. richteri* than any cyanide-puffing humans, for this Johnny-come-lately spread much more rapidly than *S. richteri*. Wherever *S. invicta* appeared, *S. richteri* disappeared. By 1949 (Wilson and Eads, 1949), *S. invicta* had eliminated

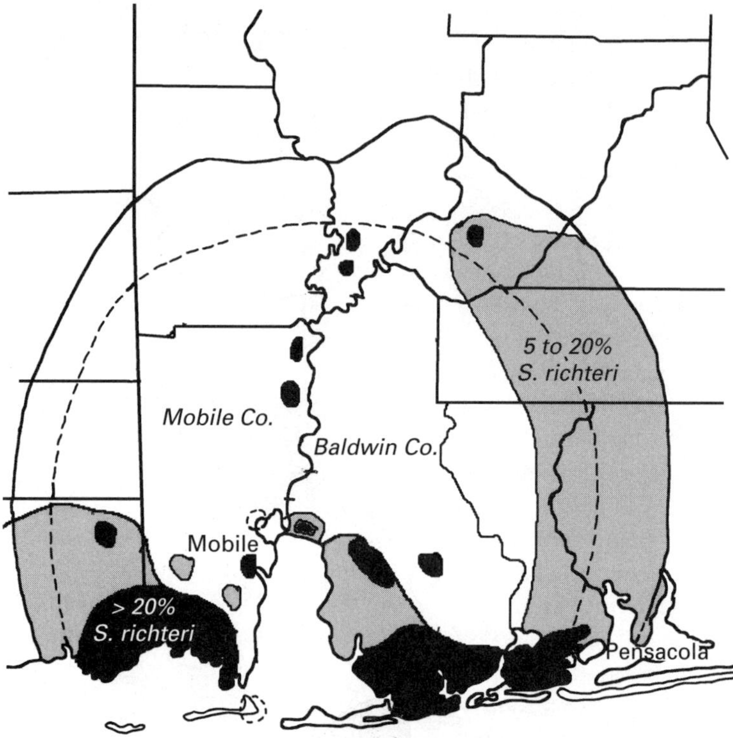

Figure 5.2. The displacement of *Solenopsis richteri* by *S. invicta*, as it appeared in 1949. Substantial populations of *S. richteri* remained only near the coast and as a few enclaves within the spreading *S. invicta* population. Adapted from Wilson (1951).

most of the *S. richteri* from the Mobile area, and *S. richteri* remained only along the gulf coast and along the eastern and western perimeter of the species' collective range (Figure 5.2). Even there it survived as a minority in a mixed population in 1949 and has since vanished from these areas. By 1953, the northern Artesia population of *S. richteri* occupied an area of about 500 km^2, but the hot breath of *S. invicta* was perceptible from the south. Although *S. invicta* eventually surrounded this northern population and reduced it in extent, it continues to persist, in part, as a pure *S. richteri* population to this day, although it has hybridized with *S. invicta* along the northeastern contact zone. Also by 1953, the Selma and Meridian populations had been completely swamped by *S. invicta*, first existing as mixed populations then being completely eliminated by *S. invicta*. Perhaps, as a result of interbreeding, *S. richteri* left behind some genes in the advancing *S. invicta* population, but as a distinct species, it vanished from most of the territory it had occupied.

After about 1945, most of the range increase of fire ants was due to *S. invicta*. Why this species, which seems so similar to *S. richteri*, spread so much more rapidly is unknown, as are why and how *S. invicta* displaces *S. richteri* or the native species of fire ant, *S. geminata* and *S. xyloni*. The answers must lie in the details of their biology, but neither comparative studies of their biology nor accounts of their interspecies interactions exist. The fire ants that reached Meridian, Mississippi, and Selma, Alabama, by 1948 were probably mostly *S. invicta* rather than *S. richteri*. In any case, the State Plant Board of Mississippi was alarmed enough by their "devastation of the hay industry" to mount the First Fire Ant War, seeking nothing less than eradication. The legislature appropriated $15,000, and the war was on. This time the antiant forces were armed with the newly developed insecticide chlordane, a chlorinated hydrocarbon, which was made available to farmers as a 5% dust to be dumped onto the mound or mixed into it. This program was begun with little or no knowledge of the biology of the ant, a pattern that was to be repeated twice more. Although chlordane may be capable of killing fire ant colonies, it proved no match for the reproductive and dispersal abilities of *S. invicta*. By 1951, the ant was found throughout the state, and the chlordane program was discontinued, having cost mere peanuts compared to those to come.

At about this time, the first fire-ant studies were begun at Mississippi State University (Lyle and Fortune, 1948; Eden and Arant, 1949), and in 1949, the USDA set up a research station at Spring Hill near Mobile. The station's duties were to determine the current range of the fire ant through surveys and to find insecticides to kill it. Studying the biology of the adversary was little mentioned. The survey found varying fire-ant population densities, from low to high, in 28 contiguous counties in Alabama, Mississippi, and west Florida (Figure 5.3) (Bruce et al., 1949).

Also during the spring and summer of 1949, the Alabama Department of Conservation hired E. O. Wilson, a student at the University of

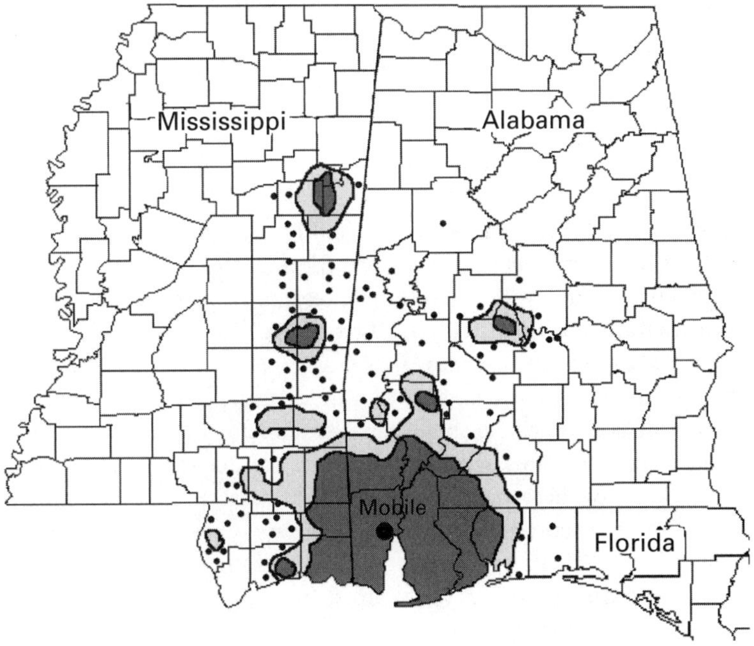

Figure 5.3. Results of a USDA fire-ant survey in 1949. Sparse populations are shown in light shading, and dense populations in dark shading. Surveyed sites from which fire ants were absent are shown as filled circles. Adapted from Lofgren (1986b).

Alabama, and J. H. Eads, who owned a car, to study the distribution of fire ants and to make a first determination of the possible economic impact of their presence. For three months, these two meandered the roads of Alabama, Mississippi, and Florida, looking for ants and interviewing farmers. Their mimeographed report (Wilson and Eads, 1949) and the publications resulting from it (Wilson, 1951, 1953b, 1952) provided the first detailed maps of the extent of the fire-ant population and included the first biological observations of both *S. invicta* and *S. richteri*, sketching the basics of colony size, nest structure, worker longevity, queen number, alate production, colony founding, and habitat preference. They also remarked on the apparent absence of parasites and significant predators of fire ants, waxed dramatic about fire-ant stings, and reported some observations on the species' food habits. Their estimates of economic losses were based only on farmer opinion in response to survey questions. In many instances, Wilson and Eads observed no fire-ant damage to the crop in question, even though farmers reported losses. Of course, no control was available for comparison. Damage to wildlife, they felt, was "a matter of opinion." Interestingly, Wilson reported that fire ants like to gnaw on peanuts and that farmers were of the opinion that the ants helped control boll weevil, a serious pest of cotton. All in all, it is interesting to note how much of the outlines of both future fire-ant research and future controversy over the ecological and economic effects of the fire ant are already visible in this early report.

Between 1949 and 1953, the USDA laboratory at Spring Hill carried out a more complete survey of the southern states. At first, they searched generally in areas likely to support fire ants and found few. After noticing that several isolated populations were located in nurseries, they switched to inspecting only nurseries and began finding large numbers of isolated populations throughout the Southeast. By the end of the survey, *S. invicta* had been found in 102 counties in 10 southern states (Alabama, Arkansas, Florida, Georgia, Louisiana, Mississippi, North Carolina, South Carolina, Tennessee, and Texas; Figure 5.4) (Culpepper, 1953). Most important, this survey firmly established a link between the shipment of nursery stock and the spread of the fire ant. During the post-war building boom of the 1950s, nurseries did a brisk trade providing the plantings for the many new subdivisions. As a result, by 1953, the fire ant's range consisted of a continuous core population centered on Mobile and almost 50 small, separated, incipient populations centered on nurseries throughout the southeast, from Texas to North Carolina. Some of these incipient populations were as much as 800 km from the main core population. Such jumps were of immeasurable value to *S. invicta* in its conquest, for most of the range increase after 1953 was the result of outward spread from these population foci to form a single continuous population. Many of these foci were separated

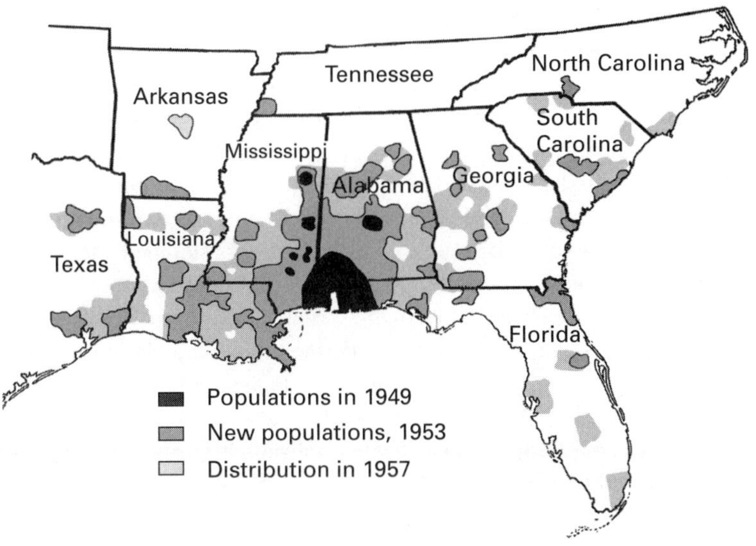

Figure 5.4. Distribution of *Solenopsis invicta* in 1953 as revealed by an intensive survey of plant nurseries throughout the southeastern states. By 1957, many of these scattered populations had coalesced. Adapted from Culpepper (1953) and Lofgren (1988).

from each other by only 100–200 km, a distance the fire ant could cover without human help in about five years. A comparison of the range in 1953 with that in 1957 shows this coalescence effect clearly (Figure 5.4).

The means by which the scattered foci were established seems clear enough—they resulted when obliging nurserymen unwittingly gave rides to hitchhiking fire ants. The mechanism of "natural spread" is less well understood, partly because experiments on the flight range of newly mated fire-ant queens are not easily reconciled with the observed rate of spread. On the average, over large areas and many years, the fire-ant population front has advanced at about 8 km per year, but several studies on postmating flight range have indicated that the vast majority of queens fly less than 1–2 km from their home colonies. The most generous estimate is that fewer than 10–15% of newly mated queens fly 1.6 km (1 mile) before settling, and the least generous is that only 3% fly that far.

The discrepancy might be explained in a number of ways. First, we know that fire ants are *capable* of flying 10–15 km and do so over water (where they might be expected to land only unwillingly), so perhaps a tiny minority of newly mated queens elects to fly very far, even over land (flight distance is known to have a genetic and physical basis). Because such long-distance fliers would be very few, they would be very difficult to find by searching. If their chances of successful founding are about 10%, they could produce densities between one and 50 young colonies per square kilometer 10–20 km ahead of the main population front. Wilson's maps show a zone of "light infestation" 10–15 km wide along the spreading front of the main population. Upon reaching maturity, these extreme outliers could then help to "backfill" the habitat lying between themselves and the main population front.

A second possibility leaves the fire ant out of the formula. Human traffic is mostly quite local, flowing to and fro daily on the scale of a few kilometers. Commuters travel to and from their jobs, goods are delivered limited distances from distribution centers, relatives often live in nearby towns, and so on. The likelihood of transportation of newly mated queens or small colonies over moderate distances in houseplants, soil, earthmoving machinery, fill dirt, farm products, sod, trash, and so on is probably high (U.S. Department of Agriculture, 1958). The tendency of fire-ant populations to spread along roads has been noted repeatedly. Net movement would always be from high density to low, much as molecules diffuse down a concentration gradient.

Local human transport cannot be easily distinguished from spread by natural mating flights. Both tend to establish low-density populations at the margins of the main population. The USDA quarantine described below was applied to counties and would not have slowed local spread of this type, which operated on a within-county scale, quickly spreading the ant across counties, a distance of 30–50 km in several southeastern states. Perhaps this type of spread can account for the 45 km of annual range increase the ant achieved in Texas. Everyone knows Texans drive way too much.

Fire ants can also be spread by water. Many observers have seen entire fire-ant colonies—workers, brood, sexuals, queen, and all—floating as mats after high water flooded them from their nests. Such colonies can drift on rivers, lakes, or ponds for weeks and can reestablish a nest after the flood subsides or they drift ashore (Morrill, 1974a). Indeed, floating colonies, complete with their abundant myrmecophiles, were first reported in South American species of *Solenopsis* as early as 1894 (Ihering, 1894; Reichensperger, 1927b). In addition, newly mated queens sometimes settle on water or on floating debris and can be similarly transported. These modes are not likely to be very important, however. From its Mobile beachhead, the fire ant spread mostly northward, but nearly all the rivers in the Southeast flow southward toward the Gulf of Mexico. Only on the Atlantic Piedmont of Georgia and the Carolinas would the rivers and the ants have been going in the same direction; colonies could have rafted down the French Broad, the Savannah, or the Great Pee Dee.

A perplexing aspect of the rate of spread is seen in comparisons between *S. richteri*, and *S. invicta*. The range of *S. richteri* increased by about 2–3 km per year, but when *S. invicta* had largely replaced *S. richteri*, the population front moved north at 8–12 km per year. Surely a great deal of the difference resulted from transport by humans, but what critical differences in the biology of these two species caused one to be transported so much faster than the other? Their biology seems very similar, so perhaps the difference simply resulted from increased abundance and shipping among nurseries.

Another puzzling point is that, although it became clear in 1953 that fire ants spread primarily through the movement of nursery stock and soil-containing products, such movement was not quarantined until 1958. The agricultural authorities instead wagered their resources on a draconian attempt to eradicate fire ants from North America.

The Second Fire Ant War: Heptachlor and Dieldrin

Every failed program can teach a lesson. The lesson agricultural officials and researchers seem to have learned from the experience with chlordane might be caricatured as, "If it doesn't work, use a bigger hammer." In 1957, the USDA petitioned the U.S. Congress to provide money for a fire-ant control program, ostensibly because of pressure from the Southern Association of Commissioners of Agriculture and from various organizations and members of the public. Of these entreaties, little record exists. In any case, the House Appropriations Committee obliged in July 1957, allocating $2.4 million for "control and eradication" of fire ants (Canter, 1981) even though only two of the 10 affected states listed fire ants among their top 20 pests, and neither ranked it high. This expensive program was mounted because "[*S. invicta*] is a destroyer of plant, animal, and bird life; it builds mounds that interfere with farm operations and *mar landscapes* [emphasis mine]; it [has] a fiery sting, which . . . is painful and can cause serious illness." The federal money was to be matched by state, local, and individual funds in a typical "cooperative program," but in practice over the next three years, contributions by local authorities and individuals were sometimes not forthcoming and were often waived. When you are fighting the good fight, what difference does money make?

In October 1957 the USDA and the Southern Plant Board met in Memphis to draw up plans for a control and eradication campaign. First, the USDA finally proposed a quarantine on shipping soil-containing products from areas with fire ants to areas without them. The quarantine went into effect in May of 1958 and prohibited the shipment of soil-containing nursery stock, grass sod, stumpwood, sand, gravel, or timber with soil attached, unless the shipment was first treated with specified insecticides or washed free of soil. The appearance of new fire-ant populations decreased sharply after this quarantine took effect but, of course, it was already much too late.

At the same meeting, the USDA and Southern Plant Board developed a plan to spray 8–12 million ha of fire ant–infested land in 10 states from the air or ground, laying down over 22 kg of clay granules containing 10% heptachlor per hectare, over 2.2 kg of active ingredient per hectare. The *Mobile Press Register* reported that "Uncle Sam is ready to use a fleet of 60 planes to go to war against the dreaded imported fire ant" (Dunning, 1957). After rain disintegrated the clay granules, the heptachlor was expected to persist in the soil, remaining lethal to fire ants (and who knows what else) for three to five years (Blake et al., 1959)! Populations of the ant were to be treated from their edges inward until they were extinguished and North America was fire ant–free. The new organic insecticides had been so effective that many agencies and individuals became completely reliant on their use, neglecting other approaches to insect control and retarding the study of insect ecology by decades. Such was the atmosphere in 1957. Spokesmen for the USDA said "only the modern airplane, dropping insecticides on the [8 million hectares] . . . , can hope to stop the menace" (Dunning, 1957).

The research basis of this plan was minimal, to put it mildly. Neither the effectiveness of the treatment nor its impact on human health, wildlife, and other nontarget organisms was assessed, nor was any money provided for research or posttreatment monitoring. When the plan was announced, conservationists, entomologists, and wildlife officials objected vociferously. All worried about toxicity to mammals and birds and other environmental side effects of such a massive spray program. Both heptachlor and dieldrin were known to be much more toxic to wildlife, fish, and other aquatic life than that old favorite, DDT (Tarzwell, 1958). One USDA bulletin (U.S. Department of Agriculture, 1958) addressed this issue by stating, "Wildlife stands to gain from the eradication of this pest." Other officials answered with bland assurances that everything had been carefully planned, and all concerns taken into account (Brown, 1961). The record shows otherwise.

In spite of all appeals, spraying began almost immediately in October and November of 1957, as mandated by Congress and presumably advised by USDA. Despite a last-minute rush, wildlife-monitoring efforts by various agencies were not in place until after large areas had been treated with poison. The USDA first met with officials from the U.S. Department of Fish and Wildlife one month after spraying began. Enthusiasm for the cooperative program among the states seems not to have been very high, and only four (Alabama, Florida, Georgia, and Louisiana) appropriated the money expected of them during the first year. The USDA therefore organized an "education campaign" to show southerners that the spray program was needed. Eventually, Texas also ponied up.

The environmental effects of the spraying were disastrous. Wildlife mortality was widespread, domestic livestock and pets died, and birds were eradicated from large areas. This disaster energized the environmental movement and helped to move Rachel Carson to write her classic book, *Silent Spring* (1962). One USDA author noted that the first applications were carried out at the beginning of an unusually cold, wet winter and wrote, "[T]he actual truth of how much damage was due to insecticide and how much to severe environmental conditions will never be known" (Lofgren, 1986b). In fact, plenty of adequate, careful studies put the blame mostly on the insecticide, although interaction between the insecticide and weather conditions cannot be ruled out. Without the insecticide, bird and mammal mortality would have been low. A few examples follow.

Bobwhite quail is a much-prized species on the hunting plantations of south Georgia and north Florida. There, prescribed fire and other land-use practices are used to boost quail populations for the sport of rich people. During the winter of 1957–58, large areas of these plantations were treated with heptachlor or dieldrin. During the breeding season three months later, Rosene (1958) compared the abundance of bobwhite quail on 4000 ha of land treated with insecticide with that on 4000 ha of similar, untreated land. He counted 297 calling males at predetermined stops along six transect routes on the untreated land but only 37 on the six transects on treated land, a difference of almost 90%. Most males heard on the treated land were near untreated land. Six months after treatment, quail had not repopulated the treated areas.

Game officials in Alabama, Louisiana, and Texas searched areas treated with insecticide for dead birds and mammals and found them in abundance (Baker, 1958; Glasgow, 1958; Lay, 1958; Newsom, 1958; Byrd, 1959). Their photos of the accumulated victims laid out in windrows make sad viewing. In Alabama, 13 coveys of quail completely disappeared from the 1200 ha of treated land. Four of the five found dead contained enough insecticide to have caused their deaths. Searches recovered 187 dead vertebrates belonging to 53 species. Of the 100 specimens analyzed for pesticide, all but six contained lethal doses. Similar observations were reported from Texas, where a mere 43 man-hours of searching found 114 dead birds of 19 species and numerous mammals belonging to seven species, as well as dead fish, frogs, and crayfish in the water and 50 species of dead insects. All 41 mammal specimens analyzed contained lethal amounts of insecticide. The same study reported that bird counts declined by about 95% within two weeks after treatment. Even on the edges of the treated areas, bird counts declined by 50%. Nesting success among blackbirds dropped 50–100%, and many birds were found dead at their nests. Bird counts were still subnormal seven months later, when the report was written. The hatching rate of eggs laid by breeding quail fed only 0.001% dieldrin in their diet dropped 50%.

The experience in Louisiana was similar. A search of 280 ha of pasture and cropland three to 10 days after treatment with heptachlor found 20 dead birds and 13 dead mammals, including two dead dogs. Residents of the area lost 72 domestic ducks. Similar counts were recovered from other Louisiana sites. Nesting success of birds dropped drastically, and bird counts decreased 65–80%. Two dead rabbits and 133 dead birds, including 95 geese and 10 other species, were recovered one week after spraying. Of 17 birds analyzed, all but one contained lethal amounts of pesticide. In an 81-ha marsh treated with about 1 kg of dieldrin per hectare, an estimated one million fish and all the crabs died.

The effects were not limited to wildlife. A great deal of mortality was reported among domestic livestock and pets; many animals died suddenly in fits of convulsions. Following heptachlor treatment, the town of Monroeville, Alabama, hauled more than 700 domestic bird corpses and 60 cat and dog corpses to the city dump. A veterinarian reported the deaths of 100 head of cattle in treated areas and reproductive failure in 150 brood sows but found no similar losses in untreated areas. Angry farmers in Alabama sought payment for their losses (Cottam, 1958). In Louisiana, sugarcane farmers petitioned for relief from losses to sugarcane borer, claiming that heptachlor treatment had upset the natural balance, allowing the pests to proliferate. Beekeepers joined in as well, claiming heavy losses after aerial application of pesticide.

The disastrous consequences of this ill-conceived program were not lost on the public. Toward the end of 1958, about one year after spraying began, Alabama's Senator John Sparkman and Representative Frank Boykin cosponsored a bill to stop the spraying until the benefits and costs of the program could be determined. The bill failed to pass, but in 1960, Alabama, Florida, and eventually Texas refused to pay their share of the spraying program. Support was evaporating. Many landowners refused to participate in the cooperative program. At legislative hearings in Alabama, witnesses testified that the fire ant was merely a nuisance, not a serious pest causing losses of crops and livestock. The Plant Commissioner of Florida freely admitted defeat in the battle for eradication and opined that eradication was impossible, a rare voice of clairvoyance. The Southeastern Association of Game and Fish Commissioners went on record as opposing the spraying program. One speaker at the association's conference commented "the control procedure is so drastic and destructive that it is analogous to scalping the patient to cure dandruff!" (Cottam, 1958).

The USDA's response to this disaster, and to the storm of criticism by environmentalists and wildlife officials, was to reduce the amount of heptachlor to 1.4 kg of active ingredient per hectare in 1959. In 1960, they began testing applications of 0.27 kg per hectare, applied at three- to six-month intervals. The lesson drawn this time seems to have been, "If a sledge hammer is the wrong tool, perhaps a tack hammer is the correct one." The idea that perhaps a wrench or a screwdriver might be more appropriate seems not to have occurred to USDA officials. Faith in chlorinated-hydrocarbon insecticides remained unshaken, as subsequent events in the Fire Ant Wars confirmed.

The final blow to hopes of eradication came in early 1960 when the U.S. Food and Drug Administration reduced the tolerance for heptachlor and dieldrin residues on harvested crops to zero—heptachlor is degraded to an even more toxic epoxide, which turns up as residues in meat and milk when fed to livestock. Croplands could no longer be treated. Even use on pasture was not advised by a number of state entomologists. The heptachlor program continued sporadically for two more years, until it finally ceased in 1962, about four years after it began and three years after severe environmental damage was reported. In all, a million ha, 10% of the intended total, were treated with insecticide at a total cost of $15 million (1962 dollars).

And the fire ant? Between 1957, when spraying began, and 1962, when it ended, the fire ant increased its range from 8 million ha to about 12 million. Many of the treated areas were quickly reinvaded, and isolated populations thought to have been eradicated reappeared. Furthermore, reports began to appear that reinvasion resulted in much higher densities of colonies (Blake et al., 1959), another "preview of coming attractions."

The Third Fire Ant War: Mirex

What did the USDA and other pesticide proponents learn from their experience with heptachlor and dieldrin? Simply this: that a *different* hammer was needed, perhaps a lighter-weight one, but a hammer was still the right tool. If only one could devise a hammer that could accurately and selectively hit fire ants, everything would be fine and eradication could still be achieved. The search for that accurate hammer was the task given to the newly organized Methods Development Laboratory in Gulfport, Mississippi, and between 1960 and 1962 staff there came up with what they hailed as "the perfect pesticide," the accurate hammer that would nail fire ants and fire ants only. This miraculous hammer was a poison called mirex. A new weapon brought new hope for a final victory. In the sense that World War II was a continuation of World War I, the Third Fire Ant War was a continuation of the second. Much like the "Wunderwaffe" V-2 rocket renewed hope of German victory, a new weapon, mirex, buoyed the spirits of the USDA.

The first poisoned bait developed by the laboratory used the toxicant kepone mixed into peanut butter and dispersed in short sections of soda straws. Kepone's slow action solved two problems often encountered by poisoned baits intended for ants (Williams, 1983). First, it could be shared widely among nestmates before the poison took effect. A poison that killed foragers quickly would not be effectively transported back to the nest and shared. Second, it reduced the problem of bait-shyness, the avoidance of toxic foods after initial contact. By the time the poison took effect, avoidance no longer mattered. Two additional properties were also desirable: the toxin was not repellent when mixed with food, and it was lethal over a wide range of doses, so delivering uniformly high doses mattered less. Kepone, which will reappear as a player later in this drama, was soon replaced by mirex, which had all four desired properties. Allied Chemical Corporation had patented mirex as an insecticide in 1954. Hooker Chemical Co. (later made famous by the story of Love Canal) had patented it as a flame retardant. Here was a remarkable literary coincidence: a quencher of fire and of fire ants.

Chemically, mirex and kepone differ only slightly—where kepone contains an oxygen atom, mirex contains two chlorine atoms. Both are impervious molecular cages of 10 carbon atoms. Strictly speaking, mirex is a chlorocarbon ($C_{10}Cl_{12}$) rather than a chlorinated hydrocarbon (see Alley, 1973, for a review of the chemistry and toxicology of mirex). The bait was a solution of mirex in soybean oil, which acted as food and attractant. This poisoned oil was absorbed by ground corncobs (grits), and the result was a free-flowing, granular powder that was 85% corncob grits, 15% soy oil, and only 0.3% mirex (Lofgren et al., 1964). When development of mirex bait was announced in 1962, the same year in which Rachel Carson published *Silent Spring,* serious biological research on fire-ant biology had yet to be done. Beginning that same year, mirex replaced the discredited heptachlor and was applied by ground or air to large areas in several southern states at the rate of 11.2 kg per hectare. By 1963, the standard application rate was lowered to 2.8 kg per hectare, and in 1965 to 1.4 kg per hectare. The lowest rate deposited less than 5 g of the mirex toxicant per hectare. In a later formulation that reduced this amount still further, the corncob grits were first soaked in nontoxic oil and the toxic oil applied only as a coating on the outside. Perhaps it seemed inconceivable to USDA officials and pesticide proponents that such a small amount of toxic material could be problematic. In fact, they listed the bait's lack of toxic residue in the soil as a "disadvantage," because fire ants would be able to reinvade treated areas and treatment would have to be repeated.

Early successes with mirex rekindled the fond dream of eradicating the fire ant from North America. Once again, a coalition of large farm interests, farm bureaus, local politicians, and commissioners of agriculture went to work to convince the U.S. Congress to spend large amounts of money on the Fire Ant Wars, again through a federal-state cooperative program. Success did not come immediately or easily. The U.S. Budget Bureau pressured for a reduction in budget in 1964, but Jamie Whitten, congressman from Mississippi, chairman of the Committee on Agriculture (and thus in charge of the USDA's budget), and a stalwart, unyielding General for Eradication in the Fire Ant Wars, overruled the budget cut. His counterpart in the Senate, Spessard L. Holland of Florida, was lavishly in favor of the program, claiming that folks in his district were "screaming for relief" from the fire ant. The Government Accountability Office (GAO, Congress's watchdog agency) issued a scathing report on the fire-ant control program, accusing the Agricultural Research Service (the USDA's research arm) of ignoring scientific opinion and wasting government money. The GAO blocked funding for the program in 1965 and 1966, and it began to appear that the federal-state cooperative program might be discontinued.

Once again, the Southern Plant Board and the USDA went to work and proposed the goal of eradication to Congress. Their support came from a loose, by-now familiar coalition of large landowners, livestock and dairy organizations, farmers, truck farmers, local granges, farm bureaus, commissioners of agriculture, and local politicians courting the farm vote. Congressional hearings were stacked with commissioners of agriculture and colorful locals who trotted out a "parade of horrors" richly illustrated with photos of fire ant stings. Under this pressure, no southern member of these congressional committees had the courage to vote against the appropriations. President Lyndon Johnson cut the fire-ant program from the recommended budget, but Representative Whitten reinstated a $5 million recommendation, which was raised to $8 million by Senator Holland (Shapley, 1971b). Such favorable treatment of fire-ant appropriations did not change until Senator Holland retired in 1970 and Senator Gale McGee of the fire ant–free state of Wyoming took over his chairmanship. In 1967, however, the Agricultural Appropriations Subcommittee of the Senate ordered that money be transferred to the Agricultural Research Service's Insects Affecting Man and Animals Research Laboratory in Gainesville, Florida, to pay for large-scale tests of the feasibility of eradicating the fire ant.

The eradication-feasibility trials were to take place on three blocks of land totaling 1.2 million ha in Georgia, South Carolina, Mississippi, and Florida: 104,000 ha in the Starkville-Columbus area of Mississippi, mostly occupied by *S. richteri*; 254,000 ha in the Tampa-St. Petersburg area of Florida; and 863,000 ha centered on Savannah, Georgia. The last two were occupied by *S. invicta* (Banks et al., 1973a). Granular mirex bait was first applied in the autumn of 1967, then again the following spring and autumn. Commercial applicators under USDA supervision used aircraft—Lockheed PV-1s and PV-2s, as well as refurbished World War II B-17 bombers—that dispersed the bait in swaths 60 to 70 m wide along the flight path. Two additional contractors provided electronic guidance ensuring that adjacent swaths overlapped by about 15 m.

Six months after the last application, almost all colonies on randomly located check plots were dead, as were those in general searches, demonstrating the lethality of multiple applications of mirex. Searches one and two years later turned up small colonies up to 40 km inside the treatment boundaries, although most were within 8 km of the untreated source populations. The USDA authors conjectured that queens had flown in from the surrounding untreated areas to found these colonies, but queens do not regularly fly that far. Technical problems, they said, were surmountable, concluding that "total elimination of imported fire ants from large isolated areas may be technically feasible." The opposite conclusion was reached by an earlier National Academy of Sciences report (Mills, 1967) requested by the USDA at about the time of the eradication trials. Eradication, it said, was neither biologically nor technically feasible, nor was it justified, because only the weakest evidence indicated that fire ants did significant harm to human health, agricultural production, or land values.

The USDA went right ahead to propose the complete eradication of the fire ant from North America. USDA staff members calculated (Lofgren and Weidhaas, 1972) whether it would be theoretically possible to eradicate the fire ant from the USA and concluded (surprise!) that it was. It was, that is, if the entire then-current range of *S. invicta*, all 51 million ha of it, could be sprayed three to nine times with mirex bait over a period of 12 years, at an estimated cost of $200 million (1970 dollars). Eradication "by an effective, acceptable, and economical procedure [was] a commendable goal." Success depended primarily on availability of "a highly effective method of control that is inexpensive relative to the benefits gained by eradication." Once again, a fleet of World War II bombers was to be launched (déjà vu all over again) to distribute 213,200 metric tons of mirex bait containing over 800 metric tons of mirex toxin. Like an echo from the Second Fire Ant War, a high USDA official said, ". . . nothing can prevent the fire ant from moving westward to California unless an eradication program is carried out now" (Nordheimer, 1970). By the end of 1970, 5.7 million ha had been sprayed with mirex, as the eradication effort got under way.

This program was not the only one spraying mirex, however. Even while it was going on, mirex was being applied to huge areas in Florida, Georgia, and Mississippi by the USDA-state-local cooperative program, in spite of poor coordination and erratic state funding. In Florida alone, the Florida Department of Agriculture sprayed mirex on 7000 large farms after the owners paid a mere $0.37 per hectare. In addition, individual farmers, who could buy mirex bait at a low price, treated areas of unknown and unrecorded size. The USDA estimates that mirex was ultimately applied to over 45 million ha between 1967 and 1975. Much of this area may have been treated two or three times, so the extent of the treated area may be less, but at least 58,900 metric tons of mirex bait were nevertheless applied to the landscape during this period.

Proponents of mirex were dead wrong in their judgment that such a small amount of toxin per hectare could do no harm—several properties of mirex made even trace amounts problematic. In 1969, a scientific panel called together by the U.S. Department of Health, Education and Welfare determined that mirex was a moderate (class B) carcinogen and recommended limiting its use to situations involving human health. Other problems also surfaced—mirex was extremely persistent in the environment, was bioconcentrated and biomagnified in the food web, was very toxic to aquatic life, and was not specific to fire ants. I discuss these properties in greater detail below.

In response to these revelations, opposition to the mirex program and to eradication began to coalesce, making the years 1970 and 1971 very active ones (Nordheimer, 1970). The U.S. Department of the Interior, whose Fish and Wildlife Service had fought bloody battles against the USDA during the Second Fire Ant War, reviewed the mirex program. In 1970, Secretary Walter Hickel banned all but small-scale use of mirex on lands under his jurisdiction, citing harm to fish and other aquatic life and striking a body blow to the concept of eradication. Another blow came when a Select Study Committee on Mirex called by the governor of Florida recommended that Florida postpone its participation in the eradication effort until a list of environmental questions had been answered (Livingstone et al., 1970). The members referred to the program as "another of the endless rounds of Russian roulette with our environment." Although presidential candidate Richard Nixon had promised effective control of the fire ant during his 1968 presidential campaign, in March of 1971 his own Council on Environmental Quality reviewed the USDA's plans, recommending that the use of mirex be limited and that alternate modes of fire-ant control be sought.

In response to this storm of criticism, and to the two withdrawals from the program, the USDA stopped calling it an eradication program in 1971, referring to it thenceforth as a control program. Eradication had called for three treatments in a row, but control used only single treatments for "relief" from fire ants. In August 1970, the Environmental Defense Fund and other citizens' environmental groups filed a motion in U.S. district court in Washington, D.C., to stop the spraying (Shapley, 1971a). The new National Environmental Policy Act required that the USDA file a detailed, final environmental impact statement on the mirex program before spraying could begin, but the USDA went ahead with its spraying program in the fall of 1970 anyway, having filed only a preliminary impact statement. In court, the USDA agreed to stop spraying in November but to resume the following spring. Although the Environmental Defense Fund's motion was denied in April 1971, other such motions followed.

Another setback for mirex followed when the regulation of pesticides was transferred from the USDA to the newly formed U.S. Environmental Protection Agency (EPA). On 18 March 1971, the EPA issued a "notice of cancellation" of mirex's label (the label is essentially a use permit), citing that enough questions existed to justify a full review of the scientific foundations of the program, such as they were. These hearings did not get under way until July 1973. Meanwhile, the state support for the mirex program was faltering. By April of 1971, only Georgia and Mississippi had pledged enough funds to proceed. Other states were virtually out of the program or made deep cuts. A governor-appointed committee in Florida recommended a pause, even though the 1970 Florida legislature had already appropriated almost a million dollars to spray 7.7 million ha in 1971. Florida's Division of Health (like several similar bodies in other states) rated the fire ant below mosquitoes, sand flies, stable flies, horseflies, midges, bees, wasps, chiggers, and other mites on their list of nuisances and pests—not exactly a powerful justification for a multimillion-dollar eradication effort.

Justification of a different sort was offered by a USDA official at a meeting I attended toward the end of 1971. Describing the mirex program's plans for 1972, he said that funds were available to spray 8 million ha, and that "aerial contractors are needin' the work. Our cooperators won't be satisfied by the one-treatment status for long. They're gonna be hollerin'. . . . The program is in dire need of assistance" (personal notes, 1971 Fire Ant Workshop).

The year 1971 also finally saw a sharp increase in the research effort on fire-ant biology, much of which was funded by a one-time release by the USDA of $800,000 in cooperative-research agreements with university scientists. As a result, research on a wide range of biological topics began. Most fundamental, George Markin and his group finally published the basic outlines of the life cycle and biology of *S. invicta* in the early 1970s. At about the same time, and perhaps in response to criticism of mirex, the USDA was busy developing "improvements" to mirex baits. One of these was "microencapsulated" mirex bait, tiny droplets of poisoned soybean oil, each packaged within a thin film of gelatin or plastic (Markin and Hill, 1971; Markin et al., 1975a, b). For reasons not clear to me, National Cash Register Corporation produced these baits. Gone was the bulk of corncob grits. The capsules had a longer field life and were effective during cool weather too. The baits could be produced with as little as one-fourth as much mirex, in the hope of assuaging critics. Best of all, without the bulky grits, airplanes could spray five times as much area without reloading.

In 1972, a report by a study panel of the National Academy of Sciences said that mirex was a serious hazard to aquatic life. As a result, in 1973 the EPA prohibited the aerial application of mirex to any coastal counties, bodies of water, or heavily forested areas. In Florida, half the counties are coastal, so this ruling effectively limited the application of mirex to ground treatment on open land only. If Florida had not already abandoned the eradication effort, this prohibition would certainly have ended it.

On 4 April 1973, the EPA finally published a Notice of Intent to Cancel Mirex. The issues to be covered in the cancellation hearings were summarized by William Ruckelshaus, administrator of the EPA, in the *Federal Register* (Ruckelshaus, 1973). Using his authority under the Federal Insecticide, Fungicide and Rodenticide Act, he announced hearings to determine whether mirex presented an unreasonable hazard to the environment and to humans; whether the benefits of mirex, as used, exceeded the costs; and whether alternative control measures were available.

The promised scientific hearings began on 11 July, Judge David Harris, an administrative law judge, presiding. In the role of plaintiff was the EPA, and in the role of respondents, the USDA and Allied Chemical Corporation, the manufacturer of mirex bait. If anyone had any doubts about which the USDA served, those doubts were promptly dispelled. For two years, each side arrayed its experts and presented its case before Judge Harris, parrying and thrusting through cross-examination and redirect, piling datum upon datum. As a young assistant professor at Florida State University, I had the good fortune to play a role in this case as a result of a peculiar circumstance. Almost all of the many independent experts on pesticide chemistry and biochemistry, effects on nontarget species, movement of mirex through food chains, economic aspects, fire-ant biology, and so on either worked for the USDA or had received research funds from the USDA. I was one of the few who did not have such a conflict of interest, so the EPA signed me on as their expert on fire-ant biology. The testimony I prepared was drawn mostly from published and unpublished USDA results, as well as from my own experience. It was great fun.

The Case Against Mirex

By the time the hearings were suspended in 1975, over 100 witnesses had given testimony, creating a court record of 13,000 pages. The facts established in the case encapsulate nearly all of the controversy surrounding the fire ant and the mirex program, which I summarize here.

As an insecticide, mirex was far from perfect. Its highly symmetrical molecule is resistant to oxidative attack, so it remains in an animal's body for most of its life. Resistance to microbial degradation gives mirex a half-life in the environment of about 12 years, longer even than that of DDT, so it is one of the most environmentally stable compounds known (Hinkle, 1982; Metcalf, 1982). Eighty percent of the mirex applied to a site could be recovered unchanged from the soil after five years, and 50% of it after 10 years. Some of the breakdown products, including kepone, were themselves toxic.

Mirex is only very slightly soluble in water but quite soluble in fat. In the bodies of animals, it accumulates in their fat and resists excretion. Rats fed radioactively labeled mirex stored 82% of it in their tissues, especially in fat, in unchanged form. Such accumulation is called bioconcentration, a phenomenon already known from experience with dieldrin and DDT (Gaines, 1969; Gaines and Kimbrough, 1970). As long as an animal's diet contains mirex, mirex will continue to accumulate, reaching ever-higher concentrations in the animal's tissues, far in excess of its concentration in the food. Mirex leaves the body through processes that mobilize fat, such as egg laying in birds and lactation in mammals.

Mirex accumulates at even higher concentrations as it moves up the food chain, a process called biomagnification (already known from DDT). At each step in the food chain, predators retain most of the mirex already bioconcentrated in the fatty tissues of their prey, so even trace amounts of mirex in the environment came to reach toxic concentrations in animals near the top of the food pyramid. For example, in a mirex-treated area of Louisiana, mirex occurred at 0.01 to 0.75 parts per million (ppm) in whole animals at the bottom of the food web; 1.2 to 1.91 ppm in birds, one level up; and 74 ppm in the fat of top predators. Biomagnification was even more obvious in the aquatic food web. In an estuary, minnows, shrimp, and blue crabs carried concentrations of mirex in their bodies that were 2300- to 41,000-fold higher than its concentration in water. Mirex found its way into the bodies of a large number of fish species (Kaiser, 1974) and into the eggs of many bird species, including four species of hawks.

News flash: humans are not exempt from the laws of nature. Samples of fatty tissue from people in southeastern states contained from 0.16 to almost 6 ppm, not so different from the amounts in birds (Kutz et al., 1974). In an EPA study, a quarter to a third of southerners in eight states had mirex in their tissues; the average was 0.4 ppm and the range from a trace to 1.32 ppm. The frequencies of mirex residues were much higher in states that were heavy mirex users. Samples of food contained mirex residues, but no systematic estimates of exposure through diet were conducted. A retired USDA official went on record to say, "if [mirex] was the most dangerous thing facing us, we would be in a hell of a good shape" (Crider, 1977). Few unretired officials could afford to be so cavalier.

The risk to humans did not derive from mirex's toxicity, however, but from its carcinogenicity. Between 1969 and 1973, the National Cancer Institute carried out two high-dose feeding studies of mirex's carcinogenicity in mice and rats and found a highly significant increase in hepatomas (liver cancers), approximately equal to that in positive controls (rats fed known carcinogens). A second study of mirex confirmed the elevated levels of carcinomas and hyperplastic nodules (precancerous growths).

Although mirex's toxicity is low, that of its breakdown product, kepone, is not. Kepone is a carcinogen like mirex, but its more immediate effects became disastrously apparent some time later in Hopewell, Virginia. A plant that manufactured kepone had contaminated the James River and much of Chesapeake Bay with over 35,000 kg of kepone waste and had exposed factory workers to high levels of kepone. Many suffered severe poisoning symptoms, including weight loss and liver damage, stillbirths in women, and sterility in men, as well as tremors, memory loss, and slurred speech.

The claim that mirex posed no health threat to humans was no longer tenable. USDA officials therefore had good reason to worry when, later in 1975, kepone "in very small amounts" was found as a degradation product of mirex in the field. The discovery that it broke down into kepone under environmental conditions, however small the amounts, was the death knell for mirex. Kepone, like its mother mirex, could be expected to show up in humans in due time.

No deaths or reproductive effects were expected in birds, mammals, or fish at the rate at which mirex was applied (Alley, 1973), but at high doses, chickens, Japanese quail, and rats developed liver lesions (Davison et al., 1976). In contrast to its low toxicity to vertebrates, mirex was extraordinarily toxic to aquatic animals, especially crustaceans, even at minute concentrations (reviewed by Alley, 1973), and caused great concern for aquatic ecosystems and fisheries. Mirex entered the aquatic systems as particles readily taken up by filter feeders or from solution—adult blue crabs concentrated solutions of 0.22 parts per billion (ppb) of mirex 140-fold in their tissues (Schoor, 1974).

As if the effect on aquatic systems were not enough, mirex turned out not to be very specific on land either. Native ants disappeared from mirex-treated plots in broadleaf forests. Forests were often sprayed, even though they harbored no *S. invicta,* and the forest ant community was devastated (Lee, 1974). In open areas, mirex killed at least 14 species of the native ants, some of which probably competed with *S. invicta* (see Showler and Reagan, 1987, for a list of species and references). Mirex could in no way be considered specific to fire ants—"perfect pesticide" indeed!

Mirex was not even specific to ants. Ground beetles and rove beetles, both predators of sugarcane pests, were reduced 60 to 70% by a single application of mirex. Second applications, when fire ants were no longer present to remove the bait quickly, were devastating to populations of crickets and beetles, which fed directly on the bait (Markin et al., 1974a). Studies in sugarcane corroborated these findings. Like heptachlor (Long et al., 1958), mirex had detrimental effects on sugar yield because it reduced populations of ground beetles, rove beetles, and crickets (Reagan et al., 1972). Mirex was not a hammer that nailed only fire ants; there was never any reason to expect that it would be.

Effects of Fire Ants on Humans and Public Health

The habitat created by humans for their own use is the very habitat most congenial to fire ants, allowing them to reach high densities. Ant and human are therefore inevitably thrown together and frequently come into contact, especially when humans blunder into "ant beds." The ants surge forth from their violated nest to sting in defense of their home. Agricultural workers were considered to be particularly at risk. During the EPA hearings, various experts justified the mirex program on the grounds that fire ants could interfere with hand harvesting of citrus, cotton, strawberries, tung nuts, and pecans. No attempts to quantify such problems were made.

Although victims occasionally seek medical help, stings are rarely of much health significance. At the time of the mirex cancellation hearings, the USDA had made no serious estimates of the magnitude of the public-health problems posed by *S. invicta*. Not until the mid-1970s and later did limited surveys estimate frequencies of stings, medical attention, and costs (by extrapolation) (Clemmer and Serfling, 1975; Adams and Lofgren, 1981, 1982; Lofgren and Adams, 1982; Drees, 1995). For example, a study at Fort Stewart (Georgia) revealed that about 3% of the population per year sought medical help for fire-ant stings, at a total cost of about $5000 (1981 dollars). In a nonmilitary Georgia population of 77 families, 95 individuals were stung at least once during the year; stings were more likely among rural and young people. In 240 New Orleans households, the overall sting rate was about 30% per year; it was over 50% among children under 10 years old. Just over 4% of those stung sought medical attention. A few cases of indoor attacks have been reported. For the great majority of southerners, this ant is just a nuisance and hardly justifies a huge eradication program. During the cancellation hearings, many experts testified to the low ranking of fire ants as serious pests.

There are exceptions, however. A very small percentage of individuals suffer anaphylactic shock when stung and may even die. For such hypersensitive individuals, relying on the control of fire-ant populations for personal safety is probably ill advised. A much better strategy would be to carry an emergency kit containing injectable epinephrine.

Fire Ants and Crop Agriculture: Losses and Benefits

In the early days, the fire ant was regarded largely as a rural, agricultural problem, causing damage to many crops and livestock. Much of the evidence was anecdotal or the product of surveys of farmers (Wilson and Eads, 1949). The EPA's notice of its decision to end mirex use (Johnson, 1976) cited various experts who testified that fire ant mounds can interfere with hay harvesting; that fire ants cause injury to farm animals, pastureland, livestock, soybeans, and other crops. None of these claims was based on careful estimates of the actual economic losses caused by fire ants. Although the mirex program was justified on the basis that *S. invicta* caused large economic losses, the program was implemented in the early 1960s without much evidence that the losses existed. The horse lagged behind the cart by over a decade—the first serious attempts to quantify losses did not appear until the mid 1970s (perhaps stimulated by the mirex cancellation hearings?), and interest in the subject fizzled by the mid-1980s, after the show was over. These studies have been summarized in three reviews (Ofiara, 1983; Lofgren, 1986a;

Showler and Reagan, 1987), where the reader can also find a more exhaustive list of references. A number of papers reported direct damage to corn, citrus, okra, eggplant, cucumber, sunflower, longleaf-pine seedlings, and potatoes, but these came after the EPA's notice of decision and played no role in it. Even so, few were careful experimental studies, and few included serious estimates of economic losses, although several authors could not resist wild extrapolations.

At about the same time, reports of beneficial effects of fire ants were appearing, giving the debate a kind of good-ant, bad-ant aspect. Louisiana sugarcane was the first to weigh in. The benefits of a healthy fire-ant population had become apparent during the Second Fire Ant War, when sugar yields were lower in heptachlor-treated, fire ant–free fields. Experiments in the early 1970s showed that fire ants were by far the most important predator on sugarcane borer, reducing crop damage by this pest. In Florida sugarcane fields, cane borer is largely controlled by two species of insect parasitoids, and the impact of fire ants is less clear.

Other good-ant cases were soon added to the list as fire ants were discovered to prey on a number of agricultural pests, including velvetbean caterpillar, southern green stinkbug, tobacco-budworm eggs, pecan weevil, striped earwig, pine tip moth, soybean looper, and a number of others (listed in Sterling et al., 1979; Reagan, 1986). Some reported significant reductions of the pests; others simply reported the fact of predation. A more careful study of cotton in Texas showed that fire ants reduced the populations of boll weevil without reducing other beneficial insects. After fire ants advanced through areas of northeastern Louisiana, the lone star tick, a pest of cattle and wildlife, essentially disappeared. Ticks carry diseases such as tularemia, tick fever, Lyme disease, and Rocky Mountain spotted fever, so the fire ant invasion may even confer public-health benefits. In pastures, fire ants invade cowpies and reduce the populations of some of the pest insects that breed in them, including horn flies, stable flies, and face flies.

In the final analysis, is *S. invicta* friend or foe? No single answer is likely to be possible—it would depend on who, where, when, and what and would vary with the circumstances: geography, crop system, stage of crop development, farming practices, economic conditions, and other biotic factors. As in a ledger, one must add up all the credits and debits to obtain a meaningful balance, and this is undoubtedly a complex accounting problem. Losses in one area may be partly balanced by benefits in another area or at another time, and of course, not all effects of *S. invicta* can be assigned dollar values.

Economic Impacts

At the time of the EPA's cancellation hearings, no consensus about or even reasonable estimate of *S. invicta*'s economic impacts was available.

The situation has not improved much since then. Ofiara (1983) referred to the entire subject of fire ants in southeastern agriculture as "confusing and poorly documented," not allowing the assessment of the economic significance of the ant. The field was and still is plagued by low-quality, unreplicated, poorly designed studies and poorly chosen, small, or biased samples. Many are simply searches for evidence that fire ants are causing significant losses, rather than unbiased assessments of impacts, whereas others indulge in unjustified extrapolation across widely different geographic regions and agricultural situations. Little consideration is given to the many factors, in addition to fire ants, that affect yield and how these factors might interact with fire ants. Ofiara lamented, "limited, imperfect, biased data presently is [sic] all that is available." From such studies, no bottom line on the balance sheet can be computed or probably ever will be. The decisions of three scientific advisory committees that reviewed the subject during the 1960s and 1970s—that the ant was not economically significant in production of row crops—were therefore not surprising. A group of panels convened in 1982 came to similar conclusions and stated that the ant was primarily a nuisance and human health pest. This opinion was also voiced by the great majority of farmers in a Florida survey, which incidentally showed greater losses on mirex-treated farms than on untreated farms (Wilson, 1975).

Because direct estimation of economic impacts was so unsatisfactory, Semenov and his coworkers took an indirect approach, reasoning that, if the fire ant had a negative impact on crop production, production should have dropped after any given area was invaded (Semenov et al., 1997). Comparison of the productions of corn, cotton, sorghum, soybeans, and wheat before and after invasion by *S. invicta* (and adjustment for other variables that affect production) revealed a slight increase in cotton yield (+3%) and an equal decrease in soybean yield. Area harvested did not change significantly for any of the crops. Although this result suggests small effects of fire ants, it is still not an estimate of economic impacts. Crop yields and area might have been maintained at the cost of pesticide against fire ants, a cost that goes on the debit side of the ledger. Other costs might include changes in crop quality, equipment repair costs, and so on. No such data are available. Clearly, fire ants do not cause major losses in these crops, but whether they add to the cost of their production is still an open question. The ledger remains unbalanced.

Other Impacts of Fire Ants

Many other sins have been attributed to fire ants, some of which are regularly trotted out to head the latest parade of horrors. These include killing farm animals, causing infections of milk-cow udders, increasing the cost of fence maintenance, chewing up the silicone seals in concrete highways, damaging airport landing lights, damaging insulated cables, and my personal favorite, eating the wax seal from under a toilet in a Florida state park, causing flooding and repairs. Occasional home and nursing-home invasions have been added to the list, and a report was even published of the invasion of three motorized vehicles, all belonging to USDA fire-ant researchers (revenge?) (Collins et al., 1993). In one, the ants made themselves comfortable in a camper. In another, they nested in the organic junk that accumulates at the top of the firewall under the hood, and in the third, they nested in the moist pollution-control canister and foraged on insects caught in the radiator. Are these ants smart, or what? Should we not feel admiration, even if only grudgingly?

One interesting association was not described until 1989 (and was therefore not a factor in the mirex hearings): fire ants are attracted to electric fields and therefore sometimes accumulate in outdoor electrical equipment, such as traffic-light relays and air conditioners, causing malfunctions, especially in Texas (MacKay, 1988; MacKay et al., 1990, 1992a, b; Vander Meer et al., 2002). Bill MacKay (with a coauthor appropriately named Sparks, among others) carried out experiments revealing that the relays were attractive only when powered, that both AC and DC fields were attractive, and that attraction increased with field voltage. Contact with the conducting surface was required. For this reason, ants seem to accumulate only in equipment with bare contacts, junctions, sockets, or plugs. The attraction is not the result of electromagnetic fields, ozone, or the type of insulation. The attraction is probably general to ants, but the reason for it is unknown, although the release of various exocrine secretions upon shock has been suggested.

The Cost-Benefit Ratio of the Mirex Program

As long as the mirex program's goal was eradication, the real effectiveness of mirex bait was an important issue. Because the USDA claimed almost 100% effectiveness of mirex, making eradication seem feasible, the EPA scrutinized these issues carefully. In reality, individual treatments often resulted in kill rates well short of 100%, for reasons that were not well understood. For example, on plots treated by air between 1969 and 1972, survival rates of 10 to 50% were fairly common. In the Tampa and Starkville eradication trials, control was adequate on only half the plots, allowing rapid repopulation by survivors. Even colonies treated with mirex emitted healthy, colony-founding sexuals for up to three months after treatment (Morrill and Bass, 1976).

Several reasons are possible for this lower-than-advertised effectiveness. First, uneven application was a persistent problem caused by poor equipment calibration, errors in flight-path alignment, and wind drift. Long strips were sometimes left with no bait at all, and colonies there survived. Second, colonies show idiosyncratic preferences for different food types (Glunn et al., 1981), and some may not have foraged on the oily baits. Third, well-fed colonies often do not take bait or do not distribute it as effectively within the colony. Fourth, rain or high temperatures may have reduced foraging or made the bait less lethal through weathering. Fifth, even under the best conditions, survival of small numbers of colonies was to be expected simply as a result of the statistics of toxicology. Sixth, occasional survival of mirex-poisoned colonies could not be ruled out because little was known about mirex's mode of toxicity or how it moved through the colony. USDA scientists assumed that workers shared the poisoned oil directly with the queen, but the little evidence that was available suggested that the queen was insulated from the direct effects of mirex because she was fed mostly on glandular secretions. In my laboratory, large colonies that I reduced to a queen and five workers survived and grew back into full colonies.

Of course, when the USDA changed the program's official goal from eradication to control, the question of effectiveness became less critical, but the EPA was also obliged to consider the cost-benefit ratio, and in this, the mirex program did not fare well. It had been deployed without even a clear identification of where the costs and benefits resided, other than "giving people relief from fire ants," especially where the state and local governmental agencies had requested federal cooperation. The program was, in effect, a huge federal give-away, pork barrel of the classic kind in which politicians gained favor with their largely rural power base by providing free federal services. In one of the few cases where a cost-benefit ratio was calculated, Wilson (1975) showed that, in a Florida panhandle county, the government was spending about five dollars for every dollar of benefit to the farmer. Florida's "farmer treatment program" required the farmer to pay only a fraction of the actual cost of applying mirex. Farmer participation in fire-ant control programs declined sharply when they were required to pay a large share of the cost.

Alternative Control Methods

EPA also considered the issue of alternate control methods. At the time of the mirex cancellation hearings, the only other pesticide registered for use against fire ants was chlordane. Most of the dozens of products available today had not yet been registered. Oddly, even the EPA notice of decision did not mention any nonpesticide alternatives. Not until the early 1970s did USDA scientists seek potential biological-control agents in South America, and 15 more years elapsed before serious development work on these agents began. The allure of pesticides faded slowly. This allure was not limited to fire ants, nor has it faded much in the intervening years. Pest control remains dominated by chemicals to this day, and millions of tons of pesticides are dispensed every year.

Sundown for Mirex

In 1975, on the basis of information that came to light during the hearings, the EPA restricted aerial applications of mirex to one per year and placed other restrictions on its use. The USDA announced that, as of 30 June 1975, it would no longer fund the fire-ant program (states and private concerns were free to continue) and blamed it all on the EPA, beginning one of those petulant exchanges between government officials that occasionally make reading newspapers so rewarding (Anonymous, 1975a, b; Brody, 1975). Earl Butz, the Secretary of Agriculture, laid the blame for the fire ant program's failure squarely at the EPA's feet. The EPA's restrictions on the use of mirex had made the program "completely unworkable." Without fussy, wrong-headed interference from the EPA, the USDA could have eradicated the fire ant "with negligible effects on the environment." Instead, the department had now been "forced . . . to distribute persistent pesticides into the environment indefinitely." Butz ended with dire predictions of how the ant, loosed from its fetters, would now wreak havoc on a helpless South.

John Quarles, deputy administrator of the EPA, shot back that he was surprised that the USDA had suddenly chosen to eradicate rather than to control fire ants. He went on to point out that the USDA had abandoned the concept of eradication well before EPA had restricted the use of mirex. Twisting the knife, he also reminded the USDA that between 1962 and 1972, when use of mirex was unrestricted, the fire ant had been neither contained nor eradicated. The most memorable quote of the year came from E. O. Wilson, by then on the faculty of Harvard University. Taking note of the $148 million already spent on the fire-ant program, and its complete failure to halt the spread of the ant, Wilson pronounced the program "the Vietnam of entomology."

Butz's salvo was probably intended to shift the blame for the failed program away from the USDA at a time when the administration of President Gerald Ford was looking for ways to cut spending during a recession. It was certainly not the end of the Fire Ant Wars.

The mirex hearings were put on hold in March 1975, but negotiations between the EPA on one side and the USDA and Allied Chemical Corporation on the other dragged on. Restrictions on the use of mirex must have shrunk demand considerably. In 1976, Allied Chemical got tired of all the fuss, or perhaps simply saw the inevitable end of mirex, which was, after all, one of their minor products, and picked up their marbles and went home. Henceforth, they announced, they would no longer formulate mirex bait at the Aberdeen, Mississippi, plant. The state of Mississippi experienced severe withdrawal symptoms and bought the plant for $1.00. Using ingredients left behind by Allied Chemical, they continued formulating baits for sale to other states at $528 per metric ton (Crider, 1977). Negotiations continued but at this point between the state of Mississippi's Fire Ant Authority and the EPA, who finally settled out of court. On 29 December 1976, the EPA published its decision in the *Federal Register* (Johnson, 1976). Mirex was to be phased out over a period of 18 months, and the Mississippi authority agreed not to appeal this cancellation order. The hearings were to be suspended, and the EPA was to publish its reasons for cancellation. As a result of this agreement, no official finding was ever issued by Judge Harris, but of course the 13,000-page record of the hearings contains all the information advanced by both sides.

Meanwhile, back in Mississippi, the money from mirex sales financed research into alternate formulations of mirex. After running out of mirex, the plant shut down for five months but resumed operations after buying 11,340 kg of technical mirex from Hooker Chemical in New York. The Mississippi Commissioner of Agriculture, Jim Buck Ross, exulted, "We are back in business." He planned to make enough mirex bait to cover 7.3 million ha before the 31 December 1977 deadline prohibited aerial application.

The year 1978 was not a good one for the mirex interests. The National Cancer Institute concluded that kepone was also a carcinogen in laboratory rats and mice, and unfortunately, samples of mirex bait from the Mississippi production plant were found to contain up to 2.6 ppm of kepone, whose label had been cancelled in May. Could things get worse? You bet they could! The EPA found mirex in mothers' milk (that's *human* mothers!). Psychologically, that was hard to top. Finally, the out-of-court settlement between the EPA and the state of Mississippi took effect. As of 1 January 1978, mirex was no longer legal for aerial application, although mound and ground broadcast use was permitted through 30 June 1978. Thereafter, no uses whatsoever of mirex would be allowed. It had ended with a whimper. Adding injury to insult, the U.S. Office of Management and Budget reduced the annual appropriation for the fire ant program from $9 million to $900,000, saying it was "like pouring money down a rat hole." Congress appropriated about $1 million anyway for "methods development" and quarantine in fiscal year 1978 and $4.5 million for fire-ant control.

Ferriamicide: Mirex Tries to Rise from the Ashes

Meanwhile, chemists at the Mississippi State Chemical Laboratory were busy working behind the scenes to rescue their beloved mirex. The lesson they had learned in the previous decade was not that a hammer was the wrong tool but simply that it needed a bit of tweaking. If the persistence of mirex was the problem, why then they would make it less persistent! By adding ferrous chloride and an organic base to the mirex formulation, and reducing the amount of mirex, they claimed to have made mirex that was quickly degraded by exposure to sunlight (Alley, 1982). They called it ferriamicide; cynics immediately dubbed it "Son of

Table 5.1. ▶ History of ferriamicide, the "degradable form" of mirex.

Early 1977: Ferriamicide is announced. The manufacturer (the state of Mississippi) claims it has reduced the half-life of mirex to 0.15 year.
Late 1977: EPA grants an experimental use permit. Environmental groups claim that EPA has yielded to political pressure.
31 December 1977: Aerial application of mirex becomes illegal; Mississippi applies for an emergency exemption for aerial application of ferriamicide. The Environmental Defense Fund opposes the permit because ferriamicide is based on mirex. EPA receives 12,000 letters and 15 calls from southern congressmen. The Federal Insecticide, Fungicide and Rodenticide Act is in revision; the conference committee includes several southern members.
March 1978: EPA grants 1-year "emergency" use permit for Mississippi, Alabama, and Georgia. Many restrictions apply: it can be packaged in one-pound bags only, homeowners must own >0.41 ha or have >50 fire ant colonies and must wear rubber gloves. Enforcement is infeasible. EPA admits not having the data to justify the permit.
Early 1979: Samples 54 days old contain up to 22 ppm kepone (a banned pesticide); mirex used for formulating bait contains 3 to 6 *percent* kepone. Kepone is shown to be a degradation product in the field. The Environmental Defense Fund legally challenges the permit.
February 1979: EPA "discovers" year-old Canadian studies (Smith, 1979) demonstrating the toxicity of ferriamicide; ferriamicide breakdown products are more toxic than mirex. Environmentalists claim this late discovery as more evidence that the EPA yielded to political pressure. The finding is contested by EPA and Mississippi scientists (Alley, 1982).
September 1982: EPA grants a 1-year emergency exemption for Mississippi, Texas, and Arkansas. The Audubon Society, Environmental Defense Fund, Sierra Club, and National Wildlife Federation sue EPA, arguing that no emergency exists.
19 October 1982: District court issues a restraining order on sale of ferriamicide and reprimands EPA for not following legal permitting procedures (Marshall, 1982). The court voices the opinion that the distinction between the banned mirex and ferriamicide is without merit.
1983: EPA finds that 40% of the original mirex and 94% of the toxicant remains after three years. Ferriamicide is finished. Congressman George Brown of California accuses EPA of giving in to political pressure to "make a mockery of scientific principles and common sense" (an astonishing statement from a politician!).

Mirex," after the serial killer Son of Sam, who was very much in the news at the time. Mississippi hoped that ferriamicide would overcome the objections to mirex and soldiered on for six years, trying to make its pesticide business pay, but before long, it looked as though the dollar they had paid for the Allied Chemical plant had been a bad investment—damning facts popped up like mushrooms, and legal challenges multiplied. Table 5.1 chronicles the history of the ill-fated ferriamicide.

While the ferriamicide battle was going on, the pro-mirex faction made one more attempt to resurrect mirex. A congressman from Georgia attached an amendment to the Federal Insecticide, Fungicide and

Rodenticide Act that would lift the ban on aerial spraying of mirex for two years. Napoleon was to return from Elba. Several congressmen from non–fire ant states were most unkind during debate. The House resoundingly rejected the amendment by 224 to 167 in late November of 1979. Mirex had met its Waterloo.

The sun had set on mirex and was also setting on its kin. Did the failure of the mirex program discourage the USDA from large-scale chemical control projects? Do you need to ask? By 1982, USDA scientists had laboriously screened over 7500 chemicals as potential fire-ant poisons (Battenfield, 1982; Williams, 1983), but only one was effective enough to be a possible substitute for mirex. It was American Cyanamid's AC 217,300, trade-named Amdro. In 1979, the USDA tested an Amdro-based bait under an experimental-use permit. Like mirex, it was a slow-acting stomach poison consisting of 1% of the toxicant dissolved in soy oil and absorbed by puffed corn grits. The recommended dosage applied 10–15 g of the toxicant per hectare. It was considerably less effective than mirex bait, having average kill rates of 80 to 90%. On the other hand, it degraded rapidly in sunlight, left no toxic residues, was biodegradable by microorganisms, and was not biomagnified in the food chain. Once the USDA applied for a permanent registration, the EPA was under tremendous political pressure, including threats to its budget, to grant the registration, because mirex had been cancelled and ferriamicide was mortally wounded. Amdro was granted a conditional registration in August 1980, in record time. It could not be used on cropland, and could be aerially applied only on pastureland. Later entries in the fire ant–bait sweepstakes received no such rapid service, plodding through all the required steps over several years. Amdro was made available to the public at $125 for a twenty-five-pound bag. Questioned about the high price, an American Cyanamid spokesman remarked that, for the owner of a 0.41-ha lot, the bag was a 25-year supply (Bonner, 1980). Clearly, he did not expect Amdro to eliminate fire ants.

Scattered Guerilla Resistance

After Amdro, a crowd of chemical companies brought a long list of baits, fumigants, and mound drenches to market. The Fire Ant Wars may have been lost, but chemical manufacturers clearly perceived the retail market for weapons in the endless rearguard guerilla war that was to follow, and like arms dealers, they were ready to cash in on this lucrative business. Nowadays, walk into almost any garden-supply, hardware, building-supply, or nursery store in the Southeast, and you will face large displays of fire-ant poisons: stacks of orange bags, each with the image of a vicious-looking fire ant in full attack; pyramids of green shaker cans that dispense poison dust, promising to "kill the entire colony AND the queen"; and rows of red-and-white plastic bottles filled with toxic bait, ready to be offered as a final repast. Drenches, dusts, solutions, granular baits, aerosols, and fumigants all promise the satisfaction of a quick and complete kill. Death is promised as a result of stomach poisons, nerve poisons, suffocation, gassing, or drowning. Some products do not deliver death directly but instead interfere with larval development or egg laying by the queen so that the colony gradually dwindles away. The products have names like Affirm, Spectracide, Amdro, Logic, ProDrone, Exxant, Fire Ant Killer, and so on (Collins, 1992). Expect soon to see labels reading MegaDeath, Scorched Earth, Biocide, and the like. Every fall, as fire-ant colonies grow larger and more obvious, so do the displays. Sales are steady. It is big business—$50 million worth per year, $32 million of it spent by homeowners. The 1980s also saw a subtle change in attitudes, however, that was correlated with the suburbanization of the South and the movement of population to the Sunbelt. Killing fire ants was no longer carried out by large government programs serving mostly agricultural interests but became a pastime for individual, largely suburban southerners. In the scientific, extension, and regulatory communities, the wind had shifted too. Talk about eradication grew quieter and rarer, and words of accommodation more frequent. People were accepting that, as much as they hated it, *S. invicta* was here to stay, and they needed to learn to live with it.

Why Did the Wars Start at All?

Why did the Fire Ant Wars begin, and why did they rage for so long? They were a cultural phenomenon, but both sides needed to make the science work for them, because in these days you look like a fool if you support an issue that is clearly contradicted by science. The original choice of position, however, was based not on science but on the deeply held values and beliefs, often below the level of consciousness, that are rooted in history, family, culture, and *Weltanschauung* and that give emotional comfort. We can reduce the opponents in the fire ant wars to a simple cartoon. On the one hand are those people whose attitudes are fairly described by the metaphor in Genesis 1:26–28, in which God grants dominion over the earth to mankind and admonishes mankind to replenish it. At the other end of the spectrum are those who see that humans depend on healthy ecosystems and are part of them, both functionally and ethically.

The first point of view is broadly represented in our society, quite independently of religious beliefs, and divides the world into humans and directly useful animals and plants (us) and the rest (them). "The rest" is of little interest or value if a harmful action is "good for people." Here is a typical statement: "It is believed that the hazards evident from toxicological results [for ferriamicide] are mitigated by the low exposure expected *for humans*" (emphasis mine) (Alley, 1982). One USDA bulletin phrased it clearly: "Since before the dawn of civilization, insects . . . have wanted to occupy places chosen by man." Even the EPA's statement of decision on mirex (Johnson, 1976) devotes just nine lines of its nine small-type pages to "nontarget insects." Throughout the Fire Ant Wars, toxic baits were used without testing *even of their effects on other ants.* Spraying the entire Southeast was proposed in the absence of any knowledge of ecological side effects, and millions of hectares were poisoned with dieldrin and heptachlor.

Those at the other end of the spectrum have powerful allies in the ecological and evolutionary sciences, which provide their basic metaphor. If individuals of this persuasion read Genesis 1, they would take note that it is a message of both dominion and (in the word "replenish") stewardship.

Of course more mundane forces were at work, too. Many of the people involved in the development and marketing of mirex belonged to a generation that had recently won the most deadly war in history, largely through technology and a can-do attitude. Faith in technology was at an all-time high. Perhaps the "Us versus Them" thinking of the Cold War era spilled over into the entomological realm. After all, the official USDA common name was the *red* imported fire ant. Significantly, the war metaphor permeated the entire era of the Fire Ant Wars and must have conjured up powerful and motivating images in a generation that had experienced war. Words of war pepper news reports, the speeches of politicians, and the proclamations of USDA officials. The vocabulary included dangerous foreign invaders, beachheads, battles, attacks, advances, and retreats. There were body counts, eradications, new *Wunderwaffen,* and civilians to be rescued or caught in the cross fire. Even the equipment spoke of war—World War II bombers rescued from the scrap heap and modified to rain death on a different enemy.

But, of course, it was not a war. It was a massive effort to manipulate ecology and *much* more complicated than a war. To have any hope of succeeding, one needed a great deal of ecological knowledge, humility, and concern for nonhuman values. Victory required subtlety, sensitivity, and wisdom, not aggression and frontal attack. The generally entomophobic nature of the American public did not help bring the proper perspective to this undertaking, but ultimately, the entire fire-ant program and the public support it enjoyed were based on what one participant in a 1982 fire-ant symposium called "a pathology of the American perception of reality." The news media were responsible for much of the hysteria. Rather than inform, most news reports sensationalized, propagandized, incited, and misinformed, repeating the same litany of unsupported "factoids" over and over. Having no schooling in biology, reporters were sitting ducks for whatever wild stories came along. I spent many an interview with reporters trying to bring some balance, knowledge, and skepticism to their thinking. E. O. Wilson once quipped that I was "trying to get a fair deal for fire ants."

No doubt the balkanization of science also contributed. As a result of specialization, chemists can invent chemicals that not only are incredibly toxic to insects but have disastrous effects on ecosystems, and concern for ecosystems is not in the chemist's job description. For mirex, the lag between introduction and the recognition of undesirable side effects also played a role. It produced no pathetic windrows of dead animals as heptachlor had. Once the dangers of mirex became apparent, a great deal of denial, dismissal, and general resistance followed. Professional people tend to be defensive about their decisions. Some exhibited just plain cussedness, a kind of suicidal joy in sticking to contrary beliefs. How, otherwise, can we explain why a chemical-company executive ate a DDT-sprinkled bowl of breakfast cereal on television? Even some scientists carried this denial through the years, long after persistent pesticides had fallen out of favor. I often suspect that cultures change only as the old guard dies off.

A large measure of the dubious "credit" must also go to agricultural commissioners in the still largely rural South, who were typically elected to office. To be elected and reelected, these commissioners had to deliver noticeable goods and services, including getting rid of fire ants. So much the better if the services came for free from our nation's capital. "Forget the ant," one official told a *Washington Post* reporter (14 February 1978). "The key is the money. It lets these guys launch fleets of spray planes and get up at the Fourth of July picnics and tell everyone that without us to get rid of those ants you would not be here." Decisions that should have involved a lot of science (though not just science) were made instead by politicians responding to the horror stories presented at agricultural commission meetings, state houses, or the halls of Congress.

Thus the drama unfolded, a complex brew of science, politics, journalistic hyperbole, public hysteria, and legal maneuvering. It roiled and boiled for more than two decades, and when it finally calmed to a simmer, the fire ant was the clear victor. The moral of this tale, if it has one, might be, "Know thine enemy."

Overview of Range Expansion from 1918 to 1995

Before 1958, estimation of the range of the fire ant was somewhat hit or miss. The best range estimates are probably those of Wilson (Wilson, 1951; Wilson and Brown, 1958) and Culpepper (1953). Plotting their data shows that *S. richteri* was expanding its range geometrically and that the pace of expansion increased after about 1940, when *S. invicta* took over. Between 1930 and 1949, the range increase was almost precisely geometric and best explained by constant radial growth outward from Mobile at a rate of 8–12 km per year (area $= \pi r^2$; doubling the radius of a circle increases its area fourfold). In 1953 the estimated range stood just short of 50,000 km^2 (Figure 5.5).

Figure 5.5. The range expansion of exotic *Solenopsis* fire ants in the USA before 1954 (data from Wilson, 1951, and Wilson and Brown, 1958).

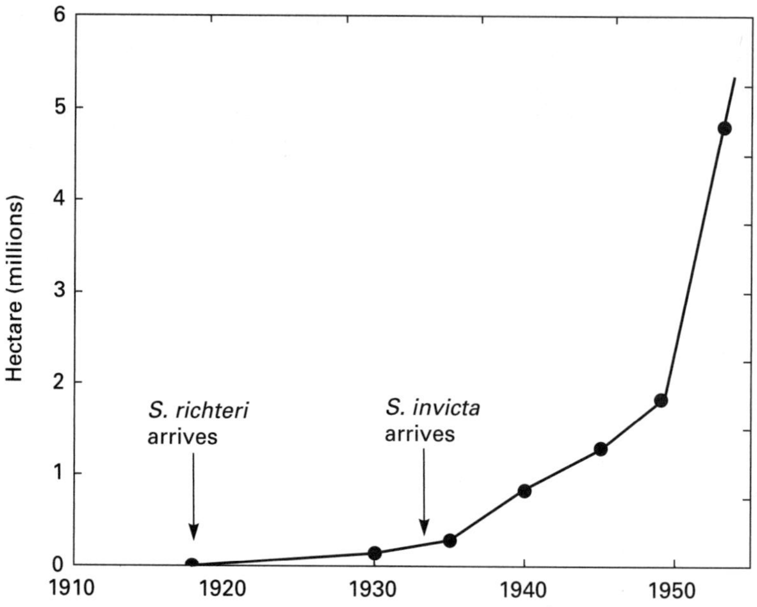

Once a quarantine was put on fire ant–infested counties in 1958, annual physical surveys were needed to determine which counties were within the range of *S. invicta* and which ones became newly occupied. Upon detection of fire ants, the county, or sometimes part of the county, was declared "infested" and placed under quarantine. Most range estimates since 1958 have used these data, although not always in the same way (George, 1958; Wilson and Brown, 1958; Adkins, 1970; Buren et al., 1974). Simply summing the total area of "infested counties or parts of counties" gives the year-by-year range of *S. invicta* in "infested acres." This measure is somewhat misleading because fire ants never occupy the total area of any county, being absent from areas of water and dense forest. The estimates are thus best considered simply to reflect the geographic range of the fire ant and to convey little or no information on the population densities within this range.

In 1996, Callcott and Collins cumulated all the data to estimate the ranges of fire ants from 1918 to 1995 (Callcott and Collins, 1996). They did not separate the two species in their analysis, but of course, after about 1940, most of the increase was of *S. invicta*. The progress of fire-ant range expansion (Figure 5.6) makes interesting viewing. The 5 million ha occupied by the ants in 1953 barely show up on the scale of the range of 112 million ha in 1995. As in the growth of a bacterial population, the period before about 1950 constitutes a lag phase, and the period after 1950 a rapid-growth phase. Twenty years (1955 to 1975) of rapid, approximately linear expansion of 3.4 million ha per year were followed by slower expansion between 1975 and 1995. By 1975, the colder climate to the north limited the ant's invasion, closing an entire frontier and slowing the expansion rate to about 1.5 million ha per year. Since about 1980 to 1985, nearly all of the range expansion has occurred along the western frontier, although a small expansion continued northward into the mild lowlands of Virginia, Maryland, and Delaware. The slowing of the westward expansion may be the result of aridity, but we cannot be sure. This question will resurface in the discussion of potential future range, below. In 1982, *S. invicta* was discovered in Puerto Rico (Buren, 1982). In 1986, it appeared in Mesa, Arizona, and was eradicated (Frank, 1988), but in 1988, it popped up again in Phoenix and reached St. Croix, U.S. Virgin Islands. By 2000 it had leapfrogged along the West Indies, taking the Turks and Caicos, Antigua, and Trinidad, just a stone's throw from South America's shores once again (Davis et al., 2001). As of 2005, it was invading El Paso, Texas (MacKay and Fagerlund, 1997), and had established populations in Oklahoma City, Oklahoma, and Lubbock, Texas. It has been discovered in Albuquerque, New Mexico. Although it appeared in Brownsville, Texas, in 1991, it honored the international border and had not crossed into Mexico 15 years later. In 1988, an infestation in a nursery in Santa Barbara, California, was eradicated, but 10 years later a large population was discovered in southern California, kicking off the Western Fire Ant War, which continued into the 21st century.

The status of the North American population in about 2005 is shown in Figure 5.7. In 2001, *S. invicta* was discovered in Brisbane, Australia, and Auckland, New Zealand. (Confusion surrounds *S. invicta*'s approved common name, the red imported fire ant, RIFA. Should it be

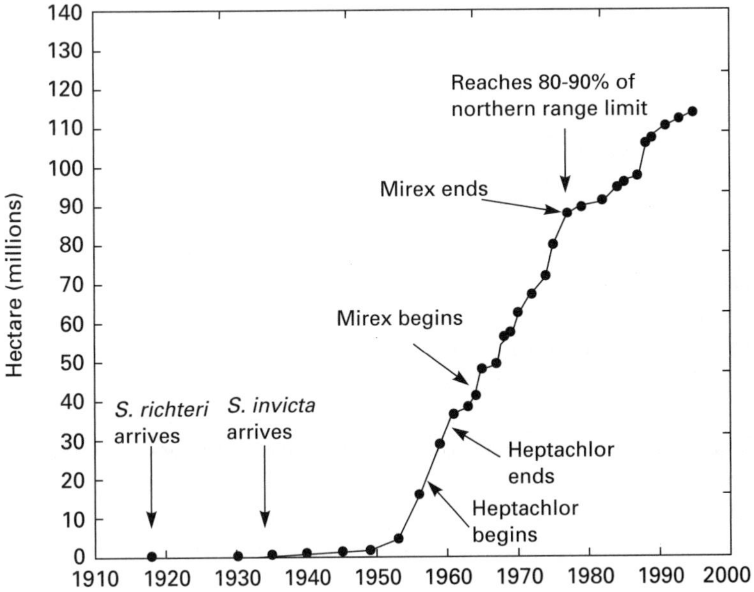

Figure 5.6. The range increases of exotic *Solenopsis* fire ants, from 1918 to 1995. Arrows indicate the points at which use of the pesticides heptachlor and mirex began and ended. Adapted from Callcott and Collins (1996). Data for 1918 to 1953 from Wilson (1951) and Wilson and Brown (1958).

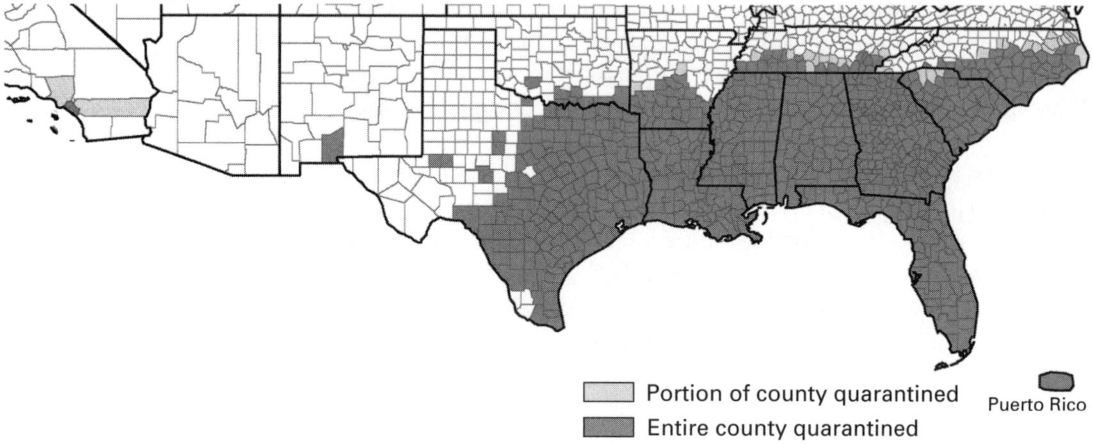

Figure 5.7. Range of *Solenopsis invicta* in May 2000, based on U.S. Department of Agriculture quarantine map of counties.

called the red *exported* fire ant, REFA, in South America? The arrival of *S. invicta* from the USA in Australia and New Zealand deepens the confusion. Should it now be called the red imported exported fire ant, RIEFA? Or the red imported^2 fire ant, RIIFA? But I digress . . .). In early 2005 *S. invicta* appeared on the shores of Taiwan, causing the Chinese to worry about a Taiwanese invasion of the mainland (the reverse of the usual relationship). Indeed, only months later, the ant was discovered in southern China. The source of all these migrants is almost certainly the huge North American population rather than the original South American homeland. As world commerce surges upward, *S. invicta* will find ever more chances to hitch rides to foreign places, repeatedly demonstrating its prowess as a traveler and invader. Thanks to humans, its heyday is far from over.

Comparing human efforts to eradicate the fire ant with the ant's range-growth curve is also instructive. Neither the Second Fire Ant War, the one based on heptachlor, nor the Third Fire Ant War, the mirex debacle, had a detectable effect on the rate at which the fire ant occupied new territory. Let us be fair, though. Heptachlor was applied to only about 10% of the then-current range before public outcry stopped the program. Without doubt, its use temporarily and locally decreased the abundance of the ant (as well as those of birds, fish, mammals, pets, and so on). Between 1964 and 1975, mirex was applied three times to about 20 million ha, close to half of the ant's total 1967 range. After 65 million kg of mirex bait was spread on these lands, at a cost estimated to be as high as $200 million, the program's effect on the fire ant's spread was undetectable (Figure 5.6) (Johnson, 1976). Even at the height of the program, the rate of spread did not even slow. Again, to be fair, the program surely reduced local fire ant densities temporarily. Also, because the range was calculated from the area of quarantined counties, it would show a decrease only if the quarantine were lifted from some areas. These data thus cannot really reflect any reduction in fire-ant population densities. Nevertheless, even with population reductions, even when the program had an explicit goal of eradication, the fire ant continued to spread as if it did not even notice the mirex program. Between the beginning and the end of the large-scale mirex program, the fire ant's range almost doubled, standing at about 90 million ha in 1975.

Here then is the final score. Between 1964 and 1975, 60 million ha (600,000 km^2, an area about six times the size of Alabama, somewhat more than half the range of the fire ant in the year 2000) were treated with mirex (Lofgren, 1986b). Airplanes flew over 15 million km, dropping over 86 million kg of mirex bait containing 320 metric tons of mirex. These numbers do not include mirex applied by state programs or by private individuals through purchase or give-away programs, all undoubtedly also large numbers.

Did Mirex Aid the Spread of Fire Ants?

The final irony of the Fire Ant Wars is that, in the long run, mirex and its predecessors and descendents probably had the opposite of the intended effect. Rather than eliminating *S. invicta,* it actually helped this invader gain a foothold in southeastern ecosystems faster and more firmly. A number of authors had taken note of the apparent increase in fire ant abundance after recolonization of mirex-treated areas. The first such reports went largely unheeded—Summerlin and his coworkers (Summerlin et al., 1977) used mirex to kill a diverse population of native ants and a few *S. invicta* in a Texas pasture. When ants recolonized the pasture, *S. invicta* had changed from a minor player to the dominant ant. Using its legendary colonizing abilities, *S. invicta* had simply beaten all the previous native ant inhabitants in reoccupying the formicine vacuum created by mirex. Unfortunately, the study lacked a control and was not replicated, as well as being on the spreading frontier of the fire ant population.

No matter. The question was specifically tested by Jerry Stimac, Lois Wood, and Bill Buren (Stimac and Alves, 1994), who treated some plots with mirex and some with Amdro and left some as untreated controls. Initially, all plots contained diverse populations of native ant species, including *S. geminata,* and *S. invicta* was only a minor presence. As expected, the ant population of the treated plots dropped sharply, but more importantly, when the plots were recolonized, all the treated plots were strongly dominated by *S. invicta,* which remained a minor component of a stable ant fauna in the control plots. Clearly, killing a diverse ant fauna with indiscriminate formicides created an empty habitat waiting for the fastest colonizer to take the lion's share, and that lion was *S. invicta.*

On the basis of this and other information, Bill Buren argued energetically in letters (pers. comm.) and print that large-scale pesticide programs were not defensible, because in the long run, they would produce higher populations of *S. invicta* through resurgence and the elimination of native ant competitors. He argued that repeated treatment with indiscriminate poisoned baits would create a situation in which *S. invicta* was the only ant present. It would be difficult for a community to recover from this situation, because inocula of native ants would be absent.

The widespread use of mirex over millions of hectares that were home to native ants, but few *S. invicta*, therefore probably sped the rate of invasion and increased the subsequent dominance by *S. invicta*. During the EPA hearings, a top USDA scientist testified that 29% of the 3.25 million ha treated in spring 1972 "contained almost no [fire] ants, and should not have been treated." Often, entire counties were treated in spite of inadequate pretreatment surveys establishing *S. invicta's* presence. Frequently, USDA workers were unable to find areas with enough fire ants in which to establish pretreatment plots so that they could later estimate the effect of the mirex treatment. At other times, areas known to harbor very few fire ants were treated in a kind of "holding action" to "prevent their spread." The unnecessary treatments probably killed many of the native ants and thereby left the areas more susceptible to invasion by fire ants.

Is this conclusion reasonable? I believe it is, because this situation is exactly parallel to that created in the studies by Stimac and Summerlin. This classic case of unintended consequences could have been anticipated with a little ecological research. The Florida official who described the purpose of his state's "farmer treatment program" as being "retarding, but by no means halting, the spread of this pest" certainly had no idea that his efforts were probably having the opposite effect. Perhaps in an era when pesticides were the answer to all insect problems, no one was very much concerned with taking a properly scientific approach.

Another Immigrant Moves West

From my airplane window, I study the deserts of west Texas, brown and tan below, dissected by shallow channels that carry water perhaps once a century, hummocky dunes piled willy-nilly by the shifting winds. To the north rise the Guadalupe Mountains, the bones of an ancient barrier reef, now capped by dark woodlands. Ocean and desert side by side, separated by vast stretches of time. To an animal needing abundant moisture, the arid land below is as much a barrier as the ocean is to land animals, and like the mirror image of watery oceans, these dry oceans contain patches and ribbons of moist habitat that are islands to moisture-loving creatures. Some of these island creatures are refugees from moister times, marooned when the waters receded, but the islands and river corridors have more recently allowed other creatures to hopscotch across the American West. The human pioneers that emigrated to the West across the great American deserts used the same strategy, following river corridors and hopscotching from island to island of moisture.

Now, summoned by the California Department of Food and Agriculture (CDFA), I am on my way to see the latest immigrant to California, the fire ant *Solenopsis invicta*. Like many thousand human immigrants from the east, it has settled in Orange County on the outskirts of Los Angeles. The fire ant is a creature of the humid subtropics. It did not make its way across the arid west without help from friends in high places. No, almost all of its spread in the United States has been achieved with help from the fire ant's best friend, mankind. Here is yet another example. Under the best of conditions, a fire-ant queen can fly up to 11 or 12 km, but in the arid plains below me, that would plunk her down right in the middle of some desiccated, creosote bush–covered gravel flat where she would curl up and cook in the sun in a few minutes, and even if she could find a hiding place, she would die of thirst in a few hours.

The fire ants that migrate across those hostile expanses use the same vehicles that humans do. The most frequent means of transport for emigrating fire ants are products containing soil, such as trucks full of sod or earth-moving machines whose owners are casual about cleanliness. Potted plants from nurseries are especially popular. Fire ants in a potted plant, say, a bromeliad, palm tree, or rooted juniper cutting, whether a solitary newly mated queen or a small colony, are traveling in comfort and style. It's like moving to California by putting your entire lot, house and all, on a truck and watching TV and having parties all the way.

But almost anything that moves can provide a ride. According to the folks at the CDFA's border inspection stations, the list includes empty trailers, the chassis of family cars, boxes of ceramic tiles, bundles of broom straw, loads of shingles, lumber by the truckful, shipments of watermelons, houseplants in travel trailers, cases of Jack Daniel's whiskey, used car parts, shipments of windows or prison beds or used tires or hay or paper towels or plumbing supplies. Fire ants can travel in truckloads of hush puppies, Heinz baked beans, frozen orange juice, pork, or school desks; in trailers full of furniture or horses or Girl Scout cookies; in video cabinets or loads of carpet, cases of beer or salad dressing; they snuggle among bags of potato chips or nestle in rolls of paper towels, hide between bricks, rest among boxes of chicken parts or cartons of books, among cans of Campbell's soup or frozen fish; they ride with fencing material or loads of Tide detergent or rice; bathroom fixtures, peanut butter, printing paper, wire, cases of hot sauce or cooking oil, among hatching chicken eggs, in potted pecan trees, among cigarettes, between potatoes or tomatoes, in boxed coffee cans, in clothing or yarn, between concrete slabs, with transformers, in empty pickup trucks or freezers, among bananas and yams, in shipments of baked goods or office supplies or corn on the cob or V-8 juice; in vehicles carrying crude synthetic rubber, jalapeño peppers, roofing materials, Gatorade, plastics, plastic bags, frozen french fries, water valves, picante sauce, catfish, sweet potatoes, mesquite wood chips for barbecues, onions, pork and beans, citric acid, school chairs, logs, lawnmowers, telephone parts, soy sauce, aluminum sheets, computer forms; in Coca Cola trucks, within fireplace inserts, with green coffee beans waiting to be roasted, between stacked planks, with fiberglass, firewood, and fertilizer. Even the military lends a hand by letting fire ants ride among tank parts.

Wherever these vagrants end up, their prospects of establishing a new home are best where the winters are mild, as in Arizona and California. They probably arrive by the thousands every year in Arizona, California, Oklahoma, Nevada, and New Mexico. Of course, people have caught on to this fire-ant express. The movement of nursery products out of fire-ant areas requires the shipper to take a number of precautions, but they sometimes comply carelessly or not at all. Although many of the travelers are intercepted by agricultural inspectors at the borders, human barriers to entry are imperfect. Products are not inspected, sequestered ants are missed, dishonest people deliberately bypass inspection, or the imagination the ant applies to hiding is greater than that the inspector applies to seeking. The system is leaky.

At the other end, the receiving states keep an especially close eye on nurseries. As a result, just as fire ants were gaining a foothold in a nursery in Santa Barbara, they were discovered. The beehive trick almost worked. Square miles of almond orchards are pollinated every year by bees brought from Texas. The pallets on which the hives are trucked support fire ants as well as beehives, but the invasion of the orchard was discovered and contained. Now beehive pallets are carefully inspected at the California borders.

Almost two hours pass at 600 miles per hour before the arid zones are east of us. White Sands, the Organ Mountains, the deserts of Arizona, the heat-kill zone of the lower Colorado River valley, the red/tan/maroon/white basins of the Mojave Desert all slide by below. Finally, outside the airplane window, the San Bernardino Mountains come into view, smog, the poison breath of Los Angeles, billowing through Cajon Pass into the western Mojave. We start our glide into LAX as the endless grid of LA begins to twinkle in the evening light.

Terminal 6 is an expansive hall, with ringing marble floors, an espresso bar, a pseudo-Mexican cantina with a touch of Hollywood, a newsstand, and a western-wear store full of sweatshirts bearing messages to inspire or place-names to envy. A dozen TV sets blare, most of them tuned to stations broadcasting live car chases filmed from helicopters. People by the dozen stand transfixed, frozen at tables, in walkways, one hand on a railing, as if paralyzed in mid step—lawyers, janitors leaning on mops, ladies with puffy hair and straw handbags, young business suits, surfers in baggy clothes and bleached hair. In LA, everything of interest happens on freeways.

Alamo Rent A Car puts me in something freeway-ready, and I negotiate the 6 or 8 lanes of I-405 South to Costa Mesa. The traffic is dense, endless, and fast. The Doubletree Hotel that is my destination could have been anywhere; it's not supposed to matter. Two gardeners are spraying pesticides on the shrubs outside the front entrance. A message from Aurelio is waiting for me at check-in—meet us at the bar at 6:30 p.m.

I spot David's curly mop from a distance; the smile and the kidding start as I close the gap. Also at hand are Homer, veteran of the mirex wars, taciturn and lugubrious as only a Mississippi boy can be, and Bart the modern, up-to-date extension specialist from Texas, cowboy boots and all. Together, we make up the Scientific Advisory Panel. Is the acronym deliberate or accidental? Our hosts from the CDFA make introductions all around. They are a large group, some from Sacramento, some LA. Plans have been laid for the next morning: a press conference, then a field trip to view several fire ant–infested sites. Then, over beers, the SAP will discuss, recommend, discuss, revise, and submit recommendations. What to do about the fire ant in California, according to the experts. Contain? Eradicate? Nothing? The next morning, another press conference, more discussions, then home.

The morning press conference is packed; not even standing room remains, and some latecomers spill into the hall. Reporters are there with notebooks and microphones. The introductions take a long time.

Several politicians get some free airtime. Yes, this is a serious threat and the (choose one) legislature, county commission, commissioner of agriculture, CDFA will set off in vigorous pursuit of victory. Money is discussed. Millions. The press asks questions. The SAP responds, individually. Poisoned baits are described, as is some basic fire ant biology. No, the state has not yet decided what it will do; that's why the SAP is here. Representatives of the nursery industry do their best to direct blame away from themselves. We're good citizens, they say; The ants could have gotten here any number of ways. The meeting ends and we pile into six vans to head for Trabuco Canyon, the center of the fire-ant infestation.

From I-5 we spill onto Alicia Parkway. Apartment complexes, one after the other, gas stations, strip malls, then eventually, the density of development thins as we approach the spreading edge of greater (greater?) Los Angeles. Here, the hillsides are covered with native chaparral, the scent of sage and *Artemisia* barely perceptible in the rising smog; here and there a spiky yucca or the fine, dark foliage of chamise contrast with the pale gray-green of the sage. Golden spines glow on prickly-pear pads basking in the open spots, and dark live oaks shade the boulders of the stream bottoms. We pass four hills that have been converted into a gigantic staircase, each step dead level, the debris from the decapitations pushed into the stream bottom, burying it. All slopes, all levels conform to exact engineering specifications. No trace remains of the complex living community that once called those hills home. Even the soil is gone. On the uppermost level stand three huge earth-moving machines, yellow and idle, their jobs done.

A little farther up the road, a half-finished apartment complex sprawls up the hillsides. Roofless frames are mixed with near-finished buildings, siding mostly in place; others are merely the promise of a poured slab with stud bolts. From the road edge, a carpet of green vegetation is spreading like a fungal colony—sod rolls, a dozen exotic species of shrubs and trees, ground covers, and grasses, not one of them native to California. Sprinklers dispense water imported from northern California as though it were in endless supply.

What we see here is not simply major ecological disturbance. This is total floral replacement, transported by truck. The replacement fauna will find its own way here—the roaches, pill bugs, Argentine ants, house mice, English sparrows, and all the other myriad little guests that follow in our wakes. And now, another little guest, the fire ant *S. invicta*, has joined this lucky crowd. If the fire ants here could join us in addressing the press, they would, like their relatives back east, undoubtedly name us as their greatest friends and benefactors—we transported them here, we made this world safe for them, we killed their enemies, we gave them water, and we will even feed them. Sure, now and then we get inexplicably angry with them and poison a few colonies, but such events are minor setbacks, of little consequence. If fire ants have a religion, humans must surely occupy the position of God, preparing a place for them with large, yellow earthmoving machines and truckloads of exotic sod. Their hell lies in the native chaparral, where certain death awaits in the form of desiccation or dismemberment by spiders, beetles, and native ants.

The open-air press news conference is scheduled at a "city park"—a couple of acres of vacant land squeezed between a housing development and a nursery. Piles of building debris and soil rise here and there, pieces of bent pipe protrude from among broken concrete blocks, and the annual growth of exotic Mediterranean weeds has been newly stimulated by the recent arrival of winter rains—fire-ant heaven, and several colonies have laid claims. Forty or 50 journalists, print and broadcast, swarm around the few flag-marked fire-ant colonies. A dozen vans full of millions of dollars worth of electronics wait to beam their discoveries to waiting satellites and into every bar and living room. Gaffers prowl around with fuzzy microphones, well-groomed anchorpersons look around hungrily for someone to interview. David and I hunch over a fire ant colony, opening the nest with the trowel I brought in my carry-on luggage. Heck, these sonsabitches look different, seconds before a fuzzy mike is lowered surreptitiously between us. No conversation is private today.

The ABC news crew cuts Bart from the herd and corrals him near one of the fire ant–infested piles of dirt. He explains carefully about fire ants, their biology and control. The *Los Angeles Times* reporter, a radio interviewer, and a journalist from a minor local newspaper each corner an expert. "Is this going to be bad?" they ask. Their tone suggests, Please, no complex answers. NBC has backed David against a bit of surviving fence, while I hover nearby. David lets himself get stung for the camera. His hand, holding a male fire ant, goes out over national TV. You could run for office, I kid him. The cameraman turns to me. Can we get a picture of you digging up a mound? Nobody has a shovel. My skinny trowel will do temporarily, as a CDFA employee races to a van, to return 20 minutes later with a square-nosed shovel, the shiny label declaring it to be a genuine Made-in-America True-Temper Razorback with a taper-forged shank and a genuine grown-in-America hickory handle. Good for shoveling coal or scraping pig manure off a concrete slab, but not so good for digging fire ants. We make do with it, though, through at least four more takes. The fire ants are confused and exhausted. The supply of experts is running low, and surplus newspeople are hovering for their chance to swoop in to capture a bite of sound or some sensational tidbit. One of them is interviewing a fat kid straight out of a Far Side cartoon who is leaning over the wall of the housing development. Have you been bitten? Do you know about these fire ants?

We leave the news crews to their broadcasts and voice-overs, and our six-van motorcade heads for other fire ant sites and sanitary facilities. The Copley News Service woman is right behind us. When we pull into a Jack in the Box restaurant a couple of miles down the road, she pulls in beside us and leaps out to chase after us, notebook open and ready, until she realizes we are heading for the men's room. Her hopes for a scoop are dashed. No, we don't have a secret agenda or destination—only more fire ants to look at. We intend to box the area in which they have been found, about 7 miles from end to end, perhaps 20 square miles, with clusters of colonies here and there. All are at least three or four years old, to judge by size. The distance between clusters suggests multiple introductions, not spread by mating flights. Who gave them a ride? When? We drive to the end of a winding road into the hills. Two straggling colonies hold territories at the base of palm trees in the trampled yard of a boys' reform school. We have to sign in and out. The uniformed boys march by in disciplined lines. A horse whinnies on the other side of the fence. We stand in a circle around the fire ant colonies, our hands in our pockets—fire-ant guys. We wonder how they got to this remote place.

Next stop is a county park with picnic shed and a piece of irrigated lawn bulldozed out of a gravel bar next to a dry creek. The ants are happy with their new home and have settled at regular distances from one another, like the good neighbors they are. We try an excavation, but the ground is too stony for square-nosed shovels. We end our survey in an upscale horsey development with white fences, big houses, and now fire ants too. It starts to rain, and we head back to the Doubletree.

The SAP's discussion over beers is mercifully brief. Without any preliminaries, Homer drawls, "The only hope they got of ee-radicat'n is to use airplanes and baits." The others nod in assent. A bit of limp discussion of details follows. We propose three alternatives, topped by an eradication attempt from the air, using baits, two applications a year, and monitoring. I acquiesce on the condition that native communities be strictly avoided (nothing native is left in the areas to be treated anyway). If the ant is not extinct in five years, give up, and focus on management. Business done. I'm not even halfway through my first beer.

Next morning, fire ants are already old news. Only a single reporter shows up for the press conference to find out what the SAP recommended. She has four scientists and ten state officials all to herself, but the audience gets restless and breaks up into small knots of conversation. I join one that has formed around a large, red-faced man with a bulbous nose. His name is Jack, and he is head of a nurserymen's association. Pat, who works for the CDFA, observes, You know, it's peculiar, he says, that we have had very few introductions of fire ants to California for all these years, and then all of a sudden, bang, we've got 'em all over. I wonder if anything's changed or what's causing it.

Jack leans back in his chair. Oh, I know what it is. It's the boom in tropical foliage trade. Bootleggers bring 'em in by the truckload. I ordered a load of *Dieffenbachia* not long ago from a guy in Arizona who got 'em from a grower in Hawaii. Only, when I was working with 'em, they were potted in clay, pearlite, and white sand. I looked at the label and sure enough, it said Hawaii, but if they were from Hawaii, they would have had volcanic ash, not sand, in the mixture. I called the guy who sold me the load and told him, now come straight with me, I know these are from Florida, and he admitted it. Got 'em a few bucks cheaper but had to get around the fire ant quarantine. This goes on all the time. I knew a guy got a truckload in on Sunday night, opened the truck and the fire ants came out. He 'fessed up and called the CDFA and said, you take care of it! But mostly, they arrive, get unloaded, and immediately sent out to 50 Home Depots or some other retailers. By Monday morning, they're for sale in 50 locations. The trade is worth millions. Knew one guy admitted turning over $1.5 million a year. About one truck in 50 gets inspected, and then the papers are often fake. They get 'em from Florida, unload 'em in Arizona or Nevada; a few days later, reload 'em and ship 'em to California. Origin Arizona, no quarantine. Or they take back roads, where there's no inspection stations.

(Eighteen months later, the SAP revisited Orange County for a briefing on the progress of anti–fire ant efforts. In a slack moment, I reminded Pat of the conversation about the tropical foliage trade. Was there anything to it? Did the CDFA follow up on Jack's comments? Yeah, sure, he said. We inspect retailers like Home Depot and Wal-mart regularly. Never found any ants there. We hear these kinds of "insider hints" all the time. It's a tool of competition—you sic the authorities on your competitors by dropping stories like this. No, the ant was spread through nurseries, no doubt about it. What I didn't know then was that these nurseries all share and trade stock. If one can't fill an order, he buys from a competitor, or they pay for one species with another they have plenty of, trading all over the place. So any fire ants that started in one nursery would soon be spread to dozens of others. That's how we think it happened.)

Yes, greed is the ally of many, even fire ants, but it's time to make the run to LAX to catch my flight. The freeway is dense with hurtling vehicles, even at 10:00 a.m. From the air, we look like ants following an odor trail to a dead roach. Just beyond the fusion of I-405 and I-5, I see an illuminated billboard for Reliable Mortgage Co. "Not all rates are created equal," it declares, and alternately flashes the rates for a 30-year mortgage and a long-term treasury bond. Yes, I think, that's the key here. Unequal rates. Until recently, the immigrating fire ants died faster in California than they arrived. They have been arriving in a steady trickle as long as the ant has been in the Southeast and people have traveled from there to California—border inspection stations make it clear—but they died even faster. Many were undoubtedly queenless groups of workers, incapable of establishing colonies. Others were probably complete or incipient colonies or newly mated queens. Most of them were probably simply unlucky and landed in places where they could not survive. A few may have become established but then died of natural causes or were discovered by CDFA inspectors and killed. For almost 50 years, the rate of extinction was higher than the rate of immigration, and most colonies vanished before they were detected.

In the 1990s, though, perhaps because of the tropical foliage trade or some other form of commerce, the rate of immigration has increased and now surpasses the rate of extinction. Suddenly, fire ants are popping up in multiple locations. The goddess of probability has rolled the dice again; California has lost and the fire ant has won. Yes, not all rates are created equal, and which of the two rates is larger is critically important.

California is still an immigration state. Every day, hundreds or thousands of people arrive from elsewhere, but in our myopia, we hardly notice that many of the immigrants are not people. They are myriad small creatures, like the fire ant, that depend on us to prepare a place for them and then even to bear them to their new land. To have such friends is a wonderful thing.

6

Predicting Future Range Limits

Models and Predictions, Past and Present

Prediction of future *Solenopsis invicta* range limits has been irresistible to authors for decades. All the crystal-ball gazers have seemed to agree that climate would ultimately limit the fire ant's range expansion, but they differed in how it would do so. The verdict on winter cold was unanimous, but not everyone agreed that aridity would limit westward expansion. Early predictions simply proposed that the ant would invade all areas with average minimum temperatures in excess of −12°C, about one-quarter of the USA. For several predictions, the rate of range expansion was simply extrapolated into the future on the assumptions of certain levels of cold and drought tolerance. Some authors suggested that *S. invicta* would ultimately occupy the entire ranges of the native fire ants, *S. geminata* and *S. xyloni*, because the cold and drought tolerance of these three did not differ greatly.

Most of these predictions were unsatisfactory, because, for one thing, they were spectacularly wrong. Hung and Vinson (1978) speculated that *S. invicta* would ultimately occupy all of Texas and much of Oklahoma and would move south into Mexico. Moody et al. (1981) and Cokendolpher and Phillips (1989) predicted that *S. invicta* would colonize 63% of Texas by the year 2000, but the actual fraction occupied even in 2005 was about 50%. Pimm and Bartell (1980) studied the correlations between range expansion since 1965 and a large number of climatic variables in the newly occupied zones and then used climatic data to predict that an additional 50 Texas counties would be invaded by fire ants by 1986. Even by 1993, however, the fire ant had occupied only a few of them. Texas is far less palatable to the fire ant than most people, especially Texans, believe.

Two more recent attempts linked daily air temperatures and rainfall to colony growth and reproduction in models to find the northernmost point at which a virtual colony could just grow and reproduce itself. In the first (Stoker et al., 1994), the model implied that *S. invicta* could do so all the way to the northernmost county in the Texas panhandle, but the authors stopped short of actually predicting this outcome, instead joining the consensus that *S. invicta* could not colonize north of the −17.8°C January isoline. The second (Killion and Grant,

1995) predicted that a colony could grow in Oklahoma City, but not in Wichita, Kansas (a bit lacking in precision, no?). In neither model did the virtual colonies have to cope with loss of workers to the cold, and in the second they did not have to produce sexuals to maintain their lines (clearly not realistic).

The only study that incorporated sufficient biological detail to approximate reality was that of Korzukhin et al. (2001). The authors reasoned that if a population is to sustain itself, let alone increase, the average fire-ant colony must produce at least one surviving female alate during its lifetime. Their model therefore computed the expected alate and worker production by individual colonies under the climatic conditions prevailing at over 4500 weather stations throughout the current and potential range of *S. invicta*, and it incorporated colony size, growth rate, worker longevity (Calabi and Porter, 1989), temperature-tracking behavior, cold coma, freeze kill (loss of workers directly to freezing), and temperature thresholds for brood production. The estimated increase or decrease in virtual colony size was cumulated in twice-daily time steps over the colony lifetime at each location, producing the colony size throughout its life. When the virtual colony had reached a threshold size, it began, seasonally, to produce alates (the model's second output), much as real colonies do. Total alate production was cumulated over the lifetime of a colony at each of the weather stations. The totals were based on the previous three to 12 years of weather data and were projected nine years into the future.

Now imagine that we are standing at the current edge of the fire ant population, peering into as yet unoccupied territory. The fire ant will occupy this territory if a colony founded there produces enough female alates to ensure that at least one of them will produce a mature colony there. No estimates of the necessary number of alates exist, but we know that colonies actually persist at the current northern edge of the fire ant's range. The lowest estimated alate production at four such locations represents a provisional minimum for successful colony reproduction. The model was thus calibrated with the lowest alate production known to exist in viable populations, an average of 3900 alates per year and a minimum of 2100.

Each currently unoccupied weather station could be invaded only if estimated alate production there exceeded these minima. Sites at which lifetime estimates exceeded 3900 alates were "certain" to be occupied. Between 2100 and 3900 invasion was "possible" because occasional freeze kill might actually cause reproductive failure. Invasion of sites below 2100 was "undemonstrated" because no populations with production lower than that are currently known to exist. Finally, when virtual alate production was zero, invasion was "improbable."

The outcome is seen in the beautiful Plate 8. It predicts that the eventual range will almost certainly encompass another 80 to 150 km northward in Arkansas, Oklahoma, and Tennessee, as well as along the

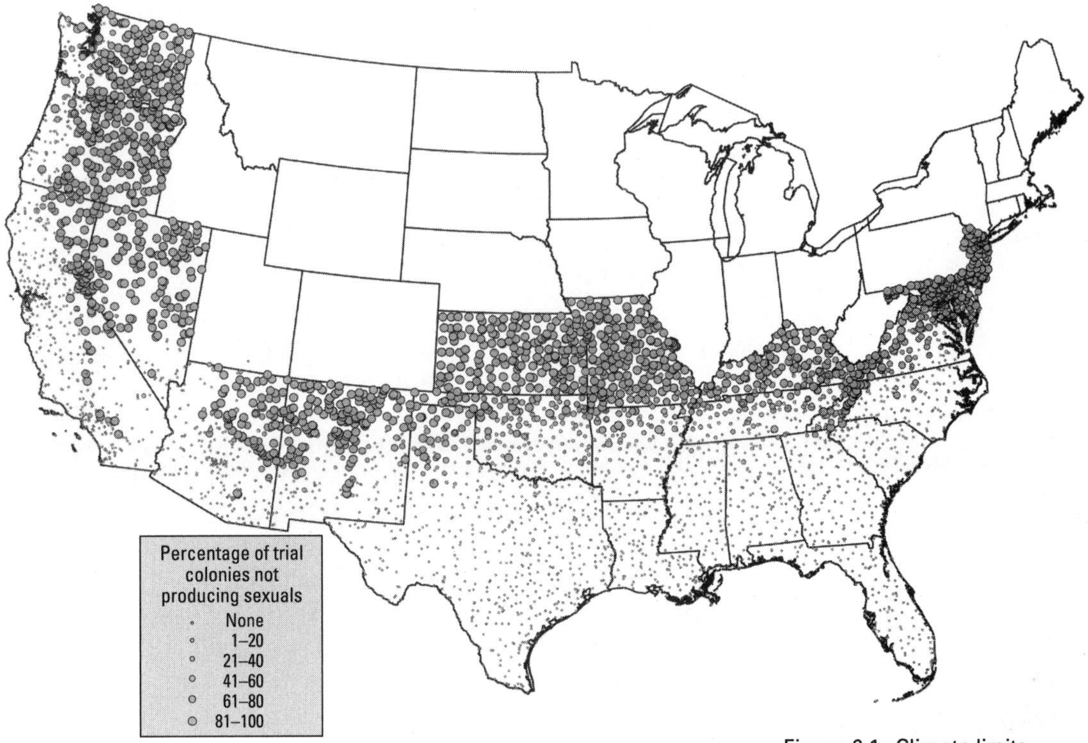

Figure 6.1. Climate limits northward expansion of the fire ant *Solenopsis invicta.* When enough workers are killed by winter cold, colony size often remains too small for production of sexuals and thereby limits further spread of the population. Adapted from Korzukhin et al. (2001).

maritime parts of Virginia. Rainfall of less than 510 mm per year (west of 102° west longitude) was deemed insufficient to support fire ant populations, although the basis for this judgment was shaky. In western Texas, southern New Mexico and Arizona, and the drier parts of California, fire ants would be limited to river margins and wherever human irrigation made life possible. For northward expansion, death of workers from cold-coma or direct freezing was a central, limiting factor. In its absence, almost all colonies at almost all sites would have been capable of growing to maturity. Colonies exposed to mean winter minimum temperatures between –3 and –7°C generally failed to produce any alates at all, because repeated reduction of colony size by cold coma prevented them from ever becoming large enough to produce alates (Fig. 6.1) (global warming could move this zone northward, as modeled by Levia and the appropriately named Frost; Levia and Frost, 2004).

Less certain is the occupation of western Oregon and Washington, but there cool summer temperatures rather than winter cold will probably be limiting. Cool summers would not allow colonies to become large enough to produce sufficient alates for self-propagation.

The predictions of Korzukhin and his colleagues are the most realistic to date. Subsequent application of the model to data from weather stations around the world produced a world map showing the warm, moist areas where fire ants might be able to invade if obliging humans transport them there (Morrison et al., 2004). The actual future range will also depend, however, on interactions with the local biota, and these are not well understood, nor were they part of the model. Especially at the range margins, the establishment of *S. invicta* may depend on competition from ants better adapted to the cold and on other local ecological effects, which might tilt the balance against *S. invicta*, keeping its population density too low for successful invasion.

Physiological Limits

Several investigators have tried to predict the maximum potential range of the fire ant by matching the limits of its temperature and moisture tolerance with the geographic distributions of these conditions. Almost all of these seem to have missed the point that the temperatures and humidities we humans measure with our fancy weather stations and satellite sensors are not the ones that fire ants actually experience, so the resulting range predictions are illusions. Even when air temperature is low enough to turn ants into Popsicles, the colony can be snug in its subterranean nest, where the temperature is well above lethal. Even when it's hot enough on the ground surface to fry an egg, the fire ants can be taking a break at comfortable temperatures below ground. The physiological limits described below characterize the tolerances of fire ants, but they do not help much in predicting where the ant will live in the future.

Temperature. Fire ants are "cold-blooded"—that is, their body temperature more or less tracks that of their environment. Because temperature is never uniform in any real environment, fire ants can choose, to some degree, temperatures that are as favorable to them as possible, but ultimately, beyond a certain limit, they cannot escape what the environment doles out. Above and below certain temperatures, ants will die. Drop a few fire ants onto dry, bare soil on a midsummer day, and they will be dead in less than five minutes. Put them in the freezer for a few hours, and they will not revive on thawing.

Upper and lower temperature tolerances are usually studied in the laboratory; simple apparatus is used to expose ants to a series of constant temperatures, revealing (1) the highest or lowest temperature at which the animal still shows normal movement or the ability to right itself (critical thermal maximum or minimum) and (2) the highest and lowest temperature that an animal can survive (heat or cold tolerance) (Francke et al., 1984, 1985, 1986; Francke and Cokendolpher, 1986; Cokendolpher and Phillips, 1990). The effect will depend on the length of exposure as well as the absolute temperature.

The critical maximum and minimum for colonies maintained at 30°C were 40.8°C and 6.6°C. Colonies maintained for an extended period at about 10°C showed minimal acclimation—their maximum decreased by only 1°C and their minimum by less than 4°C. At least part of this modest acclimation might have resulted from the gradual loss of less cold-tolerant workers.

Raising or lowering the temperature still more eventually kills the ant and defines the heat or cold tolerance. These are usually reported as the LT_{50} and LT_{95} values, the temperature at which 50% and 95% of the ants died. For four species of fire ants, the LT_{50} (one-hour exposure) was between 41 and 44°C, and the LT_{95} between 43 and 47°C (air temperatures of 47°C occur fairly commonly in deserts). Workers remained normally mobile up to the temperatures at which the first ones die and the maximum temperatures at which they normally still forage. Death from overheating does not result from desiccation—the lethal temperatures were similar at 0 and 100% relative humidity. Again, little acclimation was evident.

The consensus that the northward spread of the fire ant would eventually be limited by climate formed early. After all, *S. invicta* is a subtropical ant, unlikely to tolerate freezing. Over the years, several researchers documented freeze kill of fire-ant colonies (Green, 1959; Morrill, 1977a; Morrill et al., 1978; Thorvilson et al., 1992; Callcott et al., 2000). In State College, Mississippi, mounds froze when a cold snap followed a mild rain. Although few colonies were killed outright, so many workers froze that most colonies were too small to produce sexuals in the next spring. After a thaw, surviving workers piled up dead ants in large, funereal heaps. In the Piedmont region of central Georgia, the unusually harsh winter of 1976–77 killed 58 to 96% of the colonies, but colonies in south Georgia experienced only about 20% mortality. In the mid-1980s, 75 to 84 days below freezing during two winters in Lubbock, Texas, killed up to half of 200 marked colonies. Soil was warmer and survival higher in protected locations near structures. A similar isolated population in Tennessee suffered significant winter kill in 1993–94 (Callcott et al., 2000).

Under cooling conditions, the mound, ironically, becomes a liability. As long as the sun shines on it, it is an excellent heat collector, but it too obeys the laws of physics—a good absorber is also a good radiator, and on cold nights, the mound and the soil beneath it cool faster than the surrounding soil. Below 10–12°C., workers move the brood and themselves out of the mound to warmer zones below ground, but strangely, they are not programmed to keep moving down about 20 to 40 cm below ground level. As cooling continues, they become immobilized and are sitting ducks for death by freezing. Thus as the ant moves northward, the winter-freeze depth eventually overtakes the residence depth, and freeze kill becomes commonplace, reducing colony size (and therefore reproduction and ecological impact) or killing colonies outright. For example, the Lubbock colonies were much smaller than normal. Winter kill makes it likely that *S. invicta* is currently near its northern range limit. Interestingly, the two native species of fire ants that do not build mounds, *S. aurea* and *S. xyloni*, occur considerably farther north than the one that does, *S. geminata* (Moody et al., 1981), even though their absolute temperature tolerances do not differ. A detailed temperature study of mound and nonmound nests might be in order.

Interestingly, although *S. richteri* comes from more temperate parts of South America, its hybrid with *S. invicta* is no more cold hardy than *S. invicta* itself. Survival rates of marked colonies of *S. invicta* and of hybrids in northern Georgia were similar (Diffie et al., 1997), but this conclusion is somewhat weak because all the winters during the study were mild and survival of both forms high.

To the individual ant, death from cold either results from cell damage caused by ice formation or, for poorly understood reasons, follows a long period of immobilization by cold. As an ant is cooled experimentally, freezing begins at a point several degrees below the actual freezing point. This "supercooling point" is marked by a sudden increase in the ant's temperature, followed by a period of steady temperature (the actual freezing point). When the ant is completely frozen, the temperature falls again. Supercooling results from substances that inhibit initiation of ice formation and is considered a potential route to evolving increased cold hardiness.

The exact temperatures involved are well known, because several researchers have pursued the logic that death by freezing must set the ultimate range limit and have developed a cottage industry of freezing hundreds of fire ants of five different species and some of their hybrids (Francke et al., 1986; Taber et al., 1987; Diffie and Sheppard, 1989;). Supercooling points of pupae were much lower (–15 to –24°C) than those of larvae (–7.4 to –16°C) or workers (–6.2 to –10.3°C), but none of the five species (*S. aurea, S. geminata, S. invicta, S. richteri,* and *S. xyloni*) differed in this value. Maintaining the colonies at low temperature did not lower the supercooling point (i.e., produce acclimation). Supercooling is probably not an adaptation for surviving winter cold at all but is a side effect of unrelated physiological processes. Consider that pupae have the lowest supercooling points, yet pupae are typically absent during the winter. The more temperate *S. richteri* has the same supercooling point as *S. invicta*. Fire ant workers taken directly from the field in Lubbock, Texas, showed a gradual 1.5°C increase of their supercooling points between November and February, whereas a comparable colony maintained at 22°C in the laboratory not only had lower supercooling points but showed a gradual decrease during this period.

Cold can also kill without freezing, if it persists for long enough (James et al., 2002). Worker ants kept in a refrigerator at temperatures between 0 and 4°C for five to 10 days mostly fail to recover upon rewarming. They have succumbed to what S. D. Porter has called "cold coma" (pers. comm.). Death comes not from dehydration or starvation but from damage to unknown physiological processes. Mortality rate increases as the temperature decreases from 4° to –4°C. No workers survive seven days at –4°C, although this temperature is still above their body's freezing point. Workers of *S. invicta* die at somewhat higher rates than *S. richteri* and the hybrid.

Winters with extended spells of cold, cloudy weather eventually cool the soil in fire-ant nests to near 0°C, and cold coma ensues. Porter found that soil temperatures along the northern boundary of the fire-ant range rarely get cold enough to freeze fire ants, but they do frequently get cold enough to kill them through cold coma. This may be the reason that the length of periods of near freezing temperature was a better predictor of winter kill in a Tennessee population than was minimum temperature (Callcott et al., 2000). During the winter of 1993–94 and again in 1995–96, 80 to 90% of the colonies there died as a result of five to seven consecutive days below 1.1°C. In Lubbock, Texas, midwinter soil temperatures at a depth of 30 cm never fell below 4°C, and death probably came from cold coma. Such events limit fire-ant populations to latitudes considerably to the south of those where the ants would actually freeze.

Cold coma and freeze kill are important in setting the northern range limit, even though climate data do not pinpoint this limit (Figure 6.1). The effects of cold coma depend more on the frequency, intensity, and duration of cold snaps than on minimum temperatures. Lethal cold snaps need only occur every few years to keep populations in check, but outright killing is not the only way temperature limits the fire ant's range. Temperature extremes also restrict the seasonal and daily extent of profitable activity until the annual maintenance cost in an area exceeds the energy the colony can reap from the environment. In such areas, colonies can no longer survive. The interplay of all these physical and biological factors is complex, and infrequent, unpredictable extremes play a large role. Predictions of future range drawn only from temperature tolerances are not likely to be successful and are sometimes off the mark by half a continent or more.

Desiccation Resistance. The westward spread of *S. invicta* across Texas has taken it from a humid climate to ever more arid ones. The height of a string stretched between the tops of a 140 cm stake at the eastern boundary of Texas and a 22 cm stake in El Paso at its western tip would be a pretty good estimate of the mean annual rainfall at all points in between. Just as cold hardiness is thought to determine the ultimate northern range limit of *S. invicta*, so resistance to water loss under arid conditions has been postulated to set the western range limit. This postulate stimulated several papers on the desiccation resistance of fire ants, but once again, the humidity conditions we humans measure are not those experienced by the fire ants in their daily lives, and desiccation resistance has not proven very useful for future range prediction.

All terrestrial organisms lose water by evaporation from the outer body surface and the respiratory surfaces and by excretion of urine and/or wet feces. Insects resist evaporative loss by means of waxy, waterproof coatings on the cuticle and by means of spiracular valves that open the tracheal system only as needed (Vogt and Appel, 2000). For foraging ants under desiccating conditions, evaporative water loss is probably the most important and imposes strict limits on the activity of the ant. If water loss is not balanced often enough by water intake, physiological functions are impaired, and death may result. Some insects (e.g., clothes moths and some tenebrionid beetles) have reduced evaporative and excretory losses almost to zero and never need to touch a drop of liquid water, but fire ants are not nearly so resistant to desiccation.

Measuring the actual rate of water loss across the surfaces of an insect requires some serious equipment, so instead, Greg Hood in my laboratory (Hood and Tschinkel, 1990) and workers at Texas Tech University (Braulick, 1982; Munroe et al., 1996; Phillips et al., 1996; Li and Heinz, 1998) simply determined how long ants could survive under specific conditions of temperature and humidity, all the while hoping fervently that God is not an ant (a hope all myrmecologists share). The time it takes for half of the workers in each test group to die (LT_{50}) is positively related to the rate of water loss over the test period. Of course it is also related to water content at death, which may differ with species or experimental treatments. In a sense, however, the combined effect cuts to the heart of the matter and is more ecologically meaningful because it provides a final measure of stress tolerance.

Hood studied 36 different ant species, so his findings represent a nice sketch of some of the general principles that govern water loss in ants, including *S. invicta* and *S. geminata*. Three major factors affect the rate of water loss by evaporation. Loss rate increases and LT_{50} decreases with decreasing relative humidity, and increasing temperature. The effect of temperature has two causes: the vapor pressure of body water increases, and the relative humidity of air decreases. The third factor is body size. Within and among species, the larger the worker, the more slowly she loses water and the longer it takes her to die, even though the rate of water loss per square centimeter of cuticle is similar (Appel et al., 1991). You would expect as much simply from geometry—small animals lose a larger proportion of their body water in a given time interval because the ratio of surface area to volume is larger in smaller animals. The effect of body size was similar for *S. invicta, S. geminata, S. aurea,* and *S. xyloni. Solenopsis geminata* was the most resistant to desiccation, *S. xyloni* next, and *S. invicta* least, but the rates of water loss contradicted these rankings, perhaps because of differences in minimum viable body-water content (which was not measured). The results of Hood and Braulick both contradicted some of these findings, suggesting that firm conclusions about relative desiccation resistance of these four species are probably not yet warranted.

Braulick and Monroe both made future-range predictions based on *S. invicta's* desiccation resistance and those of the native *S. xyloni* and *S. aurea,* but basing such a prediction on a single physiological character is risky. Perhaps the most convincing argument against doing so comes from the work of Hood. After adjustment for body size, desert ants were no more desiccation resistant than ground-nesting ants from the southeastern USA. The adaptations for desert life are primarily behavioral, not physiological—avoidance of highly desiccating conditions. In contrast, the real "desert" ants are arboreal ants, who are without refuge during truly waterless periods. Under the same desiccating laboratory conditions, arboreal ants took eight times as long to die as did ground-nesting ants of similar size. Waterproof cuticular waxes greatly reduced their rates of water loss. Large differences in desiccation resistance may explain habitat differences between arboreal and terrestrial ants, but the small differences (if any) among fire-ant species are of no value in explaining or predicting geographic range.

If desiccation resistance has a genetic basis, and can therefore evolve, then survival times of sisters should be more similar than those of unrelated workers. Li and Heinz (1998) reared simple families (that is, families of full siblings) from polygyne queens and determined desiccation resistance of the workers. The "heritability" was the fraction of the total variance that was associated with families, but it included genetic, maternal, and dominance effects. It was 0.42, significantly greater than zero, but relatively low. The authors argue that polygyne fire ants are evolving increased drought resistance and that they have a competitive advantage over monogyne fire ants in western regions. Again, the authors assumed that ecological performance can be predicted from a single physiological trait, even though several studies have shown that this assumption has little basis.

A final point: Temperature and humidity are inevitably related, as already noted. The studies above tested either temperature or humidity. Testing both together gives a more realistic picture of what ants are up against. At 1% relative humidity, it took nine hours for 99% of small fire-ant workers to die at 27°C, six hours at 32°C, and four hours at 38°C (Phillips and Cokendolpher, 1988). As the maximum tolerated temperature is approached, humidity becomes irrelevant. Ants subjected for one hour to temperatures between 41 and 44°C died as rapidly at 100% relative humidity as they did at 0%. Under very hot conditions, death comes much sooner from overheating than from desiccation. Low humidity thus restricts the timing and duration of "outdoor" activity rather than being directly lethal. As with temperature, desiccation resistance, considered alone, is not really very informative about ultimate geographic limits of *S. invicta.* The ants have too many behavioral tricks for avoiding harmful conditions up their tiny little sleeves. We cannot easily match up the conditions that are lethal in the laboratory with those actually experienced by the ants in the field.

II

Basic Needs and the Monogyne Colony Cycle

An Important Note: Monogyne and Polygyne Social Forms

Solenopsis invicta exists in two social forms. In the monogyne form, each colony is a simple family composed of the offspring of a single queen. The colony defends a territory against neighboring colonies and reproduces by emitting winged sexual ants. After their nuptial flights, the mated female sexuals found new colonies without the aid of workers from their natal colonies. In contrast, polygyne colonies in the USA are aggregates of unrelated queens and their offspring. The nests are not territorial and reproduce by colony budding. In addition to the obvious differences in their life cycles, the biology and ecology of these two forms differ in less obvious but radical ways. In many ways, the two differ in biology as widely as do many pairs of separate species.

We know much less about the polygyne form of *S. invicta* than we do about the monogyne form. Although many aspects of the monogyne life cycle and colony function have been worked out in elegant detail, no comparable body of work exists for the polygyne form. Consequently, the sections that follow, unless the polygyne form is specifically mentioned, are primarily about the biology of the monogyne form. The extent to which monogyne biology can be extended to the polygyne form is often uncertain. The biology of the polygyne form is treated in detail in Part IV.

7

Shelter

Easily the most conspicuous feature of fire ants is the earthen mound they construct, which is often referred to as an "ant bed" by southerners, who seem frequently to "get into ant beds" (southerners have even been known to sleep in them, but only when dead drunk). If we think of ant colonies as superorganisms, the structure of their nests immediately springs into importance, for it organizes the colony in physical space and in time and influences interactions among the individual ants. The nest and mound of the fire ant colony is a functional adaptation and, as such, is part of the "extended phenotype" (the animal plus its constructions) of the superorganism. Because little has been published about fire ant nests, most of what follows is taken from my own published (Tschinkel, 1993b) and unpublished observations.

For most ants that nest in soil, the excavated soil is mere waste, to be disposed of with the least labor possible. Fire ants, unlike their competitors, are among the small minority of ant species that use this excavated soil to build a chambered mound in which to live. This feature places them in a class with such famous species as wood ants and thatch ants, some of whose nest mounds exceed the height of a tall human. In all of these species, mound building has important ecological consequences, including soil aeration, fertility, turnover, and nutrient cycling. Fire-ant mounds change the soil characteristics, contain more phosphate and potassium than the surrounding soil, and are less acidic. As a result, the grass growing on them contains more protein and is readily grazed by cattle and sheep (Herzog et al., 1976; Lockaby and Adams, 1985). The ecological importance of these various effects is uncertain.

Most unconstrained mounds are dome-shaped. In winter, when the mounds reach their largest size during the annual cycle, they have a smooth, rounded shape (Figure 7.1) and make nice accents to an otherwise dull, flat lawn. As the mound ages, grass may grow through it and on its flanks. During the summer, when the mound decreases in size because the colony within it is smaller, only the central region of the mound dome may show signs of active reconstruction, the flanks becoming washed down and grassy.

Depending on how many ants live in the colony, the volume of the nest mound may range from a fraction of a liter, fitting readily in the

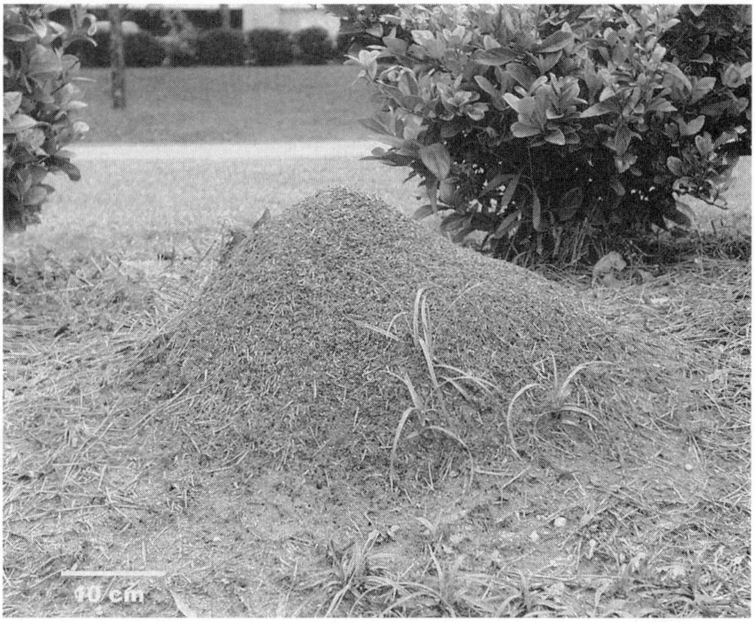

Figure 7.1. A fire-ant mound in midwinter. The grass-free dome collects heat on sunny days.

palm of a child's hand, to a colossal dome as large as 100 liters. The strength of the mound is low, especially in sandy soils, where the adherence is weak—even light contact causes ruptures from which workers pour to defend their nest. Similar-sized colonies have larger mounds in heavy clay soils than in less adhesive sandy soils. In south Florida, where it is always warm and the soil is sandy, *S. invicta* often fails to build mounds altogether.

Mound construction is possible only when the soil is damp and the grains stick together. The mound grows in episodes stimulated by rain, rather than continuously like the colony within. Immediately after a rain, or even during light rain, workers create many small holes in the surface of the existing mound, and on its surface build up serpentine crenulations, which they eventually roof over to form one or more layers of mound passages. When the soil is dry, the workers simply dump the excavated soil to the sides of the nest opening, forming craters similar to those of many other species of ants.

The best way to visualize the shape of the internal space of a fire ant nest is to fill it with dental plaster (Williams and Logfren, 1988; Cassill et al., 2002), lead (Markin, 1964), or molten zinc, then to excavate it and wash away the soil (Plate 9). These casts reveal the remarkable structures the ants create. Whereas the mound is produced by construction, the belowground nest is excavated, two fundamentally different processes. The spaces within the mound are serpentine and winding, like a mound of interconnected tubes or a pile of spaghetti without the meatballs on top (Figure 7.2). In contrast, below ground,

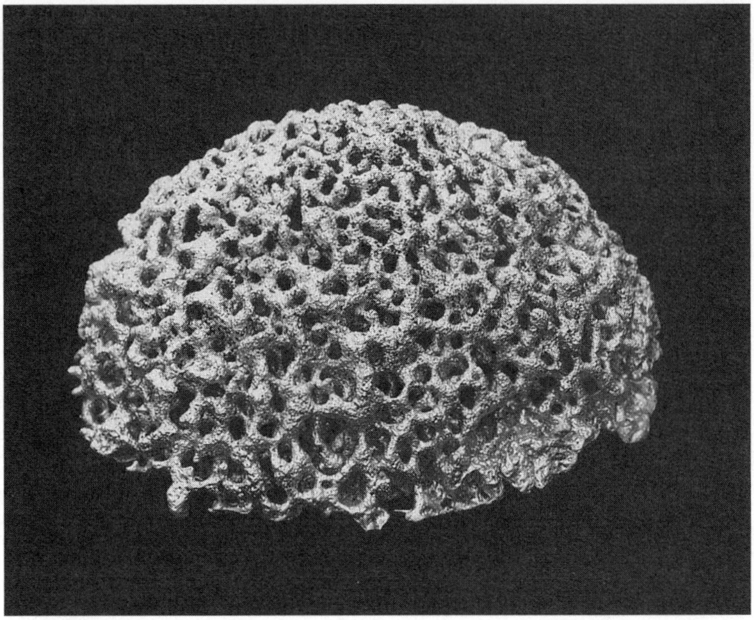

Figure 7.2. Zinc cast of the mound of a *Solenopsis invicta* nest showing the serpentine tunnels that make it up.

the ants hollow out flat chambers that are round to oval, as seen from above, and are connected to chambers above and below by narrow, vertical tunnels to form vertical series, like shish kebab on a spit. Short vertical shafts connect the winding mound tunnels with the shallowest true chambers at depths of 10 cm or more below ground level. Chambers range from three to 10 cm^2 in floor area and average about five. Their generally oval outline becomes more complex when adjacent chambers fuse to become one. This structure is already apparent in very young colonies, barely out of the incipient stage (Markin, 1964; Markin et al., 1973).

Good estimates of the total volume of living space in mature nests are nonexistent. Lead casts of the nests in the incipient stage, though hard on the ants, showed their volume to be about 1.5 ml. One month later, it was 8.4 ml, and in another six months, it had increased 50-fold to 400 ml (Markin 1964). The ants enlarge the nest in an orderly manner through four simultaneous processes—they deepen vertical tunnels, add more chambers at the nest's lower extremity and between existing chambers, enlarge existing chambers, and add new shish-kebab units (whole new vertical tunnels with their own horizontal chambers) just outside the outer perimeter of the existing ones (Figure 7.3). All these processes are regularly related to one another and to the size of the colony within, but unfortunately, the nests are so complex and difficult to map, measure, and describe that no one has yet determined the rules of construction. As a spatial description of the superorganism becomes more critical to understanding its functions, the importance of understanding the nest architecture will increase.

The size relationships of the aboveground mound have been more adequately described. The mound is first apparent at three to five months of colony age and is enlarged layer by layer by the growing colony. About 65% of the mound volume is hollow space, the rest soil. Colony size is so tightly correlated with mound volume that mound volume is used as a surrogate for colony size by many researchers (Porter et al., 1990; Tschinkel, 1993b). Typically, about 85% of the variation in mound volume can be explained by variation in colony biomass or worker number. This relationship is detailed under the heading of colony growth.

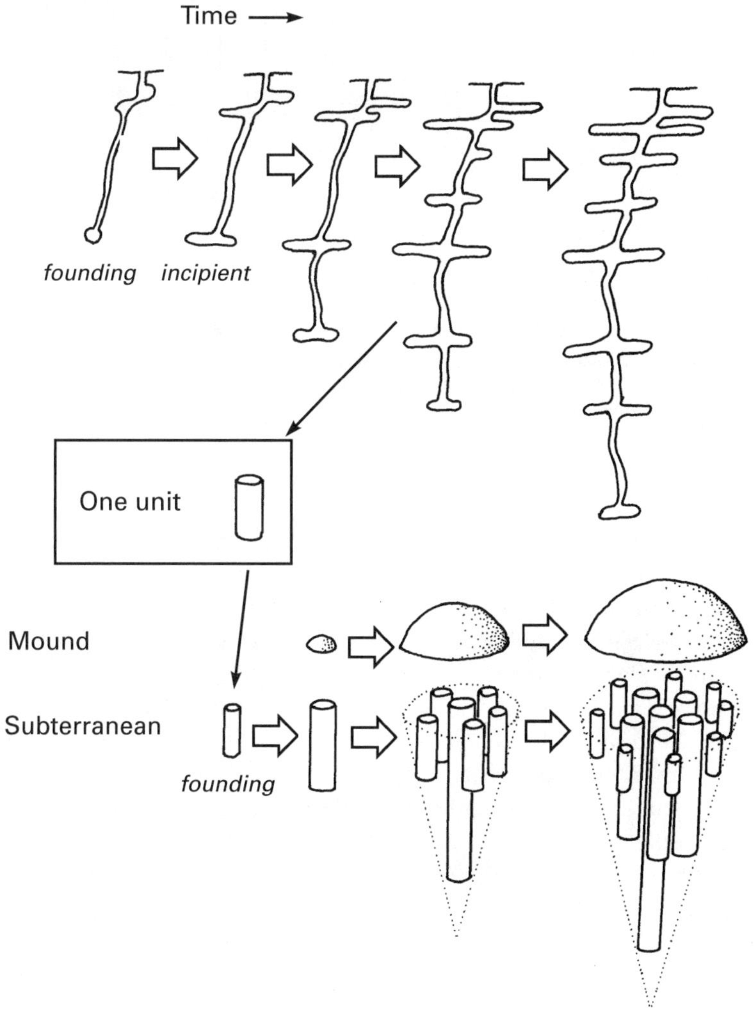

Figure 7.3. Nests grow through the simultaneous deepening of nest units, addition of more chambers, chamber enlargement, addition of more peripheral nest units, and enlargement of the mound. The upper series shows how the basic nest unit grows, and the bottom series how these units are combined in nest growth.

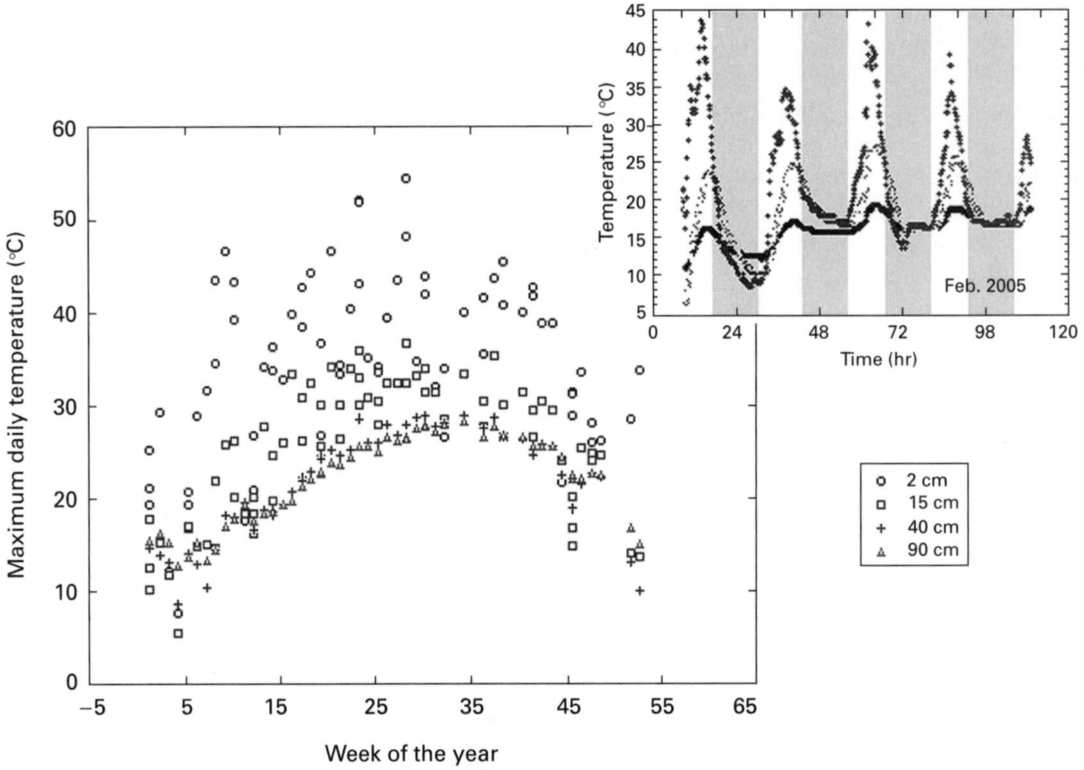

Figure 7.4. Daily and seasonal temperature cycles at various depths in *Solenopsis invicta* nests. The inset shows a five-day record of temperatures at three mound depths and in adjacent soil. Mounds heat and cool much faster than adjacent soil (inset) and always reach higher maximum daily temperatures (main graph) (annual data from Sanford D. Porter, with permission).

The function of the underground nest, as for all ants, is first of all to serve as an easily defended living space—a place to carry out the quotidian tasks of life in an environment protected from predators, desiccation, and overheating. The floors of the many chambers serve as places to keep brood, rest, receive and process food, interact with nestmates, and any number of other tasks that fill the lives of fire ants. Nests also offer a choice of environmental conditions. Rain soaking into the ground creates a gradient of soil moisture. The surface warms in the sun during the day and cools at night, creating a cycle of temperature and season that decreases in amplitude with depth (Figure 7.4).

In contrast to the range of uses of the belowground nest, the function of the aboveground mound seems to be primarily to regulate temperature and perhaps soil moisture. The domed mound dries rapidly, and its winding passages make it an efficient solar heat collector, warming in the morning sun (Figure 7.4, inset). The mound is expanded more on its warmer, southern face in winter, causing mounds to assume an elongate shape whose axis is oriented north-south (Hubbard and Cunningham, 1977). During the cool season, the colony and its brood move into and out of the mound on a regular cycle (Plate 10) tracking the temperatures that speed the development of brood and increase colony growth rate (this thermomigratory behavior also allows queens to be captured in the mounds on cool, sunny winter mornings). Mounds are little used during the height of summer, when most of the colony spends its time below mound level. The lack of need for thermoregulation may explain the absence of significant mounds in southern Florida, where the soils are always warm.

Sanford Porter estimated the amount of benefit to brood production by measuring the temperatures of fire ant nests at four depths (2, 15, 40, and 90 cm) throughout one annual cycle (Porter, 1990, 1994). On sunny days, the temperature at 2 cm was 7°C higher than that of the surrounding soil; the differential dropped to 2–3°C at 15 cm. Below 40 cm, little difference was detectable. Fire ants carry brood into regions of this thermal gradient as close to the preferred temperature (30–32°C) as are available. On cloudy days or in the shade, the mound provides no advantage.

How much does the colony benefit from this construction and transport behavior? Porter estimated the benefit by modeling how the mound affects the accumulation of degree-days (the sum of the mean hourly temperature multiplied by 1/24) over one annual cycle at different nest levels, from the mound down to 90 cm depth in the soil. He then assumed three hypothetical colonies. The first tracked the temperature gradient, the second kept all its brood at 2 cm, and the third kept all its brood at 40 cm, too deep for the mound to affect. Over one year, the tracking colony achieved 13% higher brood production than did the colony fixed at 2 cm and 30% higher than that fixed at 40 cm.

To address the benefit of building a mound, let us assume two hypothetical colonies, both of which track temperature, one in the soil, without a mound, and the other in the soil with a mound above soil level. In this case, the mound provides a 10% benefit over a year, most of which probably occurs during the spring and fall, when the surrounding soil is too cool for brood development. Taken together, mound construction and temperature-tracking behavior provided the colony with approximately a 23% annual thermal benefit over a colony that did neither. In competition among colonies, brood-production rate counts, so that 23% is probably a considerable edge. The cost of building a mound has not been estimated.

No one has tried to estimate the labor, energy, opportunity, time, and exposure costs of constructing an entire nest. Although we can assume doing so must pay dividends in fitness, we can only guess at the mechanisms of payoff, which might include protection, choice of microclimate, and organization of the work force. An understanding of the connection between nest architecture and social function is still far in the future.

◂•• *There's Nothing Like Getting Plastered*

Trying to imagine the shape of what is *not* there is not easy—I mean creating a mental image of the volume contained within any sort of walls. Because I work on ants, the volumes of greatest interest to me are the cavities that ants hollow out in the soil. Nests are formed by the removal of soil to create an unsubstantive space. Of course the space is filled with air, as every schoolchild knows, but air, at a density of 0.0013 g/ml is pretty unsubstantive, compared to soil with a density of about 2.5 g/ml. Anyway, you can see soil, but you can't see air, so it's as good as not there, practically speaking. To build their nests, therefore, ants make nothing out of something and then live in the nothing.

I used to think I had a pretty good idea of the nothing, based on having excavated dozens of fire ant nests and seen the vertical and horizontal sections through the nests. But after I turned the nothing into something by filling it with plaster, and then removing the soil-something from around the plaster-something (formerly nothing), I realized that I had known next to nothing about the shape of the nothing in the ant nest. It was nothing if not something of a surprise.

The big breakthrough in the nothing-to-something problem came when David Williams first switched from plaster of paris to orthodontal plaster. Not only is it much stronger than plaster of paris, but it comes in pleasant colors, too. You can choose from buff, yellow, blue, or green when you order your 50-pound boxes from the dental supply company. The boxes come with carrying handles so that dentists can heave them around without subsequently requiring the services of chiropractors. It works for myrmecologists, too.

Using the plaster isn't highly technical. First, you locate your victim nest and decide, through sheer guesswork, how much plaster it will take to fill it. In step two, you transfer that much dry plaster into a large container, then mix in enough water to make a thick soup. So far, this is all pretty orthodox plaster-mixing technique, right off the instructions on the box, but now we pour in enough water to at least double or triple the total volume. There is method in our madness, method learned through pure empiricism and a good deal of wasted plaster. Soil, especially the almost pure sand of Florida, sucks the water out of the plaster soup by capillarity, so the soupier the plaster, the farther it will flow down into the ant nest before becoming too thick to continue. At that point, it will dam the tunnel, backing up the plaster soup behind it.

Step three begins when you pour the thin plaster soup into the ant nest. When the nest is full, the job is not yet done, because you will notice that the soup level slowly recedes and must be repeatedly topped off. The soil is sucking the water out of the soup, bringing it to the proper thick consistency needed for good strength after setting. If you fail to top off, most of the upper levels of the final cast will be hollow and easily broken.

Step four is the easiest, at least for patient people. It is simply to wait about an hour for the plaster to set. This time can be spent in eating a delicious sandwich of Würzbrot, butter, and Lachsschinken or in lolling in the shade reading a book.

Step five is the excavation phase, when the nothing-now-turned-into-something is retrieved from the soil. Dig a pit next to the nest and carefully dig in from the side until hard substance is struck. Because fire ant nests contain large numbers of chambers that are confluent and linked by multiple tunnels, the whole volume congeals into a single structure, removable as a single piece still embedded within its soil matrix, and weighing 25 to 50 kg. Back home, the soil is washed away with water squirted from a hose, and voila!, we have made visible the nothingness of a fire-ant nest.

Less interconnected nests (that is, those of most other ant species) come out of the ground in pieces and must be reassembled after they dry. One might think reassembling a nest cast broken into a hundred pieces would be nigh impossible, but the pieces can be sorted into groups on the basis of regular changes in soil color and chamber shape with depth. In the end, one need only match the details of broken ends within a few groups of chambers. A lesson in classification has been learned.

Other lessons can be learned. Plaster is anhydrous calcium sulfate and is almost insoluble in water. Plaster soup is therefore a suspension of solid particles in water, a point extremely important to making casts in soil. Try making a cast with a true liquid, say a hardening plastic, and you will see why—the liquid infiltrates the capillary spaces in the soil, cementing the soil particles together and creating a cast much larger than the original space: in short, a mess. It is the filter action that passes liquid and retains solid particles that makes the plaster cast possible.

Lessons in geometry and physics are available, too. Our plaster soup will flow nicely through long plastic tubes as narrow as 1 or 2 mm. Yet in soil, tunnels of this diameter plug up before the soup has flowed more than a few centimeters. When the tunnels are, say, 8 mm across, an entire, 2-m–deep harvester-ant colony will fill plumb full from bottom to top. Why the difference? First, recollect that the soup stops flowing when it has lost enough water to the soil to become thick. If you do the math, you will see that, for a unit section of tunnel, the surface area to which water can be lost increases with the radius, but the volume of water to be lost increases with the square of the radius, i.e., a lot faster. Therefore, the wider the tunnel, the lower the relative loss rate. For example, water is lost eight times as fast from a 1-mm than from an 8-mm tunnel, so the soup flows eight times as far in the 8-mm tunnel. All exchanges between the environment and a living organism, be they of heat, gas, water, or nutrients, are governed by the same rules. If you understand why plaster flows so much farther in larger tunnels, you will also understand why gas and nutrient exchange takes place only through the walls of blood capillaries, the very tiniest of all the blood vessels, rather than through those of larger vessels. Do the math. Clearly, things cannot be otherwise.

8

Space

Some ants defend their food, their nests, and/or their territories against other members of their species (and often closely related species). Fire ants clearly belong to an absolutely territorial group, poetically dubbed "large-scale conquerors." Territorial defense tends to evolve where resources are uniformly distributed, as seems likely in fire-ant habitat. An ordinary animal defending its territory can be in only one place at a time, but a territory-defending colony is fundamentally different in that it can be everywhere at once by deploying workers to all boundaries simultaneously. Whereas the cost of defense by ordinary animals can be measured in effort, energy, and risk, that of social insects is probably more closely related to the number and cost of workers engaged in territorial behavior.

All the superorganism's resources derive from the territory. Food consists of insects, invertebrates, and plant exudates. Very rarely, the colony may reap a large bonanza when a bird or mammal dies within its borders, to be reduced to a skeleton in a few hours. When that mammal is large, like a cow, the feast may continue for weeks, first on the cow's flesh directly but later on the succession of decomposers that reduce the cow to other animal flesh, a skeleton, and a dark stain on the ground. Such bonanzas are too rare for ants to depend on, and fire-ant food consists mostly of the small creatures that the colony's territory produces in continual supply.

Studying the territoriality of a nonnative ant may seem unlikely to yield good results—that is, results generalizable to the natural world (similar criticisms of a range of fire-ant ecological studies have been voiced)—but it also has distinct advantages. The most important is that, although *Solenopsis invicta* shares the habitat with many ant competitors in South America, its communities in the USA are much simpler. Prime landscape is blanketed almost continuously with a mosaic of *S. invicta* territories, like tiles on an old-fashioned bathroom floor, so all territorial boundaries are with other *S. invicta* colonies. We can study the complexities of territoriality in almost pure, continuous populations of a single species, without the complicating influences of other competitors. Furthermore, the high densities of many fire-ant populations assure that the interactions among competing colonies will be intense and obvious. As researchers, what more can we ask?

Let us start with a practical question. How do we know where the

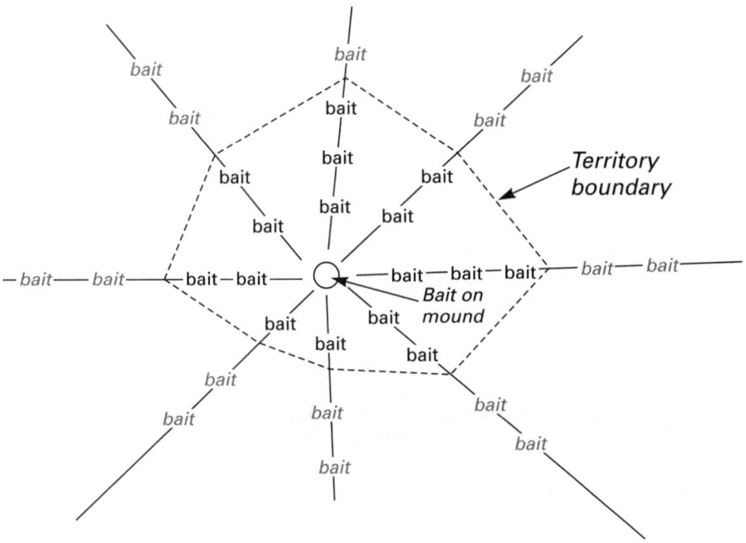

Figure 8.1. Determining territorial boundaries by the baits-on-spokes method. Workers attracted to the baits along the spokes are tested for aggression against workers coming to that on the mound. Workers at baits printed in roman type fight with those at baits printed in italics.

territory of one fire-ant colony ends and that of another begins? After all, fire ants create no visible boundaries of cleared, fenced, and mined land, patrolled by dogs and armed vehicles, and no Great Walls. Two properties allow us to determine territorial ownership: first, territories are occupied exclusively by the territory owners. Early investigators fed different colors of dyed food to colonies, then determined the colors of workers coming to a grid of bait stations (Wilson et al., 1971). Remarkably, each color of workers was found only at contiguous bait stations, identifying that color's territory. The second method takes less time and exploits the aggressive behavior of ants from different colonies to one another. Nonnestmates brought into contact typically fight, but nestmates do not. Bait, such as canned meat or tuna, is placed on a small board on top of the focal mound, the one whose territory is to be determined, soon attracting large numbers of workers. Baited test tubes are placed at approximately 1-m intervals along eight to 10 lines radiating like the spokes of a wheel from the focal mound. Scouts rapidly recruit nestmates to these tubes. Tubes from the spokes are then brought to the mound, and the ants within them are brought into contact with a few ants from the mound bait. If the ants in the tube are from a different colony, a fight will result; if not, the ants will mingle peacefully. A variant is to place a tube known to contain focal-colony workers mouth-to-mouth with each of the tubes along a spoke, one at a time, working outward from the mound. The boundary lies between the last spoke tube whose occupants mingle peacefully with the focal workers and the first where a fight results. Repeating this process on eight or 10 spokes and connecting the resulting boundary points establishes a polygon that approximates the focal colony's territory (Figure 8.1; Plate 11). We

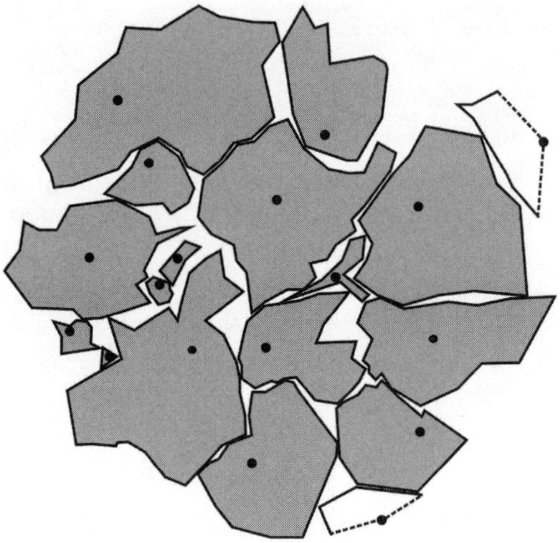

Figure 8.2. Territorial mosaic of a neighborhood of *Solenopsis invicta* colonies. Territory size increases with colony size and decreases with the number and size of neighbors (figure by Eldridge Adams, used with permission).

can map entire neighborhoods in this way, to make visible the mosaic of ownership that is so obvious to the ants yet otherwise invisible to us (Figure 8.2).

The Mechanics of Territory Defense

The fire-ant territories that blanket suitable habitat are of various sizes, proportional to colony size (Tschinkel et al., 1995). Territory size determines how many colonies can fit on a piece of land. If an average of 100 m^2 were needed for the territory of a mature colony, about 100 mature colonies could live on a hectare, a density commonly observed in pastureland. Each territory typically abuts those of five to seven neighbors (Figure 8.2). Boundaries between neighbors are usually no-ant's land, a 10- to 100-cm–wide zone into which fire ants from either colony wander only rarely. In these zones, less aggressive ant species can feed in relative peace, and newly mated fire ant queens have a somewhat better chance of survival. The mutual repulsion among colonies causes the mounds to be rather uniformly distributed on the landscape (Figure 8.3), and the uniformity is especially obvious when the ants strongly prefer to settle along an edge, such as a road or curb. Under those circumstances, the distances separating colonies are fairly constant (Figure 8.3, inset).

The territory grows with the colony, but colonies of the same size may differ by threefold or more in territory area. Factors other than colony size, most often interactions among neighbors and habitat quality, must influence territory size. First, territory is not simply a property of the resident colony but is the outcome of interactions among the workers from neighboring colonies. A colony's ability to defend territory should therefore depend on the size of its worker population.

Eldridge Adams tested this logic by removing tens of thousands of foragers that came to large baits throughout their territory (Adams, 2003). By repeating the process several times, he reduced the worker force of experimental colonies by an average of about 70,000 workers. Their territories shrank by an average of about 40% within three days or less, but those of unmanipulated controls did not. For one colony, he kept the workers alive in the laboratory and later added them back. The colony rapidly regained over 60% of the lost territory (Figure 8.4). Clearly, number of foragers and territory size are related in some numerical way, but interestingly, some buffer against loss seems to exist—a substantial number of workers had to be removed before any territory was lost.

Theory suggests (and experimentation supports) that a territorial contestant should adjust its defensive efforts in proportion to the

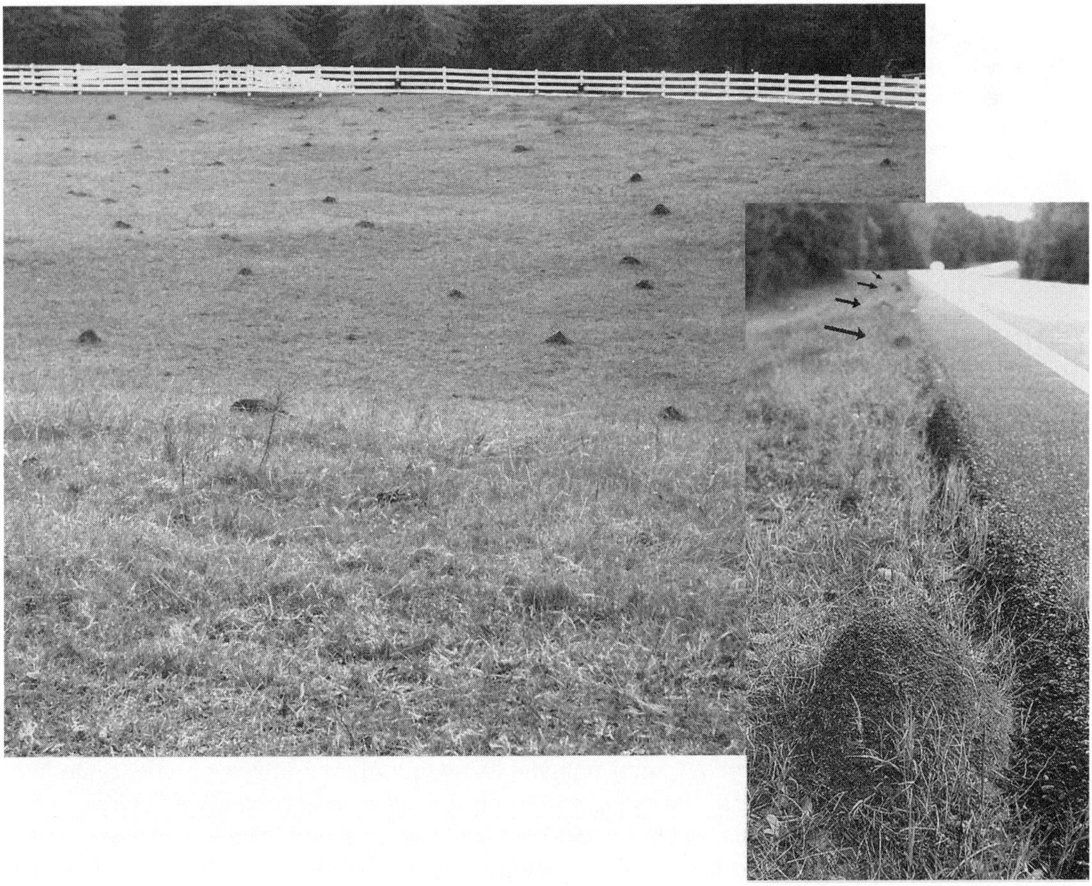

Figure 8.3. The mutual repulsion associated with territorial behavior causes colonies to be rather evenly spaced, as in this horse pasture in Georgia. When this repulsion is combined with a preference for edges, such as roads, colonies are spaced uniform distances apart (inset). Arrows point to distant colonies.

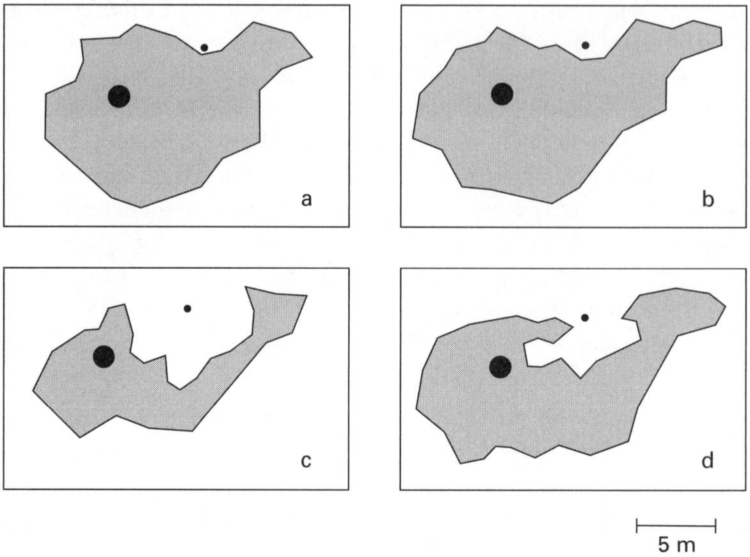

Figure 8.4. Sequential changes in territory size for a colony of *Solenopsis invicta*. The position of the nest mound is indicated by the large, filled circle and that of a neighboring nest by a smaller circle; the areas of the circles are proportional to mound volume. a. The original territory area was 117 m^2. b. Over 28 days without disturbance, the territory area increased slightly to 121 m^2. c. Six days after 17,000 workers were removed, the territory had shrunk to 58 m^2. d. Seven days after the workers were returned, the territory area had increased to 97 m^2. Reprinted from Adams (2003) with permission from Oxford University Press.

richness of its territory, and in fact, some animals do just that, defending larger territories in poorer habitat. Would fire-ant workers occupying a food-rich area defend a smaller territory than those of an equal-sized colony occupying a more meager territory? Adams simulated this situation by choosing size-matched pairs of colonies and providing one of each pair with frozen crickets and sugar water at a rate twice the minimum daily intake. Although the colonies ate this food with gusto, often until they could not stand the sight of it, they did not change their territory size any more than did the unfed ones. This result suggests that territory is not based on the economics of foraging costs and benefits. Perhaps economics would play a role if the neighbors presented no opposition to expansion, but this situation is rare in the USA.

Nevertheless, territory size changed as the summer wore on. Large colonies lost territory, and small colonies gained it, and supplementing with food made no difference. The percentage size change declined from strongly positive for the smallest colonies to strongly negative for the largest colonies (Figure 8.5), because of differences in the seasonal cycles of colonies of different sizes. As colonies grow, they invest increasingly in sexuals during the spring and early summer, reducing worker production and increasingly permitting colony size to decline until late summer. Territory size changes in parallel with these relative changes in the worker force. These seasonal patterns probably allow small colonies to grow and new colonies to enter the population much as saplings in a forest must wait for the giants around them to die or weaken in order to grow. The annual weakening of large colonies through sexual production provides that opportunity.

The importance of forager number suggests that the relative abundance of foragers from neighboring colonies at the boundary mediates gain and loss of territory. If so, then inducing greater forager densities on one side of a boundary than the other should cause the boundary to shift toward the lower density. For about a month, Adams fed one colony of matched neighbors along a two-meter section slightly inside the boundary by scattering finely divided food twice daily. The amount of food was enough to elevate foraging but not enough to satiate the colony.

Feeding did indeed induce approximately 25% higher forager densities in these zones than in unsupplemented control zones at the opposite side of the same territory. Despite the increase, the territory boundary did not move any farther at the supplemented than the unsupplemented zones, nor did territory size increase overall. In light of the dramatic territory losses that follow the overall reduction of the forager population, why did this local increase not have the opposite effect? Perhaps the density increase failed to exceed the substantial (and undetermined) buffer to territory change noted earlier.

A Territorial Model

Models can usefully guide research, provided they are based on realistic assumptions and can be modified upon testing. A good territorial model must predict not only the size but also the shape (and perhaps other qualities) of territories. If we picture a colony sending its workers into the field, not only to forage but also to expand the foraging area, two "thought models" come to mind. The first suggests that a colony optimizes the economic gain from foraging in relation to distance, the solution being the territory. Such a simple economic model seems inappropriate for fire ants because it does not include competition from

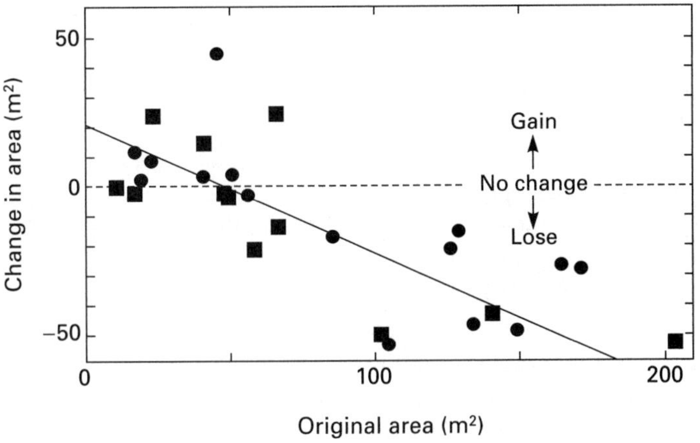

Figure 8.5. In the course of a summer, colonies with large territories tended to lose area, and those with small ones to gain. Colonies receiving food supplements (circles) did not differ from unsupplemented controls (squares). Adapted from Adams (2003).

neighbors, and we already know that territory size is not adjusted by foraging economics. A better model begins with the image of outward pressure. Workers move outward into new areas, expanding the territory until increasingly frequent contact with neighboring foragers (who are similarly expanding outward) stops them, much as a balloon ceases expanding when the elastic resistance of the rubber wall plus atmospheric pressure equal the expansive pressure of the gas within. The territory can be thought of as a two-dimensional balloon.

Eldridge Adams (1998) created a successful general territorial model based on the concept of aggressive pressure in neighborhoods. Neighboring territorial contestants apply pressure on each other to yield ground through mutual repulsion and avoidance or through fighting, depending on species. When Neighbor 1 applies relatively more pressure on Neighbor 2 than vice versa, the boundary shifts to enlarge the territory of Neighbor 1 while reducing that of Neighbor 2. Among many animals, including *S. invicta*, the ability of a territory holder to apply aggressive pressure on its neighbors depends on several factors, which are included in the model. (1) Expansive pressure decreases with distance from the core of the territory. (2) Aggressive pressure varies with fighting ability, which for a social insect colony largely means worker number, as we saw earlier. What else it may mean (e.g., larger workers, better nutrition) is not known, nor is whether some workers specialize in territorial defense. (3) Fighting ability relative to that of the neighbor is more important than absolute fighting ability. (4) As a colony expands its territory at the expense of its neighbors, it must defend an ever-longer boundary, reducing its outward aggressive pressure at any given point along that boundary. In contrast, neighbors that lose territory have less boundary to defend and can therefore increase the aggressive pressure at any point. (5) A colony's aggressive efforts therefore tend to be redirected from a neighbor with whom it is enjoying little success to one it can win against, and this tendency maximizes territory for a fixed effort. (6) Thus, the contesting colonies arrive at a point at which the aggressive pressure applied by each is equal, and a stable boundary emerges.

Adams's model uses three primary, readily measured input variables—fighting ability (mound volume/colony size), nest distance, and territory area—in a computer algorithm that solves for the positions of boundaries (points of equal pressure) between neighbors. The result is a map of the territorial boundaries of the entire neighborhood produced from the spatial locations and sizes of the colonies. To test the model, Adams mapped 27 real colonies on our Southwood Plantation plots and compared the shapes, sizes, and displacements (distance and direction of the mound from the territory's center of gravity). Agreement between the territories predicted by his model and the real, mapped territories was quite good; the correlations were 0.73 for area

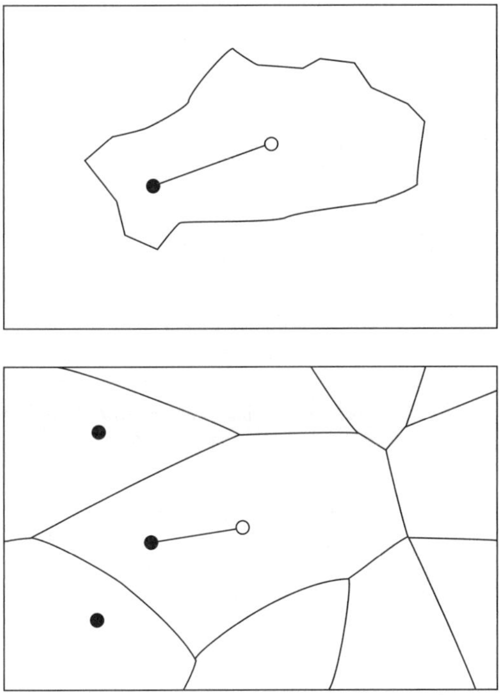

Figure 8.6. Observed territory of a fire-ant colony (upper panel) and the territory predicted for it by Adams's model (lower panel). A vector connects the nest site (solid circle) with the territory's center of gravity (open circle). Reprinted from Adams (1998) with permission from the Ecological Society of America.

and 0.63 to 0.68 for various shape variables (Figure 8.6). Remarkably, the model even predicted that, when colonies were crowded onto a narrow erosion-control ridge, their territories on the flats on each side were elongated or dumbbell-shaped.

Real colonies, however, were consistently more elongated than predicted, and their mounds were farther off center. This elongation could result because in real colonies (but not in the model) some sectors receive more workers and defensive effort than others, perhaps because they contain more resources (we know forager density responds to food abundance). Possibly the underground trail system through which most traffic moves (see below) allocates workers unevenly to territorial sectors.

A more direct test of the model's assumptions was to modify the fighting ability of a neighbor colony by simply eliminating it. Figure 8.7 shows what happened to the territory boundaries of two colonies when two of their neighbors were killed (Adams, 1998). The expansion of the two remaining colonies in the direction of their liquidated neighbors was suddenly unopposed, and boundary shifts began within days. Over two months, the surviving colonies took over much of the newly vacant zone. More significantly, their boundaries with their surviving neighbors receded because movement into the newly acquired territory

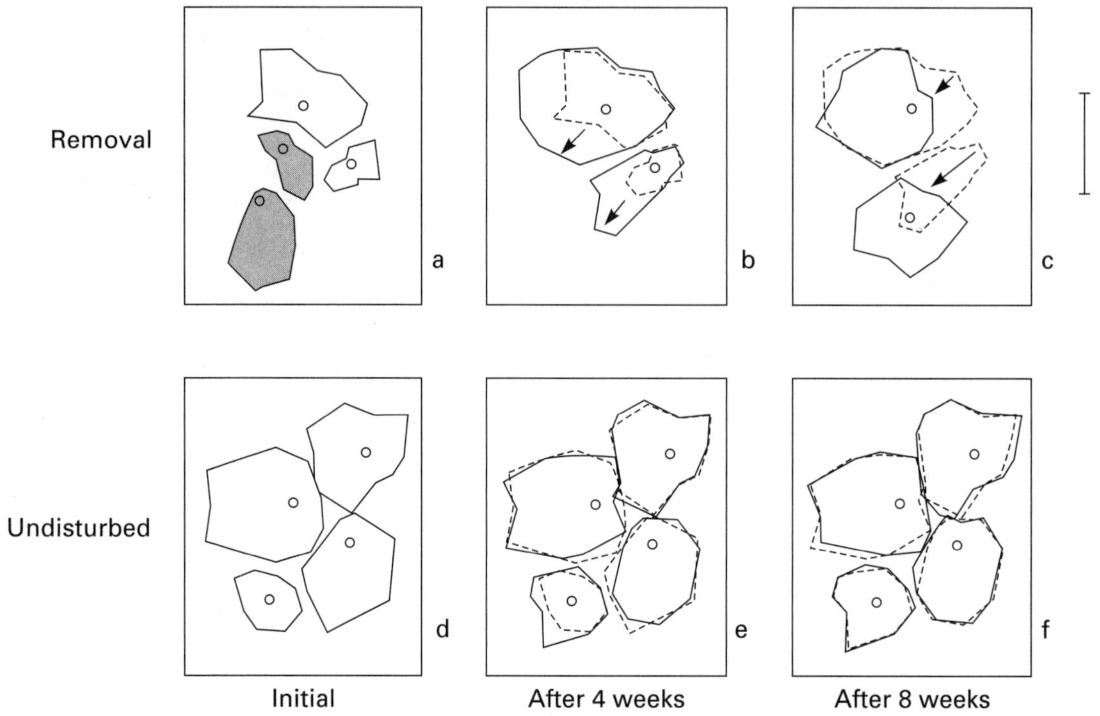

Figure 8.7. A removal experiment showed territorial boundaries to be interdependent. The shaded colonies and five others extending toward the lower left were killed, and territories were remapped after four and eight weeks. Dotted outlines show the initial territory boundaries. Colonies whose neighbors had been killed expand into the vacated space, eventually shifting their entire territories in that direction. Controls showed no such changes. Reprinted from Adams (1998) with permission from the Ecological Society of America.

weakened pressure against those neighbors. Therefore, their territories not only enlarged but also shifted toward the vacated area. No such boundary changes or shifts were apparent in a set of four control colonies.

Clearly, all the boundaries of all the neighbors were interdependent, determined not through simple two-party interactions but through simultaneous neighborhood interactions extending even to more remote neighbors. The size of each territory depended not only on the resident's characteristics but also on those of its neighbors, both near and far. The territory will be smaller if the colony has many large or aggressive neighbors, and vice versa. A colony that loses ground to a stronger neighbor may redirect more of its aggressive pressure to another boundary, gaining territory there, causing the whole territory to shift in the direction of the weaker contender, as in the removal experiment above. Sequential maps show that these effects move outward over the entire neighborhood like a wave, gradually damped by distance.

Neighborhood interactions can also explain differences in forager density. For colonies of a given size, the forager density increases as the territory size decreases (Figure 8.8). Colonies probably field a characteristic proportion of their worker force as foragers, so as the strength of neighbors increases, these foragers are compressed into smaller territories, producing higher forager densities.

A model like Adams's is mathematical and provides no information on the mechanism by which the ants actually acquire and hold the territory. How the behaviors of individual workers combine to produce a territory is not well understood. Before Adams and I began our territorial work, I expected continual boundary skirmishes, a sort of slow burn, desultory war between neighbors. Although small scuffles can sometimes be seen along boundaries, they seem too rare to explain the existence of the boundary. Large food items simultaneously discovered by both neighbors can create larger conflicts, but these occur only occasionally and haphazardly and seem unable to account for such stable boundaries. The behavior of workers at boundaries suggests that these are maintained by mutual avoidance rather than by bellicosity. Foragers from neighboring colonies soon began laying trails into the territory of a colony weakened by forager removal. These conspicuous trails were on the surface and often very long because the neighbor lacked underground tunnels in the new area. Neighbor foragers probably continually invade each other's territory until they encounter a neighbor forager and then leave that area. The colony that deploys more foragers in an area will thus gain territory because a higher proportion of its workers

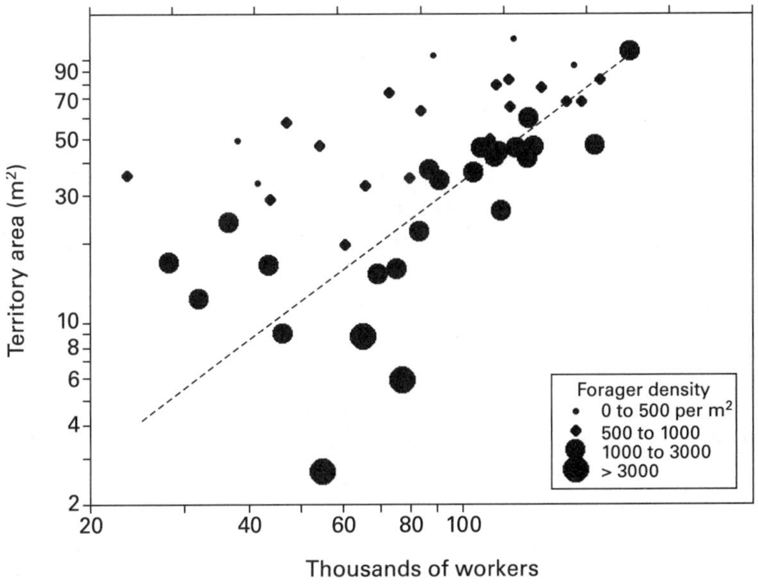

Figure 8.8. Territorial pressure from neighboring fire-ant colonies can compress territories, increasing the density of foragers within them. For colonies of a given size, the smaller the territory, the higher is the density of foragers within it.

will fail to encounter a neighbor and will remain. Reduction of foragers would allow more neighbors to enter unopposed and eventually to gain that territory. Such numerical effects would require little fighting but could account for the effects attributed to "aggressive pressure." Changes in relative density of workers at boundaries should therefore cause boundaries to shift, but in the single experiment that tested this prediction (Adams, 2003), it did not. Obviously, much remains to be understood about the behaviors that create territories and boundaries.

A Territorial Postscript. Several species of ants mark their territories with colony-specific pheromones (Hölldobler and Wilson, 1990), suggesting that a test for such pheromones in fire ants might be worthwhile. In laboratory colonies of *S. geminata,* Jaffe provided evidence that scouts mark territories with a colony-specific pheromone from their metapleural glands (Jaffe and Puche, 1984). Intriguing as that work was, it should be repeated with *S. invicta* and more rigorous methods. More recently, these glands were shown to contain various fatty acids with antibiotic effects, suggesting a possible hygienic function (Cabrera et al., 2004).

Foraging Tunnels

The way in which fire ants get around their territory is little short of fantastic. Anyone who waits by the side of a mound to see the foragers pour forth on their way to their hunting grounds waits in vain, because even as our watcher waits, under his feet a steady stream of foragers is departing from and returning to the nest in underground tunnels. (One of the consequences of this underground traffic is less nuisance contact between fire ants and humans. Such contact occurs most commonly at the mound.) Two to five such trunk foraging tunnels radiate from the mound, 2–10 cm below ground, and extend to the vicinity of the territorial boundaries (Figure 8.9). The tunnels branch repeatedly and provide exits to the surface at intervals, so no part of the territory is more than 30 to 50 cm from a tunnel exit (Markin et al., 1975c). A forager therefore makes all but the last 30 to 50 cm of both the outward and return trips underground. Successful foragers take the subway home.

Mapping out these subterranean highways is a tedious business, but Markin and his coworkers did so for two complete and three partial tunnel systems. Beginning at a trench around the mound, they repeated filled sections of the tunnels with a lead alloy and excavated them, then filled the next and so on, working outward from the mound until the entire tunnel system was mapped. Byron and Hays, by repeated trenching at intervals, made partial maps of the tunnel systems of a 16-colony neighborhood (Byron and Hays, 1986). More than three-quarters of the mounds were connected to one or more other mounds through underground tunnels. Workers from connected active mounds responded to each other with tolerance, so they were probably from

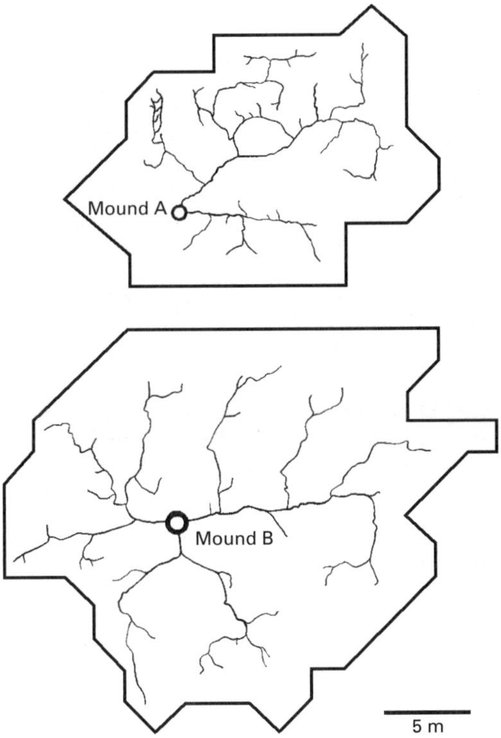

Figure 8.9. Foraging tunnels of two *Solenopsis invicta* territories. The tunnels act as distributary and tributary systems for the flow of foragers to and from the foraging areas. Adapted from Markin et al. (1975c).

polygyne colonies (although, unfortunately, the authors did not establish polygyny by other means). In contrast, colonies whose workers were aggressive toward each other were never connected by tunnels and were probably separate, monogyne colonies with individual territories and tunnel systems (Figure 8.10). Active mounds of both types were usually connected to inactive mounds as well, probably as a result of underground emigration from one mound to another, as occurs within a territory one to several times a year. These underground routes explain why emigrations are rarely observed, except when the colony must cross an expanse of concrete or other impervious material.

The tunnel system is constructed by the foragers and is a large investment in colony infrastructure. One of the colonies mapped by Markin et al. (1975c), which had a territory of 53 m^2, had constructed almost 60 m of tunnels fanning out from the mound; the other, with a territory of 94 m^2, had 84 m of tunnels. Both figures are underestimates, because very few of the smallest tunnels were cast. Fire ants do not construct tunnels in simple and linear fashion, like Swiss tunnel-boring engineers drilling through the alpine massif. Rather, they dig shallow vertical shafts every 10 to 20 cm along well-used surface trails and from their bottoms make several horizontally radiating tunnels, some of

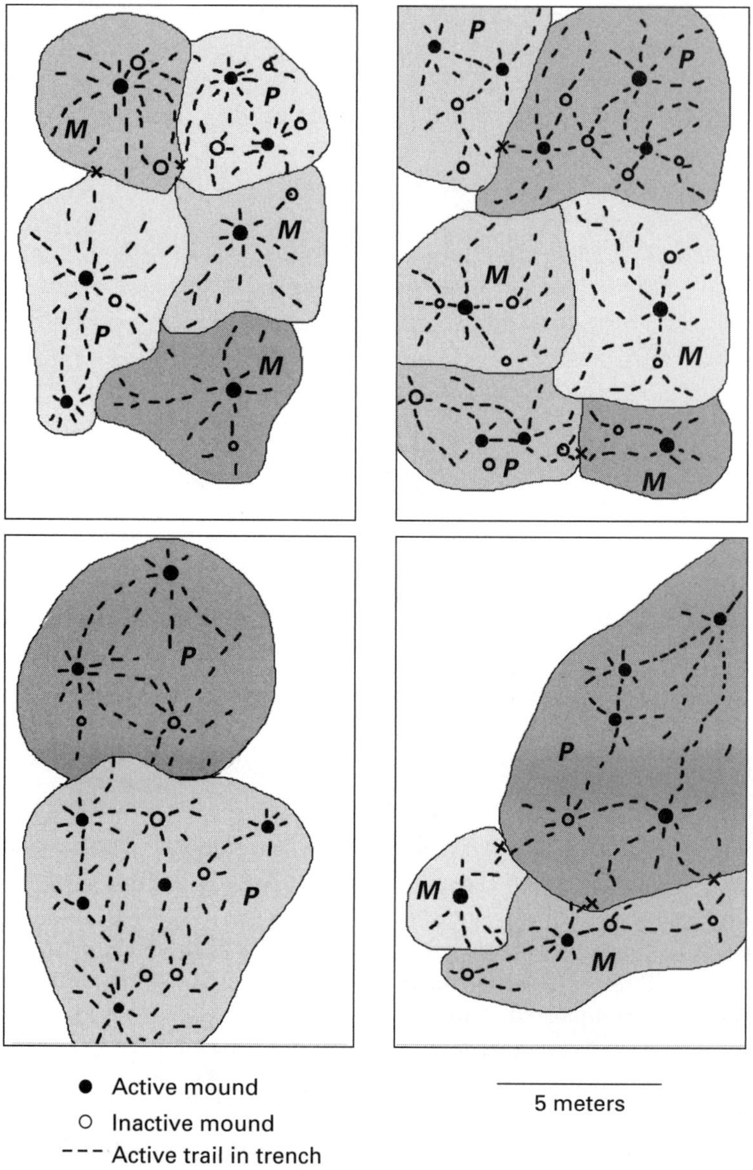

Figure 8.10. Foraging tunnel connections among mound neighborhoods. Dashes indicate the direction of foraging tunnels where these intersected with trenches. Territories containing multiple connected active mounds were probably polygyne (P), whereas those with single mounds were probably monogyne (M). Each spot where fighting was observed between neighbors is marked with "x." Adapted from Byron and Hays (1986).

which eventually connect to each other. Thereupon, all the dead ends are closed off, and the connecting tunnels are enlarged and straightened. This cycle is usually completed in 48 hours or less. In this way, segment by segment, the ants construct a subterranean transport system on an enormous scale, consisting of 100,000 to 200,000 body-lengths of tunnels or even more. On a human scale, the equivalent would be 200,000 to 400,000 km of tunnels. Individual tunnels can be as long as 20 to 30 m (Figure 8.11). Underground tunnels become especially obvious on dirt roads after rains, when the dirt dikes resulting from tunnel reconstruction run straight across roads , waiting to be flattened by rain or human traffic.

We have no way yet of estimating the costs and benefits of the tunnel system, though evolutionary theory requires that benefits (in fitness, measured as alate production) outweigh costs. Among the

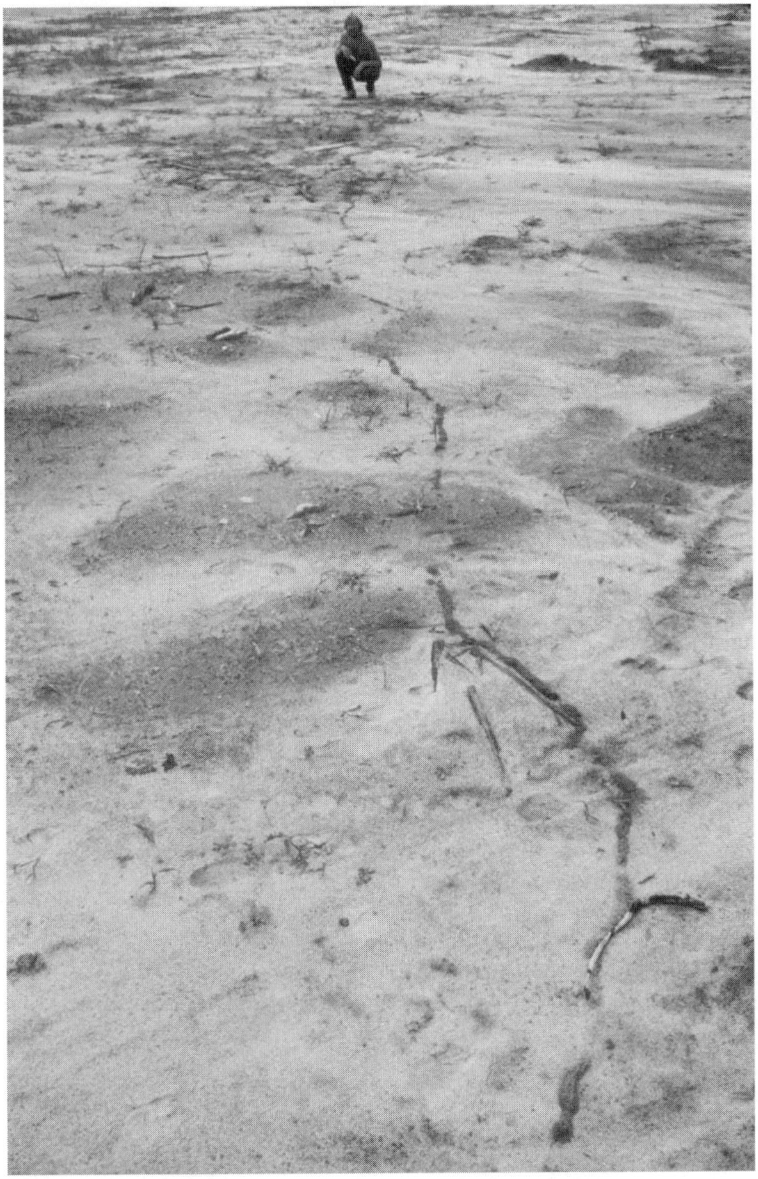

Figure 8.11. An underground foraging tunnel in the process of reconstruction after heavy rains. The bare ground makes the tunnel easily visible at this stage, but once it is completed the surface features will be erased by wind and rain.

benefits of a tunnel system are, first and most obviously, that exposure of foragers to bird, spider, beetle, and mammalian predators is minimized (but who knows what goes on underground?). Second, foragers escape from activity-restricting, possibly lethal surface temperatures. Third, fire ants have only normal insect desiccation resistance (Hood and Tschinkel, 1990; Arab and Caetano, 2002). Without the tunnel system, foragers could probably not exploit the full territory via surface traffic on hot, dry days; although the surface soil may sizzle at a lethal 45 to 50°C, the tunnels below remain at a pleasant 30 to 35°C. Porter estimated that foraging tunnels extend a colony's foraging time and range by perhaps 40% or more over that available by surface traffic only. The payoff is larger territories, larger colonies, and more alates, which probably more than repay the costs of tunnel construction and maintenance.

Foraging tunnels may have other, even less well-known benefits. To an unknown extent, the tunnel system may extend foraging into the domain of subterranean fauna, such as earthworms and soil-dwelling insect larvae, both of which are recorded as prey items. Some species of the subgenus *Diplorhoptrum* (thief ants) of the genus *Solenopsis* are subterranean predators that continuously extend fine, branching tunnels throughout the soil to search for prey (Thompson, 1980, 1989). Perhaps this system shares a common evolutionary origin with the tunnel system of *S. invicta.*

A final benefit of a tunnel system is not apparent in North America. In South America, fire ants on the ground surface are frequently harassed by parasitic phorid flies of the genus *Pseudacteon*. These parasites probably take a major toll. Moving the foraging traffic underground greatly reduces exposure to them and must confer large benefits by reducing worker losses. Such parasitism may have provided much of the selective pressure for the evolution of the tunnel system (Porter et al., 1995c).

9

Food

The mechanics of foraging and food sharing are described in later chapters. Here, we simply ask: What do fire ants eat, and how much? Ant diets range from mostly sugary liquid (e.g., that of *Formica rufa,* the wood ant; Skinner, 1980) to mostly prey protein (as in many ponerines, like *Odontomachus brunneus,* the trap-jaw ant), but most ants, including fire ants, are dietary generalists, relying on opportunistic predation, scavenging, and collection of sweet exudates from plants or homopteran insects like aphids (the meats-and-sweets diet). *Solenopsis invicta*'s diet also changes with season (Stein et al., 1990)—during the warm season, the larvae need protein, but in the cool season worker maintenance requires only sugar. Several enzyme classes needed for the digestion of a rather generalized diet have been detected, but not quantified, in major workers of *S. richteri* (Ricks and Vinson, 1972b).

What Do Fire Ants Eat?

Diet bears on the question of an ant's ecological role, but quantified studies are relatively rare. In the laboratory, most species of ants will prosper on sugar water and whatever insects or other animal matter they are offered. In the field, they are attracted by almost anything edible, even canned Vienna sausages (known to impoverished graduate students as "tube steaks"). Their actual diet is probably determined not by strong preferences but by what is present in the habitat, can be captured or found, and is not being carried off by competitors. Actual diet must be determined from the burdens carried by captured ants. Because fire ant foragers travel mostly in underground tunnels, their traffic must be exposed to view and capture. For example, a shallow trench dug around the mound forces the ants to cross an open area, or they can be captured before they enter the foraging tunnels (Hays and Hays, 1959; Wilson, 1969; Morrill, 1977b; Ali et al., 1984). Because the ants cut up most of their larger food items, less than half to three-quarters of the bits can typically be identified. Most such studies have ignored liquid food carried in the foragers' crops.

Overall, these studies confirm that fire ants have a very catholic diet, consuming a tremendous variety of arthropods and other invertebrates, which they either scavenge or kill. A list of solid food items typically contains at least 30 to 40 taxa. The opportunism of fire ants is apparent from dietary differences in different habitats (Hays and Hays, 1959; Wilson, 1969; Morrill, 1977b; Ali et al., 1984; Showler et al., 1989; Tennant and Porter, 1991; Vogt et al., 2002a). For example, termites top the list in young pine plantations, making up 16% of the items, but rank only 14th and less than 1% in nearby pasture. In central Texas, collembola (primitive wingless insects that live in soil or leaf litter) were the most frequent item and earthworms were second. In some sugarcane fields, fly larvae were the most abundant prey (12 to 14% of the total), whereas in others, crickets made up 73% of the total. Studies in grassy, weedy, and nonweedy sugarcane fields showed that foragers respond to prey abundance rather than diversity; for example, when fire ants foraged in greenhouses infested with whiteflies, 90% of the prey consisted of whiteflies. Further, immature whiteflies were taken ten times as often as adults, probably because the adults can escape by flying. Similarly, when alfalfa weevils or pea aphids were available in the greenhouse, they figured largely in the ants' diet (Morrill, 1978). In four different Oklahoma habitats, the foraged items reflected abundance in the habitat. At the lakeshore, flies and their immatures predominated; at wooded roadsides, termites made up over a fifth of the total; in grasslands and pasture, seeds formed an important component of the diet (Vogt et al., 2002a). The amount of liquid food brought home by foragers also varied with habitat and season, in parallel with the source. All in all, the answer to the question, what animal matter do fire ants eat, seems to be, anything they can get—they are essentially omnivorous. As a practical matter, the result is that by far the largest portion of their animal-matter diet is insects, although occasional vertebrate carrion may provide a temporary bonanza. In spite of reports that fire ants devour baby birds, rodents, politicians, and even cattle, they are primarily opportunistic predators and scavengers of small invertebrates.

How important is predation, as opposed to scavenging? Again, the answer is probably a matter of opportunity. Fire ants are clearly effective predators, often suppressing prey populations, and therefore sometimes control agricultural pests. Fire ants pluck boll weevils from cotton "squares," mosquito eggs from wet soil, sugarcane borers from their canes, termites from their burrows, pecan weevils from their pecans, horn flies and stable flies from their cowpats, and lone star ticks from the places where they lurk waiting for hosts (Harris and Burns, 1972; Agnew and Sterling, 1981; Dutcher and Sheppard, 1981; Reagan, 1982; Fillman and Sterling, 1983; Summerlin et al., 1984b; Wells and Henderson, 1993; Lee et al., 1994).

The importance of plant material in fire ant diets almost certainly varies greatly. Small seeds seem to be a small but consistent part of *S. invicta*'s diet (0.5 to 4% of the solid items) and a substantial part (30% of solid food) of *S. geminata*'s diet (Wheeler, 1910; Van Pelt, 1958; Wilson and Oliver, 1969; Tennant and Porter, 1991). Under some circumstances, fire ants may also damage young plants of some agricultural species, including okra, eggplant, citrus, potato, soybean, and corn. The ants may be primarily after the sugary sap. Depending on species, the ants gnaw on roots, growing tips, flowers, stems, fruits, or sap, ingesting unquantified amounts (Smittle et al., 1983, 1988; Adams, 1986). Plant parts made up only 1.3% of the solid items in the study by Tennant and Porter (1991), but plant sap may have formed part of the supply of liquid food. Hays and Hays (1959) tested 18 species of germinating seeds and young plants and reported damage only to peanut, corn, and okra seeds and to young plants of okra. Under some laboratory and field conditions, especially starvation or confinement, fire ants will gnaw a variety of seeds (Drees et al., 1991; Ready and Vinson, 1995; Morrison et al., 1997a) and germinating soybeans and their roots (Shatters and Vander Meer, 2000). *Solenopsis invicta* is probably as opportunistic when feeding on plants as it is on animals. Modern agricultural practices often leave fields very low in weeds and arthropod life, a circumstance that may favor feeding on some species of plants.

Plants attract ants with extrafloral nectaries, gaining some protection from herbivores. Many ants, including *S. invicta* and *S. geminata*, also collect the nectar they produce, preferring some plants because of the higher sugar and amino-acid content of their nectar. Presumably the plants are repaid in greater protection from herbivory (Lanza, 1991; Lanza et al., 1993).

The most careful and quantitative study of fire-ant diets is that of Tennant and Porter (1991), who compared the diets of *S. invicta* and *S. geminata* foraging simultaneously in the same areas (so the effects of habitat differences on diet were minimized). They found that earlier studies of fire-ant diet underestimated the amount of liquid food collected by fire ants of both species. Irrespective of season, they returned with liquid food five times as frequently as with solid food, so almost 80% of the successful foraging trips yielded liquid returns. This liquid contained about 25% dissolved solids (0.05 mg per average crop load). The liquid crop load of both species was about 35% of their body weight, so two to five times as much food may be returned in solution as in solid form.

These fluids contained glucose and fructose, and most also contained amino acids, but their concentrations did not clearly identify the sources as plant or animal. Tennant and Porter concluded that the most likely sources of crop fluids were plant sap and homopteran honeydew, possibly from root homopterans associated with exotic pasture grasses (Helms and Vinson, 2003), but because they pooled crop samples they could probably not have detected a minority of ants collecting insect hemolymph. I believe that insect body fluids are probably an important part of fire-ant diets. I have often observed that foragers first "dry" tuna or insect baits by sucking out the fluids and then transport the dry, solid "meatballs" and "tuna sticks" back to the nest.

Although the two fire-ant species collected the same kinds of fluids, *S. invicta* collected about 40% more than did *S. geminata*. In contrast, *S. geminata* collected eight times as many seeds as *S. invicta*, about 30% of its solid diet. Otherwise, the solid food collected by these two species largely overlaps, and the overlap may be one of the reasons why *S. invicta* and *S. geminata* cannot usually coexist. In both species, 50 to 70% of foragers returned without any detectable food, and success rates were lower in summer than in winter or spring.

In sum, fire ants are truly omnivorous, feeding on fluids derived from plants or animals, acting as both predators and scavengers and at times even primary consumers. A small caveat should also be noted: most of the colonies of both species that Tennant and Porter studied were polygyne and thus able to exist at higher densities than monogyne colonies. Because polygyne colonies crop their foraging areas so much more intensely (Porter et al., 1988), they may depend to a larger degree on plant fluids for sustenance. Similar studies should be done on monogyne fire ants.

What Are the Minimal Dietary Needs?

To what degree is this two-part natural diet of insects and sugary water, necessary? Clearly, the brood needs protein, but will any protein do? In spite of their easy acceptance of a wide range of foods, from fried chicken to peanut butter, fire ants require a diet that includes insects for normal growth. Complete substitution of noninsect protein for insect protein resulted in smaller workers whose exoskeletons never darkened (Williams et al., 1987). These workers remained pale and soft, their stingers too soft to be effective and their internal organs visible through the translucent cuticle. The path of colored food sloshing its way through their alimentary tracts could be clearly traced, a minor amusement for bored lab personnel. The return of insects to the colony's diet did not reverse the condition of these unmelanized workers, but later workers were normal. The identity of the magical insect substance that produces healthy cuticles remains unknown.

Although we have known for a long time that diets including sugary solutions in addition to protein and fats cause colonies to grow much faster (Williams et al., 1980), it remained for Porter (1989) to demonstrate the full strength and importance of this requirement by feeding colonies as much as they wanted of six combinations of sugar, insects, and artificial diet. To reduce variability, he created small, homogenous subcolonies from pooled polygyne colonies. Each colony was weighed at the beginning of the experiment and six weeks later at the end.

As might be expected from a protein-free diet, colonies fed only sugar water actually decreased in weight because they could not sustain brood production. The artificial diet, which lacked insect tissue, resulted in only enough brood production to maintain the starting size. Colonies fed only insects increased threefold in size, but this increase was only 40% of the growth achieved by colonies that received sugar water in addition to insects. Both insects and sugary solutions are therefore necessary for the full realization of colony growth potential.

Why do sugar and insects have this large, interactive effect on colony growth? The answer probably lies in the dietary differences between larvae and workers. Workers need primarily sugars as fuel for their various activities, and they need them in solution because, like most adult Hymenoptera, they cannot ingest solid food particles larger than 1 μm in diameter (Glancey et al., 1981c). In the absence of sugar water, workers get their liquid energy food from the insect prey, but most insects contain relatively little hemolymph. Use of hemolymph by workers also reduces the amount available to brood. A limited amount of solid food can be converted to liquid form by fourth-instar larvae, but this process does not seem to produce sufficient amounts for the adults. In sum, then, the effectiveness of the sugar-insect combination results largely from its satisfying the fuel needs of adult workers and therefore freeing the insects for larval consumption. It may also have direct effects on the larvae. The artificial diet, intended for use in the laboratory, provided no advantage over insects and sugar water and is probably not worth the trouble and expense of preparation.

How Much Food Does a Colony Need?

Below a certain rate of food intake, colony growth will be negative (the colony will shrink), and above it, positive, up to a limit. Oddly, despite their ecological importance, very few studies have addressed these relationships in ants. Fortunately, thanks to several feeding studies by Porter and his students, the relationship between *S. invicta*'s food intake and colony size has been characterized (Porter and Tschinkel, 1987, 1993; Porter, 1988, 1989; Tennant and Porter, 1991; Macom and Porter, 1995). Starting with standard small fire-ant colonies (1 g workers + 1 g brood), Macom and Porter (1995) fed each colony one of six diets that differed in the number of crickets per day (fixing their protein intake) and in availability of sugar water: unlimited (= unlimited energy supplement) or none. Each colony was weighed every week until it stopped growing.

When the caloric intake was fixed (crickets only, no sugar), colonies that received twice as much food reach final colony sizes twice as large. In these colonies, the crickets provided the material for both colony growth and worker energy. When unlimited sugar water was added to the diet, the availability of protein limited colony growth, but in a nonlinear fashion. That is, with unlimited sugar, a colony receiving one cricket per day increased by 2.2-fold, but a four-cricket colony increased by only an additional 1.3-fold. When colonies were offered both crickets and sugar, a colony that received two crickets reached a colony size only 1.7-fold larger than that of a colony receiving one. Doubling protein again produced an even smaller differential; a four-cricket colony reached a size only 1.4-fold that of a colony receiving two. Why does doubling protein not double colony size as in the crickets-only treatment? Macom and Porter were unable to pinpoint the reasons for this decline in effectiveness under conditions of unlimited energy.

The amount of food actually collected (not just offered) tracked colony biomass closely in all treatments. No matter what their sizes or diets, colonies take in about 1.14 kilocalories per gram of live ant per week at 29°C (the rate depends on temperature). Colonies receiving both sugar and crickets grow more because the cricket protein need not be "wasted" on providing workers with energy. The cricket-collection rate therefore drops 40% when sugar is available to meet worker energy needs. For poorly understood reasons, the value of this sugar seems to decline with the number of crickets and colony size, as I noted above.

The finding that colonies take in 1.14 Kcal per gram of ant per week, no matter what the colony size, is somewhat perplexing because it is counter to expectation. Metabolic and growth costs decline as colonies grow because brood-rearing rates decrease and the increasing size of workers results in lower respiration rates (Calabi and Porter, 1989). Together, these observations lead us to expect that the per-gram caloric intake of a colony should decline with colony size (Tschinkel, 1993b), not remain constant. I favor the explanation that Macom and Porter's experimental colonies spanned only an 18-fold size range (1–18 g), too small to detect such effects. Over the full five orders of magnitude of natural colony size, we expect only a 25% savings in the per-gram metabolic cost. The expected savings in the Macom-Porter experiment would be only about 2.5% and probably not detectable.

If we accept these inadequacies, the roughly constant per-gram caloric intake allows us to estimate food intake in the field. A single mature colony of fire ants weighing 200 g would take in 230 Kcal per week in insects and sugar water (mostly from homopterans) when the temperature averaged about 29°C. Extrapolating intake to a hectare occupied by about 90 colonies produces a figure of about 4 kg of food, valued at about 21,000 Kcal, flowing into fire-ant colonies per week. (For comparison, a 75-kg human needs about 14,000 Kcal per week). Converted to wet weight, the consumption by a hectare's worth of fire ants would be about 3.1 kg of insects and 13 liters of sugar water per week (in some cases, a large fraction of the sugar is provided by exotic mealybugs on exotic pasture grasses; Helms and Vinson, 2003). If the per-gram intake actually decreases with colony size, as expected, this estimate would be lower by about 25% but is still a large number. No one has compared these numbers with the production on a hectare of pasture, but clearly the fire ant must have a substantial ecological impact on the ecosystems in which it thrives.

Metabolic Storage

When the colony's food intake exceeds need, some of the excess is stored as metabolic reserves in the fat bodies of workers—fat, glycogen, and storage protein—to be drawn on in times of scarcity or when demand exceeds the ability of the workers to collect food fast enough. Much less is known about storage proteins than about fat, and their importance is probably underappreciated. Three types of ant storage proteins have been characterized—(1) hexamerins, each consisting of six subunits totaling between 460,000 and 580,000 in molecular weight; (2) a protein unusually high in glutamic acid and glutamine; and (3) very high-density lipoproteins. All provide amino acids when they are needed. Storage proteins and lipids are not limited to social insects; they evolved to carry insects through metamorphosis, a period of intense rebuilding and energy need, and they still serve this function in ants (for example, in *Solenopsis xyloni* [Wheeler and Buck, 1992, 1995; Wheeler and Martinez, 1995]). The function of storage for colony founding by queens and storage of metabolic reserves by workers were later add-ons evolved from this earlier adaptation. When larvae, the queen, or sexuals are absent, workers accumulate metabolic reserves, so they can synthesize them but apparently have the lowest claim on the resources to do so, storing only what is in excess of colony needs. This type of allocation has not been tested in *S. invicta*. The regulating mechanism may involve juvenile hormone.

Effects of Weather, Temperature, and Season on Foraging

In a sense, heat can be thought of as a resource, because it regulates and limits many aspects of fire-ant lives and is optimized to the degree possible; it is even collected through the mound. One very important temperature-limited function is foraging. Research of this basic type barely seems to register on the mental radar of many modern biologists, but it is just the kind of foundation project that Sanford Porter is drawn to. In a well-grazed cow pasture, Porter determined foraging rates of *S. invicta* at three times of the day once a week for an entire year (Porter and Tschinkel, 1987). Like a mink trapper, he laid baits (hot-dog pieces) out in traplines, retrieving them after 30 minutes. The sample times bracketed the daily maximum and minimum temperatures, as well as day and night. The results apply best to foraging for high-quality food that requires recruitment of nestmates. Results for small or low-quality food might have been somewhat different.

The best predictor of foraging rate was soil temperature at 2 cm

Figure 9.1. Relationship between fire-ant foraging rate and soil temperature at 2 cm depth. Reprinted from Porter and Tschinkel (1987) with permission from the Entomological Society of America.

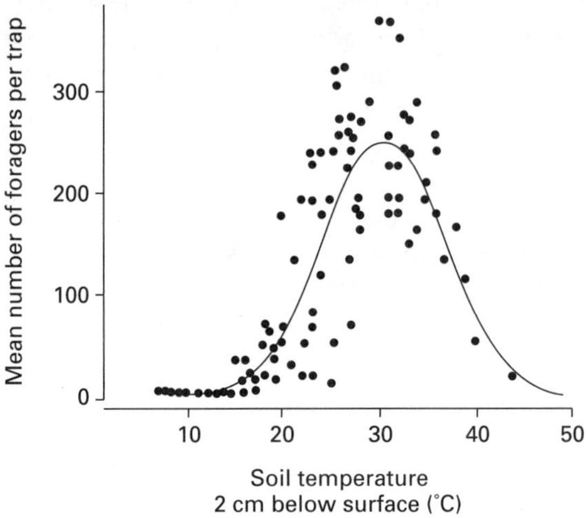

depth (Figure 9.1). It accounted for about 60% of the variation in foraging rate (no other temperature measure explained more than about 40%). Foraging is essentially zero below about 15°C, rises rapidly to a maximum at about 32°C, and drops off sharply at higher temperatures. At the highest observed temperature, 44°C, foraging rate was only about 8% of the maximum at 32°. Significant rates of foraging occurred only over a 14° range between 22 and 36°C. Because foragers travel about 90% of their round-trip distance in underground foraging tunnels, the close relationship between foraging rate and shallow soil temperature is not surprising, nor are the results of other studies, which reported some foraging even at surface temperatures up to almost 50°C!

These data can be viewed in another way. The number of ants per trap is a measure of recruitment, but the proportion of traps that contained any ants at all is a measure of scouting. Between 22 and 36°C, the rate at which baits are found was not related to temperature—essentially all the baits are found. This finding is good news to those of us who use baits to study fire ants.

As expected because traffic is mainly underground, no other single environmental measure was significantly related to foraging. Most of the effect of season was due to soil temperature, although temperature-adjusted foraging rate was lower in the late fall and higher in the early spring, perhaps because sexual brood was being produced in the early spring and that production stopped in the late fall. Adding season, rain, and cloud cover boosted the explained variation in foraging to 81%. We can probably safely conclude that all of these variables act mostly through their effect on temperature.

Because foraging occurs mostly between 22 and 36°C, the daily "foraging window" changes with the seasons (Figure 9.2). In open pasture, colonies enjoyed maximal foraging rate during 60% of the year, reduced rates during another 27%, and no foraging during only 14%. In contrast, colonies in shady woodlots enjoyed maximal foraging during only 42% of the year and reduced foraging during 35%. About 23% of the year would have been too cold for foraging. No doubt, food intake rates would depend strongly on these temperature patterns, possibly explaining in part why fire ants prefer open areas. Reduced temperatures also reduce food demands, however. Combining foraging windows, food demand, and the complexity of nest temperature regulation and tracking via the mound makes simple claims for the role of temperature begin to wobble.

Foraging windows derived from soil temperatures at several locations throughout the southeastern USA showed that high temperature was much less often limiting than was low temperature. High temperatures never limited or reduced foraging for more than 7% of the year, but low temperatures reduced or eliminated foraging during 55 to 65% of the year in the more northern locations, probably playing a role in setting the northern range limit of fire ants. In Oklahoma, maximal foraging was possible during only about 25% of the year (Vogt et al., 2003), much lower than the 40 to 60% in north Florida. In contrast, foraging conditions were maximal throughout most of the year in southern Florida and southern Texas.

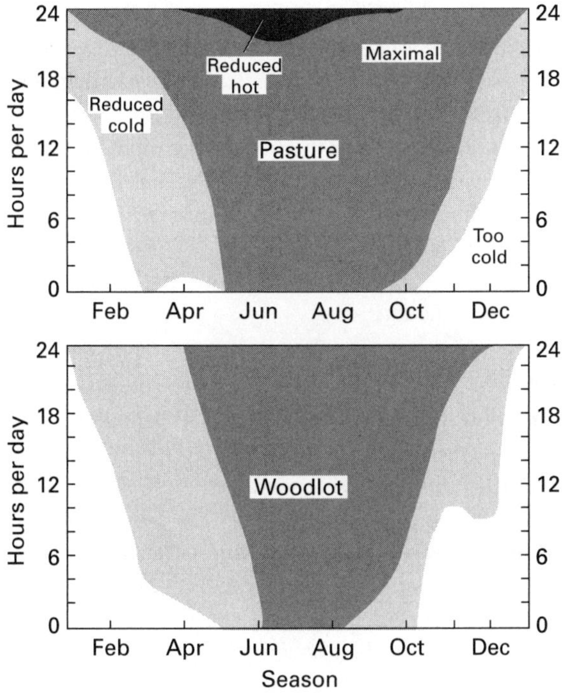

Figure 9.2. Seasonal windows of foraging activity. Optimal foraging conditions and total time during which foraging was possible were lower in the shady woodlot than in open pasture. Reprinted from Porter and Tschinkel (1987) with permission from the Entomological Society of America.

◂•• *Mundane Methods*

In science today, a lot of attention is paid to the latest technology, with its fancy machines stuffed with integrated circuits, and to procedures and products that create a major drain on the resources of large granting agencies. I do not mean to minimize the importance of such innovations, because much scientific progress occurs only when someone invents a new device or method that makes possible something that was previously impossible. I started my own research career proudly using the newly invented technique of gas-liquid chromatography to separate minute amounts of chemicals, and I identified these by using still other expensive and sophisticated machines. These technological advances created an explosion of research in chemical ecology, and they continue to benefit this field. Expensive, complex machines have sex appeal and attract the public's eye.

Research also has a mundane, but equally important, side that few people other than the researchers themselves are aware of, a world of humdrum little problems that require solutions before the fancy machines can be put to use. This work is the research equivalent of getting out of bed and putting on one's clothes—hardly worth talking about. Nevertheless, the solutions can mean the difference between being able to carry on and not. All of us who work with fire ants have had to devise a number of very unglamorous methods so we could proceed, and I find it interesting that so many of us converged on similar solutions (Banks et al., 1981). Perhaps there aren't really that many ways to skin a cat after all.

Rather than repeating standard methods every time they come up in this book, I will describe them here and then simply refer to them in the text. This process may call to mind the old joke about the sailors on shipboard, who have all heard the jokes so often that they simply refer to them by number and break up in laughter. I won't go so far as to number the methods, and you are asked merely to nod in recognition, not to burst out in laughter.

The Dig-and-Drip Method

The problems start right at the outset. Say we need a fire ant colony for a lab experiment. Well, no fire ant colony will voluntarily leave its cozy soil nest and move into your experimental laboratory nest just because you want it to. The ants show hardly any signs of such cooperation. Quite the contrary, they bend every effort to make your attempts to persuade them a hellish misery. What? you say, a big smart human can't outwit a dinky little ant? Well, he or she can, but not without some thought. It's easy enough to shovel a fire ant colony into a bucket and transport it to the lab, but unless you keep them from rushing up the shovel handle and out of the bucket, you will be doing a lot of shaking and stamping and will look for all the world as though you were involved in some primitive dance ritual. Small as they are, workers command disproportionate attention when several hundred are injecting venom into your skin.

You will be a much happier person if you first dust the handle of the shovel with talcum powder (the scented stuff is for babies and sissies) and do the same to the inside of the bucket. If you plan to wade around in boiling fire ants, a pair of tall, talc-dusted rubber boots is a nice touch. Now you can shovel to your heart's content, assured that the angry ants will fall like rain from your dusty surfaces, frustrated in their attempts to kill kill kill.

So now you have a bucket full of dirt and ants. What good is that? Ants in dirt are useless for most experiments because they cannot be seen. How do you separate 100,000 angry ants from 40 pounds of soil, though? Certainly not with tweezers.

Here, an observation made in the wilds is helpful: fire ants float. Having evolved on the flood plain of a major river system that is under water for months at a time (the Pantanal of Paraguay and Brazil), they cope with these workaday floods in a remarkable way. As the water rises in their nest chambers, the entire colony moves upward, finally setting sail from the top of the mound, a couple of hundred thousand ants holding hands to form a floating mat. Workers, brood, sexuals, the queen, all together pull up anchor to drift for weeks, living off food stored in worker crops and eventually cannibalizing their brood like seafarers cast adrift on the South Pacific. Finally the flood subsides or they drift ashore, whereupon they dig a new nest and move in. Crisis over.

We exploit their buoyancy to separate ants from dirt. A slow drip of water fills the bucket until the ants are floating on the surface of the water. From there they can be scooped up with a large serving spoon and placed in a dirt-free lab tray for further manipulations. The ants are so busy clinging to one another that the ball can be briefly held in the hands and rolled around. Not for too long though.

When a lot of nests must be thus collected, most labs set up a manifold system of yards and yards of plastic tubing so water can be dripped into many buckets at once. In the early days, we used laboratory screwcocks to regulate the drips, but plastic saline-drip valves scrounged from hospitals have long since replaced them, because they are much easier to regulate. Dripping water into a bucket isn't noticeably different from dripping saline into someone's veins.

Trap Nesting

If you haven't collected a lot of dirt with the ants, the dirt can be spread in a large tray and allowed to dry out. As it dries, the ants readily move into darkened, moist nests placed in the tray (Markin, 1968). When all have entered, one needs merely to lift out the trap nest to obtain a clean sample of ants and their brood (and queen if present).

Just Want Alates?

In some cases, you might want only the alates from the nest you just dug up and dumped into the bucket. For this rather special situation, USDA scientists built a device for automating the separation of alates from soil (Stringer et al., 1973). A 50-pound cylindrical lard can (can you still get these?) is topped with a no-return trap and a light. The soil with alates is placed in the lard can, moistened, and warmed. Those conditions induce them to fly, and when they do, they fly toward the light and end up in the trap, from which they can be retrieved.

Ant Hotels

The newly acquired soil-free ants can be housed in nests of various designs, from simple petri dishes to fancy formicine mansions. The nest is placed in a plastic tray that serves as a foraging arena for the ants. In the old days, we kept them from climbing out of these trays by dusting the sides with talcum powder, but unless we redusted regularly, the busy ants would nag and worry at the dusty barrier until they had cleared a dust-free path. The next day, you might find the entire colony in the sink drain, or under a copy of *The Insect Societies* on your bookshelf, or in the cushion of a chair (oh joy supreme). Worse yet, they might turn up in the lab next door, drowned in carefully sterilized culture medium or devouring some valuable lab animal. Your relations with your colleagues would be sure to turn frosty.

So when the miracle compound Fluon came on the market, we myrmecologists rushed to buy it in spite of the stunning price of $50 a gallon. The price has since risen to $245 per gallon, and we are still buying it. This is not our caviar; this is a sack of beans to us, the basic stuff we need for survival. Chemically, Fluon is a milky emulsion of tetrafluoroethylene (i.e., Teflon) that can be painted on smooth surfaces. When it is well applied and dry, no ant in the world can get a tarsalhold for months and months. For the ants, there is no escape, unless someone carelessly scratches the magic paint. As wonderful as Fluon is, though, I'm pretty sure it was not invented to keep ants in trays. I've never bothered to find out what the inventor had in mind. I can't imagine that it has been a greater boon to the intended beneficiaries than it has been to myrmecology.

Aspirations

Now we are ready to start some experiments, so we have to make up some experimental nests, each of which contains a specified (large) number of workers, larvae, and pupae. Give a moment's thought to shoveling workers and brood into experimental nests, and you will see that there must be a better way. Here, the aspirator, a device known to entomologists for a century, comes in handy. An aspirator is a mouth-operated vacuum cleaner with which ants are sucked into a bottle or tube. A screen on the outlet prevents the ants from being sucked into your mouth and stinging your tongue (ants are hard to get off a tongue because of all those convenient papillae they can grip). With some practice, one can aspirate ants even from loose dry soil without getting grit in one's mouth, but novices usually spit out lots of dirt-colored, gritty saliva before they master the trick.

When large numbers of ants must be aspirated, we drop a house vacuum line from the ceiling, much like the line for the air hammer in a car-repair shop. Again, with some practice and judicious adjustment of air flow, individual ants can be selected and sucked from their moorings. For collecting large numbers of ants in the field while avoiding mouthfuls of dust, the investigator is wise to modify a battery-powered vacuum cleaner to perform aspirator duty. In its various incarnations, the humble, homemade aspirator is an essential tool for many entomologists and for most myrmecologists. In use, it often draws curious looks from the lay public, to whom it must often seem that the user is breathing in lungfuls of ants.

Separating Adults from Brood

Now, say we have to count workers and brood separately. Workers are no more willing to let go of the brood than mothers are of their babies. In that case, the aspirator vial full of ants is first exposed to ether vapors until the workers just barely stop kicking. Next we pour the comatose workers and brood onto a sheet of black construction paper (OK, any color as long as the paper is rough) and wait until the workers begin waking and staggering around, like drunks with hangovers. In this stage of stupefaction, they hang onto the paper but do not pick up any brood. Now we tilt the paper and let the brood and those workers that are still comatose roll off into a tray, then shake the clinging workers into another tray (Fluoned, of course). We pour the brood with remaining workers back onto the paper and repeat the process until the brood and workers are in separate trays. Now each can be handled separately and conveniently, like pure chemicals.

Speaking of chemicals, many years ago, the fire ant researchers at the USDA lab in Gulfport, Mississippi, perhaps in consultation with Rube Goldberg, invented a winnowing machine to separate the different stages of fire ants (Stringer et al., 1972). A fan in the bottom kept a blur of ants suspended in a whirlwind inside the tube, each type lofted to a different height, depending on the details of its aerodynamic profile, I suppose. Much like a distillation tower in an oil refinery, it delivered sexuals to a tap at one level, workers to another, and brood to a third. In operation, it was a sight to behold.

I have to admit that we have never devised a way to separate pupae from larvae effectively, so we put the workers to work for us. Given four hours or so, workers will segregate pupae and larvae into separate piles. A quick lift of the lid and a precise suck on the aspirator nets you almost pure larvae or pupae (plus some workers of course); only minor cleanup is needed.

Separating Large from Small

For many experiments, we must also control the mixture of sizes of workers, as in all experiments dealing with division of labor. We must constitute the experimental colonies from pure stocks of workers of each size class. How can we separate a seething mass of workers into their constituent size classes? Well, it's not that different from the way a geologist sifts sediment into different particle sizes. In fact, we use a stack of geologist's sieves, from a number 18 to a number 35. These stack together into a nice unit with the coarsest sieve at the top, progressing to the finest at the bottom. Making sure that the inner walls of all the sieves are fluoned to prevent ants from crawling back up, we dump the seething mass of workers on the top. They tend eventually to come to rest in clumps, so every once in a while, you have to bang on the sieves to wake them up again. A light shining on the top helps too. They move down through the sieves to escape the light. When we first used this method, we borrowed a rotap machine from the Department of Geology. The image of graduate student Lois Wood eating her lunchtime sandwich in the midst of its deafening banging and thumping is still vivid in my mind. Each worker comes to rest on the first sieve through whose mesh she cannot fit her body. After a few hours, you have nicely segregated your seething mass into six or seven size classes with very little overlap, a sort of ant chromatography. Again, you can mix them, as you would pure chemicals, in whatever proportions the experimental design calls for.

Is There a Point Here?

I've presented just a few of the mundane methods upon which fire-ant research is built. The point, if there is one, is that the scientific-industrial establishment has hordes of engineers, chemists, biochemists, and biologists who labor through the night devising sophisticated solutions for difficult technical problems. The fruits of their labors are often breathtaking, both in what they can do and in their purchase prices. If you can talk a granting agency out of a pile of money, the device or product is yours, and you need only sit down and apply it to your particular problem. You don't even have to know how it works, for the expertise of an entire industry stands behind you. But when the problem is mundane, you are entirely on your own. Any solution will be the product of your own cleverness, insight, and hands. The question is, is the pleasure of the process any less? Is it any less a part of successful research? Scientists and the people who report on science rarely mention these mundane aspects of science, the little problems solved in humble ways. Perhaps they think this side of science lacks sex appeal, flash, and noise and therefore doesn't seem worth any attention. That's why I mention it here.

10

Mating and Colony Founding

The Colony Cycle

Ant biology cannot be understood without an intimate knowledge of the colony cycle. Here, I will describe how new *Solenopsis invicta* colonies are founded, how they grow and mature, how they reproduce, and finally how they die. Because ants are social, colonies go through life cycles that are analogous to the life cycles of the individuals that make them up but at a higher level of organization. In social insects, the goal of individual reproduction has been moved "upstairs" from the individual to the level of the colony—the goal of colony life is the production of new colonies through the emission of sexual males and females. All the actions of individuals within a fire-ant colony serve this communal goal, and all individuals share the fitness benefits of success. Not surprisingly, this added level of function has created some fabulously intricate and interesting biology, which is only slowly being deciphered.

When myrmecologists first began to study colony founding in the fire ant *S. invicta,* it appeared to be simple. Here was an example of an ant that practiced independent, solitary, claustral (closeted, sealed in a chamber) colony founding in the classic manner of "most" ants. Each virgin queen left the nest, mated, found or made a shelter, and raised her first brood from body reserves. Three decades of research have complicated that picture greatly. We have learned that the fire ant has several colony-founding options at its disposal, bringing each into play in the appropriate situation. Each step of colony founding has been honed and refined by unmerciful natural selection acting through extremely high mortality rates. The richness of detail with which we can now recount the story of colony founding in fire ants is largely a consequence of the tremendous abundance of this ant. Mating flights, newly mated queens, and incipient colonies are abundant. If an experiment requires 1000 newly mated queens, these are readily found. Complex experiments testing several factors simultaneously can therefore be set up and replicated, then modified and run again, until patience or time runs out. Less obviously but just as importantly, the foundress is a tiny, self-contained unit engaged in a well-defined process taking place during an exact season and stage of colony life. No one can observe a newly mated fire-ant queen for long without being impressed by how totally her tiny being is focused on a single goal—to raise that first brood. Nothing else seems to matter; everything else is outside her perceptual world. This tightly bounded and focused condition makes colony founding highly amenable and attractive to scientific study.

Production and Maturation of Sexual Alates

The production of sexuals occurs primarily in the spring. Both the haploid males and the diploid females develop through four larval stages but grow far larger (5 to 6 mg) than do worker larvae (0.75 to 1.3 mg). Their sex is not visible to us until they pupate (Plate 6). Development of the adult within the pupal skin is apparent through the darkening of the eyes, the tips of the mandibles, and other strongly hardened parts until finally the pupa looks like an adult alate trussed up like a turkey, ready for the oven.

The appearance of adult sexuals in the nests occurs as early as February in Tampa, Florida, and as late as the end of April in northern Mississippi and Alabama (Markin et al., 1974c). These dates are clearly affected by weather: during the mild winters that began in the mid-1980s and continued into the 21st century, sexuals appeared earlier than normal.

Adult sexuals are not ready to fly and mate the moment they emerge from the pupa. The females, especially, must undergo maturation for about two more weeks (Keller and Ross, 1993b; Tschinkel, 1993b). During the maturation period, the males gain little weight, but the females feed copiously and almost triple their body weight, from about 3 mg (dry weight) as pupae and new adults to about 7 to 9 mg when ready to fly. Females gain fat about 35% faster than they gain lean weight. They are about 30% body fat when they first emerge from the pupa but about 50% when ready for the mating flight. A large part of the lean weight gain is probably storage protein (Wheeler and Martinez, 1995). The sequestered fat and protein make up the metabolic stores from which females rear their first brood during the claustral period (Toom et al., 1976a, b, c) and on which they survive until the new workers can bring home significant amounts of food. A smaller store of glycogen serves as fuel for the mating flight.

The ovaries of female alates appear in the pupal stage and mature during this preflight period, so by the time of the mating flight, one to four eggs are almost ready to be laid. Each ovary consists of 80 to 100 narrow, conical tubes called ovarioles (Hermann and Blum, 1965) (Figure 10.1). Meiosis occurs in the tissue at their inner ends, and each egg cell thus formed is incorporated into a follicle that moves down the ovariole as it matures and gathers yolk protein and other materials. When finished, the eggs of all ovarioles are ovulated into a temporary storage sac (the calyx) from which they are laid at the appropriate time. Just before ovulation into the calyx, the follicle cells secrete a chorion

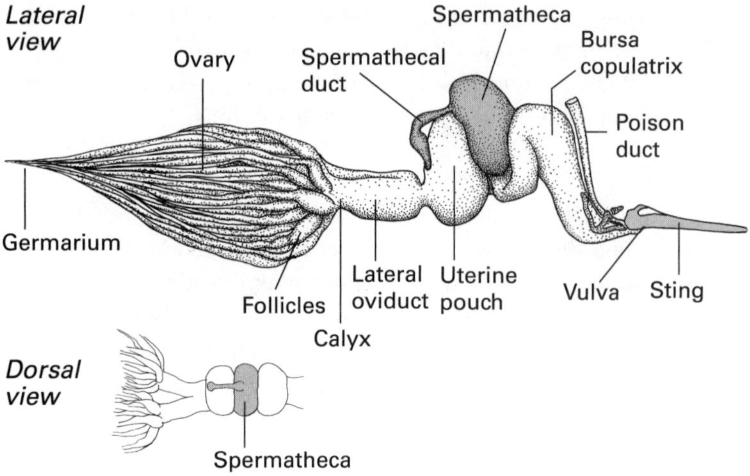

Figure 10.1. The female reproductive system. Each ovary is composed of 90 to 100 ovarioles. Meiosis takes place in the germarium. Each oocyte develops inside a cellular follicle as it moves down the ovariole. Shortly before ovulation into the calyx, the follicle cell secretes a cuticular chorion (egg shell). As the egg is laid, it passes into the uterine pouch, where sperm from the spermatheca fertilize it.

(egg shell) around each egg, leaving a tiny opening (the micropil) for entry of sperm during fertilization. The founding queen lays the first eggs two to three days after mating, but the ovaries do not really gather speed until the colony begins to grow.

In contrast to females, flight-ready males are quite lean, having only 20% body fat. Males are little more than a delivery system for sperm, a sort of formicine heat-seeking missile for inseminating females during a single mating on a single flight. Like kamikaze aircraft, they are designed with the minimum equipment they need for this job and just enough fuel (in the form of glycogen) to get it done. Each has a small head, large eyes, a large thorax with powerful wing muscles, and a tiny gaster in which he carries his supply of sperm.

The testes of the males reach their maximum size during the late pupal stage, then produce all the sperm they ever will during the first 10 days of adulthood. This ready-to-use sperm is stored in the distensible seminal vesicle (Figure 10.2). The final tally of sperm in the average male is about 8.7 million, a lot for such a small creature and enough to mate with one female but not two (Ball and Vinson, 1984; Glancey and Lofgren, 1985; Glancey et al., 1976b). Their job finished, the four-lobed testes degenerate into indistinct lumps of tissue attached to the inner ends of the seminal vesicles (Ball and Vinson, 1984; Ball et al., 1984; Glancey et al., 1976b). This syndrome of testicular degeneration before mating is almost universal among ant males (Hölldobler and Wilson, 1990). For a male who has only a single chance to mate in his lifetime, retaining functional (and heavy) testes makes no sense. Once the degeneration is complete, the male is ready to depart on his mating flight.

During mating, sperm flow from the seminal vesicles through the accessory glands, the ejaculatory duct, and the aedeagal bladder (not visible in Figure 10.2) before exiting through the aedeagus (functionally, the penis). The aedeagus is surrounded by appendage-like structures of the external genitalia whose function is to grapple with the female's genitalia to effect copulation. My student Sasha Mikheyev observed that, among ants and bumblebees, males that mate singly (as do fire ants) have large accessory glands that produce a mixture of fatty-acid esters and protein and those that mate multiply do not. In male bumblebees, these fatty acids (in particular linoleic acid, a "drying oil" used in many paint bases) oxidize to form a mating plug in the female's genital tract after copulation, preventing her from mating a second time (Baer et al., 2001; Sauter et al., 2001), a salvo in the war between the sexes that safeguards the male's reproductive investment. Fire ants possess the same fatty acids in their accessory glands, so they probably also form mating plugs (Mikheyev, 2003), possibly explaining the uniformly singly-mated condition of fire-ant queens. During this single copulation, the bursa copulatrix receives a lifetime supply of sperm, which is stored in the spermatheca to be metered out during fertilization through a musculated valve in the spermathecal duct.

Seasonality of Mating Flights

Mating flights have been observed in every month of the year at least as far north as South Carolina (Morrill, 1974b; Bass and Hays, 1979). The flights peak in frequency and size during May and June in north Florida (Figure 10.3) and a month or more later in South Carolina. Although this pattern indicates a broad mating season, it does not mean that

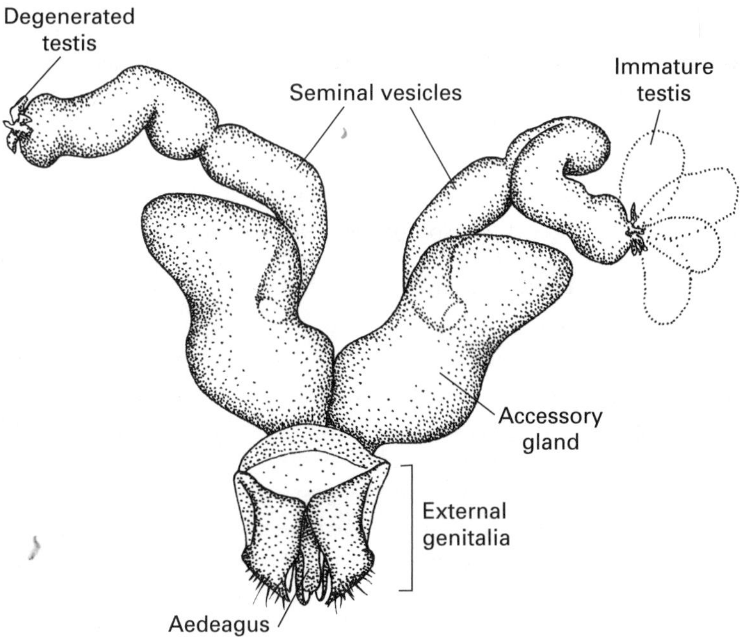

Figure 10.2. The male reproductive system. By the time the male is mature, his testes have degenerated (dotted outlines), and the sperm have been stored in the seminal vesicles. The accessory glands probably produce a substance that forms a mating plug in the female's genital tract, preventing her from mating again.

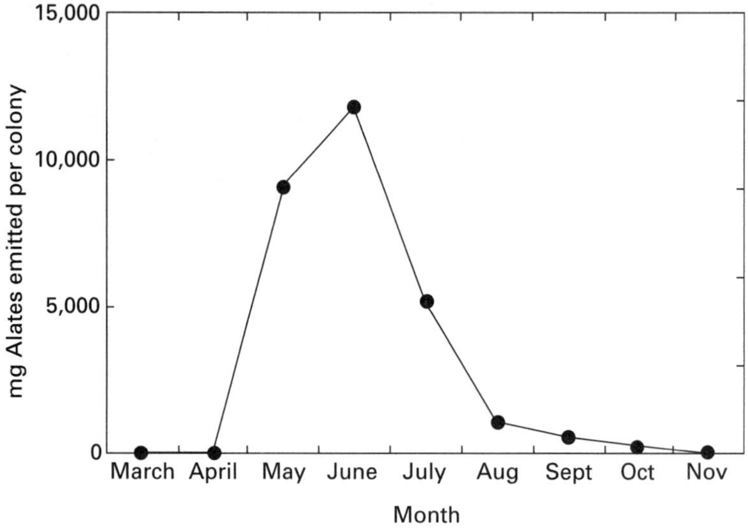

Figure 10.3. Seasonality of mating flights. Alates are produced from mid-February through June and fly from late April through July. Data from a 2004 study at Southwood Plantation.

colony founding is equally likely throughout the year. Over 90% of the sexuals fly between April and August, most having been produced between March and June. In addition, the small, early-spring mating flights of the autumn-reared sexuals that have overwintered in the nest are part of a different reproductive strategy, as explained below. The probability that a queen will succeed in founding a colony declines as the season progresses, partly because of competition from incipient colonies founded earlier that season and partly because overwinter survival, especially in the northern parts of the range, is lower for smaller colonies. Finally, average soil temperature must be above about 24°C for successful colony founding, a condition that is met for only three months of the year in the northern parts of the fire-ant's range but for over six months in mid-Florida. Altogether then, independent colony founding in fire ants is sharply seasonal, as it is in most ants, although the season is broader than that for many other species.

Timing and Characteristics of Mating Flights

Mating flights are triggered by substantial rain, especially when it follows a prolonged dry spell. Typically, the mating flight takes place on the day after the rain, during the late morning or early afternoon (Markin et al., 1971; Bass and Hays, 1979; Morrill, 1974b; Milio et al., 1988; Obin and Vander Meer, 1994). If the relative humidity is high enough, experimenters can induce mating flights by watering mounds, even in the absence of rain. The sky can be clear or cloudy, but the temperature must be over 24°C, the relative humidity high (80% or higher), and the air nearly calm. (Winter mating flights can occur under much drier, cooler conditions and will be discussed in a later chapter.) When the rains are local, like summer thunderstorms, the flights may be quite local too, covering as little as a square kilometer or so. When the rains are associated with a frontal system, the mating flights can occur synchronously over very large regions, perhaps covering a substantial part of the fire ant's range in the United States. Flights of decreasing size may occur for several days after heavy rains. Conditions that trigger fire-ant mating flights also trigger the swarming and flight of many other insects, including other ant species, termites, and many aerial predators. Mating-flight days are thus generally busy days in the insect world.

Half an hour to an hour before the mating flight, workers swarm excitedly over the mound, opening large holes in its surface. They climb vegetation and spread out for some distance from the mound, presumably to protect the alates that will soon follow. The frenzy of the workers is similar in many respects to alarm behavior, and it has been suggested that it is brought on by an alarm pheromone released by male and female alates from their mandibular glands (Obin and Vander Meer, 1994), but this conjecture has not been confirmed. The frenzy peaks as the alates exit through the mound openings, often surrounded by workers. They climb the vegetation and, without much fuss, take flight, ascending at a steep angle until they are out of sight. If wind comes up during a mating flight, or the nest is fanned with a piece of cardboard, the alates reenter the nest, and workers may even seal the holes.

A single mound may release from a few to over 1000 alates; the average during the peak flight season is 600–700 per flight. Overall, the sexes are usually almost equal in number, although one sex almost always strongly predominates in each nest. Males begin to take flight about half an hour before females begin and seem to stay aloft for at least four hours, perhaps even until dark. In contrast, most females remain aloft for only half an hour or so; few remain for as long as one to two hours. Alate departures may be spread over several hours. What brings the sexes together is unknown, but more than 95% of the descending females have mated, as confirmed by dissection of their sperm-filled spermathecae.

The males and females mate in the air, although no one has ever seen this copulation take place. Attached to laboratory flight mills, males flew 40% faster than females (1.0 m/sec vs. 0.7 m/sec; Vogt et al., 2000). This disparity makes eminent sense, for how else could males catch up to females? In addition, lighter females flew faster than heavier ones, making the heavier, more desirable ones easier to catch. What little else we know about this aerial matchmaking comes from studies in which light aircraft outfitted with nets were flown at specified altitudes during mating flights (Markin et al., 1971), the first case of "extreme" sweep netting. About 98% of all the alates captured were males, again indicating that females spend very little time in the mating swarm, mating and descending very quickly. A small number of pairs were caught *in copulo.* The highest number of males (70% of the total) was usually about 100 m above the ground, but males were not horizontally concentrated into local clouds. Rather, the males seemed to swarm in a layer at a particular altitude. No males were captured over a large swamp, so the male swarms are limited to suitable fire-ant habitat. Site selection for colony founding therefore probably begins with the locations of male mating swarms.

This description of mating flights applies primarily to monogyne fire ants. Polygyne ants are expected usually to mate locally and to seek readoption into their natal nests. The degree to which this is true of polygyne fire ants in the USA and in their native land is discussed in a later chapter.

Most mature mounds do not release sexuals at every opportunity, nor do they release all their alates in a single flight. The proportion of colonies releasing sexuals on any given mating-flight date (in north Florida) varies from a low of less than 10% in October to 75% in May and June. Each mature colony participates in several mating flights during each flight season. In 1994, the 40 colonies we sampled participated significantly in three to 16 mating flights between May and November; the average was around eight or nine (unpublished data). The average number of sexuals released per flight by a colony varied in parallel, from about 10 in October to 500–700 in April-June. Annual production in prime habitat was therefore about 470,000 alates per hectare, a mean of about 4400 per colony per year.

Dispersal and Selection of a Nest Site

After mating aerially, the newly mated queen either descends directly to the ground or flies some distance before descending. Queens in flight probably make life easier for themselves by flying downwind—a study in north Florida found almost 90% of newly founded colonies downwind from their (at that time still isolated) source population (Rhoades and Davis, 1967). The question of dispersal was especially critical during the years when mirex was being used in an attempt to control or eradicate the fire ant. Large dispersal distances would have decreased the effectiveness of large-scale fire-ant control, so several experiments on dispersal distance were carried out (Markin et al., 1974c).

To estimate postmating dispersal distance, Markin and his coworkers fed the fire-ant colonies in a 12-ha pasture with bait containing a fat-soluble blue dye; within a four-month period, they had dyed the insides of about a quarter million alates blue. After a large mating flight, they searched at increasing distances from this treated area and crushed any newly mated queens to see whether they were blue. Only about 1% of the queens were blue 400 m from the source, and at 800 m only a single queen of 314 (~0.3%) was blue. A second mating flight showed even lower rates of recovery. These results suggest only that a tiny proportion of the newly mated queens fly more than a few hundred meters from their points of origin. Unfortunately, no search for queens was made nearer the nest, so we have no background recovery rate against which to compare the recaptures. This experiment leaves open the possibility that the great majority of queens fly only very short distances.

A second experiment in Markin's study gave different results. By means of mirex bait, a square area 8 km by 8 km was cleared of 99% of its fire-ant colonies. After three large mating flights from surrounding colonies, open fields, and dirt roads at increasing distances into the cleared area were searched for newly mated queens, and the numbers were compared to those from an untreated area. The drop-out rate was about 80% in the first 400 m. Beyond this distance, half the remaining queens dropped out for every 1300 m. In the fall of the same year, after new colonies established by the arriving queens were large enough to be easily seen, the cleared area was surveyed again. At that point, every additional 320 m into the cleared area was accompanied by a 50% decrease in the number of new colonies. Beyond 1.6 km, no trend was apparent, and the colonies there could have originated from colonies surviving the mirex treatment.

These three experiments leave us uncertain about just how far newly mated queens fly before settling to the ground. The experiment with dyed queens suggests that only 3% of queens fly as far as 400 m. The young colonies in the cleared area suggest that 3% of queens fly as far as 1.6 km, whereas the collection of newly mated queens suggests that about 10–15% fly that far. Because no statistics other than totals were reported, no statistical comparisons among these experiments can be made, and the results can give us only a rough indication of flight distances. Another difficulty in interpreting the data stems from the nonpoint source of queens in the cleared-area experiments. A queen coming from a nest outside the cleared area has already flown some distance when she crosses the boundary into it, whereas a queen from a colony on the boundary has not. The dyed-queen experiment approximates a point source, and its estimate of flight distance is by far the lowest. For the time being, we can say only that queens usually fly less than 400 m before settling but sometimes fly up to 1.6 km or even more. Aerial trapping also supports the hypothesis of low dispersal distances—the female catch at all altitudes between 60 and 300 m was very low (2% total), indicating that almost all females descend quickly and directly after mating and do not fly far. If they did, they would have been caught.

Under special circumstances, fire-ant queens are capable of flying quite long distances. Markin and his coworkers used mirex bait to clear two islands of fire ants, one 11.3 and one 16 km from shore in the Gulf of Mexico. During times when major mating flights occurred on the mainland, they found 36 newly mated queens on the closer of the two islands and one on the more distant. They also enlisted the help of five charter and fishing-boat operators, who reported newly mated queens landing on their boats or on the water as far as 4.8 to 8 km offshore. Clearly, over water, queens can fly quite far. Less clear is whether the overwater distances are similar to overland distances. After all, queens may well be programmed to continue flying as long as the surface below is water, in which case these observations may tell us little about the ordinary dispersal distances over land. Indeed, all indications are that the vast majority of queens disperse much less than 1.6 km.

Queens cannot fly indefinitely, because, like all creatures, they have limited energy and water stores, both of which are depleted during flight. Perhaps if we knew what those limits were, we could at least determine the maximum *possible* flight range. When females take flight tethered to circular flight mills in respirometers, their respiration rate increases 40- to 50-fold, so flight (unsurprisingly) is a very energy-consuming activity (Vogt et al., 2000). The relative rates of oxygen consumption and carbon-dioxide production indicate that females rely on carbohydrates (glycogen) as an energy source during flight, saving their fat stores for colony founding. The size of the initial glycogen store sets a limit on flight distance. The average field-caught postflight female had used up more than 60% of this glycogen supply (Toom et al., 1976c). Females lose more than 20% of their body water for every hour they are in flight, another limit on very long flights. Combining the time it took for exhaustion to end female flight on the flight mill, female flight speed, water-loss rate, and energy depletion, Vogt calculated that females could cover a maximum distance of less than 5 km. This is the physiological flight limit of newly mated females. Downwind flights could exceed this limit in proportion to wind speed.

One final point deserves attention. With respect to dispersal distance, queens fall into two overlapping categories: long-distance dispersers and stay-at-homes who mate and settle near the natal nest. Both types are found in polygyne colonies (Wuellner, 2000), but monogyne colonies produce primarily dispersers. These differences are important for understanding polygyny.

Site Selection, Settlement, and Formation of the Claustral Nest

Newly mated queens almost certainly choose their founding nest sites in a hierarchical fashion, starting while still airborne. First, they choose a place to land, on the basis of attributes that we do not yet understand. Although no quantitative data are available, newly mated queens seem to be most abundant in partially vegetated, ecologically disturbed habitats (see Figure 31.1). Roadsides, dirt roads, recently disturbed land, and various other messed-up properties are good places to find large numbers of newly mated queens after a mating flight (although sometimes parking lots, swimming pools, and other unsuitable sites seem to attract them). Because queens are so much more visible on bare ground than among plant cover or leaf litter, an observer might overestimate the difference in abundance. Our knowledge of fire-ant biology, however, and the speed with which they colonize recently disturbed sites indicate that queens probably actually recognize and prefer such sites, which are likely to be unoccupied by fire ants. The details of site selection, however, may depend on whether the newly mated queens hail from monogyne or polygyne nests and on their body weights.

Having landed, the queen rather quickly assesses whether or not to stay, breaking off her wings within minutes of landing. To do so, she throws her leg over the wing on each side and snaps it off at a line near the wing base (Plate 12E). As with other ant queens, the short wing stubs will forever clearly identify her as a queen. When mating flights are large, and queenfall in an area heavy, the wings accumulate in gossamer drifts near small obstructions (Plate 12D), witness to the life-and-death drama that grips the tiny lives underfoot.

Sometimes, a queen seems to judge a site unsuitable and once again takes wing. I have observed this behavior occasionally when the landing site was especially harsh, such as an asphalt parking lot or a very barren, dry surface. If a queen is attacked by resident ants before she sheds her wings, she occasionally takes flight again (Nickerson et al., 1975).

Our queen hurries along the ground, under fallen leaves, poking into clumps of grass, seldom tarrying long. The scale of her search is quite small (Tschinkel, 1998) compared to the probable scale of her aerial search for a promising landing spot. Queens travel between about 2 and 50 m during their search (half of them travel 10 m or less). Within a few minutes to a couple of hours after landing, the queen either has found a hole to modify into a nest of her own or starts one from scratch. Half the queens are underground within 40 minutes of landing. The founding chamber is rarely more than 2 to 3 m from the landing spot. The queen digs with mandibles and forelegs and brings pellets of dirt to the surface and deposits them in a characteristic ring around the entrance hole. Depending on soil hardness, she digs a descending tunnel ending in a small chamber about 2 to 10 cm below ground. Well before she has reached full depth, she uses the pellets resulting from deepening to form a soil plug that closes off the nest. She continues to deepen the chamber by adding excavated soil to the underside of the plug. Often a queen builds a small side chamber just under the soil plug, perhaps to allow a choice of microclimate within the nest or to allow room to turn around.

Reasonably, our queen should avoid nesting near obvious danger. Indeed, when given a choice, queens from polygyne colonies prefer to dig their founding chambers in uncontaminated soil rather than in soil contaminated with refuse from their eternal enemy, *Pheidole dentata* (Kaspari and Vargo, 1994). They also avoid soil from other polygyne *S. invicta* colonies about 80% of the time. Both choices increase the queen's chance of survival. Whether monogyne queens make similar choices is unknown.

◂•• *Spring among the Fire Ants*

Early springtime in northern Florida is a pleasant season, with warm sunny days and cool nights. Spring weather is dominated by recurring frontal systems that make their diagonal sweep from the northwest, their leading edges heralded by powerful thunderstorms. Lightning pierces the night, and thrashing trees freeze in the momentary glare. It rains so hard that water bounces back up to waist height, then flows in sheets over the ground, piling little dams of debris behind every obstruction. Sunrise reveals a coolness that wasn't there the previous day, as though the thermostat had been creeping upward day by day until a large, wet thumb set it back a few degrees. The growing heat of summer is briefly foiled.

By mid-May, the cool morning respite after these frontal storms evaporates early in the day as the accumulating damp heat of summer begins to weigh heavily. By late in the morning, sensible Floridians seek shade. Standing stock-still in the sun, you can feel your skin film with sweat, and you know that, in their mounds, fire ants are preparing for their mating flights. A late-morning walk across a pasture reveals mound after mound with large holes on all sides. Workers busily tamp and smooth the edges until the holes look as though they have been melted rather than excavated. Every mound has holes. A big flight is imminent.

By noon, your shirt adheres to your skin, and the flight begins. On every mound, ants rush from the melt-holes, blanketing the mound and its vicinity in excited workers, ready to attack any threat. The much larger sexuals pour out too, their wings glittering in the sun. They climb the grass protruding from the mound, prodded and poked by crowds of frenetic workers. At the tip of the blade, they hesitate momentarily, waving antennae and forelegs, as if to test the breeze, then open their wings, and in a flash they are airborne—first the males; a little later, the females. As you lean over the mound, the ants rise by the hundreds, perhaps thousands, until they are pinpoints, then vanish from sight. A droplet of sweat falls from the tip of your nose, unnoticed by man or ant.

Unseen in the sky above, the males swarm in a layer, perhaps 100 or 200 m above the ground. This is their only day of freedom, their only chance to reproduce. By dusk, most of them will be dead. The heavier females tarry in the swarm only long enough for an aerial mating with a single, lucky male. How one tiny pinpoint finds another up there in the big, empty sky is still a mystery to science.

Twenty minutes after the flight begins, and continuing for one to two hours, the mated females begin to come back down to ground, like a gentle, noiseless rain. Raining cats and dogs may be a metaphor, but a rain of fire-ant queens is real. Queen after queen hits the ground, snuffles around for a few minutes, then hoists a leg to break off each wing, one at a time. With hardly a hesitation, each queen continues afoot on the urgent errand of finding a place to dig a nest, escaping from sharp-eyed predators on the surface. Here and there, one queen meets another, hesitates for a mutual head-to-tail inspection, like dogs sniffing each others' anal glands. Sometimes they decide to dig a hole together, sometimes they part ways. By late afternoon, most of the queens are below ground, each sealed in a moist, earthen chamber, ready to begin laying eggs.

The actors in this drama may be mere specks in our eyes, but the scale of the phenomenon far exceeds our parochial consciousness. Imagine a camera on a satellite in stationary orbit over the southeastern United States, one that makes visible in green the storms that accompany a spring frontal system but also detects and counts the winged male and female fire-ant sexuals as these rise from their mounds to mate aloft. Picture each individual showing up on the screen as a tiny red dot, and imagine that the events of seven days are compressed into a film lasting seven minutes, one minute for each day.

As the weather front enters the field of view during the first minute of our film, we see a narrow green line stretch diagonally from eastern Texas across northern Louisiana, clip the southeast corner of Arkansas, and vanish from view somewhere northeast of Memphis, Tennessee. As our film continues, the line sweeps to the east, majestically, relentlessly, night and day, maintaining its diagonal southwest-to-northeast orientation. During the first day it rips through Tupelo, Mississippi, bringing the snakes out of their burrows. By that evening, thunder rumbles over Demopolis, Alabama. By the fourth day, the streets of Palatka, Florida, are under 10 cm of water, and frogs are migrating by the thousands. On the fifth evening, the front cuts only across the Florida peninsula, drenching Immokalee and the water-stressed Everglades, and finally moves offshore over the Atlantic Ocean to die over Bermuda on the sixth day and in the sixth minute of our film. This is the weather as we are accustomed to seeing it daily on color weather radar.

By late morning of the second day, in the second minute of our film, considerably to the west of the green line, we see a faint red band gather strength until, by early afternoon, it is a deep, glowing strip of fire running parallel to the green band of water. This is the signal compounded of myriad fire-ant sexuals rising from their mounds, in every pasture, roadside, lawn, and field, 10 billion tiny dots on our radar screen, so many that a band 320 km wide shines with a red brilliance. By late afternoon, the band has faded, winking out before sundown, only to reform just as brightly by the middle of the next day but considerably to the east, chasing the green line of storms on their way across the South.

But now, we also see a fainter echo of the first day's red band, as a smaller mating flight forms on the second day after the rain. From our vantage in space, we see the continuous and stately sweep of the green storm line, then the red band forming at midday, growing broader day by day, faded on its western margin, glowing bright on the eastern leading edge. Every night it fades, to reappear 18 hours later. By the time the green front moves out to sea on the fifth day and out of our view, the southeastern United States, from northern Mississippi to northern Florida glows red, bright in the east, dimming in the west. On the fifth day the midday band narrows eastward, and on the sixth it stretches only across Florida and south Georgia. By the seventh day, it fails to appear.

We must not confuse size with scale. The tiny lives of fire ants dance to the tune of weather fronts on a continental scale, a point that is likely to escape us as we lean over a mound shimmering with wings and a drop of our sweat momentarily plasters a winged queen to the ground. Fire ants are more than these frenzied little creatures in our field of view, and they are more than the colony hunkering inside its mound. Fire ants also exist as a physical entity that spans a fifth of the United States and a fair chunk of South America. This entity has a life, form, and behavior of its own, arising from but not found within the creatures that make it up. We can visualize this incarnation of the fire ant only through a mental effort such as our imaginary satellite radar. The fire ants we observe at the scale of our daily lives are the atoms and molecules of the larger reality, just as we ourselves are also the atoms of a larger human existence.

11

The Claustral Period

The problem faced by the queen during the time she is sealed in her chamber is to turn the fixed amount of reserves stored in her body into the optimum number of workers. To do so, she must lay eggs, convert her metabolic reserves into a form usable by the developing larvae, and nurture these larvae until they become workers. At the end, she must have enough reserves left over to survive until her workers can provide for her. As straightforward as this process sounds, it is full of complexity and trade-offs and can fail at many points and for many reasons. Although many aspects are adjusted to mean values by evolution, the queen makes a number of important adjustments ("choices") to suit her particular situation, and these adjustments can be quite important to her final success.

Fortunately for us, founding queens are so driven that they will happily carry out the whole colony-founding process in a little test tube with a moist cotton plug in the bottom, rather than a dark earthen chamber. Colony founding is thus one of the most easily studied parts of the fire ant's life cycle and has been described in great detail.

As long as the female alate remains in her natal nest, her mother's queen pheromone prevents her from becoming reproductively active. The inhibitory pheromone is sensed by the alate's antennae, suppressing her juvenile-hormone secretion, possibly by suppressing brain dopamine, which is needed to stimulate juvenile-hormone secretion (Boulay et al., 2001). Low juvenile-hormone levels in turn prevent her ovaries from producing mature eggs (Vargo, 1998), possibly by preventing the uptake of vitellogenin (yolk precursor) (Chen et al., 2004b), and prevent her from shedding her wings and resorbing her flight muscles (Barker, 1978, 1979; Burns et al., 2002). After alates are removed from the pheromonal influence of a functional queen, their juvenile-hormone biosynthesis rates climb sharply, causing concentrations to peak after three days. This peak coincides with dealation (wing shedding) and with the activation of the gene coding for vitellogenin synthesis and several other genes (Lewis et al., 2002; Chen et al., 2004b; Tian et al., 2004) and is followed two days later by the first production of eggs (Brent and Vargo, 2003). Surgical removal of the corpora allata (the source of juvenile hormone) prevents all these effects, even in the absence of the queen's inhibitory effect (Barker, 1978, 1979), and topical application of juvenile hormone reverses the effect of removing the corpora allata. Moreover, even in the presence of the queen, alates shed their wings when dosed with juvenile hormone. Finally, in the absence of the queen, dealation can be prevented by application of precocene, a compound that inhibits the secretion of juvenile hormone by the corpora allata, and this effect can be reversed by application of juvenile hormone. No doubt, juvenile hormone plays a causal role in initiating reproduction in female alates.

As soon as the alate female leaves her nest, her former queen's inhibitory influence is left behind, and the cascade of physiological processes involved in colony founding is set into motion. Mating is followed by dealation, the initiation of rapid egg development, and the resorption of the flight muscles. The yolk precursor, vitellogenin, circulates in the female's blood even before she mates and increases steadily in concentration during the entire claustral period that follows, as she converts body stores into eggs and secretions (Lewis et al., 2001). Within one to three days, the queen begins to lay eggs and usually has a clump of 15 to 20 by the third day. Within about a week, her egg clump contains 20 to 100 eggs; 30 to 70 is typical. Queens frequently tend these eggs, licking them and moving them around, while keeping them together in a clump ((Markin et al., 1972; O'Neal and Markin, 1973). Untended eggs soon become moldy. Even after the founding period, when workers take over all brood care, the eggs are kept in clumps.

As in all insects, the rate of development depends on temperature. The temperature actually experienced by the queen in her chamber depends on many factors—time of day, sun or shade, water content of the soil, time of year, vegetation, and soil type. Most studies of larval development have been carried out in the laboratory at constant temperature (Markin et al., 1972). At the optimal temperature of about 30°C, the first larvae hatch in about a week. The larval stage and pupal stage each also last about a week, so the first workers appear in about 18 to 20 days. At temperatures below 24°C and above 35°C, queens fail to rear adult workers. Therefore, when soil temperature averages less than 24°C, colony founding stops. At the other extreme, queens can tolerate occasional daily highs of 45°C (Tschinkel, 1993a).

Upon hatching, the first-stage larvae remain stuck among the unhatched eggs and proceed to eat some of them. This is not an aberration, for about half of the eggs the queen lays do not form embryos, even when incubated for a week, and are probably laid as food for the first-stage larvae (Glancey et al., 1973b; O'Neal and Markin, 1973; Voss et al., 1986; Voss and Blum, 1987). In these eggs, meiosis proceeds normally, but the first mitotic division is blocked, arresting further development. Such eggs-to-be-eaten are common among ants and are called trophic eggs (Plate 13C). They may be produced by queens, by workers (though not in fire ants), or by both. During colony founding, they are one of the most obvious solutions to the problem of converting reserves stored in the queen's body into a form usable by the developing larvae.

This laying of trophic and embryonated eggs is precisely timed. For the first six days, almost all the eggs laid by the queen are embryonated. From the seventh to the tenth day, as these eggs begin to hatch, the queen switches quickly to laying almost exclusively trophic eggs and continues doing so until the first worker brood emerges on about the twenty-first day (Voss and Blum, 1987). That is, as soon as hungry developing larvae are present, the queen lays eggs only to provide food and does not increase the population of hungry mouths until adult workers can take over feeding and food gathering. How larvae distinguish between embryonated and trophic eggs is unknown.

Beginning with the second larval stage, the queen feeds her offspring regurgitated liquid food. This liquid may originate partly from the crop, which usually contains large amounts of oily material, or from glands that convert metabolic stores to ingestible secretions. When the feeding of these larvae commences, the queen's fat reserves begin to decline (glycogen is already depleted). Fat is either converted to carbohydrate or incorporated directly into larval food by the salivary glands of the queen. During colony founding, even young minim workers solicit liquid feeding from their mothers. After colony founding, the flow of food is always from workers to the queen.

The mating flight, followed by the production of eggs and nutritive salivary secretions, draws down the queen's body reserves. The energy required by her mating flight depleted 60% of her preflight reserves of glycogen (Toom et al., 1976c), and the rest is depleted by the twelfth day, an example of the widespread tendency among animals to use carbohydrate energy sources in preference to fat. In the course of producing the first minims, 95% of the large amount of oil she stored in her crop before leaving her natal nest is used up (Vander Meer et al., 1982a; Vinson et al., 1980). As the queen lays eggs during the first week, her protein reserves, which consist of both storage protein and the histolysing wing muscles, decline sharply (Toom et al., 1976a, b, c,). Muscle histolysis begins two to four hours after mating as changes in the muscle-cell membranes activate calcium-dependent proteases that break down the muscle fibers (Jones, 1977; Jones et al., 1978, 1982; Jones et al., 1981; Jones and Davis, 1985). The free amino acids produced by this muscle digestion undoubtedly became available for other uses, probably including the production of eggs. The space freed up by this loss of flight muscle may allow the expansion of the esophagus into a thoracic "crop" to replace the fluid-storage capacity of the regular crop that was lost as a result of the enlargement of the ovaries (Glancey et al., 1981b), or so it is conjectured. Interestingly, injecting unmated female alates with hemolymph from mated ones can bring on flight-muscle histolysis. First heating the hemolymph eliminates this effect, suggesting that a protein factor in the hemolymph initiates flight-muscle degeneration (Davis et al., 1989).

As the claustral period wears on, the heavy investment made by the queen becomes apparent even to the naked human eye. What began as a shiny, plump, sassy queen is an emaciated wraith by the end. As a result of laying eggs, producing secretions, maintaining herself, and caring for her larvae, the queen loses about half her body weight, dropping from about 14 mg at mating to about 7 or 8 mg at the end of the claustral period. Because a good deal of her stored reserves were fat, which has a higher caloric content than lean body components, she loses two-thirds of her total energy content during the claustral period, much of it as maintenance cost. About 25 to 35% of the lost energy, or 70% of the lost dry weight, is converted to offspring (between 5 and 35 minim workers; Plate 13D).

Why Minim Workers?

The first workers produced by most species of independently founding ant queens are the smallest in the whole life cycle. Their tiny size is acknowledged in such names as "minim workers," "dwarf workers," or, for the more classically inclined, "nanitic workers." Minims are assumed to have evolved as a trade-off between worker size and worker number. Because the queen starts with fixed reserves, she can produce more workers only by producing smaller ones. Another trade-off is that earlier pupation comes only at the price of smaller workers, because larval growth rates are fixed. Producing smaller workers entails costs—they are known to be short-lived (Calabi and Porter, 1989), more sensitive to desiccation (Hood and Tschinkel, 1990), and perhaps more susceptible to other stresses as well. Producing more workers and producing them earlier have at least two clear benefits. The incipient colony stage is second only to the claustral stage in vulnerability, and colonies die from many causes, including brood raiding by other incipient colonies, as we will see shortly. The more workers in the initial force, the less serious is the loss of each worker. It follows that the number of workers in the first brood and the speed with which the colony can grow out of this vulnerable colony size are both important to survival. The number of workers has long been assumed to be more important to brood rearing than their size and that most ants produce minim workers in the first brood as a compromise between the need to maximize number and some minimal functional size.

Sanford Porter, working in my lab (Porter and Tschinkel, 1986), showed these assumptions to be correct. We created experimental incipient colonies that contained a fixed weight of workers, but different colonies were given workers of different size, ranging from minims to workers about six times as heavy. This is exactly the trade-off between worker size and worker number that was hypothesized to have led to

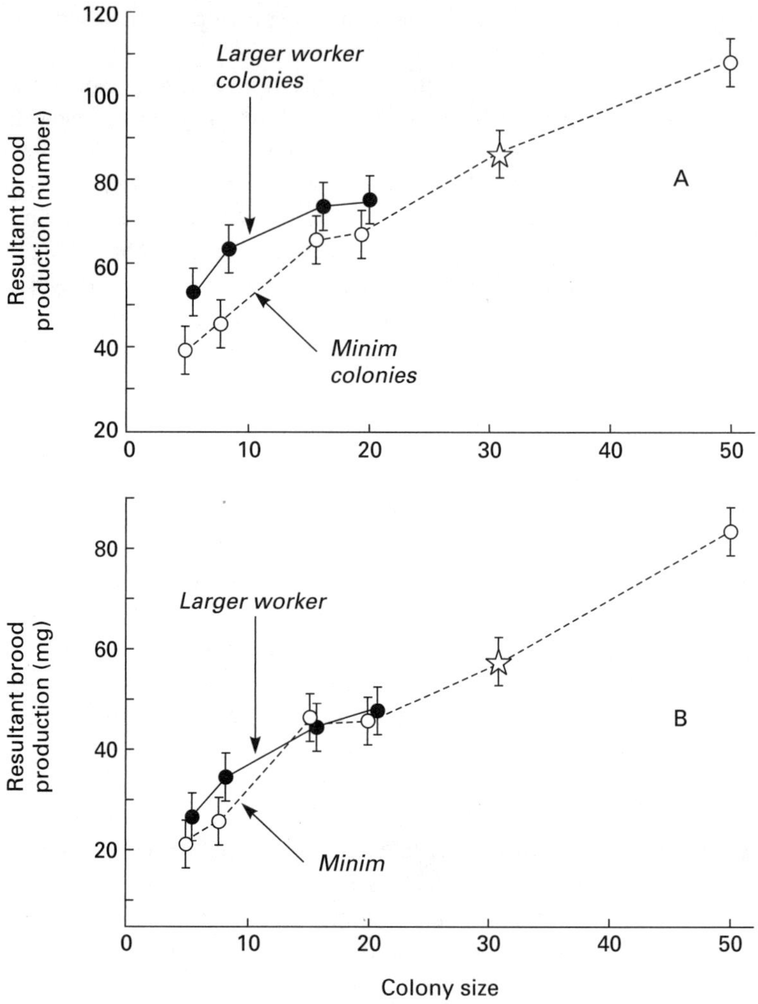

Figure 11.1. Gram for gram, minim workers are more efficient at brood production than are larger workers. Adapted from Porter and Tschinkel (1986).

the evolution of minim workers in the first place. The result: the larger (and therefore fewer) the workers in these colonies the less brood they raised in the second generation; brood production dropped 40% across the experimental range of worker size (Figure 11.1). Although minims were somewhat less effective individually, the number of workers was much more important in these experiments than was their size. In another experiment, brood production tripled as the number of minims in experimental colonies was increased from 5 to 50. Together, these two studies show that brood production, and therefore colony growth rate, depends primarily on the number, not the size, of workers.

The benefit of minims is thus that the queen can produce more of them from her fixed resources, not that they are better at rearing brood. More workers in the first generation lead directly to a higher colony growth rate. A comparison of special interest is that a given weight of minim workers produced almost a fifth more brood than did the same weight of the smallest workers from postincipient colonies. This advantage can be considered the minimum that a founding queen gains from producing such diminutive workers.

The size of workers in the second generation is already within the normal postincipient size range, indicating that the trade-off between size and number has already shifted significantly in favor of size by this stage. Average body size continues to increase throughout colony growth, as discussed in a later chapter.

The number of minims produced in the first brood is highly variable, ranging from 0 to about 35. In view of the importance of the number of minims to ultimate colony-founding success, this great variation is mysterious. Would queens that produce few minims in our experiments produce more under other circumstances (an impossible experiment)? Does the variation have adaptive meaning in the sense that the strategy for success in queens producing few workers is different from those producing many? These questions are, as yet, unanswered.

Clearly, though, minim production is a subtle and complex process and a range of strategies, or "choices," may be available. Against expectations, the number of minims a queen produces is not related to her initial weight. Although most queens weighed between 14 and 16.5 mg, these weight differences did not result in detectable differences in the number of minims produced. On the other hand, the more minims a queen produces, the more weight she loses. Also, the more offspring she produces, the cheaper they are per individual—perhaps a benefit from an economy of scale—in part because the queen's own maintenance is a fixed cost and in part because, the more minims a queen produces, the smaller they are. Minims from families of about 30 weigh only 65–80% as much as those from families of four or five. In addition, these smaller workers develop faster—when about 80% of the more numerous group have emerged as adult minims, only about 6 to 8% of the less numerous groups have (Figure 11.2). At this point, whether the advantage derives from the number of workers, their smaller size (unlikely), or their earlier appearance is not clear. The relationship among size, number, and speed of development is a typical trade-off, but we do not know whether the observed variation in minim number is part of the queen's reproductive strategy, that is, an active adjustment to her particular situation. The possible benefit of larger workers might be their greater longevity, their improved fighting ability, or their greater resistance to hunger and stress. The cost is fewer, later workers. The interpretation that best fits current knowledge is that smaller minim size is a strategy dictated by the need to produce the most minims as fast as possible, because doing so confers an important advantage in winning the brood-raiding competition that follows the claustral period. More

Figure 11.2. The number, size, and speed of development of minim workers trade off. Queens that produce more minims produce lighter ones that develop faster. Development rate was estimated as the fraction of brood that were minims after a fixed elapsed time. Different symbols indicate queens from different mating flights. Reprinted from Tschinkel (1993a) with permission from Springer.

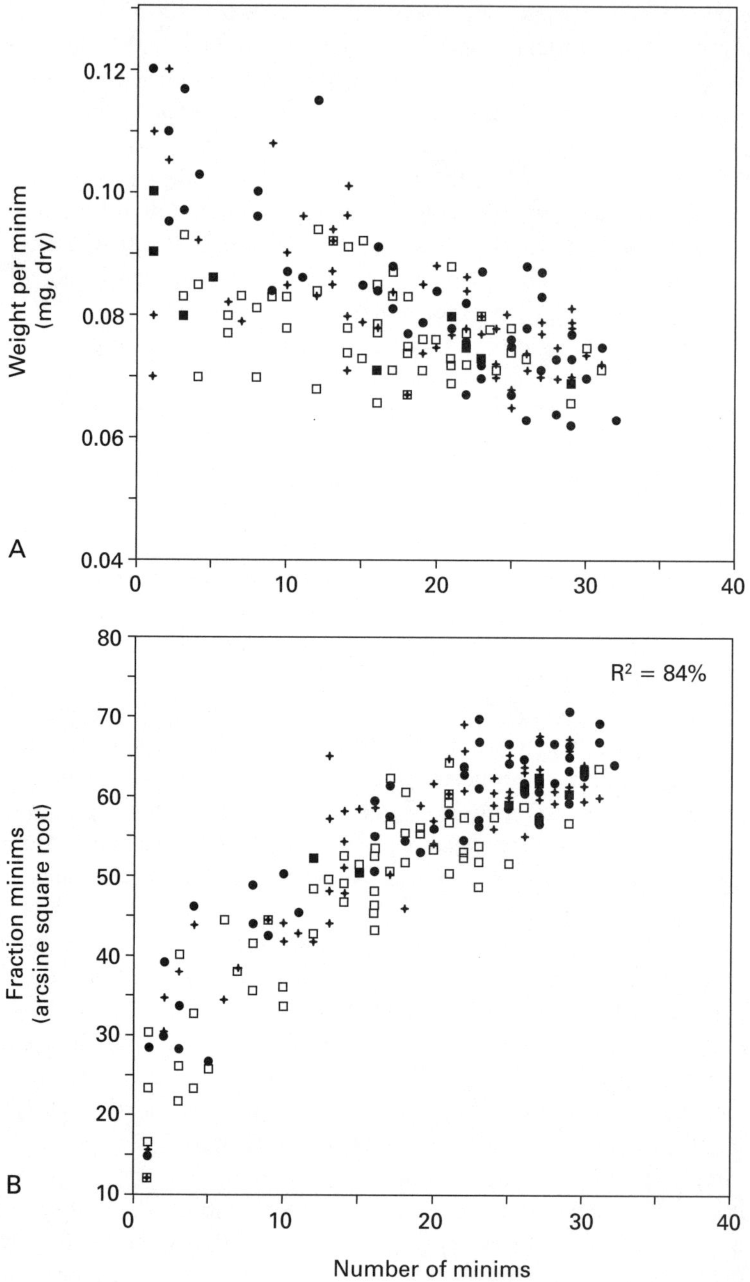

support comes from the observation that minim brood develops 35% faster than does brood from established colonies, mostly because of faster larval development (Porter, 1988). In addition, the minimum temperature for brood development is 24°C for minims, but 21°C for regular brood.

Cooperative Colony Founding (Pleometrosis)

As this tale unfolds, the reader should keep in mind that, for each queen landing in a neighborhood, the object of the "game" is to become the mother queen of a mature colony in that neighborhood. Natural selection will favor any attribute that improves the queen's chances of winning this competition. When many queens are attempting to found colonies in the same area, the game and the choices change in important ways. All the queens that settle in a neighborhood (a few hundred square meters) of unoccupied land will ultimately compete for that same space. A successful strategy thus must not only take into account what goes on during the claustral period but must "plan" for the intense period of competition among incipient colonies that follows it. Several phenomena associated with colony founding can be interpreted as responses to this future competition. Chief among these are cooperative founding by groups of queens (pleometrosis) and brood-raiding during the incipient period.

I have already noted that newly mated fire-ant queens seem to favor partly vegetated sites for founding colonies. By late afternoon on the day of a mating flight, holes surrounded by small circles of pellets identify the places where queens have dug their initial nests, and in many of these, queens will still be bringing soil pellets to the surface. In others, a soil plug is already in place, and the process of nest deepening, though still continuing, is not visible from the surface. These initial nests are easily marked for later survey and excavation. Several such studies revealed that many of these initial nests were occupied by more than one queen (the range was 1 to 18). The frequency of such cooperative founding was related to queen density on a very local level (Tschinkel and Howard, 1983). The more queens were recovered from a small patch (one to several square meters), the greater was the fraction of nests with multiple queens and the higher the average number of queens (Figure 11.3). If each queen dug her own nest, the number of nests and queens in each square meter would be equal, but rarely were more than five or six nests found per square meter, although up to 30 or 40 queens might occur in that same area. At these densities, additional queens entered existing chambers rather than excavating their own.

Are the queens simply joining other queens that they happen to run into at random, or are they making some sort of informed choice that tips the odds in their favor? If the queens in our small patch chose among the available chambers at random, the number of queens per

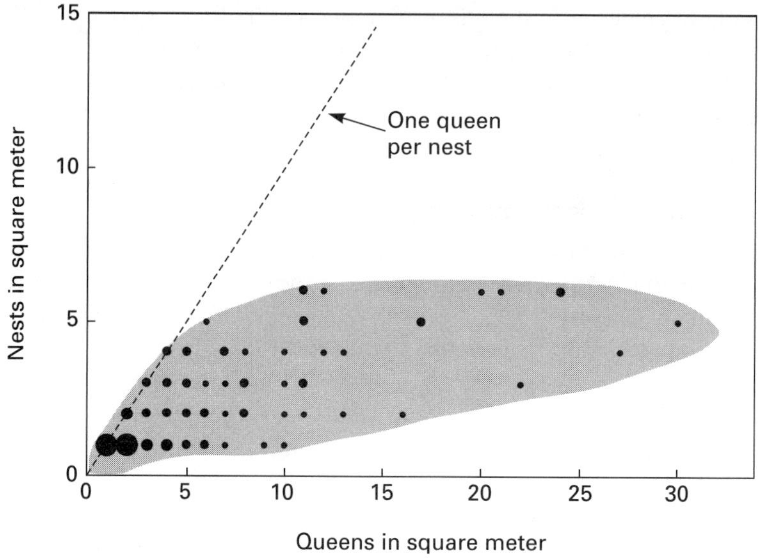

Figure 11.3. As the local queen density increases, the frequency and degree of multiple occupancy of chambers increases. No square meter ever contained more than six founding chambers, even when that square meter contained up to 30 newly mated queens. Adapted from Tschinkel and Howard (1983).

chamber would be distributed according to the Poisson function, a right-skewed distribution. When mating flights were small and queens were not abundant, the queens were indeed randomly distributed among chambers. When queen density was higher, however, queens did not assort themselves randomly but were found clumped in certain areas of the plots and in certain chambers within those areas. More low and high numbers of queens per chamber, and fewer intermediate numbers, were found than would be expected under random distribution. Furthermore, the higher the queen density, the greater was this deviation from randomness.

These observations suggested that queen density is driving the formation of associations of queens, but such a causal relationship could only be supported by experiments. Therefore, we seeded small experimental plots (5 by 5 m) with from 5 to 320 queens, marked their founding nest sites, and then excavated them over the next few days (Tschinkel and Howard, 1983). As we increased the queen density over this range, the mean number of queens per nest more than doubled, because queens tended ever more frequently to leave chambers with few queens in favor of those with higher numbers. This movement not only increased the mean number of queens per chamber but also increased their aggregation in certain chambers and certain parts of the experimental plots (remember that a queen's initial nest is seldom more than 2–3 m from her landing spot; Tschinkel, 1998). The frequency of contacts with other queens probably provides information on queen density, which in turn predicts the intensity of future competition, and therefore the value of joining. As the frequency of these contacts goes up, queens choose to share nests with increasing frequency. Any factor that increases the local queen density, such as larger mating flights, airborne site preference, microtopography, shade, and so on, drives up the clumping of nests and queens.

When queens associate, are they choosing the other *queen*, or do they simply want the associated *hole*? Even casual observation leaves little doubt that preformed holes exert a powerful attraction for newly mated queens. A queen coming upon a hole of adequate depth will often enter it and never reappear at the surface, a process we dubbed "hole-diving." Given a choice between an inadequate hole 1 cm deep and an adequate one 7 cm deep, queens chose the deeper hole, more evidence that the hole is indeed a valuable resource. When the difference in depth was less extreme, the differences in preference were also less.

The attraction of preformed holes is so strong that we were able to manipulate the dispersion of newly mated queens introduced into experimental field plots by varying the dispersion of preformed holes in them (Tschinkel, 1998). When preformed holes were distributed uniformly, founding queens were also distributed uniformly, and when the holes were clumped, so were founding queens. The queens loved our holes and used them.

Queens might still choose queens, rather than just the holes they occupy (Tschinkel, 1998). In a series of field and laboratory experiments, two queens were released within a small arena containing five identical, evenly spaced holes. The holes were in soil in the field or in a plaster-filled cup in the laboratory. If queens chose the holes randomly and had no preference for one another, then 20% of them would share holes. If they were attracted to one another, then the proportion of pairs would be significantly higher than 20%. By varying the spacing and depth of the holes, and the order of addition of queens, we showed that queens in the field paired significantly more often than expected only when the holes were of inadequate depth, close together, or both (in the absence of holes, the queens always paired). Because queens immediately began to modify holes they entered, though, their changes might have made the occupied hole more attractive than the others, confounding the effect of the hole with that of the queen. However, lab queens also paired more frequently than expected, and paired more often when the holes were shallow (52%) than when they were deep (37%). Because these lab holes were in plaster and could not be modified, pairing must have been the result of attraction between queens.

The studies revealed that, when holes are of adequate depth for colony founding, the queen chooses one at random and "hole dives," rarely coming to the surface again, thus reducing her chances of meeting and pairing with other queens. On the other hand, when the hole is not deep enough for safe founding, the queen returns frequently to the surface during her excavations or leaves to continue searching. She is thus more likely to encounter another queen or to enter an already occupied hole. Increased contact allows the attraction between queens to take effect. The depth of the available holes regulates the interaction among the queens and encourages them to make the choice most favorable for their situation. Joining is advantageous when a deep refuge is not present because it results in faster nest construction at lower individual cost. In addition, part of the attraction of preformed holes for newly mated queens is almost certainly that they offer an escape from the ground surface, where the queen is easy prey for any sharp-eyed predator, and desiccation and heat are common threats. Preformed holes, whether constructed by other queens or of another origin, probably greatly increase survival, but the advantages of joining probably trade off with the fact that only one of the queens will ultimately survive. It is one of many gambles the queen makes, as will be explained below.

Even when a queen chooses to join another, she may assess the other's individual qualities before making her choice. After all, the queen she pairs with is likely to be her competitor during the post-founding period, and ought to be chosen to tip the odds in our queen's favor. The morning after queens in field arenas chose to join or not to join a resident queen in the process of digging a nest, all queens were excavated and allowed to rear their brood individually in nest tubes. Test and resident queens did not differ in brood production, nor did paired and nonpaired ones, but queens that joined or were joined were significantly heavier at the end of the founding period than were those that had not joined or been joined. In addition, test queens that joined were significantly more likely to survive the founding period than were those that did not. No such difference was evident between resident queens that were joined and those that were not. Therefore, queens *that made the choice* to join another queen were more robust in some important way and were more capable of surviving the founding period, even by themselves. Greater weight gives queens an edge in surviving the future competition, fighting, and execution in incipient colonies. For a more robust queen, joining is the right choice.

Another question is whether queens choose to join in such a way as to optimize the performance of the association. If so, then randomizing self-assorted groups of queens should reduce their performance during the founding period, but in an experiment in which two- and three-queen natural associations were reassorted, brood production and final queen weights of the randomized groups did not differ from those of the natural groups. At least with respect to these two measures, who the queen joins does not matter. Any payoff from joining is not collectible until after the claustral period.

At this point, we have a population of claustral nests consisting of both single queens and associations of varying numbers of queens. Associated queens rear brood cooperatively and communally—they do not segregate their own offspring but keep all brood together in a single clump. Nevertheless, careful observations have revealed that, even within this claustral nest, shut off from the outside, the queens are struggling among themselves to capture a future competitive advantage (Tschinkel, 1993a). The details of colony founding are modified in subtle and important ways as a result, because queens make additional strategic choices.

Life in the claustral chamber begins peaceably enough. Usually, each of the queens in the group lays eggs at the same rate as if she were alone. The eggs are sticky, and the queens form them into a large clump, like damp gumdrops. Care of these eggs and later brood is shared by the queens. Like solitary founding queens, they lay trophic eggs (about half the total), and this fraction does not vary with the size of the founding association.

As the larvae hatch and develop, the situation changes. By the time the minims appear, the number of offspring is no longer a simple multiple of the number of queens but has changed into an optimal function in which groups of four to seven queens produce the highest amount of brood, and smaller and larger groups produce less (Figure 11.4). This optimal relationship rests on a dramatic decline in the production of brood per queen. In a typical laboratory experiment, a queen founding alone produces about 18 older larvae within two weeks, but this per-capita production drops to about 10 in groups of four queens and only four in groups of 10. This decline in per-queen contribution is absent in the egg stage, appears after larvae hatch, and continues to grow as the brood develops (Tschinkel, 1993a). In parallel with the declining contribution by each queen as group size increases, the mean final queen weight increases.

The sensitivity of brood number to queen number varies greatly with the brood stage. It is zero for eggs, rises to a maximum for late larvae, and is again low for pupae. The inescapable conclusion is that brood, especially larvae, are disappearing during development and that the rate of disappearance increases with group size. In one study, for example, 44% of the initial brood disappeared by the end of two weeks in nests of solitary queens, but this attrition rate rose steadily with group size and reached 88% in groups of 10 queens. Very large groups of queens often produce no older brood at all.

So figure it out: larvae vanish without a trace, and queens end up losing less weight—I would say that queens cannibalize larvae. They may be doing so indiscriminately in order to boost their own weight and vigor at the end of the claustral period, enhancing their competitive ability, or they may be eating the brood of other queens, thereby increasing the fraction of their daughters among the brood. Bernasconi's

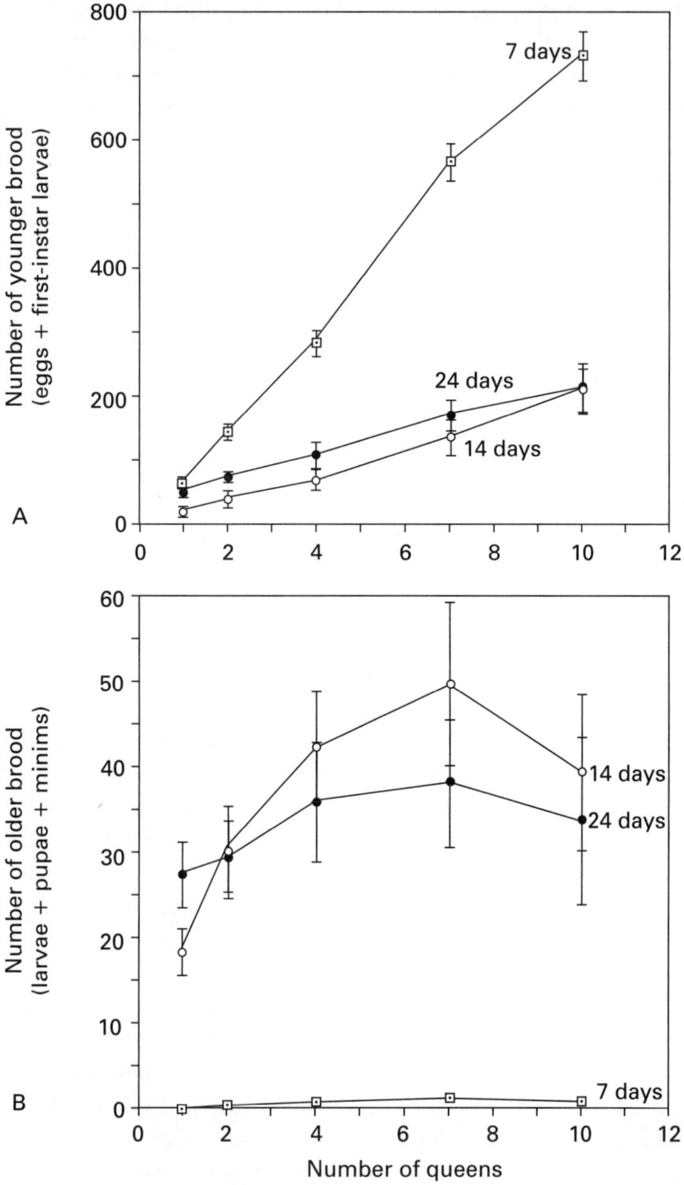

Figure 11.4. High queen density in founding chambers does not inhibit egg laying or hatching—amount of brood is a simple multiple of the number of queens—but older brood disappears with increasing frequency at higher queen numbers and as time passes, causing older-brood numbers to decline at high queen densities. Reprinted from Tschinkel (1993a) with permission from Springer.

finding that the queen losing *less* weight in two-queen associations produced more offspring (Bernasconi et al., 1997) is probably explainable by cannibalism. Kin recognition need not be invoked—even if the more productive queen cannibalized brood randomly, the surviving brood would still contain more of her offspring, and she would appear to have lost less weight while producing more brood. No matter who is eating

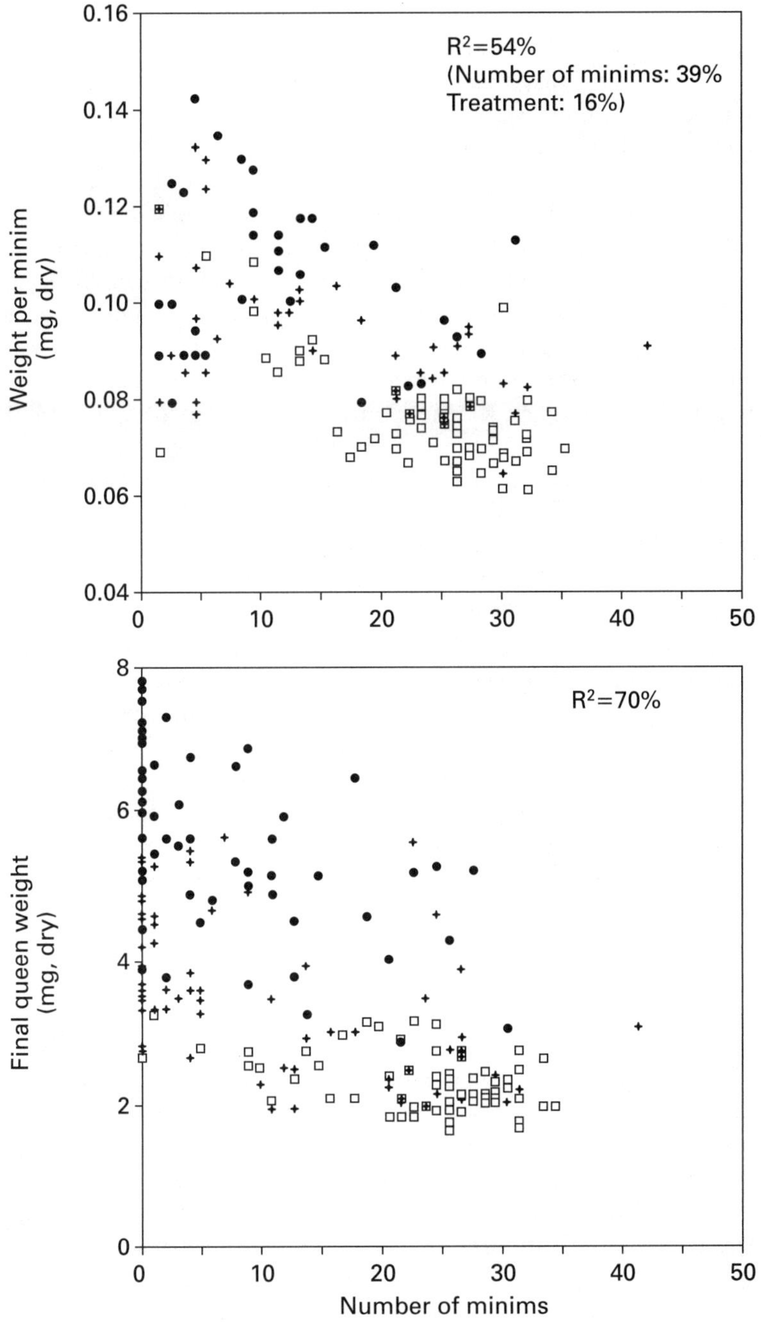

Figure 11.5. The carnivore-cannibal experiment. Solitary queens that were given the opportunity to feed on dead queens (cannibals, closed circles) were heavier at the end of the claustral period than unfed controls (open squares) or queens fed beetle larvae (carnivores, crosses) and produced heavier minims. Reprinted from Tschinkel (1993a) with permission from Springer.

whom, because the group began the founding period with fixed reserves, the effect of cannibalism is to redistribute material and energy within the founding group. We can probably assume that each queen sees to her own future, but how that future is best served involves a complicated set of trade-offs and compromises.

Queens do not stop at cannibalizing brood. If one of the queens in the founding group should die (we have no evidence that they kill each other during this period), her remains are often eaten by the survivors. A portion of this recaptured energy (as well as that from cannibalizing larvae) is probably invested in the remaining larvae, resulting in larger minims, and some goes to reducing the queen's weight loss. Minims of solitary queens weigh about 0.08 mg, dry. For a given number of minims, their mean weight increases with the number of live queens and the number of dead queen pieces eaten. Nests of 10 queens in which four or five queens had died produced minims that averaged over 40% heavier. Larger groups of queens produce less brood per queen, lose less weight, and produce heavier minims.

Notice, however, that these conclusions are based on correlations. To establish that cannibalism increases queen and minim weight, we must do an experiment. If the hypothesis is correct, then feeding solitary founding queens should have the same effects on queen and minim weight as observed in the groups of queens. The experiment we did consisted of solitary founding queens who were fed pieces of dead founding queens, pieces of beetle larvae, or nothing. In the spirit of cuteness, we will refer to these groups as cannibals, carnivores, and controls. For equal minim production, carnivore queens were 22% heavier than the unfed controls, whereas cannibal queens were a whopping 76% heavier (Figure 11.5). Carnivore queens produced minims that were 11–14% heavier than those of the unfed controls, whereas the minims of cannibal queens were 24–30% heavier. This experiment shows that at least part of the effect of group size on queen weight probably results from cannibalism.

Although the founding group is a closed system with fixed reserves, patterns of allocation can clearly be altered, altering in turn minim number, minim size, and queen weight loss. Higher body weight benefits queens during the competition of the ensuing incipient period and give them generally more resistance to stress. We can only speculate on the advantage of larger minims. Larger body size does give greater resistance to some environmental stresses, particularly desiccation (Hood and Tschinkel, 1990). Perhaps larger minims are also more effective foragers or defenders. Whatever their advantage, it is likely to be important primarily during the incipient period, because the mean worker size increases rapidly beginning with the second generation.

◂•• *Sharon's House of Beauty*

In spring in north Florida, fire ants spend their accumulated biological capital to produce large numbers of winged reproductive males and females. By mid-May, these are waiting, poised and fat, for the right conditions to fly from the nest and mate. And they like it hot and sweaty, the day after a heavy rain. By late morning, I leave my air-conditioned office to stand in the shade of a nearby oak. If my skin is damp with sweat in five minutes, I know a big mating flight is coming. The all-hands-on-deck alarm is sounded and my students, helpers, and I head for the back of the Winn-Dixie Supermarket, squatting in a strip mall on Tennessee Street alongside Eckerd Drugs, a Dollar Store, Sharon's House of Beauty, USA Video Rental, the Washeteria, and an AutoZone.

The mews behind the Winn-Dixie mall must have been designed by a fire ant researcher. For reasons that I have never understood, something about this messy, dumpster-accented, pot-holed stretch of cheap asphalt seems to be irresistible to newly mated fire-ant queens. By noon, the queens are mating and coming down on the asphalt. The smart ones take one look around, spread their wings and take off again, but most aren't that smart, and by the time they realize they have landed in fire ant hell, they have discarded their wings. Escape on foot is impossible. They run on the simmering pavement, seeking any scrap of shade or shelter, any bit of litter, loose chunk of asphalt, piece of board or damp paper they can find. As the hot sun draws the dampness out of each bit of litter, they rush to find another, gradually accumulating in ever larger numbers under the remaining shelters. A hundred queens under one wet Baby Ruth wrapper is not unusual. We cruise around this septic place with our aspirators (a myrmecologist's most important tool), flipping over bits of litter. As the queens scatter in the sudden light, the aspirator tip comes down, and a mighty suction is applied on the other end. The queens find themselves flying through a narrow tube to crash into the bottom of a plastic bottle, each head making a tiny tink as it hits the bottom. We suck them up so fast they sound like a muted typewriter—tinktinktink . . . tinktink . . . tink. When the bottle is over half full, we dump the 25 ml of queens into a larger container. Fifty ml of queens is about 1000. How many do we need today? ask my helpers. In a couple of hours, we have collected 5000 newly mated queens. At other times, even 10,000.

Such large numbers of queens are a myrmecologist's dream. They allow us to perform experiments with sample sizes as large as we wish to handle. The situation may be unfamiliar to those working on other species of ants, but in fire ant research, the limitation is laboratory labor, not queens. Today, we are collecting for a large factorial experiment that will require about 3500 queens. The extra 1500 give us a margin of safety for very little effort.

The back of the Winn-Dixie is the sort of space that people go to only if they must. Deliveries are made early in the morning, long before the queens fly. Reluctant employees emanate from the rear doors occasionally to heave bags of trash into a dumpster, then scurry back into the cool interiors. The two homeless guys who live in the woods across Ocala Street come by to rummage for half-rotten oranges in the dumpsters and to wash at the faucet on which some careless person has left the handle. This is the third year I have seen the skinny one. He has developed quite an interest in fire ant biology and remembers what he learns from year to year. Somewhere in his happier past are two years of college. The one with the red beard and grime-shiny clothes moves slowly, his eyes so turned inward that he pays scant attention to us as we bend over, aspirator tube in mouth, sucking stuff off the ground. His world is dark and full of ghosts.

The queens are especially drawn to the shadows of the buildings to the south and accumulate in crowds under any shelter along the bases of their walls. Just as I discover a large cache of queens under a drainstone that diverts the water from the roof gutters, the back door of Sharon's House of Beauty opens. A young woman wearing a white cover-up shakes out a hairy cloth before noticing me behind the open door. She wants to ask, but her eyes say, Don't talk to strangers. After a moment or two, she goes back in, looking back over her shoulder. Five minutes later, we have moved down a couple of doors. The back door of Sharon's opens again, and a young man walks toward us. It looks as though he's been delegated. It also looks as though Sharon's specializes in unisex poodle cuts. Hands in pockets, he asks, Wha' ch'all doin'?

Collecting fire ants, we say.

Fire ants! Sh____t, you come on over't my house. I got all the fire ants you want!

As often as I've heard this phrase, I have yet to formulate a zingy response. What am I supposed to say? That's nice, dear? Oh, I'm so sorry? What a lucky person you are? Really, if you live in the South, you've got fire ants. This is hardly news. Most often I take the staid professor route and explain that the usual home fire-ant supply consists of workers but that we are collecting queens that have just had sex. The mention of sex sharpens my listener's attention, so I say, Yeah, they only do it once in their lives, but it's probably a great experience because they do it up in the sky, and the male's genitalia literally explode into the female's vagina.

Eyebrows go up. No kiddin'?

I detect a teachable moment in those raised eyebrows, and find it hard to keep quiet (hey, I'm a professor). Yeah, and she stores a lifetime supply of sperm in a little sac in her body. For about seven years.

I sketch a brief account of the fire ant colony cycle, punctuated by sucking up more newly mated queens.

Eventually, he asks the question I know must come—So, how do I kill 'em? We have arrived at a sort of conversational crossroads. If my audience has been unpleasant, or snide, I would now launch into standard advice no. 1—Get yourself a brick and a hammer—but after his initial suspicious challenge, this guy has been friendly and genuinely curious. So I give him standard advice no. 2, about poisoned baits and where to buy them. I can't resist slipping in the observation that having a lawn means you are providing fire ants with an irresistible opportunity. We chat some more, as I continue sucking up queens.

I gotta get back to work, he says. That was real in'erestin'. As he heaves open the door and vanishes into the cool cavern of Sharon's House of Beauty, I hear him yell, Them guys is suckin' up fire ants for . . . . The door shuts behind him. Another collectible experience comes to an end.

On the asphalt, the queens are destined to die sooner, huddled in their shelters, or later, in my lab. The space behind the Winn-Dixie is a black hole in the universe of the fire ant, sucking queens in and swallowing them like a giant aspirator. Even its vicinity must be dangerous, for it seems to emanate some fatal attraction. For us humans, it is one of those places we try to pretend doesn't exist, places that don't fit our suburban concept of "nice," places where nasty but necessary things happen, the dumps and sumps, the slaughterhouses and scrap yards that support our existence. But for those who look closely, even here there is life alongside death, beauty struggling with ugliness, treasure among the trash, order battling confusion. You don't have to look far to see the beauty in the newly mated queens, refined by thousands of generations, the beauty of function, each queen a perfect treasure, and that in them life struggles with death, struggles against odds, struggles to bring the order of colony life into existence.

The back door of Sharon's opens again. Another young woman steps out and shakes out a hairy cloth onto some discarded shelves from Eckerd Drugs. She glances at us and smiles. Snatches of a Merle Haggard song float out through the open door, to be snuffed out as it swings shut. Deep in the scented and air-conditioned dimness of Sharon's House of Beauty, beauty and ugliness are arbitrary, as fickle as a teenage fad. Out here among the trash, the beauty is timeless, and the sunshine is lethal.

12

The Incipient Phase and Brood Raiding

The emergence of the minim workers marks the beginning of one of the most dynamic and confusing phases of the colony life cycle. During this period, nests may shift from place to place, brood is transported like sack goods on the backs of slaves, worker and queen numbers rise and fall. When it is over, entire neighborhoods have consolidated into a single nest, and queens have fought to the death or been executed by workers until only one is left. The major engine of these rearrangements is a process named "brood raiding" by Bartz and Hölldobler (1982), who first described it in laboratory colonies of *Myrmecocystus*. At the end of the claustral period, the minim workers open the nest and begin to explore its vicinity. If during their explorations they come upon other incipient nests, they enter these nests and engage them in reciprocal brood-stealing contests. I made the first field descriptions of brood raiding on fire ants in the early 1990s (Tschinkel, 1992a, b). The dense, monospecific populations of fire ants and their preference for founding in largely ant-free habitats are a boon to the study of brood raiding.

Because the nest interior is not visible in natural nests, the initiation and progress of brood raids must be studied in the laboratory. In a typical experiment, two glass nest tubes, each containing a founding colony with newly eclosed minim workers, are placed in a tray and the nests opened. Workers from two different mature monogyne (single-queen) colonies are invariably hostile to one another when they meet, but this is decidedly not the case for the minim workers from incipient colonies (Tschinkel, 1992b; Balas and Adams, 1996b). When these workers enter each others' colonies, they encounter little or no aggression in over three-quarters of cases. A host worker may occasionally bite and hold an intruder's antenna or petiole for a while. Very rarely, a worker may flex her gaster as if to sting, but stinging is never observed. On the contrary, hosts and intruders often engage in mutual grooming and/or food exchange (even with the host queen), very nonaggressive behaviors indeed.

The reduced aggression within incipient colonies seems to be largely a characteristic of the minim workers themselves and is not the result of a failure to detect nestmate recognition cues. Minims usually attack and kill minor workers from mature colonies (Balas and Adams,

1996b), but minor workers that have been fostered as pupae in incipient colonies are attacked less vigorously. Rearing environment therefore probably contributes to the nestmate recognition cue. In any case, minims are clearly capable of discriminating between nestmate and nonnestmate workers.

Eventually, the intruder returns to her natal nest, and when she does, she usually lays an odor trail. In most cases, these initial trails were laid by workers from the larger nest. Brood raiding begins when a worker (usually from the larger nest) simply picks up a pupa and carries it back to her own nest (Tschinkel, 1992b). This theft meets with no active resistance. Occasionally, a "traitor" worker may start the raid by carrying brood from her own nest into the larger one. Returning workers sometimes alert nestmates by excited antennation, but often they simply deposit the stolen brood and return to collect more. The number of raiding workers varies from one to all but is usually a minority of the nest. Fairly commonly, one of the nests provides no raiders at all. In the majority however, workers from the raided nest soon begin to carry brood back from the raiding nest, so a two-way traffic in brood develops. The tempo of raids is highly variable. When a nest fields multiple raiders, only one or two do most of the carrying. When brood is scarce, a larva or pupa might be carried into a nest only to be deposited directly in the mandibles of a waiting rival worker who immediately carries it back. This merry reciprocation may continue for a few hours to several days.

Brood raiding seems to be carried out primarily by young minims and is one of the first orders of business after they open the claustral nest. By the time the first cohort of minims has finished emerging, 60–95% of the nests have raided (Figure 12.1). A minority of laboratory nests fail to raid for unknown reasons. Raiding ends when all the brood remains in one nest, and the workers from both colonies assemble in the nest with the brood. This nest can be regarded as the winner, though the competition is far from over. Why raids end and how the winner is decided are not clear.

A moment's reflection will reveal that, at this point, the workers from the losing nest have abandoned their own mother, who is pining in their now-empty natal nest. How could natural selection have produced a behavior in which workers (who are completely sterile) abandon the individual in whom their entire reproductive future resides in order to aid unrelated individuals? We will take up this apparently non-Darwinian behavior under the heading of usurpation, below. Another peculiar phenomenon that hints at mysteries still to be unraveled is the existence of "traitor raiders," workers that carry brood from their own nest into another. How does this behavior serve the interests of the traitors? Are they simply hurrying this competitive process to its inevitable end? Is it a mistake, an imperfection of nature?

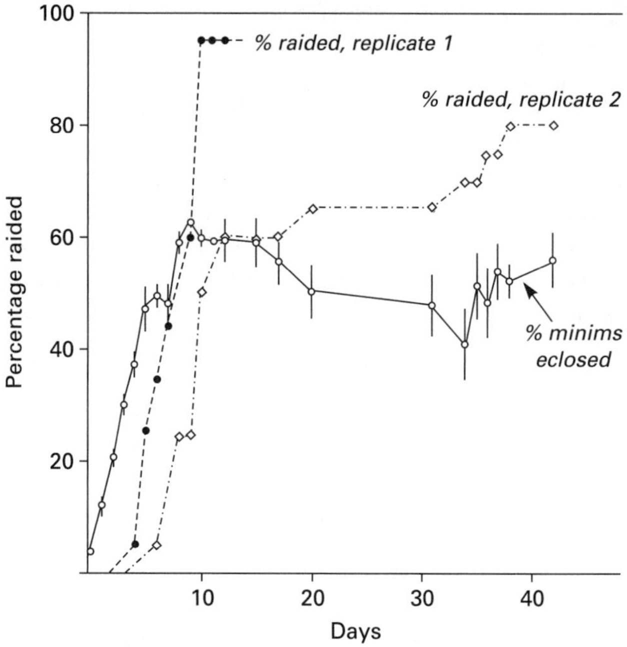

Figure 12.1. Brood raiding is one of the first behaviors in which newly emerged minims engage. By the time the first brood of pupae had eclosed, most nests had been raided. A second wave of raiding coincided with eclosion of a second brood of pupae. Adapted from Tschinkel (1992b).

Winning a Brood Raid

Only one important factor affects the outcome of raids in lab colonies—the number of workers (Tschinkel, 1992b). Larger nests usually win the contest, and the likelihood of winning increases with the difference in the initial number of workers (Figure 12.2). Victory does not depend on the number of larvae or pupae in either nest, the ratio of larvae to pupae, the number of queens, or whether the workers in the two nests are related to one another (or to the queen). To win, a nest must have more workers (a bigger army, if you will), and the more it outnumbers its opponent, the more likely it is to win. The effect of number does not seem to depend on devotion of more workers to raiding by the larger nest; nests usually commit about the same number of raiders. The contest seems therefore to be settled by some sort of "negotiation" or "agreement." How the workers finally agree on the winner remains to be determined.

Although queen number itself has no direct effect on probability of victory, incipient nests with multiple queens tend to produce more workers initially, and these larger nests have a better chance of winning. This advantage of worker number brings into focus one of the likely reasons why cooperative colony founding evolved in fire ants—it confers an advantage in the next phase of colony development by favoring victory in brood raids.

Brood Raiding in the Field

Although events in the nest chambers cannot be seen in a field study, the laboratory studies have already given us a good idea of what transpires underground (Adams and Tschinkel, 1991, 1995a, b, c; Tschinkel, 1992a, b). Still, the laboratory studies do not prepare us for the occasional scale and scope of this phenomenon in the field (Tschinkel, 1992a). Most of my field studies were made at the Tharpe Street site (Plate 16B), an area that had recently been cleared of forest and was in the early stages of invasion by weeds—in other words, prime habitat for newly mated queens. Observations of natural raids were followed by experiments designed to reveal causal links. Brood raiding was limited to the first few weeks after worker emergence and was suppressed when soil was dry.

In naturally settled cohorts, the closer together the incipient nests, the more likely they were to raid each other. Of course, these nests and queens were self-assorted, responding to each other and to small features of their landscape when deciding where to locate their nests. To make the claim that the distance between nests controls the raid frequency, we had to perform an experiment in which queens were assorted randomly into nests that differed only in how far they were from their neighbors. We did so by planting incipient nests in regular arrays of 25, varying only the distance between the nests. The raiding proportion varied from none when nests were 120 cm apart to almost

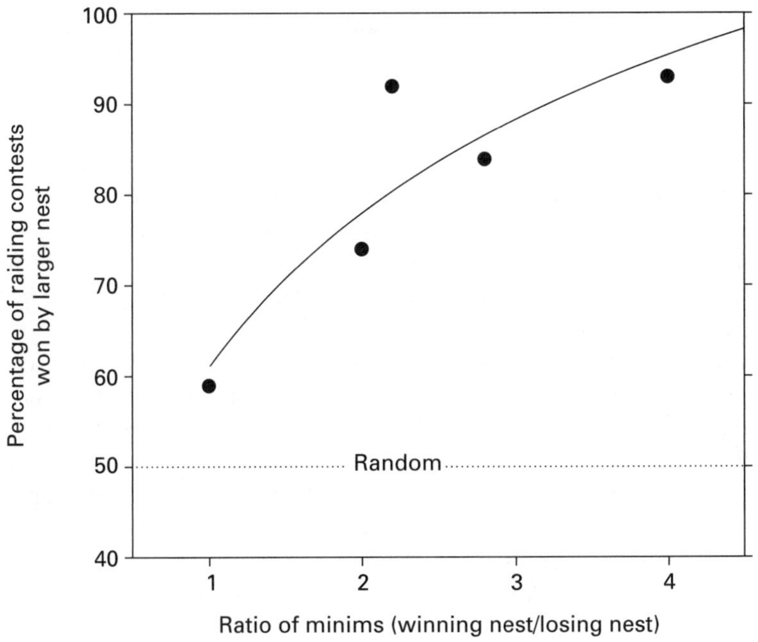

Figure 12.2. The greater the colony-size difference between mutually raiding nests, the more likely is the larger nest to win the contest. Winning enlarges the winning colony, making it more likely to win the next series of raids. Winning is thus a self-catalytic process.

70% when they were 15 cm apart. The closer together the nests, the more likely they were to raid one another, the more complex were the raids (the more nests were involved in mutually raiding groups), and the more likely they were to result in colony mortality and queen migration.

Natural nest chambers are not arranged in regular arrays but tend to be clumped for a variety of reasons, as discussed earlier. In contrast to those in a regular array, some nests are closer together and some farther apart than the average. The greater proximity ought to facilitate raiding. Indeed, in clumped arrays, over twice as many nests participated in raiding than did so in regular arrays of the same average nest density. The great majority of raided colonies were less than 50 cm away, but some were as distant as 132 cm.

Queen number might also be expected to affect raiding, because larger groups of queens (up to a limit) produce more workers, and nests with more workers might be more likely to engage in raiding and win the contests. This expectation was borne out when nests founded by multiple queens were shown to be more likely to initiate raids. In experimental arrays of nests with different numbers of queens, queen number had no effect on the probability of raiding or of surviving raids, independent of worker number.

A major effect of brood raiding was to increase the size of some nests at the expense of others, that is, to make worker number more variable. The variability of worker number tripled as a consequence of raiding at high nest densities. This result confirms what we saw in laboratory experiments—nests that were larger to begin with were more successful in amassing further brood and workers—competition was asymmetric.

Brood raids took place on the soil surface on odor trails connecting the colonies involved. During sunny weather, raids were limited to the morning and evening because midday soil temperatures were forbiddingly high. Because the raiding workers used precisely the same trails even after such lapses, the odor trail must have contained a long-lived component that survived the high soil temperatures. Worker density on the trails and trail length were highly variable but ranged up to 100 per meter and many meters, respectively. Usually, fewer than 10% of the workers carried any brood, causing rational humans to scratch their heads in puzzlement. Brood was also usually carried in the two directions with equal frequency, but gradually one direction became dominant, and eventually traffic in the opposite direction ceased. When the nests involved were excavated, the nest showing the net loss was empty, whereas the other contained workers, queen(s), and brood (Figure 12.3). Sometimes, such successfully raiding nests contained many times the number of minims and brood that could have been produced by any group of queens (up to 1700 minims and 11 queens). These nests could

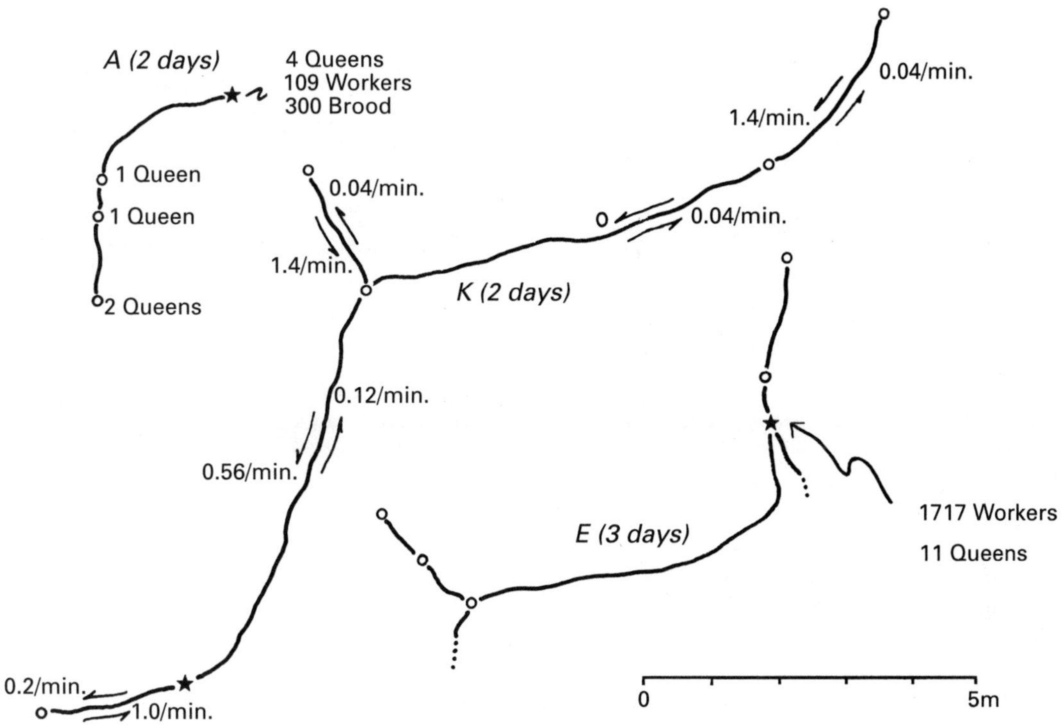

Figure 12.3. Examples of brood raiding in the field, showing the duration of the contests and the numbers of queens, workers, and brood in the participating nests. Arrows show the rate of brood transport in each direction. Only one of the 11 queens in E had fully functional ovaries and had won the competition. Reprinted from Tschinkel (1992b) with permission from the Entomological Society of America.

only have been the result of mergers of several incipient nests through brood raiding.

Because success in a brood-raiding contest increases the number of workers in an incipient colony, the successful colony is more likely to win the next contest and can probably raid farther as well. The distance between raiding colonies, the duration of contests, and the number of colonies involved in them all increased from June through August (Tschinkel, 1992a). Brood raiding is therefore a self-catalytic, cascading process that expands across the local founding population as the season progresses, aggregating incipient colonies into ever-larger units.

I observed one extreme example—"The Megaraid." This raiding contest lasted more than a month and involved at least 80 incipient nests, which at the end had aggregated into two units (Figure 12.4). At its peak, this contest involved more than 70 m of active raiding trail spread over 1500 m^2 and connecting up to 15 colonies at one time. The raiding

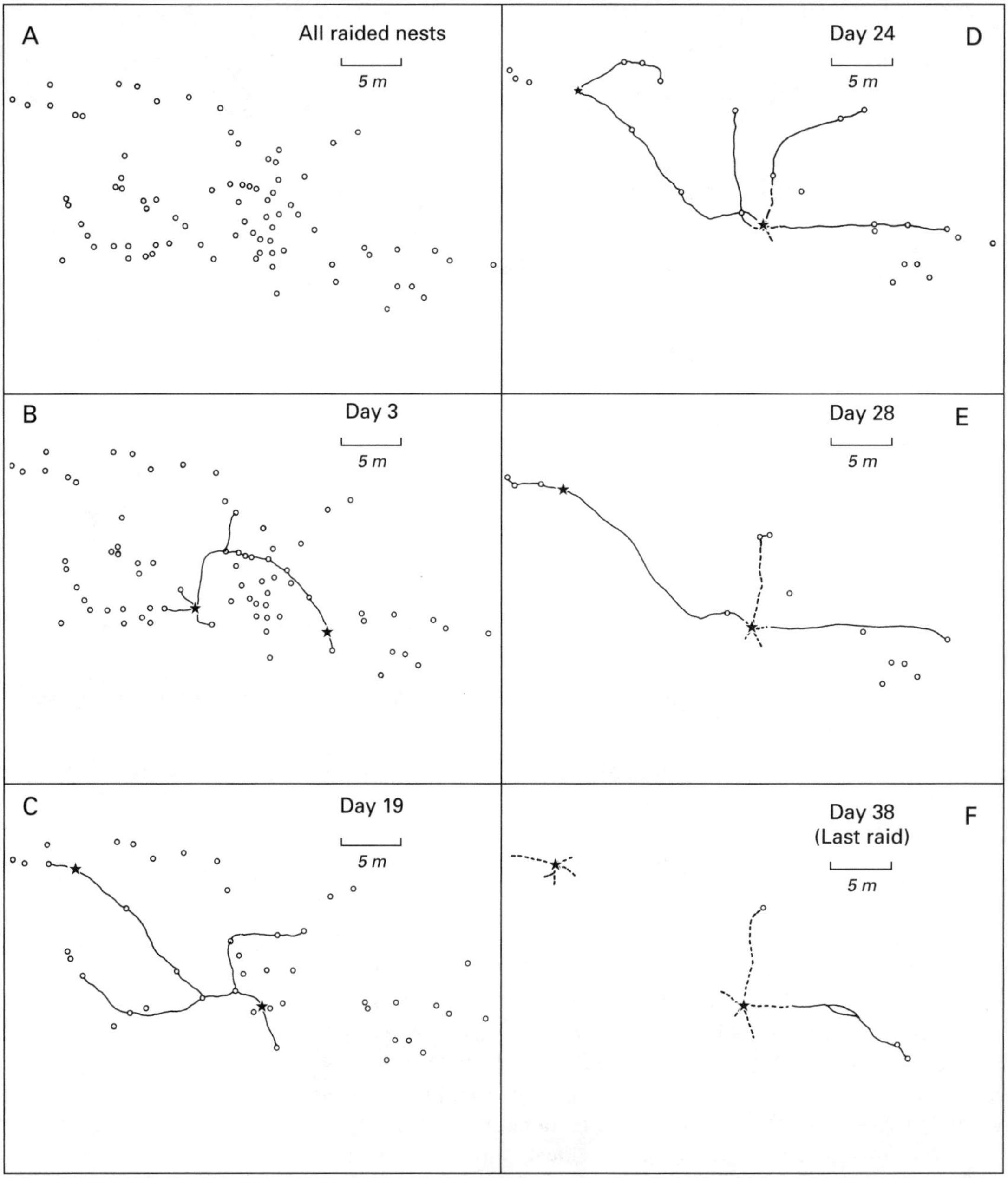

Figure 12.4. The "Megaraid." This complex raiding contest lasted more than a month and ultimately involved more than 80 nests, condensing them into two winning nests, each with hundreds or even thousands of workers. All participating nests are shown in the first panel, but these were not all present initially. Nests that lost raiding contests are omitted from subsequent panels. Stars show the dominant colonies, and dotted lines the eventual movement of the raiding trails underground into foraging tunnels. Reprinted from Tschinkel (1992b) with permission from the Entomological Society of America.

expanded for the first 24 days, catalyzed by the raid-generated growth of the winners. Both nests that won initial bouts eventually lost to other nests. As the winning colonies grew by raiding, the contests with additional colonies became ever more asymmetric, so eventually brood was carried only toward the winning nest, never the reverse. What had started as a contest became simple plundering by a vastly superior force. Newly discovered incipient nests were emptied of brood in just a few minutes. New colonies were discovered by what appeared to be groups of scouts moving in a loose cluster, much like the slave-raid parties of *Polyergus breviceps* that steal brood from *Formica sanguinea* (Hölldobler and Wilson, 1990).

Mentally reducing ourselves to the scale of ants gives an impression of what an amazing undertaking such a raiding contest is. Tiny workers weighing only 0.25 mg carried loads as large as themselves and ran at full speed in the hot sun for distances of up to 20 m (7000 body lengths; multiply your height in meters by seven for the equivalent distance in kilometers) between individual nests. From our height, we are apt to overlook the magnitude of the undertaking for the ants themselves, at soil level.

The "Megaraid" ended very suddenly. One day, workers from the winning nest attacked workers from a distant nest in typical territorial-defense fashion. No more raiding was seen after that day, although the workers continued to use the same routes as trunk foraging trails. The new territory acquired by raiding is probably organized around these trunk trails; the area between the arms of the trail is gradually brought under the colony's control. All observations have revealed that raiding lasts only a few weeks—thereafter colonies defend food and territory in a typical aggressive manner.

At this point, we pause for perspective on what has transpired. We began with a rain of newly mated queens onto a likely piece of real estate and aggregation of the queens into founding nests. The advantage of such cofounding can now be understood to include the benefit it confers during brood raiding, because groups of queens produce more workers initially, and more workers mean a better chance of winning the upcoming brood-raiding contests. Raiding condenses the local incipient population into a few winning nests, which thus begin "normal" colony growth with an enormous boost in size. Because the winner of the struggle for space is the colony that grows the fastest, the value of this boost for a colony is obvious. In fact, the two colonies that won the Megaraid produced sexuals and were of mature size within one year of founding.

The Fate of Queens: Migration, Usurpation

Workers from laboratory nests that lost brood-raiding contests abandoned their mothers to join the winning team. In the field, excavation

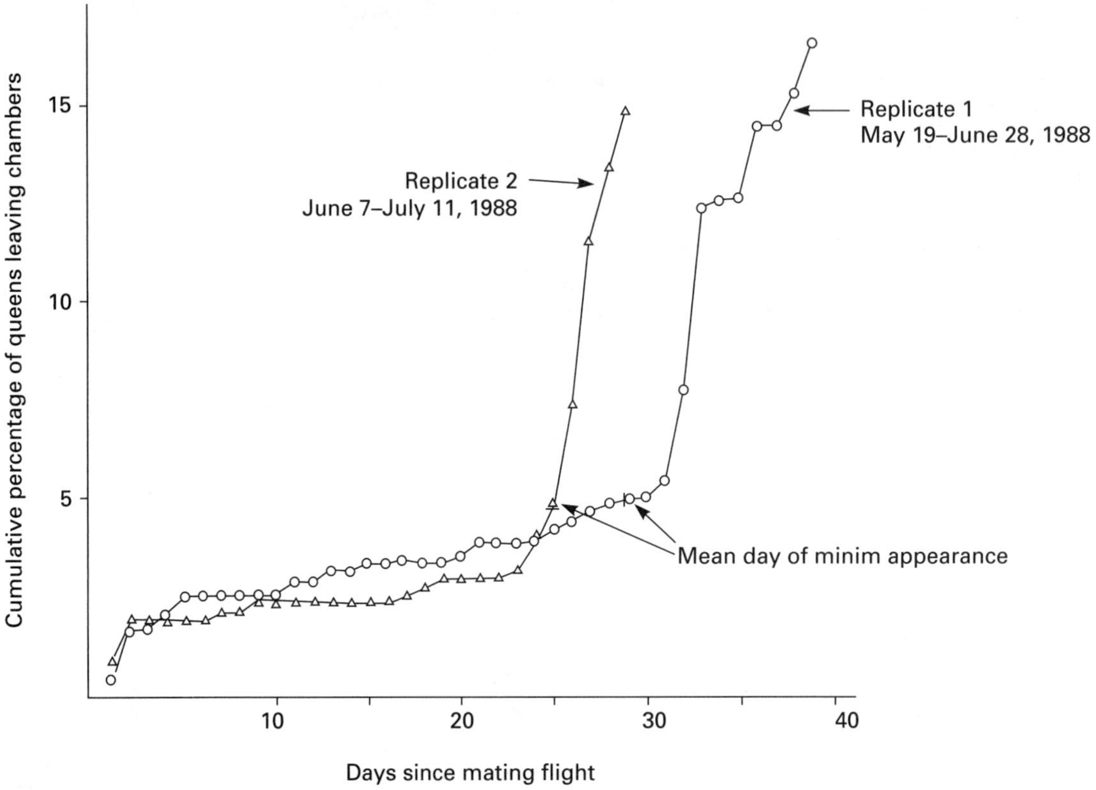

Figure 12.5. The departure of postclaustral queens from nests coincides with the appearance of minims and usurpation. Warmer soils caused the later cohort to develop faster than the earlier one. Adapted from Tschinkel (1992a).

of losing nests often uncovered only a queen; the brood and workers had joined the winning nest. Lest the reader shed a tear over these abandoned mothers, this is far from the end of the road for these queens. Queens are commonly seen running on the ground surface on days when no mating flights have occurred, and close inspection shows these to be queens that have already passed the claustral period—they are inseminated and have a few developed ova but are light in weight, have thin gasters, and lack large fat bodies, having invested their reserves in brood. An experiment in which queens were trapped as they left the small area containing their founding nests showed that the number of queens leaving founding nests was very low before minims appeared and that it increased manyfold thereafter (Figure 12.5). Departure was associated with nest failure, and a major cause of that failure was brood raiding (Tschinkel, 1992a; Adams and Tschinkel, 1995a, b, c).

Departing queens often follow raiding trails and attempt to enter occupied nests. Doing so usually requires digging, because minim workers make the nest entrance just large enough for their tiny bodies to pass. Resident workers and queens often resist the entry of foreign queens and attack, expel, or kill the majority of them, but many such queens were not attacked, so that death or expulsion is not the invariable outcome of intrusion. In some cases, the resident queen was killed or expelled. In a somewhat constrained experiment, I found that the chances of successful colony founding were about the same for queens that remained in their original nests and those that became intruders in other nests.

In unmanipulated laboratory raiding contests, a losing queen's chances of usurping the winning nest are about 25%, provided some of the workers from her own nest are present in the winning nest. Replacing the usurping queen's workers with workers that were both unrelated and unfamiliar (i.e., raised by another queen) caused usurpation success to drop by half. It returned to the unmanipulated rate as long as the winning nest contained brood *familiar* to the usurping queen—they

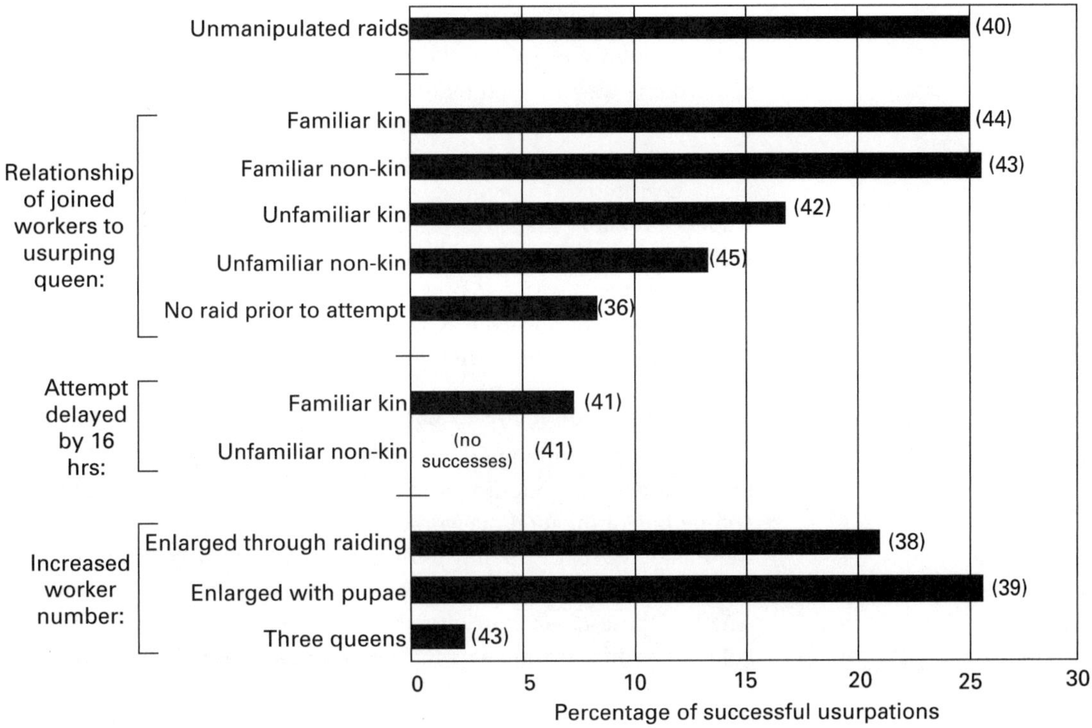

Figure 12.6. Outcome of usurpation experiments. Conditions of the experiments are noted to the left. Presence in the nest of familiar kin increases a usurping queen's chances. Reprinted from Balas and Adams (1997).

did not need to be related (Adams and Tschinkel, 1995b; Balas and Adams, 1996b, 1997). In fact, even a queen's own daughters did not help her success if these were unfamiliar to her because another queen had reared them. A losing queen's chances are greatly improved if workers from her own nest (not necessarily her own daughters) are present in the nest she is trying to enter.

When the post-raiding-contest usurpation attempt was delayed by 16 hours, success was halved again with familiar workers and was zero with unfamiliar (Figure 12.6). Nestmate recognition cues probably play a role in this discrimination, and these cues, or the workers' assessment of them, probably change rapidly in the amalgamated nest. To some degree, this recognition also extends to cofoundresses, who are less aggressive toward a reintroduced former nestmate than to an unfamiliar queen.

The most likely interpretation is therefore that, in migrating from the natal nest to the winning nest, the workers are simply trying to make the best of a competitive loss and are infiltrating the winning nest in order to help their mother enter and continue the struggle for primacy. The queen's chances of finding the nest to which her workers have migrated are probably fairly good—the odor trail along which the raids took place guides her there.

The number of workers in the winning colonies had no influence on usurpation success, but the number of queens did. When the winning colony was founded by three queens, successful usurpation was very rare. Resistance to usurpation may therefore be another advantage of cooperative colony founding and another choice a queen must weigh at the beginning. Cooperating with a few other queens to exclude intruders probably improves a queen's chances of surviving the final culling. We should also be cautious in extending the findings of usurpation experiments with two queens (most of the above) to higher numbers of queens, because the dynamics of raiding and usurpation may be somewhat different for higher queen numbers.

Reduction to Monogyny, Queen Execution

The moment the minim workers begin to forage, and as brood raiding begins, queens in founding associations cease cooperating and begin competing for the food the workers bring into the colony. At this stage, the advantages of competing outweigh the advantages of (more or less) cooperating. If an incipient colony contains more than one queen, whether these queens are members of the group that founded that colony or additional queens that migrated in, the reduction to a single queen soon begins (reduction, usurpation, and brood raiding may all occur simultaneously). Untangling the means by which queens live or die, and whether the losers are killed by the workers, the other queens, or both, has been challenging and has produced a few surprises.

Several reasonable armchair scenarios might explain how workers, queens, or both determine which queen survives. First, we might expect that the queen producing the most workers would be favored, because a greater proportion of the workers would be her daughters. This scenario would imply a sort of democratic vote on who survives and assumes that workers can distinguish their own mother from the other queens. Alternately, workers might ignore motherhood and choose the queen on the basis of her fecundity, because high fecundity is favorable to future colony success. In either case, queens ought to produce as much brood in cooperative founding groups as they can. Second, queens might fight and attempt to kill each other. If so, queens ought to behave in ways that maximize their fighting ability at the end of the claustral period, probably, at least in part, investing less in brood production in order to retain more vigor and reserves for the fighting to come. Third, workers might select queens on the basis of a dominance hierarchy that queens establish among themselves by fighting or other forms of competition. In a number of ant species, the more dominant workers assume positions on the brood pile. Thus, queens might struggle for positions on the brood pile, and workers might kill those not on the pile.

These details have been teased apart in a number of studies (Adams and Tschinkel, 1995b; Balas and Adams, 1996a, b, 1997; Bernasconi and Keller, 1996, 1998, 1999; Bernasconi et al., 1997; Adams and Balas, 1999). To begin with, in two-queen colonies, a queen's chances of survival are increased with every increment by which her initial weight exceeds that of her partner and decreased by every increment by which her weight loss (initial minus final) exceeds that of her partner. (In this context, recall that queens that were more robust at the end of the claustral period were more likely to join another queen at the beginning.) Weight loss is associated with investment in brood, but the correlation is broad, accounting for only about a third of variation in brood production. Factors that account for the rest probably include brood cannibalism, already mentioned, through which a queen might manipulate her weight loss in her own favor. Natural selection acts in opposite directions on the association and the queens within it. Thus, although initial weight and weight loss are correlated with queen survival, the reason for this correlation must still be clarified.

Does kinship play a role? Newly mated queens were allowed to rear brood alone until each nest contained pupae. Nests were then combined into two-queen nests and the brood of one of the queens discarded, so that all of the soon-to-emerge minims were the daughters of one of the queens and unrelated to the other. Dismayingly, daughters did not favor the survival of their own mothers—although they were less likely to attack their own mother, survivals of the mothers and the unrelated queens were about the same. Did the workers choose the more productive queen? By means of genetic markers, we could determine which of the workers in an unmanipulated incipient nest were the daughters of each of the two cofounding queens. Because we already know that relatedness is not a factor, this experiment tested fertility. The less fertile queen was as likely to be the survivor as the more fertile one (Balas and Adams, 1996a).

A hint about what was going on came from observations of aggressive postures and fighting between queens. Many surviving queens bore injuries and were more likely to be attacked by workers. When queens fought, the heavier queen was more likely to survive, explaining why the surviving queen in earlier experiments was the one that was initially heavier and/or lost less weight during the claustral period. Some evidence suggests that queens can adjust their weight loss according to the social situation and the characteristics of their nestmates (Bernasconi and Keller, 1998). Why greater weight confers an advantage during queen fights is not clear, but it suggests that physiology, not kinship, is the basis of worker discrimination between queens. In the absence of fighting (e.g., if minims are prevented from emerging), weight conferred no advantage, and the heavier queen had no survival advantage. Heavier queens also did not survive longer if their brood was removed at the end of the claustral period (unpublished data). The advantage conferred by weight therefore derived from improved fighting ability.

Queen aggression and fighting are stimulated by the appearance of the minim workers. If pupae were repeatedly removed from two-queen nests so that no workers appeared, or if all the brood was removed, queen aggression and mortality remained low (about 10%). In nests with workers and queens only, this mortality increased to about 40%, whereas in nests with brood, workers, and queens, it reached over 80% (Adams and Balas, 1999). The timing of this outbreak of hostilities coincided with an exodus of queens from natural founding chambers in the field and in the exodus-trapping experiment described above. When the colonies are not imprisoned in escape-proof laboratory nests, exodus from a losing situation seems to be an option. The wandering queens mentioned earlier are at least partly the result of this exodus.

Fighting among queens is probably not always an attempt to kill or injure an opponent, nor is the workers' role limited to executing *injured* queens. When queens were separated by screens through which workers could freely pass but that prevented queens from fighting with each other, workers attacked and killed one of the queens. Remarkably, they were virtually unanimous in their choice, rarely attacking both queens (Balas and Adams, 1996a; Adams and Balas, 1999). When the experiment was arranged so that all the workers were the daughters of one of the queens, the workers attacked the unrelated queen more frequently, but their mother was not more likely to survive (as in experiments noted above). Accelerating a queen's development by keeping her at a higher temperature increased her chances of dying at the hands of the workers. So did removing one of the queens from the brood and workers for several days in isolation and then returning her to the two-queen, screened nest. Workers usually moved the brood to the queen they would soon select to survive. In most experiments, by the day before the death of one queen, the brood was with the surviving queen in 95% of the cases. In some cases, workers made this choice within a short time after eclosing.

In preferring queens that have been in the recent presence of brood, workers tend to reinforce the outcome of queen fights, because the winning queen assumes a position on the brood pile. Queen fighting seems to result in a dominance hierarchy in which one of the queens monopolizes the position close to the brood and the other is banished to more remote parts of the nest, where workers are more likely to attack and kill her. Monopolization of the brood pile may also give the dominant queen access to more feeding by workers and to the fertility-stimulating material produced by fourth-instar larvae, making her still more attractive to workers and further improving her chances of winning the struggle for survival. In contrast, excluded from access to the necessary resources, the queen isolated from brood continues to decline in attractiveness to workers. We concluded therefore that workers prefer queens that are in better physiological condition rather than queens to which they are related, but treatment of one of the queens with juvenile hormone, which stimulates fertility, does not affect worker choice of which queen to attack.

The relative roles of queen and worker aggression in reducing colonies to monogyny can be seen in the following comparison. When both queen and worker aggression were prevented by removal of pupae so that no minims appeared in the nests, 59% of the colonies still had two queens after three weeks. When queens were prevented from fighting by screens separating them, and workers therefore performed all the executions, both queens survived in 34% of the colonies. When colonies were unmanipulated, so that both worker and queen aggression had free rein, both queens survived in only 8% of the colonies (Adams and Balas, 1999). Elimination of supernumerary queens thus results from the interplay of brood development, queen fighting, and worker aggression under the influence of the queens' physiological condition. Finally, a caution—all these experiments have been done in two-queen nests. Strategies such as queen fighting may differ somewhat in larger nests.

Strangely, these behaviors seem to reduce rather than increase the inclusive fitness of workers in incipient colonies. On average, by killing the physiologically more depleted queen, which workers do even in the absence of queen-queen fighting, they kill the queen who has invested more in reproduction and is therefore more likely to be their own mother. The explanation of this puzzle is currently unknown.

The Advantages of Cooperative Colony Founding (Balance Sheet)

Calculating a balance sheet for solitary and cooperative colony founding is not a simple matter. Overall, the increase in founding success through cooperation must exceed the chances of execution in the group. For example, joining a group already containing three queens must improve a queen's chances of success by more than fourfold over those founding nests alone. After all the rearrangements and usurpations in a field experiment were accounted for, this conclusion was borne out for four-queen associations but not for two-queen ones (Adams and Tschinkel, 1995a). Although queens seem to prefer the largest available groups, we do not know how they would, evolutionarily speaking, "leap over" the reduced success of small groups.

On the other hand, cooperative founding offers other benefits too. After the incipient period, colony growth is exponential. An incipient colony that begins the growth period with 100 workers as a result of brood raiding will always be four times as large as a colony that began with 25 workers, as long as both colonies are growing exponentially. Because competition among fire-ant colonies is largely a race to occupy the most space at the earliest date, this size advantage probably leads directly to a larger territory (Tschinkel et al., 1995) and earlier and more reproduction. Vargo (1988b) showed that cooperatively founded colonies produce sexuals earlier, enjoying the fitness benefit of reduced generation time, and larger colonies are known to produce more sexuals (have greater fitness) (Tschinkel, 1993b).

Colony size may also be important to winter or drought survival. Larger colonies can dig deeper nests, possibly helping them escape dry or cold conditions. They can probably also suffer greater worker losses and still recover. Another possible advantage of cooperative founding, especially when queens land in occupied territory, is that groups of cofounding queens may be more successful in defending against invasion by mature colonies (Jerome et al., 1998).

The Chances of Successful Colony Founding: Ledger of Death

The fate of a male departing on his mating flight is certain—he will be dead by the end of the day. If luck is with him, he will have mated and therefore have a chance, in the body of his mate, of successful reproduction, but the chances of his mate are not much better. A female alate embarking on her mating flight faces formidable odds against success. No one would blame her if, poised on the tip of a grass blade, ready to take flight, she pondered the road ahead and then simply turned on her tarsi to reenter the nest, where she would live out her life in safety and comfort. But take flight she does, subjecting herself to predation by birds and insects and to the chance of desiccation, heat death, starvation, execution, and usurpation. As soon as the alates rise from the nest, dragonflies and a variety of aerially feeding birds swoop in to scoop them out of the air. Alates must run this gauntlet again as they descend after mating (Whitcomb et al., 1973). Unfortunately, we have no estimates of the toll this takes on the alates.

As the queens settle to the ground, a further fraction is taken by ground-based predators, especially other ants, more than 20 species of which have been recorded to attack newly settled queens both in the United States (Whitcomb et al., 1973; Nichols and Sites, 1991) and in South America (Williams, 1980). When the ant *Conomyrma insana* is abundant, the fate of 97% or more of the fire ant queens is to be overwhelmed by *C. insana* workers, to have their appendages cut off, and to be dragged, still alive, into the attackers' nest to be eaten at their convenience (Nickerson et al., 1975). Hollowed out bits of queen body parts are cast out of the *C. insana* nests over the next day or two. An assortment of spiders, earwigs, and tiger beetles also prey on the queens. When predators are abundant, as few as 5% of queens may succeed in entering the ground (Whitcomb et al., 1973).

Existing fire-ant colonies are probably the most frequent killers because of their abundance (Nichols and Sites, 1991). Should a queen land in territory already occupied by a fire-ant colony, her chances of survival are very small even after she seals herself into a subterranean chamber (Whitcomb et al., 1973). Founding in groups seems to provide some protection—groups of two or four queens planted in fire-ant territories survive better (56%) than do solitary queens (34%) (Jerome et al., 1998). Usually, both queens are alive, having defended successfully, or both are dead, having failed. Rarely does only one queen in the nest survive the attack.

Survival through the claustral period does not put the incipient colony in the clear, for such survivors continue to be attacked by workers from mature colony upon discovery. Nor does having more workers help. In the laboratory, having more workers did not confer an advantage in nest defense. In similar experiments in the field, dead nests contained evidence of invasion—fragments of invading workers, warriors fallen in battle, alongside the remains of the queen and minim workers. Three weeks into the incipient period, survival in occupied territory is between 1 and 10%, but the survivors are probably still being attacked, and mortality continues.

Subterranean dangers other than fire ants tend to be highly variable.

The tiny, subterranean thief ants (members of the old *Solenopsis* subgenus *Diplorhoptrum*) prey on queens in their chambers, but their effect varies greatly among sites and trials (Lammers, 1987). Ants are probably generally important enemies of founding queens in most areas, although their impact also seems quite variable. In a Texas study, queens buried in ant-accessible vials survived better in monogyne than in polygyne areas, perhaps because of the lower density of scouting ants in monogyne populations (MacKay et al., 1991).

In the recently cleared sites that I favor for research on colony founding, the mortality is much lower, because predators, especially mature fire-ant colonies, are less abundant or absent. About 25% of the founding nests survive the claustral period (Tschinkel, 1992a). By the end of the raiding period, only 1 to 3% of these are still alive. Attractive sites are colonized by 8–16 queens per square meter during each summer, many founding cooperatively. I estimated that 10,000 to 20,000 queens settled on my plots during one summer, but only 0.1% survived to become queens of mature colonies.

As a comparison, consider founding success in an area supporting stable populations of mature fire-ant colonies, such as a pasture. If all the queens produced on an acre of such pasture mated and returned to that same pasture, about 100,000 queens would settle on each acre. Because the annual turnover in stable populations of mature colonies is about 13% and stable density about 50 mounds per acre, only about six or seven new colonies can survive per acre every year. Therefore, all but six or seven of the 100,000 queens that settled on this acre of pasture would die without producing a mature colony, a survival rate of 0.007%. Even if this estimate is accurate only to within an order of magnitude, a queen's chances of success are clearly (1) mighty poor under the best of circumstances and (2) about 10- to 15-fold better if she lands in unoccupied territory.

Such high levels of mortality imply very intense natural selection. Indeed, this intense selection is the force that has produced the many precise, complex, and special adaptations just described. Evolution has seen to it that the queen is not a passive player, relying only on luck. Facing each situation, she makes choices that tip the odds slightly in her favor—should she join or not? If yes, whom? How much to invest in her brood? How to balance her own needs against those of the group? At no time does a choice result in a certainty, for information is always short and the future hard to predict. During the hurly-burly of brood raiding and usurpation she continues, through her choices and behavior, to try to tip the odds in her favor. The "right" choices, along with a huge measure of luck, might place her with the few queens that survive. Evolution takes note of what it took to do so. In addition, if the selective forces in the nonnative range and South America were different, comparison of the two populations would reveal how fire ants have responded to selection. Finally, the density of newly mated queens in South America, and therefore encounters among them, may rarely be as high as in the USA, and some of the queen responses in the USA therefore not necessarily adaptive.

13

Dependent Colony Founding

In the face of such tiny chances of successful independent colony founding, we can reasonably ask why monogyne fire ants have not evolved dependent colony founding without becoming polygyne. The answer is that they have but that its existence was only recognized in the 1990s and represents a previously unknown phenomenon in ants (Tschinkel and Howard, 1980; Tschinkel, 1996; DeHeer and Tschinkel, 1998). The discovery linked two previously disparate phenomena, the mating flights of overwintered female alates (Fletcher and Blum, 1983a) and the replacement of queens in orphaned colonies (Tschinkel and Howard, 1980; Tschinkel, 1996).

In the late 1970s, my laboratory showed that about a third of orphaned colonies could replace their queens with perfectly normal, inseminated queens. We assumed that the replacement queens were already members of their colonies but not reproductively active until the active queens were removed. An attempt to repeat this work in the late spring of 1992 failed to detect *any* queen replacement, leading me to fret about a ruined reputation. The reason soon became apparent; queen replacement was quite seasonal. When I orphaned colonies in north Florida between January and mid-February, 17% had inseminated replacement queens after six to nine weeks. For those orphaned between mid-February and late March, the number jumped to 39%, and after March it dropped to zero. Because the mating flights of the overwintered sexuals were taking place during the time of peak replacement, it seemed likely that the replacement queens were actually overwintered female alates that had entered the orphaned colonies.

A direct experimental test is always better than correlation, so six colonies were orphaned, and 10 newly mated, overwintered queens, each marked with a fine wire, were released on the mounds, two each day for five days. Seven weeks later, the colonies were excavated and searched for a queen. Two of these colonies contained fully functional queens, each of which, *mirabile dictu,* bore the tell-tale wire mark (Plate 13B). My test queens had succeeded in being accepted by their host colonies, to whom they were not related (the overwintered queens were originally collected about 15 km from the experimental nests). Subsequent experiments and observations gave a clear picture of what happens.

After midsummer, as colonies grow back to their full midwinter size, they also produce small numbers of male and female sexuals, perhaps 10% of the number produced in spring. Given the opportunity, these sexuals participate in mating flights and colony founding just as the spring-reared ones do. Newly mated queens collected in November produce as many minim workers as queens collected in June, but as cool weather sets in to inhibit mating flights, a change comes over the female alates that remain in the nest—for unknown reasons, they gradually lose weight, so by February, they are 30% lighter and their fat reserves have declined from 50% of their weight to well under 40%. Moreover, their blood levels of circulating vitellogenin (yolk precursor protein) have steadily increased (Lewis et al., 2001), their ovaries have developed, and they are laying a few eggs in their natal nests. These changes are similar to those that occur during colony founding, a similarity that is anything but coincidental.

When the first warm spring days arrive, these overwintered sexuals participate in mating flights, but these early spring mating flights (late February through the end of March) are quite different from the late-spring ones (April to July). They take place in much cooler weather (20–22°C), are triggered more by temperature than by rain and therefore take place even under fairly dry conditions, and involve much smaller numbers of sexuals (10's to 100's per mound per flight). The overwintered female alates that participate in these flights are different from spring-reared ones in important ways. We have mentioned that they are lighter and that their ovarian development is advanced. After mating (they may mate near the nest) and settling back to the ground, they do not attempt to find or dig a founding chamber into which to seal themselves. Rather, they land on and try to enter the mounds of mature colonies and can regularly be seen there. Although they are usually attacked and killed by resident workers, the experiment with marked queens shows that, when the host colony is orphaned, the chances of being accepted are quite substantial (possibly as high as one out of six). Once past the workers' defenses, the queen is treated as if she were the workers' own mother. Their feeding and care soon stimulate her ovaries to pump out eggs at a high rate (Plate 13B). For a time, as the workers of the host colony rear these eggs into workers, the colony consists of two families, but as the original queen's workers die of old age, the new queen's workers gradually replace the original queen's. When this process is complete, the means by which the queen founded her colony can no longer be determined.

Overwintered *S. invicta* queens are wedded to this form of temporary social parasitism for starting their own colonies. First, they lack the body reserves needed for producing brood on their own, having lost them during overwintering, but they also lack the necessary behaviors: when forced to attempt independent founding in the laboratory, they

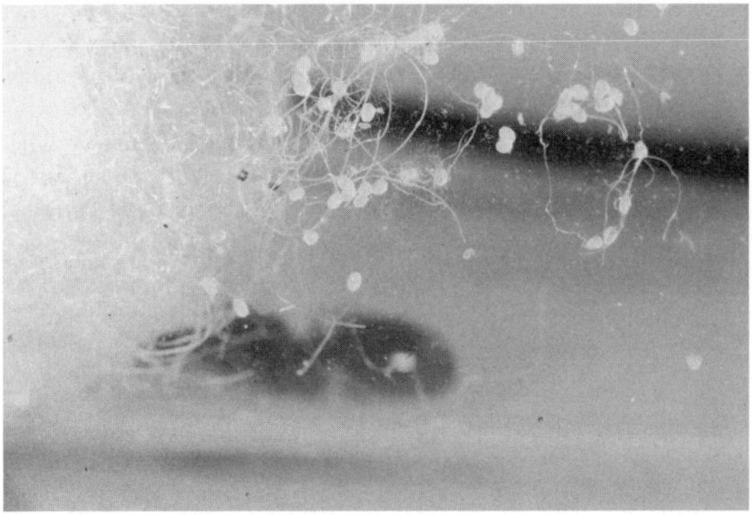

Figure 13.1. A newly mated overwintered queen in a nest tube, showing the absence of egg-clustering and care. A spring queen would have gathered all the eggs into a single cluster and tended them. The fibrous material to the left is cotton.

lay eggs, but they do not cluster them together (Figure 13.1) and do not tend the larvae that hatch from them. Rarely do they produce any minim workers, and when they do they produce too few for successful founding. In these characteristics, in their fertility, and in their attractiveness to workers, overwintered queens resemble colony queens rather than newly mated ones. All the evidence indicates that this strategy is an alternate mode of colony founding, one in which a queen that has developed physiologically beyond the claustral stage exploits the worker force of an unrelated colony to aid her in founding her own.

Does the orphaned host colony have any defense, and does it matter, because their mother is already dead anyway? To answer the second question first, yes, it matters. If a colony contains overwintering female alates at the time of orphaning, one or more of these females will shed her wings within hours and begin to lay eggs. Being unfertilized, these eggs can only develop into males, but these (often very large) broods of males give the colony a last chance at reproductive success during the upcoming late-spring mating flights. The entry of an unrelated replacement queen robs the colony of this final reproductive gasp and is the feature that exposes the socially parasitic nature of this phenomenon. The workers of the host colony gain no fitness from helping the unrelated replacement queen. Any sexuals they may help to rear are not relatives of theirs.

What about defense? The same overwintered alates that are the agents of this exploitative colony founding are also the defense against it. When a colony is orphaned, the level of queen pheromone drops quickly, and some of these alates shed their wings and begin to lay eggs. The assumption of this queen-like state by family members returns the colony to a more or less normal state (even though the queens are not mated) and protects against intrusion by unrelated newly mated queens. The field results make clear that, when colonies contain female alates just before orphaning, 14% of the replacement queens are inseminated parasitic queens and 86% are uninseminated sisters of the workers (Figure 13.2). In contrast, when female alates are lacking at orphaning, 62% of replacements are parasitic and only 38% are sisters. This defensive function of alates may explain their strongly female-biased sex ratio during the winter (Morrill, 1974b; Tschinkel, 1996).

As colonies release their overwintered sexuals, the ability to invade colonies and to resist such invasion decline together. By the end of March (in north Florida), most of the alates have flown. By the end of April, the spring-reared alates are ready to fly and await the triggering effect of rain.

Independent and dependent colony founding in fire ants are most successful under different population conditions. When the habitat is new and free of mature fire-ant colonies, newly mated queens enjoy relatively high rates of success because one of their chief causes of death (workers from mature colonies) is absent. As the habitat becomes stocked with mature colonies, the success of queens trying to found independently drops to very low levels. The situation for the dependently founding queens is the opposite. Because they need mature but orphaned colonies to employ their strategy, they have no chance of success in habitat that has not yet been colonized by fire ants. In contrast, fully stocked populations, especially after at least one colony life span has elapsed, offer regular founding opportunities in the form of orphaned colonies. The relative success of these alternate reproductive

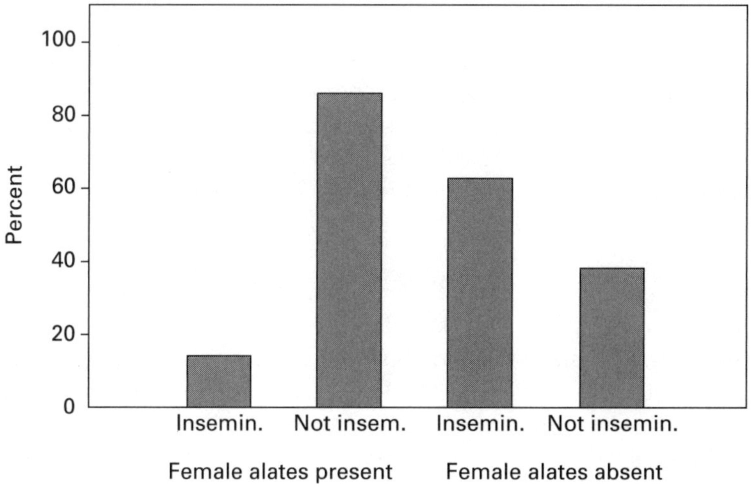

Figure 13.2. Presence of female alates in the nest reduces social parasitism by unrelated overwintered queens by more than 75% compared to that in nests lacking such female alates.

strategies therefore depends on what proportion of the habitat is vacant and what proportion is stocked. *Solenopsis invicta* colonies seem to invest 8–10% of their total annual female alate production in dependently founding queens. I calculate that, when only fully stocked habitat is available to colonies, the dependent queens will contribute about 35% of the reproductive success. When only 80% of the habitat is occupied, this figure drops to 5%, because the vacant habitat contributes so much more to success. This calculation emphasizes that the balance between independent and dependent founding is determined by the typical degree of habitat saturation encountered by a species. In *S. invicta,* the low investment (8–10%) in dependently founding queens suggests that it is primarily a weedy species that seeks out new habitat to colonize and that it depends on a continuing supply of such habitat.

Obviously, because the queen is replaced with another of the same species, one cannot simply inspect queens to determine which are original queens and which replacements. A clever way of determining the success rate of dependent, parasitic founding relies on the knowledge that, in the USA, about 13 to 17% of newly mated queens are match-mated (that is, mated by chance with males that matched them genetically at the sex-determining locus) and produce diploid male offspring. Match-mated queens are readily accepted into orphaned colonies, but when they are, they produce large numbers of diploid males and few workers. Diploid males are noticeably larger than haploid males (DeHeer and Tschinkel, 1998), so male size and abundance can serve as field signs. Because a match-mated queen cannot keep a colony going (she does not produce enough workers), such queens must have entered the colony in the previous few months. The replacement nature of a queen can then be confirmed by tests showing that the genotypes of the workers and the males are different (they have different mothers) and that the males have two sets of genes instead of one. This method provides over 99% certainty of correct classification. One can estimate the proportion of the colonies headed by any type of replacement queen by dividing the proportion headed by a match-mated replacement by 0.13 (the frequency of matched mating). This procedure, applied to over 3000 colonies tested for match-mated queens, revealed that about 1% of queens are replaced by dependently founding queens each year and that about 3% of any population of colonies is headed by such queens.

The case of the tropical fire ant, *Solenopsis geminata,* provides an illuminating contrast with that of *S. invicta.* Like those of *S. invicta,* the sexuals it rears in the spring participate in summer mating flights and found new colonies independently. Like those of *S. invicta,* after a hiatus in sexual production, many colonies produce fall sexuals, but unlike those of *S. invicta,* these females (microgynes) are both smaller and lighter than the spring sexuals (macrogynes) (McInnes and

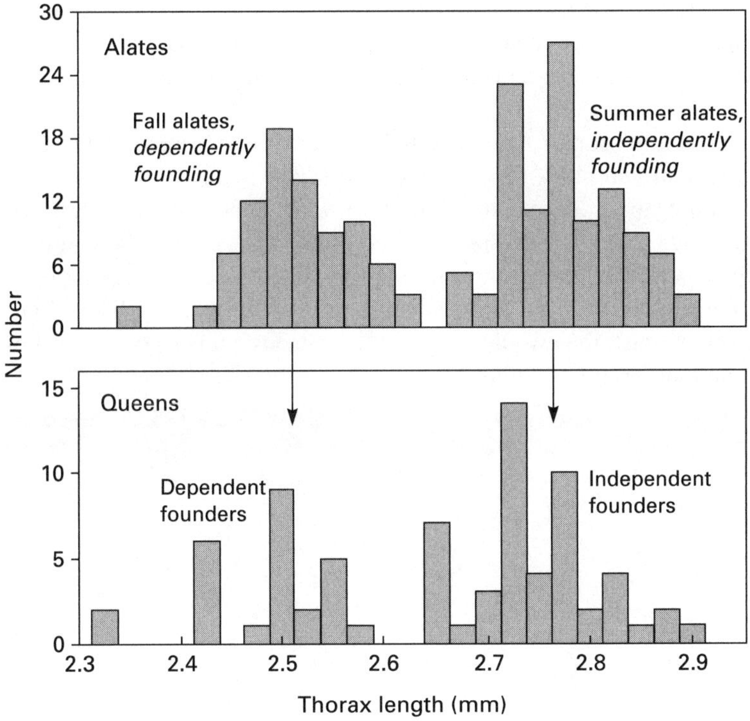

Figure 13.3. Dependent (parasitic) colony founding in *Solenopsis geminata*. Dependently founding queens are smaller and are thus easily recognized when they become replacement queens in mature colonies. Dependently founding queens fly in the fall, whereas independently founding queens fly in the summer. Adapted from McInnes and Tschinkel (1995).

Tschinkel, 1995). The microgynes of *S. geminata* correspond to overwintered queens of *S. invicta* and, like them, lack the reserves to found independently. They fly and mate in the fall and enter (probably orphaned) mature colonies of *S. geminata*. That they succeed is apparent from the approximately one-third of queens in mature *S. geminata* colonies that are microgynes, about 10 times the fraction of colonies headed by overwintered parasitic queens in *S. invicta* (Figure 13.3).

We therefore know that *S. geminata* also practices both independent and dependent parasitic colony founding but with much greater emphasis on and success through the dependent mode. This pattern is in keeping with the much less "weedy" nature of *S. geminata*. Its habitat is less ephemeral, its populations more stable over long periods, and its colonies more likely to find themselves under near-saturation conditions. In this circumstance, claustral founding is difficult and investment in dependent founding pays off relatively well. Colonies then should, and do, invest more of their effort in this form of reproduction, about one-third of their annual effort. The selective environment has adjusted the balance point between these two strategies until their relative payoff is the same and shifting the balance no longer improves total fitness, a so-called evolutionarily stable strategy. In addition, dependent founding does not require alates with a lot of reserves, so the microgynes are smaller, less fat, and therefore cheaper to make. Because a colony can produce more such individuals from the same amount of material, its chances of successful founding are increased—it gets more bang for the buck.

An interesting but unanswered question is whether *S. invicta* and *S. geminata* can alter their relative investment in these two modes of founding in response to the situation in which they find themselves. Do *S. invicta* colonies in saturated populations produce more overwintered queens than do those in unsaturated ones? Do *S. geminata* colonies in new habitats produce fewer fall and more spring queens? If these capacities exist, they would represent a previously unsuspected level of adaptability and flexibility.

14

Colony Growth

When the founding period is over, brood raiding is finished, and queen execution has run its course, each surviving colony finds itself with a single queen (not necessarily the mother of all the workers), a handful to thousands of minim workers, and a substantial number of larvae and pupae, some of which may have been pilfered from other incipient colonies. The queen will never again rear brood through her own labor (Cassill, 2002; Stringer et al., 1976) but will depend on her daughters to do so. The nest is a vertical tunnel with one to a few small chambers. The workers have recently stopped tolerating intrusions from unfamiliar workers and now attack them with the typical aggression of territory-holding ants, managing to hang onto a desperately small territory wedged between two warring superpower colonies.

In contrast, a mature colony consists of up to a quarter of a million workers of diverse sizes, a nest with hundreds of chambers surmounted by an elegant mound, and an exclusive territory of 50 to 200 m^2. This chapter tells how the tiny incipient colony, teetering on the edge of extinction, grows into the proverbial 800-pound gorilla, how an astoundingly complex, coordinated family of a couple of hundred thousand daughters arises from a single queen.

This transition is much more than simply the addition of more workers. The mature colony is no more like the incipient colony than an adult human is like a baby. Like adult organisms, the mature superorganism arises from the founding queen through a complex and subtle process of development called "sociogenesis" (Wilson, 1985). As the name implies, sociogenesis is analogous to embryogenesis, the development of an individual from a fertilized egg, but it occurs at the next higher level of biological organization, a society. I will describe sociogenesis of the fire ant colony from three angles: the mathematics of the growth of the worker population, the economics of converting energy and materials into the components of the superorganism, and the qualitative changes in the superorganism brought about by differences in the relative growth of its components.

The Basis of Colony Growth: Brood Development

The basis of sociogenesis is the production of new ants, primarily workers, the essential modular units of a colony. For each ant, the process

Figure 14.1. All brood stages develop faster at higher temperatures. Adapted from Porter (1988).

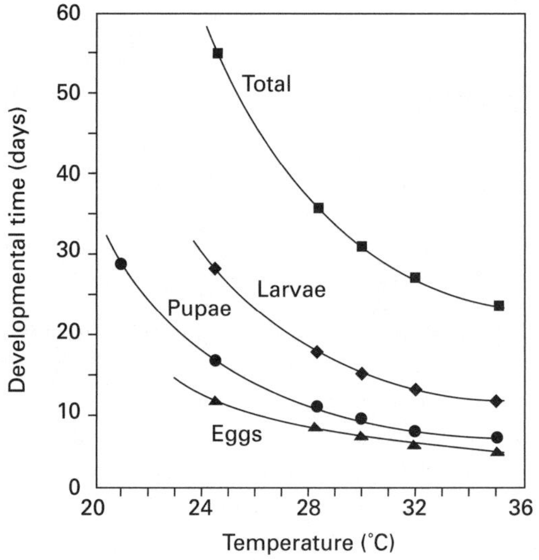

begins with an egg laid by the queen. This egg is only a promise, the foundation on which workers build a new ant. Through ceaseless feeding and care, they convert this 4-μg egg into a worker 250 to 1000 times as large or a sexual female up to 3500 times as large. This conversion, not egg laying, represents the vast bulk of production and growth in an ant colony (the coordination of the queen's egg-laying rate with the available labor force is discussed in a later chapter).

The development rate from egg to adult ant depends strongly on temperature, an unremarkable fact. This rate seems to be fairly uniform across several species of ants, including *Solenopsis invicta* (Porter, 1988). Stage durations are usually determined by laboratory rearing, so small groups of queenless workers can rear brood at several constant temperatures (note: the ways in which the queen, and more realistic fluctuating temperatures of the same average value, affect development have not yet been determined). The average stage duration is the time required for half the test ants to pass through the stage. The strong temperature dependence of stage durations can be seen in (Figure 14.1). Total development time declines from 55 days at 24.5°C to 23 days at 35°C.

Total development time also varies with body size, but growth rate does not. Therefore larger ants result from longer development (O'Neal and Markin, 1975a). At 32°C, the tiny minim workers require 18 days to reach adulthood, minor workers 24 days, and the large majors 28 days. Sexuals, which are larger still, take about 34 days. All of these pass through four larval instars, which can be differentiated on the basis of their mouthparts (O'Neal and Markin, 1975b). Invariance of growth rate implies that the colony's daily production rate of new tissue depends only on the number of individuals, not their type.

Interestingly, the proportion of the total development time spent in each life stage does not change with temperature. Thus 21% of each ant's total development time is spent as an egg, 41% as a larva, and 30% as a pupa (in founding colonies, 25% as an egg, 35% as a larva, and 40% as a pupa). The development rate, expressed as percentage of the total development time passed per day, increases about 0.4% per day for every degree increase in temperature (Figure 14.2). At 35°C, the rate is 4.3% per day for brood (5.8% for founding brood). Brood development stops below about 24°C (21°C for founding nests) or below 1.5% per day. Brood development thus occurs only within a rather narrow window of temperature, about 12 to 14°C, imposing absolute seasons for brood production. Brood production ceases during the winter in the northern parts of *S. invicta*'s range in the USA but not in the southern parts (Markin et al., 1974c).

Seasonal temperature changes produce changes in colony growth rates, in part through their effect on brood development rate, but colony growth rates cannot be precisely predicted from temperature alone, because they depend not only on individual development rates but also on the number of larvae a colony nurtures. In laboratory colonies fed *ad libitum* (Figure 14.3), growth was zero below 24°C and increased sharply to peak at 32°C. Shorter development times squeeze 2.2 generations into the two-month experimental period at 32°C but only 1.1 at 24°C. At 35°C, production declines sharply as a result of some general interference with colony function at this high temperature. The colony growth "window" was therefore a mere 12°C (similar in founding

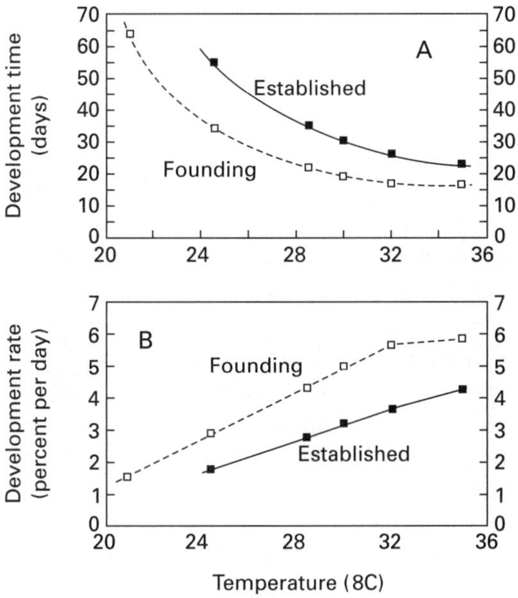

Figure 14.2. A. Total development time for workers from established and founding colonies decreases nonlinearly as temperature increases. B. Daily development rates as a percentage of total development time has an approximately linear relationship to temperature. Adapted from Porter (1988).

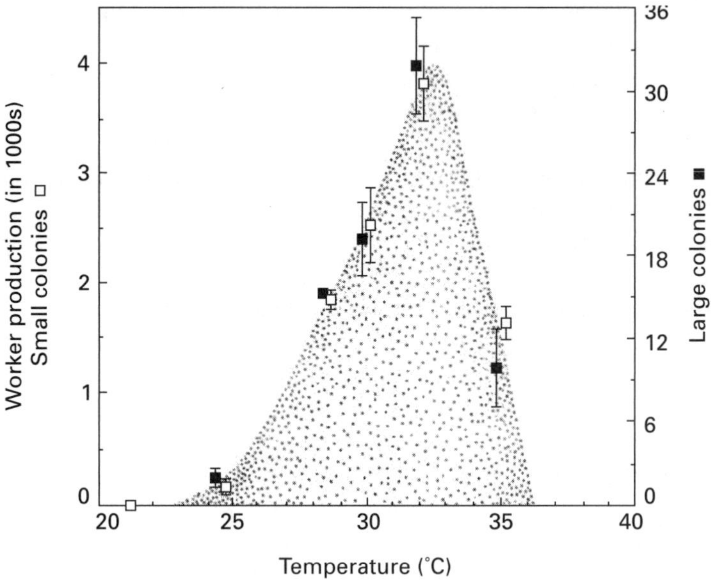

Figure 14.3. Net worker production in relation to temperature in small and large colonies two months old. Note that large and small colonies have different scales on the y-axis. Adapted from Porter (1988).

colonies). The optimal temperature for growth, 32°C, is also the preferred temperature of well-fed colonies.

Field colonies employ several tricks to increase the availability of this favorable temperature window. Chief among these is the construction of a solar collector mound. The temperatures actually experienced by fire ants outside the laboratory are not well characterized, in part because temperature in nature varies so much on several time scales, and fire ants have temperature choices within their nest.

Growth of the Colony as a Population of Workers

An ant colony is obviously a population of individual organisms, each capable of sustaining life. The colony starts with a single queen and a lifetime supply of sperm, grows when its birth rate exceeds its death rate, declines when the reverse is true, and ultimately meets a limit to further growth. Also like a population, the individuals that make up the colony live out their lives and are replaced by new ones, in a process called turnover. Although convenient, laboratory studies of colony growth in population terms are of limited usefulness because conditions in nature are so different from those in the laboratory and conspire to make field growth rates different from laboratory growth rates. Truly relevant growth data must be taken from natural colonies. Fortunately, *S. invicta* is one of the few ant species for which data on the age and size of colonies are available.

For *S. invicta*, estimation of colony age capitalizes on the even-aged cohorts of colonies it establishes on recently cleared forestland. These even-aged cohorts retain their integrity until colonies begin to die of old age (5–8 years), to be replaced by younger colonies (Tschinkel, 1988a). Markin created an even-aged cohort by killing existing colonies on several large blocks of pastureland with the poisoned bait mirex, then followed the cohort for three years (Markin et al., 1973).

However the even-aged cohort was created, measuring colony size over several years produces a description of colony growth. Many researchers have struggled with determining how many ants are in a colony. The small number of species for which the answer is known is testimony to the difficulty of the task. A mature fire-ant colony contains hundreds of thousands of ants spread through hundreds of chambers extending from the mound down to 2 m below ground. In principle, several methods might be useful, but each makes incorrect assumptions about how ants behave. In practice, the problem has been solved through two rather crude, brute-force methods. Markin and his coworkers drove a wide metal cylinder, open at both ends, into the soil to surround a mound, then slowly flooded the nest until the ants floated in the water-filled ring (Markin et al., 1973). Small colonies were simply excavated and brought into the laboratory, where the soil was allowed to dry until the ants moved into a moist trap nest. Both methods recovered about 90% of the workers.

The second method is based on excavation and sampling in the field (Tschinkel, 1988a); it obviates the need to carry large amounts of water into the field, and the investigator gets a good deal of exercise in the bargain. The entire colony is excavated, and the dirt and ants are mixed and weighed. Several small, weighed subsamples are returned to the laboratory, where the various types of ants they contain are separated and counted. Their number is the same fraction of the total population that the weight of dirt in the subsample bag is of the entire excavated nest. Neither this method nor the previous one includes the foragers that were out in the field at the time of the census. As I will show below, the result is a considerable underestimate of the total population of workers. For now, these estimates will suffice to reveal the essentials of population growth.

When colony size is estimated on approximately the same date of each year, fire-ant colony growth meets the expectation of being approximately logistic, with an inflection point at about half the maximum size. In addition a complex seasonal pattern exists whose real nature was not recognized until my student Sanford Porter found that mature colonies were largest in midwinter and smallest in midsummer (Tschinkel, 1993b). The size fluctuation is associated with the production of sexuals in the spring and workers in the fall. The larger the colony, the greater is the annual size fluctuation. That is, the difference between the annual maximum and minimum also grows logistically as colonies grow. Superimposition of this seasonal fluctuation on the annual-mean growth curve results in a wave function whose annual

Figure 14.4. Growth of fire-ant colonies is logistic, with logistically increasing seasonal size fluctuation. Reprinted from Tschinkel (1988a) with permission from Springer.

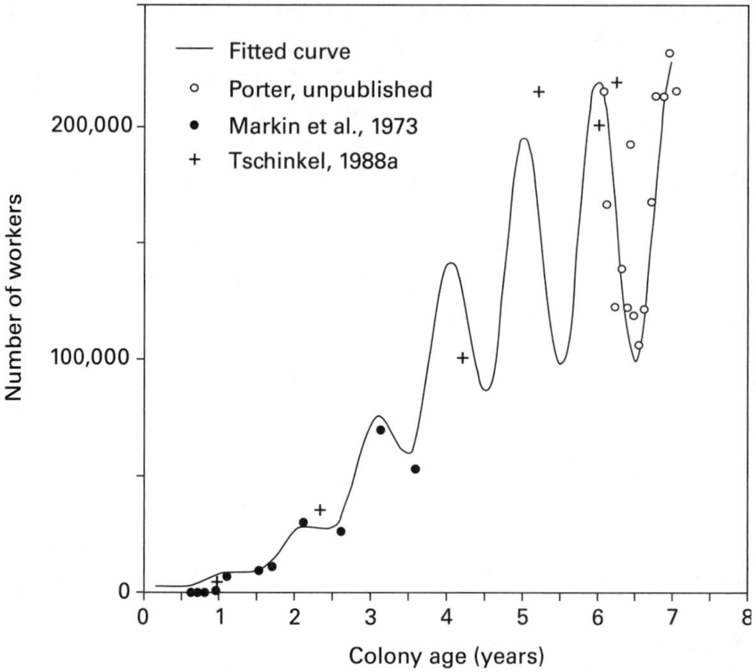

mean and annual variability both increase logistically (Figure 14.4). My own data, Porter's, and those of Markin's group can all be fit well by this complex logistic function. Recognition of this seasonal fluctuation of colony size, predicted on economic grounds by Oster and Wilson (1978) and by Brian and coworkers (Brian, 1978; Brian et al., 1981), was an important insight and will be discussed in the context of colony reproduction.

After four to six years, colonies no longer increase in size but simply fluctuate annually around a mean size. Totaled over the annual cycle, birth rate and death rate are then equal. Several growth-limiting factors suggest themselves, including an upper limit on the queen's egg laying rate, an increasing proportion of sexuals with a parallel reduction of worker birth rates, size-related inefficiencies, food and foraging limitations, and others. In truth, however, we do not know what finally puts the brakes on colony growth.

Ant Economics and the Factory-Fortress

Social-insect colonies are obviously producing output from input, and much of what has been written about their growth is dominated by economic thinking, much of it of a rather theoretical nature. Oster and Wilson (1978) recognized this analogy when they wrote that "viewed from within, the colony is like a factory enclosed in a fortress." In this formulation, popular with many myrmecologists, we see the colony as an entity in the business of production, subject to economic rules much like a human manufacturing firm (Sudd and Franks, 1987). Unlike most human firms, the colony is under constant physical threat by enemies and protects its investment within a defended nest, as anyone who has blundered into a fire ant nest will attest.

For an ant colony, *production* means the conversion of food into new biomass, be it workers, reproductives, or stored reserves. Because insects grow only as larvae, most production occurs when workers feed larvae and make them grow. The production of stored fat and protein represents the setting aside of *capital reserves* or *metabolic reserves* for later use. Workers are the unit machines of the factory-fortress. Their cost does not end with their production, because simply keeping them alive entails a *maintenance cost,* and the work they do takes energy and therefore entails an *activity cost.* Reproductives also incur both kinds of postproduction costs, though in different amounts. When the colony eventually draws on its metabolic reserves, it must again make most of the same investment choices it made with the food itself—production of workers, production of reproductives, maintenance costs, or activity costs. The corporate ledger would show worker production to be a form of reinvestment, either to replace worn-out machinery (dead workers) or to increase the size of the factory. The ledger would show sexual alates as the profitable output of the factory, the object of the entire enterprise. These sexuals are shipped out during mating flights and attempt to start new factory-fortresses. Available data do not allow a full economic analysis of colony growth, but at least we can apply a few economic ideas, beginning with food input and its allocation within the factory-fortress.

Food Is Always Limiting. A factory can have all the productive capacity in the world, but if the rate at which raw material enters the factory is low, output will be limited by input. Similarly, colony growth rate can be expected to be food limited until food is available in excess, a circumstance that is probably rare in nature. Below this limit, one can manipulate laboratory-colony growth rate by varying the feeding rate (Macom and Porter, 1995). Above this limit, only the intrinsic capacity for growth limits colony increase. We found that some incipient colonies fed in excess produced about half a million workers in the first year. In comparison, field colonies produced only 3 to 4% as many during this period, probably because of limited food. Porter also found very high growth rates associated with *ad libitum* feeding (Porter and Tschinkel, 1993).

Another observation supports food limitation of growth in nature. One of our study sites had been partially stripped of topsoil as well as forest. Although the better half supported a dense stand of weeds 1 to 2 m high, the poor half supported only small weeds, and a great deal of ground remained bare. Low plant biomass means low availability of insects for food, and vice versa. By midwinter of the first year, colonies on

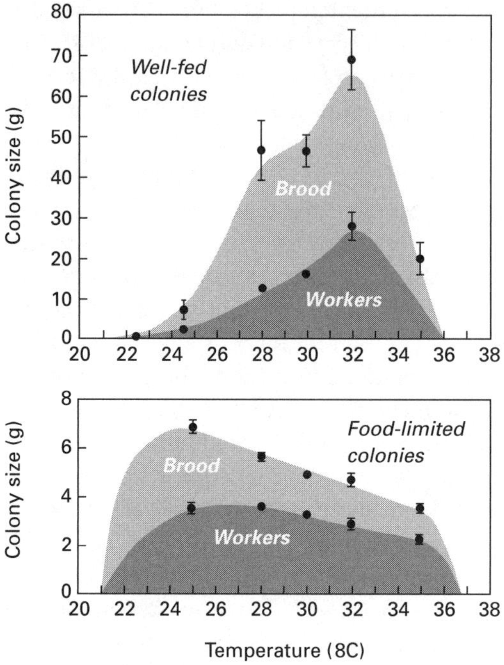

Figure 14.5. Well-fed colonies keep their brood and workers at higher temperatures than do food-limited colonies. Reprinted from Porter and Tschinkel (1993) with permission from the Entomological Society of America.

the better half were 8 to 50 times as large as those on the poor half. Such general food limitation implies the existence of a family of logistic curves generated by differences in growth rates under different food, density, and competitive conditions. These may also generate differences in maximum size.

The food intake rate may fluctuate in the short term as a result of weather and is therefore not highly predictable. When a colony finds itself with far more larvae than it can feed, workers reduce the number of mouths by killing some of the larvae and feeding them to their sisters, conserving energy and material, while at the same time reducing demand. Waste is a sin in which ants do not indulge, and apparently, so is sentimentality. As expected, larval cannibalism is more common under conditions of starvation or poor diet. Its occurrence at rates of 8–13% even in well-fed laboratory colonies suggests a general regulatory role as well (Sorensen et al., 1983b).

Buffering Growth Against Food Shortage. Interestingly, a fire ant colony can stretch the effectiveness of available food by choosing the temperature, from the available range, at which to maintain brood and workers. Sanford Porter, working in my laboratory, found that the maximum growth rate of small colonies occurred at 32°C when they were well fed but 25°C when they were food limited (Figure 14.5). Of course, the absolute value of the growth rate at each temperature was much lower for the food-limited colonies, but it is the shape of the rate-temperature curve that is important here (Porter and Tschinkel, 1993). This result implied that food-limited colonies should prefer lower brood-rearing temperatures than well-fed ones because such behavior would maximize colony growth under each circumstance. Also, if the ants are smart (and we know they are), workers not actively engaged in brood care should choose lower temperatures for themselves because doing so saves the colony energy that can be invested in brood.

Porter pushed two incubators together, side by side, set them at different temperatures, and set up 20 experimental colonies in them. Each colony was given two brood chambers/foraging areas, one in each incubator, connected through the incubator walls with plastic tubing. The temperatures in the incubators were then changed every 12 hours, so that the ants had to make two temperature choices every day for 22 days. Ten of the experimental colonies were fed an excess of food, and 10 received insufficient food. Every 12 hours, Porter estimated the percent of brood and workers in each colony at each of the two temperatures (Porter and Tschinkel, 1993). As the colonies grew, Porter removed workers and brood to keep each at a roughly constant size.

Choices for brood were tempered by the nature of the alternative. When the cool option was too low (say, 25°C or 28°C), a substantial fraction of the brood was moved to a temperature higher than the preferred one (say 36°C), but temperatures as high as 38°C were not chosen under any circumstance.

Workers are capable of fairly complex adjustments. Food-limited colonies preferred 30°C for brood rather than the 31°C preferred by well-fed colonies, and they avoided 32°C under all circumstances. Given the choice between 30 and 32°C, workers in food-limited colonies moved 90% of the brood to the lower temperature, whereas well-fed colonies distributed them 50:50. The balance between energy expended on growth and maintenance by larvae is probably more favorable to growth at the lower temperature. Well-fed workers chose similar temperatures for themselves and their brood, except that roughly 30% chose the lower temperature even when almost all the brood was placed at the higher one. In contrast, 50–60% of food-limited workers under similar conditions were always found at the lower temperatures. Lower temperature reduces worker metabolism and extends life span, saving energy that can be invested in brood (each 2°C decrease extends life span 14% and reduces metabolism 7–10%). These temperature choices allow food-limited colonies to stretch their food resources and maximize colony growth to the extent possible. Workers under both food regimes always chose the alternative that maximized colony growth, but food-limited colonies could not quite compensate for the higher respiratory costs of workers at higher temperatures, so they declined somewhat in size over the two-month experimental period. They tried their best.

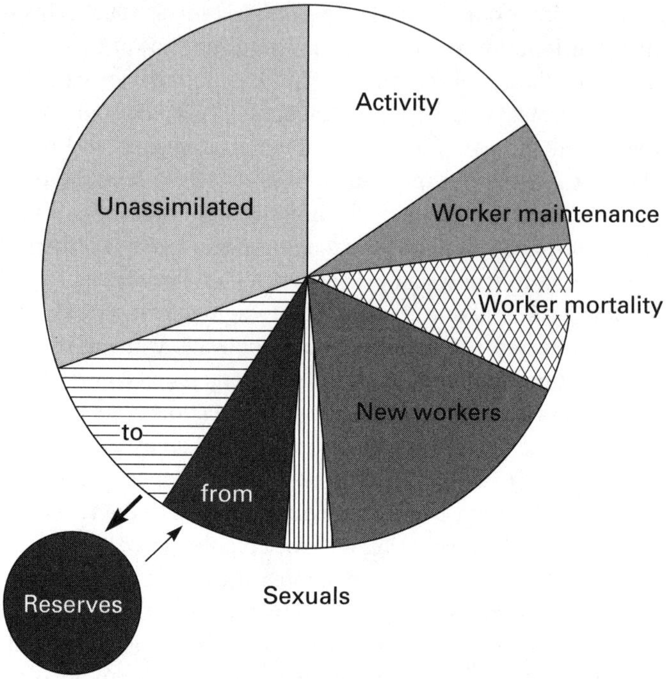

Figure 14.6. Relative flow of energy (calories/day) to colony functions in August–September. Sexual production and worker mortality are low, worker production and activity are high, and net flow to reserves slight.

When food-limited colonies were suddenly switched to an excess of food, their temperature choices changed to those characteristic of well-fed colonies within 12 hours. The opposite switch took one to two weeks and was more gradual. Switching shows fire ants are indeed smart, choosing temperatures that maximize colony growth, but how smart are they in nature, given the very complex food and temperature regimes there? What temperature choices do they make when food is limited, as it usually is? Perhaps someone will someday harness the power of modern electronics to study this phenomenon in greater detail.

Energy and Material Flow in the Colony Economy. Because food flow cannot be increased much, if at all, in the short term, increasing the food intake to increase fitness is usually not an option. Managing a colony is a fixed-sum game in which the details of how food and labor are allocated will determine the success of the entire enterprise, i.e., the conversion of resources to sexuals. We can represent this allocation schematically as in Figure 14.6. The total area of the circle represents the total food input rate in calories per day, and the sectors represent the amounts of energy allocated to the indicated functions. These data are derived from my census of 90 colonies of all sizes through one annual cycle (Tschinkel, 1993b) and are reasonably representative of allocation in a typical mature colony in August–September.

A large portion of the food that enters the colony, perhaps half, cannot be assimilated into ant tissue because of inefficiencies in digestion and absorption and costs of processing and metabolic conversions. Some materials, such as chitin, cellulose, and some plant materials, are simply indigestible. Only the assimilated portion can be allocated. Whatever the colony invests in one stream is not available for others, and herein lies a key to understanding colony development. Increasing allocation to one stream can be achieved only by reduction of one or more others. Even so, a colony's worker maintenance, mortality, and activity costs may be relatively fixed, narrowing options for others. The major allocation flexibility in the short term is among (1) the streams into and out of the storage reserves, (2) the production of sexuals, and (3) the production of new workers. These more flexible allocation rates trade off against one another. Most of the seasonal and life-history patterns of fire-ant colonies result from shifts in relative allocation to these three streams, in conjunction with worker mortality rate.

In our example in Figure 14.6, allocation to storage reserves (worker body fat) is only slightly higher than the reverse, so the reserves accumulate slowly. I have included stored reserves as colony growth because reserves can later be converted to any other allocation stream, albeit with conversion losses. All these rates vary with temperature, which I have ignored for the time being. In August–September, allocation to sexual production (output) is low, but that to new workers is high. Because much of the worker population is young, workers die at moderate rates, and their death rate is lower than the production of new workers. As a result, this colony is growing rapidly and getting fatter but is producing very few sexuals. Other seasonal patterns will be discussed in the next chapter in the context of reproduction.

Seasonal Variation and the Effect of Temperature. All of these rates, including the input rate, are temperature dependent. As a result, for a colony of a given size, the absolute values of inputs and allocation streams vary seasonally. Total production rates are high in summer and drop dramatically when the weather turns cold in late fall. In January, it is a mere 0.02% per day; it exceeds 2% per day by late March, peaks at 3.8% per day in June, and declines to below 2% again by late September (Figure 14.7). However, seasonal production rate differences are not driven by temperatures alone but also by changes in colony size. Colonies get smaller during the spring as they divert much of their productive capacity to sexuals, and a smaller colony means fewer calories per day. The reverse occurs in the fall; colonies produce mostly workers, causing colony size to increase, thereby gradually increasing production rate in joules per day.

The Importance of Efficiency. As I argued above, increasing food intake is usually not a feasible way to increase the rate of sexual production. Instead, colonies can increase their profitable output of sexuals in three major ways: (1) become more efficient (produce more output per unit input) in gaining and using food; (2) exploit increasing economies

Figure 14.7. Production rate through the year, in energetic terms, relative to the production rate in January. The spring peak of production is fueled by withdrawal from metabolic reserves to produce sexuals. Adapted from Tschinkel (1993c).

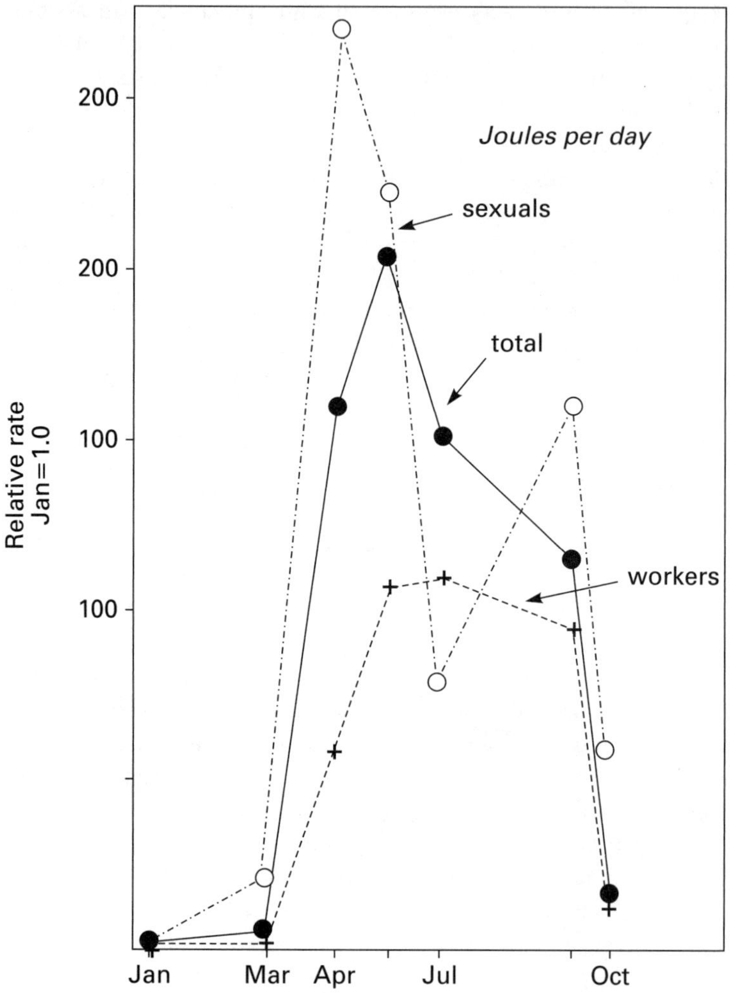

of scale, producing output at lower unit cost; and (3) grow to larger sizes in order to produce more sexuals, trading off earlier reproduction against greater output. Do real ant colonies benefit in any of these three ways as they grow? Fire ants are among the very few species about which we know enough to test any of these ideas. I will take up this discussion in the section on sociogenesis.

15

Relative Growth and Sociogenesis

Body size is one of the most significant characteristics of all animals, including ants, and has profound consequences for an animal's physiology, life history, and ecology (Calder, 1984) as well as the physical laws that dominate its life. In social insects, colony size is the counterpart of body size and probably has equally great consequences for colony biology. Once we think of the colony as a superorganism, we see that simply describing its increase in size during growth misses a host of profound changes. Just as the study of embryogenesis has provided tremendous insight into how organisms arise from single cells, so we can expect that a study of sociogenesis will (eventually) provide similar insight into social biology. The account of fire-ant sociogenesis I provide below is among the first and, like early studies of embryology, is mostly descriptive and based largely on an analysis of the relative growth of the superorganism's component parts in relation to the whole. As the real and metaphoric factories get larger, their floor plans may change, their component parts may grow at different rates. The mix of machines may change. They may change functionally too—the relative magnitude of inputs, fixed costs, reinvestment, production costs, and maintenance costs may change.

Relative growth and changing proportions are best analyzed with the methods of morphometry. Most of the hundreds of possible measures and components of a colony can be analyzed in this way, from incipiency to maturity. When these measures grow at different rates, their proportions to each other and to measures of the whole colony change, and the colony changes *qualitatively*. As a result, large colonies are not simply inflated small ones, any more than adult humans are magnified babies. Relative growth relationships may also change through the annual cycle, imposing a regular fluctuation over the effects of increasing colony size. This seasonality was dealt with in greater detail in an earlier chapter.

Sociogenesis: The Consequence of Changing Allocation Patterns

Allocation of new biomass by the growing colony is probably a sequence of hierarchical, quantitative decisions, like a decision tree. First, the colony must "decide" how much to allocate to sexuals (if any) and how much to workers. Many subsidiary decisions follow: How large should workers be? How much fat and protein should be stored in their bodies? These decisions have consequences for division of labor, maintenance costs, worker longevity and turnover, and many more subtle effects, as discussed below. Sexual production also requires many subsidiary allocation decisions. What proportion of each sex? How many individuals? How large? How much storage reserve should each carry? How much for maintenance?

The best allocation decisions will depend greatly on the size of the colony and the season, patterning sexual production, worker production, and worker size in relation to season and colony size. My data reveal only average patterns, but in real life, even colonies of the same size will vary, often widely, in all the measures of production, investment, and seasonality. To the extent that this variation is based on genetic differences, it is the raw material on which natural selection acts to adjust the fire-ant population's life history to its environment. Identification of selective pressures would require linking investment variants to fitness in a range of environments. I doubt I will live long enough to see such data.

Allocation Decision 1: Workers or Alates?

Picture a colony with a unit of new ant biomass, ready to invest in alates or more workers or a mixture of the two. The choice has great consequences for the colony's life history and fitness. Reinvestment in workers causes the productive capacity (future fitness) to grow, because each worker will produce more workers. An investment in alates, on the other hand, although a direct investment in current colony fitness, does not increase the standing workforce and productive capacity of the colony because alates do not produce more workers. If the colony invests all its production in workers, it achieves indefinite growth without colony fitness, and if it invests only in alates, it achieves a single burst of fitness and dwindles away to nothing as its workers die of old age (as do, for example, paper wasps and bumblebees).

The solution that has evolved in *Solenopsis invicta* is a common one among temperate species—small colonies invest only in workers and therefore grow rapidly (the *ergonomic* phase), and large colonies alternate investing in alates and workers (the *reproductive* phase). Why alternate rather than produce a sustainable mixture? Continuous production would not work because successful colony founding occurs only during the warm season, the earlier the better (Markin et al., 1972). Therefore, as soon as mound temperatures permit, colonies begin to invest about half their productive capacity in sexual brood. Such investment reduces the capacity to make workers, which are not replaced as fast as they die, so colonies get smaller. Smaller colonies mean lower production capacity, and so on it goes, until alate production is switched off in July by an unknown mechanism. At this time the colony has roughly half as many workers, and therefore about half the production capacity, that it did in November.

After July, almost all of the capacity is once again applied to worker production, and the colony returns to its midwinter maximum size, ready for the next year of birth, sex, and death.

Allocation Decision 2: If Workers, What Size Workers?

Given a certain amount of material to invest in workers, the colony can make a lot of small workers or fewer large ones or some mixture. In a mature fire-ant colony, the largest workers weigh about 15 to 20 times as much as the smallest and have heads almost three times as wide. This great size variability is called *polymorphism* and occurs in about 15% of ant genera. The majority of studies use the head width across the eyes as a convenient measure of worker size, but most body parts are similarly affected. Differences in size are often accompanied by differences in shape and are always associated with division of labor.

The minim workers of incipient colonies are monomorphic and have a mean head width of 0.51 mm, but one month later, the sizes of the second brood of workers are both distinctly larger and slightly asymmetrically distributed. The size distribution becomes ever more skewed as larger workers increase in frequency (Figure 15.1). Mean individual worker weight therefore increases 60% for every 10-fold increase in the number of workers, until workers in the largest colonies weigh three or four times as much, on average, as those in small colonies.

"Normal score" analysis of worker head widths reveals that this increase in mean worker size is the result of the differential growth of two slightly overlapping, normally distributed worker subpopulations (normal distributions yield straight lines in normal-score plots; Figure 15.2). The modal subpopulation consists of smaller, modestly size-variable workers (the minors), whereas the "tail" consists of a class of larger, much more variable workers (the majors) ("media," although sometimes used to refer to medium-sized workers, has no biological basis). Major workers from mature colonies weigh about three or four times as much as do minors, on average, and have a weight range about 20 times as great (Figure 15.3).

The existence of two distinct groups suggests that majors and minors differ by only a single, binary developmental event, some kind of developmental switch that changes a larva's destiny from minor to major. As in other insects, metamorphosis is triggered when the larva reaches a critical size, shutting off the secretion of juvenile hormone. Indeed, if a larva attains a certain threshold size in the third instar, metamorphosis is reprogrammed to a larger critical size and later time (Wheeler, 1990). The default critical size results in minor workers, and the reprogrammed critical size in major workers. This switch also extends the larval period by two to four days and may involve juvenile hormone. Treatment of all

Figure 15.1. The development of worker polymorphism during colony growth. The skewness of the worker-size distribution increases most rapidly in young, small colonies, slowing with maturity. Adapted from Tschinkel (1988a).

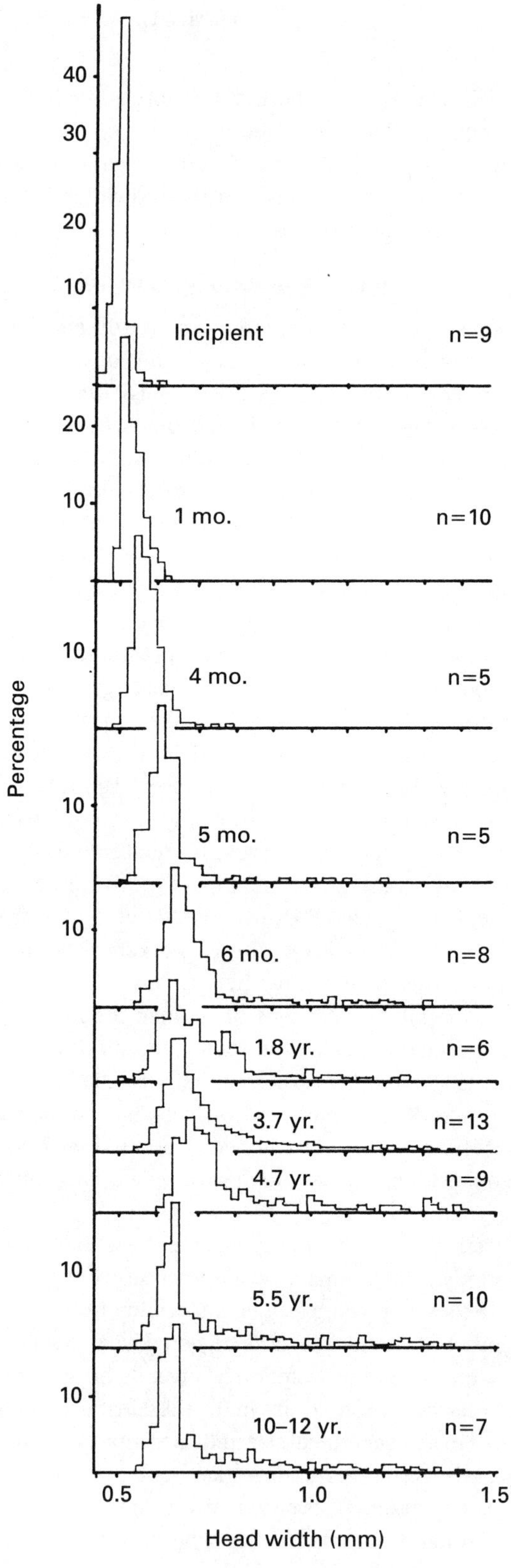

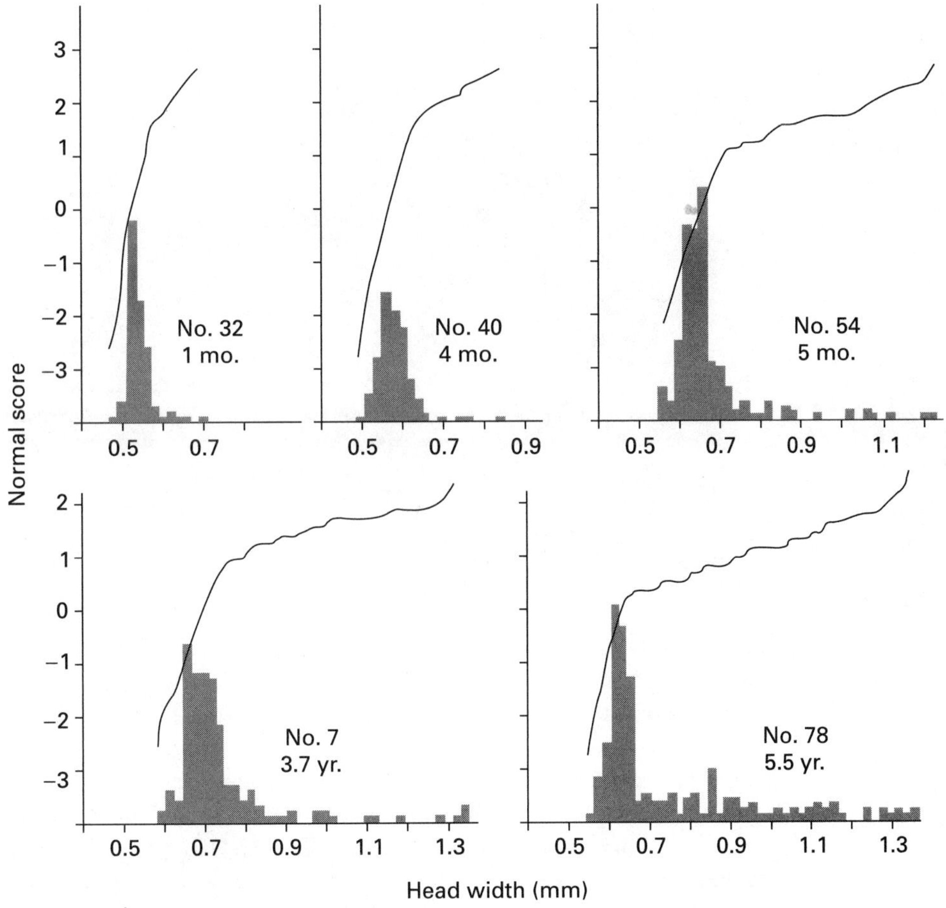

Figure 15.2. Worker size distribution and normal-score plots for colonies of several ages. The kinked linear normal-score plots show that the size distribution is composed of two adjacent normal distributions, minors to the left (the modal class) and majors to the right (the skew or tail). Adapted from Tschinkel (1988a).

larval instars with an analog of juvenile hormone resulted in pupae larger than controls. The mechanism of this effect is not yet clear.

Analyzing the major and minor workers separately leads to a more precise understanding of how worker size changes during colony growth. When colonies are less than six months old, two simultaneous processes increase the mean worker size (Tschinkel, 1988a; Wheeler, 1990). On the one hand, colonies produce a few major workers and gradually increase their number. On the other, the average sizes of both majors and minors increase until the colony is about six months old and then remain unchanged for the rest of the colony's life.

For the remaining 95% of colony growth, the entire increase in average worker size results from a continual increase in the proportion of

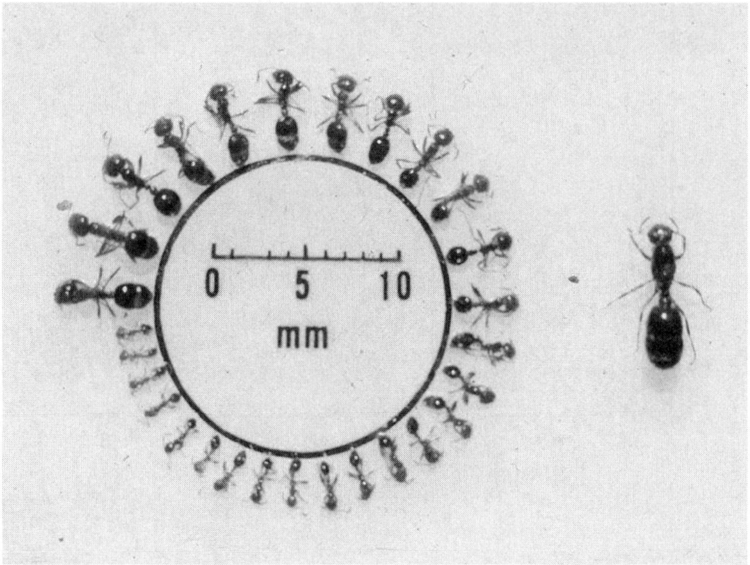

Figure 15.3. The range of worker sizes in a mature fire ant colony. The largest workers weigh 15 to 20 times as much as the smallest. The queen is shown to the right. (Photo by Sanford Porter, used with permission.)

major workers. Although both subpopulations grow logistically, the major subpopulation grows faster than the minor (Figure 15.4). As a result, the proportion of majors approximately doubles for every 10-fold increase in colony size, stabilizing only when colony growth stops. At that point, the worker force averages about 35% major workers by number but about 70% by weight (Porter and Tschinkel, 1985a; Tschinkel, 1988a). Biomass allocation has shifted from all minors to 70% majors and 30% minors. Small workers account disproportionately for the numbers and large workers for the biomass of a colony. Each of these measures of colony size is more meaningful in some contexts than others.

The existence of an early size threshold that reprograms minors to majors suggests a colony-level mechanism that drives the increase in the proportion of majors. It is rather hypothetical, though testable, and goes like this: as colonies grow, they become better nourished, as evidenced by increasing fat storage. At the same time, the ratio of workers to brood increases, so that more workers care for each larva, which therefore grows larger (Porter and Tschinkel, 1985a). Together, these changes produce ever-faster early larval growth, causing an ever-larger proportion of minors to be reprogrammed into majors. Thus, the proportion of majors increases with colony size.

The Changing Body of the Superorganism

During its two to six years of growth, the colony thus undergoes dramatic changes in its worker force—from a small population of uniform, small workers, it becomes a huge population of much larger workers of much more variable size—but worker size has profound effects on

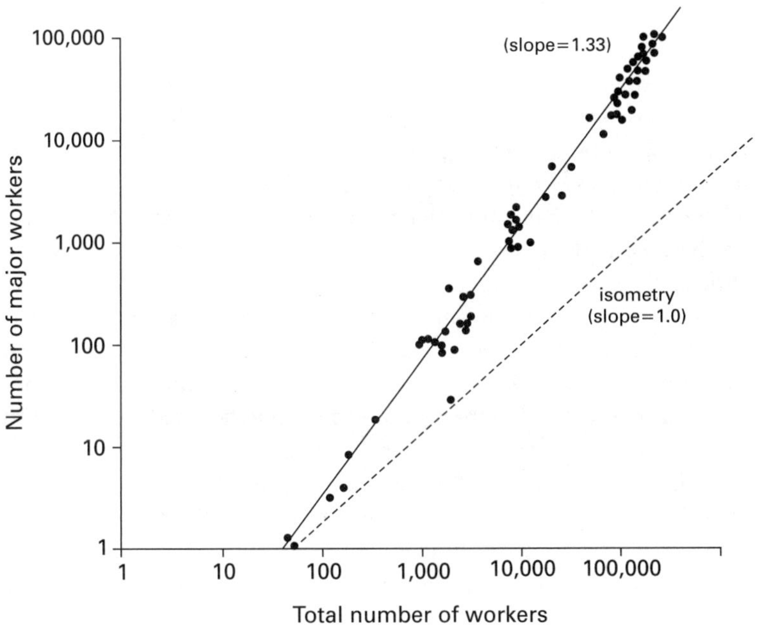

Figure 15.4. The number of majors in a colony increases faster than the total worker population. Points reflecting equal rates of increase would fall on the line of isometry. Adapted from Tschinkel (1988a).

worker characteristics that cascade upward to affect the nature of the superorganism. Some of these changes are simply the aggregated worker properties, such as biomass, but many are more subtle shifts in how the workers and the colony function. Several are discussed below.

Metabolic Reserves. Being major depots of metabolic reserves, fat and storage protein accumulate when food income exceeds need and are used in times when food need exceeds income, but they may also be depleted when production rate exceeds the capacity of the workers to collect and process food fast enough. Fat storage tells us a good deal about the physical condition, well-being, and nutritional status of an individual and a colony. Storage proteins may be equally indicative, but relevant data are not available. Estimates of body fat usually rely on the weight lost upon extraction, with ether, of fat from dried, weighed bodies of the ants. Thousands of workers had to be weighed, extracted, and reweighed for my study. Fortunately, all that tedium paid off.

Fat storage increases faster than does either worker size or colony size. Therefore, the percentage of body fat was highest in the largest workers of the largest colonies and the lowest in the smallest workers of the smallest colonies. At the collective level, total colony fat reserve increases 25% faster than worker biomass. As colonies grow, they store an ever-greater excess of energy as fat in the bodies of their workers, and a higher proportion of this fat is in the larger workers, who function in part as energy depots (proportionally larger gasters and slower metabolism suit them for this function). This pattern suggests that colony nutrition improves with colony size.

The improved nutrition as colony size increases may have multiple origins. First, larger workers may be more effective at capturing insect prey, producing a higher-quality diet. Second, the respiration costs of larger workers are lower for their size, leaving an increasing excess that can be stored as fat as colonies grow. Third, the ratio of foragers to brood increases with colony size, producing a greater excess of food above the needs of the brood. Fourth, at the individual level, large workers, because of their lower metabolic rates, would have more left over to store as fat after their maintenance costs are met.

Fat storage also varies with season in important ways, but this variation is similar for all colony sizes. Fat content reaches a maximum in July and then declines to a minimum through the winter until about June. Fat decreases in winter, when metabolic demand exceeds food intake—foraging is greatly curtailed in winter, but worker maintenance cost only decreases about 50%. During the spring, the fat is drawn down greatly during production of sexuals, more in large colonies than in small ones, so during the spring, the relationship between percent fat and colony size disappears. These patterns are discussed in more detail in the next chapter.

Colony Maintenance Cost. All living things require energy simply to stay alive, because homeostasis, uptake, turnover, excretion, and other necessary processes all require energy. These energy costs are usually estimated as the basal metabolic rate and have been measured for individual ants. The basal metabolic rates of entire fire-ant colonies have not been measured. Estimation of this measure as the product of individual basal metabolic rate multiplied by the number of ants requires the assumption that the respiration of ants alone does not differ from their respiration in a whole colony. The assumption remains to be verified.

As expected, the per-milligram maintenance cost decreases by two-thirds as the size of the colony increases by five orders of magnitude from incipience to maturity, because the proportion of larger, more slowly respiring workers increases. A 10-fold increase in colony size results in only an 8-fold increase in maintenance costs. The share of the food input that goes to maintenance therefore decreases as colonies grow, and the savings are available for other investments, including sexuals.

Metabolic rate and colony maintenance cost depend on temperature and therefore vary about twofold seasonally. This variation is small compared to the five orders of magnitude of variation associated with colony size. Additional savings result from the behavioral strategies discussed above that use temperature differences to achieve savings in worker maintenance costs.

No fire-ant data exist on the cost of doing work, that is, movement and muscular activity. Among other species, an active ant respires at several times its resting rate (Lighton et al., 1987; Lighton, 1989; Fewell et al., 1996). This difference would be most important to foraging and less important to nest workers, a large fraction of which is usually inactive and whose respiratory rate is essentially at base (K. Mason, unpublished data). If the per-milligram cost of activity decreased with worker size, large colonies would spend proportionately less on worker activity than small. We do not yet know whether this is the case.

Worker Birth Rate, Mortality Rate, and Turnover. Larger workers live longer, and their longevity has consequences for colony birth and death rates and therefore for the turnover rate, i.e., the rate at which workers die and are replaced. When the birth rate is lower than or equal to the mortality rate, all the new workers are simply replacing workers that die, and the turnover rate is simply the birth rate. When the birth rate exceeds the mortality rate, turnover rate is equal to the mortality rate, and the excess is the colony growth rate. Each of these rates is usually estimated as percentage of the colony turning over per unit time.

Worker birth rate has been estimated from the pupal census and the temperature. The numerical worker birth rate increases much more slowly than colony size—4-fold for every 10-fold increase in worker biomass. Partly, this result stems from production of larger, fatter workers. Thus, the energy represented in the birth rate increases about 8-fold in the same interval. Although this declining worker birth rate should put serious brakes on colony growth, the greater life span of the larger workers reduces its effect dramatically. These birth rates rise and fall with seasonal temperature variation, although their relationship to colony size remains the same. When the total annual births equal the total annual deaths, the colony has reached its maximum size, averaged over the annual cycle.

No method exists for direct estimation of worker mortality rate within individual colonies, but we can estimate it as the difference between the rate of colony size change (positive or negative) and the birth rate. Like the birth rate, mortality rate is high when temperatures are high and low when they are low. Here, we are interested in the *cumulative annual totals* for birth and deaths, which we obtain by summing the weekly percentage values of each over the entire year.

The results are startling. Colonies in the smallest size class have an annual turnover rate of over 600%, an average of 50% per month! Their entire worker forces die and are replaced six times during the year. Most of this turnover takes place during the warm months, when the monthly rate may be between 75% and 100%. As colonies grow, turnover rate declines rapidly, dropping to 350% in the second size class and 270% in the largest size class (Figure 15.5). Most of this change must be the result of producing larger workers, which live longer.

Two points emerge from Fig. 15.5. First, except in the very largest size class, the total number of ants born every year exceeds the number that dies. Only in the largest size class are these values equal, so that growth from year to year ceases. Second, a strikingly small fraction of

Figure 15.5. Cumulative annual mortality and natality in relation to colony size (colony size classes differing by increments of 50,000 workers). In all but the largest size class, natality exceeds mortality, causing a net annual size increase. Adapted from Tschinkel (1993b).

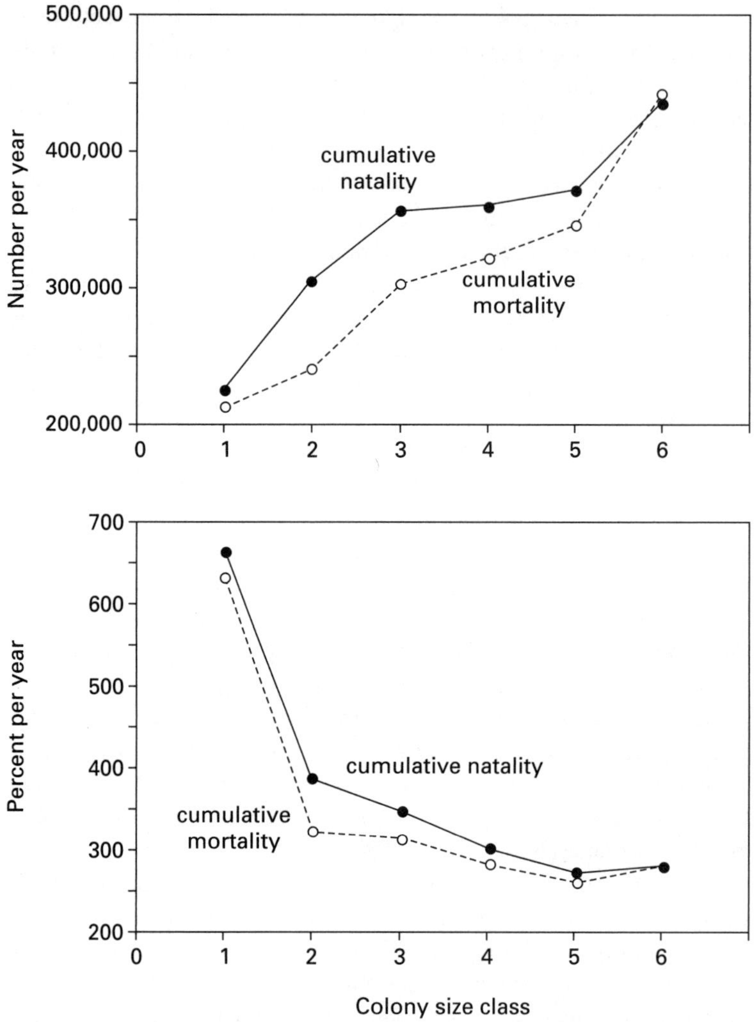

the total worker production goes toward colony growth—most of it is turnover. For fire ants, turnover is therefore a very large share of the cost of doing business.

What happens to colony size within each annual cycle, discussed in the following chapter, is dramatically different from this apparently simple annual increase.

Shifting Labor Sectors

Colony growth has still another consequence, and this one also derives from the increase in worker size. In a later chapter, I show that larger workers do little brood care and more outside tasks and that smaller workers spend a larger proportion of their lives at tasks within the nest.

Overlap in the jobs that they do well, or do at all, is limited. The small and medium-sized workers are the generalists, capable of rearing brood at normal rates (Porter and Tschinkel, 1985b), but large workers seem incompetent when it comes to rearing brood. Even when they must, because no small workers are present, they reared brood, fed larvae, and groomed other workers at less than 5 to 10% the rate of small workers (Figure 15.6), suggesting less engagement in duties within the nest. On the other hand, large workers seem to be much better at capturing insect prey, being recruited to food, moving large objects, and possibly defense. These tendencies of large and small workers are relatively inflexible, no matter what the situation. The total available amount of a type of labor can thus be expressed in "worker days"—the product of the number of workers, their longevity, and the fraction of their lives spent in a specified type of labor. I have called these totals "labor sectors." Examples of possible labor sectors include nurses, reserves, and foragers/defenders, but others may exist as well. The absolute and relative sizes of these sectors should change as the colony grows and the worker population becomes ever more dominated by majors, characterized by an aversion to brood care and a propensity for outside work. Perhaps new sectors appear as well.

Are these conclusions all theoretical, or does evidence exist for shifting labor sectors in relation to colony size? So far, I have estimated

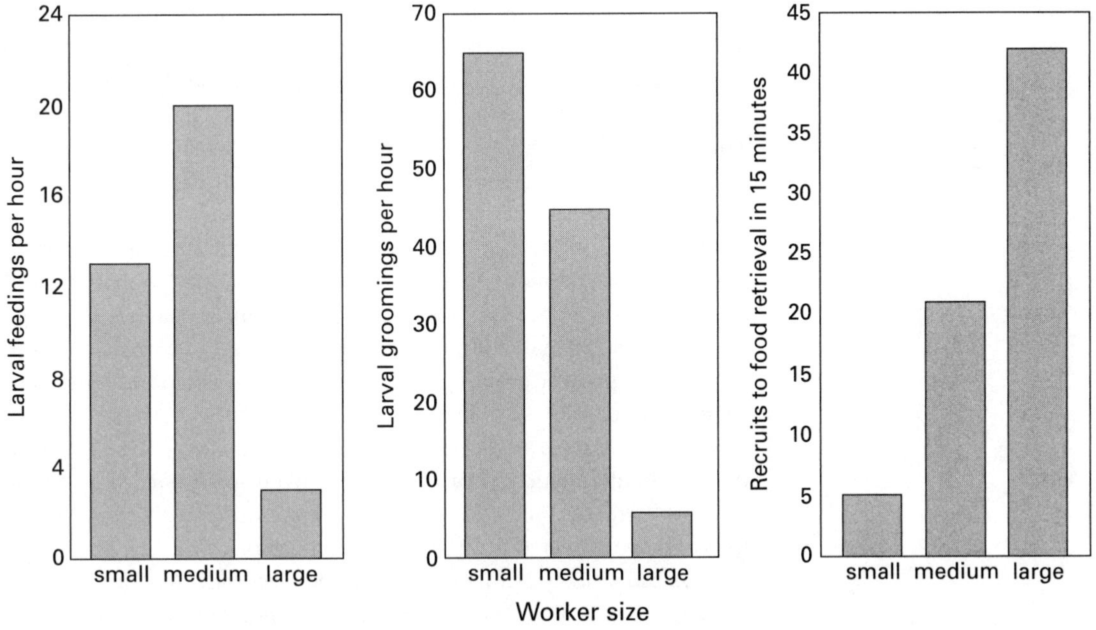

Figure 15.6. The performance of small, medium-sized, and large workers at brood rearing and response to recruitment. Large workers do little brood care but are much more likely to be recruited for the retrieval of food. Data from Cassill and Tschinkel (1999b).

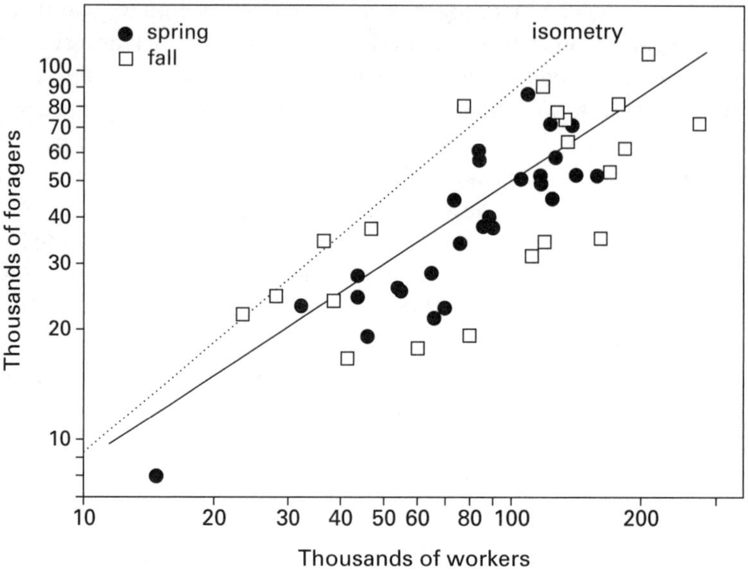

Figure 15.7. The number of foragers in a colony increases more slowly than the number of workers, so the proportion foraging decreases with colony size. Equal rates of increase are shown by the line of isometry. Filled circles represent colonies sampled in fall, open squares those sampled in spring.

only the size of the forager sector and must rely on indirect methods to estimate those of nurses and reserves (unpublished data). The foraging sector can be sharply defined as those workers that search for or can be recruited to retrieve food.

The fraction of the total worker population that foraged decreased as colonies increased in size. (In a so far unpublished study, I estimated the forager population of a number of noncontiguous territories using mark-recapture estimation, then censused each colony by excavation. The result was an estimate of the entire ant population, both in the nest and in the foraging territory.) For every 10-fold increase in colony size, the foraging/recruitment sector increased only 7-fold (Figure 15.7). Therefore, the proportion of the worker force that forages drops from about 70% in colonies of 20,000 workers to about 35% in colonies of 300,000. Nonforagers ("nest workers") make up the minority of the worker force in small colonies, half of the force in colonies of about 85,000, and an ever-greater majority in still larger colonies. The labor sectors of colonies smaller than 85,000 are tipped toward foraging, whereas larger ones are tipped toward nest labor. Oddly, this labor shift occurs simultaneously with a declining need for brood care—brood number increases only half as fast as worker number. This is not exactly the result we expected from the increase in worker size.

We can crudely estimate another labor sector by assuming that the nursing sector is roughly proportional to the brood they care for (we will see in a later chapter that the nursing population is limited by the area of brood piles). The size of the nest-worker sector increases 16-fold for every 10-fold increase in colony size, but that of brood increases only 5-fold.

Therefore the relative size of the sector needed for nursing also decreases dramatically with colony size (we have no reason to believe that nurses become less efficient). The only sector to gain is reserve labor (Figure 15.8). To be sure, reserves and nurses can probably do each other's jobs to some extent, and thus perhaps ought not to be considered distinct sectors. A direct method for estimating the size and overlap of these sectors would clarify this issue. Whatever the details, major shifts in labor availability clearly occur as colonies grow, and these must have important consequences for sociogenesis and colony function.

In view of the relative decline of the forager population, we might well ask whether they make up for their declining proportion by their greater foraging effectiveness. Direct estimates of forager efficiency (e.g. calories harvested per forager hour) do not exist, but one can argue that their individual effectiveness does indeed increase. If it did not, food input per gram of ants would drop by half as the colony grew. A constant per-gram intake would require foragers from large colonies to be twice as effective as those from small. The increasing fatness, i.e. storage of excess food, of larger colonies suggests that this compensation (and more) occurs, but the lower maintenance costs of larger workers probably also contributes. Until someone untangles this multivariate knot, the question of foraging efficiency will remain unresolved.

Efficiency of Converting Food to New Ants

Theory suggests that a colony might benefit from an economy of scale as it grows, doing more with less. Is the production rate (grams of new ants per gram worker per day) in large colonies greater than, less than, or equal to that of small colonies? The only useful data are from my own census of 90 fire ant colonies (Tschinkel, 1993b). The birth rate (individuals per day) is calculated as the number of pupae divided by the pupal period in days. Multiplying by the weight or energy content of a pupa converts this rate to weight or energy terms.

The colony's production rate increases at exactly the same rate as its total worker force (slope = 1.0 on log-log plots). Doubling colony size doubles production rate. The daily production rate is simply a fixed

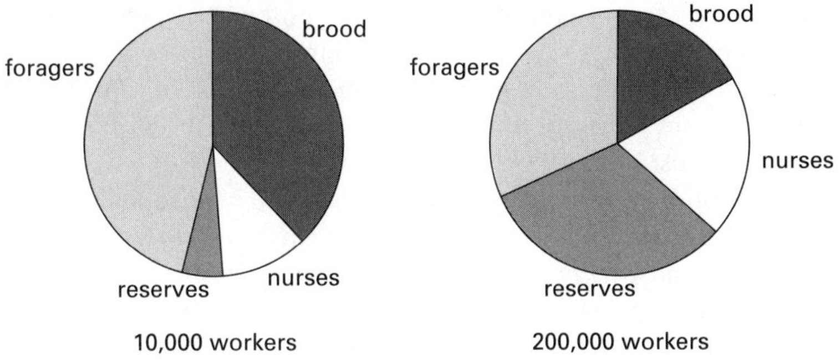

Figure 15.8. The labor sectors of a small colony contrasted with those of a large one.

percentage of standing worker biomass or energy content. Contrary to theory, no economy of scale is apparent, at least not by these measures. The total worker force of larger colonies is no better at producing biomass than that of smaller colonies.

This is not the final word on efficiency, however. This estimate suffers from a number of problems and is sometimes contradicted by information from laboratory experiments. For many species of social insects, the workforce is generally accepted to become less efficient as the colony grows (Michener, 1964). In laboratory colonies of fire ants, the weight of new ants produced per gram of workers also decreases with colony size (Porter and Tschinkel, 1985b), a result in direct contradiction to my findings in field colonies.

What might account for these contradictions? One glaring difference is that, in field colonies, the size of working groups of ants within the colony is effectively limited by the multiple-chambered nest structure. Chamber area averages about 5 cm^2, and working group size about 200 (Cassill et al., 2002). In chambers with brood, brood and workers were about equal in number. In contrast, each laboratory nest consists of a single chamber shared by all the workers and brood. Larger groups are less efficient at brood production (Brian, 1956), possibly accounting for the differences we observed. Would subdividing laboratory nests increase the efficiency of the worker force? The only test of this possibility was not run for long enough to reveal efficiency differences (Cassill et al., 2002). This subject deserves some careful experimentation.

What is more, the above efficiency is not the total efficiency with which colonies convert food to ants but only a portion of this complex process. In addition, the calculation did not include foragers out in the field. Because the forager population increases more slowly than that of nest workers, including foragers would cause the total worker force to grow more slowly than the rate of production, making colonies more efficient as they grow. Likewise, if we add the declining cost of worker maintenance, efficiency increases with colony size. The ways in which the rates of food collection and assimilation change as colonies grow are also unknown. If these are not isometric with colony size, efficiency will change during growth. Further, the estimates do not include the amount of energy expended in activity. In three species of formicine ants, and possibly ants in general, the metabolic cost of running was about 7.2 times the cost of rest (Jensen and Holm-Jensen, 1980). The metabolic cost of transport in *Camponotus herculeanus* was a linear function of the combined weight of the ant and its load (600 to 2200 J/g-km), no matter what the weight of the ant (Nielsen et al., 1982). Such costs must be incorporated into our calculations. We still have a long way to go before we can estimate all these aspects of fire ant efficiency.

Growth of the Nest

As the worker force grows, it excavates more chambers below ground and constructs a larger mound at the surface, raising questions of costs and benefits, as well as relative growth. The quantitative relationship between chamber volume and colony size is not known, except for very small nests. Between one month of colony age and seven months, chamber volume increases from about 8 ml to about 400 ml (Markin, 1964), but filling a nest with molten lead makes it more than a little difficult to count the ants within it.

On the other hand, the relationship between the volume of a mound and the size of the colony is known rather precisely and has been put to use in many studies as an excellent surrogate for colony size. Colonies begin to build mounds when they contain between 5000 and 10,000 workers. The mound volume and colony size are tightly correlated—about 85% of the variation in mound volume is explained by variation in colony biomass. Mound volume grows about 20% faster than colony biomass (Figure 15.9), so as the colony grows by five orders of magnitude, space per milligram of ants increases 2-fold. The advantage

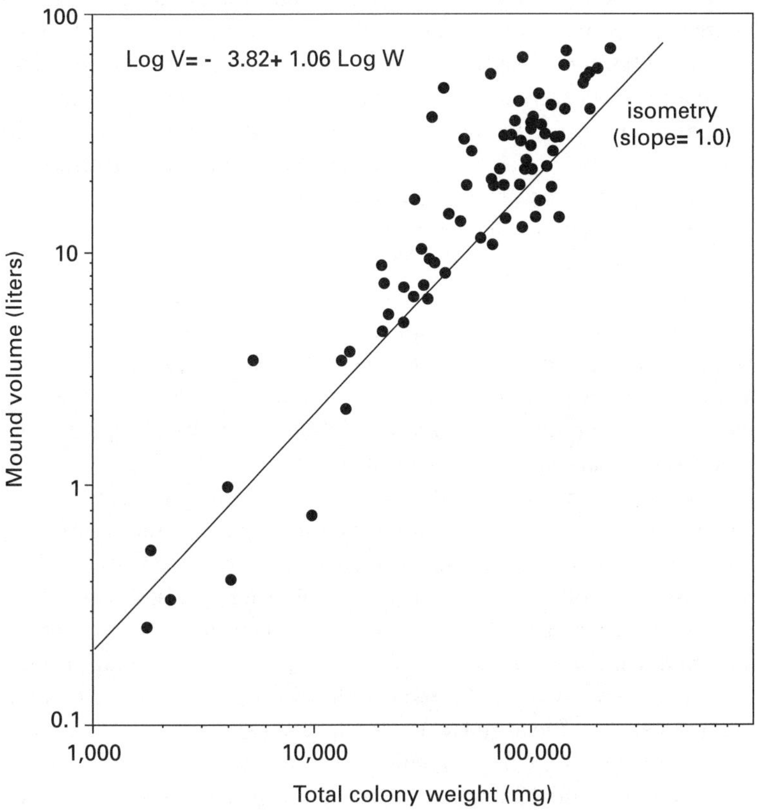

Figure 15.9. Relationship of mound volume (V) to total colony weight (W). Adapted from Tschinkel (1993b).

of this extra room remains unknown. Perhaps it is simply a consequence of the increase in average worker size. We can presume that the mound and the underground chambers are in a constant proportion to one another, but what this relationship might be is not known.

Through the seasons, mound volume changes as the colony within the nest changes size, but the volume change lags behind that of colony size. For a given size of colony, the mound is 60% larger (about 3.5 ml mound/mg ants) in the spring and summer, when colony size is decreasing, than it is in the fall and winter (2.1 ml/mg), when colony size is increasing. The intercept of the regression of colony size on mound volume is therefore slightly higher in spring than in fall. These differences are small compared to those resulting from the changing size of the inhabiting colony.

Territorial Growth

Territorial behavior appears suddenly at the end of the brood-raiding period. Almost overnight, workers shift from tolerating intruders from other colonies to attacking them. When the incipient colony is located in an unoccupied area, it need only stake its claim. When it is one of the few survivors of newly founded nests in a fully stocked area, its future is much more tenuous. What little we know about this phase comes from a territory neighborhood mapped by Eldridge Adams. By chance, this map included one very tiny colony, too small to produce a mound, which lived a fugitive existence in the no-ant zone between its large neighbors. From day to day, its territorial boundaries shifted as its neighbors squeezed it into contorted shapes, first long and thin then contracted or dumbbell-shaped. Eventually, the colony grew large enough to stabilize its territorial boundaries, carving out a small space between its giant neighbors.

Because the territory is the source of all the colony's resources, its size and that of the colony are inevitably linked, although probably not as tightly as expected, as a result of neighborhood interactions. Because colonies undergo such large annual size fluctuations, the relationship between territory and colony size must be tested during at least two seasons—say, late spring and late fall.

In the late spring, when the colonies are merrily producing sexuals and are near their annual minimum size, territory area and colony biomass are isometric—a 10-fold increase in one corresponds to a 10-fold increase in the other (Tschinkel et al., 1995). Every gram of ants is supported by an average of 1.5 m^2 of territory, but because a gram of workers from a small colony includes more ants than does a gram from a large one, a colony of 1000 workers holds about 3.4 cm^2 of territory per worker, whereas one of 100,000 holds 6.4 cm^2 per worker.

In the late fall, when colonies are ending worker production and are near their annual maximum colony size, a 10-fold increase in colony biomass results in only about a 4-fold increase in territory area; i.e. larger colonies have proportionally less area in the fall and support more grams of ants with every unit of territory area. Do these differences represent changes in territory size or in worker density? A large part is a real increase of worker density, because even when colonies produce gobs of workers, neighbors greatly limit their ability to expand their territories. The patterns are therefore best understood in this way: in the spring, colonies hold territory in proportion to their weight of ants. In fact, during sexual production, they have lost territory in proportion to their size. When they switch back to worker production in midsummer, they begin to grow in biomass without growing in territory (or perhaps they gain back some of what they lost), thus increasing worker density. At the annual maximum, the larger the colony, the more ants it supports from each unit of territory area. Are they magicians? Are they really doing more with less? We know too little about the complex relationship among season, food production, and colony life to formulate a detailed answer.

Here is a working hypothesis: the function of the ants produced in the fall is not to increase the territorial empire, but to be converted, through their labor and their stored fat, into sexuals in the following spring. The more of these "excess" workers a colony produces, the more sexuals it will produce the following spring (Figure 15.10). By the middle of the next summer, the colony returns to nearly the same size it was in the previous year, with nearly the same size territory. Significantly, the difference between the annual maximum and minimum colony size is strongly related to the total number of sexuals produced. This difference represents the

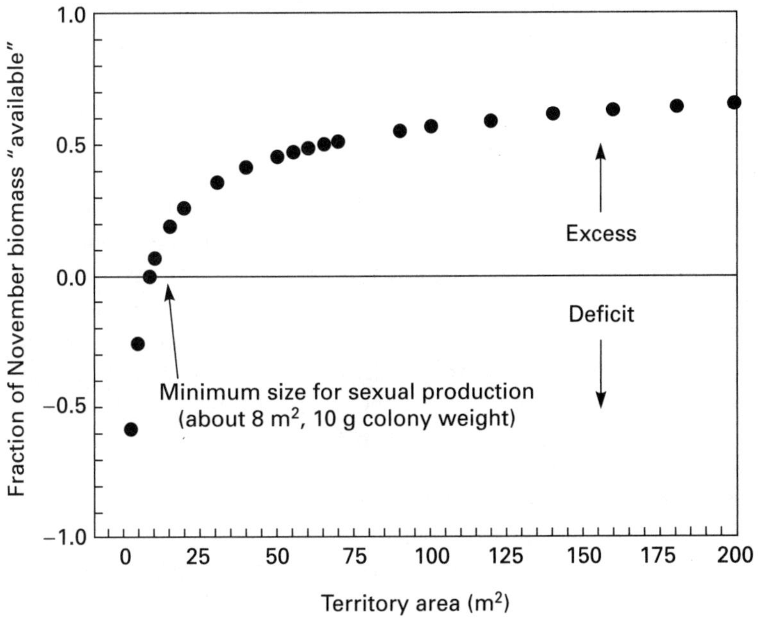

Figure 15.10. The difference between the maximum and minimum annual biomass (calculated from regression equations) represents the excess production available for producing alates.

excess workers cashed in to produce sexuals (see below), which left the nest. Also significantly, the density of foragers is not higher in the fall, so these excess workers are not much involved in territorial matters.

Does the cost of territory change during colony growth? In the spring, every 10-fold increase in the number of workers results in a 20-fold increase in territory size. The colony is therefore netting more territory per worker as it grows larger, but the total weight of workers increases at the same rate as territory size. In other words, larger colonies produce larger workers, and large workers are more effective at territorial conquest and defense.

Territory geometry probably plays a major role in the per-unit defense cost. Contact with neighbors, and therefore territory defense, occurs only at the territory boundaries. The defended perimeter, and therefore the cost of defense, increases with the radius of the territory, whereas the area, and therefore the benefit of defense, increases with the square of the radius. The cost of defense therefore becomes an ever-smaller fraction of the food benefit wrested from the territory, creating an economy of scale.

Of course, only the foragers are involved in territorial matters. The rest of the workers are indoors taking care of the home front. The effect of foragers on territory size is much greater than that of total workers, at least in the spring. A 10-fold increase in foragers results in a 25-fold increase in the size of the territory. Thus, each fielded forager nets a bigger increment of territory for a large colony than it does for a small colony. As a result, forager density generally declines with colony size, but territorial compression by competing neighbors increases forager density greatly (see Figure 8.7), causing it to differ by 10-fold or more for colonies of a given size. Forager density also increases as the percentage of a colony foraging increases. The combination of these factors means that, in the largest territories, only about 400 to 500 workers comb each square meter of territory, whereas in the smallest, this number is over 6000, 12 to 15 times as high. Absent neighborhood effects, the unit cost of territory defense should therefore decrease with territory size.

Foragers reach all parts of their territory through underground tunnels, but how this system grows in relation to colony size is uncertain. The two territories for which Markin and coworkers measured both area and tunnel length differed by almost twofold in area but had the same ratio of tunnel length to territory area (Markin et al., 1975c). More such measurements are needed, but because of the enormous labor involved, we shall probably not see them soon.

Reviewing Growth and Sociogenesis

We have now traced the full journey from the one or two dozen workers of the founding queen's first brood to the 200,000- to 300,000-worker behemoth at colony maturity. The changes have been many and profound, including physical characteristics, relative size of colony sectors/components, the scale of colony life and geography, colony functions, and efficiency. The average worker size quadrupled as major workers gradually increased in proportion until they made up 70% of colony biomass. Larger worker size reduced maintenance cost, increased the average worker life span, decreased the birth and turnover rates. They also shifted the labor economy from one dominated by brood care to one dominated by outdoor labor, holding more territory with fewer foragers. From this territory, an ever-decreasing proportion of the work force wrested the food for an ever-better-nourished colony that stored ever-more fat and protein reserves for overwintering and spring alate production (see next chapter). Finally, after three to five years of seasonally fluctuating sigmoidal growth, the annual worker birth and death rates reached equality, and the colony grew no more, although its seasonal fluctuation was at its maximum.

Death of the Superorganism

In keeping with the sexual nature of their reproduction, queens do not live forever. When the queen's demise ultimately comes, she probably does not go belly-up, kicking her tarsi feebly as she gasps her last breath. Rather, her capacity to produce workers approaches zero as her supply of sperm, stored in the spermatheca from her long-past mating, dwindles to exhaustion. Without sperm, the queen can no longer fertilize eggs and therefore no longer produces workers to replace those that die, but male production probably increases. As her fecundity declines, so does her pheromone production, until perhaps she can no longer prevent a few of the female alates still in the nest from laying eggs (unfertilized and therefore destined to produce males). The colony's final months are probably a time of male production and gradual decline, the coup de grace perhaps delivered by a vigorous neighbor.

Even so, male production contributes to lifetime colony fitness. Lifetime production of males can probably be boosted by perhaps 10 to 20% in dying colonies if these are produced during the normal reproductive season (otherwise they would find no females to mate with). So far, however, this phenomenon has not been established by direct observation.

However it ends, the maximum functional life span of a fire-ant queen is determined by the size of her sperm supply and the rate at which she uses it up. Determining the sperm supplies of queens of known ages, and then extrapolating to zero sperm, would estimate queen life span. The even-aged cohorts that colonize newly cleared land provide colonies of known age. We capture their queens, dissect out the pearly spermathecae, and estimate the number of sperm using a hemocytometer (Tschinkel, 1987; Tschinkel and Porter, 1988).

A newly mated fire-ant queen holds, on average, 7.3 million sperm in her spermatheca (standard deviation = 1.0 million; Glancey and

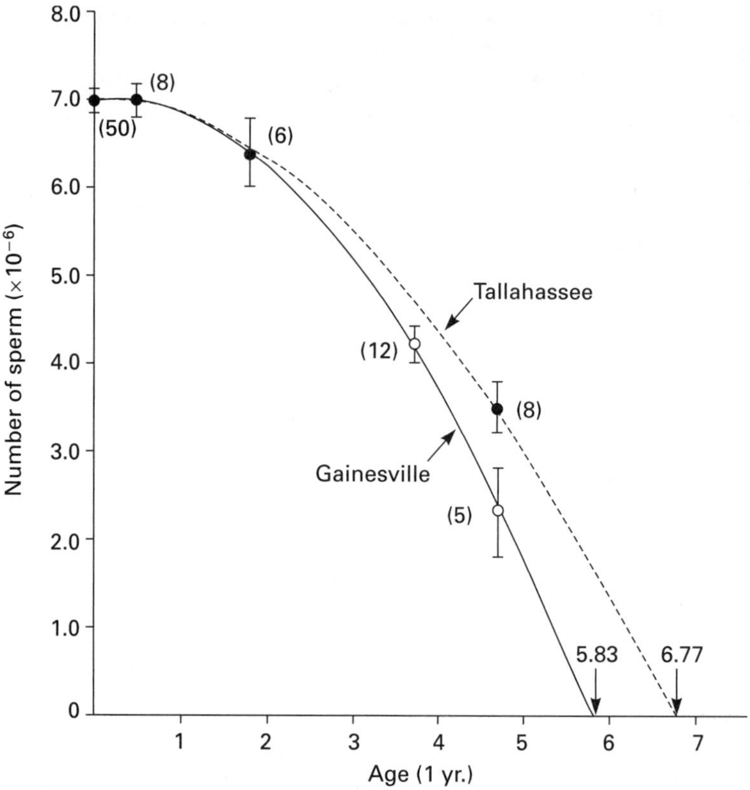

Figure 15.11. Extrapolation to zero of sperm count of queens of known ages gives an average life span estimate of about seven years. A queen that has run out of sperm cannot produce workers, so her colony is doomed. Each point is based on the number of queens given in parentheses. Adapted from Tschinkel (1987).

Lofgren, 1985; Tschinkel, 1987). This supply begins to dwindle with increasing speed as the colony grows and the rate of worker and female-sexual production increases (Figure 15.11). By extrapolating the sperm count curve to zero, we find that Tallahassee queens run out of sperm in about 6.8 years. With the slightly longer growing season and warmer latitude of Gainesville, Florida, they last 5.8 years. The 13% turnover rate in the stable, natural population at Southwood Plantation corroborates these longevities. The reciprocal of this turnover rate, about 7.7 years, is the mean colony life span. This estimate agrees remarkably well with that derived from sperm counts.

A little thought suggests that fire-ant queens must be very parsimonious with their sperm and, in fact, that sperm parsimony is the key to a longer life. After all, starting with a supply of 7 million, the queen rears and constantly replaces a colony of about a quarter million workers for about seven years. How many sperm does she expend for each female offspring? We determined the initial sperm counts of part of a group of newly mated queens and allowed the other part to rear colonies (Tschinkel and Porter, 1988). We fed them to excess, so that the queens produced the largest number of workers in the shortest possible time (we did not want to hang around longer than necessary), and produce workers they did! In a little under a year, the four surviving queens produced an average of over half a million workers each (most of whom were no longer alive at the end of the year, of course), using 1.7 million sperm of their initial supply of 7.3 million to do so. Dividing the number of sperm used by the number of workers yielded a rate of 3.2 sperm per adult worker—very frugal indeed!

We estimated this rate by a second, independent method, using worker turnover rates. For colonies of a constant size from year to year, births and deaths are equal, and both are equal to the turnover rate. Cumulating birth rates every two weeks over the entire year showed that the entire colony is replaced 3.85 times (385% turnover). We estimated worker mortality rate from the relationship between worker longevity, temperature, and the number of workers. Again, cumulated over the entire year, this procedure yielded an annual turnover of 3.20 fold (320%). The two estimates average to 353%. We then applied this value to the growth curve (Figure 14.4) to estimate that a queen produces 2.64 million workers in her average life of 6.8 years. Because she expended 7 million sperm producing them, the queen used 2.64 sperm for every adult worker she produced. Of course, if a worker larva dies before reaching adulthood, a sperm has still been expended. Sperm probably also die during storage. The number of sperm expended per fertilization is thus even lower than these already astoundingly low rates, perhaps close to one sperm released for every egg fertilized. How the queen accomplishes this frugality is unknown, but it probably involves the muscle that controls the spermathecal duct. For comparison, consider that humans expend hundreds of millions of sperm for each successful fertilization, horses billions.

This extreme sperm parsimony is food for reflection, for it dramatizes the value of sociality. Life-history theory predicts that juvenile mortality rates will shape maternal reproduction. When most offspring die early, a mother must produce more to overcome this mortality. Alternately, she can provide care to the offspring to improve their survival and trade off the cost of care with a reduction in fecundity. Within a fire-ant colony, parental care is extreme, juvenile mortality is very low (at least 30% survive to adulthood), and the queen's fecundity is only just high enough to supply the colony's needs for workers. Compare this situation with reproduction at the colony level. The founding queen receives no parental care (except for the metabolic reserves donated by the natal colony), and juvenile mortality of colonies is extreme; only a fraction of a percent survive to maturity. To overcome this mortality, the colony (mother) must produce thousands of founding queens. Reproduction within the colony reaps the benefits of sociality, but reproduction of new colonies is nonsocial and garners no such benefits.

So after six to eight years, the curtain comes down on the fire ant colony, its sperm spent, its worker force dwindling and largely idle. For a small percentage of colonies, insult is added to injury when an unrelated, newly mated queen enters the colony and parasitizes the labor of the worker force to found her own colony. Most colonies simply disappear, the process of dwindling and dying never having been directly observed.

The Porter Wedge Micrometer: Mental Health for Myrmecologists

In the bad old days, when you wanted to know how big an ant was, you would put it on the stage of a stereomicroscope so that it was staring up at you, and then you would superimpose a transparent scale placed in the microscope's ocular over the ant in the field of view and note how many scale units the head measured across the eyes. The ocular scale had been previously calibrated, so that the measurement could be converted to millimeters. You might measure some body part other than the head, but the procedure was the same, as were the frustrations. Just as you were about to make the reading, the ant would pop out of your forceps to vanish somewhere in the detritus layer on your work table. Or maybe you couldn't get the specimen to sit at a good angle, so the measurement was wrong. God help you if you had indulged in several cups of coffee that morning, because then you couldn't hold those little buggers still enough to get reliable readings. True, with practice, you might get pretty adept, but even under the best of circumstances, it was a lengthy, tedious, and trying procedure.

Imagine (or, in my case, remember) that, in your relentless quest for truth and the Nobel Prize, you have committed yourself and your lab to a research project that requires measuring the heads of thousands of workers. You have now entered a dangerous domain, in which you cannot remain for long without suffering serious brain damage. One of my master's students quit graduate school after the 4000th head-width measurement. Her own head slightly unstable on her neck, she bumped into the doorjamb on the way out.

Why was ant size estimated this way? Because it had been done this way forever, and nobody had ever come up with a better way. That is, not until my student Sanford Porter invented the *wedge micrometer* (Porter, 1983). This device is so laughably simple that the most remarkable thing about it is that no one had thought of it before. It consists of two straight-edged pieces of material, say a couple of microscope slides, glued to a base with a narrow space between them. The slides are at a slight angle to each other so that the space is a little larger at one end than the other. The exact distance between the two edges is measured at several points along the length and marked on the slides. This procedure creates a scale from, say, 0.4 mm at one end to, say, 1.6 mm at the other. Any small object, when slid down this narrowing space from the wide end, will jam and stop when its dimension is equal to the distance between the slides. Reading the scale at the stopping point will then reveal the dimension.

Now, instead of fumbling each ant into position under a microscope only to have it pop across the stage, we put all the dried ants to be measured into a vial and unceremoniously crumble them with the eraser end of a pencil. From the resulting ant-part ragout, the heads are sorted and placed to one side of Sanford's magical micrometer. With the fingertip, each head is then slid into the wedge micrometer until it stops, the head width is read off the scale, and the head slid back out and to the other side of the micrometer. Even after a training period of only five minutes, a beginner can measure 200 heads an hour or more.

The value of such a humble invention is easy to underestimate—it includes no high-tech concepts, no special materials, no computer programs, no difficult manufacturing, no lengthy training for use, and practically no cost—only a simple device based on a simple geometric principle already well known to the ancient Greeks. Yet the wedge micrometer is to myrmecology what the steam engine was to manufacturing. Each replaced brute labor with an efficient device that increased the output of the laborer manyfold. Each also reduced the physical toll that the work exacted from the laborer.

And like the steam engine among early 19th-century manufacturers, the wedge micrometer has still not been fully accepted by the head-measuring community, even though it has been among us for 22 years now. Several times in the last few years, I have had a colleague tell me about measuring heads with an ocular micrometer, or I have seen this method described in a manuscript I was reviewing. Is it ignorance? Sometimes, perhaps, but often I think not, for most of these colleagues know about the wedge micrometer. In those cases, I believe it is simple cultural viscosity. Oh, I hear excuses, like "I wanted more accurate measurements," but these are just rationalizations to hide the emotional basis of the decision. Scientists, in spite of their pretense to total rationality, are traditionalists who often don't like to let go of the familiar. Like most normal people, they too need a balance between the old and the new, between the foundation in the past and the titillation of the future. Some, perhaps, feel that the wedge micrometer robs them of the heroic achievement of measuring thousands of heads the old way, a way that requires an ability to persevere, to endure, to prevail over boredom, sore back, and eyestrain. Perhaps these myrmecologists feel as did John Henry, pitted against that steam drill there at the Big Bend Tunnel on the C & O Line, but the outcome of a contest between the ocular and the wedge is not in doubt. When the contest is over, the loser will get up from the microscope in a daze and, on the way out, will bump into the doorjamb.

16

Colony Reproduction and the Seasonal Cycle

Age or Size at First Reproduction

For a wide range of both organisms and colonies, reproductive output increases with body or colony size. As a result, young organisms or colonies do not reproduce, because they cannot produce enough offspring to assure the continuation of their line. Worker production is maximal in young colonies as "workers beget more workers," increasing colony size rapidly. In this *ergonomic* phase, new-worker production exceeds the rate at which workers die, the flow into and out of storage is small, reserves are low, maintenance costs are a large part of total costs, and activity is even larger (Figure 16.1, left). These patterns persist through the year as long as temperature permits.

Having grown larger than a certain threshold size, the colony begins to produce winged sexuals, passing from the ergonomic to the *reproductive* phase of the life cycle. In both the laboratory and the field, this change occurs when colonies contain between 23,000 and 33,000 workers (Markin et al., 1973; Vargo, 1988b). This size is only about 10% of the maximum, so colonies achieve 90% of their growth concurrently with producing sexuals, i.e., during the reproductive phase of the life cycle. Clearly, the ants are pursuing a mixed strategy, investing in both growth and reproduction.

Colonies grow at different rates for a variety of reasons, some taking longer to reach reproductive size than others. Is the switch to sexual production regulated by colony size, age, or a compromise between them? Laboratory colonies founded by groups of queens, or colonies condensed during brood raids, maintain their size advantage as long as growth is exponential (Tschinkel and Howard, 1983). In the laboratory, cooperatively founded colonies produce their first sexual pupae about 100 days earlier than colonies founded by solitary queens (Vargo, 1988b), but their size at this transition is about the same—between 23,000 and 33,000 workers. The colony's size, rather than its age, therefore seems to determine when it will become reproductive, but some caution is advisable. Under other circumstances, for example if differences in growth rate resulted from differences in feeding or from defensive expenditure, the outcome might be different. Many organisms compromise age and size at first reproduction—as it

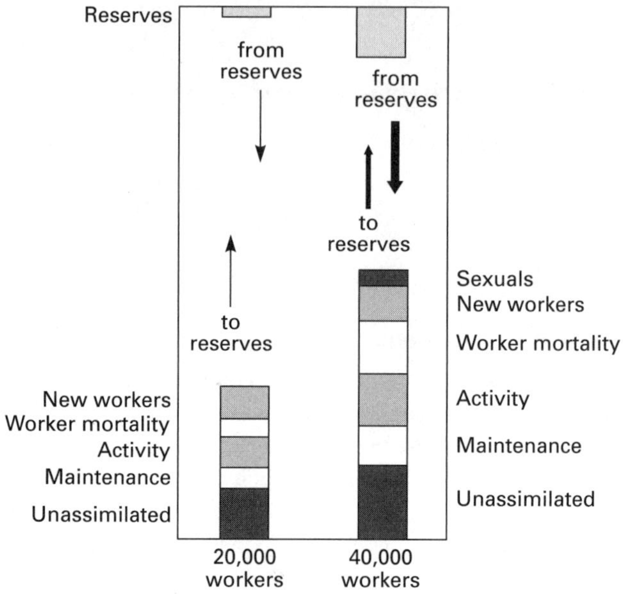

Figure 16.1. Energy flow (schematic, relative calories per day) to major colony functions in April in a colony that is too small to produce sexuals (left) and in one just large enough to produce a few (right). Metabolic reserves play a large role in sexual production.

takes longer and longer to achieve reproductive size, threshold size decreases.

The Annual Cycle and Colony Reproduction

The annual cycle of colony life is the consequence of four basic facts. First, for a colony of a given size, the production of workers and sexuals is a zero-sum game—an increase in one results in a decrease in the other. Second, sexuals produced early in the year have the best chance of successful colony founding. Third, the "weedy" life history of *Solenopsis invicta* requires the production of large numbers of sexuals. Fourth, workers live only an average of one-sixth to one-third of a year. Through natural selection, these four facts lead to an annual pattern in which colonies decline in worker population in the spring as sexuals are born and increase in worker population in the fall when few sexuals are produced. Production ceases during midwinter wherever temperatures are too low. The fire ant is a tropical, not a temperate, ant and lacks true hibernation.

Figures 16.2 through 16.6 show the energy allocation patterns that result from these four facts at key points in the annual cycle. (In the absence of hard data, we will assume that food intake is roughly proportional to colony size and that activity costs are about twice maintenance costs.) In the immortal words of my undergraduate chemistry professor, of whom more later, one can immediately see what a useful tool these figures are for visualizing energy flow in colonies. Figure 16.1 shows these patterns in April for a colony just large enough to be producing a few sexuals. This colony had about 40,000 workers in midwinter.

It is now drawing on a small reserve of fat and protein, as well as part of the food intake, to produce a small number of sexuals by mid-May. Worker production rate falls slightly below the worker mortality rate, so that the worker population falls slightly. By midsummer, the colony switches back to producing only workers, and the colony grows exponentially until winter. By midwinter, the colony's size, and therefore the rate at which it can produce new ant biomass, is almost twice what it was a year before.

This basic seasonal variation increases in amplitude every year as the colony grows to full maturity at about 220,000. For four to five years, the worker population is larger at the end of each year than at the beginning. Thereafter, it simply returns to its previous size or declines.

Midsummer. Figure 16.2 shows energy allocation patterns for such a full-grown colony at four key points during the annual cycle, beginning just after sexual production has ceased in July. The worker population has declined to a mere 125,000 workers, about half that at the beginning of the year. Productive capacity is therefore also about half, and the spring capital reserves of fat and protein have been depleted. Food intake exceeds energy expenditure on worker brood, causing the net buildup of fat and protein depots in worker bodies (Ricks and Vinson, 1972a; Tschinkel, 1993b). The relative youth of the workers may increase their tendency toward fatness. Worker birth rate has increased from its spring low and approximately equals the death rate, causing colony size to change little for a while. Sexual production is too low to

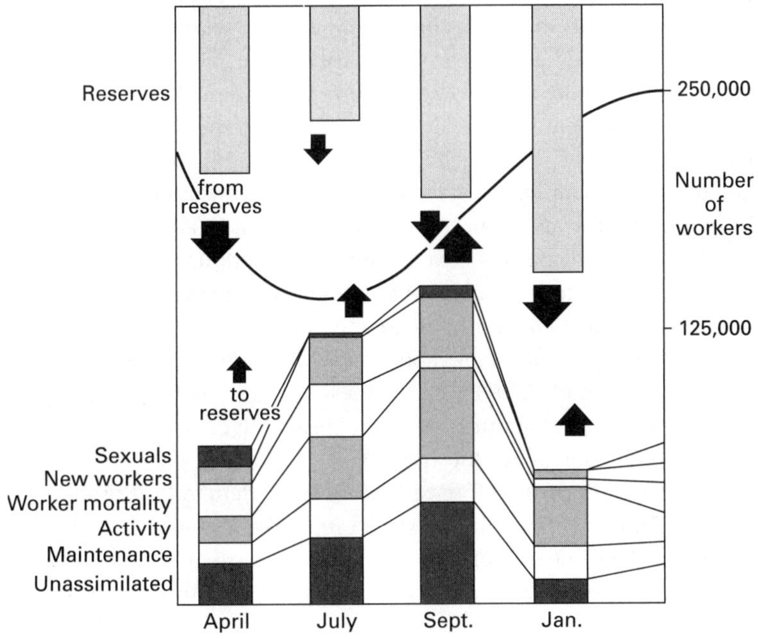

Figure 16.2. Energy flow (schematic, relative calories per day) in a mature colony throughout an annual cycle. Changes in the metabolic reserves result from changes in fat storage combined with changes in colony size. Although fat content is highest in September, total reserves are greater in January because there are more workers. Colony size is shown as the curve in the upper half of the diagram, right scale. Sexual production draws heavily on metabolic reserves. Flow to and from reserves is indicated by the width of the arrows.

depress worker birth rate. The costs of maintenance and activity are relatively high because temperatures are high.

Fall. By September–October, the worker birth rate exceeds the death rate, because most of the workers born in the spring have died, and the majority of workers are young. The worker population grows more rapidly than at any other time of the year and is composed largely of young and relatively small workers. Because metabolic storage exceeds use, these workers are also being fattened, and the larger workers are fattening proportionally more (see above). Therefore, whereas late-spring workers average about 42% fat, fall workers gain fat (Figure 16.2, September) to about 52%. At the onset of winter, the worker population is at its largest, fattest, and most protein-rich of the year. Sexual production is less than 15% of production costs.

Midwinter. By the time falling temperatures reduce brood-rearing rates, the colony has returned to its maximum size of 250,000 workers. Low temperatures curtail brood production, and food demand is low. Temperature has a threshold-like effect on brood production, shutting it down almost completely, but it reduces maintenance energy much less, so most of the food income in the winter is spent on the costs of maintenance and activity (Figure 16.2, January). It is supplemented with withdrawals from metabolic reserves. Meanwhile, worker deaths and births are low but about equal, so colony size does not change much during midwinter.

The factory is now at peak capacity, about twice what it was in August (or would be at the same temperature). This capacity is composed of the labor of the worker population and the metabolic reserves stored in their bodies. A midwinter colony is thus poised and fully equipped not only to survive the winter on stored worker fat but to start producing sexuals at the earliest possible date. The factory floor is swept, the equipment sparkling, the inventory at an all-time high, and the savings accounts flush. Only the winter temperature prevents it from forging ahead. Now the importance of the mound becomes apparent, for it warms rapidly in the morning sun, allowing brood production in spite of low prevailing air temperatures (Porter, 1994).

Spring. The disparity between mound temperature and general soil temperature creates a problem in addition to an opportunity. When soil temperatures (at 2 cm depth) are less than about 15°C, fire ants cannot forage effectively, and below about 24°C, they cannot produce brood (Porter and Tschinkel, 1987; Porter, 1988). In April in Tallahassee, production is already at about 60 to 75% of peak, but cool soils limit foraging to about half of its July rate (Porter and Tschinkel, 1987). The food-collection rate therefore cannot meet the demands of brood production. Drawing down the fat stored in worker bodies makes up the difference (Figure 16.2, April).

But why not simply wait until foraging rate and brood production capacity are more in line with each other? The trade-off here is that early sexuals have a better chance of founding successful colonies. The later in the season a sexual flies, the smaller her chances of successful colony founding and the lower her reproductive value to the colony. The optimal pattern is therefore to produce as large a burst of sexuals as early as possible, holding back some as a hedge against early failure.

A further point is important. The May peak of total production rate (in energy) precedes the midsummer peak of foraging rates and temperatures (see Figure 14.7). The May peak is almost entirely the result of the addition of sexual production on top of a fairly ordinary, temperature-tracking curve of worker production. The use of stored fat allowed production rates that far exceeded what could have been sustained by foraging. How do we know? One indication is that, in midsummer, the workers have the lowest fat reserves of the year. The production capacity at this time is not restricted by temperatures and probably indicates what a colony can sustain primarily on foraged food. This rate is 25% lower than the May peak, about 150 times the January minimum (the May peak is 200 times).

Now we understand the value of the fat reserves—they allow abundant and early sexual production, as well as tiding the colony over the winter. During peak sexual production in April, the colony spends lives, labor, and metabolic reserves as though there were no tomorrow. The rate of withdrawal from metabolic reserves far exceeds input, driving them to their annual low. Sexual production consumes half of the production capacity. As a result, the worker birth rate is near its annual low, well below worker mortality rate, causing the worker population, already down to 175,000, to decline faster than at any other time of the year. The spring workers were mostly born in the previous fall and are near the ends of their lives. The smaller workers among them die faster, so the average worker size reaches its annual maximum during this period.

By midsummer, the colony is back to its smallest size and has the lowest reserves of protein and fat of the year (Ricks and Vinson, 1972a; Tschinkel, 1993b; Figure 16.2, July) and the annual cycle begins again.

Overview of the Annual Cycle and Its Variation

Although sexuals are uncommon in colonies from August to February (in the Tallahassee latitudes), it would be wrong to conclude that sexual production begins in March. The production really begins in the middle of the previous summer as colonies switch from mixed sexual/worker broods to almost exclusive worker ones and begin to grow again. Colonies grow to their maximum possible size and fat content by late fall, and then "convert" this reserve of material and labor into sexuals in the spring. Unpublished data from my field study show that the more the worker population of 23 colonies decreased between winter and summer, the greater the total weight of alates emitted from these colonies.

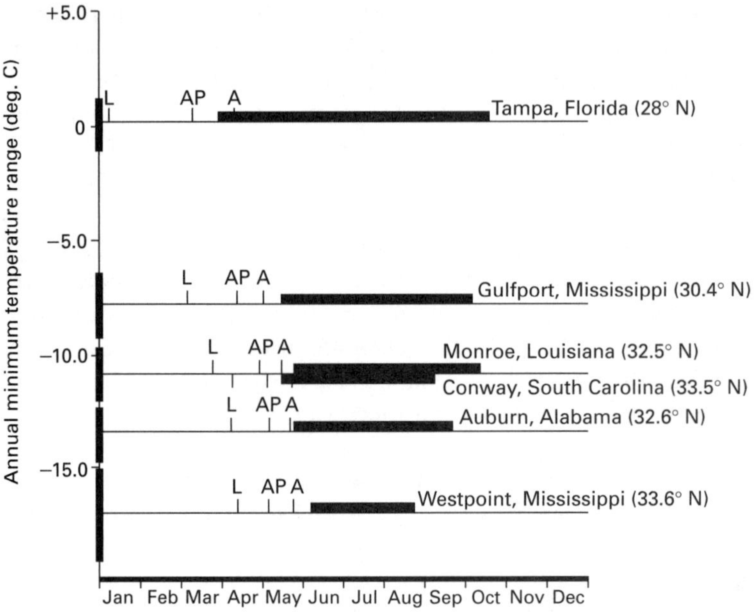

Figure 16.3. Brood-production and colony-founding patterns for sites at different latitudes, arranged in relation to their annual mean minimum temperature range (heavy bars on vertical axis). Heavy horizontal bars represent the period of the year when soil is warm enough (>24°C) for colony founding. L, AP, and A are the approximate times when larvae, alate pupae, and adult alates first appear in mature colonies. Adapted from Markin et al. (1974c).

Every liter of mound-volume decrease (a surrogate for decrease in worker number) was associated with an additional 1.3 g of sexuals (about 160 female alates or 480 males). Many ant species rely on such strategies to produce early sexuals.

The capture of heat by the mound has its limits and ultimately cannot prevent low temperatures from reducing or shutting down brood production. In the USA, as in South America, the range of *S. invicta* spans enough latitude to experience year-round reproductive conditions at one end of its range and complete shutdown and occasional killing temperatures at the other. In the southern parts of this gradient, brood production begins earlier and lasts longer than in the northern parts, as expected from the annual temperature variation (Markin et al., 1974c; Figure 16.3). Sexuals are produced more or less year-round in Tampa, Florida (although at lower rates in midwinter). These patterns agree with our understanding of *S. invicta* as a tropical insect lacking effective adaptations to winter, whose physiological processes merely track temperature. No matter what the latitude, the average temperature at which the first worker larvae appear is 15.1°C; that at which pupae appear is 20.4°C, sexual pupae 19.7°C, and alates 22.9°C. Temperature combines with length of brood-rearing season to assure that southern colonies can produce much more brood in a year than northern ones, allowing greater maximum colony size and alate production. Not surprisingly, foraging intensity also tracks temperature.

Colony founding is even more restricted seasonally because founding nests have no capacity for capturing heat with a mound. Soil temperature limits successful founding to June–August (80 days) in northern Mississippi and to April–September (200 days) in central Florida (Figure 16.3). Combining alate-production season with the season for colony founding, we expect that the capacity for population growth should decrease northward, finally reaching zero at the northern range limit.

Seasonal Worker Turnover

Workers' birth rate, death rate, and turnover all decrease greatly as colonies grow (see above) but are not evenly distributed throughout the year (Figure 16.4), being low in winter and peaking during the summer (Tschinkel and Porter, 1988; Tschinkel, 1993b). (Birth rate and turnover were computed from the pupal census and prevailing temperatures. Mortality rate was calculated indirectly from the difference between the rate of colony size change and the birth rate.) The annual cycle reveals several interesting patterns (Tschinkel, 1993b). Seasonal temperature has by far the strongest effect—no surprise there for an exothermic ("cold-blooded") animal. More interestingly, in all but the smallest colonies, spring is predominately a time of dying, whereas fall is a time of birth. In the four largest colony size classes, 60% of the workers are born in the fall and only 40% in the spring. Conversely, 60% die in the spring

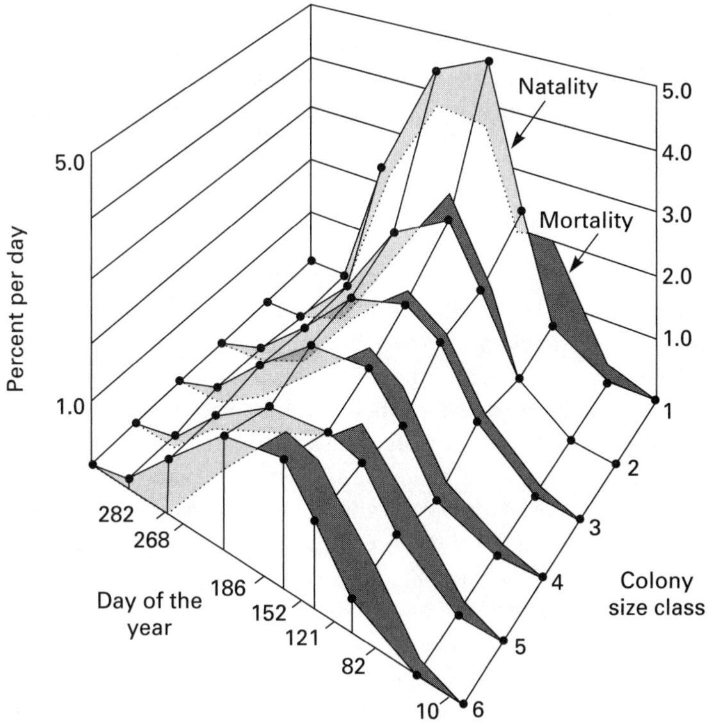

Figure 16.4. The annual cycle of worker birth and death rates in colonies of increasing size (grouped into size classes separated by increments of 50,000 workers). Both rates are highest in the smallest colonies and decrease with colony size. Mortality exceeds natality in the spring, but the reverse is true in the fall. Note reversed horizontal scales. Adapted from Tschinkel (1993b).

and only 40% in the fall. This asymmetry derives from reduced worker production during spring and causes the mean age of fall workers to be lower than that of spring workers (which is why mortality rates are higher in the spring than in the fall). Because larger workers live longer, the mean body weight of spring workers gradually increases 50 to 80% in comparison to that of fall workers.

What Do Colonies Profit per Year?

Colony fitness is most meaningfully estimated as total lifetime alate output. This lifetime output cannot currently be determined, but total annual alate output in relation to colony size can be estimated from my data. For each midwinter size class, the mean daily production of workers and sexuals is summed over the entire following year, and interpolation is used between sample dates. Because we track size *classes* through their seasonal changes of size, the sums are not confounded by these size changes. Although approximate, the results show some interesting patterns.

The total annual production of colonies increases more slowly than colony size, no matter what the units (Figure 16.5), i.e. the per-capita or per-gram output of large colonies is less than that of small ones. For example, a January colony of 25,000 workers (size class 1) had produced 225,000 new workers by the end of the following December, about 9 new workers per original worker. Larger colonies produced more workers but fewer per capita. Thus, a class 5 or 6 colony produced only about 1.5 to 1.6 new workers per original worker during the year. Workers change size, but the pattern is basically the same for weight or energy. A class 1 colony starting with 10.5 g of workers in January produces about 88 g of workers and sexuals by the end of the year, or 8.8 grams of new ants per original gram. This value drops to about 2.2 g/g in class 6. The production efficiency of a class 6 colony is only about 25% of that of a class 1 colony.

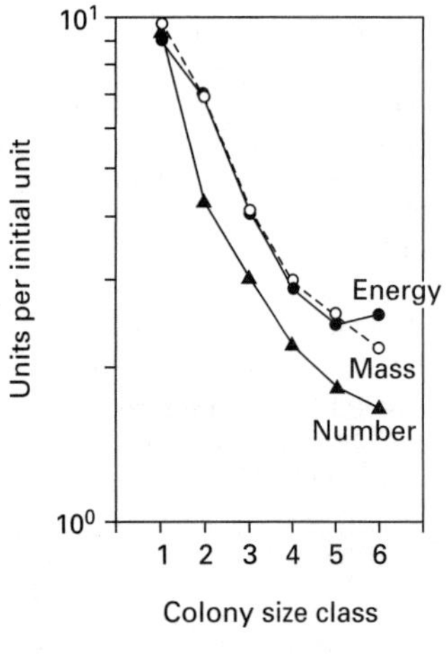

Figure 16.5. No matter what units are used, the production efficiency of *Solenopsis invicta* colonies decreases as colonies grow, in part because sexuals are exported and do not contribute to future production. Colony sizes classes as in Figure 16.4. Adapted from Tschinkel (1993b).

But did we not say earlier that production rate is a simple multiple of colony size? Double the colony size, double the production rate, right? These two observations can be reconciled as follows. Production of workers leads to higher production rate in the next generation, but production of sexuals does not. Therefore, the larger the colony at the beginning of the year and the greater the sexual production during the year, the more colony size and therefore *instantaneous* production rate decline. Therefore, even though the efficiency at any sample time does not decrease in relation to colony size, the falling size and production rate cause the *annual* efficiency (in units per unit per year) to decrease with increasing colony size. The price of producing sexuals is a loss of productive capacity and efficiency.

Particularly interesting is the fraction of the total annual production a colony manages to convert to sexuals. Class 1 colonies export about

16% of their total annual production as sexuals, whereas all larger size classes export about 33%. Larger colonies produce more sexuals but not because of increased efficiency. Including the cost of worker maintenance lowers this profit to 10–14%. Including activity costs would further decrease it by an unknown amount. This nearly constant profit after maturity is somewhat surprising. The expectation from life-history theory and my own skimpy alate data was that reproductive effort should increase with colony size. Perhaps the estimate of the annual total is too crude or subject to systematic errors, but it is currently all we have.

Variation of Alate Weights with Season and Colony Size

Why do alates evolve particular characteristics and not others? Queen number, weight, and size are typical trade-offs, balanced against each other to optimize founding success. Investment in individual alates should be correlated with the reproductive value of those alates, that is, the expectation of their future reproductive success. Might this reproductive value change with colony size, season, or both?

For monogyne colony size, the answer is a clear no. Whether female alates hail from a small or a large colony, their mean weight is similar. It might appear that a large colony could "afford" to produce more robust sexuals, but apparently the constraints on successful colony founding are too great. On the other hand, the seasonal variation in the weight of female alates is substantial (male variation is small and probably unimportant). Female alates are lightest in March when they first appear and their weights peak in June as a result of increases in both lean and fat weight (Figure 16.6). Average minim production during founding increases, so late-June queens produce an average of five more minims than late-May queens (Tschinkel, 1993a). After June, female weight declines gradually, and by November female alates are 15% lighter than their June counterparts (Tschinkel, 1993b). The consequences of these later weight trends are not known.

The trend in number and quality of alates through the seasons parallels their reproductive value to the colony. As successful colony founding becomes ever less likely, the expectation of future reproductive success dwindles. The story is quite different for alates that spend the winter in the nest, for these participate in a different, dependent type of colony founding. Such queens have no need for metabolic reserves or maternal behaviors.

Queen Pheromones and the Annual Cycle

Fertile queens of *S. invicta* produce a pheromone (or pheromones) that regulates several important aspects of colony life. The evidence for this regulation was derived largely from research with the polygyne form. Briefly, a queen pheromone(s) (1) attracts workers to the queen, assuring her constant care and grooming, and probably facilitating

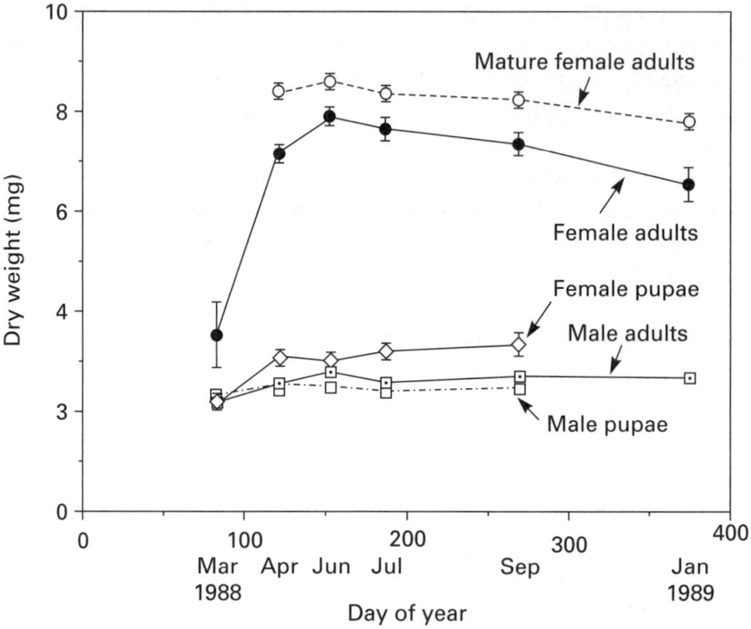

Figure 16.6. Trends in the body weights of male and female alates through the seasons. Investment in individual females peaks in June and declines gradually through the remainder of the year. Investment in individual males varies less. The lack of seasonal trend in pupal weight shows that differential investment in alates occurs after adult eclosion. Adapted from Tschinkel (1993b).

pheromone dispersal within the colony; (2) prevents female alates from dealating and initiating reproduction in their natal nest; (3) inhibits the sexualization of larvae; (4) causes monogyne workers to execute all but one female secreting queen pheromone. The pheromone is secreted by at least two glands—the poison sac and the postpharyngeal glands—and possibly more. Each egg is coated with the secretion from the poison sac as it is laid, so pheromone release rate is strongly and positively related to egg-laying rate.

With this knowledge, we can postulate how the pheromone regulates the annual cycle. Large numbers of spring sexuals would be induced as follows: During the winter, low temperatures result in low queen fecundity and therefore low populations of fourth-instar larvae. When spring temperatures once again allow brood production, low pheromone levels allow sexualization (development of female larvae into alates rather than workers) of large numbers of larvae. Because individual fourth-instar larvae of sexuals and workers stimulate queen fertility about equally, the relatively small number of sexual larvae in the spring keeps queen fecundity relatively low, and larvae continue to be sexualized until colony size declines below some threshold.

Pheromone levels might also explain why colonies smaller than about 25,000 workers do not produce sexuals—they never experience a level of queen pheromone low enough to sexualize larvae. Perhaps queen-pheromone production lags increasingly behind colony growth, leading to greater sexual production in larger colonies. This pheromonal picture has not been specifically tested, and other factors are almost sure to influence sexual production as well—nutritional status, temperature, photoperiod, fat stores, worker:brood ratio, worker size and age distributions, and colony genotype are only a few.

To Fertilize or Not to Fertilize? Sex Ratio

In "ordinary" animals, sex is chromosomally determined, and the ratio of males to females is about 1:1. Hymenoptera are not "ordinary" when it comes to sex determination, and their sex ratio is not thus constrained. Because males develop from unfertilized, haploid eggs, female hymenopterans can control the sex of their offspring by fertilizing each egg or not and can produce broods ranging from almost no males to a lot of males, depending on the situation. In the honeybee queen, control over sex is complete. The honeybee queen lays only haploid, unfertilized eggs in drone cells (which are larger), and only diploid, fertilized ones in worker cells (Ratnieks and Keller, 1998). The fire ant queen's intentions are not so easily read, but her extreme efficiency of sperm use suggests that she has tight control over the release of sperm for fertilization and therefore also over sex determination. Her control over a tiny muscle at the end of her spermathecal duct has far-reaching consequences.

Because the proportion of sexual males and females in the Hymenoptera can be adjusted through fertilization control, sex allocation has become a testing ground for much evolutionary theory. Bourke and Franks (1995) have reviewed the subject. A paper by Trivers and Hare (1976) began the inquiry by pointing out that the sex ratio that maximized worker fitness required that three times as much energy be invested in females as in males but the strategy that maximized the queen's fitness was equal investment in the two. The asymmetry arises because workers are three times as related to their sisters as to their brothers, whereas queens are equally related to their sons and daughters. All parties maximize their fitness by investing in proportion to their relatedness to each sex. The interests of the workers thus appear to conflict with the interests of the queen. By comparing the sex ratios of populations of real colonies with the theoretical ratios, we might be able to tell which party controls the sex ratio, but the task is complicated by additional effects—factors such as dispersal patterns, multiple mating, worker reproduction, and cooperative colony founding by relatives and by others all affect relatedness and therefore the predicted sex ratio. Deviations in relatedness from 3:1 will lead to parallel deviations in the theoretical preferred ratio.

Queens control sex ratio through fertilization of eggs, and workers through selective brood rearing or killing. Destruction of males bears a cost in lost energy and material, which, above a certain limit, exceeds the fitness benefit gained from manipulating the sex ratio. We suspect a struggle: the earlier in development workers can recognize males, the lower the cost of killing them; the longer a queen can "hide" the identity of her sons, the more control resides with her. Many possibilities can be tested with theoretical models (Bourke and Franks, 1995).

A credible analysis of sex ratio data therefore requires a great deal of knowledge about the social structure, genetic structure, breeding biology, and dispersal patterns of the test species, both for populations and for colonies. Ed Vargo, then in Austin, Texas, recognized that knowledge of *S. invicta* met these conditions better than that of most ant species. Moreover, *S. invicta* exists in two social forms that are predicted to prefer different sex ratios, promising a sort of intraspecific experiment. Predicting the sex-investment ratio for the monogyne form is rather straightforward. Each colony has a single singly mated queen, and all workers are her offspring. Workers are therefore 75% related to female brood and 25% to male brood, as verified by allozyme data. Monogyne populations are predicted to produce a female:male energy-investment ratio of 3:1.

Vargo collected 25 entire, mature monogyne colonies in north Georgia. To his own samples, he added sex-investment ratios calculated from published numerical male:female ratios (Markin et al., 1971; Morrill, 1974b; Bass and Hays, 1979). After adjusting for submature alate weights and sexual differences in preflight maintenance costs (Boomsma, 1989), he arrived at an annual average investment ratio of 0.64 (female:total sexuals) over all populations. This value was significantly higher than 0.50 but was not significantly different from 0.75.

To Vargo's values, I added my own data from 1989. Because dry weight underestimates the energetic sex ratio by 4–5%, I included the disproportionate cost of female fat, a refinement not used by Vargo. Each mating flight–ready male represented a total investment of about 81 joules and each female 275 joules, including two weeks of maintenance in the nest. These data produce an annual sex ratio of 0.70 (95% confidence interval, 0.62–0.78) for the Tallahassee population, not significantly different from 0.75. In a separate study at Southwood Plantation in 1994, we trapped sexuals from 23 colonies during an entire season of mating flights. The population energetic sex-investment ratio was 0.63, close to Vargo's value. Next, I recalculated all Vargo's values using individual sexual costs and then averaged them with my own (Table 16.1). The result proved to be 0.752, indistinguishable from the theoretical 0.75. With my fat correction, the seven individual populations ranged from 0.70 to 0.79, plus an unusually low ratio of 0.57 from South Carolina.

All together, the sex-investment ratio of 0.75 across populations of *S. invicta* suggested that workers have strong control over the sex of the alates, although the ratio deviated significantly for some populations and in some seasons. Perhaps under some circumstances the sex ratio is actually a compromise between queen and workers, or each party gains temporary advantage in an ongoing tug-of-war. In addition, the

Table 16.1. ▶ Estimates of sex ratios in *Solenopsis invicta* colonies at various times and places. Together, these figures indicate that ratios in the field more closely approximate 0.75 (the ratio predicted if workers control sex ratio through selective brood care) than 0.5 (that predicted if the queen controls sex ratio by sperm allocation).

Source	***Ratios calculated from numbers of individuals (male/female)***	***Ratios calculated from total energy investment in alates (female/total)***
Markin and Dillier (1971)	0.917	0.79
Morrill (1974b)	0.923	0.79
Bass and Hays (1979)	2.6	0.57
Tschinkel (unpublished, from Tharpe Street, 1989)	0.82 (adults) 0.49 (pupae)	0.70
Tschinkel (unpublished, from Southwood Plantation, 1994)	2.04	0.63
Vargo (1996)	1.39	0.71
Porter (unpublished)	0.996	0.77
Mean	1.38	0.752 (arcsine derived)
	(42% female, 58% male)	

value of one sex or the other may be increased by local conditions or climate. Reasons may also exist for male bias to be greater than theoretical. Males are more exposed to predation and physical stress during mating flights than are females, and they are usually less robust. In order to end up with a 3:1 sex-investment ratio in the mating swarm, populations may therefore overproduce males. Possibilities like these remain to be investigated.

The skewed investment is not apparent in male and female pupae, into which energy is poured at about equal rates (per-day energy ratio, 56% female, not significantly different from 50%). Because male and female pupae weigh about the same, this investment produces an approximately equal numerical sex ratio. Five of the seven populations were 45–50% males, assuring a female mating frequency of at least 95%. Skewing of the investment takes place through the deposition of large metabolic reserves in females during their adult maturation. A flight-ready female represents 3.4 times as much energy investment as a male. This final cost is the basis of the sex-investment ratio.

Sex Ratio in Polygyne Colonies

The predicted sex ratio in polygyne *S. invicta* must take into account the lower average relatedness among workers within the nest. Relatedness declines in relation to the "effective number" of queens (that is, the number of queens weighted by their relative contributions to colony reproduction) and their relatedness to each other. In the north Georgia population, queens were unrelated, the effective queen number was between 8 and 10, and worker relatedness asymmetry predicted an investment ratio between 0.64 and 0.74, but comparison was complicated because 78% of the males were sterile diploid males that did not contribute to the breeding population. How they should be included in the sex ratio depends on how the workers regard them—as males, as females, or not as sexuals at all. We do not yet know if workers regard diploid males as males (which seems likely), and we adjust for the energetic cost differences, the observed annual mean sex ratio was 0.62. This value is not significantly different from the lower value bracketing worker control but is significantly higher than 0.50, the ratio expected under queen control. This result is consistent with some level of worker control of sex investment, but inbreeding and other factors may also play a role.

Strictly speaking, the workers that depart with queens during colony budding should also be included in the investment calculations, but such data are not available.

Knowledge of the sex ratio in the native Argentine polygyne population, without the complication of diploid male production, would be helpful. There, effective queen number per colony is lower, and queen–queen relatedness much higher. Uncertainty also remains whether the workers that accompany a group of queens during nest budding should be counted as part of the investment in females, but Vargo argued that the high rate of exchange between mother and daughter colonies mean that it should not. Clearly, some of these decisions, and whether diploid males are males, females, or neither, verge on the philosophical, so we can be sure that knowledge of polygyne sex-ratio patterns will be some time in coming.

Split Sex Ratios in Monogyne Colonies

Monogyne colonies tend to produce mostly or entirely one sex or the other, rather than the population-wide average of 0.60–0.75. Such so-called split sex ratios are widespread among ants. The theory on this subject relies on variation in patterns of relatedness and local resources. Our data from Southwood Plantation, based on the total energy content of a year's production of alates, are typical. Colonies that produce almost entirely one sex are more likely to produce females (Figure 16.7). Mixed-sex broods are more likely to be male biased than female biased. This female bias is more spread out in the polygyne form, and mixed-sex broods are more likely to be female biased (Figure 16.8). None of the theoretical factors that have been proposed to explain split sex ratios were unequivocally supported by Vargo's analysis, and many simply do not apply to *S. invicta*. Although a good hypothesis explaining split sex ratios in fire ants is currently lacking, some progress has been made in understanding the mechanism producing them.

If we could determine what the queen intended to invest in the two sexes and compare that ratio with the one finally reared by workers, we would see whether the queen and workers had different "intentions." Serge Aron and his colleagues (Aron et al., 1995) tried to do just that. The queen is likely to manipulate the sex ratio by varying the proportion and/or number of unfertilized (male) eggs she lays. Queens were collected from 12 field colonies that produced either male or female alates (not mixed broods). The sexes of samples of their eggs was determined from chromosome squashes—16 chromosomes meant the egg was male, and 32 meant female. Squashes of 50 eggs from each queen gave the primary sex ratio and indicated the queen's "intention." Each colony's secondary sex ratio was revealed by the sexuals reared in the laboratory from the same cohort of eggs tested for the primary ratio and from the sexuals present at the time of collection.

Queens from the five colonies chosen because they produced 90% male alates laid 18 to 22% haploid (male) eggs (mean 19%) immediately after capture, whereas those from the seven colonies that produced female alates laid 8 to 14% haploid eggs (mean 11%). In the laboratory, the male-producing colonies continued to produce males, and female-producing ones produced females, so the sex ratio did not change (97% males, 99.9% females).

Attrition of haploid eggs during development was high. In male-producing colonies, 19% of the eggs were haploid, but only 7.5% of the pupae were male. In female-producing colonies, these numbers were 11% and 0%. Losses were greater among male than worker eggs, either because of differential viability (unlikely) or because workers eliminated males. For every 100 eggs, 11 more male than worker eggs vanished before pupation. In male-producing nests, about eight males remain per 100, but in female-producing nests, none remained.

Is this a battle between queens trying to produce equal numbers of the two sexes and workers shooting for a 3:1 ratio of females to males? If we back off from the freighted concept of "conflict" for the moment, we can ask simply whether the sex ratio is a property of the workers or of the queen. Exchanging the queens of male- and female-producing colonies should provide the answer. Under worker control, the sex ratio would remain unchanged, but under queen control, it would accompany the queen. When queens were thus exchanged, male-producing colonies quickly began producing only female sexuals and vice versa (Passera et al., 2001). Exchanging queens of colonies producing the same sex did not change the sex produced.

At least at the extremes of sex specialization, queens seem to be able to impose the sex investment ratio on workers. How the game plays out in mixed-sex broods is unknown. If the proportion of haploid

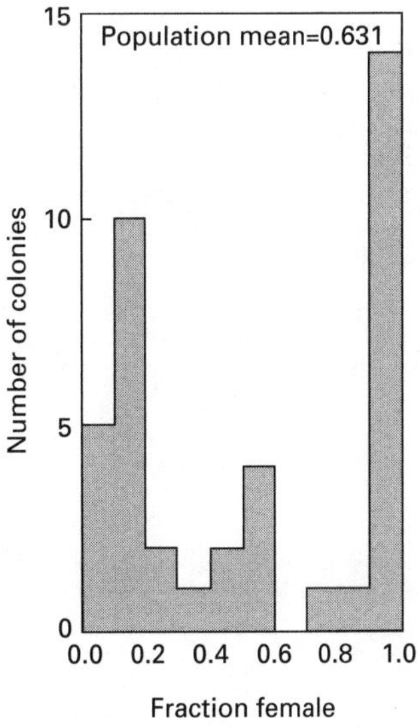

Figure 16.7. The distribution of mating-flight sex ratios (based on dry weight) among colonies of *Solenopsis invicta* at Southwood Plantation in 1994, showing the phenomenon of split sex-ratios. Colonies that produce almost entirely one sex are more likely to produce females, but those that produce mixed-sex flights are more likely to be male biased than female biased.

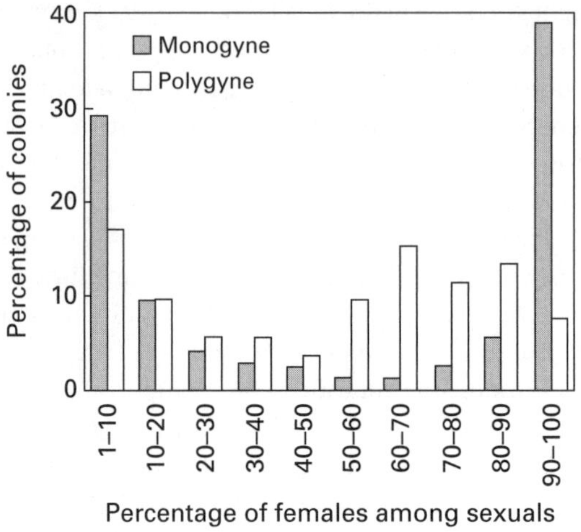

Figure 16.8. Comparison of the distribution of mating-flight sex ratios in individual monogyne and polygyne colonies. Polygyne sex ratios are less female biased. Adapted from Vargo et al. (1996).

eggs is the causal link, we would expect to find a correlation between sex ratio and number of haploid eggs in mixed-sex broods, but this possibility has not been investigated.

The following (hypothetical) scenario could produce split sex ratios. The total resources to be allocated to sexual production is set by season, colony size, and colony condition. At the beginning of the spring, the queen begins to lay haploid eggs, and a short time later, a nutritional switch causes some diploid larvae to begin female sexual development. If the queen lays few haploid eggs, the sex ratio will inevitably be strongly female biased. When haploid egg production is higher, a tipping point must exist above which the colony produces a strongly male-biased ratio. Whether this mechanism involves workers' killing males is unknown, but the higher attrition of male than of diploid eggs suggests that it might (lethal mutations, which are unopposed in haploid eggs, might also explain higher male attrition). Perhaps sexuals inhibit the production of more sexuals of either sex, and one can "get the jump" on the other. Queen age could conceivably play a role; older queens might produce more males as their sperm supply dwindles, but this possibility is another that has not been tested.

Is the queen trumping the workers by manipulating the production of haploid eggs? Probably not. The queen control of sex ratio within the colony is an illusion—workers get their way at the level that counts, the population. It may look like a struggle between the workers and their mothers, but at the population level, the aggregate of these family fights produces the 3:1 ratio favored by the workers, not the 1:1 ratio favored by the queen. The "contest" between queens and workers within the monogyne colony is revealed as merely a mechanism for producing the worker-preferred sex ratio. Its control lever is the laying of more or fewer haploid eggs by queens, which adjusts the tipping point between female and male alate broods within colonies so that the population ratio conforms to the worker-preferred 3:1 ratio. Even if this apparent fight is not quite what it seems, however, queen control over male production is clearly important in contexts other than control of sex ratio. For example, during the development of a colony, the queen adjusts the primary sex ratio from no males at all in founding colonies to various proportions of haploid eggs in mature ones. Egg fertilization is therefore definitely under some sort of regulation; natural selection has a grip on it.

III

Family Life

◂•• *Deby Discovers Ants*

Very few people have epiphanies, blinded by a bright light on the road to Damascus. In today's blasé world, many of us wear both physical and perceptual shades. But then perhaps an epiphany is more likely if you have the kind of enthusiastic personality that is predisposed to seeing bright lights and to yelling out Wow! without embarrassment. Enthusiastic certainly described (and still does) my student Deby Cassill. After 18 years as a bureaucrat in the health field, Deby decided to go back to college for a second bachelor's degree, this time in biology. Older students are always a pleasure to have in a class, and Deby single-handedly made Insect Biology a pleasure to teach. Toward the end of the semester, she indicated that graduate school was her goal, but she was still undecided about the field. After some weeks of coaxing and a collegial tug of war between a cell-biologist colleague and myself, she decided to work with me. Something entomological, was the agreement. No sharp focus yet.

I suggested that Deby set up an observation colony of fire ants and spend a few hours watching them. This exercise, I said, would make her future readings in entomology a lot more meaningful. Deby agreed with a look that said, *You're* the major professor. Whatever you say. I mean, what's there to know about ants except that they get in your food and sting your ankles now and then?

She did as she was told, apparently without much expectation, but what she saw surprised her, as it has many people before and since. Later that year, she wrote up her impressions of this moment, and I present them here as they are, in her own rather purple prose.

> Nonetheless, the following day found me sitting on a cold metal stool trying to focus the dissecting microscope on a colony of lab ants. Then it happened. Unexpectedly, a whole new world exploded into view. It was as though I had been yanked off the stool, sucked through the scope, and plunked smack-dab in the middle of a city teeming with the most beautiful glass-like creatures. All I could do was gape, wide-eyed and speechless, at the exquisite beings hurrying about their daily work. Their colors and shapes were like nothing I'd even seen before. Some were glowing curves of amber, some blocks of white marble, and some translucent raindrops freckled in white.
>
> Later . . . I realized that here, within this scope, lay the kind of adventure for which I'd been yearning . . . . Each day thereafter, I sat for hours watching these astonishing creatures until hunger or a tired back forced me to quit. Slowly, patterns of behavior began to emerge out of the lovely chaos . . . . Soon, watching wasn't enough for me. So many "Whats?" and "Whys?" were

emerging that I sometimes didn't even have a chance to write down one question before another popped up. [I came to understand that] some of the most exciting discoveries in science have come, not from finding something new, but from seeing something old in a new way. And in a moment of true inspiration, I saw something old in a new way."

Deby went on to get her master's degree and Ph.D. with me, turning in a groundbreaking dissertation based mostly on thousands of hours of watching ants. Most of these observations were made from videotape recordings played over and over on fast forward. After a couple of years as a postdoctoral associate with me, Deby went on to Arizona, then Texas, and finally Florida, where she is still watching ants with undiminished enthusiasm. She is now a competent, sophisticated biologist. But it all started with a single look through a dissecting microscope at the beautiful, magical world of fire ants.

17

Nestmate and Brood Recognition

A discussion of fire-ant family life must begin with how family members recognize one another. Many a young boy has observed that ants deal very roughly with a worker from another colony dropped into their territory, like an urban gang that catches a rival gang member on its turf. In both ants and gangs, the intruder is accosted, threatened, and attacked, and at least among ants, often killed (Plate 14). Such aggressive behavior is the basis for most nestmate-recognition research, because it is the overt sign that ants have encountered an individual they do not recognize as a nestmate—it is therefore a reliable assay for such recognition.

Nestmate recognition exists because, to reap the benefits of cooperating within an organization, the members must be able to tell who belongs and who does not. This rule applies as much to an individual organism or superorganism as it does to many human organizations. The cells that make up an organism have evolved a multitude of molecular and physiological mechanisms for recognizing, rejecting, or isolating things that do not belong among them. The cells can then work harmoniously together, sharing the products of their cooperation only among themselves. Similarly, when ant-colony members reject nonmembers, they limit the benefits of superorganismal cooperation to their own members and keep the shared goal of the colony intact. These "others" may be from other colonies of the same ant species, other ant species, or other species of insect. Because the overwhelming majority of insect societies are nuclear or extended families, rejection of nonmembers limits the benefits of social life to relatives, whose fitness is improved through the increase in the indirect component of inclusive fitness (i.e., the fitness of relatives discounted by the degree of relatedness) that results from cooperation. Therefore, kin ought to be able to recognize kin, to make sure that the colony consists only of relatives, but in most cases, social insects merely discriminate between nestmates and nonnestmates. Under ordinary conditions, all the members of a colony are related to some degree, and to the degree that they are, nestmate recognition is equivalent to kinship recognition. In fact, in the monogyne form of *Solenopsis invicta,* all the workers are the daughters of a single mother and a single father, and nestmate and kinship recognition are the same thing. In the polygyne form, nestmate recognition is weak, and efforts to demonstrate kinship recognition have failed.

Vander Meer and Morel (1998) provide a comprehensive review of nestmate recognition. Using aggression assays, Robert Vander Meer, Martin Obin, and Laurence Morel explored the nature of nestmate recognition in fire ants in a series of papers published between 1985 and 1993 (Obin and Vander Meer, 1988, 1989a, b, c; Vander Meer et al., 1989, 1990b, c; Lavine et al., 1990a, b; Morel et al., 1990; Obin et al., 1993; Vander Meer and Morel, 1998). In nature, most intrusions, and therefore most cases in which ants encounter individuals they do not recognize as nestmates, take place in the foraging territory, so responses of foragers in laboratory arenas represent what happens in the field better than would those of ants within the nest. Each of Obin's nestmate-recognition tests began when a worker in the foraging arena of one colony walked undisturbed onto a pair of clean, odor-free forceps and then walked off the forceps to become an "intruder" in the foraging arena of a test colony (the "residents"). This method of introducing intruders assures minimal disturbance and reduces possible bias from any alarm signals. As a control, workers were also allowed to walk off the forceps in other parts of their own foraging arenas.

Each test was a bit like a wrestling match, in which a round lasted a minute or less. The responses of the first five residents to encounter the intruder (or the first 20 encounters) were classified into three levels of aggression. In encounters at the "inspection" level, a resident worker simply antennated and inspected the intruder, then ignored it. The "challenge" level included rapid antennation, inspection, following and mandibular gaping, and possibly holding and later release. At the "attack/kill" level, the intruder was held and dismembered by nestmates or was immediately attacked, grabbed, and stung. These last always resulted in the death of the intruder.

Colony Odor and Its Creation

No one has seriously challenged the hypothesis that nestmate recognition in *S. invicta* begins with an odor on the body surface. The degree to which the behavior of the confronting workers may intensify the aggression is not clear (Vander Meer and Morel, 1998). A great deal of circumstantial evidence from various social insects supports the odor hypothesis, the first piece of which is that recognition is almost always preceded by a quick sweep of the antennae over the other ant, as if for an olfactory inspection. Contact with any part of the intruder's body suffices to release the aggressive behavior of the resident, implying that the chemical cue is nonvolatile and generally distributed over the body. We can accept a number of facts/working hypotheses for ants in general: (1) Each colony has a characteristic odor synthesized by the ants or accumulated from their environment or both (Hölldobler and Wilson, 1990). (2) The extreme frequency of contact among workers in the nest transfers surface materials among them, much as we transfer lipid material to whatever we touch in the form of fingerprints. Frequent licking and grooming further speeds the rate of transfer and reduces the odor differences between workers, until nestmates generally share the same blend of chemicals. This blend is termed the colony odor. Saturated hydrocarbons are thought to be particularly important to colony odor. (3) The cuticular lipids trap nest, food, and other environmental odors, incorporating them into the colony blend. (4) The cues that ants use for nestmate recognition are probably always a subset (the "label") of the total colony odor, but direct evidence about the chemical nature of these cues is rare and weak. (5) In at least some species of ants, and possibly in ants generally, a glove-shaped gland, the postpharyngeal gland, opens into the mouth cavity in the head and contains the same blend of chemicals found on the cuticle. Application of the contents of a nestmate's postpharyngeal gland to an alien ant reduces aggression to the alien, and conversely, application of an alien's gland contents to a nestmate increases it (Soroker et al., 1994; Hefetz et al., 1996; Lahav et al., 1999). (6) Ants acquire the cuticular lipids of nestmates through grooming and food sharing and sequester them in the postpharyngeal gland, creating a colony blend. The ants then apply this blend to their own cuticles through self-grooming, admixing their own, personal cuticular lipids in the process. Although mutual grooming alone may spread and blend cuticular lipids sufficiently fast in small colonies, exchange of gland contents by food sharing is probably required in large colonies. (7) The sensory half of the recognition system is understood only in general terms. Presumably, the ants compare the chemical cues on the body of another ant with a memory (usually called the template) of their own colony-specific odor stored in their nervous system. To understand nestmate recognition systems, we must understand the nature of the specific cues and the neural recognition process leading to aggression or acceptance.

The transfer of cuticular materials in a colony is especially clear when the recipient is not an ant. *Solenopsis invicta* is host to two myrmecophiles, a scarab beetle, *Martinezia dutertrei* (Vander Meer and Wojcik, 1982), and a eucharitid wasp, *Orasema* sp. (Vander Meer et al., 1989). The beetle simply barges into the ant colony, then, protected by its armored cuticle, it plays dead until it has acquired its host colony's label. After that, it feasts on the host's brood, dead workers, and food. The tiny, first-instar larva (planidia) of the wasp lies in wait on plants and enters the colony by hitching a ride on a passing worker. Once in the ants' nest, it is left in peace to devour ant brood, having also acquired the colony's label. Label material is physically transferred during this acclimation. Both the beetle and the wasp larva acquire the host-colony-specific pattern of cuticular hydrocarbon compounds after they enter the colony, probably by physical transfer. Furthermore, both lose these hydrocarbons once they are no longer in the host nest. If transfer can produce a colony-characteristic mixture on the surface of nonants, it can certainly do so for the ants themselves.

The key nestmate-recognition role played by the postpharyngeal gland in two species of ants from different subfamilies suggests that this system is widespread among ants, including fire ants. Historically, the functions attributed to the postpharyngeal gland have ranged widely, to include the sequestration and digestion of fats, lipid secretion and absorption, pheromone production, and hydrocarbon secretion. In fire ants, the postpharyngeal glands lie in front of the brain and open into the posterior region of the pharynx. Each gland is composed of four lobes in small workers, 24 lobes in majors, 85 in the queen, and only nine in males. This difference in absolute and relative size suggests great variation in the level of function. Newly emerged adult workers and female sexuals have colorless, flaccid glands, and these remain so unless lipid-containing food is offered. Thereupon, the glands fill very quickly with an oily liquid. Workers fed on dyed oil and then starved for long periods lose the color from the crop but not the postpharyngeal gland, suggesting that these glands are not storing food. Radioactively labeled fatty acids injected into the hemolymph of queens do not appear in the postpharyngeal glands in appreciable amounts, so most materials in the gland are probably sequestered from food rather than secreted. If this is the case, then it is not, in fact, a gland. Indeed, absorption of lipids has been proposed as a function of this mysterious organ (Phillips and Vinson, 1980a, b; Vinson et al., 1980; Vander Meer et al., 1982a).

The larger postpharyngeal glands of female alates and queens of *S. invicta* have been more intensely studied. In young queens, the contents of the glands are not derived from food and are chemically different from crop contents in that they contain large amounts of the major cuticular hydrocarbons, suggesting a role in nestmate recognition (Nelson et al., 1980; Thompson et al., 1981). Removal of the gland from queens has no discernible effect on queen behavior or function and no effect on the behavior of workers toward them. This finding may not rule out a role in colony odor formation, because mutual grooming, especially in small colonies, can also spread such odors, but similar analyses of workers have not been done, nor have the appropriate experimental tests, so the role of the postpharyngeal gland in colony odor of fire ants remains uncertain. The postpharyngeal glands of the queen have also been shown to produce and dispense queen pheromones. We cannot yet fully resolve the role(s) of this gland.

Environmental and Heritable Contributions to Colony Label

Most fire ant researchers have noticed that laboratory colonies seem less aggressive than colonies fresh from the field, so colony odor may include an environmental component (Obin, 1986). Compared to that of field colonies, the odor environment of laboratory colonies is much simpler and less variable: they grow up without soil, in a shared air space with other colonies, and all eat the same, much simpler diet. When workers were transferred between large, laboratory-reared colonies from two geographic locations, they were received with moderate aggression. About 20% were ignored, 60% were challenged, and only 19% were killed. Intruders from field colonies were received much less cordially—74% of them were immediately attacked and killed and the rest challenged and detained. They were never simply inspected or ignored. The aggression was as likely to be initiated by the intruder as by the resident, suggesting mutual nonrecognition. The geographic origin of the colonies did not affect these results. All differences were associated with lab rearing, not geography. The murderousness of lab residents toward field intruders showed that these residents are capable of strong aggression and suggests that their relative "cordiality" toward lab intruders results from weaker ("half-empty") recognition cues on the intruders' bodies. Unfortunately, Obin did not test field colonies as residents, so we cannot say that lab and field residents are equally aggressive or that they differed in both cue strength and aggressiveness, but the recognition cues are "half full" too—even though they were reared under identical conditions for months, the colonies still retained individually recognizable colony labels. Obin referred to these as "heritable" cues, assuming that they were the products of enzymes and metabolic processes. Of course, the differences in these cues are not necessarily the result of genetic differences. We can say only that these cues resist obliteration through rearing in a shared environment (i.e., are heritable in the broad sense).

The results discussed so far suggest a role for diet in determining the environmentally derived nestmate-recognition cues. From large lab colonies, Obin and Vander Meer created several subcolonies, then fed one of them the same diet as the home colony and others a different diet. The diets consisted of honey water and/or sugar water as the sugar source and moth pupae and/or roaches as protein. The diets of treatment subcolonies (1) lacked a dietary item that the controls and source colonies received or (2) contained an item they did not or (3) differed from the controls and sources only in the proportion of the diet made up by an item. After one or two months, foragers from these subcolonies were used as intruders in their own colony of origin. The presence of unfamiliar dietary items, the absence of familiar ones, and change in their proportions all caused significant increases in aggression from former nestmates, showing that diet contributes to nestmate recognition and suggesting that the ants are judging the perceived odor by its *overall similarity* to their remembered template. In nature, the diet and therefore the colony odor must change constantly as old sources of food dwindle and the ants find new ones. Workers must therefore continually update this template to match the changing recognition cues.

In some species of ants, the queen contributes importantly to the colony odor. Fire ants are attracted to the odors of their own queens more strongly than to those of other queens (Jouvenaz et al., 1974), suggesting that in fire ants, too, the queen might play a role in colony odor. To determine whether she does, Obin and Vander Meer created small broodless, queenless subcolonies and fed these the same diet as their source colonies for six months, until they had reared eggs to adulthood and the original workers died. None of these newly reared workers had had any contact with their own queens since they were eggs. Groups of workers from these queenless subcolonies were then exposed to their own or an alien queen, or kept queenless, and were then used as intruders in the original source colonies.

Intruders from the subcolonies kept queenless were met with little aggression in the original colony of their own kin, so the immediate suggestion is that the queen contributes little to colony odor. Transfer to nonkin complete colonies produced higher aggression; about 70% were challenged and 2% killed. The response to kin workers that had been housed with an alien queen depended on the location from which the intruder was selected. Only kin workers taken from the queen's retinue (i.e., those that had had direct contact with the queen) were attacked, and about 30% were killed. Fifteen minutes of exposure to the alien queen sufficed. Those taken from the foraging arena (i.e., those who had had no queen contact), met with no more aggression than did a queenless-control intruder. In a real colony, nearly all of the aggressive interactions with intruders are limited to the foragers out in the territory. Although the queen's odor is transferable by contact (Sorensen et al., 1985c), and can be extracted from tending workers, this transfer does not reach as far as the foragers.

More puzzling are the responses of kin to intruders housed with their own mothers for 24 hours. When these were tested on their kin colony (now queenless for 24 hours), they elicited significantly more aggression than did the queenless controls, although it rarely exceeded the challenge level. Presumably, no foreign odor had been added. Perhaps 24 queenless hours suffice to bring the queen odor in the source colony to such low levels that intruders with queen odor are recognized as foreign.

These experiments test the effects of queens on the colony odor of intruders—they do not test the effect of the queen on the residents, which turns out to be quite dramatic. Polygyne workers, whether housed with a queen or not, respond with little aggression to intruders from either monogyne or polygyne colonies. In contrast, monogyne colonies respond with lethal aggression to such intruders, but when the queen is removed the aggressiveness declines until, after two weeks, it does not differ from that of polygyne colonies (Vander Meer and Alonso, 2002; Figure 17.1). The presence of brood makes no difference to this decline. When *S. richteri* intruders are used, they are met with lethal aggression, so it is not the general recognition ability that has declined but the specific response to nonnestmates. Such queenless colonies of either form are also less aggressive toward newly mated queens and readily adopt them—a single one for monogyne workers and sometimes multiple queens for polygyne ones. When originally monogyne or polygyne queenless colonies have adopted newly mated queens from monogyne colonies, the aggressive response toward nonnestmates reappears. Moreover, the aggressiveness of polygyne workers is as high under these circumstances as that of monogyne workers. They are also equally and highly aggressive toward additional newly mated monogyne queens. Newly mated polygyne queens were not tested.

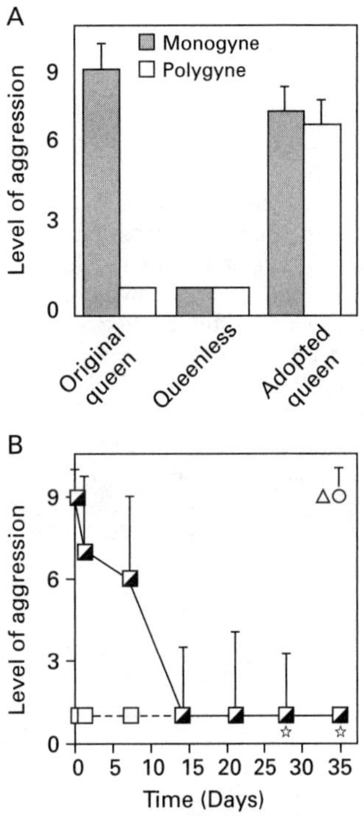

Figure 17.1. Effect of queen removal and replacement on worker aggressiveness. A. Monogyne colonies treat intruding workers much more aggressively than do polygyne ones, unless they have been queenless for a period. When either monogyne or polygyne queenless colonies adopt newly mated queens from monogyne colonies, their aggressiveness reappears. B. After 15 queenless days, the aggressiveness of monogyne colonies (half-filled squares) decreases to that of polygyne colonies (open squares). Symbols in upper right corner show level of aggression in monogyne control colonies with functional queens. Adapted from Vander Meer and Alonso (2002).

The first experiments showed that the colony odor is modified only a little by queenlessness, so the later ones seem to show that the response to colony odor is expressed only in the presence of a queen. How might such a reponse be regulated? Vander Meer and Alonso proposed that the queen secretes a pheromone that increases the sensory ability of workers to distinguish the small differences among colony odors within their own species. Unfortunately, no direct evidence supports such an effect of queen pheromone or such a change in sensory thresholds. Whatever the mechanism, the effect of the queen is to make the workers aggressive toward nonnestmates, and this aggression in turn creates territorial behavior, which in turn, monopolizes all the resources within the territory for the use of that queen and her colony. By extension, the declining aggressiveness of workers after the death of their queen puts the colony at risk of invasion by an unrelated replacement queen.

What might the heritable components of colony odor be? Since about 1985, the favorite hypothesis has focused on hydrocarbons on the surface of the ants' cuticle. These compounds are universal among insects because, along with other lipids and waxes, their primary purpose is to reduce water loss through the cuticle (Blomquist and Dillworth, 1985). In social insects, cuticular hydrocarbons are thus available to evolve the additional function of nestmate-recognition cue, partly because of their exposure and partly because they are complex mixtures of 30 to 200 compounds that could be recombined as infinitely varied bouquets. According to this popular hypothesis, ants recognize nonnestmates by differences in hydrocarbon patterns—the greater the difference, the more hostile should be the reception.

Extracting hydrocarbons from ant cuticle is simple (a seven-minute soak in hexane does the job), but their separation and identification requires many thousand dollars worth of fancy equipment, especially gas chromatographs, which yield hydrocarbon profiles, like

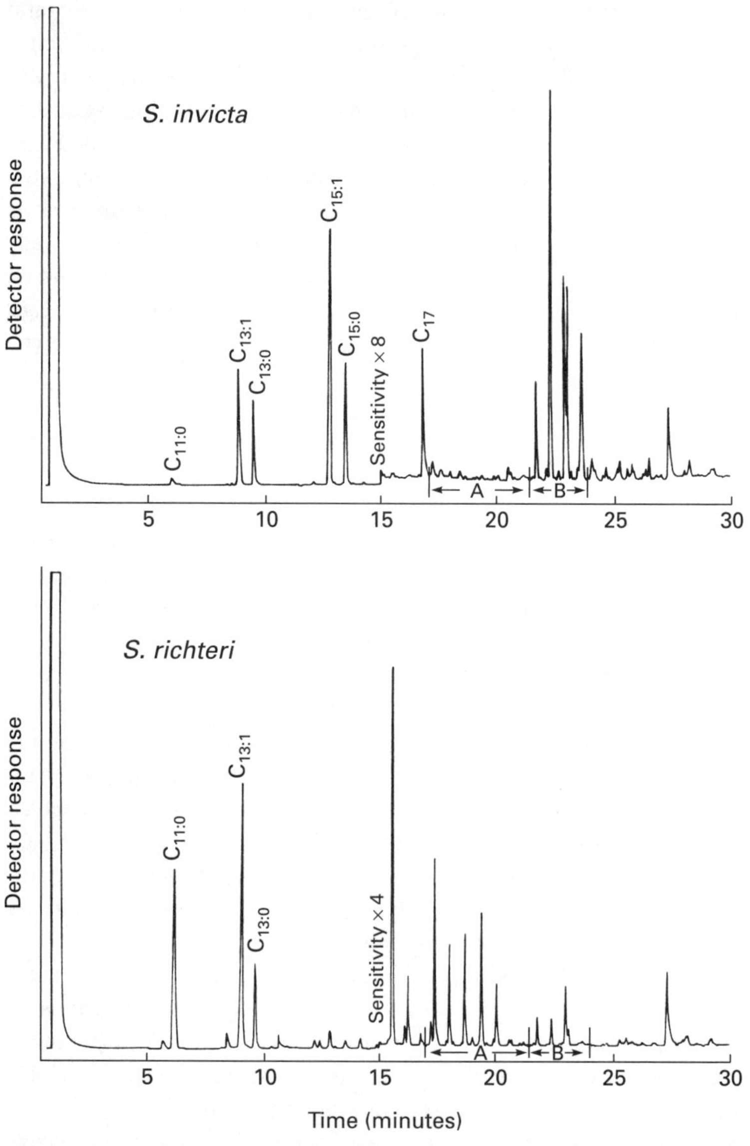

Figure 17.2. Gas-chromatographic cuticular-hydrocarbon profiles for *Solenopsis invicta* and *S. richteri*, showing the species-typical patterns. The peaks labeled A are hydrocarbons characteristic of *S. richteri*, and those labeled B are much more prevalent in *S. invicta*. Although these profiles are from pooled individuals, differences also occur at the individual level. Adapted from Vander Meer and Lofgren (1990).

fingerprints, of individual workers or groups of workers (Figure 17.2). Fire ant cuticular hydrocarbons consist of a large number of both linear and branched hydrocarbons containing from 20 to 30 carbons (Lok et al., 1975; Nelson et al., 1980). Although correlations between the hydrocarbon profiles of the five most abundant compounds and the colony's origin (lab or field) were significant, these differences were not associated with differences in the proportion of intruders killed. For example, of the 21 field intruders that were not killed, only 9 had lab-colony-like hydrocarbon profiles. These results do not support a strong role for cuticular hydrocarbons in nestmate recognition.

Heritability implies a certain constancy, just as environmental origin implies variability. The heritable component, like hydrocarbons, is expected to be a product of enzymes, so Vander Meer and his associates used hydrocarbons as a "model" for how the heritable component might behave in time. Workers from each of several colonies had colony-specific profiles of the five major hydrocarbon peaks at each sample time over four to eight months, but the profiles of all colonies changed, so the profile for a colony was as distinct from its own profile at an earlier time as it was from those of other colonies at each sample time. No temporal patterns in the profiles were apparent, and variation seemed random and unsynchronized. The cause of these changing profiles is unknown.

Not only, therefore, are dietary contributions to the nestmate recognition cues variable, but the hydrocarbon profile, a possible heritable component, also changes continually. Where does this variability leave the poor worker who must discriminate her own colony label from those of others? If the hydrocarbons are indeed part of the colony label, they include no constant elements in the recognition cues, and the memory template is also without constant elements. This inconstancy emphasizes the importance of its continual revision.

Perhaps part of the problem lies in our understanding (or misunderstanding) of "heritable." Does heritable mean "completely determined by the genes"? Ross et al. (1987) showed that the hydrocarbon profiles of hybrids between *S. richteri* and *S. invicta* are intermediate between those of the two parent species and closely track the genetic differences based on allozymes, but that result does not necessarily mean that the differences between colonies are also fixed by genetic differences. To the extent that they are secreted, rather than adsorbed, the cuticular hydrocarbons are presumably the products of enzymes, which are the products of genes, yet the hydrocarbon profiles are not stable over time. The actual mixture of compounds may be under nongenetic regulation, causing genetic differences not to be expressed one to one as differences in hydrocarbon profiles.

The lack of correlation between aggression and profile differences and the temporal variability of the profiles argue that the hydrocarbons are not part of the nestmate recognition cues. In cases where subcolonies were fed the same diet as the source colony, intruders from them were accepted without aggression by the source colony, even after several months. Both the environmental and heritable (nonenvironmental) components of the label therefore matched—the first as a result of identical diets and the second because it is stable over time—but the "heritable" hydrocarbon profiles changed quite drastically over time. If these profiles are indeed the heritable component of the colony label, they would have to change exactly in parallel in the source colony and the subcolony during several months of separation. Is this expectation reasonable?

A third difficulty with hydrocarbons as nestmate labels is that fire-ant brood bears hydrocarbon profiles like those of adults (Vander Meer et al., 1989), yet brood is readily accepted by foreign colonies whereas adult workers are not. Morel suggested (Morel and Vander Meer, 1988) that the reason is that larvae lack agonistic behavior and therefore do not escalate resident-worker aggressive behavior. To what extent the behavior of each worker contributes to the aggressiveness of the next is currently unresolved.

Only recently have ant hydrocarbons been directly, experimentally tested as nestmate recognition cues (Lahav et al., 1999), but in the Old World ant *Cataglyphis niger* rather than our own *S. invicta.* The hydrocarbon fraction of postpharyngeal gland contents, but not their other lipids, was capable of modifying resident responses as expected for a colony label—resident hydrocarbons applied to an intruder reduced resident aggression, whereas intruder hydrocarbons on a resident increased resident aggression. Treatments increased aggression more than they decreased it, suggesting that the ants are detecting odor differences rather than similarity. Similar direct experiments with fire ants have not been done. The rather distant phylogenetic relationship between *Cataglyphis* and *Solenopsis* makes extrapolation risky. For example, some correlation studies indicate that different classes of compounds may be used as labels in different species.

Modification of the Template

What is the nature of the memory "template" used to compare odors? The existence of both polygyne and monogyne *S. invicta* provides an opportunity to determine some of the template's properties. Morel and her associates (Morel et al., 1990) tested monogyne and polygyne colonies from several separate locations in Florida as both intruder and resident against all others. To avoid lab-induced reduction of aggression, they carried out the tests within two days on small subcolonies that were housed in their own soil. Whereas an intruder from a monogyne colony in another monogyne colony was met with high levels of aggression (mean = 6.9 on a scale of 8), the same intruder in a polygyne colony met only investigative interest (mean = 2.4). Intruders from a polygyne colony were met with high aggression by monogyne residents (mean = 7.1) but were only inspected by polygyne residents (mean = 1.5). Vigorous attack of *S. richteri* intruders showed that polygyne residents were not intrinsically less aggressive. Although monogyne and polygyne workers are similarly aggressive, the latter do not seem to recognize conspecific foreigners as readily.

A worker emerges from the pupa neither bearing the colony-identifying cues nor "knowing" the colony label; she acquires both during her first hours of life, the odor by a combination of secretion and transfer and "knowledge" of the label by learning. Because polygyne colonies consist of many family lines, each with its own heritable cues, and neighboring mounds may exchange workers, brood, and food, the polygyne colony label should contain a much broader and more homogenized range of labels than that of a comparable monogyne colony, so a newly eclosed worker in a polygyne colony probably learns a much more inclusive template. Indeed, as we saw, polygyne workers accept a much wider range of labels as familiar and therefore do not respond to them with hostility. What if we were to narrow the range of environmental cues available to a polygyne colony? Such colonies should tolerate a narrower range of intruder odors. For this experiment, monogyne and polygyne colonies were maintained in the lab for several months in soil-free nests on one of two diets that differed in the type of insects and sugar source (Obin et al., 1993). This situation should decrease the environmentally derived differences in colony labels and narrow the range of cues. When both the diet and the collection locality were different, the polygyne intruders were attacked, and half of them were killed (mean = 6.7). When the diet was the same, intruders were met with little aggression, whether they came from the same collection locality or not. This result suggested that geographic, heritable contributions to colony label were small and that differences of diet caused the increased recognition. Unfortunately, Obin and his coworkers did not test colonies from the same collection locality fed different diets, so an interactive effect of diet and locality cannot be ruled out for fire ants.

Monogyne intruders fed the same diet as the resident polygyne colony were challenged and restrained but not killed, but if they had received a different diet, over 80% were killed. Even when the diet was the same, the polygyne residents were much more aggressive toward monogyne intruders than they would have been if tested as residents straight from the field. Finally, all the residents were capable of full aggressive responses, because intruders from *S. richteri* colonies were invariably killed.

We interpret these findings thus. In the laboratory, the variety of environmental inputs to the colony odor is greatly reduced, whereas the heritable components are presumably little affected. A worker born into a polygyne lab colony therefore learns a narrower range of environmental components than a worker born in the field, and should have a template accepting a narrower range of cues as familiar. Any intruder is therefore more likely not to match the template and to be attacked. This pattern is consistent with the great acceptance of intruders by polygyne colonies fresh from the field, because their templates were formed from exposure to a wider range of environmental and heritable cues.

Perhaps one of the simplest hypotheses is nestmates recognize not nestmates but nonnestmates. Their senses might simply become habituated to the odor of their own colony, so that they respond only to unfamiliar odors. In such a situation, any random set of odors could serve as a colony label (Carlin, 1989). Different workers might even fix on different sets of odors. Habituation easily accounts for much of the experimental data and solves the problem of the continually changing label/template, but it raises other issues that are not easily resolved (Vander Meer and Morel, 1998). Perhaps different species employ different sensory mechanisms, among which habituation plays a role.

Finally, we should note that the recognition and acceptance of sexual females differs from the worker nestmate recognition system (Vander Meer and Porter, 2001). Queenright monogyne colonies (that is, colonies with functional queens in residence) invariably attack nonnestmate workers, as we saw above, and they also kill all nonnestmate sexual females offered to them, whether these are alates, newly mated queens, or ovipositing queens. Queenright polygyne colonies generally accept nonnestmate workers, but they also accept nonnestmate virgin alates, newly mated queens, and ovipositing polygyne queens of a particular genotype (discussed in a later chapter). They kill all other sexual females. Whether this pattern is related to the nestmate recognition system or to queen pheromone characters is not clear.

The Present Situation and Future Needs

All these recognition cues and templates are our mental constructs of how the system *probably* operates in rather general terms. As of this writing, we cannot choose confidently among a large number of possible explanations. For example, the chemical nature of the colony label has not been rigorously demonstrated, although it is almost universally assumed. In fire ants, no particular compounds or even classes of compounds have been fingered, nor has their location on the cuticle been experimentally established. No physiological or metabolic processes have been linked to nestmate recognition. The manner in which diet affects the cues is unknown. Nothing is known about how, where, or in what form the experience with an ambient colony odor is stored as a template in the nervous system or about how olfactory input is processed and compared to the stored label memory. The best we can do is to state that the attack on a nonnestmate is (probably) a response to differences in a (probably) chemical label that is composed of materials from both the environment and (probably) physiological processes. At least the environmentally derived part of the label changes continually and must (probably) be continually relearned. The perceived odor on another ant is (probably) compared directly with the current colony label or indirectly with a (presumed) stored memory of this label.

In the study of nestmate recognition, methods of chemical analysis are decades ahead of the necessary biological experiments. Thirty years after the invention of the gas chromatograph made analysis of cuticular compounds easy, almost all of the supporting evidence for their role as recognition cues is still circumstantial, and their importance in nestmate recognition remains primarily a hypothesis. Work by Lahav and others (Lahav et al., 1999) serves as a suitable model of what needs to be done with *S. invicta*. The tools and concepts for a reductionist analysis of nestmate recognition are available, but too few attempts have been made to apply them with rigor.

Other Levels of Recognition

Nestmate recognition is but one manifestation of recognition in general. Because workers behave differently toward other workers, larvae, pupae, and the queen, they can obviously discriminate among these, though how they do so is mostly unknown. Workers may also recognize members of classes of workers within the colony, on the basis of age, task, location, and so on. The possibility of kinship recognition within polygyne colonies is discussed in a later chapter. At the finest level of discrimination lies individual recognition, a capacity that is astoundingly well developed in humans and some other mammals. In an ant colony with tens or hundreds of thousands of workers, the tiny worker brain could not conceivably cope with remembering so many individual features, even if so many unique features could be generated. Most ants are therefore highly unlikely to have individual names, Disney movies notwithstanding. Perhaps individual recognition occurs in the tiny colonies of some primitive ants, but it has not been tested.

Brood Recognition

If our understanding of nestmate recognition seems less than completely satisfying, that of brood recognition is even less so. Part of the reason is that the subject has received much less attention, but a larger part is that the research has rarely produced clear answers. The problem is this. Imagine yourself to be an ant worker patrolling the underground chambers of your colony's nest. In complete darkness, you must recognize objects that are brood and distinguish them from anything else that might be present. Moreover, you must be able to discriminate among eggs, young larvae, fourth-instar larvae, pupae, and callow adults, for each of these must be treated differently. How do you do it? Vision is of no use in the darkness, and because they are soft and relatively immobile, most brood stages are unlikely to produce sounds. Tactile and chemical cues remain, and indeed, most research has quickly gravitated to those.

Oddly—but perhaps importantly—although most ant species attack adult workers from other colonies of the same species, they just as commonly readily accept brood from other colonies. In some species, if their own brood is removed, they will even rear the brood of other species to adulthood. These adults may even be accepted as normal colony members, because adults of many ant species acquire and learn the identifying odors of their colonies as young adults. Such learning and acquisition occur in slave-making species and may occur among colonies within a species as well (as in brood raiding). Whatever cues emanate from ant brood, they say "brood," and probably "stage," without signaling colony identity or even species.

How can we get the ants to share with us the brood cues and signals to which they are responding? Workers that normally respond to brood (perhaps only a subset of workers) should respond similarly to objects giving off "brood" cues at the appropriate intensities. Tests of this form work best when the response the investigator watches for is very specific and distinct from other responses. Unfortunately, in the study of brood cues, response specificity is not the rule. A list of possible worker responses to brood includes attraction, aggregation, grooming, transport, feeding, and retrieval during nest disruption. None of these is specific to brood. Studies on brood recognition must all deal with the problem of possible confounding of multiple cues that trigger unspecific behavior.

Work on fire ant brood recognition began at the USDA Methods Development Laboratory in Gulfport, Mississippi (Glancey et al., 1970). The investigators ground up large numbers of unsorted ant brood and extracted this mess with hexane. When corncob grits (what else?) were coated with this extract in unspecified amounts, workers retrieved these bits into the colony, kept them "with the brood," and carried them around "like brood" when alarmed. Soybean-oil coated grits, as a food control, were not treated in this manner, so the USDA researchers concluded that they had extracted a brood-tending pheromone. Oddly, only larval extracts, especially those of sexual larvae, were active, although workers retrieve all brood to the nest.

A follow-up publication claimed the existence of a volatile brood pheromone (Glancey and Dickens, 1988). When 300 larvae were placed in a two-choice olfactometer (an apparatus in which ants can "choose" one entrance or another in response to odors), workers chose the arm containing the larvae significantly more often. Worker ants are also attracted to carbon dioxide, however (Hangartner, 1969a), and 300 larvae would produce a lot. Further, the olfactometer was not cleaned between trials, so later workers could simply have been following trails or other cues laid down by those tested earlier. Workers usually respond to individual larvae rather than to groups, and these cues are those of greatest interest to ant biologists. Attraction to larval extract required the equivalent of 500 larvae, and the improper food control compounded the error. The electroantennogram results included in the study were not credible either, having employed an improper control and demonstrating only very weak responses to the putative larval cue.

In a rush to isolate this purported brood pheromone, Bigley and Vinson (1975) used an even less specific bioassay, namely, retrieval to the nest followed by palpation and antennation of treated pieces of filter paper, and compared this response with that to a soybean oil control. They used this bioassay to trace activity through a grinding/extraction/separation procedure, finally isolating and identifying triolein as the brood pheromone of sexual pharate pupae (young pupae, still contained within their unshed larval skins). Even on the face of it, this result is unlikely. Triolein is a storage lipid extractable from body fats, rather than from the cuticle surface, and it probably elicits a strong food response. Attempts by Vander Meer (Vander Meer and Morel, 1988) to repeat this work showed that, not only did the thin-layer-chromatography spot claimed by Bigley and Vinson to be the brood pheromone contain at least eight different triglycerides, but triolein was not one of them. Nor was triolein active in causing brood retrieval. Clearly, a poor bioassay, lack of proper dosing of tests, and weak chemical evidence had led to erroneous identification of the brood cue.

To my very first graduate student, John Walsh, and to me, these earlier studies seemed to violate so many protocols that we set out to do the test for ourselves. Unable to repeat the published work, we began our own study, which still carries a certain sentimental value for me because not only was Walsh my first graduate student, but the resulting paper was my first publication on fire ants (Walsh and Tschinkel, 1974). Like the USDA group, we used brood retrieval as an assay, but we took care to reduce the problem of confounding a brood with a food response. Our experimental nests had four chambers, the innermost of which was darkened by red cellophane and was used by the ants as their main brood chamber. We placed the test objects, two at a time, in the foraging arena. Workers quickly retrieved brood into the nest, as they did objects containing brood cues. Some brood was kept in the outer chambers, interspersed with items of food, but in the panic that ensued when we blew into the nest entrance, the ants quickly moved all brood into the inner, darkened chamber and the food items out. This difference in treatment of objects was our evidence that the workers considered the test object to be brood or at least that brood-like cues emanated from it. Although the response was still not an entirely unambiguous, brood-specific one, it reduced the likelihood of confounding brood and food responses. To date, no one has done any better.

We used only sexual larvae and pupae, whose large size made various manipulations easier, but as a consequence, our results cannot be directly applied to other brood stages and types. These might all have cues in common, but because workers treat them all differently, each must also produce its own type of cue. Preliminary tests showed that none of a number of environmental and social conditions affected our bioassay. Adding insects to the diet eliminated retrieval of bits of tenebrionid larvae. In retrospect, perhaps we should have added body contents of fire-ant sexual larvae to the diet in order to head off the criticism that retrieval of some of our sexual-larva-derived test objects was a response to novel food. Ah! The value of hindsight!

Sound is not involved in brood retrieval, for freeze-killed sexual larvae were retrieved as readily as live ones. Holding the dead larvae for various periods before testing showed that retrieval by workers gradually dwindled over two and a half days. Whether the cue loses strength or is masked by the products of decomposition we cannot say. Next, we sought to pinpoint the pheromone's source on the individual larva. We reasoned that the cue would be stronger near its release point. By dipping larvae or sexual pupae into molten paraffin wax to different depths, we found that the reduction in retrieval was related simply to the percentage of the cuticle "hidden" by wax and was the same whether we dipped from the head end or the hind end. The brood cues seemed to be more or less evenly spread over the larval and pupal cuticle. The brood cue was not present in adult sexuals—70 to 90% of sexual pupae were retrieved up to the time of adult eclosion, but after eclosion retrieval dropped to 20 or 30%, and most of the sexuals (which were from unrelated colonies) were attacked and killed. One can reason that colonies should accept worker brood from other colonies because these are sterile and can contribute to the retrieving colony's fitness, but sexuals from other colonies represent the fitness of those colonies and reduce the fitness of the host colony by drawing on its resources. Killing them is a wise move.

The top brood cue candidate was clearly chemical, so we tested chemical hypotheses. To begin with, sexual larvae were not attractive in a two-choice olfactometer, so the cue was not volatile. The cue could not be rubbed onto filter paper and was not transferred by long exposure, so it was probably tightly bound. Ideally, demonstration of a chemical cue or signal requires that, when the chemical is removed from the animal, the response disappears and that, when it is added back at the same dosage, the response reappears (the "removal-therapy" procedure). Soaking sexual larvae in water did not reduce the retrieval rate, but soaking for an hour in hexane reduced it from about 85% to 8%, without any visible alteration in the appearance of the larva. A chemical cue had probably been removed from the cuticle. Unfortunately, we were unable to get the retrieval response back by applying the extract either to a dummy larva or the extracted larva. The failure of the "therapy" experiment leaves half the equation unsolved.

Further evidence that the retrieval cues emanate from the cuticle came from a rather gruesome experiment in which we poked a hole in the hind end of each larva and pressed its contents out to leave a thin wafer of cuticle, a sort of "ant chip." These flat, empty cuticles bore no physical resemblance to sexual larvae, yet workers retrieved them quite readily (50%, opposed to 83% for whole, killed larvae). The body contents were also retrieved, at the same rate, but represented much more material than the cuticles. To be sure the response was not to food, we offered tenebrionid larval tissue as a control, which was not retrieved. Sandwiching the cuticles between bits of tissue paper prevented contact without preventing the diffusion of volatile cues. Such sandwiches were not retrieved at all, providing support for the earlier evidence that the cues resided on the cuticle and were nonvolatile.

Less direct tests for other properties of chemicals helped build a case. We exposed larvae to several vaporous chemical reagents on the assumption that changing a chemical cue would eliminate its activity. Of the various reagents, only a one-minute exposure to bromine vapor, followed by a water rinse, eliminated the cue (control, 83% retrieval; bromine-treated, 8%). Bromine reacts readily with double bonds, implying that the chemical cue contains such bonds. Workers no longer retrieved even *live* larvae exposed to bromine. The result was not as clean as we would have liked because two other reagents that react with double bonds, ozone and hydrogen bromide, did not reduce retrieval (although ozone came close), adding some uncertainty.

The final score for our study was that the cues that cause workers to retrieve sexual larvae are probably contact chemicals on the cuticle, but the failure of the therapy experiment makes this conclusion less than airtight. Perhaps the responsible molecules must be presented in a particular organization whose disruption renders them ineffective. A contact chemical seems to make sense, given the context of its function. For purposes of feeding and grooming, larvae must be recognized one at a time in the close, dark quarters of underground nest chambers. The cue from each larva must therefore be distinct from those of its neighbors, who may even be touching it. A contact cue evenly spread on the cuticle would seem to fill this bill very well.

Still, not everyone was satisfied with this interpretation. Morel and Vander Meer (Morel and Vander Meer, 1988) published a criticism of the brood pheromone literature based on a close reading (something all too rare in science). After noting all the gaps, flaws, and errors in the existing literature, they proposed that the retrieval response of workers is not based on a pheromone at all. Taking note of information from diverse species, they proposed that most behavior of workers toward brood could be explained by a combination of nestmate-recognition cues, body form, and behavior and that brood-specific pheromones do not exist. Personally, I am unable to follow their argument. At least in fire ants, body form is unlikely to play an important role in recognition (retrieval)—workers retrieve "ant chips" that lack the larval form but do not retrieve bromine-treated larvae that have it. Next, nestmate-recognition cues release aggressive behaviors toward members of other colonies, not retrieval. Still further, solvent extraction eliminates retrieval without visible changes in body form. Finally, unless "larval behavior" means lack of behavior, it cannot play a role because dead larvae elicit retrieval as effectively as live ones. Besides, if lack of behavior is a cue, why do workers not retrieve all kinds of objects that lack behavior?

Morel and Vander Meer also argued that the worker response to brood may be learned through the reward that workers get when larvae regurgitate to them. This may indeed be true, but it does not address the question of how workers recognize the larvae in the first place. It also fails to explain why workers also treat eggs and pupae, from which they get no reward, as brood.

One additional possible brood pheromone requires mention. During eclosion, worker pupae require the assistance of adult workers to remove and consume the pupal cuticle (Lamon and Topoff, 1985). Absent worker help, they often fail to survive. Apparently, worker helping behavior is cued by a chemical emitted by pupae about to eclose. Objects treated with extracts of eclosing pupae, but not early pupae, were treated like brood by workers. Little more is known about this possible pheromone.

We still know little about the cues that direct and regulate fire-ant worker behavior toward brood. Studies using worker brood are in error, and our work was carried out only with sexual larvae. Although workers retrieve all forms of brood, this behavior may or may not all result from a common cue. We studied only retrieval, although the behaviors of workers toward brood are many and complex, including grooming, feeding, sorting, spacing/piling, and more. Brood cues are obviously an important, challenging, and understudied area.

◂•• *Ant ID Systems*

Ants don't have individual names, like Georgina Lukowski or BettyLou Deever or the like. Sure, there's *Solenopsis invicta,* but that's like *Homo sapiens*—not too useful in identifying the guy you're trying to serve with a subpoena. Ants have no phone book, no birth certificates, no social security cards. Of course, through much of their history, myrmecologists have assumed that ants don't have given names, don't have individuality, and are pretty much interchangeable, but how can we make that claim if we can't tell one ant from another? For that matter, if we can't tell them apart, at least by group, we can't even claim that older ants do different jobs than younger ones or that some ants spend most of their time in the brood area. The trouble is that not a myrmecologist alive can look at an undamaged worker ant and pick that same individual out one minute later, unless that worker has a unique mark. After that, it's easy.

Marking ants has therefore long been popular among people who want to know individual behavior patterns, or who need to find a particular ant after some time has elapsed, or who just want to develop a highly personal attachment to one favorite ant. The most obvious way to mark ants is to use paint, and this method has a long and glorious history. Myrmecologists have been dabbing, squirting, and spraying ants with colored paints almost since the field was conceived. Who knows, perhaps some Canaanite in the land that was promised to other people, bored with watching his sheep munch bunchgrass, turned his attention to the seed-harvesting ants, marking a few individuals to see how often each returned with a seed. That would seem a welcome relief from watching a bunch of bunchgrass-munching sheep.

Nowadays, of course, we use standard paints. You soon learn that the paint and the ant must match, not in color of course (what would be the point?), but in chemistry. Some ants retain the color as long as they live; others seem to shed almost any paint you throw at them. Unfortunately, fire ants fall into the second category. Their cuticle is smooth, shiny, and waxy. When painting a fire ant, you are painting wax. Too often the dabs of paint appear on the garbage heap outside the nest, and the ants are as clean and shiny as ever. So, like many other myrmecologists, we have, over the years, been steady buyers of a wide variety of paints—spray paints, lacquers, temperas, quick-dry oils, urethanes, epoxies, stains, metal flake, opaques, lakes, and inks. We always think there ought to be a better paint. That must have been on Dan Wojcik's mind when he tested over 150 of them (Wojcik et al., 2000).

How you mark the ants depends on how you answer a number of questions. Must each ant be recognizable individually? Or would it suffice merely to identify it as a member of one of a small number of different groups? Must the mark last a long time or just a few days? Are the ants to be marked large or small? How many ants are to be marked?

For marks needed for a few days, and small to modest numbers of ants, several paints and manual application will do quite well. A toothpick or insect pin is dipped into the vial of paint, and the tiny bleb that adheres is quickly touched to the desired part of the anatomy of the ant held in a pair of forceps. Small ants are dabbed under a microscope. If we only need to distinguish membership in one of several groups, we use one color for each group and dab away, one ant after the other.

When many workers must be marked, but only their group membership must be identifiable, spray painting is best. In the 1980s, Sanford Porter capitalized on the waxy nature of ant cuticle by using oil-based fluorescent printer's ink, greatly diluted in ether. Twenty years later, we are still using the same ink, from the same container. The lightly etherized ants are sprayed with an ink-ether suspension from a perfume bottle. My favorite color is Day-Glo orange, though signal yellow runs a close second. The marks are not visible unless the workers are placed under an ultraviolet light; then the light show is tremendous—a chaos of moving glowing orange or yellow polka dots. These marks last many days to several weeks. So do the unavoidable paint spots on your own skin.

Individual recognition requires a lot more effort. For example, in my study of brood-raiding behavior, each minim worker had to be uniquely marked, and each of the two opposing nests included up to 35. Four dot positions on the gaster coded for 1, 2, 4, and 7, which, singly or in pairs, add up to all the numbers from 1 to 9. A color dot on the thorax coded the 10's. I was quite proud of being able to apply these marks to ants less than 3 mm in length. Neat, clean dots were my aim, a little reminiscent of the "tighter bomb patterns" of *Catch-22*.

When the marks must be individual and really permanent, the choices are few, and paint is unreliable. John Mirenda first used fine wire tied around the worker's waist, but fine enough wires are hard to find. You wouldn't necessarily think that the ant-marking problem could be solved by a visit to Leon Scrap Metal and Recycling, but it was. Leon Metal is the type of place that buys unneeded steel scaffolding, aluminum cans gathered from roadsides by poor folks, discarded cannon-control cables from the Army, stainless steel pipe with three-inch-thick walls left over from Cape Canaveral rocket launches, old automobile engines, I-beams from demolished buildings, barrels of oily metal shavings from machine shops, junked IBM typewriters, and superseded telephone exchanges. The ultimate fate of all of these is to be reprocessed for the metal they contain, but before that any can be bought from the junkman, by the pound. A pound of telephone exchange runs about 20 cents, but all I had use for was the tiny solenoid coils wound with 38-gauge copper wire, as thin as a hair. The scale does very well on a stack of car radiators, but my purchase wouldn't budge it off zero unless I got on the scale with it. Mr. Fishbein and I settled on a dollar for eight solenoid coils, four red and four green, just one handful and a lifetime supply of ant-marking wire.

Tying a wire around an ant's waist is simple, at least in principle. First, a four- or five-inch piece is tied into a simple overhand knot, no fancy Boy Scout skills needed. The wire is held in the left hand so that one end is between thumb and forefinger; the other is clamped against the palm with the little finger. The ant, in forceps, is placed within the loop. Extending the end held by thumb and forefinger tightens it around her waist. The final tightening calls for both hands, and the loose ends are cut off with a scalpel under the dissecting scope. Voila! An ant with a colored copper belt, guaranteed not removable. Sanford Porter used this method to find ants even after they died.

Individually marking larger numbers of ants is more of a big deal. Two colors of wire belts around the petiole allow five categories, and bending/not bending the wire ends doubles the count. Putting bracelets around the legs (near the body, where the thickening of the femur keeps them from sliding off) is a major marking opportunity—two bracelets combined with six legs gets into a fairly respectable number of choices, which can be increased still more by use in conjunction with petiole belts. All this involves a lot of tense tying under microscopes but can be well worth the stiff neck that results. Besides, there's no other way. If you need to be able to distinguish MaryLou from MaryLee, as long as they both shall live, wire is the only choice.

18

Division of Labor

Labor as a Resource

Labor is a central resource in an ant colony. Through their labor, ants acquire a territory, defend it, and scour it for food and, within the safety of the nest (also a product of their toil), convert the food into reproductives to carry their genes into the next generation of colonies. Sit down to observe a colony of ants under a microscope, and you will mostly see ants working—carrying, digging, rearranging, feeding and being fed, grooming and being groomed, scouting and foraging, transporting, preparing or distributing food, culling and collecting, and many more common activities. Of course that is why we call them *workers*. Physical labor is so valuable to ants that many species steal it from others; they have evolved the ability to enslave workers of other species to carry out their own work. In the most extreme cases, ants lose their own worker caste and become parasites on the labor of other species.

Division of Labor Increases Efficiency

Myrmecologists discovered early that colony members specialize in different kinds of work, that a *division of labor* exists. This division is probably the key to the phenomenal success of social insects because it forms the colony into a team whose performance is much better than the summed performances of its members acting individually. In other words, the group is more efficient when team members are specialized.

Imagine two ant colonies, each a family. In one of these colonies, the cohabiting sisters all do the same set of jobs, and none does more of any task than of any other. The family is communal but not social. The second colony is social; a single ant is specialized to lay all the eggs, and the rest to perform particular sets of duties but not others.

At least as a first approximation, the total range of behaviors of the two groups need not be different for the second to enjoy an evolutionary advantage. Specialization does not necessarily exert its beneficial effect as a result of new abilities, although it can. It is effective because some individuals are a lot better at some tasks as a result of neural development, practice, learning, evolution of a better-adapted body form, changes in physiology, and combinations of these. The price of getting better at some tasks is to become worse at others. Once the individuals who are much better at those tasks they do, and who avoid tasks for which they are not specialized, cover the entire set of tasks needed to operate a colony, they are more efficient than the nonsocial family, in which all the ants were equally good at the entire colony repertory. This efficiency is believed to lead to increased reproductive success, that is, passing of more genes to the next generation (but it should be noted that this hypothesis has rarely been tested). When members of the family become specialists in different functions, they become the parts of a larger machine, integrated into a new and higher level of organization that achieves increased fitness, in which all share. Absent that specialization, they are just a collection of similar, independent machines occupying the same space.

For humans, too, it is effective division of labor within groups that makes us so scary—we call these groups corporations, armies, teams, hunting parties, posses, agencies, assembly lines, or factories. In human groups, cooperation contributes importantly to group effectiveness. The degree to which this is true for ants is generally not well documented.

Fire-ant Castes

Queen Caste and Worker Caste. The requirements of egg laying and brood rearing are clearly different. In those nonsocial Hymenoptera that practice parental care, the adaptations that serve egg laying must compromise, evolutionarily speaking, with those that serve brood rearing. In social Hymenoptera, these functions are separated, and each is thus free to be shaped by natural selection into the particular form and function that best serve its function. The physical consequences of this specialization separate the two basic castes of ant life, the queen and the workers. This division of reproductive labor is fundamental to and universal in all ant societies (Hölldobler and Wilson, 1990), although ant species differ greatly in the degree of divergence between queens and worker.

Fire ants illustrate a fairly advanced state of the difference between the queen and the worker castes (Plates 2A, 2B, 5A). At peak fecundity, the queen is more than 25 times the weight of her average worker. Together, her ovaries consist of about 200 ovarioles whose activity is regulated over a huge range in relation to colony size. She also has a sperm-storage organ, the spermatheca. Her large fat body stores fat and protein and provides the metabolic machinery for making yolk and other components of eggs. To accommodate the tremendous range of activity and size of these organs, the exoskeleton of the gaster forms telescoping rings. As a result, the queen's body weight can vary from about six milligrams to almost thirty. The gaster of the queen's body obviously serves the function of egg laying, which is the part of reproduction in which the queen specializes.

The other features that distinguish the queen from her workers are associated with colony founding. Her head bears three ocelli, simple eyes. Her large and robust thorax bears two pair of wings and has the typical sutured, hymenopteran structure associated with flight. Its interior is packed with the large muscles that power the wings. After the nuptial flight, wings and wing muscles are of no further use. Wings are shed and the muscles are resorbed and used to nourish the first clutch of developing larvae. The queen's average life span is about six to eight years, much longer than that of her workers. A number of exocrine glands (glands that secrete their products to the outside) permit production of copious amounts of pheromone.

In contrast, the typical fire-ant worker is smaller, ranging in weight from about one-third of a milligram to about five or six milligrams on the tarsus. Workers live only about two to six months, depending on their size and the temperature. Their thoracic structure is greatly simplified, lacking wings, wing muscles, and complex suturing. Fire-ant workers lack functional ovaries and the spermatheca. Without the constraints of reproduction, natural selection is free to hone workers into the parts of an integrated machine through further division of labor. One path to increased specialization is the evolution of a wide range of worker sizes, as discussed below.

Worker Polymorphism in Solenopsis invicta. A conspicuous feature of fire ant workers is the large variation in body size, a phenomenon referred to as worker polymorphism. Wilson defined polymorphism as substantial variation of both size and shape (Wilson, 1953a), but I, and, I believe, the majority of myrmecologists, use it simply to describe any situation where worker size variation greatly exceeds "ordinary" size variation. Like the differences between the queen and the workers, it is believed to evolve because it makes the worker force more efficient. The body size of worker ants is most commonly measured either as body weight (live or dry) or as the size of some convenient body part, most often the width of the head. The head is particularly convenient because its modifications are often associated with division of labor. Size of the head, like those of other sclerotized (hardened) body parts, is invariant throughout adult life.

The polymorphism of *S. invicta* workers is pretty typical. Workers range from diminutive in newly founded nests, barely noticeable at about a sixth of a milligram live weight and half a millimeter in head width, to the behemoths in mature colonies, making the ground quake at five to six milligrams and spreading terror with their 1.4-mm heads. The range in worker body weight during the colony life cycle is thus more than 30-fold, and of head width, almost threefold. Even within a single mature colony, workers typically range from about 0.4 to 6 mg live weight, a 15-fold range. The head width ranges from about 0.55 to 1.4 mm, about 2.5-fold.

The size-frequency distribution of workers in all but the smallest colonies is strongly skewed toward larger workers; head widths between 0.60 and 0.75 are most common. This size range is characteristic of all of the polymorphic species of *Solenopsis*—the largest workers have head widths between about 1.20 and 2.25 mm, and the modal class is much smaller. The various species of the *S. geminata* group vary in size and size range of the workers (Trager, 1991).

Many animals change shape as they grow. By shape, we mean the proportions among the various dimensions of body parts. For example, the ratio of head width to head length is a measure of head shape. If dimensions grow at different rates, their ratio changes, and we perceive that the shape of the body part changes with size. The large range of body sizes among workers of *S. invicta* invites us to ask whether body shape depends on body size. To answer, we plot the logarithms of selected pairs of dimensions against each other. A slope of 1.0 indicates *isometry* (no shape change), a slope greater than 1.0 indicates *positive allometry,* and a slope less than 1.0, *negative allometry.* The slope expected for isometric growth is 3.0 when the log of body weight is plotted against a dimension, because the volume (or weight) is the product of its three dimensions, i.e. the third power. To facilitate visualization, I will refer to the increase of the focal part associated with a doubling of body size, for example, a 2.4-fold increase.

Porter (1984) found a slope of 2.98 for body weight versus head width, essentially isometric growth, but simple inspection of fire-ant workers readily reveals several changes in proportion that Wilson described as "weakly allometric" (Wilson, 1978). We undertook a more detailed study of multiple measurements (Tschinkel et al., 2003). As an estimate of whole-body size, we used the sum of the lengths of the head, alinotum ("thorax"), petiole, and first gaster segment (the remaining gaster segments could not be reliably measured because they telescope). The total body length is therefore not quite the equivalent of the dorsal silhouette of a worker. It ranged from 2.65 to 6.16 mm.

Allometric analysis revealed a number of mild allometries, that is, changes of shape with size. Most of these are visible in the scanning electron micrographs in Figures 18.1 to 18.4. The most conspicuous changes are in the head. Head length is isometric with body length, but its widths, proceeding ventral to dorsal, are increasingly allometric—isometric just above the mandibles, growing 2.2-fold across the eyes with a doubling of body length and 2.4-fold above the eyes. Seen from the front, the head of a small worker is rather barrel-shaped and that of a large one more heart-shaped because the width of the upper regions of the head grow more rapidly (Fig. 18.1).

What might these shape changes mean functionally? The relative increase of head size, especially dorsally, may be associated with allometric increase in the mandibular muscles, in turn increasing relative

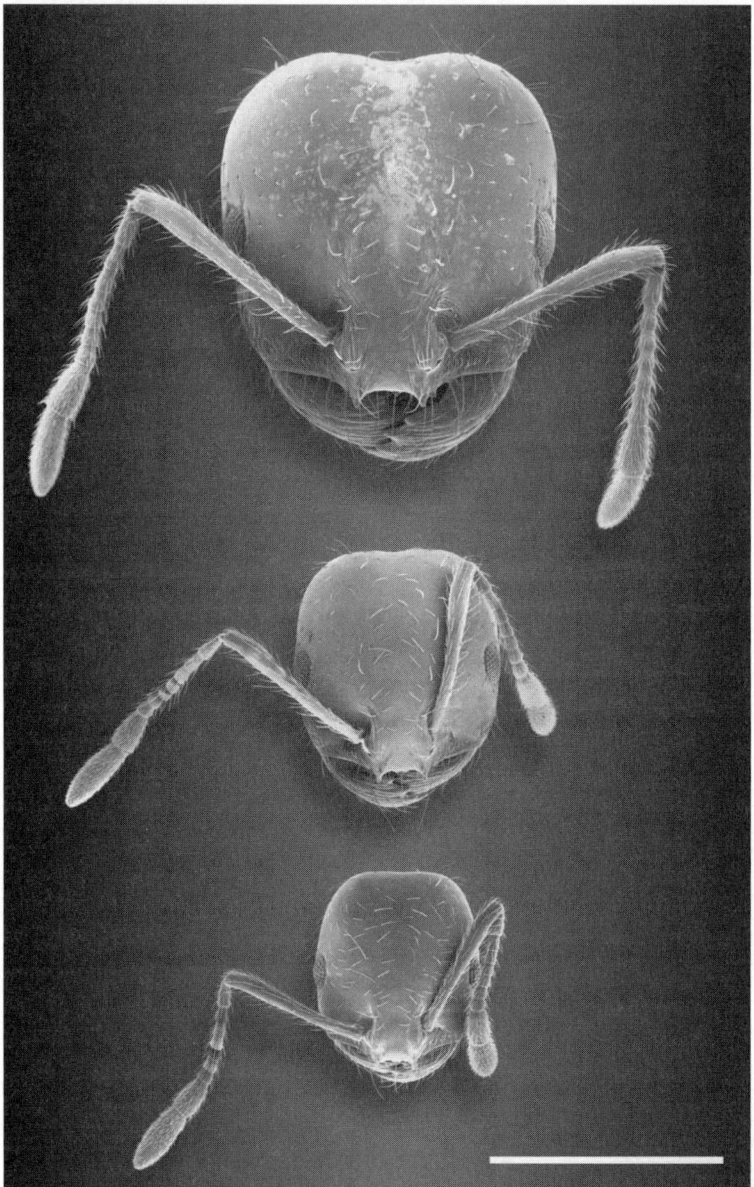

Figure 18.1. Frontal view of the heads of large, medium, and small *Solenopsis invicta* workers. Shape changes as size increases—larger heads are more heart shaped, smaller more barrel-shaped.

jaw strength of large workers or at least maintaining it against the relative weakening associated with isometric growth (Calder, 1984). Interestingly, mandible length is isometric with body length.

The antennae also change shape as workers increase in size. The entire antenna is relatively shorter in large than in small workers—doubling body length results in only a 60% increase in antennal length and decreases the antenna:body ratio by 20%. The three antennal parts do not share equally in this change. The club (the endmost portion) makes up 10% less of larger antennae than of small ones. Scaling the antenna of a large worker to the same scape (first segment) length as a smaller one (Figure 18.2) makes the relative club size and thickness easy to see.

All three pairs of legs retain a constant proportion to body length and are therefore also isometric with each other, but they are not all the same length: the legs of the first two pairs (the prothoracic and mesothoracic legs) each make up about 30% of the sum of the three lengths, and the third, the metathoracic leg, 40%, considerably longer (Figure 18.3). These relationships do not change with body size.

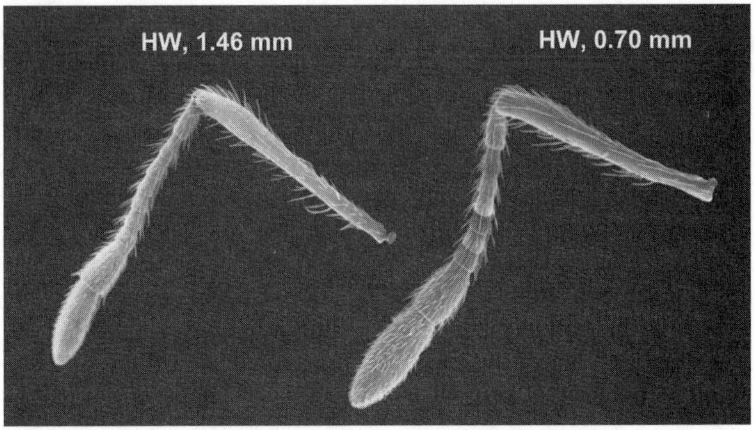

Figure 18.2. Antennae of small and large workers, scaled to the same scape length (length of the straight segment). HW, head width. The club of large workers is relatively smaller and thinner.

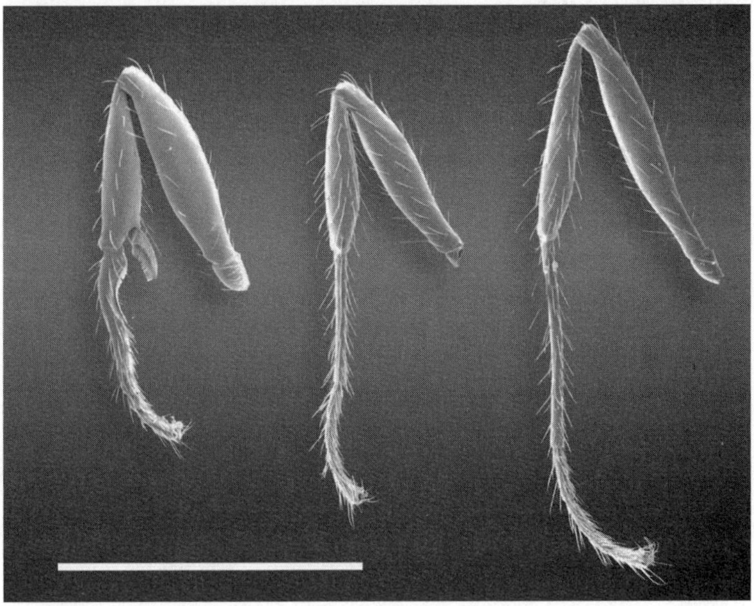

Figure 18.3. Legs of a medium worker. The three segments make up different proportions of the three legs, but all legs are isometric with body length.

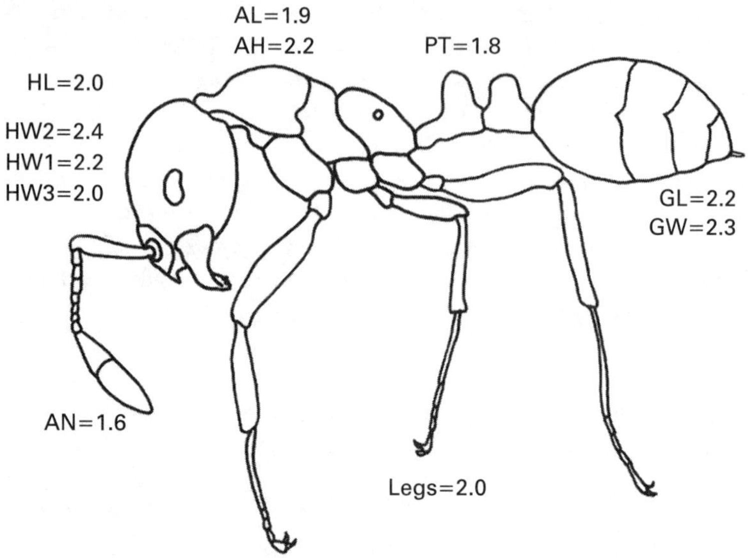

Figure 18.4. Allometric constants of the body parts of a worker for a total length increase of 2.0-fold. When the constant is 2.0, the body part in question increases isometrically with body length as a whole. HL, head length; HW1, head width at eye level; HW2 head width between the eyes and the top of the head; HW3, head width at the level of mandibular insertions; AN, antenna length; AL, alinotum length; AH, alinotum height; PT, petiole length; GL, length of the first segment of the gaster; GW, gaster width.

The proportion of each leg made up by its three segments—in order from the body to the tip, femur, tibia, and tarsus—changes differently for the three pairs of legs (Figure 18.3). The parts make up different but constant proportions of the entire leg for the pro- and mesolegs, but in the metaleg, the femur becomes relatively shorter and the tibia longer as size increases. The tarsus, although relatively long, is isometric. These reciprocal internal changes conserve the isometry of the metaleg with body length. Leg length:body isometry suggests that running speed ought to increase isometrically with body size. In army ants, leg length is positively associated with running speed (Franks, 1985), but in three species of *Pogonomyrmex*, running speed did not increase with body size (Morehead and Feener, 1998), reminding us that the functional consequences of morphological differences are often difficult to predict.

Mild allometries also characterize the major body sections (Figure 18.4). The head length grows isometrically with the body, but alinotum length and petiole length increase somewhat more slowly than body length. The alinotum itself also changes shape (Figure 18.5), because its height grows more rapidly than its length, giving large workers a distinctly more humped, robust alinotum than small workers (Figure 18.6). As with the head, the relatively more humped alinotum of large workers may accommodate a larger mass of muscle to conserve relative strength.

In contrast, both length and width of the first segment of the gaster increase considerably more rapidly than body length, causing gaster volume to increase disproportionately. If gaster dimensions had merely doubled with body length, the increase in gaster volume would have been eightfold. The actual increase is about 13-fold, and gaster volume of large workers is proportionally almost two-thirds greater than that of small ones.

A relatively larger gaster suggests a relatively greater maximum crop volume, which in turn suggests greater crop storage of food. Glancey et al. (1973c) claimed that large workers stored oil for long periods, but this finding is doubtful. Evidence that large workers imbibe more liquid food per unit weight than do small ones is rather equivocal, as discussed below. Perhaps the greater relative gaster size of large workers is for accommodating relatively greater fat stores, as large workers do.

The worker body is a mosaic of negative and positive allometry and isometry (Figure 18.4); neighboring parts and even subparts seem unlinked. This lack of pattern suggests that parts of the body can be modified by evolution separately and independently. Fred Nijhout and Diane Wheeler (Nijhout and Wheeler, 1996) pointed out that the imaginal-disc

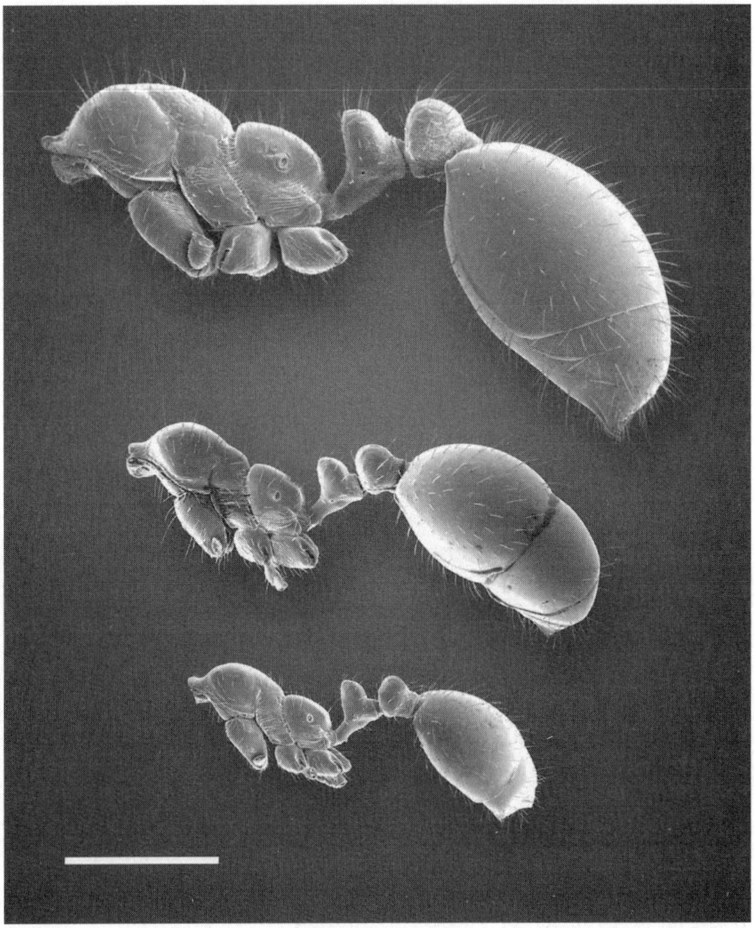

Figure 18.5. Mesosoma and gaster. As worker size increases, the trunk becomes more robust, and the gaster relatively larger.

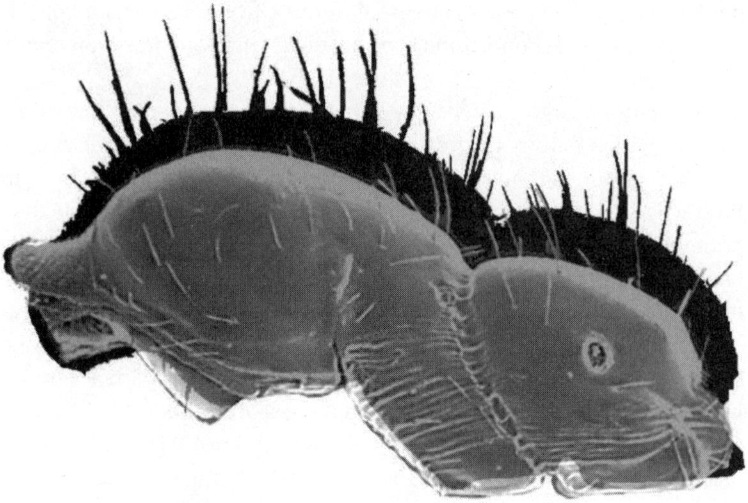

Figure 18.6. Mesosomas ("thoracic" segments) of a large (black) and a small (gray) worker scaled to the same size and superimposed, showing the relatively greater height of the former.

growth that produces the adult body in ants takes place during metamorphosis, after the larvae have stopped feeding. The imaginal discs (small clusters of embryonic cells, each of which grows into a different adult body part) therefore compete for fixed resources, and increased growth rate of one must be balanced by decreased growth by another. Such competition is expected to lead to just the kind of mosaic distribution of allometry we observed in *S. invicta*. Which particular discs grow at the expense of which others cannot be determined without experimental analysis. The imaginal discs in majors can grow much more because they are in a larger body, which has more resources. Greater growth magnifies the allometric differences intrinsic to the growth constants of the imaginal discs, but these constants do not seem to be different for major and minor *S. invicta* workers. When allometries are very large, the sapping of resources can damp growth of some parts at extreme sizes, but the mild allometries of *S. invicta* do not exact such costs. Finally, Nijhout and Wheeler's model deals only with overall disc growth, but it is apparent that the several dimensions of an imaginal disc need not grow at the same rate. Such rate differences lead to the changes in the shapes of parts.

Fire-ant colonies begin life with tiny, monomorphic workers. The patterns according to which the sizes and range of sizes of workers increase as colonies grow were discussed above, but only the question of final worker size was addressed, not that of shape. The question of how the allometries appear and develop during colony growth is still open. Do worker growth rules change with colony size, or is worker shape a function only of worker size, not of colony size? In *Cataglyphis niger*, the characteristic allometry is only established at colony maturity (Nowbahari et al., 2000). Few comparable studies exist for other species.

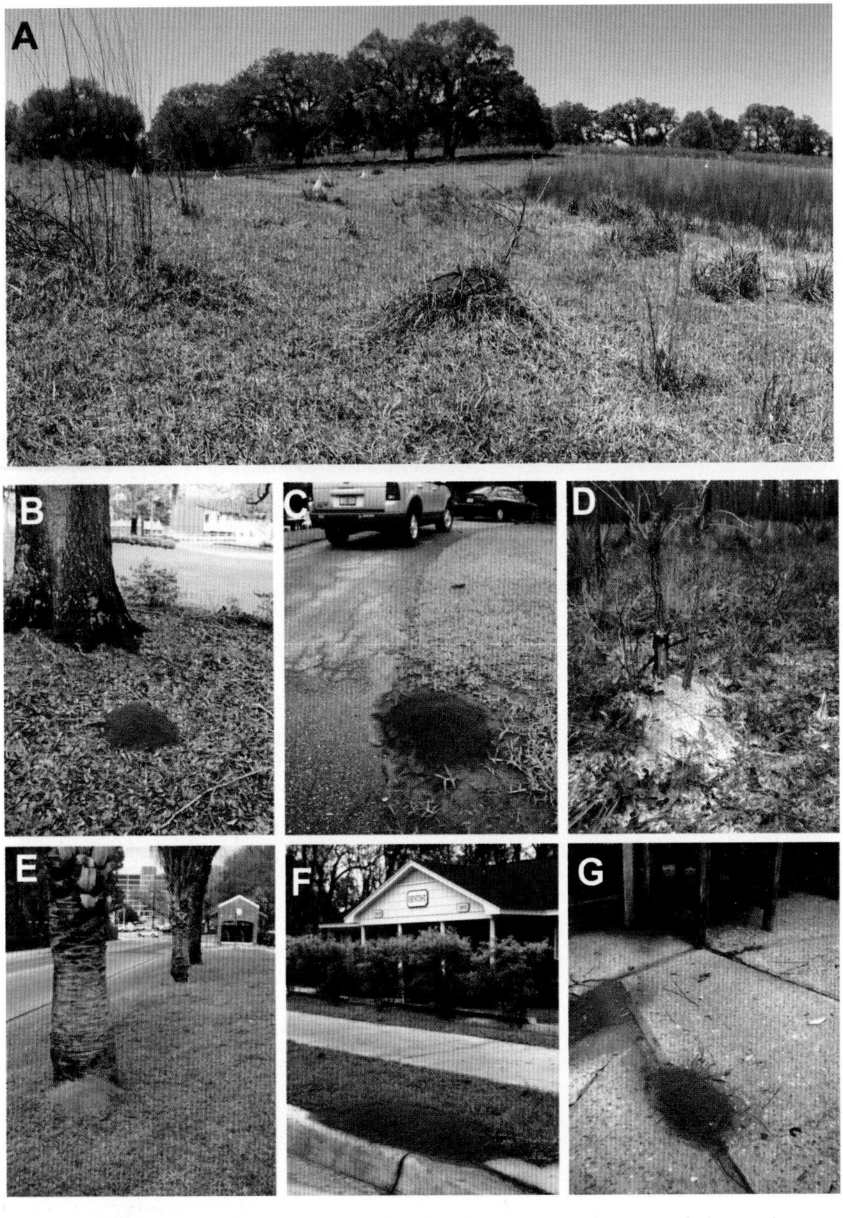

Plate 1. A. Southwood Plantation, the site of much of my research on *Solenopsis invicta* populations. **B–G.** Fire-ant mounds in various locations in both urban/suburban areas and natural pine forest of the Florida coastal plain **(D)**. Mounds are often located along edges and at the bases of trees or bushes. All photos were taken in March–April, when the mounds are large.

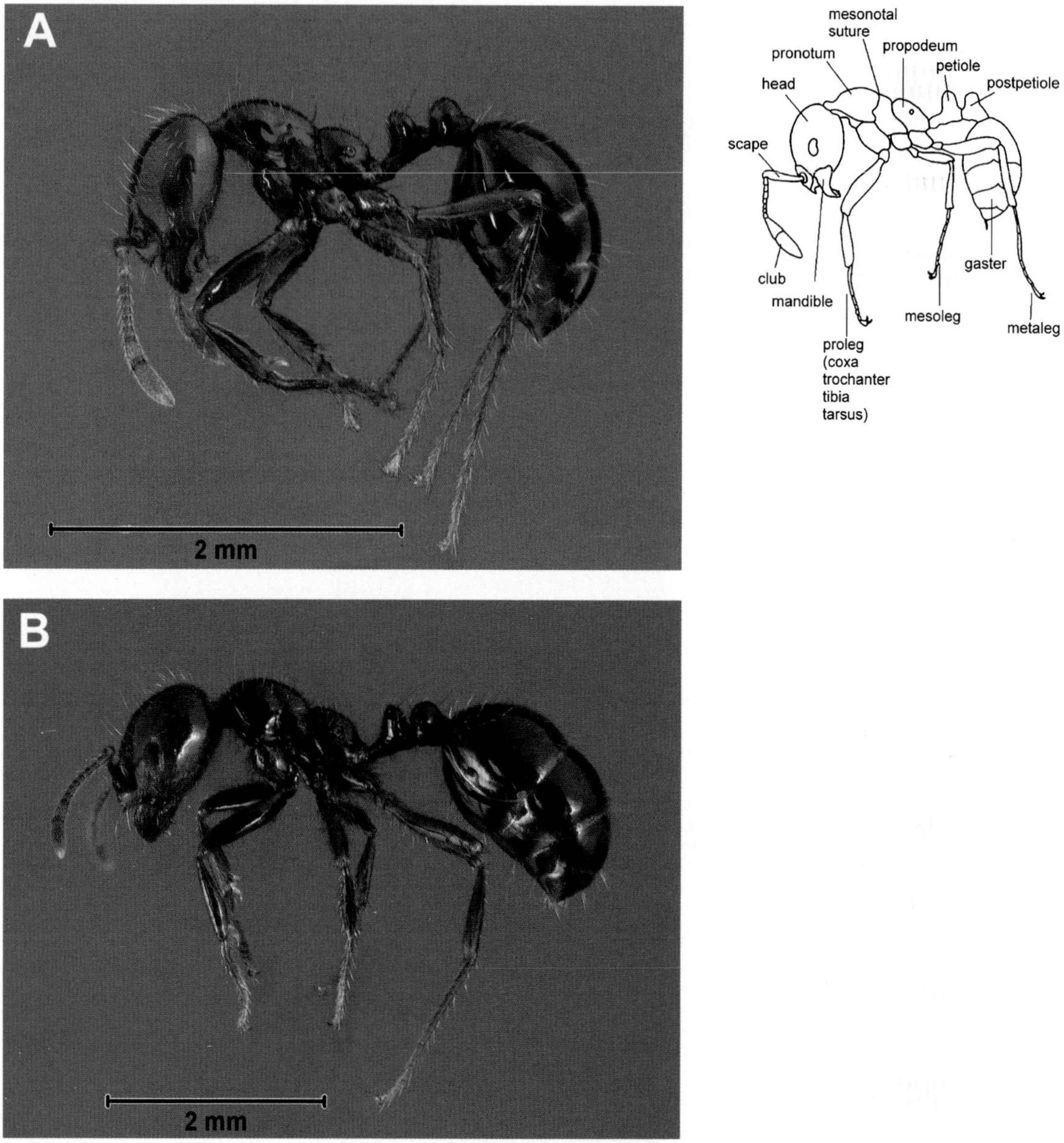

Plate 2. A small and a large *Solenopsis invicta* worker in lateral view. Major body parts are labeled on the outline drawing of the small worker.

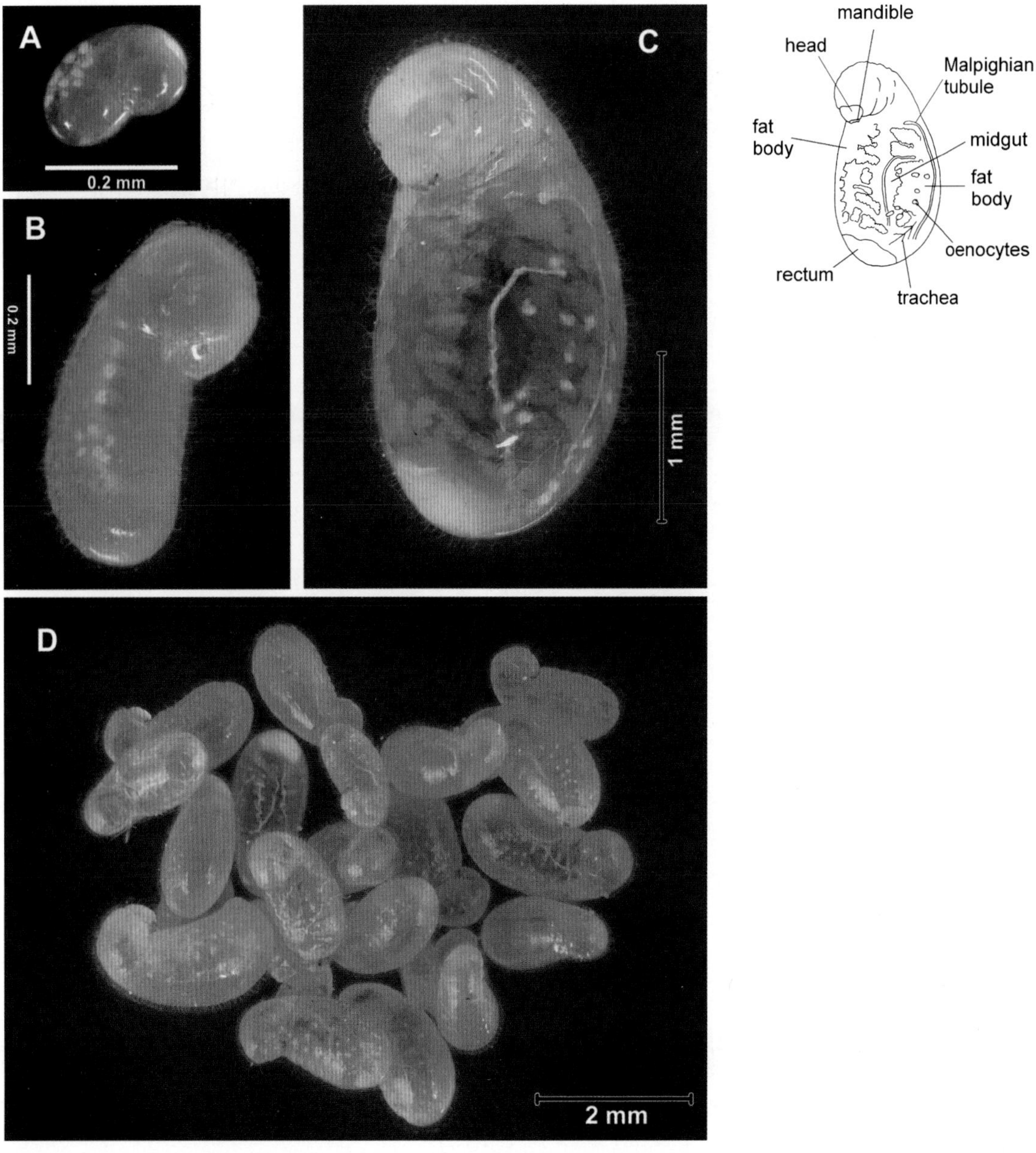

Plate 3. Larvae of *Solenopsis invicta.* **A.** A first-instar worker larva, also called a "microlarva." First-instar larvae are not segregated from the egg pile by workers. **B.** A second-instar larva. Instars 1 to 3 have unhardened mandibles and feed only on liquids. **C.** A single fourth-instar worker larva of *S. invicta.* Internal structures, including fat body (translucent clumps), tracheae (thin, branching tubes), midgut (dark region), oenocytes (round, white objects) and Malpighian tubules (long, white tubes) are visible through the transparent cuticle. The whitish mass at the posterior end is uric acid accumulated in the rectum. **D.** A cluster of *S. invicta* worker larvae. Larvae stick to each other by means of forked and recurved hairs, which can be seen in C.

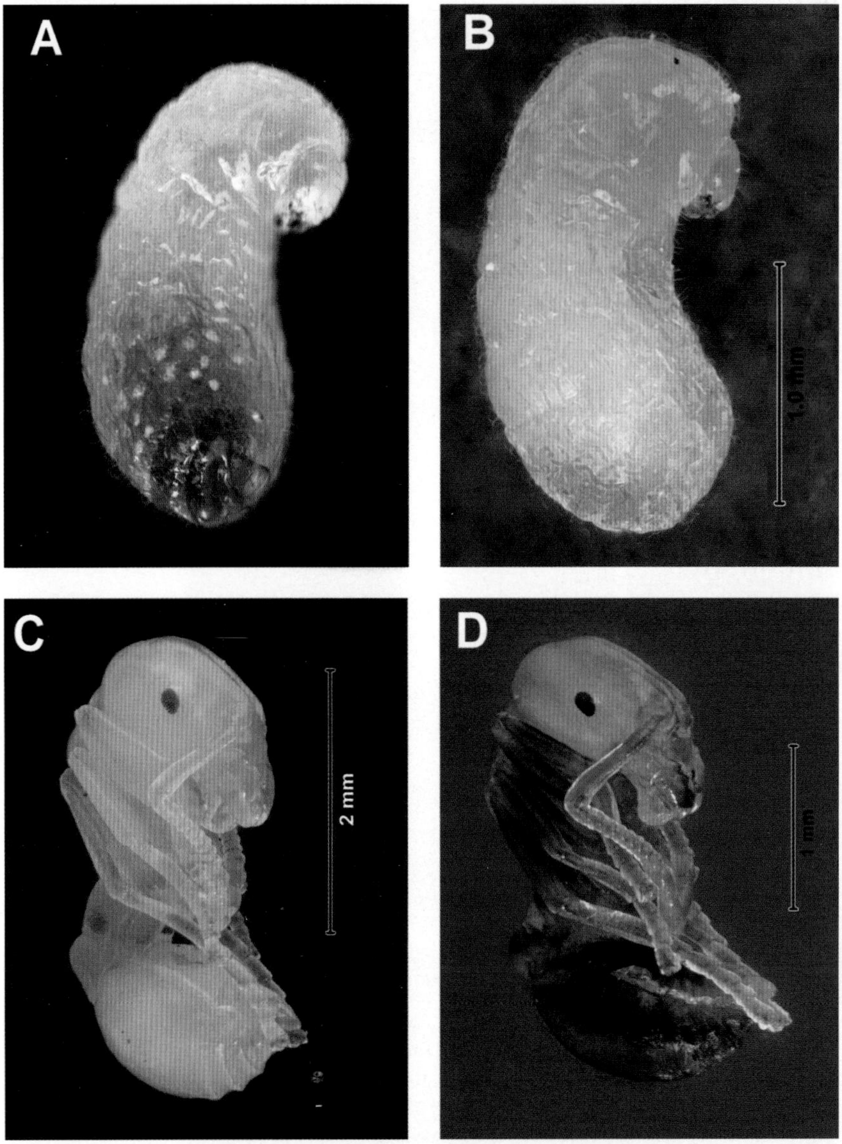

Plate 4. Metamorphosis of worker brood in *Solenopsis invicta.* **A.** A larva in the meconial stage, just after apolysis (separation of the hypodermis from the old, larval cuticle) and the formation of the new pupal cuticle. At this time, the midgut and hindgut connect, and all undigested material is moved into the hindgut to be voided as a meconium. This stage lasts only about two hours. **B.** After passage of the meconium, the pharate pupa appears shriveled, having lost so much volume. **C.** The pupa of a large worker. **D.** A pharate adult—the tanned adult structures can easily be seen through the transparent pupal cuticle. This individual is almost ready to molt into an adult.

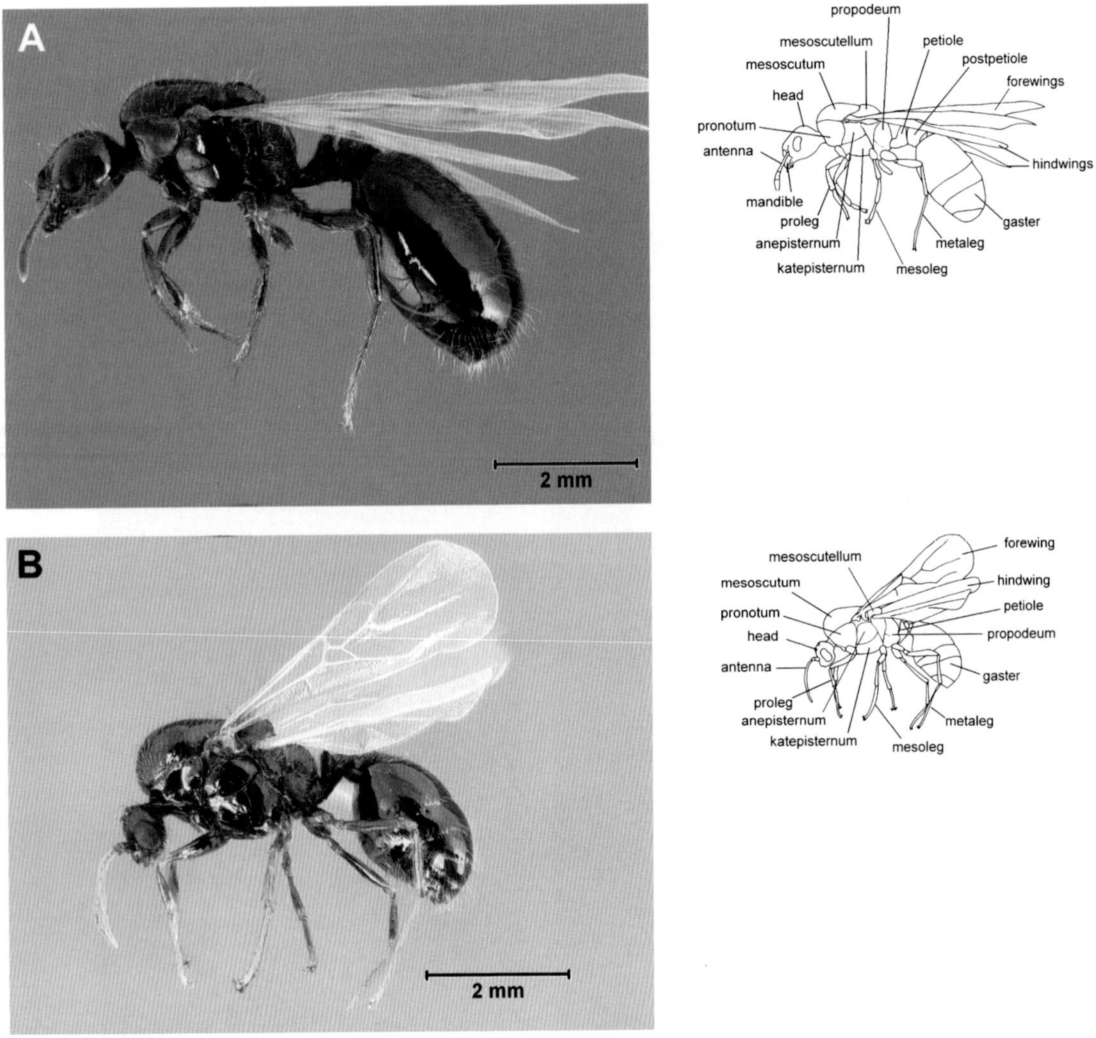

Plate 5. A. Female sexual, also called an alate or a virgin queen. **B.** Male alate. Because alates are capable of flight, the suturing of their mesosomas is much more complex than that of workers.

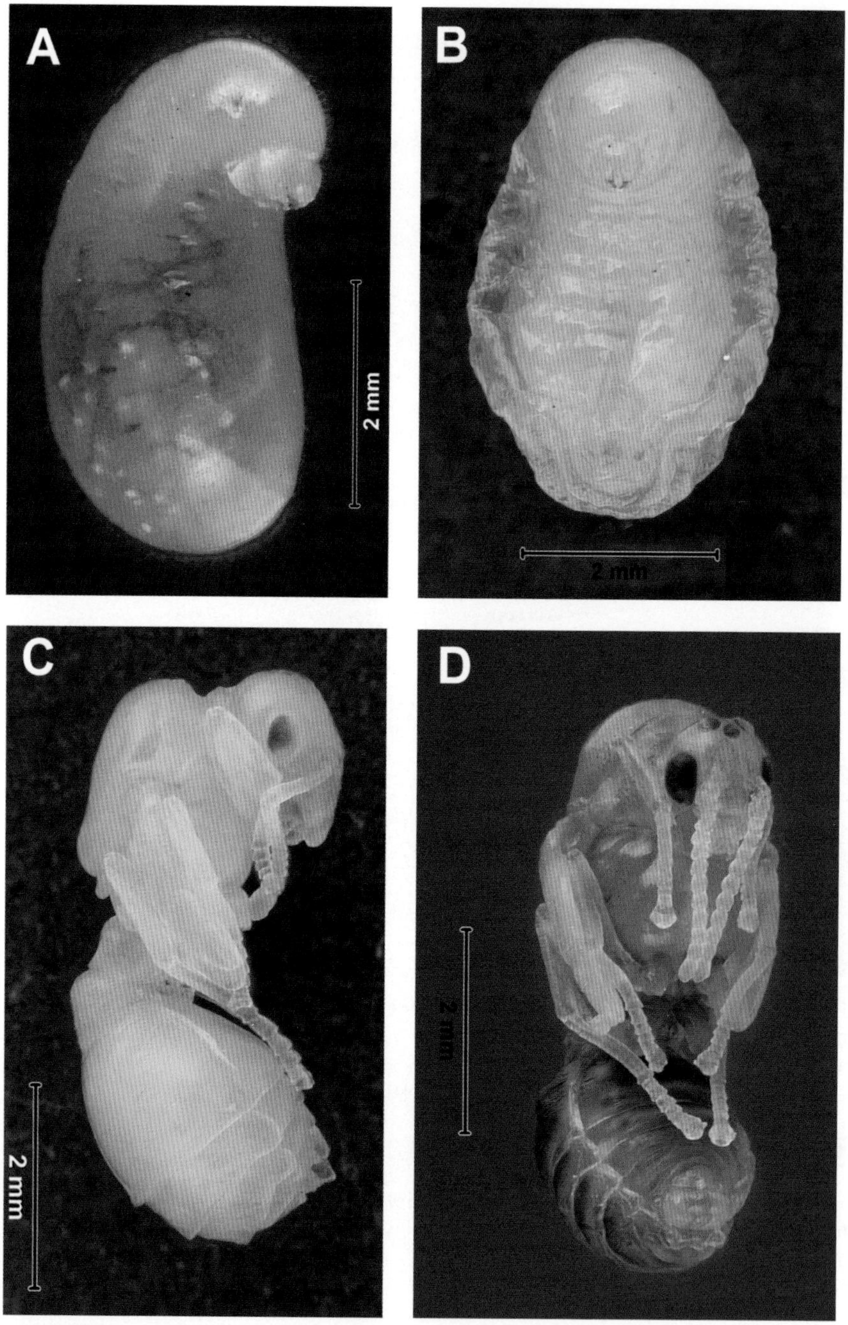

Plate 6. Sexual brood. **A.** Fourth-instar sexual larvae during the feeding period. **B.** Pharate sexual pupa in ventral view. After passing the huge meconium, the pharate pupa has a wrinkled appearance. Sex of the larva is not externally apparent until the larval-pupal molt. **C.** Female pupa. **D.** Male pharate adult.

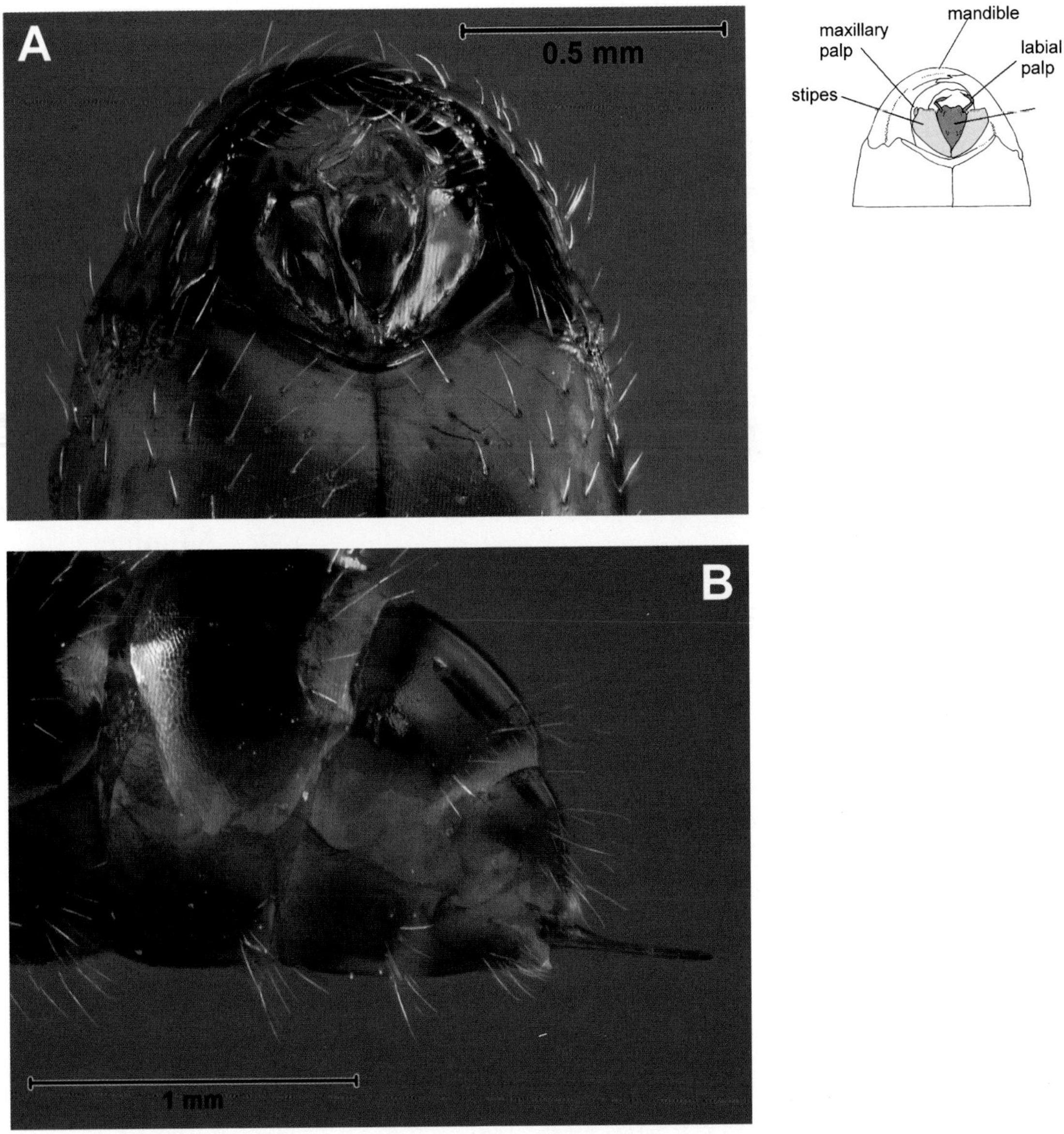

Plate 7. A. The underside of the head of a worker of *Solenopsis invicta,* showing the mouthparts. The labium and maxillae form an extrusible "tongue" through which liquids can be imbibed or regurgitated. **B.** The gaster of an *S. invicta* worker, showing the extruded sting.

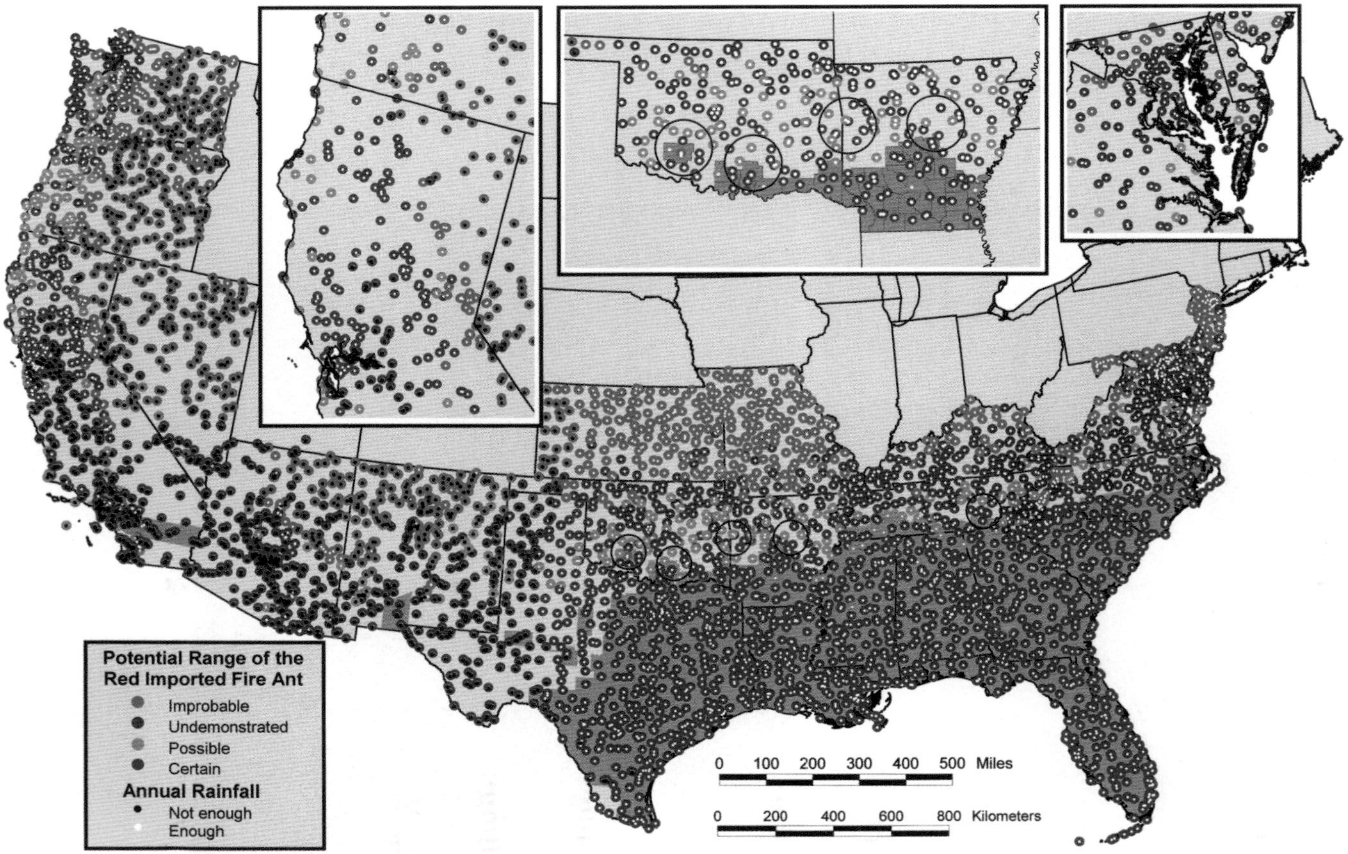

Plate 8. Potential future range of *Solenopsis invicta,* based on calculations of the effect of climate on *S. invicta*'s ability to grow and reproduce. Reprinted from Korzukhin et al. (2002).

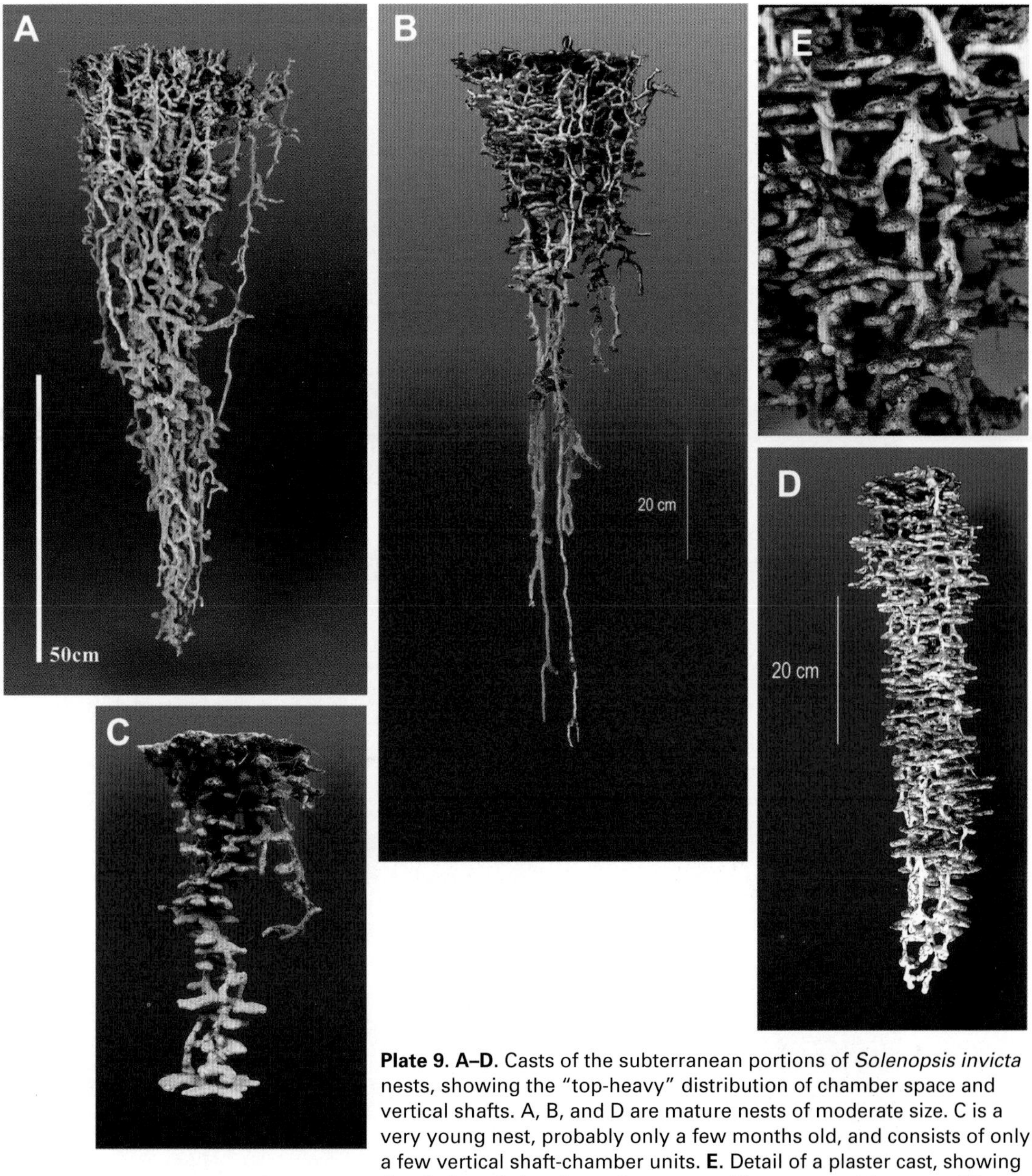

Plate 9. A–D. Casts of the subterranean portions of *Solenopsis invicta* nests, showing the "top-heavy" distribution of chamber space and vertical shafts. A, B, and D are mature nests of moderate size. C is a very young nest, probably only a few months old, and consists of only a few vertical shaft-chamber units. **E.** Detail of a plaster cast, showing the shaft-and-chamber unit structure. The nests are shown at different scales (note scale bars). A and B are zinc casts, C is cast in dental plaster, and D is an aluminum cast.

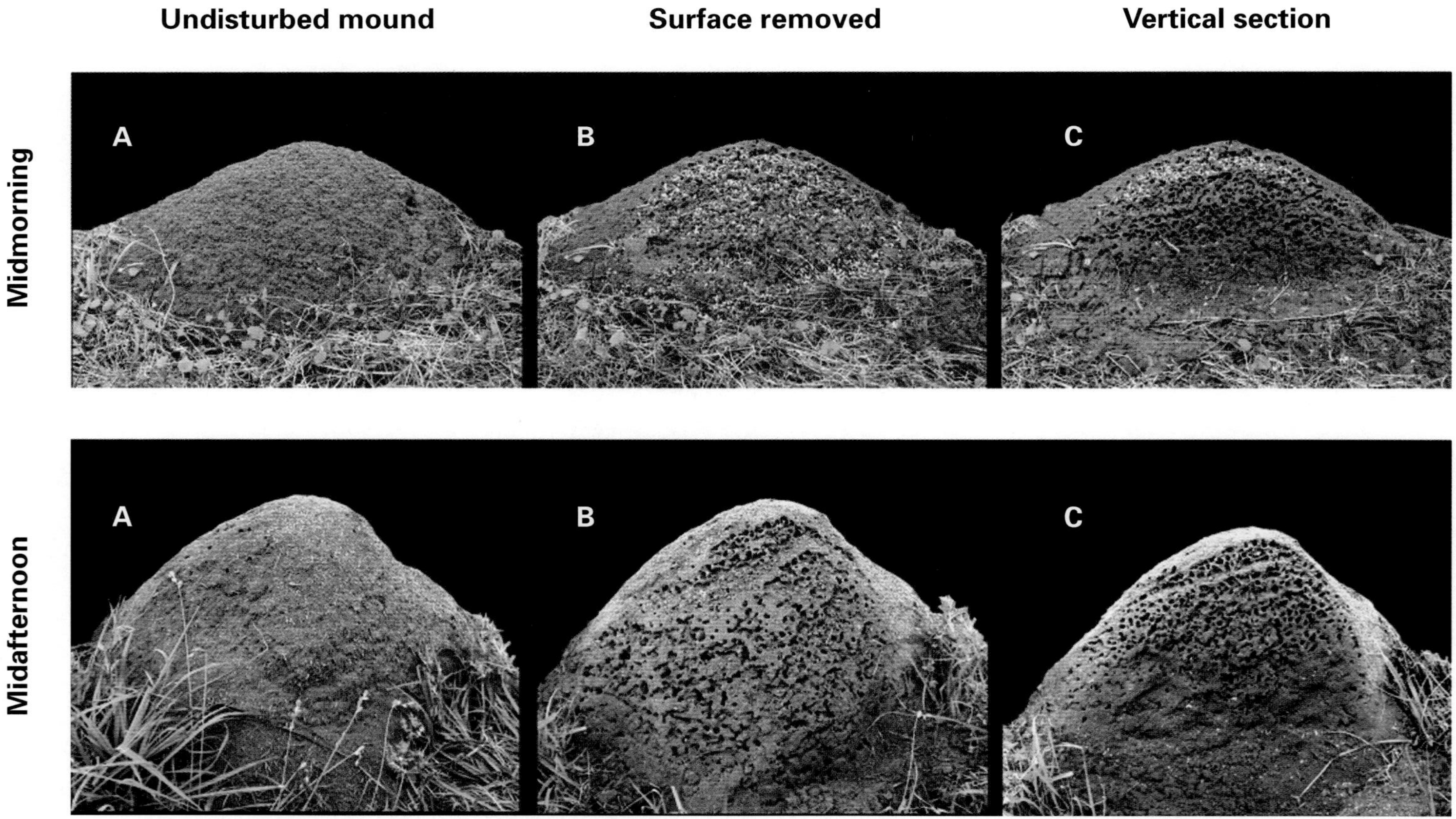

Plate 10. Two *Solenopsis invicta* mounds on a sunny spring day opened on the sunny side to show the precise placement of brood in relation to temperature. Mounds were chloroformed so that workers would not move brood during photography. Top row: a mound in midmorning; bottom row: midafternoon. **A.** Intact mound. **B.** Sunny-side mound surface removed. **C.** Mound cut in vertical section, after surface removal. In midmorning, temperature just under the mound surface was about 30°C, while in the mound core it was 23°C. By midafternoon, mound surface temperature was about 45°C and core temperature 36°C. Note the layer of brood immediately below the mound surface in the morning and the total absence of brood from the mound in midafternoon.

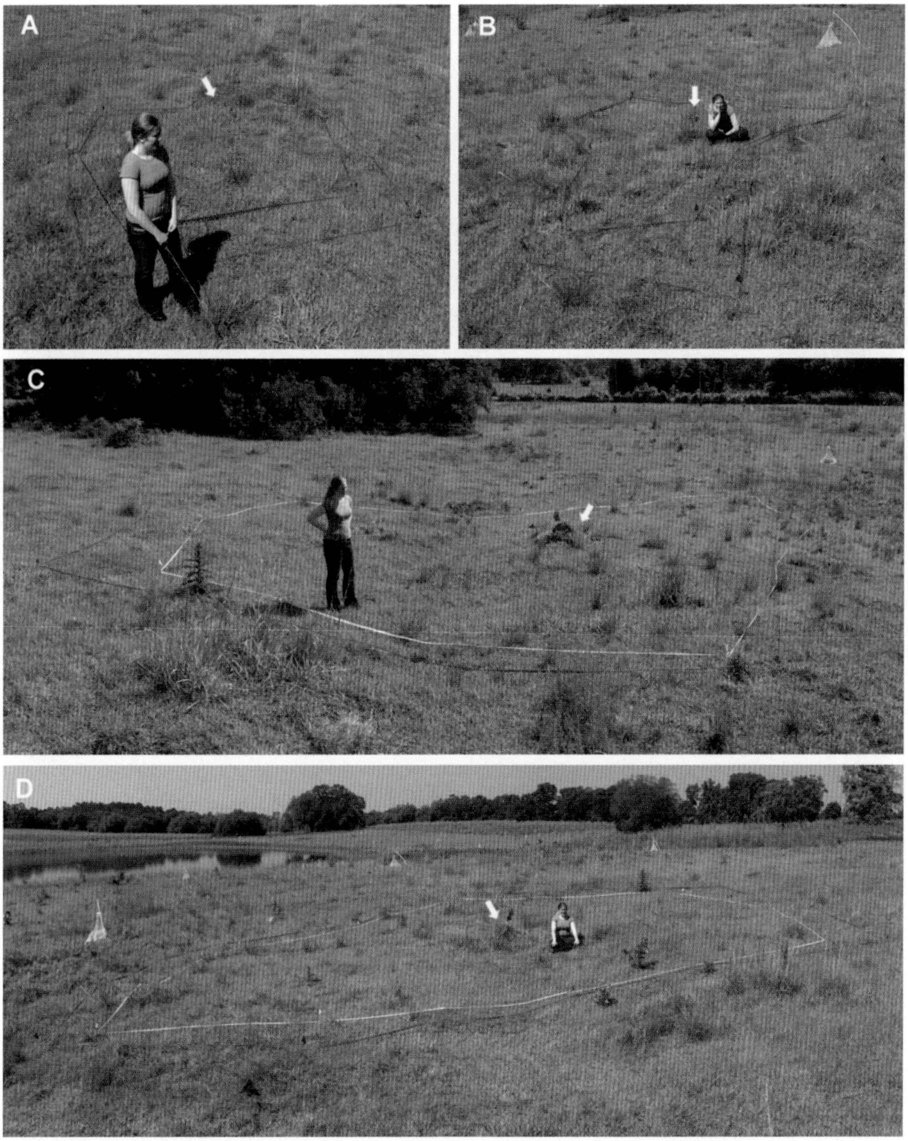

Plate 11. Fire-ant territories with their borders marked with colored tape, with Erika Tschinkel for scale. The inner tape shows the outermost positions with workers from the focal colony, and the outer tape the innermost positions with workers from neighboring colonies. The focal colonies are marked with red or blue flags and indicated with white arrows. **A.** A small territory. **B.** A middle-sized territory. **C–D.** Large territories.

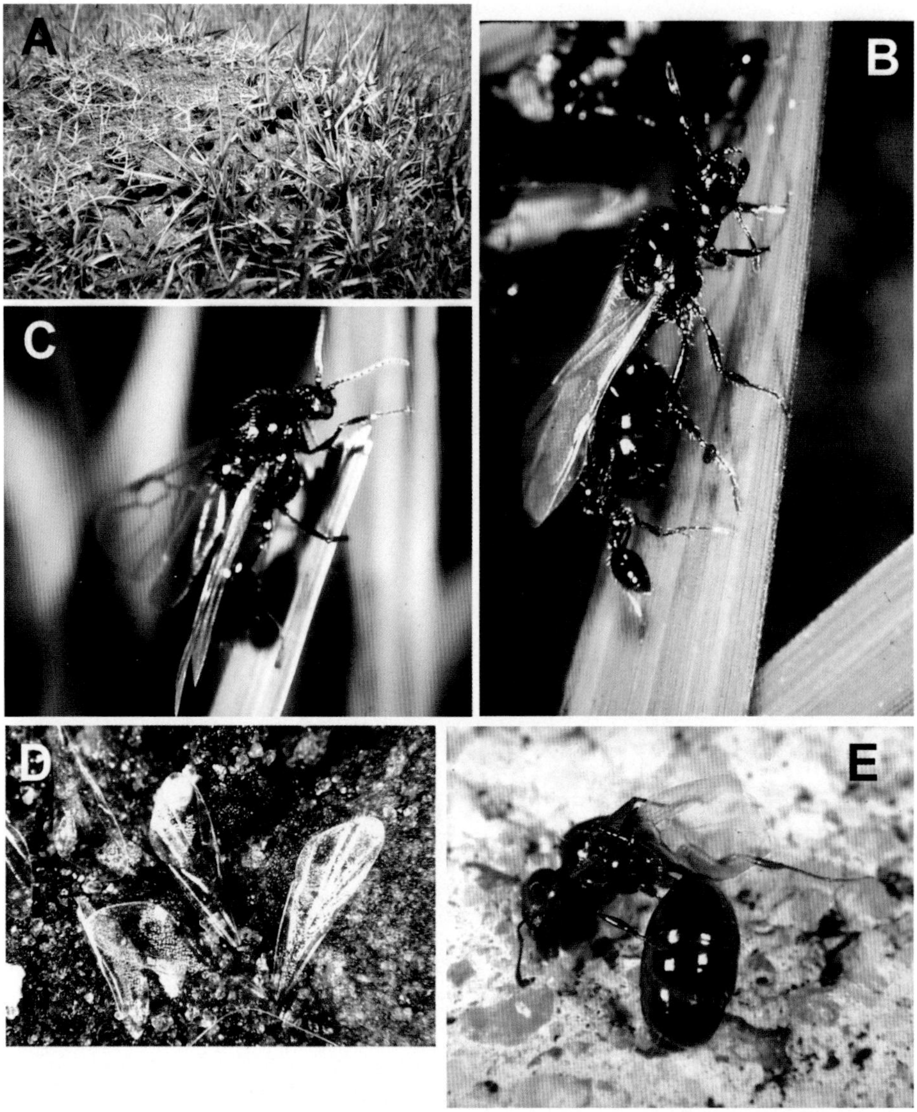

Plate 12. The mating flight of *Solenopsis invicta.* **A.** On the day of a mating flight, workers create large holes in the mound surface. **B.** A female alate preparing to take flight. **C.** A male alate preparing to take flight. **D.** When mating flights involve many thousands of queens, their discarded wings may collect in drifts. **E.** A newly mated queen breaking off a wing with her hind leg.

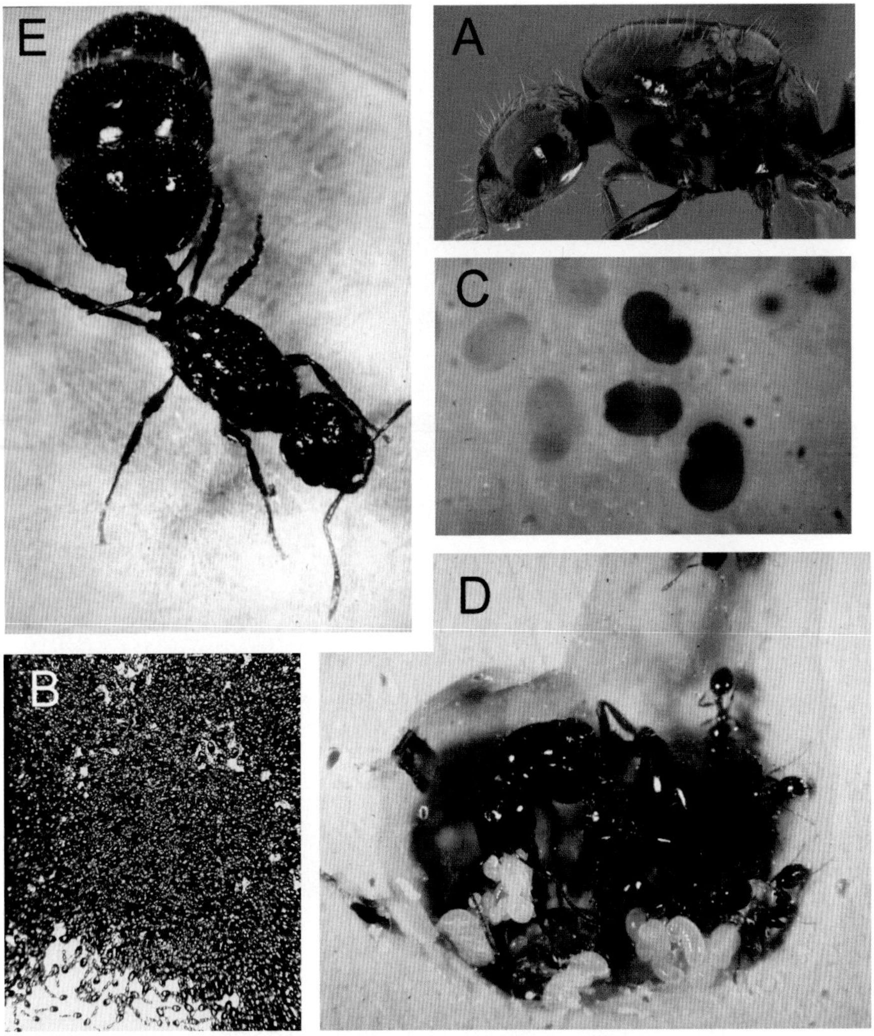

Plate 13. A. Colony founding. Close-up of a queen, showing the thoracic sutures and the wing stubs remaining after dealation. **B.** A small portion of 8000 newly mated queens collected in about two hours along a roadside in Tallahassee after a large mating flight. **C.** Trophic (light pink) and embryonated (dark pink) eggs after staining with a Feulgen DNA stain. **D.** An incipient nest of *Solenopsis invicta.* All the minim workers and younger brood were produced from body reserves stored in the queen before she left on the mating flight. **E.** A parasitically-founding queen recovered 6 weeks after being wire-marked (note the wire around the petiole) and released on the mound of an orphaned fire ant colony. She was adopted, then parasitized the labor of the host workers to begin actively laying eggs (note the swollen gaster), so that her offspring will gradually replace the host workers.

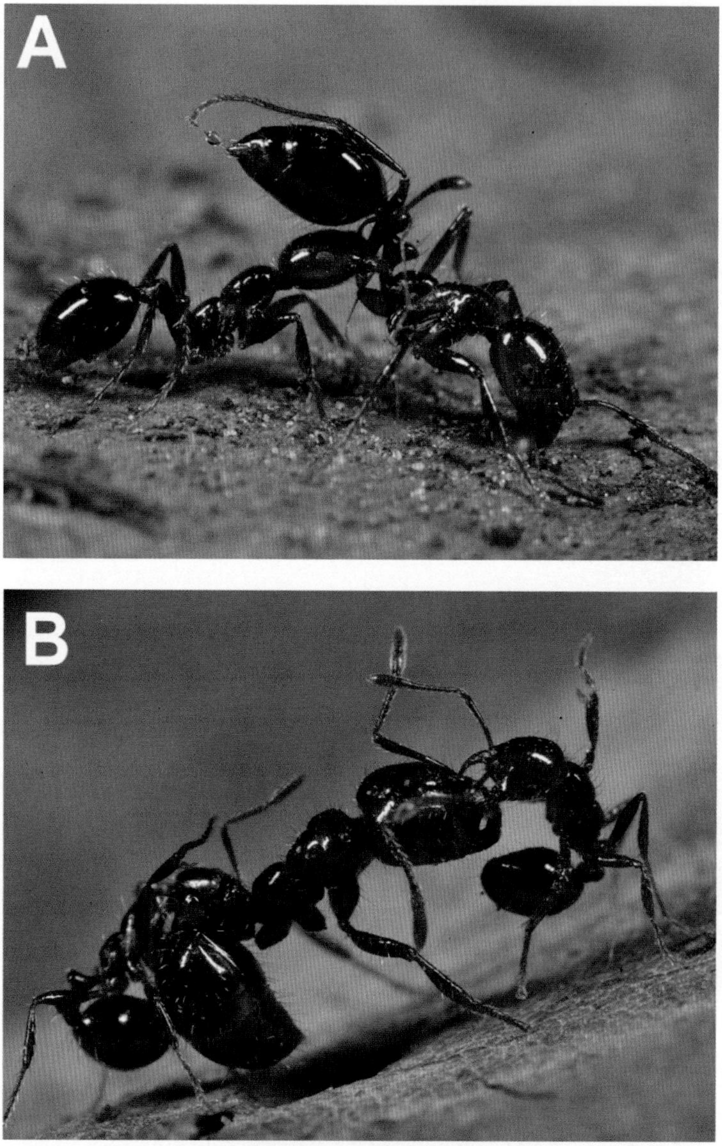

Plate 14. *Solenopsis invicta* workers fighting. Note the extruded sting with a venom droplet in **A.** The large worker in **B** has already lost several legs. Photos courtesy of Piotr Naskrecki.

Plate 15. An egg-laying queen with her retinue. Retinue workers constantly feed and groom the queen, and take away the eggs she lays. This queen was from a middle-sized colony and laid an egg every minute or two.

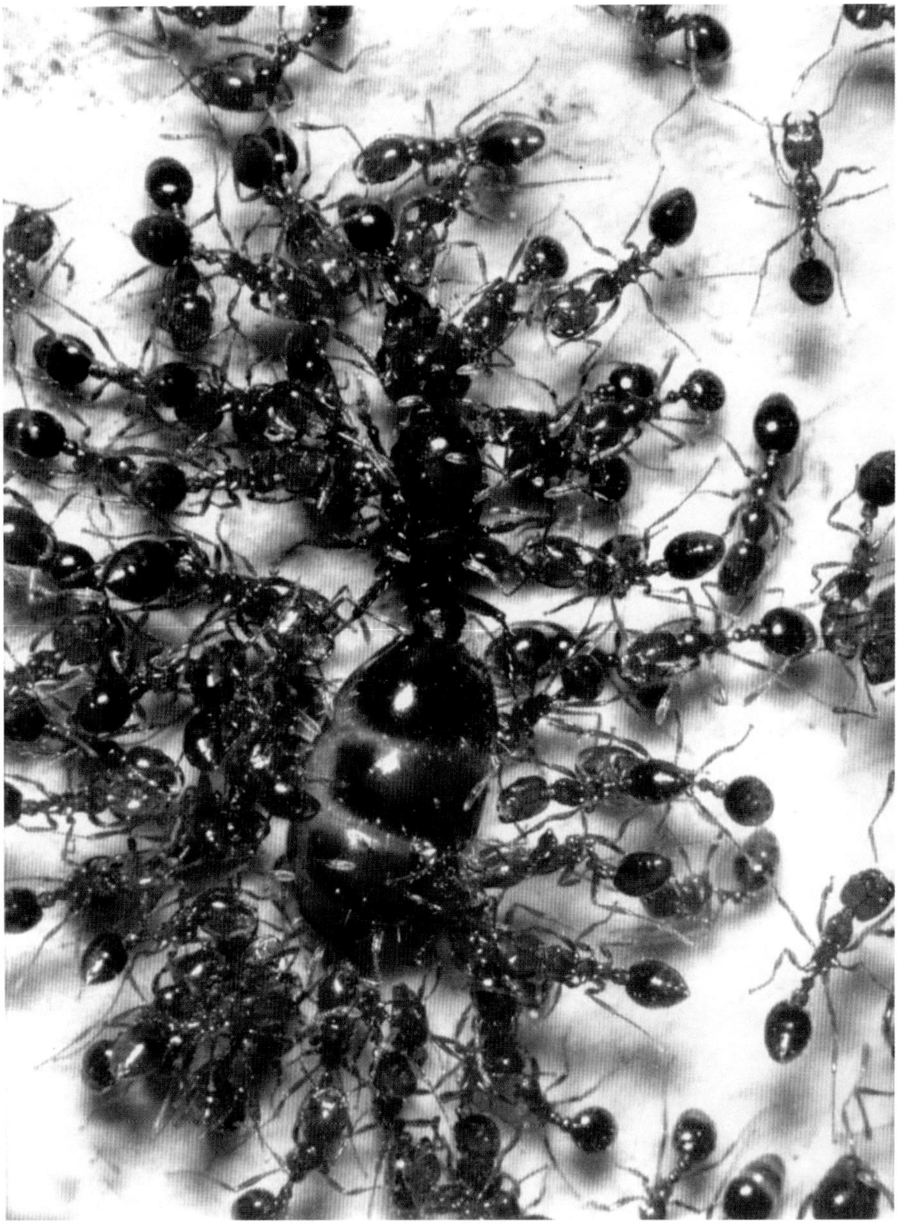

Plate 16. A. A colony of *Solenopsis invicta* afloat after heavy rains have raised the nearby pond and flooded out their nest. **B.** Cleared pasture land in the Mato Grosso, Brazil. Such disturbed habitat favors *S. invicta* in South America, just as it does in North America. Note the forest on the horizon (photo courtesy of Sanford D. Porter). **C.** The "mega" brood raid on day 3 after its discovery. The silver paint dots mark the raiding trails, and the two distant persons show the other ends of the trails.

Allometry and Polymorphism in Other Fire Ants. The workers of all but three of the 20 species of the *geminata* group of *Solenopsis* are polymorphic (Buren, 1972; Snelling, 1963; Trager, 1991). Only *S. tridens, S. virulens,* and *S. substituta* are monomorphic (i.e., vary little in size). Many species are strongly allometric as well. In particular, the shape of the head and its proportion to the rest of the body changes greatly with worker size. The more extreme the worker size, the larger is the head in proportion to the body (Trager, 1991; Figure 18.7). This trend reaches its extreme in the tropical fire ant, *S. geminata,* in which the largest major workers have huge heads that are wider than the thorax is long, the result of an allometric constant of about 1.4 to 1.5 (head width to thorax width) (Wilson, 1953a). This feature makes *S. invicta* and *S. geminata* readily distinguishable in the field. Across fire ant species, maximum head width grows more than twice as fast as maximum thoracic length. In other words, head size increased allometrically with body size during the evolution of the species of fire ants. Within each species, the relationship between head and body sizes may be either allometric, as in *S. geminata,* or nearly isometric, as in *S. invicta.*

Division of Labor Among Workers: Methods of Study. Division and organization of labor are center stage in many ant studies. Most such studies begin with the same premise, namely, that certain classes of worker are more likely to perform a specified task than others. Workers must thus be classified by some physical or physiological characteristic and possibly marked. The vast majority of studies have been observational ones,

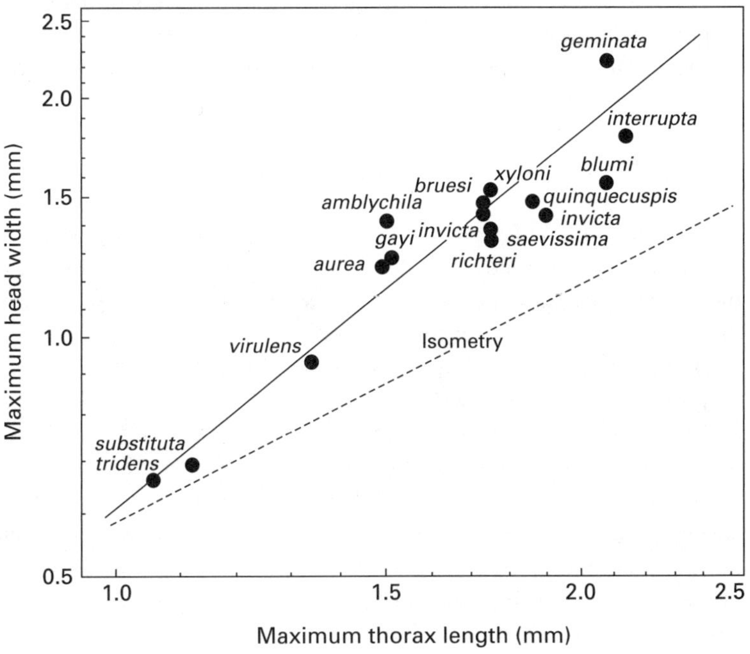

Figure 18.7. As the majors of *Solenopsis* species evolve to larger sizes, their heads become disproportionately larger. The allometric growth rules that operate within a species also seem to operate among them. The line labeled "isometry" indicates equal size-increase rates. Data from Trager (1991).

based on watching the ants during their quotidian rounds. Nowadays, videotaping followed by detailed analysis is very popular. Such studies demand time and *Sitzfleisch*, but as compensation they reward the observer with a glimpse into an alien and beautiful world.

No one has ever watched an ant for its entire life, or even continuously for long periods of time. Doing so would be a passport to the loony bin. All studies are samples, both of the ants and through time. In the "focal worker" approach, the behavior and location of sets of marked individuals are observed at specified times. Alternatively, the workers carrying out a specified "focal task" can be determined at specified times. For example, one could observe a patch of larvae to see who feeds them.

Whatever the method, the data consist of the frequency and/or duration (and sometimes the temporal sequence) with which individual workers or worker classes perform each behavior. Analysis tests for associations between the classes of workers and the performance of specified tasks. It also clusters tasks into *sets* or *roles* (for example, brood care), as well as associating tasks with locations in the nest. A significant correlation between a group of workers and a task is evidence of division of labor.

Division of Labor among Fire Ant Workers: Effect of Body Size. The foundations of our description of division of labor in fire ants were laid by Ed Wilson at Harvard University (Wilson, 1978) and John Mirenda (Mirenda and Vinson, 1981) at Texas A&M University. Early in his career, Wilson published a seminal paper on the evolution of worker polymorphism in ants (Wilson, 1953a). As a follow-up, he compared polymorphism and division of labor in the weakly allometric *S. invicta,* with that in the strongly allometric *S. geminata* in order to determine, first, whether these differences in size are associated with division of labor within each of these species and, second, whether the differences in allometry are expressed as differences in the division of labor between the species.

Wilson classified the workers in small laboratory nests of both species as minors, media (medium-sized workers), and majors, and then inventoried all of their behaviors (in rather broad categories) and their frequencies within each size class (Fagen and Goldman, 1977). *Solenopsis invicta* minors had repertoires of about 20 behaviors, similar to those of most ant species (Hölldobler and Wilson, 1990), but medium and major workers performed only 16 and 15, respectively. Smaller workers spent more time near the nest entrance and engaged more often in brood care, whereas majors were more frequent among those chewing seeds, attacking paper (a surrogate for nest defense) or alien ants, and foraging or excavating. Generally, larger workers carried larger burdens. The smallest and the largest size classes of *S. invicta* were half as likely to tend brood as the middle size class. The smallest workers did not chew seeds at all, and the largest were only 20% as likely to chew seeds as the middle group. After adjustment for their different frequencies, the types of tasks were thus unequally performed by workers of different sizes. Therefore, division of labor by body size exists, but is not sharply defined.

Solenopsis geminata was quite different. Majors of *S. geminata* performed only two acts (self-grooming and milling seeds), fewer by 13 than majors of *S. invicta*. In *S. geminata*, the minor repertory was 17 acts, and minors were the most likely to take care of brood (as were the media in *S. invicta*), the media (12 acts) were only 5% as likely to do so, and the majors never engaged in brood care. Conversely, majors were most likely to chew seeds; media were only slightly behind at 95%, and minors lagged at only 10%. In *S. invicta*, as worker size increased, the shift in the probability of performing particular tasks was gradual, but in *S. geminata*, this shift was quite sharp and much greater. The virtual dedication of the largest workers to processing seeds for consumption makes possible the substantial fraction of seeds in *S. geminata's* diet. Their exaggeratedly large, broad heads are packed with powerful jaw muscle capable of cracking even tough seeds. The price of this extreme specialization is that majors are good for little else. Such specialization was not apparent for the minors and media, whose repertories were rather similar in both species and who came closest to being the jacks-(or jills)-of-all-trades. The lesson: greater specialization is accompanied by greater differences in body size and shape and greater reductions in behavioral repertoires. The heads of workers of *S. invicta* are only mildly allometric with the body, and the behavioral specializations are smaller than in *S. geminata* and more gradually related to body size. Similar observations from many species of ants lead to the generalization that size and shape differences are always reflected in division of labor and are often amplified by differences in allometric growth (Hölldobler and Wilson, 1990).

Division of Labor in Fire Ants: Worker Age and Size. John Mirenda sought to describe division of labor in relation to both body size and age (Mirenda and Vinson, 1981) by following the lifetime careers of identified individuals in small polygyne nests. He selected 33 minor, 18 medium, and 12 major workers newly emerged from the pupa and marked each individual uniquely with a fine, colored wire (Mirenda and Vinson, 1979). Using a random sampling regime, he recorded eight different behaviors and five locations in the nest for as long as the marked workers lived. As in Wilson's study, the behaviors were sliced rather thick; all brood care was lumped into one category and all queen care into another, but donation and receipt of liquid food between adults and between adults and larvae were separated.

The most obvious outcome was that ants started their adult lives on the brood pile and generally and gradually moved from the brood to the nest periphery to entrance tube and finally to foraging as they aged.

This outward movement from the interior duties (*Innendienst*, to use the German term) to exterior ones (*Aussendienst*) is practically universal among ants, but the fire-ant workers varied widely in the rate at which they made the transition, which appeared to be closely related to differences in their size and life span. During the first 30 days of life, minors and media were more likely to be on the centrally located brood pile, and majors more likely to be in the brood-free nest periphery. Minors also first appeared in foraging areas at an earlier age (25 days) than did media (31 to 35 days) and majors (about 60 days). Having appeared once in the foraging area, a worker was seen there regularly until it died. During the last 24 hours of their lives, all workers spent all their time in the foraging arena, often dying on or near the refuse pile, a kind of self-burial. This is a level of consideration inconceivable among humans.

Further analysis showed that the proportions of time the workers spent in the various categories of behavior were highly unequal. Averaged over the entire experiment, workers spent only about 10 minutes of each half hour doing anything at all. During the remaining 20 minutes they were inactive (the idea of the industrious ant is an artifact of human culture, not a fact of ant culture). Of the time they were active, they spent 6.3% attending brood, 0.3% attending the queen, 82% grooming either themselves or other adults, and 11.3% either donating liquid food to other adults or receiving it from them.

More important, age accounted for 22 to 38% of the variation in time spent in six of the eight behaviors. Most brood care occurred before the workers were 10 days old and declined thereafter. For example, before day 10, six of 10 minor workers spent a fifth of their time at brood care but by 31 to 50 days, only two of 12 still participated in brood care, and then only for 1% of the time. Both frequency and time spent in brood care thus declined with age; the same was true for being groomed by nestmates and for queen care. These trends were correlated with the movement away from the brood to the nest periphery and eventually into the foraging arena, as already noted.

At ages greater than seven days, marked workers made the transition from the nest periphery to the foraging arena, where they foraged, and/or were recruited to help retrieve food. About 100 to 150 ants appeared at the honey water or a crushed caterpillar within 10 to 20 minutes. Minor workers mostly sucked liquid from prey, whereas media and majors cut up prey with mandibles. All sizes fed to repletion on honey water. Foragers never fed larvae directly, but donated their food to ants in the nest periphery. Within the nest, a few ants, including callows, placed solid food on bellies of larvae.

The first appearance of minors at the food (23 days) was earlier than that of media (44 days), and that of media earlier than that of majors (62 days). Age explains 56% of the difference in day of first appearance, and size 23%, but individuals were highly variable and the differences were not significant. At 50 days of age, 80% of minors still alive had appeared at food, whereas only 40% of media and 15% of majors had.

The tendency to share liquid food also changed with age. Young workers were more likely to receive food and older ones to donate it. As we shall see in a later chapter, food therefore enters the nest with the oldest workers (foragers) and moves from the middle-aged workers receiving it at the nest periphery to the younger ones engaged in brood care on the brood pile. Worker age is spatially organized in the nest as a result of the outward movement of aging workers. In addition, workers became generally less active with age and even actively avoided some behaviors, e.g., by moving away while being groomed.

Sorensen exploited the association between worker location and worker role (Sorensen et al., 1981, 1983a, 1984, 1985a, b; Sorensen and Vinson, 1985). In starved colonies, she separated and marked the three functional labor groups—she captured foragers at baits in the arena, nurses in the inner brood chamber, and the remainder (reserves) in the peripheral nest chamber. Sorensen combined these marked workers in small experimental colonies (20 to 25 workers of each type) and subjected them to experiments that often failed to yield clear-cut results. One finding was invariable, however—foragers, reserves, and nurses remained consistent during the experiments. Foragers never fed larvae, and nurses never foraged. Reserves both foraged and fed larvae, but only a minority of them did either. Reserves were therefore confirmed as the transitional group (Mirenda and Vinson, 1981). Although the results look like evidence for sharper age specialization than Mirenda described, workers were chosen on the basis of behavior, not age (which was unknown). Workers simply kept doing what they were doing, making their behavior less variable than it would have been had they been chosen for their age.

Why is it that in fire ants, as in all species of ants, the work of foraging falls primarily to the oldest workers? Foraging is the most dangerous work in the life of a worker because that worker is exposed to predators, desiccation, accident, and disorientation. The probability that a worker will die while foraging is quite substantial and can result in loss of 1 to 5% of the workforce per day (Tschinkel and Porter, 1988). In comparison, work within the factory-fortress is safe, and the probability of dying is low. In a colony that selected foragers randomly, without regard to age, mortality would act randomly, striking young and old foragers alike, and the average worker life span would be lower than that in a colony that sent out only the oldest workers. The greater worker production needed to make up for the reduced life span of random-aged foragers would directly reduce the colony's sexual production (fitness). For this reason, the oldest workers forage. They constitute a "disposable caste" (Porter and Jorgensen, 1981) and maximize the use the colony gets from its labor force. Because only the oldest workers forage, most workers must not leave the nest but have food brought to them by those that do.

Although Wilson detected significant division of labor by body size, Mirenda found only one significant difference, and even that explained only 4% of the variation. Mirenda's failure to find strong size-based division of labor is perplexing and conflicts with results of other studies, including several discussed below. As in many such cases, the studies differ in some details of material and design, so the likely reasons are difficult to select. Mirenda's colonies were very small and polygyne, and his three size classes differed in mean size and frequency from those used in several other studies. Differences in how behaviors were categorized and small sample sizes may have contributed as well. A major contributor was probably the high variation among individuals, which could be quantified in Mirenda's study but was not measured in other studies. Individual variation will be discussed further below.

In contrast, Cassill's experiments on brood care in the monogyne form found an extreme division of labor by body size. Briefly, even-aged monomorphic experimental nests of small, medium, and large workers were challenged with hungry larvae every two weeks throughout their lives. Small workers fed larvae less than half as frequently as did medium workers, and large workers hardly fed them at all. Large workers also groomed larvae much less frequently, and small workers did so more frequently than did medium workers.

Foraging tells a different story. Small workers are almost absent among workers recruited to food, medium workers are present in moderate numbers, and large workers appear in large numbers at about twice the rate of medium workers. Large workers seem to make up the recruitable food-retrieval force, and because of their greater crop storage and body fat, a food storage and processing force as well. Oddly, this food is passed on to the larvae only when smaller workers are also present in the nest. In their absence, large workers are heavily recruited to food, bring gobs of it back to the nest, but fail to feed larvae at significant rates. In the language of bureaucracy, it seems not to be part of their job description.

The tendency of workers to feed or groom larvae generally declines with age, but patterns are somewhat complex. In any case, the effect of body size on these behaviors is about six times as strong as the effect of age. Perhaps we should not be surprised—when large differences in body size evolve in ants, they are driven by greater work specialization and come at the cost of less flexibility of labor. Porter found similar effects of body size on brood care (Porter and Tschinkel, 1985b).

Let us designate the brood-tending period as a worker's "nursing career" and the foraging period as its "foraging career" (Table 18.1). The

Table 18.1. ▶ Times spent in different "careers," ages at first foraging recruitment, and life spans of three sizes of classes of *Solenopsis invicta* workers. The corresponding percentages of each class's mean life span is shown in parentheses. Mean minor head width, 0.55 mm; mean medium head width, 0.78 mm; mean major head width 1.14 mm. The larger the worker, the shorter its nursing career, but length of foraging career was much less affected by size class.

	Age or duration in days (percent of life)		
	Worker class		
	Minor	**Medium**	**Major**
Length of nursing career	10 (20%)	6 (8%)	1 (1%)
Age at first foraging recruitment	23 (45%)	44 (55%)	62 (51%)
Length of foraging career	27 (53%)	39 (49%)	33 (27%)
Total of foraging and nursing careers	37 (73%)	45 (57%)	34 (28%)
Length of reserve career	14 (27%)	35 (43%)	87 (72)
Age at death	51	80	121

larger the worker, the shorter its nursing career, but length of foraging career was much less affected.

These careers must be seen in relation to the longevity of workers, which is positively correlated with body size ($r = 0.65$). The average life span of a minor worker was 51 days, that of a medium 80 days, and that of a major 121 days, but the variability was high. Longer-lived, larger workers had longer careers. Body size and life span together explained 66% of the variation in career length. Seen as proportions of life, career lengths show some striking differences. Minors spent much greater percentages of their lives nursing than did media, and for practical purposes, majors had no nursing careers. Minors and media spend greater proportions of their lives foraging than do majors. Readers astute in math will have noticed, however, that career durations in Table 18.1 do not sum to the life spans, nor do the percentages sum to 100%. The time between the nursing and foraging careers is spent in a transitional, "reserve" career that increased dramatically with size (Figure 18.8). The highly recruitable nature of majors now comes into focus—their function is not primarily to find food but to wait in reserve (probably in foraging tunnels) until called to help retrieve it. All workers, no matter what their size, first appeared (on average) in the foraging arena when they had lived about half their lives.

A Causal Interpretation: Caste. The correlations among tasks, worker groups, roles, and locations are simply descriptive of division of labor. Naturally, we are also interested in what causes it. The caste hypothesis, as eloquently described by Oster and Wilson (1978) and in *The Ants* (Hölldobler and Wilson, 1990) implies that developmental programs drive

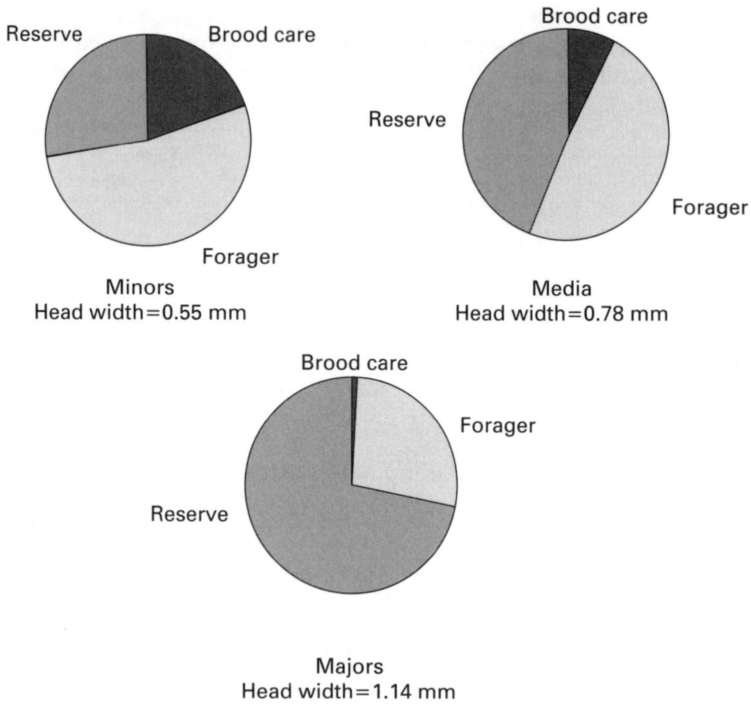

Figure 18.8. The larger the worker, the greater the proportion of its life is spent in reserve status and the less in brood care and foraging. Data from Mirenda and Vinson (1981).

role changes with age, causally linking age and body size with behavior. As a result "[evolution] can adjust the program of role changes during adult life to approach optimum ratios of temporal castes" (Oster and Wilson, 1978, p. 144). These causal links were assumed to be fairly flexible, but combinations of age and size effects nevertheless allow workers to be sorted into fuzzy-edged categories of age-based temporal castes, and anatomical differences sort them into physical castes. Fire ants include two distinct temporal castes—nurses and foragers—and the transitional reserves overlap with both. Even though the data are only correlations, this interpretation implies a necessary relationship between age and duties, however loose.

Now comes the bad news. Age and body size explained only a small proportion of division of labor. By far the largest share of variation (62 to 81%) was associated with individuals. In other words, most of what workers do and when they do it is individualistic and is not explainable by their age or body size. Age only explained 22 to 38% of variation in behavior, and size was barely significant at all (we will revisit the size issue later). Mirenda pooled data from individuals and even colonies to look for average patterns, so any insight he might have gained from following individuals through their lives was lost. Perhaps he did so because the only hypothesis on the table at the time was the caste hypothesis, which predicted that pattern should emerge from the means of grouped workers. Competing hypotheses were not offered until the early 1990's.

An Alternate Hypothesis of Causation. Similarly weak correlation between age and task in a number of ant species led in the 1990s to an alternate hypothetical mechanism, termed foraging for work, that generates division of labor (Bourke and Franks, 1995). Attributing role changes to advancing age was particularly criticized because it is based only on a correlation, and as we all learned in school, correlation does not necessarily imply causation. In addition, worker age leaves a lot to be desired as an experimental factor because it cannot be manipulated. A worker's "clock-o'-life" cannot be cranked backward or forward at will, so age and behavior will remain only correlations.

In this context, Tofts (1993) proposed that the correlation between age and task is not causal at all, but a side effect—workers emerge from the pupa in the brood pile or brood chambers and soon look for work. The closest work they find is, of course, brood care. In the course of repeatedly looking for work in a spatially structured nest, they gradually move out from the brood area, therefore changing tasks to whatever is happening in the new area. This process eventually brings them outside the nest, where they forage. Each worker is behaviorally flexible, responding to undone tasks where she is currently located. The gradual, time-related outward movement of work-seeking ants generates the correlation between age and task. The undone tasks continuously redistribute and organize labor, and the system is said to be self-organizing.

Foraging for work can even explain the task specialization that is universal among ants. Workers fixate on a set of tasks for some time, improving through practice and learning. When tasks are locally saturated, workers move to other areas, most commonly outward, because new workers are appearing in the brood area all the time. Pooling data to generate caste patterns causes individual information on the role of experience to be lost. Perhaps prior experience is important to subsequent behavior, and we should focus more attention on the "intellectual" development of workers than has been traditional. Caste, with its intrinsic programs, implies something rather robotic, whereas the role of experience suggested by foraging for work makes ants appear more intellectual, more like us. To contrast the two proposals most sharply, in the foraging-for-work view, behaviorally flexible workers interact with their environment to create both the age correlation and the spatial organization, whereas in the temporal-castes view, worker age broadly allocates labor and produces spatial organization.

I feel somewhat Solomonic in awarding each hypothesis half a baby. Are these views really incompatible? Might they both be applicable, but to different degrees, depending on the species? In situations where flexibility is low, caste may provide a more satisfying explanation;

where it is high, foraging for work would be more realistic. It also seems likely that even workers best described by their caste might forage for work within their limited behavioral flexibility. In science, point of view is often colored by the observed phenomenon and the kind of data collected, much like the blind men's view of the nature of the elephant. For example, Wilson marveled over the astoundingly huge, complex colonies of leaf-cutter ants with their sharply distinct forms of labor, whereas Tofts and Franks focused on the tiny, one-room nests of *Leptothorax unifasciatus,* where a worker can move only millimeters before bumping into the wall. Perhaps these differences might naturally predispose the one to emphasize rigidity and the other flexibility. Caste emphasizes the variation *explained* by age and anatomy, whereas foraging for work is more concerned with why so much variance is *not explained* by age and anatomy, but the amount explained or not explained varies greatly from species to species.

A third factor, however, is usually underappreciated—the centrifugal movement of workers from the brood area. Is the movement the cause of changing jobs, or is the change of jobs the cause of the movement? Under the caste hypothesis, this outward movement is part of the ant's intrinsic, age-based behavior. Under the foraging-for-work hypothesis, it is a side effect of underemployed workers looking for work. In either case, behavior and location in the nest are only correlated, just as are age and behavior. The observed temporal sequence of tasks would be generated by any mechanism that caused behaviorally flexible workers to move outward from the brood as they aged. A hint that such movement might occupy a causal position is given by the recent observations by my student Scott Powell on *Odontomachus.* In this ant, division of labor is created when dominant ants *physically drive* subordinate ants outward in the nest.

Perhaps the roles of movement and physical location are greater when the nest is large and complex, as suggested by Oster and Wilson (1978). When the entire colony lives within a single chamber a couple of centimeters across, the relationship between location, age, and division of labor may be weak. On the other hand, consider a species whose nest consists of hundreds of subterranean chambers spread through two or three vertical meters and is used for several distinct, physically separated purposes. During its lifetime, a worker born deep in such a nest can drift relatively vast distances. The correlation among worker age, location in the nest, and division of labor is likely to be quite strong.

With respect to physical castes, the two interpretations are not so far apart. The foraging-for-work hypothesis proposes that larger relative size or other anatomical differences more or less fix the worker on a subset of tasks, reducing its behavioral flexibility. This notion is not so different from that of a physical caste. When the correlation between a worker's body size and the jobs it does and does not carry out is very high, we have the essence of a caste. If the duties of a worker are the products of its work history, experience, practice, or learning, then large workers forced to perform unaccustomed labor should improve at those tasks. Fire-ant workers decidedly fail to do so, as we shall see shortly. For practical purposes, physical castes remain an experimentally useful concept.

Other hypotheses about mechanisms of division of labor can be found in the literature. The "response threshold model" has a number of adherents (e.g., Detrain and Pasteels, 1991; Bonabeau et al., 1996). It assumes that individual workers vary in their response thresholds to the cues emanating from various tasks. When the cue strength exceeds a particular worker's threshold, the worker responds by carrying out the task, which in turn reduces the strength of the cue. Colony-wide division of labor would result from patterned distribution of threshold values among the various ages and sizes of workers. Although this model can account for some observations, it has difficulty explaining the rapid alternation of workers among many very different tasks, each task presumably emitting a different cue. Moreover, the model has not been tested in fire ants.

In summary, the picture of division of labor in ants, including fire ants, is emerging in ever more detail and color. More detail will appear in chapters that follow. On the other hand, the question of how the association of labor, age, and place comes about does not yet have a satisfactory answer. The social and individual mechanisms that generate division of labor are still a mystery.

◂•• *Moving Up in a Harvester Ant Colony*

A walk in the piney woods of Florida often leads to crossing paths, literally, with the Florida harvester ant, *Pogonomyrmex badius*. Around the entrance to its nest is a conspicuous disc up to a couple of feet across. It consists of the sand excavated from the nest below, but it is often covered with a dense layer of bits of charcoal. At the outer margin of this disc is a berm of cast-off seed husks. Most casual observers of harvester ants fail to understand that the ants collect and eat a lot of seeds and lead a sort of formicine hunter-gatherer existence. The foragers hustle out into the terrain along trunk trails, which are kept clear of plants and debris, and then fan out to scour the landscape for seeds.

In spite of its name, the Florida harvester ant is an oddity in these woods, an immigrant from the western prairies that came during the cool, dry period of the last ice age and stayed after the climate became wetter and warmer. It shares that origin with yucca, prickly pear, bluestem, and wiregrass, as well as the gopher tortoise and scrub jay (Folkerts et al., 1993). It was through a study of these harvester ants in the late 1980s that I came to see the nest not just as something the ants built to live in but as the very skeleton of the superorganism, the form that organizes its functions. I wasn't the first to conclude that the architecture of the nest is an integral part of the superorganism, but the notion remained a pale ghost in my mind until I sweated through 32 excavations and saw it with my own eyes, 32 times.

Harvester-ant colonies are best excavated from a pit to one side of the nest; one exposes each chamber horizontally in turn, sucking up all the seeds, ants, and brood it may contain with an aspirator. Chamber outlines are traced on acetate film and note made of their depth and orientation. Analysis of these data yields a remarkable three-dimensional picture of the average harvester-ant colony (Tschinkel, 1999a, b, 2004). A plaster cast reveals a nest up to 3 m deep, with large chambers close together near the surface and connected by helical tunnels to lower chambers of decreasing size and increasing spacing. Chambers of intermediate depth are packed so full of stored seeds that the workers barely fit between the top of the seed stores and the chamber ceiling. A large colony may store 300,000 seeds, probably representing at least half a million foraging trips. Amazingly, the ants built this nest without a blueprint, without foremen (forewomen?), and probably even without much communication among the excavators—but that's another story.

While digging, we also noticed that some of the workers were lighter in color, ranging from pale yellow to reddish, grading finally into dark red-brown. Color is a key to age, because workers just emerged from the pupa are almost white, gradually changing to dark brown over a period of weeks. At least for a few weeks, a less than fully dark-brown color thus politely tells us that the worker is young (callow). Now we see that in the deepest chambers, which are packed full of larvae and workers, almost all the workers are callow. As we ascend toward the surface, the proportion of young workers decreases very regularly, until finally in the upper chambers, callows are rare, as are larvae and pupae. From this pattern, we deduce that the average age of workers increases steadily from the bottom to the top of the nest. The gradient from young to old stretches over 1.5 to 2.5 vertical meters.

How come? In keeping with a centrifugal pattern found throughout most ant species, workers drift gradually away from the brood as they age. In harvester ants, most brood are in the bottom third of the nest, giving the workers a long way to drift—up to 2.5 vertical meters. This association of age and location is not accidental. Young workers experimentally displaced upward in the nest move back down, and old workers displaced downward move up. Their distribution is not a meaningless consequence of having been born in the bottom of the nest. They "know" where they are supposed to be.

This vertical distribution also assures that young workers engage primarily in brood care and that, as they age and drift upward out of the brood zone, their work will change appropriately to transport, seed care, and maintenance. Simultaneously, workers also get a lot more interested in digging, until finally as they approach the uppermost chambers, they can't seem to stop digging. This change in behavior with age and location contributes to the top-heavy shape of the nest, as the most enthusiastic diggers excavate the large chambers near the surface. This phenomenon is easy to demonstrate. Capture 200 workers each from the top, middle, and bottom chambers of a nest, and pen each group to let them dig a new nest. After five days, the group from the top will have dug a nest that is three or four times as large as that dug by the group from the bottom. Callows almost never dig.

If you were trying to form a theory, or even an opinion, about whether centrifugal movement, age, and location in the nest play a role in creating division of labor in an ant colony, it makes a difference whether your study ant is so tiny that the entire colony fits in a single chamber no bigger than your thumbnail or a colossus that builds subterranean skyscrapers 500 to 1000 times as tall as the ants themselves, with 100 to 200 separate chambers and one to four helical elevators. Travel distance alone would favor a division of labor based largely on age, location, or centrifugal movement under these circumstances. Might the roles of these factors differ in different species, and should we expect to find that no universal law governs how division of labor is generated? Once you have excavated a harvester-ant nest, the scale and distances of their enterprise haunt you, and the connections between the social structure and the nest structure seem obvious. Might not evolution have used a fairly large hammer to organize their labor, more in keeping with their scale and structure? Would random wanderings of bored, unemployed workers be sufficient to create both the nest and the social structure of *Pogonomyrmex badius*? When one is trying to decide what makes ants tick, perhaps the choice of study species does matter.

19

Adaptive Demography

At the beginning of the previous chapter, I proposed that division of labor evolved because it made colonies more efficient and that the energy thus saved could be invested in increased survival and/or sexual production (fitness). Although the behavioral flexibility of workers has limits, this proposal would imply that the frequency distribution of worker ages and sizes (colony demography) should make a difference to colony success and should be adaptive. It also implies that natural selection should move the mixture of labor types toward some optimum set of proportions. The adaptive nature of demography in social insect colonies is in contrast to the demography of populations of nonsocial animals. In the latter, the age and size structure is a meaningless epiphenomenon that results from selection at the individual level. Adaptive demography is widely accepted by myrmecologists, but unfortunately it is essentially still a working hypothesis because it has rarely been directly tested (Schmid-Hempel, 1992). Undeterred, and perhaps operating on intuition, most myrmecologists report that division of labor *seems* to make colonies more efficient, but experimental studies are few. Testing the hypothesis requires manipulating the proportions of the demographic classes and measuring some output related to fitness, ideally alate production directly. Published tests are indirect, and some failed to demonstrate the fitness value of adaptive demography.

Fortunately, ant-colony demography is easily altered. Unlike ordinary organisms, which tend to respond to dismemberment and reconstitution by dying, the workers that make up the superorganism can be recombined in any proportion of size or age, as though we were mixing chemicals. Hölldobler and Wilson (1990, p. 335) charmingly described this "pseudomutant technique": "The procedure . . . is roughly comparable to an imaginary study of the efficiency of various conceivable forms of the human hand in the employment of a tool. In the morning we painlessly pull off a finger and measure performance of the four-fingered hand, then restore the missing finger at the end of the day; the next morning we cut off the terminal digits (again painlessly) and measure the performance of a stubby-fingered hand, restoring all the fingers after the experiment; and so on through a wide range of variations and combinations of hand form. Finally, we are able to decide whether the natural hand form is near the optimum."

Testing adaptive demography in brood-rearing

Adaptive demography was tested in my lab by Sanford Porter, as part of his Ph.D. research. Like others before him, and in the interests of finishing his Ph.D. in less than a decade, Porter bowed to practicality and chose worker brood production as a proxy for fitness, reasoning that higher rates of worker brood production eventually lead to larger colonies, which, in turn, produce more sexuals (Markin et al., 1973; Tschinkel, 1993b) and produce them earlier (Vargo and Fletcher, 1987). Use of this measure also permits use of laboratory experimental colonies, which are too small to produce sexuals.

Mature field colonies were separated from the soil and sieved into multiple size classes whose extremes differed by twofold in average dimension and eightfold in weight. The four experimental treatments were: (1) a standard polymorphic colony in which the size distribution approximated that of a mature field colony and three monomorphic, "pseudomutant" colonies containing the same weight of (2) small workers (head width 0.68 mm), (3) medium-sized workers (0.86 mm), and (4) large workers (1.24 mm). Running each of the four treatments with colonies containing 0.75 grams or 3.0 grams of workers separated the effects of polymorphism and colony size, but these effects are independent, and I will discuss only the results for polymorphism here. Each of the eight experimental nests was given a foreign queen. This design was replicated seven times with workers from seven different "source colonies" from the field. This "blocked" design allows inter-colony variability to be separated from the variation caused by the treatment variables.

Workers that died were replaced with similar-sized workers from the original source colony so that experimental colony size remained constant. For two months, about two brood cycles, the experimental colonies then fed as much as their tubular little hearts desired and reared brood. Because Porter could only do one or two replicate sets at a time, it took two years to complete the experiment. For a given weight and demography of workers, the amount of new biomass produced in the second month (g brood/g worker/month or g brood/individual worker/month) was an estimate of efficiency and, down the line (we presume), fitness.

Brood Rearing Efficiency by Weight and Number. The results? Polymorphic colonies produced brood at about the same rate (g/g/month) as monomorphic colonies of small workers. Monomorphic colonies of medium-sized workers produced about 30% less brood, and large workers produced almost none at all (Figure 19.1), confirming that large workers participate little in brood care. Rarely

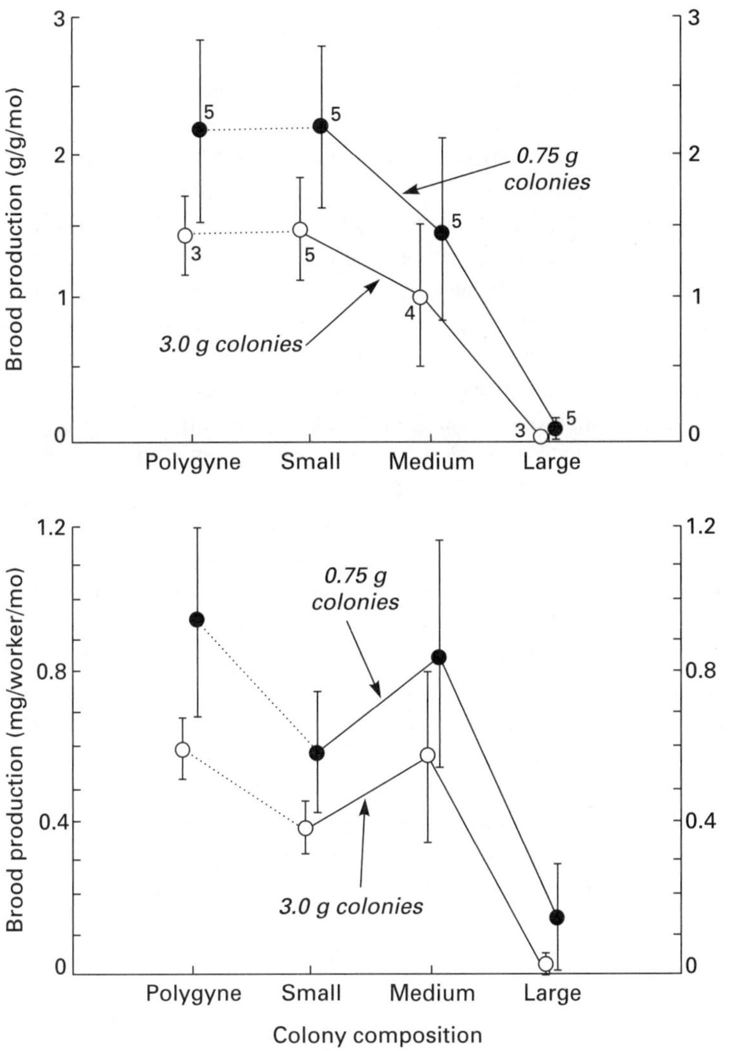

Figure 19.1. Brood rearing by polymorphic and monomorphic experimental colonies of *Solenopsis invicta.* Numbers beside points indicate the number of colonies on which each point is based. Brood-rearing competence decreases with worker size. Per unit weight, small workers are as effective as polymorphic mixtures. Large workers are almost incapable of rearing brood successfully. Rearing efficiency decreases with colony size (compare 0.75g colonies with 3.0 g ones). Adapted from Porter (1984).

did large workers rear larvae past the third instar, and their housekeeping left much to be desired—moldy food and waste littered their nests. Even after two months of "opportunity" to practice brood care, the performance of medium and especially large workers remained below the levels of small. The association between anatomy and labor seems to be inflexible, at least over a large range of body size. Physical castes are real entities.

Here then is an important question: Is the sum of the separate production of the sizes of workers (proportional to their representation in polymorphic colonies) different from their production when together? Worker size classes produced about 40% more (g/g/month) when together than when apart. The polymorphic worker groups are indeed more efficient when they are working together. This is an important insight.

Greater efficiency is even more apparent on a per-worker basis. In this approach, small, medium, and large workers are counted individually rather than by weight, and adjustment is made for the differing numbers of workers in the colonies. Under these conditions, polymorphic colonies produce significantly more brood per worker than do monomorphic colonies of small and about the same as those of medium workers. Large workers produced little brood, even on a per-worker basis. It would be desirable to determine efficiency in colonies with equal numbers (rather than weights) of workers in all four treatments, because we saw earlier that production rate decreased with larger numbers of workers, but this experiment has not been done.

Where Is the Problem? Were the differences in brood production described above associated with particular steps in the complex rearing process? Porter set up experiments with the same four worker-size treatments but challenged them with brood in different stages. In the first experiment each experimental colony was given a new queen and one gram of fourth-instar larvae (which are needed to stimulate egg laying in the queen). In this situation, the queen's egg-laying rate was determined by how effectively the workers transmitted the larval fertility-stimulating factor to the queen. Four days after addition of the larvae, the fecundity of all queens was assessed.

The queen's ovarian development estimates egg-laying rates integrated over a longer time. Dissection showed that queens in polymorphic nests had the most vitellogenic (yolk-depositing) egg follicles, followed by queens in monomorphic small-worker colonies. Large-worker colonies brought up the rear, as usual. Queens in polymorphic colonies averaged almost two and a half times as many follicles as those in large-worker colonies. Egg-laying rate declined with worker size but not quite significantly. I confirmed these patterns in later, unrelated experiments (Tschinkel, 1988c). Clearly, the brood-rearing differences described above begin at the very beginning, with egg laying.

The next two experiments tested egg-rearing and larva-rearing ability. The same four treatment groups were kept queenless, and each nest was provided with 50 mg of fire-ant eggs in the first experiment or a fixed weight of early fourth-instar larvae (the stage that does most of the growing) in the second. Workers reared both of these to the pupal stage as best they could. Whether beginning with eggs or larvae, large workers once again lagged behind polymorphic, small, and medium workers in the production of pupae. The differences were not as great when colonies started with larvae, as might be expected because the larvae were already midway through their development.

Taken together, these results make clear that the effect of worker size on brood rearing is not limited to one step of the process but seems to act on all aspects, from stimulating egg laying to rearing the last larval stage. Limited brood-rearing ability is particularly characteristic of the largest class of workers. Simply put, rearing brood is not part of their "job description." These workers must be specialized for tasks other than rearing worker brood. In contrast, the small workers and, to a lesser degree, the medium-sized workers are brood-rearing generalists, managing just fine over the whole range of tasks. We shall see below that even in this size range, preferences for different parts of brood care are apparent.

Calculating Energetic Efficiency. So far, we have looked only at the production side of efficiency, not the cost. Workers represent an energetic investment in the machinery of production. A more complete accounting of efficiency must include the production, maintenance, and replacement costs of the worker force during the month-long brood cycle. Several diverse measurements of the same three sizes of workers at several temperatures must be combined to produce this estimate: (1) their energy content (production cost); (2) their average life span, and (3) their average respiration rate (Calabi and Porter, 1989). Multiplying life span by respiration rate yields the lifetime maintenance cost. Multiplying the daily respiration rate by 30 days gives the monthly maintenance cost by size class. The cost of activity was not estimated.

Worker life spans determined in broodless nests are not reliable because such workers carry out few normal functions. Calabi and Porter therefore marked newly emerged workers of three size classes with wire belts around the petiole and made regular searches of the dead piles for the marked ants. Because ants are ectotherms, these observations were also made at three different temperatures (17, 24, and 30°C.). As expected, longevity declined almost 80% between 17 and 30°C, but this point is less interesting here than the effects of worker size. The large workers lived between 1.5 and 2.5 times as long as the small workers, depending on temperature; medium workers fell in between (Figure 19.2). The larger the worker, the longer it lived, on average, but individual variation was considerable (coefficient of variation, 40%).

The production cost of workers is simply the energetic content of their bodies, which we determine by burning a pellet of dried, weighed ants in a calorimeter and measuring the heat released. From this value, the calories per gram and the calories per ant can be calculated. The cost of a milligram of large worker is slightly higher than that of smaller workers because larger workers have a higher fat content.

The third ingredient in the total cost of a worker is her lifetime maintenance. This cost is most readily (and conservatively) estimated from her basal metabolic rate, the energy-use rate it takes to stay alive. Basal metabolic rate is usually estimated from the rate at which an animal uses oxygen or emits carbon dioxide, or both, while at rest. By assuming a mixed diet, we can make a reasonable estimate of metabolic rate from

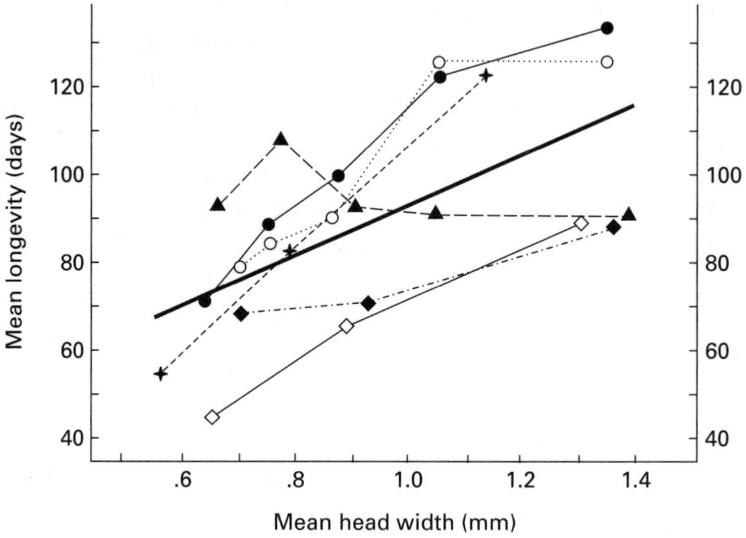

Figure 19.2. Worker longevity at 30°C as a function of body size, shown separately for each of five colonies. Longevity increases with body size. Crosses are from Mirenda and Vinson (1981). Heavy line shows the relationship after adjustment for intercolony differences. Adapted from Porter (1984).

either gas. Oxygen uptake can be determined with a simple homemade device in which a droplet of colored fluid in a capillary tube moves in response to the depletion of oxygen (or you can spend $30,000 on a fancy setup that does the same thing). These measurements show that, although each milligram of large workers is more costly to make, it is less costly to maintain. That is, the amount of oxygen respired per milligram of tissue decreases with the size of the worker. This relationship holds universally across all animal groups and is based on general physical laws that govern transport, surface-to-volume relationships, and so on.

By adding the lifetime cost of producing a milligram of worker prorated to 30 days and the cost of maintaining that worker, we have finally derived the total cost per gram of workers per month. Multiplying this value by the mean weight of each size class of workers produces the total cost per ant per month (Figure 19.3; Calabi and Porter, 1989). At both temperatures, the larger the worker, the cheaper it is per milligram per month, because larger workers both live longer and respire less (these two are connected). Services to the colony, however, probably come in units of individual ants, not milligrams of ants. On an individual basis, large workers are about six times as heavy, but their lower unit-weight cost makes them only four times as costly. Similarly, medium workers cost about 1.6 times as much as small ones.

Crude as these economic estimates are, they suggest that the services that a large worker renders to the colony in terms of fitness must be at least four times as great as those of small workers, and a medium worker must render services about 1.6 times as great as small ones. If this were not true, the colony would benefit more from converting the same amount of energy into small workers, the size class most efficient at rearing brood. What these valuable services might be is currently unknown. Candidates might include the execution of some crucial part of sexual production, the procurement of large amounts of insect prey, or the conquest of additional territory. Of course, larger workers may provide several services at once, in keeping with the rather broad division-of-labor patterns in fire ants.

Energetic Efficiency. With the "full" costs of each size class of worker finally in hand, we calculate the rate of return on these investments (efficiency) by dividing the monthly brood production of each size class of worker by their monthly total cost. The result is much closer to the "true" efficiency, but remember that it still does not include the cost of doing work. Here it becomes clear that the efficiency of polymorphic groups is about 10% higher than that of the monomorphic colonies of small workers (Figure 19.4), which were a good deal more efficient than medium workers. As usual, large workers hugged zero. So this, finally, is our answer: although the brood-rearing efficiency of workers decreases with body size, a *mixture of these workers is more efficient than any size class of workers by itself.* The whole is greater than the sum of its parts. Put somewhat differently, measuring the services of worker size classes in monomorphic colonies underestimates these services. Presumably polymorphism pays because it creates a machine that is more efficient as a result of division of labor among its members, each of whom specializes in part of the whole process, and of course, the benefits of having workers of different size can be reaped only when these workers are together.

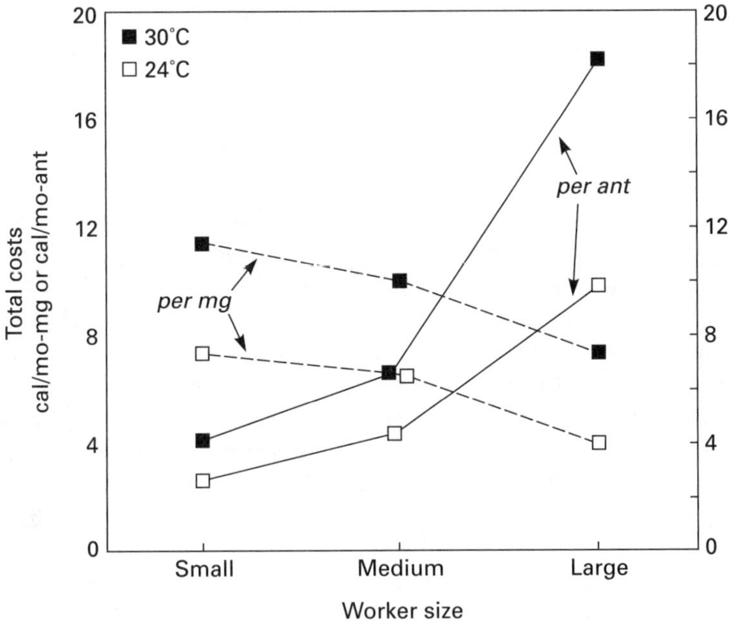

Figure 19.3. Total monthly costs per worker and per milligram of workers at two temperatures. Cost per ant increases with body size, but cost per milligram decreases. Adapted from Calabi and Porter (1989).

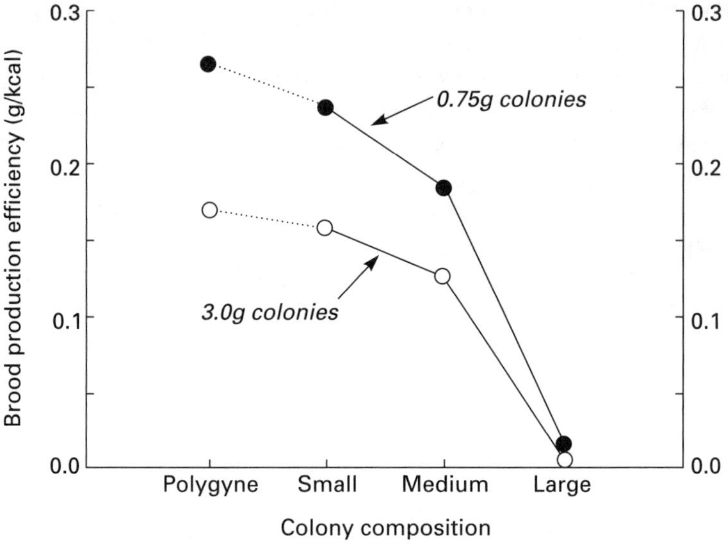

Figure 19.4. When the lifetime costs of workers are included, polymorphic mixtures are more efficient at brood production than are any monomorphic group. Adapted from Porter (1984).

Concluding Remarks

In the spirit of full disclosure, I must add other cautions. Both worker-brood production rate and colony size are imperfect proxies for sexual production. Furthermore, the experiments were deliberately designed to reduce the impact of other colony activities on brood production so that the impact of worker polymorphism would be more apparent. In their dirt-free plaster laboratory nests, colonies had no need for excavating chambers, building mounds, defending territory, migrating within the nest in response to temperature, or expending much effort in foraging because food was supplied in excess. Competition between these energy and labor choices and brood production was therefore reduced so that effects of polymorphism on brood production could be seen more clearly.

Porter's experiments tested only one standard polymorphic mixture and therefore say nothing about the optimal ratio of worker sizes. In theory, for each size class, some proportion between zero and 100% of the worker population would maximize the services rendered to the colony by that size class. Natural selection would tend to move the worker size distribution in the average natural colony toward this optimal composition, but for two reasons, the observed composition is probably not optimal. First, in the USA, selection has only had 10 to 20 generations to act, and second, fluctuating environments may assure that no single optimum ratio exists. Whether an optimum exists or not, the experiments suggest that the services of larger workers are generally maximized when they are a minority of the workers, whereas those of small workers are maximized when they are in the majority. Optimality of the worker size distribution under specified conditions is testable, suggesting many interesting if difficult experiments. Optima may differ in different ecological situations or for different colony sizes. Presumably, these differences account for the changing worker size distribution in growing colonies, but large variation in worker size distribution is also apparent among colonies of similar size. Whether it represents adaptive adjustments of the worker populations to particular situations remains to be determined. In light of the mechanism that produces the worker size distribution, we should perhaps not expect very fine adjustments. Polymorphism may represent broad compromises rather than high-precision adaptation.

Many questions, even fundamental ones, remain to be answered. No experiments to date have directly estimated the benefits of a mixture of worker body sizes for sexual production. All have been carried out in the laboratory in which many of the tasks that would compete for resources and labor in a field colony are essentially absent. In addition, large workers may appear incompetent and idle in the laboratory because the tasks they are specialized for are found only in the field. The adaptiveness of age distributions has not been tested—does a colony composed of, say, only young ants differ in fitness from one with normal age demography? The mechanism that directs fire-ant workers to move from nursing to foraging in the course of their lives is entirely unknown (but much experimental work has addressed this question in honeybees; Winston, 1987). Similarly, what mechanism programs workers so differently depending on their body size? Size distribution of workers changes as colonies grow and, combined with the behavioral inflexibility of larger workers, causes shifts in the total amount of each type of labor available. Of what importance are these ergonomic shifts?

Although we have reached the end of our discussion of division of labor, the reader will see that division of labor permeates every aspect of colony life and will inevitably affect most of the topics that follow.

◂•• *Driving to Work with* Odontomachus

The first time I saw Scott Powell in my Animal Diversity course, leaning forward in his front-row seat, eager as a pointer dog, I knew that this was not a run-of-the-mill student. The first time he asked a question, I knew that he was a natural-born biologist, not some college convert but the kind who was destined from the cradle to love biology and knew it as far back as his memory extended.

So it was gratifying two years later to have Scott elect to do a senior honors research project with me. Scott, it turned out, was ga-ga over ants, matching my own condition pretty well. While exploring our local forest, he discovered *Odontomachus brunneus*, the trap-jaw ant. It was love at first sight. This ant is classified in the primitive subfamily Ponerinae and once seen is not easily forgotten. The elongated, monomorphic workers emerge singly from its small colonies to prowl steadily and deliberately through their domain among the moldering leaf litter and the fallen twigs of the forest floor. The distinguishing feature that imparts their aura of menace is their jaws—these resemble two crowbars terminated with teeth and are carried cocked 90° to each side of the head. Between the two jaws are several tiny hairs that, if touched, cause the trap-jaws to snap shut in less than half a millisecond (Gronenberg, 1995). Any hapless insect touching these hairs is dead before the nerve impulse from its eyes has traveled halfway to its brain.

Scott came back from the forest with a dozen colonies, which he installed in dirt-free lab nests, fussing to meet their every demand. He spent enjoyable hours simply watching the ants at their daily rounds and soon noticed something common to many primitive ants—the workers actually dueled with one another! Their long jaws were closed and pointing forward, and the interaction looked like swordplay among pirates. What was going on here? Were these really fights? If so, what were they fighting over?

To answer these questions, Scott installed his colonies in nests that had three zones—an inner brood chamber, an outer antechamber, and a foraging arena. He captured workers from each of the three zones and marked them individually using fine, colored wires tied around legs and petioles. Between his classes, Scott enjoyed weeks of bliss as he watched and recorded the actions and interactions of his marked ants.

Life in these colonies turned out to be rather dramatic, full of social strife. Often, two ants met and began rapidly fencing with their antennae. Within a few seconds, one seemed to dominate, rising high on her toes to drum her antennae on the head and body of her opponent, occasionally biting as well. The loser crouched low and folded her antennae as if trying to make herself appear smaller. It wasn't that different from a dominance interaction between wolves or dogs, only these were ants (and a lot less trouble to observe). Under continued drumming from the winner, the submissive loser would creep away while shivering her antennae. Ants have no tails to tuck between their legs, but recognizing the loser of a fight was no great feat. Many times, even without any contact, submissive ants crouched in the presence of dominant ones, and everyone crouched in the presence of the queen. Here finally was an ant queen to whom the ants responded as though she were really a queen, to be feared like Bloody Mary, not just a machine for laying eggs.

Sometimes the duels would end there, but often the winner showed no mercy, and literally drummed the loser from the chamber into the next outer chamber, biting her from behind to assure that she kept moving. If she tried to return, she was driven out again. When Scott sorted through all the paired dominance duels, he found that rank and location were strongly associated; the most dominant workers were princesses of the brood pile (not queens—that job was taken). These workers were able to drive somewhat less dominant workers off the brood pile into the brood periphery. The second-rank workers in turn beat up on still less dominant workers, pushing them into the outer chamber, and you guessed it, these in turn found still less dominant workers to drive into the foraging arena. The further from the brood, the less dominant.

As in most ant nests, location allocated tasks. Workers on the brood pile took care of the brood, and gave what little attention the queen demands in this species. Workers in the brood periphery prepared food, those in the outer chamber received and butchered prey, and those in the arena foraged. Everyone busied themselves with the jobs found in the zone into which they had been banished. Their locations did not seem voluntary. Often, a worker sneaked closer to the inner sanctum, doing the tasks found there, until she was discovered and driven, under severe antennal lashing, back into her allocated zone.

After his observations were complete, Scott dissected the workers to look at the condition of their ovaries. The youngest *Odontomachus* workers had the ripest ovaries and dominated the central brood pile. The farther from this brood pile, the more withered were the ovaries and the older the workers.

So here we have task allocation and location by age, much as it has been described in many ant species. Young workers perform brood care, intermediate-aged workers do general nest duties in the nest periphery, and the oldest workers do the foraging. The difference between Scott's observations and those in other ants is that the machinery that creates the division of labor is clearly visible. Its gears, pistons, and wheels are the dominance duels that force workers to move outward and thus to change jobs as they age. Finally, when they are at the bottom of the totem pole, they leave the safety of the nest to search for food among a thousand dangers.

The system even regulates tasks in a flexible way—when the colony is hungry, duels increase and more workers are driven out of the nest to forage. When workers are not numerous enough to meet the demands of the brood, the dominant workers are too busy to drive out the subordinate workers who sneak into the brood chamber, and they all share the job.

Are these workers struggling for personal gain? For the opportunity perhaps to lay eggs and produce male offspring? For the safety of the nest? It certainly looks that way; everyone vies to be as near the brood as possible. When we humans struggle for status, don't we do so for personal gain? The trouble is, from a different point of view (West-Eberhard, 1981), the outcome of this sibling rivalry and strife benefits the colony. It assures that those workers that are in reproductive condition are near the brood, ready to lay eggs if the queen dies or perhaps to lay trophic eggs to feed the larvae while she is still alive. As their ovaries wither and their usefulness for this job declines, younger workers with fatter ovaries drive them outward until, when they have given most of what they can give to the colony and have little life left to live, they take on the most dangerous task. Foragers are disposable because they have been wrung dry (Porter and Jorgensen, 1981).

So what are the workers doing, fighting for personal advantage or serving their own needs by serving the needs of the superorganism? Winning fights may serve their own interests at two levels, directly through a chance at personal reproduction and indirectly because it organizes the colony into a more efficient superorganism, thereby raising the success of all the workers, winners and losers alike. So is the fighting really competition, or is it just the machinery that organizes the colony? Or is it both simultaneously? Maybe it all depends on how you want to see it. It is all an interpretation, after all. The only facts are those that Scott observed.

We might like to imagine that, in the course of evolution, ants became more civilized, moving outward from the brood in response to more subtle cues, cues that were nevertheless related to their ages, even in species without ovaries. Gone would be the duels, replaced by a friendly cooperation, by good formicine manners, and especially by a sense of one's place and duty. Is that what happened in the evolution of the ants? Does *Odontomachus* represent a holdover from the Wild West of ant evolution, when arguments were settled with antennal whippings, when life in the nest was raw, violent, and scary? Was this forced-labor system too wasteful because the energy expended in duels could be put to better use?

Scott himself has since left his undergraduate brood chamber, but unlike *Odontomachus* workers, who go downhill once they leave the brood chamber, he has risen to the status of graduate student. His major professor does not lash him with his antennae, nor does he drive him from the room. On the contrary, they have a very cordial relationship from which only a keen observer, like Scott, would detect that their status in the hierarchy is not the same. Whom does this hierarchy serve? Scott? His major professor? Both? A larger society? Do conflict and competition bring benefits to all contestants or just to the winners? Are these questions any harder to answer for humans than for ants?

20

The Organization of Foraging

Nonsocial animals may forage to meet their own food needs, but among ants, workers forage largely to meet the needs of the colony. When ants go a-foraging, they face many of the same problems faced by human hunting and foraging groups for much of human history. A single human or ant can bag small items and bring them back home to share, but if the food object is large, both man and ant may need help in capturing, killing, and transporting it. Many species of ants limit their foraging to items that can be carried back to the nest by an individual, but many others have evolved various kinds of communication systems that recruit help to deal with food finds that are too large for an individual to handle.

Having arrived at the food, the recruited workers may simply cooperate to pick up the whole food item and carry it back to the nest as a group. Alternately, like a family of !Kung at a giraffe kill, they may cut it up and carry the pieces back to the nest individually. Different ant species emphasize different foraging/recruiting strategies. We can presume all to have evolved under conditions of intense competition with other ants and other animals. Fire ants take the Roman approach, building roads out to the food and deploying forces that both defend it against all interlopers and cut it up for transport back to the nest. Of course, their foragers are also capable of simply carrying small items back to the nest without much fuss. It is this foraging and recruitment system that I describe here.

The Physical Context of Foraging and Recruitment

Most studies of how this recruitment is organized have been done in the laboratory, but understanding the context of this behavior in nature is important. In the field, foraging and recruitment are inextricably connected to territoriality, which I describe in other chapters. Nearly all foragers travel within the territory through a system of branching underground foraging tunnels that radiate from the central mound and act like a bidirectional river system. On the outward journey, the tunnels act as distributaries, spreading foragers into all parts of the territory, and on the return trip they become tributaries, coalescing to swell the flow of food until it finally bursts into the open spaces of the nest. There the food is transferred to waiting receivers, and the recruit returns to the field. No point in the foraging territory is more than about a meter from the opening of such a foraging tunnel, so recruits travel only short distances above ground. This organization must surely increase foraging efficiency.

In addition to the physical organization provided by the tunnel system, the foraging population is divided into scouts and recruits. On any sufficiently warm day, a few workers can be seen ambling along the ground, climbing vegetation and antennating the objects they happen upon. These are the scouts. Scouts explore areas in diverse ways (Gordon, 1988) and, at least in the laboratory, are stimulated to explore unexplored territory. In *S. geminata*, scouts may chemically mark these new areas as they explore them and therefore expend less effort going over already-explored areas (Jaffe and Puche, 1984). Scouts probably represent a small portion of the total foraging force, as becomes apparent when one finds a morsel of food too large to retrieve alone. The scout then lays an odor trail back to the entrance of a foraging tunnel, and within a short time, numerous recruits follow this trail out to the food to begin retrieval. The scout may make several such recruiting trips, making clear that these two aspects of foraging are carried out by distinct groups of workers.

Foragers recruited to food by scouts do not start from the nest mound, even when it is as close as 15 cm to the food (Horton et al., 1975). If they did, they should appear more slowly on the more distant baits because of the greater travel time, but the distance of the baits from the nest (from 0.5 to 10 m or more) is not correlated to the rate at which recruits appear on them. During the warmer months, recruits appear on most baits in less than five minutes, so much of the recruitable population must be decentralized, "stationed" out in the field in foraging tunnels, and recruits must not travel far from where they wait. This system saves a great deal of recruitment time and allows the colony to claim food quickly, before it escapes or falls into the tarsi of a competitor.

This conclusion is supported by my mark-recapture studies in the field. Recruits to large baits were marked with fluorescent ink and released. One or two days later, their colonies were excavated and censused. Only about 10% (s.d. 6.5%) of the foragers were taken in the nest. Either these foragers spent only about 10% of their time in the nest, simply entering the nest briefly to hand over their burden before returning to their posts, or they did not make the entire trip back to the nest, having transferred their burdens to a member of some kind of "transportation corps" worker (as, for example, in *Formica obscuripes;* McIver and Loomis, 1993). These possibilities are currently unexplored, although, according to one report, fire-ant foragers often hand their burdens over at the entrance to foraging tunnels (Horton et al., 1975).

Decentralizing the forager force has still another advantage. The food that fire ants exploit is often ephemeral, and recruits are brought

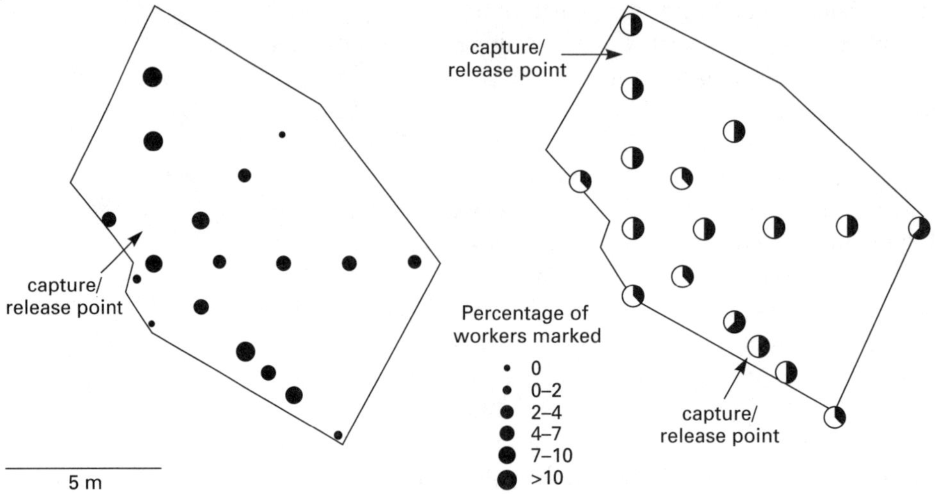

Figure 20.1. Lack of route fidelity in *Solenopsis invicta.* Left: Percentage of marked workers in recapture samples one day after marking and release. Right: Half of workers at each capture point were marked yellow, half orange. Those marked yellow were released at their capture point; those marked orange at the opposite capture point. The fraction of each circle blackened shows the proportion of recaptures that was marked orange. In the absence of route fidelity, the expected percentage orange is 50%, as observed. Lines show territory boundaries.

to this food by means of odor trails. How would an outward-bound recruit starting from the nest determine which is the correct branch to take it to the food of the moment? A system of recruitment from a forager population already in the field is a simple solution.

Interestingly, after delivering their burdens, recruits do not necessarily return to the post they just came from. In one set of experiments, I set up two bait stations in opposite sectors of the foraging territory and marked the recruits with different colors before releasing them at the bait where they were captured. Over the next several days, I baited throughout the territory to recapture foragers. Even after only one day, workers of both colors were found fairly evenly throughout the territory, even in sectors opposite the one in which they were marked (Figure 20.1). *Ortstreue* (route fidelity) does not seem to exist in fire ants. How outward-bound foragers decide which tunnel/route to take is an interesting but totally unexplored question.

The Mechanism of Recruitment

As disappointing as our knowledge of foraging at the colony level may be, we know a good deal more about the details of recruitment at the individual level. Most of this work has been done in the laboratory, beginning with the pioneering work of Edward O. Wilson in the early 1960s (Wilson, 1962a, b). When food is absent, the foraging arena is nevertheless continuously patrolled by a few scouts that cover the area in looping, irregular paths. If a juicy, dead roach is then plopped down, the scout that finds it excitedly inspects it, creeping slowly over it for 10 to 30 seconds. It does not feed. If the food is sugar water, the scout feeds until her crop is distended. In either case, she then heads back to the nest laying an odor trail. The gait of a trail-laying ant is slower and her posture more crouched. By repeatedly dragging and lifting the tip of her extruded sting like a ballpoint pen (Figure 20.2), the worker applies streaks of a minute amount of trail pheromone to the substrate. She often doubles back on her own trail briefly to reinforce it. As she approaches the nest and encounters nestmates, she often rushes toward them, partly mounting them and shaking them briefly. Eventually, she may make the entire trip back to the food, laying a second complete trail over her first, and rarely, a third. A single to triple streak of trail pheromone now connects the nest and the food.

If this scene had been played out in the field, the odor trail would lead to the entrance of a foraging tunnel. How does a tiny scout wandering about in a dense jungle of gigantic grass stems find her way back to the opening of a foraging tunnel up to a meter away? Most probably, she uses the sun as a directional cue. The forager remembers the heading of the foraging tunnel entrance in relation to the sun's position (compensated for time), and calculates the changes in that heading during her wanderings. Although this seems like a difficult task for an ant with a dot-sized brain, it has been elegantly shown that the desert ant *Cataglyphis* sp. does exactly that (Wehner et al., 1983). Two experiments suggest that fire ants also use the sun at least as a simple orientation cue. Wilson illuminated the laboratory foraging arena with a single bright light (Hölldobler and Wilson, 1990). After a scout had fed and begun her homeward journey, laying a trail, he switched off the one light and switched on another on the opposite side of the arena, shifting the light's position by 180° from the ant's point of view. The ant invariably made an approximately 180° turn. Repeated switching marched the poor ant back and forth like a soldier on a parade ground. As much fun as this may be, it demonstrates that the ant is using the single bright light as a directional

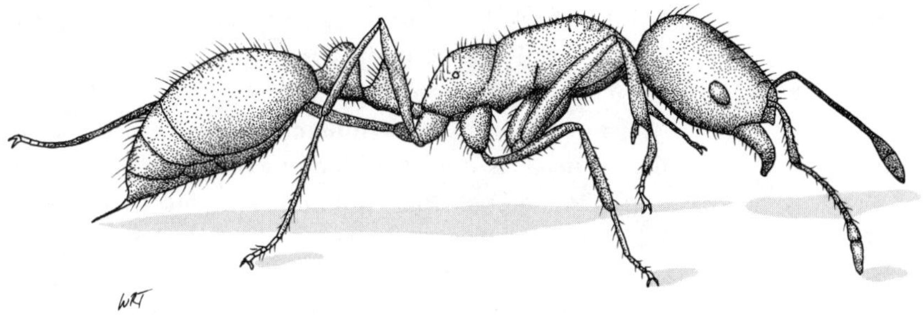

Figure 20.2. A fire-ant worker laying an odor trail with her extruded sting. Adapted from Wilson (1962a).

cue for finding its nest. I have used this demonstration many times in animal-behavior courses, to the delight of many students. The use of a sun-compass has also been demonstrated in the field, but in the context of the removal of dead ants from the colony rather than foraging. Magnetic fields have been proposed as another orientation cue, but the evidence is not convincing (Anderson and Vander Meer, 1993).

Workers in the nest are attracted to the trail pheromone and follow the trail outward toward the food. Responding workers are recruits, by definition, and are a subset of the foragers, themselves mostly drawn from older workers idling in the laboratory nest periphery. The trail does not distinguish the nestward from the outward direction, as was decisively shown by the following experiment. A trail on a strip of paper was arranged at right angles across an outward-bound trail from the nest, so that trail-following ants came upon a T-shaped intersection and had to choose which direction to take, the original outward or the original inward direction of the trail on paper. Almost exactly half of the ants chose each direction, demonstrating that the trail lacked "outward-inward" information. In addition, when a piece of a trail was reversed, the ants showed no confusion, in contrast with their confusion when the piece is simply removed. When ants follow odor trails in the presence of a strongly directional light, however, and the direction of that light is reversed, they reverse their direction of travel on the trail, indicating again that orientation cues do not emanate from the trail. (By making the second light weaker and weaker until ants no longer reverse when the light is switched, and by repeating the procedure with a range of colored lights, one can create a visual action spectrum. Like most insects, fire ants are most responsive to ultraviolet light (360 nm) and show lesser peaks of responsiveness to green (505 nm) and near-red (620 nm) (Marak and Wolken, 1965.)

Outward-bound recruits proceed down the odor trail, casting from side to side like bloodhounds on a convict's trail. Working with the ant *Lasius fuliginosus*, Hangartner (1967) showed that the trail follower compares the sensory inputs from the two antennae and balances them to stay within the "odor tunnel." Gluing the antennae in crossed-over position is a mean trick and causes the ant to have great difficulty in staying on the odor trail. Fire ants probably use a similar method to stay within the active space of the odor trail.

In this way, the recruit, guided by the odor trail, arrives at the food. It must hurry, because the trail of a single worker lasts only about two minutes on glass, effectively limiting the recruitment distance to about 20 cm. On materials that bind more strongly, such as soil or paper, the trail lasts about 10–20 minutes, making the maximum recruiting distance 1–2 m, more than the distance to the nearest foraging-tunnel exit. Whether trails are laid inside the foraging tunnels is not known.

Because the trail launches recruits toward food distant from the nest, Wilson (1962b) analyzed trail laying as though the ants were an artillery unit lobbing shells (recruits) toward a distant target. In both cases, accuracy is less than perfect because of limitations in the amount and accuracy of information and because of error (range and deflection deviation) intrinsic to the launched unit (artillery shell or trail-following worker). Ants often do not follow the complete trail but begin to cast about (search?) short of the food. If the point at which the ant leaves the trail is considered the "impact point," both ants and artillerymen scatter the "impacts" around the target. As trails become longer, impacts fall increasingly short of the target, but direction remains quite accurate. Perfect accuracy may not be optimal for either ants or artillery: a moving target (like live prey) is more likely to be hit when the shots are scattered.

What is the source of this magical trail material? Because the material seems to be applied through the sting, Wilson dissected out and crushed the hindgut, the accessory (Dufour's) gland, and the venom gland on separate fine cork tips and drew trails with them (Wilson, 1959). The hindgut and venom sac materials drew few workers to follow, but the response to that from Dufour's gland was spectacular, drawing an average of 108 recruits.

Dufour's gland is a small, narrow, cuticle-lined gland lying ventral and to one side of the poison sac (Figure 20.3; Callahan et al., 1959). A single layer of cells surrounds the reservoir. Both the poison sac and Dufour's gland are associated with the sting apparatus, as is true of Hymenoptera in general. The ducts from both the poison sac and Dufour's gland empty into the exit channel. The duct from the poison sac is opened by the contraction of a special muscle, but Dufour's gland has no obvious valve or muscle, and it is not apparent how the release of its secretion is controlled or whether venom can be released without Dufour's secretion or whether the secretion is dispensed by the sting tip or more proximally.

Regulation of Recruitment by the Trail Pheromone

An individual trail does not inform nestmates about the amount or quality of the food, simply that it is present. Wilson claimed that trail laying was an all-or-none response and did not differ in relation to food quality. Each worker, he said, simply votes "yes" or "no," as in an election. Walter Hangartner, working in Wilson's lab, looked more deeply into this question, using *S. geminata* (Hangartner, 1969b), and trail laying in *S. invicta* seems likely to be very similar. He reasoned that, if foragers could regulate the strength of the trail, they should do so in relation to the profitability of the food. He therefore tested several factors related to profitability: hunger (7 or 14 days starvation), food quality (0.01 or 1.0 M sucrose solution), food quantity (1 mm^3 or 30 mm^3), and distance (1.5 cm or 6.0 cm from nest). To make the trails visible, he contrived that foragers returned to the nest over smoked microscope slides, leaving

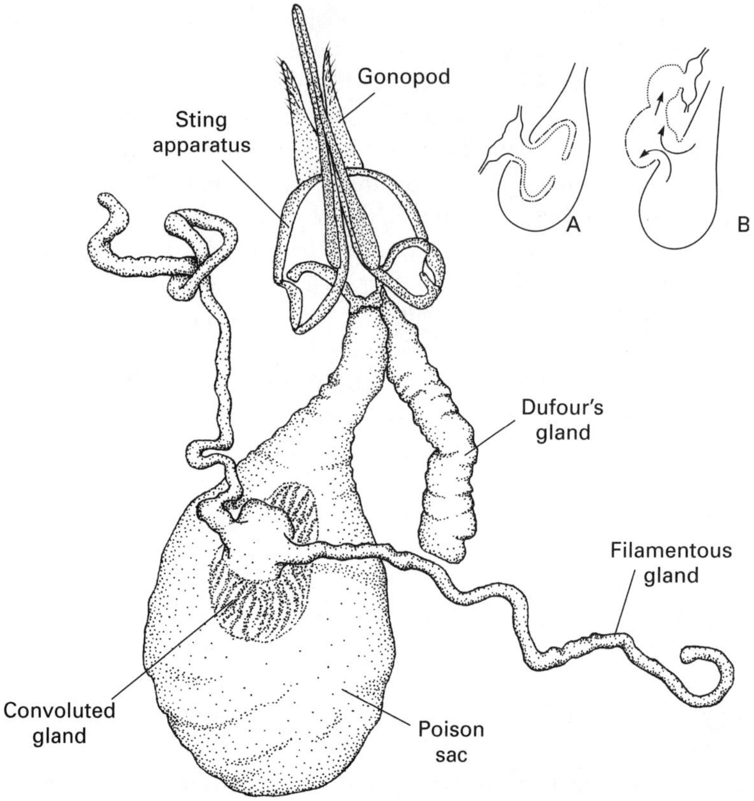

Figure 20.3. Sting, poison sac, and Dufour's gland of a fire-ant worker. Venom is secreted by the poison gland and stored in the poison sac. Dufour's gland secretes the trail pheromone. The manner in which the convoluted gland is invaginated into the poison sac is shown in insets A and B. Applying pressure on the poison sac (A) causes the convoluted gland to pop out (B). Cuticular parts of the sting apparatus are shown. Additional details are shown in Figure 22.1.

marks, which were photographically enlarged for examination. The ants made several different kinds of marks in the soot, ranging from footprints only to footprints and gaster-drag marks to the addition of partially extruded sting-drag marks and finally to the fully extruded sting marks without gaster-drag marks. The sting-drag marks indicated that the sting had been used like a fountain pen to lay trails.

Sting marks made up 55% of the trails of foragers that had been starved for seven days, but 72% of those starved for 14 days (Hangartner, 1969b). They made up 60% of the trails of ants offered 0.01 M sucrose but 75% when ants were offered 1 M sucrose. The proportions were 70% when the food was 6 cm from the nest and 94% when it was 1.5 cm. This is not the whole story, however. The sting marks made by ants fed the higher sucrose concentration were generally more definite. Poorer-quality food resulted in lighter marks, often just series of dots rather than bold lines like the trails of ants fed better food. Workers that fed longer laid more intense trails, suggesting that both hunger and food quality exert their effect through length of feeding. Together, these results show that the vigor and quantity of trail marking and, by implication, the quantity of trail pheromone, varies in accord with the food quality, retrieval distance, and hunger of the ants. In other words, ants did not simply vote "yes" or "no" on the food, as Wilson had suggested, but also had the option of voting "no" or "yes" or "yes!" or "YES!" (Figure 20.4).

Interestingly, no relationship was apparent between food *quantity*, feeding time, and trail-laying intensity. The proportion of sting marks in trails laid by ants feeding on 1-mm^3 droplets of sucrose solution were not significantly different from those laid by ants fed 30-mm^3 droplets. Foragers comment on several aspects of the food but not on how much is available.

Oddly, workers at different distances fed for similar times but differed in trail-laying intensity. Analysis of the entire trails showed,

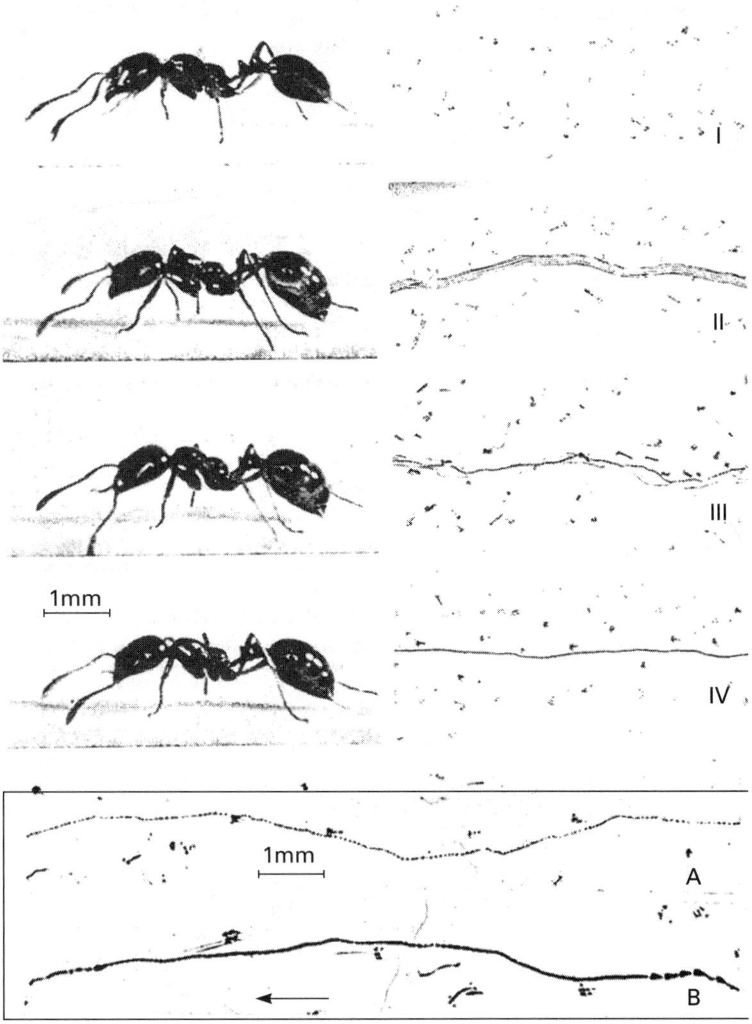

Figure 20.4. Trail-laying fire-ant workers, showing different intensities of trail laying as revealed by marks made on soot-covered slides. I. Footprint marks only. II. Marks left by hairs on the tip of the gaster. III. Combined hair and sting marks. IV. Sting marks. Panel at the bottom shows (A) a weak trail made by low sting pressure and (B) a strong trail resulting from high sting pressure. Reprinted from Hangartner (1969b) with permission from Springer.

however, that the initial intensities of long and short trails were the same but that, the farther the workers dragged their stings, the less enthusiastically and continuously they laid trails, so the overall intensity of a 6-cm trail was significantly lower than that of a 1.5-cm trail. Perhaps the worker cannot judge how far she is from the nest until she makes the trip back and therefore reserves her comment on distance for the trip home. The length of feeding is probably not the direct cause of trail-laying intensity but may respond to similar motivations.

The single trail laid by a single forager dissipates in a few minutes. It remains "alive" only if recruits that follow the trail to the food also vote "yes" by laying their own trails on top of the original one, increasing its strength and extending its life. Before doing so, however, a worker must "register to vote" by inspecting or feeding on the food. Workers who do not do so never lay trails. Thereupon, each recruited worker makes an individual decision whether to lay or not. As a result of this formicine opinion polling and these accumulating "yes" votes, the concentration of pheromone in the trail increases at a rate determined by the excess of renewal rate ("yes" votes per minute) over evaporative loss. Trail strength builds up faster for higher-quality foods, or hungrier colonies, because a greater proportion of the recruits choose to add to the trail, and their trail increment is greater. Shorter trails are stronger because the rate of loss is lower for a given reinforcement rate. When loss exceeds renewal, the trail dwindles.

Stronger trails in turn attract more workers to follow them outward. Wilson tested this hypothesis somewhat obliquely by showing that more workers were attracted to a glass rod treated with increasing amounts of the trail pheromone. If this pheromone had emanated from a trail (rather than a rod), the attracted workers would presumably have followed it out, regulating recruitment through trail strength (pheromone concentration). Both quantity and quality are communicated and responded to at the group level, a phenomenon termed "mass communication" by Wilson, who also recognized the "electoral" nature of this recruitment system. It is a dispersed, collective decision-making system based on the democratic principle of one ant, one vote, with the exception that a vote counts more if the ant feels strongly about it.

What limits recruitment? Remember that workers who do not come into contact with or feed on the food never lay trails. When the food is completely covered with gnawing and sucking recruits, new recruits fail to come into contact with it, and, after wandering about aimlessly, return to the nest without adding to the trail. As more workers return unrewarded, the strength of the trail approaches a maximum, and this maximum will be directly proportional to the area of the food. Indeed, doubling the area of a bait, providing new, unoccupied access, causes new waves of recruits to appear, presumably because the increased access leads to increased numbers of trail layers. Both the rate of appearance of workers at the food and the maximum number are regulated through the buildup rate and maximum level of pheromone in the trail. The pheromone level in turn is regulated by the probability that a recruit will lay a trail.

As the area of the food find dwindles through removal, the number of workers who can get access and thereafter lay trails dwindles as well. Trail strength gradually dwindles in parallel to food area, reaching zero shortly after all of the food has been removed. When it is gone, consider what this simple system has accomplished. Over the lifetime of each food discovery, the pheromone trail has allocated labor commensurate with the size and value of the food, and it has done so without the "knowledge" of any individual ant. No individual trail contains the information on food quality or quantity. From the combination of relatively simple units obeying simple rules, a new level of organization has been created.

This system of voting by recruits suggests that the accumulated "votes" of all the recruits could tell us about the food preference of the colony, much as the results of an election tell us about the preference for a candidate. John Glunn, an energetic undergraduate in my lab, decided to test preferences by offering colonies the three "basic ant food groups," sugar water, oil, and rat serum (protein). All were in liquid form to eliminate the ambiguity arising from differences in foraging for liquids and solids. A constant droplet area of each food was presented in each trial on the empty trailer pads of the trailer park where Glunn lived. To assure that he was measuring the preference of the entire colony, he presented the foods one at a time, three times each. Because the colony's response is not complicated by the necessity of choice, the measured preference is absolute, not relative. Once a scout discovered the bait, Glunn simply counted the number of ants at the bait repeatedly (Glunn et al., 1981). The more recruits "like" a food, the more quickly trail pheromone builds up and draws workers out to the food. The rate of worker buildup on the food therefore represents the cumulative "votes" of the recruits and is a proxy for the preference for that food.

The outcome was remarkable for the large and consistent differences among the colonies that it revealed. The 10 colonies generated seven different preference patterns for the three foods. These ranged from recruiting equally and rapidly to all three foods, to recruiting rapidly to one or two of the three, to recruiting rapidly to one, moderately to another, and not at all to a third (Figure 20.5). In last place was a colony that did not recruit much to any of the three foods. When we screened another 31 standardized laboratory colonies (10,000 workers, 1500 brood), we found 16 significantly different preference patterns. The trail recruitment system is thus capable of remarkably precise expression of the colony's preference for particular foods.

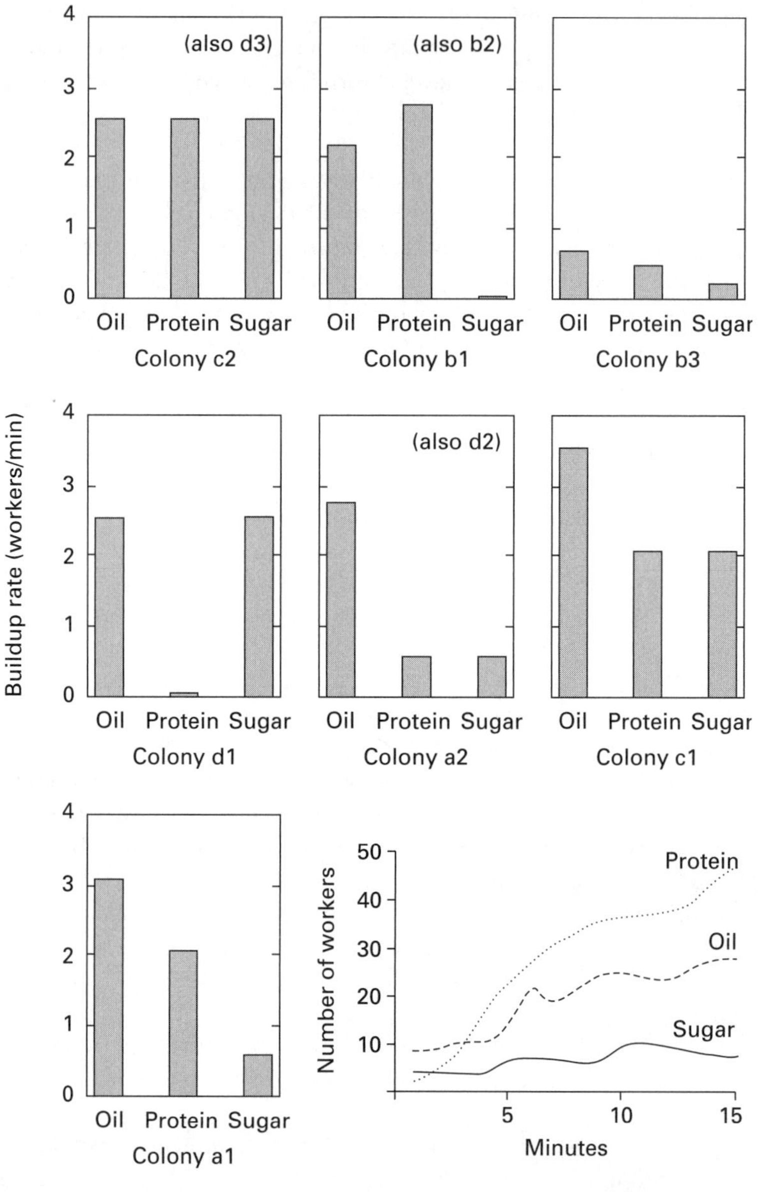

Figure 20.5. Patterns of food preferences of different fire-ant colonies. Sugar was offered in solution, and the protein solution was rat serum. The panel at the lower right shows the rate of recruit buildup at three baits for colony a1. The bar graphs show the buildup rates derived from the slopes of such graphs for 10 different colonies. Colonies showing similar patterns are noted. Data from Glunn et al. (1981).

What is the meaning of these distinct preferences? Are they idiosyncratic, or do they reflect some kind of colony-level nutritional wisdom? We have a couple of minimal hints. First, 31 laboratory colonies fed a standard laboratory diet of sugar water and chopped mealworms nevertheless expressed 16 significantly different patterns of preference, suggesting idiosyncrasy. Second, field colonies that were allowed to feed on soy oil *ad libitum* for 14 hours showed radically different preferences the next day, including the suppression of feeding on other foods in addition to oil. Preferences seem to be interdependent and to depend on feeding history. Third, the only time that colonies foraged heavily on sugar water, the primary energy food for workers, was the day after a heavy rain had caused the ants to labor mightily at repairing their mounds. The demands of heavy work may have created the preference.

Although many particulars remain to be worked out, the trail recruitment system clearly regulates food retrieval. In addition to getting food back to the colony, though, how are the benefits of particular foods, the costs of retrieving them, and time it takes balanced against each other (Taylor, 1977)? If foraging is optimal, does trail pheromone play a role in optimizing it?

Isolation and Identification of the Trail Pheromones

The trail pheromone is a wonderful chemical, mediating subtle adjustment of the colony's response to food. Interest in identifying this chemical has been quite intense. Its first partial purification for *S. invicta* found it to be unstable (Walsh et al., 1965), but in the late 1970s through the 1980s, three groups actively pursued its isolation and identification. These were the associates of Murray Blum at the University of Georgia (Barlin et al., 1976), Bob Vander Meer's USDA team in Gainesville, Florida (Jouvenaz et al., 1978; Vander Meer et al., 1981, 1982b, 1988, 1990a; Vander Meer, 1986b), and the group with Brad Vinson at Texas A&M University (Williams et al., 1981a, b).

Identifying a trail pheromone is no small enterprise—a single Dufour's gland of *S. richteri* contains only 1–10 ng, considerably less than a microwhiff (Barlin et al., 1976). Any sensible person would drop the notion of isolating the trail pheromone. Hungry ants responded to as little as 10 fg of Dufour's gland secretion per centimeter, and sated ants to 80 fg per cm. These figures suggest that hunger regulates the worker's detection threshold and thus recruitment to food, but they also emphasize the astounding olfactory acuity of worker ants. The nanogram contained in a single gland is sufficient to make 100 m of trail if the recruits are hungry or about 12 m if they are not. So sensitive are the workers that they can detect an old trail's odor imprint even when it is covered with a fresh sheet of paper.

Although simple in principle, isolation and identification of any pheromone is a difficult process, as much art as science, and requires exacting chemical techniques and expensive machinery. If one starts with extracts of whole insects or their parts, the problem becomes one of separating a tiny amount of active pheromone from a huge amount of inactive "stuff" that is also extracted. When the pheromone is contained in a reservoir that can be excised, as are fire-ant trail pheromones, the chemist has already passed GO and collected $200, for the amount of inactive material in the extract is likely to be much, much less.

The first requirement is for a good, unambiguous biological assay. The first of several bioassays was developed in Blum's laboratory and consisted of replacing the middle section of a Dufour's-secretion trail on paper with a trail drawn with a specified dose of the test substance (Barlin et al., 1976). If the ants continued along the artificial trail without hesitation, the extract contained the trail pheromone. Colony responses varied widely, allowing designation of only four approximate response levels. Barlin used this bioassay and gas chromatography to determine the species specificity of the naturally laid trails of four species of fire ants. Table 20.1 shows the results as well as the species' systematic relationships (Trager, 1991). Tests with Dufour's-gland secretion showed more cross-reactivity. The gas-chromatographic trail peak of *S. richteri* eluted later than that of *S. invicta*, showing that these are different compounds, whereas those of *S. geminata* and *S. xyloni* eluted at the same time and evoked equally strong trail following in both these species, indicating they might be the same compound. The picture is not really neat, and other authors (Wilson, 1962a; Jouvenaz et al., 1978) found different patterns of specificity. Trail pheromones are therefore probably composed of several compounds, which have varying functions and concentrations in the different species.

Vander Meer and his group found the dissection of Dufour's glands too tedious, difficult, and impractical a method. To obtain one or two milligrams of pure material for chemical identification, one would need to excise about 10,000 Dufour's glands. Knowing that the trail pheromones were unsaturated hydrocarbons, they extracted that class of compounds from almost a kilogram of blended *S. invicta* workers (almost a million). Gas chromatography and bioassays identified four fractions with trail activity. The first and most abundant peak was a hydrocarbon with four double bonds (MW=204, $C_{15}H_{24}$), identified by chemical and bioassay comparison as Z,E-α-farnesene. In the bioassay,

Table 20.1. ▶ Species specificity of naturally laid trails of fire ants. The exotic species, *Solenopsis invicta* and *S. richteri,* both belong to the *saevissima* complex, and the native species, *S. xyloni* and *S. geminata,* to the *geminata* complex (Trager, 1991). Strength of following was rated strong (XXX), moderate (XX), weak (X),or absent (—) and was clearly influenced by evolutionary relatedness.

	Strength of trail following by			
Trail source	***S. invicta***	***S. richteri***	***S. xyloni***	***S. geminata***
S. invicta	XXX	X	—	—
S. richteri	XXX	XXX	—	—
S. xyloni	—	—	XXX	XXX
S. geminata	X	X	XXX	XXX

it evoked trail following at 0.1 pg/cm (note the shift in units when trails are tested). Each worker contains about 6 ng of this material, considerably higher than Barlin's estimate of 1 ng and enough to lay 60 m of trail.

The two next most abundant compounds (both MW=218) were present at about 0.4 and 0.1 ng per worker and also had four double bonds but one more carbon. Upon hydrogenation, both yielded the same saturated, branched hydrocarbon, indicating that the unsaturated hydrocarbons were geometric isomers. Other spectral and chemical analysis, as well as comparison with synthetic compounds (Alvarez et al., 1987), showed the two compounds to be Z,E- and E,E-homofarnesene. These resulted in trail following at a dose of 40 pg/cm. A fourth compound occurred at only 0.03 ng per worker and was identified as E,E-α-farnesene. A mere 1 pg/cm of this compound resulted in trail-following. At about the same time, Vinson's group at Texas A&M identified Z,Z,Z-allofarnesene as the trail pheromone of *S. invicta*. This compound had optimal activity at 100–500 pg/cm, much higher doses than the compounds identified by the Vander Meer lab. This identification was later shown to be in error when bioassay of the authentic synthesized compound produced no trailing activity (Vander Meer et al., 1988).

The Dufour's gland thus secretes a family of α-farnesenes that differ by one methyl group as well as the configuration around their double bonds. Imagine the investigators' disappointment when the identified compounds failed to bring on the entire sequence of recruitment behaviors, from attraction to induction of trail following and trail-following. Ants already following a trail continued along trails of these four compounds, but nontrailing ants were not induced to do so. Clearly, something else was at work here, and finding out what required two additional bioassays.

In the first, called the "point bioassay," up to eight small blotter papers, each treated with a different test compound, were arranged in a test arena containing large numbers of foragers. Samples always included a Dufour's-gland-extract standard and a solvent blank. The assay was based on several responses: attraction, aggregation, settling, and chewing on the papers. Scores were normalized such that the Dufour's standard equaled 100 and the solvent blank 0 and allowed for scores greater than 100 and less than 0.

In the second, the "choice" bioassay, one worker-equivalent of a test sample on a piece of filter paper was placed in one arm of a glass Y-tube, and a solvent blank in the other. An equal flow of air passed over the two samples to converge and vent at the base of the Y, where a group of foragers waited. The arm of the Y initially chosen by the first 20 workers walking up from the base constituted the test. The apparatus was rinsed and the positions reversed between trials.

Note that these studies were careful to test all compounds at a dose of one worker-equivalent and to compare all bioassay activity against a

Dufour's-gland standard of one worker-equivalent. The object was to fulfill the requirement that the proposed pheromones substitute quantitatively for the Dufour's standard, that is, account for all of its activity. Only then could the claim be made that the entire trail pheromone blend had been successfully identified.

With these two new bioassays in hand, Vander Meer and his associates set about isolating the active compounds of what they termed the "recruitment pheromone." Preliminary studies once again suggested a hydrocarbon. One-minute fractions collected from gas chromatography of the hydrocarbons were all inactive, but recombining the first 21 fractions resulted in a score of 164 in the bioassay. Clearly, multiple compounds synergize to stimulate recruitment.

Once again, 900,000 ants met their doom in an extract, and once again, multiple fractions emerged that were active only when recombined. The combination of the synergistic fractions 9 and 13 of the initial 20 showed full activity equal to the Dufour's-gland extract. All the activity of fraction 9 was eventually shown to be a 16-carbon compound (MW = 218) with a single double bond and three rings. Combined with fraction 13, the pure compound scored 130 in the bioassay. Each worker contained 75 pg. Fraction 13 was identified as Z,E-α-farnesene, the major trail pheromone. In combination with fraction 9, it accounted for most of the activity, but not the full activity of the Dufour's-gland standard (scoring 55 of 100). Why the activity of 13 + 9 was not the same as that of 9 + 13 was not explained.

Neither compound was active by itself in the Y-tube choice assay, which measures only attraction, but together in a ratio 80:1, Z,E-α-farnesene to fraction 9, they accounted for all of the attractive (recruitment) activity of the Dufour's extract. In the point bioassay, however, this combination only accounted for 85% of the activity of the Dufour's extract and still did not induce nontrailing workers to follow trails. The hunt for additional compounds was clearly not over. Still another compound must cause attracted workers to begin following the trail.

The scene now shifts to still another bioassay, one that measured the ability of test compounds to cause workers to *begin* following a trail. Workers and brood were placed in a petri dish with openings on opposite sides. From each opening projected a wedge of paper with a 10-cm trail drawn with Z,E-α-farnesene. To the first centimeter of one of these papers was added the compound to be tested, matched in worker-equivalents. To the other, as a simultaneous control, was added more Z,E-α-farnesene. The number of ants that traveled the complete 10-cm trail in two minutes measured the ability of the test compound to induce trail following. In one series, the attractant pheromone was compared to Dufour's extract.

Addition of Dufour's extract to the first centimeter of trail caused the number of workers following a Z,E-α-farnesene trail to increase in a

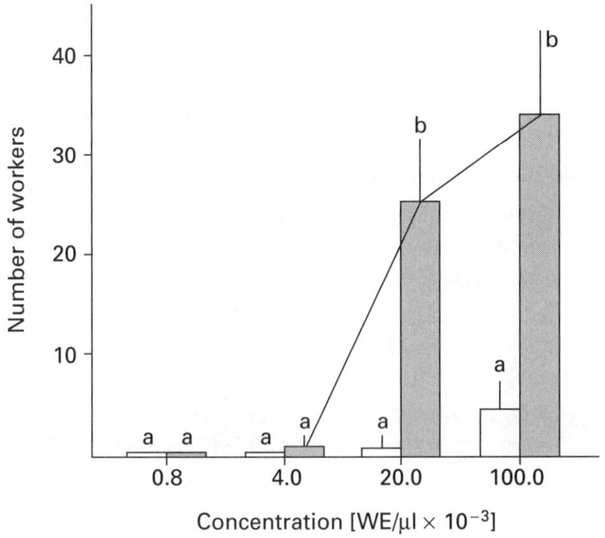

Figure 20.6. Effect of adding Dufour's-gland extract to the first centimeter of trail on the number of worker ants induced to follow a 10-cm trail of Z,E-α-farnesene. The white bars show the effect of adding an equivalent amount of extra Z,E-α-farnesene (control). Bars labeled with different letters are significantly different. WE, worker equivalents. Adapted from Vander Meer et al. (1990a).

classic sigmoidal dose-dependent fashion from fewer than 2 followers at 0.004 worker-equivalents per cm to 33 at 0.1 equivalents (Figure 20.6). Adding an equivalent amount of the attractant pheromone (Z,E-α-farnesene and fraction 9) did not increase the number following the Z,E-α-farnesene trail (both added fewer than five workers). Clearly, attraction and induction of following are not behaviorally equivalent and are not induced by the same pheromone component. When the attractant pheromone was paired with Dufour's extract, it was far less active in inducing following (about five rather than 23).

Looking over what we have now learned about the trail pheromone, we see a rather peculiar phenomenon. The trail pheromone produced by the Dufour's gland is a mixture of compounds all released together, but the individual compounds it contains induce different behaviors. As little as a few hundred-thousandths of a worker equivalent of the trail-orientation pheromone (Z,E-α-farnesene) will allow a trail-following worker to continue on the trail, but about 250 times as much material is required to induce a worker to follow this trail in the first place. Because these two compounds are mixed together, the scout must lay a very miserly trail outside the nest, one that would not induce any worker stumbling upon it to follow, but upon entering the nest, she must release a whopping dose of the same trail pheromone to start her nestmates on their way. By varying the dose with the context, a scout thus smoothly recruits nestmates and shows them the way, but no direct evidence indicates that she actually does so.

The meaning of the attractant in fraction 9 to trailing workers is less clear, because activity required doses approaching one worker-equivalent, an emission rate that does not occur in a trail-laying context.

The attractant may have other as yet undetermined functions. In any case, Dufour's secretion is also used in nontrailing contexts, including alarm and nest emigration (Wilson, 1962c). Perhaps this secretion simply heightens worker responsiveness to other stimuli in a variety of contexts.

Most of those who worked on the fire-ant trail pheromone assumed or explicitly claimed that the pheromone is the sole stimulus that organizes foraging recruitment, even though several other species of ants, indeed even fire ants, are known to supplement the trail pheromone by activating recruits through the use of sound, antennation, jerky displays, or shaking. Vander Meer and his coworkers concluded that the release of large amounts of the trail pheromone by the recruiter at the nest entrance activated recruits, but this conclusion was based only on the doses needed to get waiting reserves to follow trails and not on the amount actually released by returning scouts. The door remains open for alternate modes of activation.

Nonpheromonal Recruitment Behaviors

Just such a mode was recently suggested by the work of Deby Cassill (Cassill, 2000, 2003). She varied the quality of food (3, 9, and 27% sugar solutions) offered to experimental colonies and analyzed videotapes of the behavior of returning scouts. As expected, she found that, as food quality increased, more scouts chose to recruit nestmates, more nestmates left the nest for the food, and more nestmates engaged in sharing the returned food, but the details of the process told a more nuanced and interesting tale. They suggested that other signals help motivate recruits. In effect, she discovered the importance of the behaviors Wilson described and chose to deemphasize 40 years earlier.

Recruits that had just reentered the nest upon returning from a food find showed one to all of six characteristic behaviors: (1) trail laying; (2) rapid walking within the nest and more frequent contacts with nestmates; (3) antennal stroking of a nestmate's body (which often caused the nestmate to turn toward the scout); (4) advertising, in which a scout stopped, opened her mandibles, and exposed a droplet of food to be tested by nearby nestmates (never resulting in food exchange so probably only as a display of the goods); (5) waggle-walking, lasting from 8 to 150 seconds, in which the scout alternated walking with shaking her body from side to side for about one second, sometimes so violently that she fell over (similar behavior was observed by Obin and Vander Meer during alarm recruitment for mating flights; Obin and Vander Meer, 1994); and (6) group recruitment, in which the scout led a small group of activated nestmates out to the food, laying a fresh trail over her old one.

The occurrence of these behaviors was related to food quality. No matter what the food quality, all scouts used trail laying, followed by antennal stroking, followed by advertising, giving each alerted nestmate a taste of the food. If these are recruitment signals (we know trail laying is), their message is simply "food present" and contains no information on food quality. In contrast, the sweeter the sugar water, the more likely the scout was to walk rapidly, to waggle, and to group recruit. After finding 3% sugar water, 40% of scouts walked rapidly and 10% waggled, but after finding 27% sugar water, 70% walked rapidly and 90% waggled (Figure 20.7). Group recruitment followed in 40% of cases with 3% sugar water, but in 80% with 27% sugar. The median number of signals scouts used to recruit to 3% sugar water was three (the minimum), to 9% was five, and to 27% was six (the maximum). The mean number of nestmates recruited rose steadily with the number of signals employed by the scout, from about one recruit for one to three signals, to 6.5 for six signals (Figure 20.7). Activated workers milled around near the nest entrance until led outward by the scout. Sometimes a single scout led several groups of recruits out to the food. In short, higher quality resulted in more signals and more recruits.

Only apparently idle workers waiting in the nest periphery near the nest entrance were ever recruited. As noted above, scouts and recruits are distinct groups. For a given number of signals by a scout, hungry

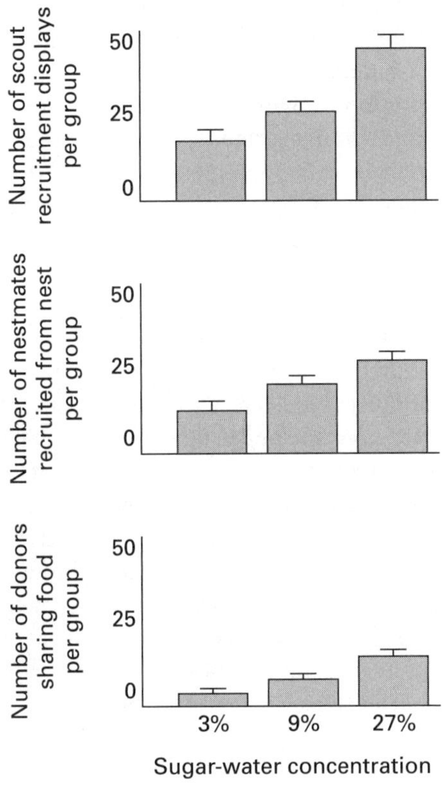

Figure 20.7. Effectiveness of nonpheromonal recruitment behaviors. The number of recruitment displays, recruitment effectiveness, and multiple sharing all increased with food quality (sugar-water concentration). Adapted from Cassill (2003).

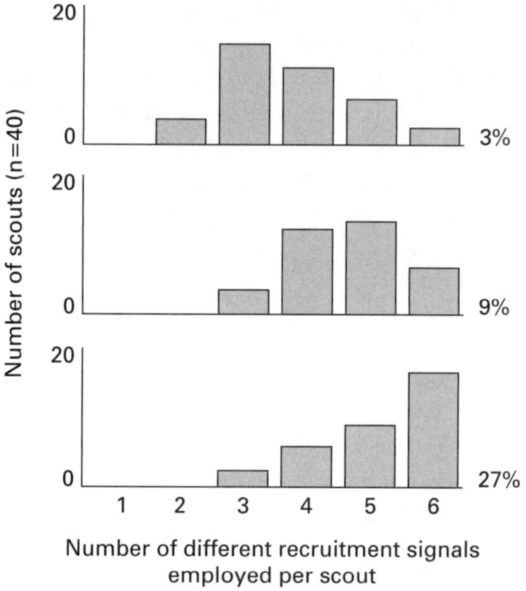

Figure 20.8. As food quality increased (from 3% to 9% to 27% sugar-water concentration), the diversity of recruitment signals employed by scouts also increased. Adapted from Cassill (2003).

nestmates were more likely to be recruited than were sated ones. This conditional response is clearly a component of how the colony regulates the allocation of labor to food collection. For a given number of signals, more nestmates were recruited to high-quality than to low-quality food, but vigorous signaling could overcome the low attractiveness of low-quality food (Figure 20.8). Finally, "showing" was much more effective than "telling"—scouts that led their recruited nestmates out in groups recruited six times as many workers as those that did not. Apparently fire ants, like human buyers, need to be shown the goods and are unwilling to buy a pig in a poke.

Interestingly, upon arrival, the proportion of the recruits that actually fed on the food was about 10% for 3% sugar water, 30% for 9%, and 60% for 27%. The remaining recruits simply patrol the area around the food. The lower numbers at the weaker sugar solutions assured that about the same number of workers patrolled around each bait, whether "by design" for the purpose of food defense or by accident we cannot say.

The effect of all this behavior is to create a peak of recruitment efficiency (recruits per scout) during the first two minutes after the scout's return. If the trail pheromone were the sole recruitment signal, the number of recruits would build up gradually through the "polling" process. Alternatively, one could invoke the suggestion of the Vander Meer group that the returning scout releases a large dose of trail pheromone upon reentering the nest, which substitutes for a live recruiter, but in real life, the process seems not to work that way. Or perhaps pheromone release is part of the recruiting behaviors, or recruiting behavior is important only in the early phases of recruitment, and the trail pheromone gradually takes over as the sole recruitment cue as its level builds on the trail.

As we have seen, recruitment behavior is complex and subtle and involves considerable individual choice. It becomes more and more difficult to view workers as tiny little robots, responding automatically to one cue at a time. More and more attractive is viewing them, in currently fashionable parlance, as *competent*, making appropriate decisions after collecting diverse information from multiple sources. Science tends to keep the complexity of an explanation to the minimum needed to explain the available facts (a principle known as Occam's razor), but one should not confuse this simplest explanation with reality. It is merely a station along the way to ever more complex realities.

◂•• *Who's in Charge Here?*

Sometimes I have the kind of thoughts you expect when you spend too much of your life socializing with ants. Take this example. While trying to decide what the subject of my next grant application should be, it comes to me: the scientific community doesn't operate all that differently from an ant colony. Neither has a boss. Both consist of a lot of workers, each laboring according to a set of shared rules, which were not created by a higher authority. Each applies these rules on its own, so that all decisions are dispersed to the individual workers themselves. No top-boss ant sends down the order that the nest needs expanding or the larvae need feeding or that, today, we need protein rather than sugar. At least in academia, no Boss-Scientist with universal vision gives the order to do more research on nest architecture today or on colony founding tomorrow. Rather, each worker decides what to do next after scoping out the local environment, factoring in recent experience, and adding a dab of more "global" (buzzword!) information gleaned from occasional contact with colleagues or sister workers.

A work task or research topic is thus chosen, and the work begins. Because of habit and improved proficiency, workers are more likely to fixate on a task they have just done. After a period of task fixation during which similar procedures are carried out repeatedly, the scientist and the ant either get bored or cannot make further progress and therefore drop what they are doing—usually well before the initial question has been answered or the job finished. There the question or the job rests until, at some later time, another worker scientist or worker ant spots it, picks it up, and completes the next part of the job or answers the next part of the question but then also fails to complete the entire project, leaving it for still another in the future. This system works because neither ant nor human is endowed with unique abilities—labor is highly redundant, especially in ant colonies. Plenty of workers are available to come along to pick up wherever the first left off, so that multistep tasks get completed by a series of workers who shuttle from one partially completed sequence to another. Myrmecologists call this the "series-parallel organization" of work. It governs a large share of the work of ants and scientists alike.

Despite this apparent lack of coordination, colonies and the scientific establishment are both capable of making "intelligent" decisions, that is, decisions that are "right" for the colony and for science. These are the collective result of the thousands of individual decisions, a sort of grassroots balloting. The outcome of this combination of dispersed, collective decision making coupled with task fixation and series-parallel operations is that labor, both formicine and scientific, flows to the activities that are most profitable, however defined by ant or man, at the moment. As these activities decline in profitability, labor flows elsewhere. None of this ebb and flow was decided by a higher, wiser authority with horizon-to-horizon vision. No, it was the collective outcome of many individual decisions made with limited vision in the damp chamber of an ant nest or the cold fluorescent light of a laboratory.

Of course counterarguments can be made. What about the National Science Foundation and other large granting agencies? Don't they decide where the money, and therefore the research, flows? They might seem to at first glance, and perhaps for the Big Science, Superconducting Supercollider crowd, they might, but in biology, their control is an illusion. Consider how most of the funding decisions are made. A panel of working scientists gets together to read *unsolicited* proposals from their own peers. Each applicant made an individual decision when choosing the topic for his or her proposal. The proposals before the panel are therefore already the collective products of a dispersed set of decisions, each made in the hope of getting some of that federal largesse. The panel, which is drawn from the same group that submitted the proposals and is therefore steeped in the same information, can only choose from among the proposals that have been submitted. Not even the program director can command the scientific community to put its efforts into this or that project. Oh, sure, the program director can announce that so many million dollars are available for work on x, y or z, but that is just bait, not much different from the baits of Hormel Vienna sausage or dry crumbs that myrmecologists use to attract ants. In either case, each worker makes his or her own decision to take the bait or not. Each worker, whether ant or human, responds to his or her own senses, motivation, and hunger. The flow of ants and scientific effort will move toward the more profitable, juicier bait, but the flow is the result of decisions by workers who bow to no central authority. They can be baited, but they cannot be ordered.

21

Food Sharing within the Colony

Basic Attributes and Their Origins

We have now seen how the foraging system is organized by division of labor, by the underground trail system, and through use of trail pheromone for recruitment. Here we turn to the fate of the food returned to the colony. A forager that enters the colony with a load of food commits that food to an intricate food traffic that ultimately allocates food of the right type and amount to each individual. As in many other species of ants (Hölldobler and Wilson, 1990), this system has several broad attributes: (1) only a small portion of the worker population is engaged in foraging at any time; (2) the foragers are drawn from the oldest members of the workforce; (3) food is brought by adults to the larvae; (4) the end recipients are reached through multiple acts of sharing among workers, never directly by foragers; (5) protein food is primarily allocated to the larvae and the queen, sugar to the workers; (6) solid foods are carried back to the nest in pieces or entire, and liquid foods are returned in a large, distensible crop.

In searching for the roots of this system, we must remember that evolution tends to build on traits that already exist, modifying, amplifying, and bending them to new purposes. Although this complex web of sharing is clearly a social trait, several of its features predate sociality. For example, the bringing of food to larvae by adults is an accommodation to the immobility of holometabolous larvae (the helpless larvae of species that undergo "complete" metamorphosis) and was already practiced by the nonsocial, wasp-like ancestor of the ants, whose female provisioned a burrow with prey for her larvae. The same is true of allocation of different food types—in the general insect life cycle, the restriction of all growth to the larval stage and all reproduction to the adult stage has dietary consequences—larval growth and queen egg production both require dietary protein (sperm production requires less), of which adult workers need little. Conversely, adults require a ready source of energy, primarily sugars, to fuel their activity (they are *workers*, after all). Therefore, food traffic in those ants that have been studied directs protein primarily to larvae and the queen and reserves sugar largely for the adult workers.

The third characteristic of food traffic is the product of division of labor, a social trait. What makes the food-sharing system necessary is that a small minority brings back the food for the majority who remain in the nest, as well as for the larvae and the queen. Direct feeding by foragers is probably impractical for a number of reasons. The evolution of a web of sharing and resharing enables a small fraction of the worker force to bring back sufficient food for the rest of the colony. The evolution of a large, distensible crop for the storage of far more liquid food than an individual can use is another necessary development for this system.

Studying Trophallaxis

Under a stereomicroscope, solid food can be seen as it is subdivided and doled out among colony members. Liquid foods, on the other hand, are shared by regurgitation from the crop of a donor worker to that of a recipient, a process named "trophallaxis" by early, classically inclined myrmecologists. In most cases, no droplets of liquid are ever visible; the two ants link their plumbing much like tanker planes refueling B-52s in flight. One obvious method of tracing liquid food is to offer dyed liquid food to foragers. Any dye in the crops of ants that have never left the nest must then have come from these foragers. The amounts exchanged can even be determined, as Vinson (1968) did. A higher-tech method uses radioisotopes, first applied to ants by Eisner (Eisner and Wilson, 1958). Radioisotopes can trace food through much greater dilution than can dyes, but they are expensive and dangerous to work with. Only radioisotopes emitting gamma rays, such as phosphorous (P^{32}) and iodine (I^{125} or I^{131}), are really useful, because their radiation is readily measurable through the bodies of the ants, their decay rates are high enough to give good sensitivity and low enough not to require adjustment for decay. Typically, they are supplied to the ants as one of their soluble salts, such as sodium phosphate or sodium iodide. Ants must be rinsed of surface radioactive contamination before their internal radioactivity is counted.

Neither radiolabel nor dye is chemically bound to the food, so oncc absorbed the label may move on a pathway separate from that of the food. Uncertainty increases with time, so we can be less and less sure that the label still represents the bulk transfer of the food rather than absorption and metabolism of the label. This problem can be temporarily overcome if the radioactive element is chemically bound to the food molecules, for example as iodinated albumin (Sorensen et al., 1980) or iodinated oil (Howard and Tschinkel, 1980), but the separation is merely postponed. After 16 days, only 5% of the unbound label (sodium iodide) was still in the ants, compared to 40% of the albumin-bound iodine label (Sorensen et al., 1980). When protein-bound iodine label was fed to larvae, it was incorporated into their tissue and remained there still longer.

Alimentary Equipment: Crop and Infrabuccal Pocket

The liquid-food-sharing system in fire ants is based on the ability to store and regurgitate liquids from a distensible crop. Many insects possess a crop, which is a cuticle-lined region of the foregut modified to store food before digestion. What little digestion occurs there is usually limited to the action of salivary enzymes. In ants, two major modifications have converted the crop into a "social stomach" whose stores can be drawn upon by the entire colony (Figure 21.1). First, it has become extremely distensible, capable of holding enough fluid to stretch the gaster to expose the intersegmental membranes. Second, the proventriculus, a grinding organ/valve at the inner end of the crop of many insects, has evolved into what was initially believed to be a passive valve separating the "social stomach" from the ant's midgut, that is, separating the social food supply from the individual food supply (Figure 21.1; Eisner, 1957). The proventriculus acts as a regulatory pump, with an intake and an exhaust valve; the bulb between them contracts rhythmically to meter a trickle of liquid food into the midgut, thus keeping the worker in adequate nutritional health.

Adult ants feed only on liquids, probably because their narrow waist and rigid proventriculus do not allow passage of larger food particles. Indeed, to prevent blockage of the proventriculus, ants have evolved a filter mechanism, the infrabuccal pocket, just behind the labium of the mouth (Figure 21.2; Wheeler, 1910). There fire ants filter out all particles greater than 0.9 μm in diameter and press them into a pellet, which they either discard or feed to the fourth-instar larvae (Glancey et al., 1981c). Thus, solids and liquids are very efficiently separated before swallowing. In contrast, larvae swallow particles as large as the diameter of their esophagus, about 45 μm. They probably act as a digestive caste. Bacteria in their midguts may play a role in digestion, although this hypothesis remains to be tested (Peloquin and Greenberg, 2003).

Food Traffic

Capacity of Foragers to Hold Food. The social-stomach function of the crop evolved from nonsocial meal feeding, serving both functions in ants. To understand the balance between these two functions, we must determine how much liquid a forager can hold in her crop and at what rate it leaks from the social into the nonsocial domain. Together, these determine how long workers can share food and how much they can share.

To find out, we fed groups of workers to repletion (four hours), then isolated them without food. Loss of food from the crop was measured at intervals from the volume of dissected crops (Cassill and Tschinkel, 1999b), from the weight of severed gasters (Cassill and Tschinkel, 1999b),

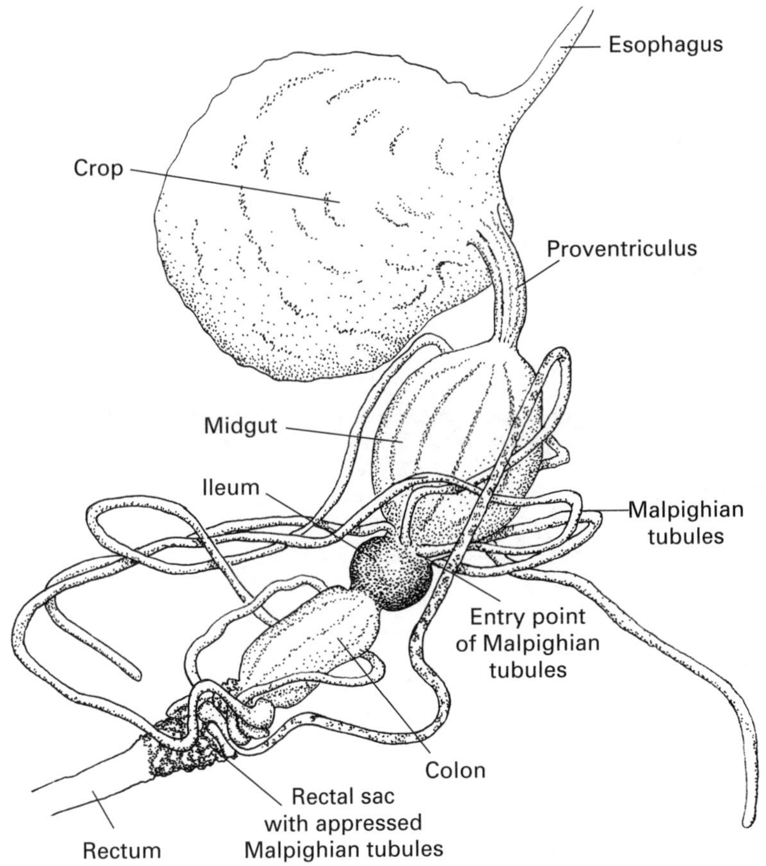

Figure 21.1. The alimentary canal of *S. invicta*, showing major gut subdivisions and Malpighian tubules. The distendible crop serves as the "social stomach" from which food can be shared with nestmates. The proventriculus functions as a valve regulating the movement of food from the crop to the midgut. The Malpighian tubules associated with the rectum probably function in water conservation.

or from the distribution of radiolabel in separated body parts (Howard and Tschinkel, 1981b). Immediately after feeding, workers' crops averaged about 85 to 90% of maximum capacity (0.25 to 0.5 μl for medium-sized workers). In isolation, one day later, this figure was down to 50%, and by three days, it was about 12 to 15% (Figure 21.3). It then dropped more slowly to 3 to 5% on the 10th day (Cassill and Tschinkel, 1999b). The slower decline after three days explains why workers starved for three days imbibed as much as those starved for seven or 14 days (Howard and Tschinkel, 1981b). All were essentially empty.

The relative ability and/or tendency to fill up with and hold liquid food is related to worker body size. Upon feeding, the gaster weights of small workers increased 23%, those of medium sized workers 73%, and those of large workers 88%, but Howard found that the volume ingested per milligram of ant was the same for all three sizes, as expected for similar relative capacities. For the time being, this question is unresolved, but the strong effect of body size on absolute crop capacity (not surprisingly) is clear.

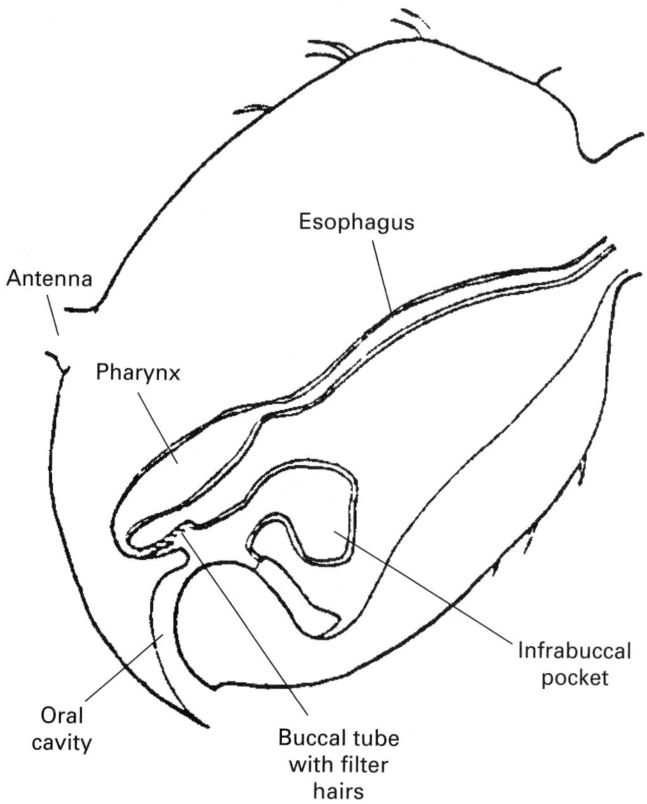

Figure 21.2. Geometry of the infrabuccal pocket of a typical ant. Arrows show the movement of food. Solid particles are captured in the infrabuccal pocket, from which they are later ejected and either discarded or fed to larvae. Adapted from Glancey et al. (1981c).

Workers vary greatly in crop volume, both before and after feeding, suggesting that food is always unevenly distributed, no matter what the colony's state of satiation. In fact, colonies are never truly satiated, in the sense that all members are full. Even with continuous excess of a single food, worker crop contents differed by five- to 15-fold even after 48 hours (Sorensen and Vinson, 1981). This unevenness is probably an important design feature of food traffic. It makes food move.

Use Rate is Food Specific. Radiolabel experiments show that, in workers of all sizes, sugar solution, the primary food of workers, passes from the crop to the midgut at a higher rate than does casamino acid solution (a partially hydrolyzed milk protein), primarily a food for larvae. Small workers fed sugar lost 70% of the label from the crop to the body in 24 hours, but those fed casamino acids lost only 40 to 50%. Moreover, sugar leaves the crop to the body at a rate appropriate to the metabolic rate of the worker—70%/day in small workers, 50%/day in medium workers (confirming the estimate by volume), and 30% in large workers. Larger workers also excrete label more slowly (20% in four days, as opposed to 40% for small workers), suggesting they use

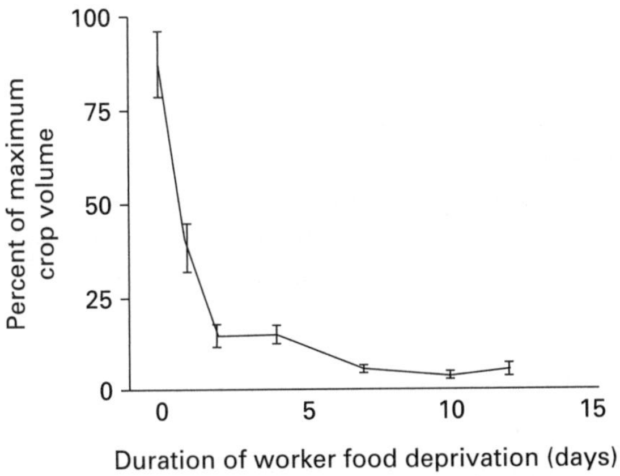

Figure 21.3. Decrease in worker crop volume with food deprivation. Workers were fed to repletion on day 0, then starved for 12 days. Adapted from Cassill and Tschinkel (1999b).

food more slowly, as expected from their lower metabolic rate, and may be able to store it longer. Initial movement from the crop can be quite fast—small workers sampled immediately after feeding on sugar had already passed 40% of the total label from the crop to the body, compared to 20% for amino acids. The starved worker must release food into the midgut within seconds, more rapidly for sugar than for amino acids. All these results suggest that the proventricular valve is not as passive a device as supposed and that nutrition is regulated within the body of the individual worker.

The loss of oil from the crop was much lower, and a large proportion of the label settled in the head, probably in the postpharyngeal gland, which is known to sequester lipids. After four days, small workers had excreted about 40% of the label, and large workers 20%. Some was regurgitated onto the nest floor, as also observed by Vinson (1968). Oil-fed workers retained essentially 100%.

Researchers at the USDA laboratory in Gulfport claimed that major workers act as repletes for oil, storing it for up to 18 months and giving it up to minor workers under starvation conditions (Glancey et al., 1973c). These claims are not credible. First, the life span of majors is only three to four months at 24°C and eight months at 17°C. Second, the mobilization experiment lacked an unstarved control. The increase in the proportion of minor workers containing oil could easily result from differential mortality of minor workers that did not contain oil. Third, no evidence supports the claim that the dye used to trace oil was actually in the crop rather than having been assimilated into the tissues. Dye in alate wing veins shows that assimilation occurs. Therefore, even though Glancey and his colleagues wrote, "[o]nly 2 logical explanations are apparent," long-term storage of oil by replete majors is greatly in doubt.

Trophallactic Behavior. Most of what we know about food sharing in *Solenopsis invicta* comes from the detailed studies by Dennis Howard in my laboratory, Ann Sorensen in Vinson's laboratory in Texas, and later, Deby Cassill, also in my lab. If we allow replete foragers to stagger back to their colony, a series of rapid and dramatic events unfolds. Upon entering the nest, the forager walks rapidly around the chamber, stroking nestmates with her antennae. If the stroked worker is hungry, she will turn toward the donor and, with mandibles always closed, join "tongues" (glossae) with the donor (whose mandibles are always open) and begin to feed (Figure 21.4). Alternatively, the full worker stands still with open mandibles, a kind of donor display, sometimes holding

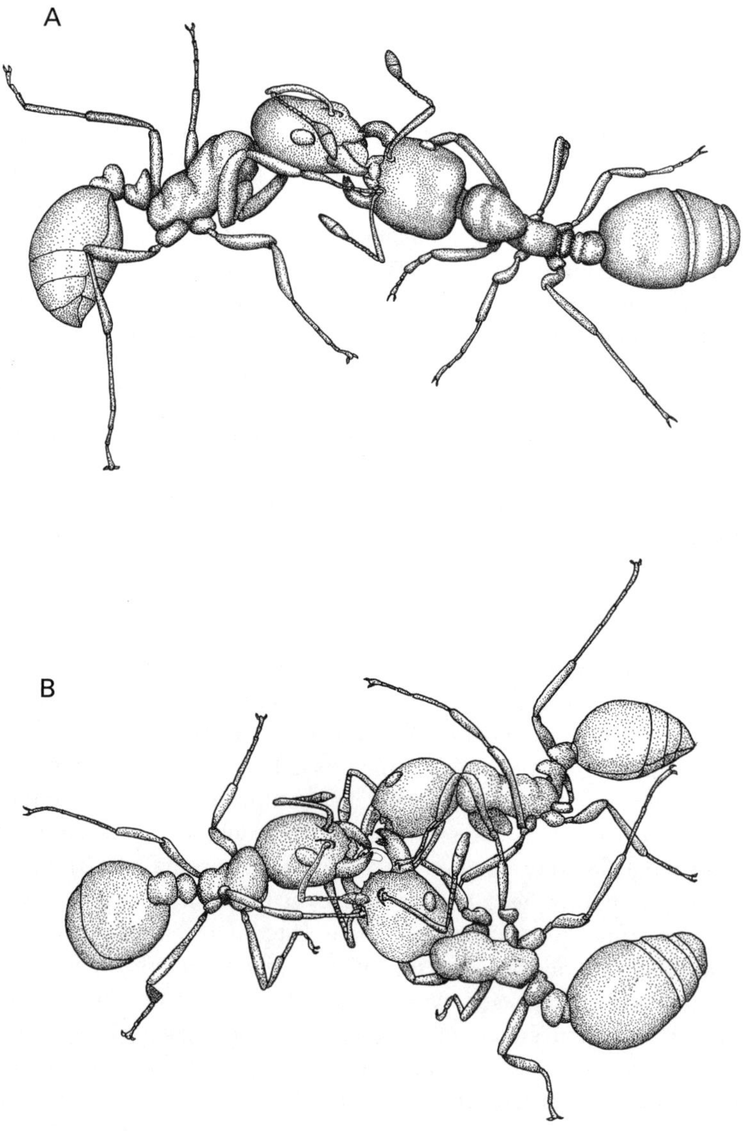

Figure 21.4. A. Trophallaxis (liquid-food sharing) in *Solenopsis invicta*. The donor's mandibles are always open, and the recipient's are closed. Food flows from the crop of the donor to that of the recipient. B. When colony hunger is extreme, groups of workers often feed from a droplet of food extruded by the donor.

a droplet of food at the tip of her glossae. A hungry worker approaches her, antennates her head, and then, with closed mandibles, engages glossae. Fluid then flows from the full crop of the donor to the less full one of the recipient. The pair stands quietly, the only movement slow antennation by the recipient, for seconds to many minutes. Either worker can disengage. Under most circumstances, no fluid is visible because it passes directly from one mouth to the other, but when the nest is extremely hungry and the foragers extremely full, the donors may "blow bubbles," extruding droplets of fluid between their mandibles, allowing several nestmates to feed at once (Howard and Tschinkel, 1980; Sorensen et al., 1985b).

A nest worker that has received fluid from a forager can in turn act as a donor, creating a spreading cascade of exchanges, like a chain reaction. In this way, food brought in by one or a few foragers can spread very rapidly throughout the colony. For example, in a small experimental colony starved for four days, fewer than 5% of the workers engage in trophallaxis, and these are always pairs. Minutes after being fed sugar water, over 80% of the workers were either donating or receiving food through trophallaxis, and the exchange groups ranged from two to eight ants, averaging four. An hour later, only 10% of the workers were still engaged in trophallaxis, and the group size had declined to two or three (Howard and Tschinkel, 1980; Sorensen et al., 1985b). Not surprisingly, the length of the starvation period affects food consumption and sharing strongly for all three food types.

Distribution of Protein. Just how rapidly and in what pattern this cascade distributes food depends on the type of food and on whether it is continuously available or consists of a single pulse. When solid radioactive egg yolk is continuously available, 70–85% of it flows to the larvae and only 15–30% to workers. Larvae end up with five to eight times as much yolk as workers, largely because the fourth-instar larvae are the only colony members that ingest solid food (see below). When the egg yolk (mixed with radioactive albumin) is suspended in water, flow is less biased toward the larvae. After 48 hours, workers contained 45% of the label and larvae 55%. Radioactive albumin is water soluble, however, and yolk is not, so the two were probably physically separated at the outset of feeding as workers pressed the yolk solids into infrabuccal pellets, which they fed to larvae, while imbibing and sharing the radioactive albumin in crop fluids.

Less uncertainty surrounds a pulse of casamino acid solution. This material is shared only through trophallaxis and is completely distributed within less than six hours (Figure 21.5). As with solid protein, it moves disproportionately toward larvae, increasingly so with longer starvation of the colony. In unstarved colonies, workers and larvae receive casamino acids in similar proportions (20% of workers to 25% of larvae). After seven days of starvation, the proportions are 40 and 75%,

Figure 21.5. Distribution, six hours after the onset of sharing, of three types of food to fire-ant larvae and workers starved for different lengths of time. Distribution is more or less complete after six hours. Casamino acid solution (a protein hydrolysate) moves preferentially to larvae. The nonlinear scales on the y-axes represent the arcsin transformation of the data. Data from Howard and Tschinkel (1981a).

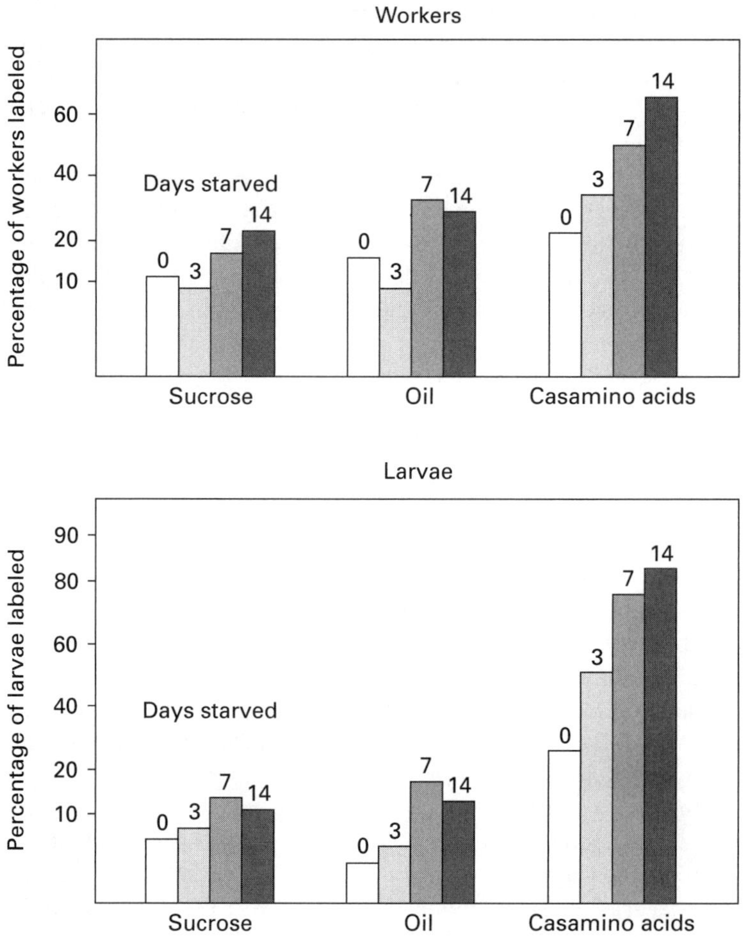

and after 14 days of starvation, they are 60 and 80%. Under the shortage conditions prevailing in these experiments, workers probably act more as pass-through agents than as consumers, perhaps because the food spends less time in their own crops. As a result, by the end of the 24-hour experiment, larvae contained on average three to four times as much as workers. Whether it is solid or in solution, workers recognize protein and feed it preferentially to larvae, who need it for growth.

Distribution of Sugar. When the colonies are fed honey or sugar water, the pattern of distribution is almost the inverse of that for protein. To begin with, consumption of sugar water increased with starvation up to seven days. Distribution of this pulse of sugar was very rapid, and the workers contained an average of five to 10 times as much as larvae, whether the food was offered as a pulse or continuously. After 24 hours, larvae contained only 5% of the recoverable total radioactivity, much less than expected on the basis of their proportion of the colony. Even though workers retain sugar for themselves, the variability in the amount they contained decreased over 24 hours because workers continue to share among themselves. This retention is much less prevalent under *ad libitum* conditions.

Starvation increases the speed of sharing and the proportion of nestmates who receive labeled food. In unstarved colonies fed a pulse of sugar, no more than 30% of workers become radioactive, even after 24 hours, but in those starved for one or two weeks, the majority is radioactive after only a few hours. *Ad libitum* feeding of honey increases the share of label in larvae to 65% after 48 hours, but at such long elapsed times, the travels of the tracer are unlikely still to represent the travels of the sugar. The cause of the difference between the *ad libitum* and pulse-feeding experiments is not clear. Perhaps shortage increases the strength of priorities, so that when sugar is plentiful, more is fed to larvae, especially in the absence of other food (all of these experiments offered only one food).

Tellingly, the amounts of amino acids and sugar ingested by foragers were similar, but a pulse of amino acids reached a much larger proportion of the colony's workers and larvae than did sugar. Clearly, workers "know" what they have and handle these two foods differently.

Distribution of Oil. Foragers took only one-third to one-tenth as much oil as sugar or casamino acid solutions, even though they fed longer on it, perhaps because of its high viscosity. Consumption of oil was little affected by starvation. Oil was shared among the workers and larvae much more evenly than were amino acids or sugar, suggesting that neither workers nor larvae have a strong priority for oil. Under *ad libitum* conditions, oil levels in workers and larvae gradually equalize over 48 hours, but even then, total volumes of oil per individual were small. Starvation increases the proportion of workers and larvae that receive oil.

Colony Size and Worker Size. Colony size, between 1000 and 20,000 workers, has at most a weak effect on food-sharing rates and patterns. Even so, size might account for some of the differences between Sorensen's experiments with colonies of 200 workers and Howard's with 10,000. Patterns of distribution among workers of different sizes were complex. We will come back to worker size later.

Larval Feeding

Video-playback Methods. The methods described above do not illuminate the behavioral dynamics of food sharing. Beginning in the early 1990s, Deby Cassill, working in my lab, studied the food-sharing acts themselves through video recording and playback analysis. The method is simple enough once the equipment is in hand—a fire-ant colony is installed in a glass-topped nest, the experimental conditions are applied, and one or more sample vignettes of the colony are recorded on videotape for an hour. The recorded scene contains 50 to 100 larvae and as many or more workers, a random sample of the whole. Nest size can thus be varied at will without compromising the ability to analyze the outcome. Another advantage is that, unlike that for dye or radioisotope methods, the sampling does not disturb the colony.

Once a recording is in hand, the tedious process of data extraction begins. The tape is played over and over, and a different larva is watched each time until an adequate sample size has been obtained or the observer runs frothing from the room. Data include the number and duration of feedings by workers. With practice, the observer can work from tapes played back at four to eight times normal speed, though doing so requires considerable concentration. Deby Cassill's capacity for playback analysis, over 2000 hours of it, is legendary. She emerged from this epic effort not only sane but cheerful (of course, her capacity for cheerfulness is also legendary).

Experiments using dyed or radiolabeled food tell us only how much each individual contains at the moment of sampling, not how it got there or what role individual hunger plays. In contrast, the videotape record yields rich detail of the behaviors involved, individual by individual, and emphasize the degree to which a very local phenomenon, the individual act of food sharing by two ants, is the basis of the food traffic in the colony.

Larval Feeding Behavior. Whereas the studies of Howard and Sorensen traced the flow of food inward from the forager in the field, Cassill's began with one of the end users, the fourth-instar larvae, and moved outward. When a worker feeds a larva, the worker must do all the orienting, because the larva is largely immobile. The worker lines up her body with that of the larva, facing in either the same or the opposite direction, then presses her glossae onto the larva's labium and mouth so that food regurgitated from the worker's crop can flow directly into the larva's mouth (Figure 21.6; Cassill and Tschinkel, 1996). During this transfer, the worker holds quite still. Her mandibles are open wide, and her antennae are bent so that the tips almost touch the larva's mouth, often making slow caressing movements. At eight times normal playback speed, workers rush and jerk over the brood pile at dizzying velocity, so those in the act of feeding larvae are conspicuous by their sudden meditative stillness. A second or two before termination, the worker raises her antennae slightly and then withdraws her glossae.

The analysis of video recordings reveals a most remarkable and important characteristic of worker-larval trophallaxis. In contrast to the duration of worker-worker trophallaxis, which varied from 1 to 240 seconds, larval feeding by workers was both brief and of constant duration. Workers fed each larva for 10–11 seconds (67% of feedings fell between 9 and 13 seconds). This interval remained constant whether the larva

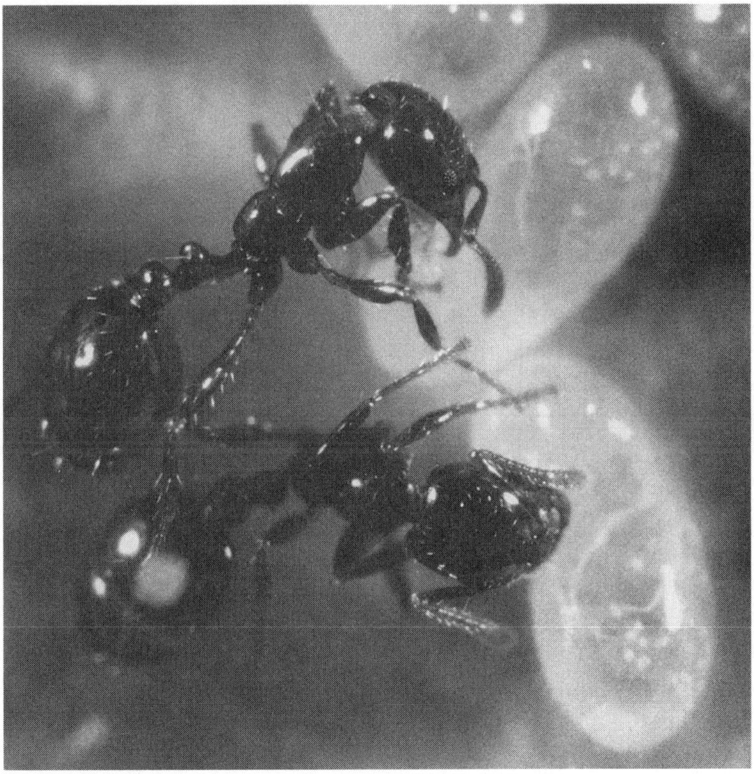

Figure 21.6. Trophallaxis between workers and larvae. The upper worker has aligned her body with that of the larva, but facing in opposite directions. Workers also align facing in the same direction. The lower worker's alignment is not typical. Reprinted from Cassill and Tschinkel (1995) with permission from Elsevier Ltd.

was sated or had been starved for up to two days, whether the larva was large or small or even a minim larva, whether it was located on the top or the bottom of the brood pile, whether its mouth was directed upward or downward, whether food had been available for less than an hour or for more than 12 hours, or whether the feeding worker herself was small, medium, or large. The duration of the feeding act was always the same.

Is the same amount of food therefore transferred during each feeding act? Cross-sections show that, no matter what the size of a fourth-instar larva, the size of the esophagus through which it swallows food is the same (it is smaller for earlier instars). Blue-dyed food is visible as a series of boluses as it is swallowed, like blue pearls on a string. All fourth-instar larvae, no matter what their size, swallowed about two boluses per second. If the swallowing rate, the bolus size, and the duration of each feeding act are all constant, then, inescapably, each worker shares a fixed amount of liquid food with a larva during each feeding, whether that larva is large or small, hungry or sated. The volume is about 1.5 nl—not a whole lot, even for an ant.

How then are the varying food needs of the larvae met? Should a large larva or a hungry one not receive more food than a small or sated

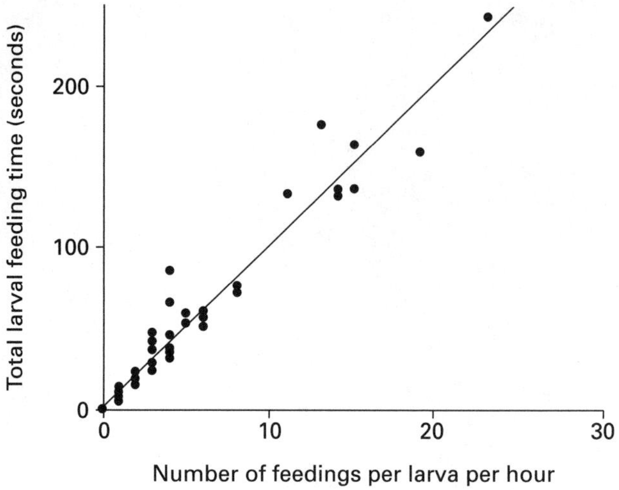

Figure 21.7. The amount of food a larva receives is simply proportional to the number of feedings. Adapted from Cassill and Tschinkel (1996).

one? The answer is remarkable for its simplicity—greater food needs are met by more frequent feeding (Cassill and Tschinkel, 1996). A larva fed twice as often receives twice as much food (Figure 21.7). Because intake rate is constant, the total feeding time (sum of the durations of all feedings) determines the total amount of food consumed (but these rules do not apply to earlier instars and sexual larvae). A worker cruising the brood pile need make only a simple decision—should this larva be fed or not?

How do workers decide whether or not a larva should be fed? To begin with, workers group larvae into "brood piles," and these piles attract workers until crowding on the brood pile reaches a steady state and as many workers leave the pile as enter it. No matter whether the larvae are hungry or sated, large or small, whether the workers are large, medium, or small, 85 to 95% of the brood pile is always covered by constantly moving workers. Nothing that larvae do, no change in larval condition or mood, could possibly go undetected for more than a few seconds. Indeed, patrolling nurse workers touch a typical larva 200 to 800 times an hour, approximately every 5–15 seconds. This rate is similar no matter what the larval condition or size. The intensity of larval care is overpowering; during its larval life, an individual larva is touched and assessed 50,000 to 120,000 times by nurse workers. The initial contact is with the antennae and, in a small minority of cases, is followed by a lick with the worker's glossae. Licking seems to be a second, more serious, level of assessment and is often followed by feeding. Despite the very frequent contact, larvae are fed only 2–50 times per hour.

Effect of Larval Size and Hunger. Workers feed larvae at frequencies appropriate to their individual size and hunger, even when larvae of different size and hunger are mixed. Service is strictly individual. How do we know? We created two sets of stock nests containing a range of worker and larval sizes. Two days before the experiment, we fed one group food dyed green, and starved them thereafter. This was our starved group, and their guts were visibly green through their translucent bodies even two days later. The sated group received red-dyed food up to the time of the experiment and appeared red. The red and green larvae were sifted into four size classes, then remixed in the desired ratio of size and degree of hunger and added to a nest of workers to be taken care of and fed.

When presented with equal proportions of the four larval size classes, half of each starved and half sated, workers fed them according to their individual size and hunger (Figure 21.8)—the larger a hungry larva, the more often it was fed. A 16-fold difference in larval volume resulted in an increase in feeding frequency from four or five per hour to 35 per hour. Hungry larvae of the same size were fed with similar frequency whether mixed or in separate nests.

Does this individualized service have limits? Can workers find and service only a few hungry larvae in a sea of satiated ones, or vice versa? The answer is yes. Even when starved larvae comprised only 5% of the larval population, the frequency with which they were fed was the same as when they comprised 30% or 100% (Figure 21.9). Workers are not confused by neighboring larvae, responding instead to an individual hunger cue at close range or upon contact, feeding each larva at a rate appropriate to its size and hunger. This efficiency no doubt rests on the very high, one might almost say excessive, frequency of worker-larva contact. When the larvae are sated, size has no significant effect on the frequency of feeding—workers simply keep them "topped up" by occasional feedings. When larvae are starved, the frequency of feeding does not increase further after 12 hours of starvation.

Time to Satiation of Larvae. At the end of the one-hour filming period, starved larvae were still being fed at almost the same rate as

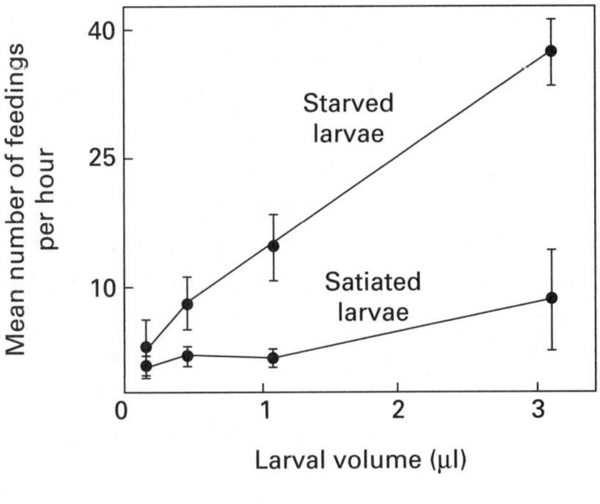

Figure 21.8. The rate at which a hungry larva is fed is simply proportional to it size. Adapted from Cassill and Tschinkel (1995).

Figure 21.9. The rate at which a larva is fed depends only upon its own size and hunger, not the condition of its neighboring larvae, even when the hungry larva is surrounded by sated ones. Adapted from Cassill and Tschinkel (1995).

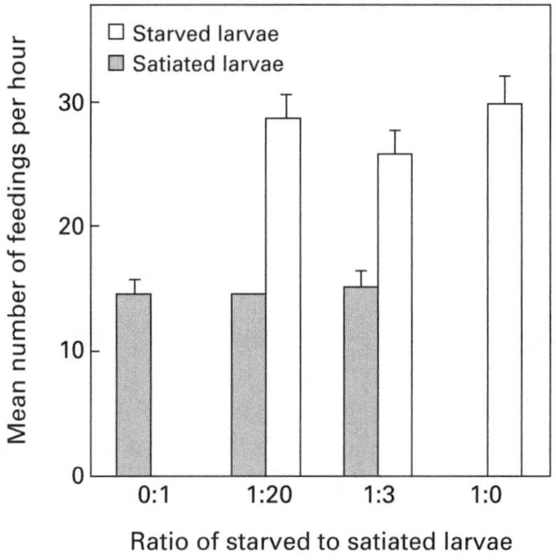

initially and were far from satiated. Initially, the feeding frequency of large, medium, and small larvae was directly related to their size (Figure 21.10). The feeding frequencies of all size classes decreased with time, most rapidly for the largest and least rapidly for the smallest larvae. All larval sizes reached satiation at the same time (after eight hours), even though they were fed at different rates. After satiation, all sizes were fed at the same rate. I believe this is still another refined feature of the larval feeding system, not mere coincidence. A feeding increment is a larger fraction of the total needs of a small than of a larger larva, and the higher feeding rate for large larvae seems likely to have evolved to accommodate their higher food needs. Over 125 feedings are needed to satiate a starved large larva, 80 for a medium, and 50 for a small. The constancy of the increment for all sizes of larva simplifies the rules that regulate behavior in workers feeding larvae.

Observing a nurse worker "plugged into" a larva raises the question of who terminates the feeding. Cassill noticed that, in their excessive enthusiasm for feeding sexual larvae, workers often missed their targets and pressed their glossae next to or above the mouths of the larvae, leaving a sticky film at the spot (she called this process pseudotrophallaxis). Significantly, its duration was the same as that of the real thing. Because no larval ingestion occurs, the worker, and not the larva, must terminate pseudotrophallaxis. Furthermore, hand-fed larvae continued ingesting for up to eight times as long as worker-fed larvae, suggesting that, given the choice, larvae would feed much longer. Nevertheless, the rate of swallowing in hand-fed larvae dropped abruptly after about ten seconds, probably because larvae become conditioned to the feeding period imposed by workers.

The Feeding Rules of Thumb

Here is the job description for nurse workers: (1) Be attracted to the brood pile and patrol it for extended periods. (2) While patrolling, constantly test the larvae encountered, looking for the larval hunger cue. (3) Having assessed a larva, make only a single, yes-or-no decision whether or not to feed it. (4) When feeding, feed the larva at a constant rate for a fixed period of time, assuring that the larva is fed only a tiny amount of food, perhaps only one-seven-hundredth or one one-thousandth of the amount it needs to complete larval development. (5) After about 10 seconds, disengage. (6) Recommence patrolling and assessing.

Feeding is probably activated when some cue from the larva surpasses a worker's threshold. Feeding briefly suppresses the larval cue. The larger or hungrier the larva, the more brief is the suppression and the more frequently do workers feed the larva. The adjustment of feeding rate by nurse workers is therefore achieved by a population response rather than by the response of individual workers. When the cue appears faster or more frequently, a higher proportion of the nurse workers detect it and thus feed larvae, and individual nurse workers will feed with a higher frequency. Adjusting the mean worker threshold upward would decrease feeding rates; downward adjustment would

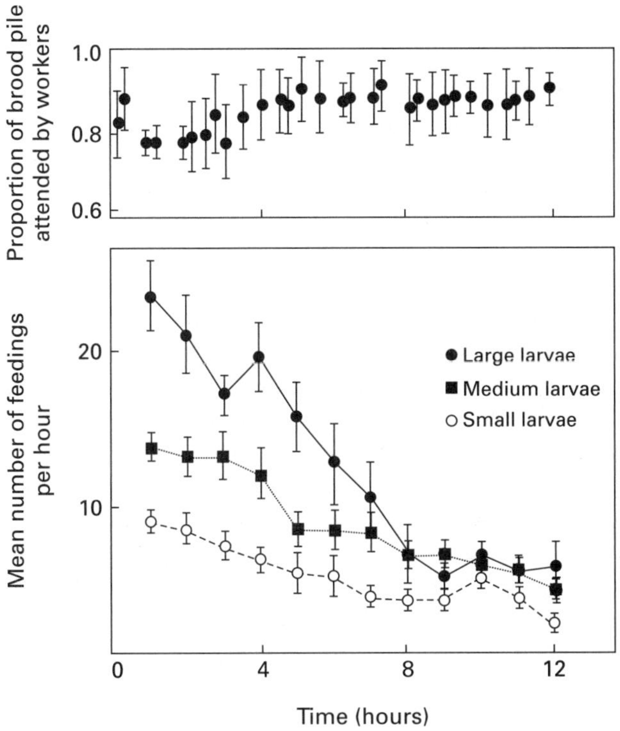

Figure 21.10. Workers fed hungry larvae at rates proportional to their size, so all reach satiation at about the same time (lower panel). The number of workers patrolling the brood pile did not change during this time (upper panel). Adapted from Cassill and Tschinkel (1995).

increase them. Recruitment of workers to food by trail pheromone is regulated in this way at the population level and is also based on a binary response. This system requires only that a nurse worker walk around on the brood pile, have a threshold for the larval cue that triggers feeding, and once it begins feeding, that an internal interval timer stop it and then be reset to zero. An alternative control system in which workers varied the amount fed according to larval needs would require considerably more sensory and neural circuitry.

The nature of the larval hunger cue is unknown, but some possibilities can be ruled out. In the complete darkness of natural nests, a visual hunger cue is unlikely, as are soliciting movements by the larvae, which almost never moved in the 15,000 feedings observed by Cassill. An auditory cue or a chemical cue detectable at even small distances from the larva would quickly lead to confusion under the cheek-by-jowl conditions of the brood chamber. As we have come to expect, the cue is likely to be a chemical one, perhaps a by-product of larval metabolism, detectable only upon contact or at very close range, like the larval recognition pheromone.

Food Flow by Type and Quality

So far, we have seen that larvae determine, through their interaction with nurse workers, how often and how much they will be fed, but the diet of a fire-ant colony consists of sugary solutions, proteinaceous insect prey and their juices, and liquid or semiliquid fats and oils, each of which is shared with some but not other ants. The behaviors that regulate this distribution once again emerge from the analysis of video recordings (Cassill and Tschinkel, 1999a).

If *only* the volume of food were regulated, then larvae should feed on different food types at similar rates. If the volume of each food type were regulated independently, then larvae should feed on each type at a rate that is not affected by how much of the other it has already received. A switching experiment tested these alternatives (Cassill and Tschinkel, 1999a). Larvae were first completely satiated on either an undyed sugar solution or an undyed casamino acid solution. Then the undyed sugar solution was replaced with dyed casamino acids, and the undyed casamino acid with dyed sugar. The dye revealed any ingestion taking place after the food switch.

The results were dramatic. Larvae fed on the "novel" food as though they had not been fed at all (Figure 21.11). When the switch was to amino acids, the rate of feeding in the sugar-first group jumped from four per hour to 25 per hour, similar to the initial rate in the amino acids–first group on amino acids. Likewise, when the switch was to sugar, the amino acids–first group dropped from 19 per hour to 13 per hour, matching the initial feeding rate of the sugar-first group on sugars. The response to each food is independent of the response to the other.

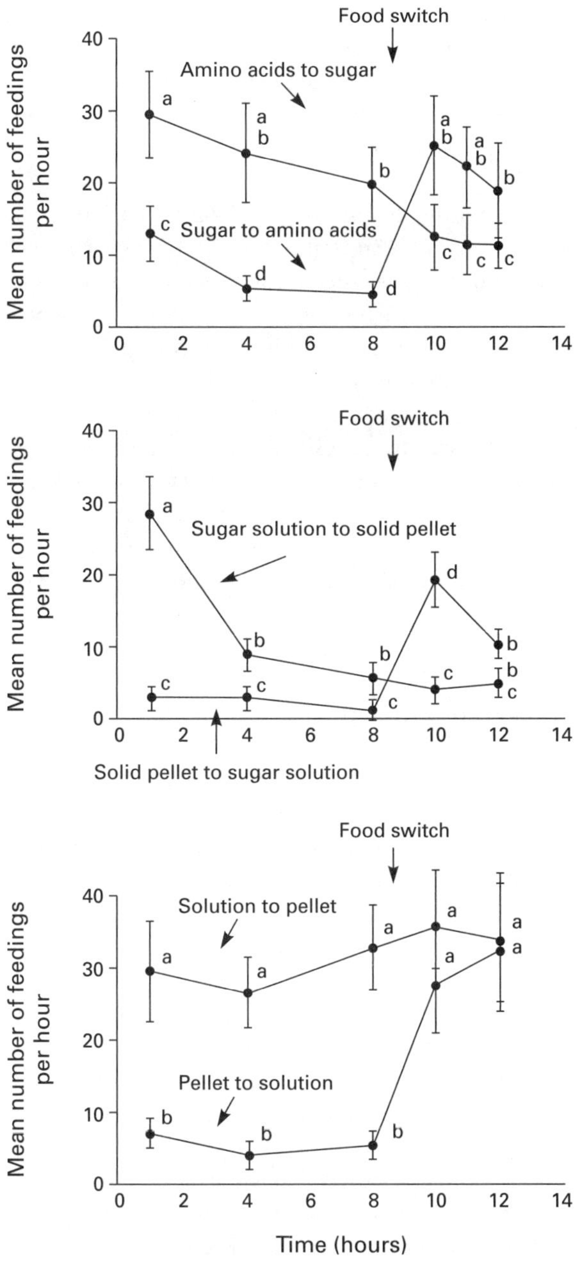

Figure 21.11. Food-switching experiments show that hunger for each food is independent of the amount of another food eaten, i.e., that larvae regulate what they eat. Points with different letters are significantly different. Adapted from Cassill and Tschinkel (1999a).

Larvae are beginning to seem pretty smart, right? Can they also discriminate the quality of food? Yes, they can. As a measure of quality, let us test sugar and casamino acids in three different concentrations, 1, 10, and 30%. Larvae take concentrated solutions of both sucrose and casamino acids at higher rates than more dilute ones. Feeding rate depends much more strongly on concentration for sugar than for casamino acids. Larvae show little interest in 1% sugar but sustain feeding at moderate rates on even 1% casamino acids. After eight hours of feeding, the weight of casamino acids (not volume of solution) taken in the form of 30% solutions is 30-fold greater than that taken in the form of 1% solutions, whereas this difference when they feed on sugar solutions is 130-fold.

In all the experiments just described, larvae were offered only one food at a time. Would they show the same preference if actually given a choice? One cannot answer this question simply by offering a colony both types of food, because the choices made by workers would bias the choices made by the larvae. In order to separate larval and worker choice, we must combine equal numbers of workers previously satiated on sugar and on casamino acids (actual values in two replicates were 55:45 and 50:50), each dyed a different color, and then allow these to tend hungry larvae. The color of the workers' crop contents can be seen easily through their stretched intersegmental membranes. The larvae were periodically scanned under the microscope for determination of the proportions that contained each color (and therefore food type) or a mixture.

Workers bearing the two food types behave quite differently. Those with casamino acids proceed much more rapidly to the brood pile, where they satisfy the larvae's preference for this food. During the first half hour, workers with casamino acids feed larvae almost three times as frequently as do workers with sugar. Casamino acids disappear from worker crops (54%) more than twice as fast as sugar (25%), because larvae prefer casamino acids about two-to-one. Workers with sugar linger away from the brood pile and share more frequently with other workers, delaying their appearance on the brood pile and emphasizing the food preferences of workers and larvae. In other experiments in which workers were allowed to forage on both food types simultaneously, twice as many workers recruited to sugar as to casamino acids. Despite this bias, one hour later 85% of larvae had received casamino acids and only 50% sugar (35% had received both). At two hours, 97% had received casamino acids, 65% sugar, and 60% both.

Both foods ultimately reach each larva, although at different rates. Where in the chain of transfer does the mixing of foods occur? Foragers invariably fed on only one kind of food. Even after three hours of sharing, when most larvae had received some food, almost 70% of workers still contained only one type of food. Almost all of the 30% that contained a mixture were reserve workers within the brood chamber that had received their crop contents from others. Moderate mixing therefore occurs through repeated food sharing as food moves toward the brood chambers. In comparison, over 60% of larvae had received both food types after two hours, more than twice the rate of mixing in workers.

Together these data reveal independent food streams entering the nest in the distended crops of foragers and gradually overlapping and mixing with each transfer, but still moving in different proportions to larvae and workers.

A look at the relative times spent donating and receiving during worker-worker trophallaxis is illuminating (Sorensen et al., 1985a, b). Foragers spent six times as much time donating food as receiving, reserves spent over twice as long, and nurses about half as long. Food mixing is thus most likely during transfer from reserve workers to nurses, who receive more than they donate because they service the end users, the larvae. If food is transferred at a constant rate during trophallaxis, it clearly undergoes net movement from the foragers to the reserves to the nurses and finally to the larvae. Altogether, 60% of all trophallactic time involves reserves, about twice as much as expected from their frequency. This finding emphasizes that one of the important functions of reserves is food sharing. The path of movement is not unidirectional but is web-like even within each functional group. Food also moves from larvae to workers, as will be discussed in a later chapter in the context of control of the queen's egg-laying rate.

Distribution of Solid Foods. So far, I have described only the movement of liquid foods, but colonies also consume solid foods such as cut-up insect prey, earthworms, and at least in the lab, powdered egg yolk (Sorensen et al., 1981, 1985b). Foragers expend a great deal of time and energy reducing prey animals to small pieces that can be transported back to the nest on their foraging trails. Workers gnaw holes in the exoskeletons of large insects and crawl inside to cut and remove pieces of tissue, like miners hauling ore from a mine (Cassill and Tschinkel, 1999a). In the laboratory, this sudden bonanza of solid food is soon stockpiled on top of the nest as dry "meatballs" or "meat wafers." Its placement is a great convenience for the experimenter: one can collect food in the form in which it is consumed, dye it with food coloring, redry it, and feed it to experimental colonies.

These "meatballs" are shared quite differently from liquid food. Of all the ants in a colony, only the fourth-instar larvae are able to feed on solid foods because only they have the sclerotized (hardened) mandibles needed for chewing solid food (Petralia and Vinson, 1978, 1979a, b), and they lack the narrow waist of workers. A worker places a chunk of solid food in the "food basket," a bare patch on the larva's belly encircled by inward-pointing hairs that hold the food in place (Figure 21.12). Earlier instars lack this food basket. If the food is wafer-like, the larva simply chews it up in about 10 minutes, like a cow eating grass. If the food is a round pellet, the larva spits salivary secretions containing high levels of four protein-digesting enzymes onto it (Petralia et al., 1980; Sorensen et al., 1983c). These proteases are absent from adults and present in much lower amounts in younger larvae

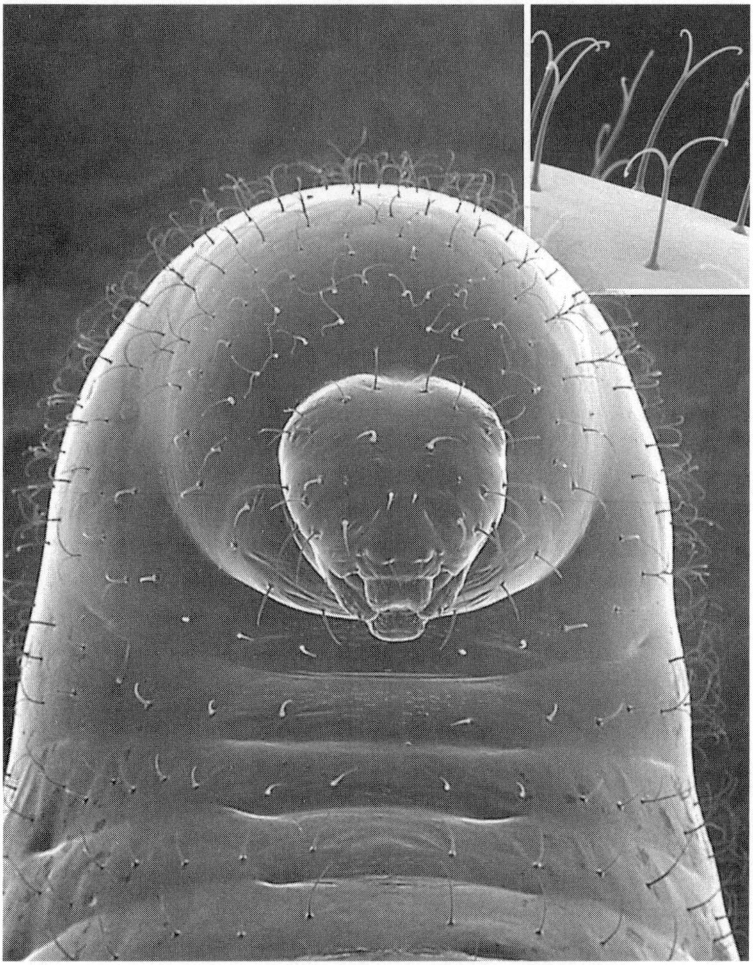

Figure 21.12. Fourth-instar fire-ant larva, showing the "food basket" in which the hairs are simple and point inward (in contrast with the forked hairs on the remainder of the body, inset).

(Whitworth et al., 1998). The forked larval salivary glands are almost as long as the body and have a reservoir near the opening (Figure 21.13). Their large size indicates the importance of these glands and their secretion.

The salivary secretion converts the pellet into a viscous mass that is generally eaten in about 30 minutes. Workers imbibe this viscous liquid from the larva's food basket and distribute it among other larvae. One hour after dyed pellets were offered to a colony, all larvae had received food, 80% directly as pellets and 20% indirectly by trophallaxis of pellet liquid. Fourth-instar larvae therefore mobilize solid food for inclusion in the trophallactic distribution system, through which it can potentially reach all colony members. Another effect is that workers that tend fourth-instar larvae have higher levels of proteases in their midguts (Sorensen et al., 1983c), presumably because they ingest larval salivary enzymes that supplement their own, inducible midgut enzymes. The importance of this extra enzyme to the digestive physiology of the colony is unknown.

One peculiar aspect of solid-food traffic is that, even when pellets are available in excess, 20% of larvae received no pellets, the majority (55%) receive only a single pellet, and only 25% received more than one (maximum three). The meaning of this frugality is not clear, but eating solid food is apparently a rare event in the life of a larva. Similarly, only a small portion of the egg yolk collected by workers was fed to larvae (Sorensen et al., 1981).

Relation of Solid to Liquid Feeding. The question of how solid feeding and liquid feeding are related was posed through another switching experiment. Half of a group of experimental colonies were first fed either sugar solution or prey pellets and then switched after eight hours; the others were fed either casamino acids solution or pellets, and then switched after eight hours.

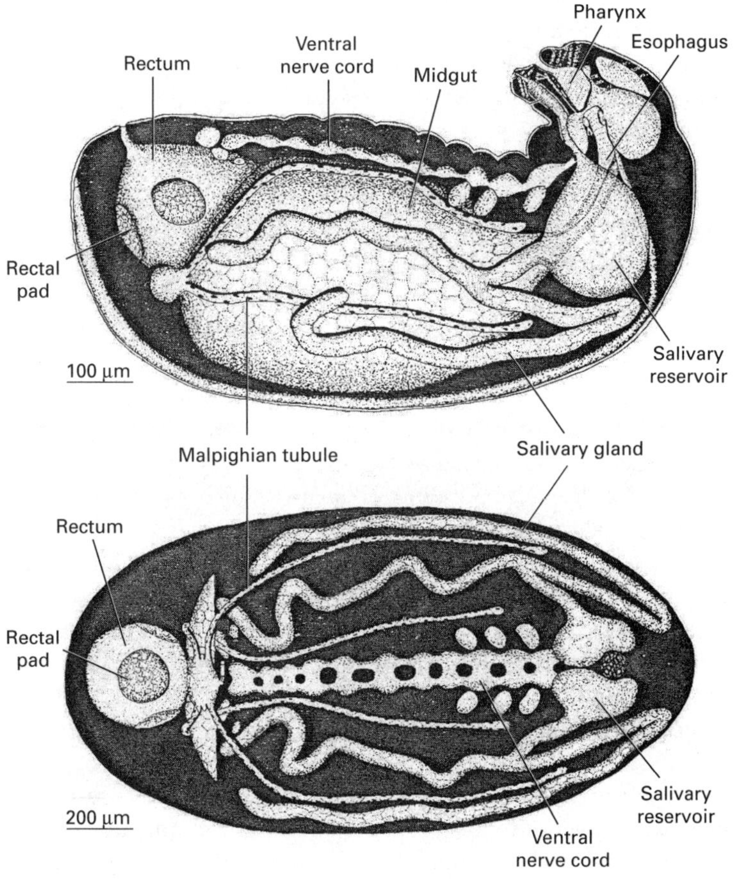

Figure 21.13. Internal anatomy of fourth-instar fire-ant larva, showing midgut, salivary glands, and other structures. Adapted from Petralia and Vinson (1980).

The rate of larval feeding in the sugar-first nests declined from an initial 30 per hour to a low level by eight hours, as expected. The rate of trophallaxis in pellets-first colonies was low, one to seven feedings per hour, and jumped when the switch was made to sugar solution but was still significantly lower than the rate of colonies fed sugar first. Prior feeding on pellets therefore seems to depress the larval appetite for sugar. This was not the case for larvae fed first on casamino acids. Switching them to pellets did not change the rate of feeding, which remained between 30 and 35 feedings per hour throughout the 12-hour experiment. More importantly, prior feeding of the casamino acid solution led to a complete absence of pellet delivery to larvae after the switch. Is a solution of amino acids an adequate replacement for solid protein but not vice versa? At the very least, the larval diet seems to show a strong preference for foods in solution.

How are we to understand these rather complex experiments? First, casamino acid solution is fed to larvae much more frequently than the pellet soup, even though these surely belong to the same food group, perhaps because the volume of soup is so much less than that of the protein solution supplied by the experimenters. If fluid volume were limiting, then a switching experiment with water and pellets should increase the rate of trophallaxis after water becomes available, but it does not. Perhaps the larva-produced soup is so much more nutritious and concentrated that a little goes a long way, but the observation that higher concentrations of casamino acids and sugar result in higher rates of feeding, not lower, argues against this explanation. We have no reason to believe that pellet soup is somehow magically different, and an additional problem is that we do not actually know the composition of the pellet soup. It could be watery; it could be rich and thick.

In summary, fire ants strongly prefer sharing liquids among all colony members, even the fourth-instar larvae. The appetite for solid food pellets is mostly independent of that for liquid foods. Furthermore, pellets are distributed parsimoniously, even when they are present in excess, and few larvae receive more than one or two pellets. The reason is not clear. Also puzzling is that larvae full of casamino acids do not receive pellets when these are made available.

Role of Workers in Food Traffic. We now need to overlay our understanding of worker division of labor on that of food traffic. How is food traffic organized within the worker force? As before, this section is based heavily on the work of Deby Cassill and Dennis Howard in my lab (Cassill and Tschinkel, 1999a; Howard and Tschinkel, 1980, 1981a, b), supplemented with the work of Ann Sorensen and her colleagues (Sorensen et al., 1981, 1983a, 1984, 1985a, b; Sorensen and Vinson, 1985).

In an experiment designed to reveal whether foragers respond to their own hunger or to the hunger of larvae, six experimental colonies were starved, three containing larvae and the other three pupae. One

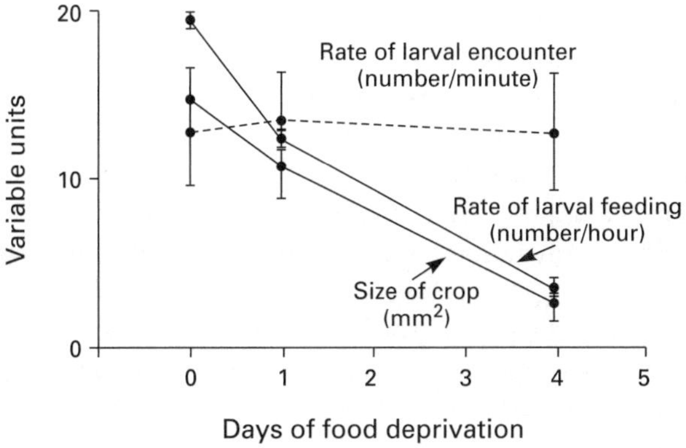

Figure 21.14. Workers feed larvae at rates proportional to their own crop fullness. Rate of feeding declines almost exactly in parallel to the decline of crop volume. Crop volume was estimated as the area of the smear resulting from crushing the gaster. Adapted from Cassill and Tschinkel (1999b).

colony of each type was then fed for two hours on sugar, one of each type received casamino acids, and one of each type was not fed. The feeding period was long enough to satiate the workers but not the larvae. After two hours, both types of food were offered to all three pairs of colonies. Pupae do not feed, so their hunger could not influence the outcome.

The results showed that, whether the nests contained hungry larvae or nonhungry pupae, feeding preferences and rates were the same. The feeding patterns of workers are not responses to larval hunger but reflect their own. Workers recruited more strongly to the food type that was novel to them—the workers first fed sugar more strongly to casamino acids (equal to previously unfed controls) and those fed casamino acids first more vigorously to sugar (also equal to previously unfed controls). The results do not make clear whether individual workers were responsible for the observed patterns or whether each nest contained groups of workers that differed in their appetites for the food types, one satiated during the first feeding and the other during the second.

Because workers feed in response to their own hunger, the fullness of a worker's crop should affect its response to larval hunger, and it does. Workers were fed dyed food (Cassill and Tschinkel, 1999b); then starved for zero, one, or four days; and then given hungry larvae but no food. Samples of workers crushed on white paper revealed the relative amount of food in their crops in the sizes of the smears they formed. The outcome was quite dramatic (Figure 21.14). Without any decrease in patrolling of the brood pile, the rate at which workers fed larvae was almost exactly parallel with the volume of crop liquid. If we connect this result with the uneven distribution of crop contents among workers (see above), we see a possible reason why individual workers feed larvae at different rates.

Does the "memory" of recent starvation cue workers to feed larvae at a higher rate than would workers without this experience? The answer is no—filling up just before feeding larvae wipes out all memory of the prior period of starvation, and the feeding rate is related only to the larval hunger. The workers live only in the present.

The poor showing of large workers in brood rearing raises the suspicion that worker size is related to larva-feeding tendencies. Monomorphic experimental colonies of small, medium, and large workers and a polymorphic mixture all show similar coverage of the brood pile, but large workers feed and groom larvae almost not at all, whereas small and medium workers feed them at roughly similar, high rates. No wonder large workers in monomorphic groups fail to rear brood (Porter and Tschinkel, 1985b)—the brood starves.

In the polymorphic group, workers could assort themselves on or away from the brood, according to their inclination. The total coverage of the brood pile was not different from those in the monomorphic groups, but despite the equal proportions (33%) of the three sizes of workers in the nest, medium workers made up about 65% of those on the brood pile, small workers about 30% (but they groomed larvae more often than they fed them), and large workers a mere 3 to 5%. Roles in larval feeding may be more sharply defined than this experiment initially suggested; as noted in a previous chapter, feeding larvae seems not to be part of the large-worker job description.

Also as mentioned in an earlier chapter, workers change jobs as they age, but pattern of change depends on their size. The tendency of workers to feed or groom larvae generally declines with age, but the effect of age on these behaviors is only about one-sixth as strong as the effect of body size.

Picturing task selection by workers as highly rigid and deterministic would be a mistake—following individual workers in their daily rounds for even short periods shows the variety and variability of what workers do. The individual is the largest source of variation. The next studies that Cassill and I conducted differ from the previous ones in an important way. All previous studies were focused on larvae. We could not tell whether the worker feeding our focal larva had just fed a hundred larvae and would go on to feed a hundred more or whether this was the only larva it fed all day and it was about to depart, exhausted from its exertions, for a long nap in the corner of the nest. It was the world seen through the "eyes" of the larva (which has none), in which the workers are averaged and have no individuality. We therefore went on to videotape 53 individual nurse workers, selected as they fed a larva, for 15 to 60 minutes each to see how they spent their time. This was the world through the eyes of the worker, where larvae are simply averaged and have no individuality.

One is immediately struck by how busy nurse workers are. Ninety percent of them were active more than 75% of the time, patrolling on the brood pile, grooming and feeding larvae, but the frequency of nurse behaviors showed great variation. The majority of the nurses fed larvae 30 to 60 times an hour, but about one in 10 fed only about three times an hour, and an equal number fed more than 90 times an hour. Some of those that fed infrequently spent their time on the brood pile grooming larvae, but others left the brood pile for other parts of the nest chamber, where they did tasks unrelated to brood care or merely strolled about. Significantly, the great majority of nurse workers did not often donate food to other workers, and they never did so on the brood pile (perhaps this pattern assures that food flows primarily from workers to larvae). The activity of nurse workers contrasts strongly with observations of entire nests, in which, over their lifetimes, workers were inactive two-thirds of the time (Mirenda and Vinson, 1981), literally "reserve" workers.

Nurses did not refill their crops frequently—when they began with full crops, only half had refilled them by the end of the second hour. Clearly, a worker crop can hold a large number of the miserly food increments fed to larvae. When nurses did refill, they left the brood pile and engaged a donating worker in trophallaxis for a highly variable period, 1 to 247 seconds (mean=35 sec, median=14 sec). Half of the workers that refilled did so only once, the remainder, from two to six times. Nurses thus seem to shuttle on and off the brood pile as they fill and empty their crops, causing net flow of food from reserve workers to nurses to the larvae. The variability of refilling may create a nurse worker population with highly variable crop fullness and may in turn be partly responsible for the variability of larval feeding rates among nurses. Recall that crop fullness is tightly related to the average rate of larval feeding.

These studies expose the patterns of food traffic, but they also provide us with a case study of how labor is organized in a fire-ant colony. If we imagine all of the tasks within a colony to make up a sort of "task space," then each worker occupies only a portion of that space, a sort of "task paddock" determined by its personal characteristics. The most important of these is its body size, which limits so strongly that, even under dire circumstances, workers cannot cross them very effectively. The limits imposed by age are much weaker and therefore more variable. Perhaps each worker's limits slowly shift with age, but these changes in limits are small compared to the effects of body size. These limits "coarse tune" worker task selection. (In an earlier chapter, I developed the idea that these constraints on flexibility create "labor sectors" that determine the total amount of a type of labor available to a colony. Because the constraints change with worker body size, the labor sectors change with colony growth.)

Within its paddock, shorter-term conditions affect the tasks a worker elects to perform, "fine-tuning" task selection. Most obvious is the effect of crop fullness. The highly variable duration of worker-worker trophallaxis and larval feeding rates conspire to keep crop fullness highly variable (see also Howard and Tschinkel, 1981a; Sorensen and Vinson, 1981) and constantly changing on a scale of minutes, with the result that the motivation to forage, feed, and/or be fed also remain quite variable (room for higher-quality food is therefore always available, should it be found). The crop fullness of the moment, interacting with worker size and age, thus organizes social feeding through its influence on whether a worker leaves the nest to forage, seeks food within the nest, donates food, stores food, or feeds larvae. Furthermore, the identity of the food in the crop influences whether the worker shares with other workers or feeds larvae. Food thus flows from fuller crops to emptier crops, biased by food type, in a chain of demand, a system driven by personal, relative hunger (Brian and Abbott, 1977, used the term "gradient of hunger"). Dennis Howard (Howard and Tschinkel, 1980) showed that, when engorged foragers had been previously starved but their colony had not, food flowed rather briskly to reach 50 to 75% of the colony, but when neither the foragers nor their colony were starved, food flow was sluggish, reaching less than 20%. The engorged foragers have relatively much fuller crops than their fed nestmates, causing food to be transferred from fuller to less full and on down the line.

Social feeding, like division of labor generally, is not absolute but probabilistic. Workers are more likely to perform some tasks than others, whether they are in mixed groups or segregated by age and/or size. Workers grooming larvae in a normal nest are therefore likely to be predominately small, to be of any age, and to have empty crops. Workers away from the brood pile and recruiting to food are likely to be predominately large, to be of any age, and to have empty or partially empty crops. Workers patrolling the brood pile and feeding larvae are likely to be predominately medium-sized, to be young to middle-aged, and to have full crops.

We can picture the sequence of events through which workers and larvae, deep in the nest, communicate their needs to the foragers. Deep in the nest, a few nurse workers sense that their particular crop loads are being gobbled fast by larvae, and as these nurses empty their crops rapidly, they solicit more of the same food from the tankers, who in turn solicit this food from the next wave outward until the pulse of specific solicitation reaches the foragers coming into the colony periphery. Those foragers carrying the food that is in high demand unload quickly and therefore recruit others, through behavioral alerting and trail laying, to collect the same food. This outward communication of specific hunger can be seen in experimental findings. That it can explain specific colony tastes is still conjecture.

The food-sharing system is decentralized, lacking any higher directing authority or guiding hand, and is the outcome active task selection by workers on the basis of many competing local cues. Food flow is the outcome of thousands, perhaps millions, of individual, independent decisions, each worker responding by simple rules to very local information, to its own condition and motivation, and to those of the larvae and workers around it.

"Accidents don't just happen, they are caused." So reads a popular safety poster, and to an awful lot of people, other thoughts extend seamlessly from it: "If accidents are caused, then someone is to blame," followed by, "if someone is to blame, then someone should pay," and finally, "When they've paid, everything will be all right." That these logical bridges are missing pylons or large chunks of the roadbed, or have collapsed, or are entirely absent doesn't seem to slow the traffic across them, traffic that leads Americans more or less straight into court in record numbers, the desire for revenge and payback burning in their souls. No surprise then, that court is also where many people get blamed for an attack of fire ants, but such cases rarely go to jury because, in American case law, property owners are responsible only for the actions of their *domestic* animals—their dogs, cats, bulls, llamas, and yaks.

So it was a big surprise when Mark called me on Monday afternoon to say that the case I had been advising him on was going to court. The very next morning. In Miami. Could I take the time to testify as expert witness? Six jurors need an education about fire-ant biology. The plaintiff 's case is that the owner of some low-income apartments, whom I'll call Mr. Laster, is a slum landlord and that the garbage, junk, uneaten food, and poor sanitation around his apartment complex have caused fire ants to view it as paradise. The landlord is therefore to blame for the stings, followed by medical complications, suffered by the plaintiff, Michelle. And he should pay. A lot. On the phone, Mark says, Assuming for the moment that the place is filthy, messy, and that there is food lying around, would that make it more likely for fire ants to settle there? No, I answer. Thank you, he says.

The Dash 8 is in the air just as dawn begins to bleach the eastern sky from indigo to gray. The two engines are running at slightly different speeds, making the cabin thrum to the slow beat of the difference. Near Perry the smoke from a small forest fire hugs the ground under an inversion, the narrow plume straight as a stick, driven the 25 miles to the mouth of the Steinhatchee River by a north wind. On paper with blue squares, I outline a set of facts about fire-ant biology, trying to link them in a logical sequence—fact and consequence, observation and implication. Then I write it out in detail. As we glide into Miami, I am not quite finished. Below, the Everglades stretch to the southwestern horizon, the invisibly flowing sheet of water parting around scattered elongate hardwood hammocks, looking like green nunataks, and redissolving into grass at their downstream ends.

Upon arrival at the Sheraton Biscayne Bay Hotel, I take a few minutes to explore the waterfront lawn behind the hotel and to sketch a map of the fire-ant colonies I find there. Fire ants are cleaning up a dead blue crab. It goes on the map too. Next door in Brickell Park, two homeless men have laid out grass mats. One is reading a novel, paying me scant attention. I find plenty of fire ants colonies, but none has a mound. I map them too.

The next cabbie speaks Jamaican English. We make our way through the heart of Miami at Flagler and Miami Avenues to the Dade County Courthouse. The guard at the door insists on keeping my Old Timer pocketknife, just in case I want to run amok. Ask for it with the number on this card, she says in heavily accented English. The elevator doors are carved brass in Art Deco style. They open into the echoing waiting room on the fifth floor, where long, dark wooden pews face each other two by two, awaiting lawyers and witnesses. The sign in front of Judge Sheedy's (again, not his real name) courtroom door says, Silence, Court in Session, and the bailiff makes it clear that I must wait outside. It is a waiting room. I wait. I chat with the nicely dressed black woman also waiting. She is a forensic nurse. After 30 minutes, Carmen comes out, followed by Mark. This case should never have gone before a jury, he grumbles. That's what he said the day before on the phone too. The jury straggles toward the elevators, on the way to lunch, so we take time to go over my testimony in the case. We get it down to the bare essentials. Juries bore easily, and less is better. Smaller target for the plaintiff 's lawyers.

The jury returns and I take my place in the waiting room again. Mark guesses that my turn will come about 3:00 p.m. It is 1:30. I head for the Latin House Restaurant down Flagler Street for a cup of *café con leche* and a piece of *flan*. The waitress spots me as an Anglo and addresses me in English. Most of the conversations are in Spanish. Back in the waiting room, I move to the more comfortable leather chair at the end of the long, glass-covered table. I wait. I doze. The four elevators take turns dinging, taking in and releasing their cargoes. Red light means up, green down. The Art Deco brass doors open and close. An immensely fat woman, tightly wrapped in blue, waddles across the waiting room, her buttocks making a little jump with each step. A full elevator load. Two lawyers confer with their client in the pews, something about cars. One of the lawyers says to the client, I'm a lawyer and I *know* everybody does it! I did it, and I'm a *lawyer*! You can't be blamed. He has gold buckles on his black shoes—a picture of Ben Franklin's feet floats into the viewing space of my mind. His belly makes his suspenders look like red parentheses. A copy of *Motor Trend* magazine lies on the bench next to the client. I doze off again.

When I awake, a handsome black woman with a wide brown hair band is eating lunch at the other end of the table. She lays everything out carefully before she begins, dissects an orange with painstaking precision, and puts the pieces to one side. Unfolds a napkin carefully and places it in her lap, then begins, with great leisure and studied pleasure, to eat. I start up a conversation. She works on the first floor in family court. It is bedlam down there. Up here it is quiet. Her daughter is a lawyer, Duke, and got a private office even as an intern. Nice office, too. She (the mother) has worked here for 17 years and still doesn't have a private office. Young people don't know how good they have it. After a while, she finishes her lunch and carefully packs everything up. Just as carefully, she works on her makeup, mirror in hand. Nice talking to you, she says, heading for the elevator. The waiting room is silent. I doze some more.

Inside the courtroom, the jury has heard from the plaintiff, how six years ago, she had been talking with her friend Monica, just about dusk, at the end of the walk in front of her apartment. I jus' stan'in under tha' tree, an' I feel this pain. My leg, like burn. I run in the house. My Dad rub my leg wit' alcohol. She describes how the next morning, her leg hurt and how her Momma say, you not stayin' home from school. But she did anyway, spending the day with her friend, Shakila, playing hookie together. By afternoon, she called the medics and her mother. It swell up real big, my leg, she says.

Her father is next. She say she standin' down under the tree, talkin', when they bite 'er. What kind of tree? Black olive. The photograph shows two cabbage palms and a myrtle.

After the father comes the defendant/landlord, Mr. Laster. No surprises. The medical doctor has detailed the condition of Michelle's leg and given an admirable description of the secondary infection and other medical complications that left Michelle's leg scarred and sent her to the hospital several times. Mark has no questions for the doctor. Nobody doubts the harm done to Michelle. Stretch marks, says Michelle, because my leg blew up so big. I had a beautiful leg.

Back in the waiting room, time has been chloroformed. Gaggles of lawyers with cell phones drift through, talking importantly to the air. The elevators go ding ding, the lights over them blink from left to right, then from right to left. Runners wheel carts full of papers through the waiting room. A handsome blonde bailiff in a white uniform shirt wanders aimlessly out of the courtroom behind me, his badge shining gold and blue. He looks around, yawns and goes back in.

At 4:00 p.m. I am finally called into the courtroom and take my place in the witness box, the jury of six to my left. Do I swear to tell the whole truth and nothing but . . . ? I do. The clerk returns to his table. Adrenaline is making my brain buzz. At the table straight ahead sits the plaintiff, Michelle, in an orange wig, flanked by her two lawyers. Her father sits at one end of the table, a maroon Banlon shirt stretched over his belly. Michelle is 100 pounds heavier than in her last photo.

Mark takes his place opposite me. We establish my credentials, then cover some basic fire-ant biology. Finally, we get to the critical question. Assume for the moment that everything the plaintiff claims about the landlord is true, does this make it any more likely that fire ants will be found there? No, I say. Why not? Because fire ants choose their habitat on the basis of openness and ecological disturbance and don't care either way about filth. Disturbance predicts a bounty of insect food for them and an absence of other competing ant species. So they are found everywhere? says Mark. Yes, wherever people spend time, because people create the disturbed habitat preferred by the ants. Even in Miami? Even where it is not filthy? asks Mark. I reach into my right pocket to pull out four vials containing fire ants in alcohol. Yes, I answer, I collected two of these samples on the well-kept grounds of the Sheraton Biscayne Bay Hotel and two next door in Brickell Park. I reach into my left pocket to pull out three more vials. These two were collected just across First Street from the courthouse, and this one—I hold the vial aloft, pause a moment for dramatic effect—this one came from the flower bed just to the left of the entrance to this courthouse. The jury stirs and someone produces the hint of a chuckle. They relax.

Your Honor, intones Mark, I would like to enter this vial into evidence. The judge looks bemused and orders the clerk to enter it into evidence. May we show this to the jury, Your Honor? asks Mark. The jury peers at the vial as it is passed around, some holding it aloft to silhouette the ants against the white ceiling. The old black woman wakes up from her doze to wonder in a whisper what is going on. After a whispered explanation from the woman next to her, she says out loud, Oh, antses!

Mark and I get back to fire-ant biology. The points follow one another like soldiers: control is imperfect and temporary; sterilization, even if it were possible, bears a high environmental cost; prophylactic treatment is neither effective nor recommended. We review the failed mirex program and its disastrous predecessor, the fire-ant eradication trials heptachlor and dieldrin. I mention the sainted name of Rachel Carson, and out of the corner of my eye, catch a juror nodding. We cover how colonies reproduce, how queens disperse and choose their nest sites, how colonies move (they just pick up all their larvae and pupae and head out on foot; several jurors laugh outright at the image) and under what conditions they move (killing a colony merely makes an opportunity for a neighboring colony). The jury learns that the exact location of colonies cannot be predicted and that, in south Florida, with its warm, sandy soils, fire ants do not construct mounds and are therefore not very conspicuous. Fire ants have minds of their own and don't obey our orders. No one is to blame for fire ants.

The opposing lawyer, I'll call him Mr. Janson, rises for cross-examination, his greenish suit jacket opening to expose his large belly. He stares at me for several seconds, then says, how long does it take a colony to grow to where you can see it? I describe the variation in rates and conditions. So there might be quite a while during which a colony might be spotted, and something could be done about it? Yes.

Now he has a circus act of his own. From a large, rustling plastic bag, he pulls out a bottle of Ortho and holds it aloft. Could this help to get rid of fire ants? he says. He pulls out a container of Amdro, then Logic, then seven more fire-ant killers in a row. Each time, he thumps it down on the table and asks, would this help? And I answer, yes. I wait for him to come to the predictable point, the one I would make now, but he doesn't. Instead, he says, what do you recommend people to do about fire ants? Do you mean the general public? Yes, he answers. I say, First I ask if they have a compelling reason to control fire ants, such as a hypersensitive child. If not, I recommend they leave them alone. They do no harm. He looks irritated.

He tries a nastier tack. Who is paying for your hotel? Nice hotel, is it? Your plane fare? How much? Cab rides? How much? It adds up to a lot of money, doesn't it? You make it sound like a paid vacation, I say. It's business; people's expenses get covered. I am thinking, and you're doing this for charity. He continues, are you getting paid for appearing here? How much per hour? I tell him, adding that I might charge a lower rate for the waiting time. How much? I haven't decided. He persists through several exchanges, how much? Several jurors are fidgeting and looking unhappy. They like me and don't respond well to seeing me bullied. I give the judge a beseeching look. I think he's told you everything he can, says Judge Sheedy, why don't you go on to something else? As Mr. Janson turns, I call him back, saying, I make a reasonable salary, my wife works, we have enough money. I don't do this for the money. I do it because I can help out with my specialized expertise. I charge because it is customary. Mr. Janson looks like he wishes he hadn't turned around. Judge Sheedy is too easygoing to object to my parting volley.

On redirect, Mark asks, If you had been hired by the plaintiff, what would you have charged? The same, I answer. And would your testimony have been different? Absolutely not, I say. Thank you, says Mark. I step down, my part done. Final arguments and jury instructions will take place the next morning. The jury files out. A couple of jurors give me a sidelong glance and smile. I hang around a while, chatting with Mark and Carmen.

On the way out, I retrieve my Old Timer pocketknife. I can again feel safe on the streets of Miami.

Back in my office the next day, Mark calls to tell me, We won! A jury of six peers and one alternate has decided that Mr. Laster is not to blame for fire ants being under that clump of trees at the end of the apartment block, not to blame for Michelle's standing on the nest, not to blame for her being stung a dozen times or for the festering sores that followed, and not to blame for the discolored scarring of her leg. The landlord is blameless, at least for the fire ants. Michelle in her orange wig, nineteen years old, black and without prospects, her out-of-wedlock child already three years old, does not know whom to blame or what to blame them for. Does she imagine that, without the fire ants, her life would be all right? Does she blame fire ants for the dead end she is in? Does she believe that if that slick lawyer and that smart-ass professor hadn't messed it up, the money would have made her future bright?

The fire ant has few defenders. Most people hate fire ants without reservation, without reflection. Perhaps that is what the fire ant has to offer us—something we can all agree to hate, something about whose reprehensibility no argument can be made, something we can blame and that won't argue back. That way we can hide, just a little while longer, the uncomfortable fact that, for most of the bad things that happen to us, no one is to blame. It is fate, bad luck, misfortune, the wrong place, the wrong time. Life catches us, all too often, standing on a fire-ant mound. The consequences are painful, and sometimes disastrous, but in a landscape full of fire-ant mounds, also occasionally inevitable.

22

Venom and Its Uses

The Hymenopteran Heritage

When ants evolved from a waspish ancestor, they inherited a stinger as part of their basic equipment. The stinger is essentially a modified ovipositor and defines the aculeate Hymenoptera (ants, wasps, and bees; Hermann and Blum, 1981). In the ancestral condition, the tube-like structure was formed from elongated appendage bases (valvulae) pressed together to serve as a conduit for eggs on their way from the female's ovary to their resting place outside the body or inside the body of a parasite's host. Two accessory glands assisted her in placing or aiding the eggs through, for example, gall formation or host paralysis. Many uses of secretions from ovipositor-associated glands probably remain to be discovered.

Once the social Hymenoptera evolved, their colonies offered tempting concentrations of juicy, nutritious larvae or stores of delicious honey to vertebrate predators. In response, worker ovipositors, their egg-conduit function no longer needed, evolved into hypodermic syringes for injecting venoms produced by accessory glands. The purpose was to make vertebrate enemies sit up and pay attention (see Schmidt, 1990, for a review). In vertebrates, pain is a sort of early-warning system for imminent physical or physiological damage. Venoms are therefore often diabolical brews of substances that cause immediate pain, that inflame and swell tissues, unglue cells from one another, or rupture blood cells, toxins that mess with nerve cells to cause lethargy, sickness, and in the extreme, death. When a single wasp can deter a human three to four thousand times its size, one can hardly doubt that a step has been taken toward rebalancing the odds in confrontations with vertebrates. Abundant (harmless) mimics of stinging Hymenoptera are further testimony to the effectiveness of these defenses. Schmidt goes so far as to suggest that without sting and venom systems, the social Hymenoptera would be unlikely to exist because vertebrate predation would make life impossible.

Having inherited this versatile tool, two entire subfamilies of ants lost it in the course of further evolution, evolving various alternative chemical defenses. Even among those subfamilies that generally have stings, scattered genera no longer use it as a sting. Nevertheless, the glands that came with the sting remain functional, becoming little chemical factories that produce a variety of compounds for communication or defense.

Anatomy of the Venom Gland

Obviously, fire ants are not among those ants that have lost the sting. Their very name derives from the effect of the sting and its venom. As in other ants, the venom of all fire-ant species is secreted by the venom gland and stored in the venom, or poison, sac. The whole arrangement is geometrically complicated (see Figure 20.3; Callahan et al., 1959). The venom gland consists of two distinct parts. The first consists of two long filaments of glandular cells that lie free in the hemocoel, suggesting the uptake of materials from the hemolymph. These join and enter the second, a bulbous part called the convoluted gland, which rests within a deep invagination of the poison sac, so that a double wall of sac completely surrounds it. Venom secreted by the enormously long, tangled convoluted gland passes into the poison-sac reservoir through a cuticular sac called the "filter" by Callahan, although it is unlikely actually to act as one. Venom is stored in the poison sac and is then ejected through the main duct during use. The capacity is between 20 and 40 nl, depending on worker size.

All of these ducts and sacs are derived from epidermal tissue and are therefore lined with cuticle produced by two types of epidermal cells. Specialized exocrine-gland cells produce the chemicals and export them across membranes dense with microvilli into the upper end of a cuticular duct (end-organ) around which the secretory cell is wrapped. This and all the downstream ducts are produced by another type of specialized epidermal cells, the duct cells, which create a collecting system that carries the products of the gland cells ultimately to a main duct and the poison sac. The main drain also receives the products of the filamentous gland at its other end. Because the venom is toxic, the cell exports it in a blocked form (usually a sugar is bound to the active group), along with an enzyme to unblock it. The final step in toxin synthesis then takes place outside the cells in tubules impervious to the toxin, eliminating self-poisoning. The tubules' great length allows the reaction to run to completion before the venom enters the reservoir. This venom apparatus is found in both worker and queen fire ants but differs in the nature of its contents, its structural details, and its uses.

Venom is delivered to the sting apparatus through the main exit duct, which first receives the duct from the Dufour's gland, the source of trail pheromone. Nevertheless, the trail pheromone and venom can be delivered separately. The worker's sting can be used to inject venom into a victim or to spatter it around the nest, or it can act as a quill to draw a trail of pheromone on the ground. The structure of this versatile tool is similar to that in aculeate Hymenoptera in general (Figure 22.1).

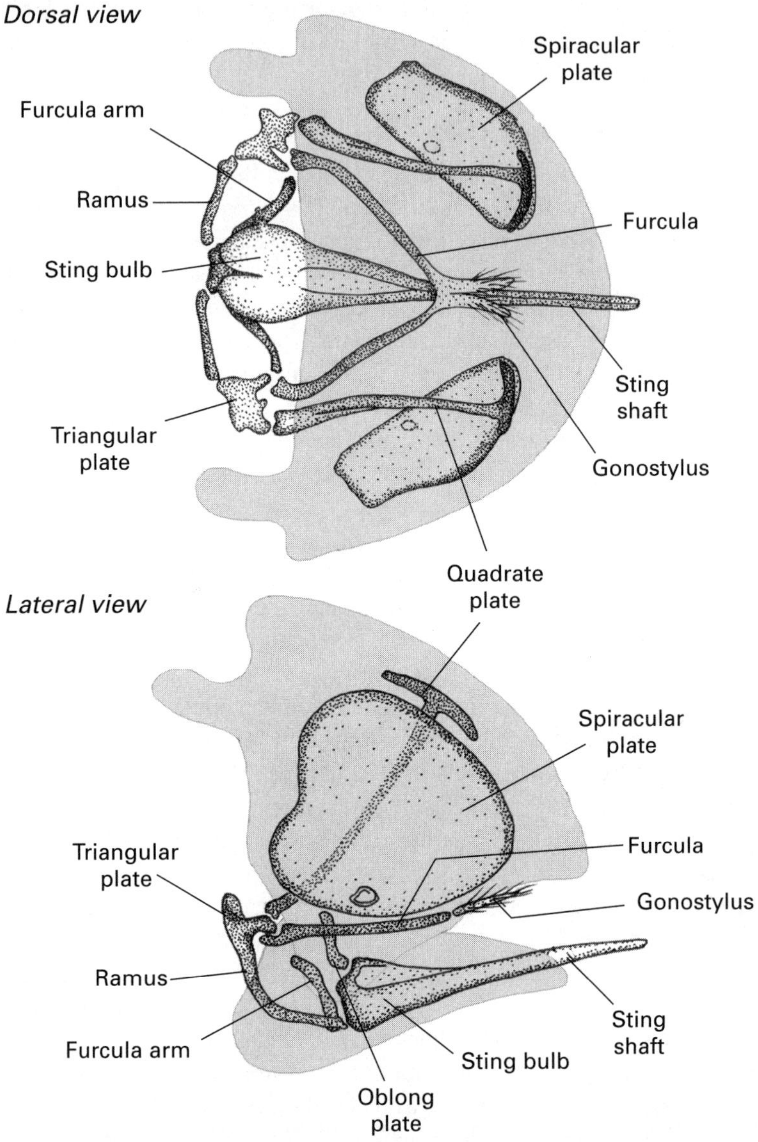

Figure 22.1. Sting apparatus of a large *Solenopsis invicta* worker. The sting is a modified ovipositor, which itself evolved from modified leg bases. Major cuticular plates are labeled.

By means of a tongue-and-groove arrangement, the valvulae form a hollow, pointed lancet, which receives the main exit duct of the poison sac reservoir at its inner end. The expanded inner end of these valvulae (the sting bulb) is articulated with several levers and arms of cuticle. The muscles attached to these allow the lancet to be protracted from the tip of the gaster, ready to be rammed into a victim or to pen a trail. Stinging ants, including fire ants, have "pump stings"—as alternate movements of the lancets bore the sting deeper, this same action sucks venom from the reservoir like a piston and injects it into the victim (Maschwitz and Kloft, 1971). Another set of muscles allows retraction of the apparatus. Hair sensillae and other sensory structures provide feedback for control.

Not surprisingly, the size of the sting apparatus and reservoir increase with worker size but less than proportionally, so major workers have 25% less venom per milligram body weight than minor workers—venom volume increases about 1.6-fold for every doubling of worker weight. Minor workers usually contain between 10 and 50 nl of venom and majors between 50 and 150 nl but sometimes more. Variation is substantial.

Stinging Behavior and Venom Dose

Stinging begins when a worker grasps the victim's skin, cuticle, or hair in order to anchor her body, then swivels her gaster forward between her legs, repeatedly inserting the weakly serrated stinger at an angle forward of vertical in slightly different locations. In my laboratory, Kevin Haight induced fire ants to sting small pieces of orange pH paper through a skin-like film of silicone cement (Haight and Tschinkel, 2003), creating green spots whose areas were proportional to the amount of venom injected. Small workers were much more vigorous stingers, averaging almost seven stings per bout, contrasting with four for large workers (Haight, 2002). As both injected about 0.66 nl (0.56 μg) venom per sting on average, small workers deliver 75% more venom during each stinging bout, in spite of weighing one-fourth as much as large workers and having less venom. Occasional workers delivered six to 17 times the average dose. To the extent that stinging is defensive, small workers do a great deal more of the defensive work, suggesting that large workers are not specialists in defense and probably should not be called soldiers. Averaged over all workers, the venom sacs contain about 18 μg of venom, 32 average doses, per worker (but variation was high). The venom dose per sting delivered by nest defenders was about 50% greater than that of foragers and was 55% higher in the spring than in other seasons.

Chemical Composition of the Venom

Fire-ant venom is highly unusual among hymenopterous venoms, although not quite as one author put it, "extremely unique." The effective ingredients of wasp and ant venoms are usually biologically active proteins, but all fire ant venoms consist of alkaloids with only a trace (0.1%) of protein. Alkaloids of animal origin are rare—among ant venoms, only those of *Solenopsis* and *Monomorium* contain alkaloids (Blum, 1988).

In the 1950s and 1960s, natural-products chemists began to collaborate with biologists to identify the many interesting pheromones and defensive secretions produced by arthropods, using new methods for

Figure 22.2. Gas chromatographic trace of the venom of *S. invicta*, showing the major alkaloid peaks. The peaks all correspond to alkylated piperidine alkaloids differing only in the long side chain (alkyl group) at the 6-position on the ring, which could consist of 11, 13, or 15 carbons (C11, C13, C15) with or without a double bond (:1). Adapted from Vander Meer et al. (1985).

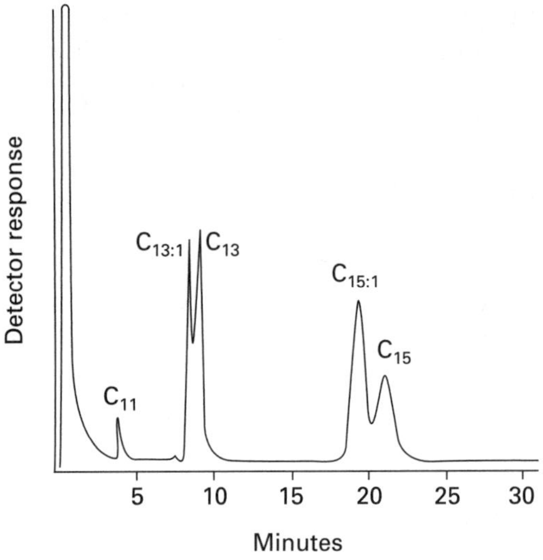

working with small quantities. Initially, fistfuls of fire-ant workers were dumped into a blender full of solvent and the switch flipped to "purée." G. A. Adrouny of Tulane University used the powerful hemolytic (red-blood-cell-rupturing) action of the venom as a biological assay to isolate 160 mg of pure material from about 30,000 ground-up ants (Adrouny et al., 1959). At a mere 30 parts per million, this material destroyed all the red blood cells in a suspension in 15 seconds. Unfortunately, he misidentified the compounds (Sonnet, 1967). At about the same time, Blum and Callahan (1960) ground up 1.2 million fire ants and isolated the insecticidal compounds from them. These compounds turned out to be the same as the hemolytic compounds isolated by Adrouny. This result stimulated Murray Blum and his collaborators John Brand and J. G. MacConnell to reinvestigate the chemistry of fire-ant venom and eventually resulted in the correct identification of all of the venom's major and many minor compounds (Blum et al., 1958, 1961, 1973; Blum and Callahan, 1960; Sonnet, 1967; MacConnell et al., 1970, 1971, 1974, 1976; Brand et al., 1972, 1973a, b; Brand, 1978; Blum, 1984, 1985).

Later venom studies relied either on dissecting out venom sacs of hundreds of fire ants or on collecting uncontaminated exuded droplets of venom in capillary tubes from the stingers of workers held with forceps. Someone tolerant of tedium could be paid a menial wage to do this collection. Venom yields from both methods ranged from 10 to 80 μg per worker, depending on worker size.

Gas chromatography separates the venom milked from *S. invicta* into five major and several minor components (Figure 22.2). Spectroscopic and chemical analysis, as well as chemical synthesis, showed these to be alkylated piperidine alkaloids. The five alkaloids, named solenopsins, differed from one another only in the nature of the side-groups—a methyl group always occupied the 2-position on the ring, whereas the 6-position was occupied by a long side chain (alkyl group) of 11, 13, or 15 carbons with or without a double bond. In addition, the molecules could differ in their three-dimensional geometry. The methyl and alkyl groups were either *cis* or the *trans* with respect to the plane of the ring. The double bond in the alkyl side chain was always in the *cis* configuration. Combinations of these characteristics create a family of 12 different compounds, each designated by an abbreviation. For example, *cis*-C_{13} designates the alkaloid with a 13-carbon side chain *cis* to the ring methyl group; *cis*-$C_{13:1}$ designates the same compound but with one double bond in the 13-carbon side chain, and so on. No 2-methyl-6-alkylpiperidines had ever before been isolated from animals, although several plant species produce them.

The relative amounts of the five main venom alkaloids change with both age and body size (Figure 22.3; Deslippe and Guo, 2000). The unsaturated C_{13} and C_{15} alkaloids decline greatly in their dominance over the saturated ones as workers age. Their dominance also declines with increasing body size and is below unity in workers of 1.2 mm head width and greater. Some of the five alkaloids are positively allometric with body size, some negatively.

Minim workers, the first tiny workers produced by founding queens, have venom that is about 94% $C_{13:1}$ alkaloid, very different from that expected from a downward extrapolation of the size-composition relationship for percent $C_{13:1}$ alkaloids. On this basis, Vander Meer (1986a)

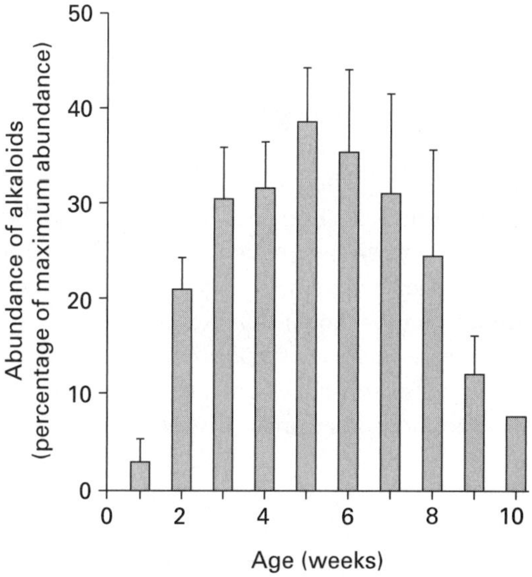

Figure 22.3. Venom alkaloid abundance peaks in midlife and declines with age. Abundance was measured relative to highest peak. Adapted from Deslippe and Guo (2000).

suggested that minims are a distinct worker caste, not merely the smallest workers. Minims are not only small, however, but also young (and short-lived). Their venom contains levels of $C_{13:1}$ alkaloids comparable to those of young minor workers (Deslippe and Guo, 2000). Vander Meer was probably comparing older minors to young minims.

Biosynthesis of the Venom

The cells that synthesize the venom have smooth endoplasmic reticulum and few ribosomes, in keeping with the nonproteinaceous, alkaloidal nature of their secretion (Billen, 1990). On the other hand, the cells have abundant mitochondria to provide the energy for the alkaloid synthesis. A good deal is known about the biochemical synthesis of these alkaloids, mostly as a result of the application of radioisotopic tracers. Because this information is biochemistry, rather than social biology, I will give but the briefest of overviews. Acetate compounds radiolabeled at either the first or second carbon position were fed to a colony of *Solenopsis geminata,* and the radioactive solenopsins were isolated. A series of specific chemical degradations cleaved them into known fragments whose radioactivity was measured. Overall, the pattern of radioactive atoms in the degradation products was consistent with biosynthesis from acetate units, unusual for insect alkaloids. Unfortunately, the authors were not able to answer the question of just how the ring is formed or how the *cis* and *trans* isomers come into being.

Concentrations of the alkaloids in the poison sac build up beyond saturation, causing phase separation (many arthropod venoms contain two phases). The alkaloids make up most of the organic phase, 95% of the secretion. The remaining 5–7% consists of tiny droplets of a suspended watery phase containing the unblocking enzyme and blocking group, glucose. The proteins in the watery phase of *Solenopsis invicta* venom total only 0.1% of the venom, making their identification very challenging. Nevertheless, separation of these proteins by molecular size resulted in four major peaks. The first peak contained the enzyme phospholipase, which cleaves phosphate bonds. The enzyme hyaluronidase (dissolves the "glue" holding cells together) was detected in the whole venom but was not localized to one peak (Baer et al., 1977, 1979). Other studies identified the enzymes N-acetyl-β-glucosaminidase and glucosidase, whose function is probably to catalyze the final unblocking of the alkaloids. At least three of these minor proteins cause the allergic response experienced by some humans upon being stung by fire ants (see below). Thus, in spite of making up only a tiny fraction of the venom, these proteins cause a great deal of trouble for some humans and possibly for other mammals.

Taxonomy of the Venom

The species of fire ants differ in which particular members of this family of compounds they secrete and in the proportions in which they secrete them. *Solenopsis invicta*'s venom contains five of these possible compounds, all in the *trans* configuration, with only traces of the *cis; S. richteri* lacks the C_{15} and $C_{15:1}$. Like those of *S. invicta,* its C_{11}, C_{13}, and $C_{13:1}$ are all in the *trans* configuration. Other species of fire ants have much simpler venoms that include alkaloids in both configurations. *Solenopsis geminata, S. aurea,* and *S. xyloni* (all in the *geminata* species complex) have major amounts of *cis-* and *trans-*C_{11}, as well as traces of *cis-*C_{13} and *cis-*$C_{13:1}$ (Figure 22.4). Assuming that the ancestral venom was mostly proteinaceous with a trace of alkaloid, we can postulate an evolutionary sequence. First, the protein was replaced by a single, thermodynamically favored *cis-*C_{11} alkaloid, as in *S. geminata* and *S. xyloni.* Next, small amounts of a ring-unsaturated C_{11} alkaloid convertible to either *cis-*C_{11} or *trans-*C_{11} were added, as in *S. xyloni.* The *trans* isomers probably evolved because they are more toxic than the *cis.* Next, alkaloids with longer side chains (C_{13}, $C_{13:1}$) were added, as in *S. richteri.* Finally, the C_{15} alkaloids appeared, as in *S. invicta* (*S. richteri* and *S. invicta* are in the *saevissima* species complex). These changes made the venom more effective—the stings of *S. invicta* and *S. richteri* have more severe and longer-term effects than those of *S. geminata* and *S. xyloni.*

The differences between species are large and useful as taxonomic characters for species of *Solenopsis* (MacConnell et al., 1976). Composition was mostly consistent with the taxonomy at the species-group or species-complex level. Within *S. invicta* or *S. richteri,* secretions from different North American and South American populations differed only moderately in composition. This pattern is probably typical of most species. The general similarity extended to species within species complexes. In contrast, large differences distinguished members of different species complexes and were generally correlated with morphological characters. Venom composition is not itself a sufficient character for sorting out taxonomy, but it can be useful in combination with more traditional morphological characters (Trager, 1991).

In the mid-1980s, Vander Meer and his collaborators discovered, rather by chance, that certain fire ants from northern Mississippi that were morphologically *S. richteri* had venom alkaloid mixtures intermediate between those of *S. invicta* and *S. richteri* (Vander Meer and Lofgren, 1988, 1989, 1990). This was the first evidence of widespread hybridization along the northern margin of the fire-ant range. A more complete discussion of this hybrid population appears in a later chapter.

Venom in Queens: Composition and Function

We saw above that the venom of *S. invicta* workers contains four alkaloids (*trans-*C_{15}, *trans-*$C_{15:1}$, *trans-*C_{13}, and *trans-*$C_{13:1}$) that are absent from the venom of *S. xyloni* and *S. geminata* workers. Interestingly, these alkaloids are also absent from the venom of their own alates. In

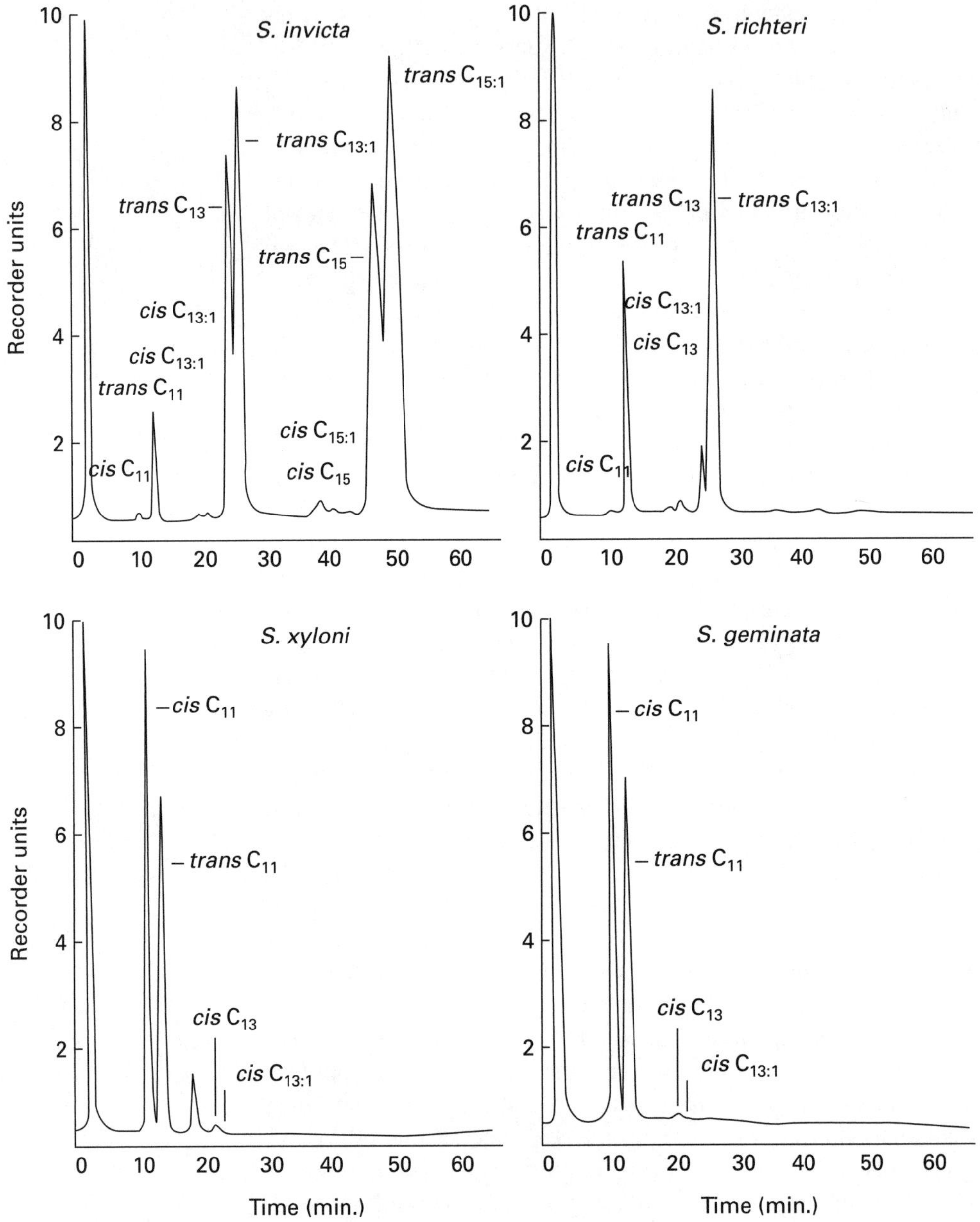

Figure 22.4. Venom alkaloid compositions of four species of *Solenopsis*. Qualitative and quantitative differences distinguish the species. Peaks are labeled as in Figure 22.2. Adapted from Brand et al. (1972).

fact, the venom of queens of all four species contains only *cis* and *trans* C_{11} alkaloids, in somewhat variable ratios (Figure 22.5). The function of queen venom is not defensive—alate and dealate fire ants rarely even attempt to sting. Instead, in addition to alkaloids, the poison sac contains components of the pheromones that regulate reproduction in fire-ant colonies, as described in the next chapter.

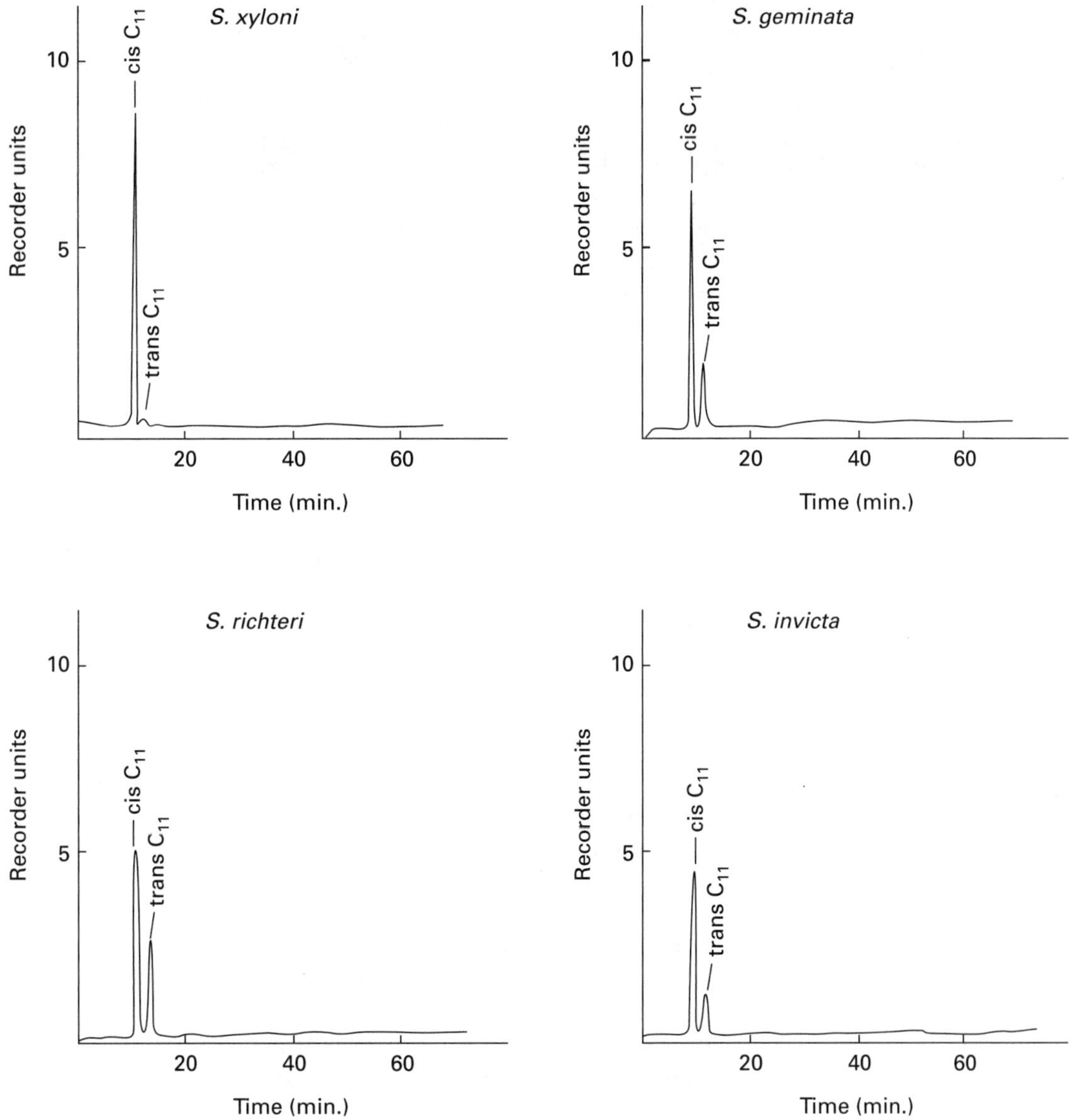

Figure 22.5. The venom alkaloid composition of the alate queens of four species of *Solenopsis* is simpler than that of workers. Adapted from Brand et al. (1973b).

Uses of Venom and Venom Action

Clearly, the venom and sting system of fire ants is of great importance to their lives. Well before the chemical identity of the venom was known, it was shown to be cytotoxic, insecticidal, repellent, bactericidal, fungicidal, herbicidal, allergenic, hemolytic, and necrotoxic, in short, a biocide. Venom also inhibits several membrane-dependent processes, including the responses of motor end plates and ATP production. Practically all of these effects are attributable to the alkaloids that compose 95% of the venom. All around, this is pretty nasty stuff. The wide range of toxicity suggests that it may have functions other than the defense against predation, and research has revealed that it does.

Nest and Self-defense. When a person blunders into a fire ant nest, the stinging he receives is obviously defense of the nest, although the human victims often take the attacks very personally. Most of what we know of fire-ant stings derives from studies on humans, even though the degree to which the venom evolved as a defense against large vertebrates is not clear.

When a fire ant stings, she injects a tiny smidgen of venom into the victim. It sets off a complex sequence of physiological events that quickly produces a reddened wheal at the site of the sting and causes a burning sensation that lasts 5–10 minutes and gives the fire ant its name. Synthetic, purified alkaloid compounds injected into human skin can produce all of the venom effects, including pain, in a dose-related manner, showing that these alkaloids are the active ingredients. The early stages of this process—local swelling, redness, itch, burning—appear to be a response to histamine, a common agent of tissue inflammation and pain (Read et al., 1978; Lind, 1982), but more than 99% of this histamine is released from the victim's cells in response to the venom rather than being a component of the venom. Antihistamine drugs injected together with fire ant venom greatly reduce both inflammation and pain. The most likely source of histamines is mast cells, a special kind of connective tissue cell that contains granules of histamine. Suspensions of these cells release all their histamines when treated with venom alkaloids diluted as much as 50,000-fold. The alkaloids interfere with cell-membrane function, leading ultimately to cell rupture (lysis).

At the sting site, the released histamines cause blood capillaries to dilate and plasma to enter the tissue along with several types of white blood cells. As events continue apace, the alkaloids rupture and kill the cells around the sting site, filling a space with their remains and forming a pustule with a thin epidermal roof as a succession of different types of white blood cells scavenge the debris. The pustule is not an infection—no bacteria can be cultured from its contents. Indeed, the venom is as toxic to bacteria as it is to mammalian cells. Many people,

I among them, cannot resist popping these little blisters and listening for the tiny "zick" sound. Rarely, opening them leads to secondary infection, but the pleasure of popping is worth the risk. Popped or not, after five to seven days, all that remains of the pustule is a discolored mark on the skin.

People often sustain multiple stings, reaping crops of multiple pustules, but for the great majority of people, even hundreds of stings present no danger of general, whole-body toxic effects. Inebriated persons using a fire-ant "bed" have sustained over 5000 fire-ant stings without signs of general toxicity (other than that of alcohol). The effects of the venom remain local to the site of injection. The consensus among physicians is that complications are rare.

In contrast to the great majority of humans, a small minority of hypersensitive people finds being stung very serious indeed; it can send them into a life-threatening anaphylactic shock (an extreme allergic reaction). A great deal of literature addresses this important topic, but because it is primarily medical, I will summarize it only briefly here (Triplett, 1973; Lockey, 1974, 1980; Rhoades et al., 1975, 1977, 1978, 1989; James et al., 1976a, b; Fox et al., 1982; Paull et al., 1983, 1984; deShazo et al., 1984; Johansson et al., 1985; Stafford, 1996; Freeman, 1997). Anaphylaxis is scary. Within minutes of the sting, generalized itching, rash, skin swelling, and shortness of breath progress to extreme difficulty in breathing, dizziness, and unconsciousness. Other symptoms include chest pain, nausea, vomiting, falling blood pressure, collapse, cyanosis, sweating, convulsions, and blurred vision. In the extreme, and in the absence of treatment, death can result. Of course, such extreme responses are not universal even among hypersensitive people, whose responses may range from localized swelling to whole-limb responses to whole-body responses. The onset of anaphylaxis usually results in a fast trip to the nearest emergency room, where injections of epinephrine and steroids, and sometimes oxygen, usually relieve the symptoms. Extremely hypersensitive individuals may be instructed to carry an epinephrine injection kit at all times, so that they can give themselves first aid if stung. Somewhat less sensitive individuals may respond well to Benadryl tablets.

Allergies, including hypersensitivity, generally require prior exposure to the antigen. In the great majority of cases, hypersensitive persons report having been previously stung, usually without whole-body consequences. The first exposure initiates the development of the immune response that is subsequently triggered by the next sting. The severity of the reaction increases with each stinging incident and, in the absence of stings, often fades with time. Bill Buren, the fire-ant taxonomist who named *S. invicta*, suddenly developed hypersensitivity after having been stung many times over many years by numerous fire-ant species.

As with a variety of allergies, hypersensitivity can be reduced through desensitization therapy, which is about 98% effective. Tiny but increasing doses of antigen are injected over several months, gradually reducing hypersensitivity. Less frequent injections maintain the patient indefinitely. Only about 2% of desensitized patients who were subsequently naturally restung experienced anaphylaxis (Hylander et al., 1989). In contrast, all hypersensitive but untreated patients that were naturally restung experienced anaphylaxis, showing that fire-ant allergies persist if untreated.

Despite the seriousness of hypersensitivity, actual numbers of cases are strangely difficult to determine, except that they are low. Studies differ in their methods, and the sampling basis is often unknown. Unchallenged allergies often fade, and repeated stings may act like desensitization treatments to reduce hypersensitivity in others. The hypersensitive population is therefore a moving target. Estimates have been made from surveys of general medical and allergy practices in various parts of the Southeast. A study in Jacksonville, Florida, estimated less than four new cases of hypersensitivity for every 100,000 population per year, 0.004% per year, of which one-fifth were at risk of anaphylaxis. Cumulated over 20 years, we would guess that perhaps 0.08% of the population should experience an anaphylactic reaction to fire ant stings. A survey of doctors in Mississippi, Georgia, and Alabama showed that medical cases increased as fire-ant populations increased (unpublished survey by Triplett cited by Rhoades, 1977). About 10% showed signs of systemic reaction, and 154 cases of anaphylaxis were reported, including 17 deaths (but few of these cases are well documented). Between 0.25% and 1.3% of the population suffered some level of systemic reaction, and 1–2% of these progressed to anaphylaxis, but these calculations depend on large assumptions and are not reliable. Other studies were not adequately documented, and still others estimated the frequency of hypersensitivity to be about 0.6% (Paull, 1984). A study of a suburb of New Orleans combined the sting rate with the hypersensitivity rate to estimate that about 0.2–0.4% of southerners, about 100,000 people, experience whole-body responses to fire ant stings annually, but a survey of 5300 physicians in 1988 revealed about 2600 patients under desensitization treatment across all southern states, far short of 100,000. Still, the real scale of the hypersensitivity problem remains fuzzy. Probably the best we can do is to say that something on the order of 1% of the population suffers at least some general allergic reaction to fire ant stings, and about 0.01% is in danger of anaphylaxis. These figures agree reasonably with the 0.08% estimated for Jacksonville above.

As mentioned above, the allergenic components of the venom are not the alkaloids but the trace of protein that it contains. Many insect venoms are highly allergenic, but then, they typically contain 10 to 40% protein, and doses are between 10 and 100 μg per sting. At only 0.1%

of the total venom, and an average injected venom dose of only 0.56 μg, only a fraction of a nanogram of protein is injected, a few ten-thousandths as much as during a typical bee or wasp sting. The protein in fire-ant venom is allergenic far out of proportion to the amount present.

The five proteins isolated from the watery phase of the venom all bind antibodies from allergic but not from normal individuals, i.e., they are all allergens (Baer et al., 1977, 1979; Hoffman, 1987; Hoffman et al., 1988a, b). The sera of individual patients showed strong responses to some (or one) of these proteins, but not to others, resulting in many patterns of reactivity to the four purified allergens—the antigenic action of each of these proteins is independent of the others. In addition, allergenicity varied seasonally (Hannan et al., 1984), peaking in May–June and possibly explaining why, although people are more likely to be stung by fire ants in the spring, they are more likely to suffer allergic reactions in the summer. The meaning of this seasonal variation is unknown, but it roughly parallels seasonal venom use, discussed below.

Some of the venom antigens of *S. invicta, S. xyloni, S. richteri,* and *S. geminata* are chemically almost identical, but others differ strongly and have parallel cross-reactivity to antibodies. In spite of these differences, enough antigens are shared to create broad cross-reactivity among fire ant species. All but one of 69 patients allergic to one species were also allergic to others.

One aspect of fire-ant hypersensitivity puzzled allergists for years. Many individuals who had never lived in fire ant regions proved to be allergic to their first fire ant stings, often within days of arrival in a fire-ant area. How could this be? Allergies require at least two well-separated exposures to the antigen. Could fire-ant venom share antigens with the venom of other Hymenoptera? A study by Hoffman and others (Hoffman et al., 1988c) found that 51% of individuals who had never been exposed to fire ants, but were allergic to stings of other Hymenoptera, cross-reacted to fire-ant antigens. These venoms contain a single antigen protein that is similar enough to one in fire-ant venom to cross-react. Cross-reactions were also obtained in about half the reverse cases, that is, those allergic to fire ants proved also to be allergic to other hymenopterous venoms.

What is the meaning of the widespread allergenicity of insect venoms? Is it just an unfortunate accident of nature, or is allergenicity the product of natural selection? This question has no clear answer, but Justin O. Schmidt, the Guru of Venom, suggests that, because the development of an allergic response requires at least two doses that are separated in time, allergenicity punishes repeat offenders, perhaps beginning with Africans who first raided beehives for their honey hundreds of thousands of years ago and proceeding to modern beekeepers. Hypersensitivity to bee venom affects 1–4% of Americans but up to 42%

of beekeepers. The allergenicity of fire-ant venom seems less likely to have evolved in response to humans, who have coexisted with the ant for only 10,000 years. The target (if one exists) is probably generally mammalian, but we do not know enough about either the tendency of other mammals toward hyperallergenicity or which mammals regularly raid fire-ant nests.

Repellency and Use in Confrontations. Venom has uses other than deterring small and large humans from sitting on fire-ant nests. For most of these, surprisingly, the ants dispense the venom not by stinging but by splattering, spattering, or dabbing droplets from the extruded sting while shaking the gaster up and down vigorously, a behavior dubbed "gaster flagging." Outside the nest, this conspicuous behavior occurs during confrontations with competitor ants. For example in a confrontation with *S. geminata* workers, the *S. invicta* worker appears to stand on her head, raises her gaster to vertical, pointing its extruded stinger (usually bearing a droplet of venom) at the zenith, then vigorously vibrates it back and forth, often while stamping her tiny little front feet (Figure 22.6; Obin and Vander Meer, 1985). Headstand flagging often followed initial contact and did not seem particularly oriented. When workers within 2–3 cm grappled, nearby workers of both species often gaster flagged. In a second type of gaster flagging, the gaster was oriented toward the object of hostility and appeared directly defensive. In such cases, the gaster is usually not raised above 45°, venom is rarely extruded, and gaster vibrations are much more vigorous.

Iodoplatinate applied as a coating to small glass plates turns from red to white or blue in response to alkaloids. Plates held near the nondirected headstand-flagging workers remained blank, but most of those held near prodded defensive-flagging workers showed small spots caused by droplets of venom thrown by the workers. Workers of *S. geminata* were occasionally repelled by gaster-flagging workers from distances up to 1.5 cm, withdrawing rapidly, wiping the antennae, and grooming. When repelled (but uncontacted) *S. geminata* workers were surface-washed in hexane, an average of 0.5 μg of *S. invicta* venom (which contains compounds not present in *S. geminata* venom) was recovered from them. Therefore, *S. invicta* workers literally hurl venom at their adversaries, and *S. geminata* probably do the same. Of course, a few control washes of *S. geminata* that had not been thus repelled would have made the case stronger. Gaster vibration may disperse up to 500 ng of venom in small, flying droplets during defensive flagging, about 2.5% of the venom sac's total content. Gaster vibration is also associated with sound production—the postpetiole–gaster joint is rubbed across a series of roughened ridges to produce a very faint squeaking sound. The possibility that sound produced by gaster-flagging workers helps spatter the venom has not been tested.

To an unknown degree, contests between *Solenopsis* species appear

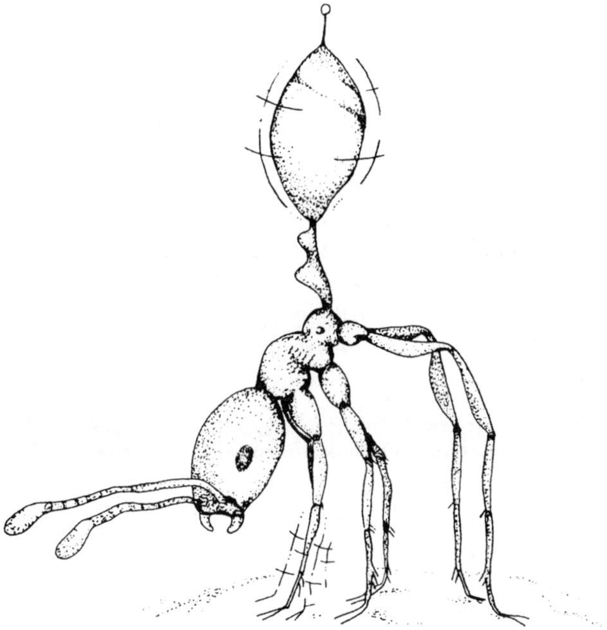

Figure 22.6. Headstand gaster-flagging behavior during encounters with other ant species. Note the extruded venom droplet at the tip of the sting. Reprinted from Obin and Vander Meer (1985) with permission from Robert K. Vander Meer.

to be chemical wars, each side trying to splatter opponents with noxious chemicals, but how venom properties and amounts affect the outcome of these contests is unknown. Use in territorial contests within species is worth testing, as is any role in competition between fire ants and other species of ants.

Subduing or killing prey is probably one of the most frequent uses of venom in the everyday lives of fire ants. Larger insects are usually quickly covered with stinging, biting workers and are quickly paralyzed or killed. Once the insect is motionless, it is cut up and transported back to the nest. No one has attempted to determine how much venom is expended in a typical colony's daily life.

In addition to being an injected poison, the venom of *S. invicta* is also a contact poison to a range of insects (Blum and Callahan, 1960). A dab on fruit flies produced instant paralysis and was about two to three times as toxic as DDT, weight for weight. For the C_{11} and C_{13} alkaloids, a topical dose of 0.6 µg/mg of body weight on termites killed half of them. The C_{15} alkaloids were about half as toxic. These toxicities were similar to that of nicotine, a powerful contact poison to insects (and humans, too, for that matter). Fire ants are much less sensitive to these alkaloidal contact poisons than are other insects. Strangely, no one has directly tested whether venom is used as a contact poison to subdue insect prey (researchers have missed a bet), and few quantitative studies have addressed the use of stinging during predation or during territorial encounters. Bhatkar and his colleagues found that *S. invicta* and a competitor ant, *Lasius neoniger*, both used chemical warfare during confrontations (Bhatkar et al., 1972). Fire-ant workers repeatedly sprayed with formic acid by *L. neoniger* usually curled up and died within 20 minutes. For their part, *L. neoniger* workers spattered with venom by gaster-flagging *S. invicta* workers retreated, grooming rapidly and repeatedly. Multiple contacts caused them to keel over and die within a few minutes. Lethal as these chemical weapons are, their importance to larger-scale competition between fire ants and *L. neoniger*, as well as other ant species, is not known.

Antimicrobial Use. Gaster flagging also occurred among workers tending brood. The gaster was never raised above 45°, the vigor of vibration highly variable, and the sting rarely extruded. In the brood chamber, gaster flagging produced sound, but showing that it also disperses venom was more difficult. First, no droplets of venom were ever visible. Second, surface-rinses of brood can be contaminated by venom extracted from the poison sacs of pupae or venom rubbed off excited adults during the brood separation. Quantifying these sources of contamination showed, however, that they made up less than 2% of the venom rinsed from the brood. After exclusion of the maximum possible contamination, the total venom on the body surface of each larva or pupa was approximately 1 ng. Unfortunately, the experiment does not exclude the possibility that the venom was transferred to larvae by grooming rather than by gaster flagging. In any case, it gets there one way or the other.

Why do workers coat their own sisters with venom, a topical irritant and contact poison? The broad toxicity of the venom to a range of microorganisms suggests that the venom acts as an antiseptic, much like Chlorox on a kitchen counter. Ants that nest in soil must cope with the huge range of bacteria, fungi, and other microorganisms for which soil is a fecund medium. The relationship of microbes to fire ants probably ranges from neutral to competitor to pathogen. In most ants, the metapleural gland produces an antiseptic, but in fire ants it does not, and the poison sac has assumed this role. Much of the evidence for antimicrobial use of venom is circumstantial and derives from the toxicity of venom to a wide range of fungi. Blum (1988) tested fire-ant venom alkaloids on cultures of 13 different species of fungus from "animals," plants, or ant brood, including fire-ant brood. Although variation was high, all of the venom alkaloids, especially the C_9 ones, caused complete or nearly complete growth inhibition among the fungi of plant and "animal" origin. More interesting is that several of the alkaloids were also highly toxic (more than 70% inhibition) to five of the eight fungi from ant brood. Interestingly, whereas fire-ant workers are quite resistant to the topical toxicity of their own venom, the venom resistance of their pupae and larvae has not been specifically tested. Still, application seems to do them no serious harm.

Another study (Storey et al., 1991) tested the effect of fire-ant venom alkaloids on the germination of four strains/species of entomopathogenic fungal spores, including one isolated from *S. invicta*. In all tests, spore germination was inhibited in proportion to alkaloid concentration. At 10 μg per cm^2 germination was less than 50%, and at 60 μg, it was 5%. The inhibition was only temporary, however. No matter what the alkaloid concentration, after 48 hours, most of the spores of three of the four strains had germinated. The strain isolated from *S. invicta* was of intermediate sensitivity. The venom alkaloids seem to affect germination by interfering with the synthesis of fungal cell walls. Three of these four strains were one-eighth as sensitive to alkaloids as were those tested by Blum.

All three venom alkaloids (C_{11}, C_{13}, C_{15}) were also toxic to eight gram-positive bacteria, including *Streptococcus, Staphylococcus,* and *Bacillus*, in paper disc tests on agar plates or in liquid culture (Jouvenaz et al., 1972). Sensitivity to alkaloids differed by no more than twofold. Toxicity tended to decrease from C_{11} to C_{15}. In liquid culture, two to eight parts per million, depending on the compound, killed 99.9% of *Staphylococcus aureus*, a gram-positive bacterium and the agent of a common skin infection. *Escherichia coli*, a gram-negative bacterium from the colon of mammals, was less sensitive, requiring 20 to 40 μg per ml to kill 99.9%. Gram-negative bacteria are little affected at concentrations that seriously affect gram-positive bacteria. Different species of another microbial group, the yeasts, occur in mound soil and in adjacent nonmound soil, but whether the difference is due in any part to the use of venom is unknown. The authors suggest that mound construction creates microhabitats that are suitable for some but not other soil yeast species (Ba et al., 2000).

Tempting as it is to believe that the ants spray their venom inside the nest in order to suppress microorganismal growth, and thus provide a healthier environment for themselves and their brood, the evidence is not conclusive. Demonstration of such a phenomenon must meet several conditions. First, the alkaloids must be present on larvae and/or nest soil at effective concentrations. In nests of workers in clean sand, alkaloids accumulate to about 0.7 to 0.8 μg per gram of sand (Storey, 1990), still far below the levels that inhibit several species of fungus. Even when venom alkaloids were experimentally quadrupled, the treated sand was no more inhibitory to fungus than the control. Similarly, in culture, inhibition of fungal-spore germination required concentrations of venom alkaloids about 1000 times greater than those on larval surfaces. The fungicidal activity reported by Blum occurred at 100 to 200-fold the amounts found on larvae. The observed alkaloid levels are also well below toxic levels for most bacteria.

We cannot currently reconcile these observations. First, disease susceptibility of larvae coated with varying concentrations of alkaloids or lacking such coatings must be tested. Second, suppression of microbial growth of "public health" importance to fire ants on nest soil and nonnest soil must be tested. Third, when pure alkaloids are added to nonnest soil in natural amounts, they must suppress microbial growth to a similar extent and specificity. Finally, although the quantification of venom on larvae and soil is difficult, it should be revisited.

The venom is also toxic to plants. It might be tempting to speculate that venom inhibits plant growth on fire-ant mounds, thus keeping them in better condition as solar collectors, but in fact, grass is often greener and taller on fire ant mounds than nearby.

Venom Synthesis Rates and Venom Economy

Clearly, venom is an everyday, workaday substance with an amazing array of uses, of great importance to colony health and success, and continuously dispensed by workers. Workers should therefore be able to replenish their venom supplies through biosynthesis, and venom production, or "the venom economy," might be a substantial part of the total colony energy budget.

Recent studies by Deslippe and Guo and my graduate student Kevin Haight quantified the venom content of the poison sacs of workers of different ages by gas chromatography. Workers start their adult lives with very little venom. Venom increases 13-fold by the fifth week and gradually declines until, in 10-week old workers, it is only about double what it was in one-week-old workers (Deslippe and Guo, 2000).

Such a rise and fall could have several causes. Haight sought to identify these causes in an experiment that sounds simple but was a nail-biter to manage (Haight and Tschinkel, 2003). He set up even-aged worker cohorts. At intervals of two weeks, he measured the venom-sac contents of one third of the workers by dissection and milked the remaining workers of as much venom as they would give. Half of these milked workers were then returned to the nest, and the venom remaining in the venom sacs of the other half was measured. Finally, when the milked workers had had two weeks to synthesize replacement venom, their poison sac contents were also measured, and their venom synthesis rate during the previous two weeks was determined by comparison with the amount present in their cohort sisters that had been milked immediately before venom measurement.

Remarkably, the rate of synthesis was highest during the first week of adult life, declined 50% by the second week, and was not significantly higher than zero after 20 days. In other words, workers make their entire lifetime supply of venom during the first three weeks of adult life. Because the broodless workers in Haight's experiment had little opportunity to use venom, their supply remained roughly steady after about three weeks. Deslippe sampled complete colonies in which workers presumably dispense venom for various uses, possibly accounting for the decline in venom after midlife. Because synthesis is limited to early life, even a constant venom-use rate would result in a midlife peak.

On the basis of the energy content of venom, less than 6% of the energy required to produce a worker was allocated to venom production. If defense of sexuals is one of the potential benefits of stinging, then venom is extremely cost-efficient. A single sexual female has the same caloric cost as roughly 1900 to 5600 (spring) doses of venom. A colony must deliver more than 5600 doses, a vigorous defense indeed, before it is no longer energetically worth defending a single female sexual. At 4700 sexuals per colony per year, that would come to 26 million doses, 23 million more than an average colony even contains. Clearly, the energetic value of sexuals far exceeds the amount of venom *S. invicta* colonies contain and the amount an effective defense is likely to require.

Total venom content of colonies obviously increases during growth, but whether it does so isometrically or allometrically is not clear. An average mature *S. invicta* colony of 160,000 workers, 35% of whom are majors, contains roughly 3.2 g of venom. During this colony's seven-year life span, it produces 2.4 million workers (2 kg), who produce roughly 43 g of venom, about 2.2% of the total weight of ants. The amount of this venom dispensed during quotidian tasks or defense and the amount present at colony death are unknown. Indeed, much remains mysterious about fire-ant venom and its uses.

You Call That Pain!?

Many southerners grow hyperbolic when describing the pain of fire ant stings, but to the connoisseur of pain, the fire ant is less than ordinary, outranked by a wide variety of ants, wasps, and bees. For example, have yourself stung by the Florida harvester ant (*Pogonomyrmex badius*) for comparison purposes. The worker will painlessly slip her stinger into your skin and, before you feel anything, will inject a proteinaceous venom that is, weight for weight, 20 times as toxic as rattlesnake venom. By the time you notice that you have been stung, it is too late. For at least 24 hours the sting will throb with a dull, chronic pain. The dampness you feel on your skin at the site of the sting is not sweat brought on by the inflammation; it is plasma leaking out through the skin as a result of an enzyme that unglues your cells from one another. The lymph nodes nearest the sting become painful, and the victim may suffer flu-like symptoms. Even so, the harvester ant is a mere tap on the forehead compared to a Central American ant fondly known as "the bullet" (*Paraponera* spp.). Following a sting from one of these giant ants, a person can enjoy blinding pain as the knees go weak and the hands tremble. Lying down is recommended. Now that's a sting!

Not many people systematically study the pain associated with insect stings, but Justin O. Schmidt, whose specialty is venom, has kept careful notes on the pain caused by the inevitable stings he receives in the course of his work. Putting together his years of notes, he produced the following pain-rating scale for stinging Hymenoptera (reprinted with permission from *Outdoor Magazine*). It remains only to be determined whether venoms come in vintage years.

Justin O. Schmidt's Pain Rating Scale

Sweat bee. Light and ephemeral, almost fruity. A tiny spark has singed a single hair on your arm. Pain rating: 1.

Fire ant. Sharp, sudden, mildly alarming. Like walking across a shag carpet and reaching for the light switch. Pain rating: 1.2.

Bull-horn acacia ant. A rare, piercing, elevated sort of pain. Someone has fired a staple into your cheek. Pain rating: 1.8.

Bald-faced hornet. Rich, hearty, slightly crunchy. Similar to getting your hand mashed in a revolving door. Pain rating: 2.

Yellow jacket. Hot and smoky, almost irreverent. Imagine W. C. Fields extinguishing a cigar on your tongue. Pain rating: 2.

Harvester ant. Bold and unrelenting. Somebody is using a power drill to excavate your ingrown toenail. Pain rating: 3.

Southern paper wasp. Caustic and burning, with distinctly bitter aftertaste. Like spilling a beaker of hydrochloric acid on a paper cut. Pain rating: 3.

***Pepsis* wasp.** Blinding, fierce, shockingly electric. A running hair dryer has just dropped into your bubble bath. Pain rating: 4.

Bullet ant. Pure, intense, brilliant pain. Like walking over flaming charcoal with a three-inch nail imbedded in your heel. Pain rating: 4+.

Rattlesnake. Deep, penetrating agony. Unmistakably full-bodied. Analogous to shooting a hot slug into your arm. Pain rating: well off the chart.

23

Social Control of the Queen's Egg-laying Rate

Control of reproduction is a central feature of ant colonies. Optimal fitness requires coordination of the egg-laying rate with the capacity of workers to rear brood so that the brood-rearing capacity of the worker force is neither exceeded nor left idle. Queens from large colonies are easily observed to be very corpulent and to have huge ovaries, whereas queens from small colonies weigh considerably less and have less developed ovaries; their egg-laying rates differ accordingly. Somehow, the colony translates information about its own size accumulated from several hundred nest chambers and two vertical meters into control of the queen's ovaries. Finding out how tiny ants with tiny nervous systems accomplish this huge task consumed four years of my life.

Egg laying is continuous. A queen in a large colony lays an egg or two every minute, around the clock. She is highly attractive to her workers, who face her to form an excited retinue of 40–60 workers around her (Plate 15). These workers lick and groom her cuticle, probably receiving a dose of pheromone for their trouble. Those at the front of the queen feed her at intervals by trophallaxis; those at her hind end receive the eggs as they are laid.

The imminent appearance of an egg is signaled by the increased excitement of the workers at the queen's hind end. One to several of them begin to antennate the tip of the queen's gaster, leaning forward expectantly, and within moments, the queen's sting begins to protrude slightly and the tip of her gaster to pulse rhythmically. A few seconds later, a glistening egg appears from her vulva underneath the base of her sting. The stinger is retracted across the egg, coating it with material from the poison sac. A worker avidly grabs the egg (workers often jostle each other for the honor) and immediately leaves the retinue (the "halo" of workers surrounding the queen), always pausing just outside the retinue, as if to show the observer her prize. She then proceeds to a nearby egg pile and adds the egg to this sticky clump. In an undisturbed nest, a queen is usually surrounded by a halo of egg clumps a short distance outside her retinue.

Most of the tiny dose (1.5 to 4 ng) of poison-sac secretion that the queen deposits on each egg consists of venom alkaloids unique to the queen (*cis*- and *trans*-C_{11} alkaloids). These compounds can be rinsed off freshly laid eggs, not yet handled by workers (Vander Meer and Morel, 1995), and are known to be powerful antifungal and antibacterial agents (Blum et al., 1958). They probably prevent eggs from succumbing to disease or mold in their microbe-rich soil nests (Storey et al., 1991), but this conjecture has not been experimentally tested.

The queen's poison sac also contains a minor amount of a worker attractant (about 4 ng), one or more primer pheromones, and minor amounts of sesquiterpenoids of unknown function. Because these all occur together in solution, all are deposited on each egg as it is laid. This process ties the pheromone-release rate to queen fecundity through the simplest mechanism possible—the more eggs, the more pheromone. An egg is therefore not only a future ant but also a vehicle for a chemical communication system that regulates several aspects of reproduction, including queen attractiveness, egg care, and suppression of reproduction in female sexuals. Whether the queen has voluntary control over this anointment or it is an automatic part of laying eggs is unknown. One suspects the latter. The function and control of the associated Dufour's gland in the queen is unknown.

Although some researchers measured oviposition rate by isolating queens with a few workers for five hours (Fletcher and Blum, 1983a), I watched the undisturbed queen for half an hour under a movable dissecting microscope and counted the eggs as she laid them (Tschinkel, 1988c). Egg-laying rate was determined on multiple replicates four days after the experimental conditions were applied.

Because queens in larger colonies are heavier and lay more eggs, we might naturally expect that more workers would lead to a higher the egg-laying rate. This mechanism would coordinate the number of workers with the number of larvae needing care. The prediction is dead wrong. No matter whether the (broodless) experimental colony had 10 workers or 10,000, the queens laid very few eggs, and the egg-laying rate was unrelated to the number of workers. On the other hand, addition of larvae to half of the broodless experimental nests caused the queen's egg-laying rate to surge to a peak in four days, while it remains low in the broodless controls. Just as clearly as workers alone do not control queen fecundity, this experiment shows—totally unexpectedly—that larvae do. Every once in a while, it is refreshing to be jerked awake by an unexpected outcome.

This discovery opens a world of labor-intensive experiments. First, how is the number of larvae related to the egg-laying rate, and does the number of workers make a difference when larvae are present? For six combinations of larva and worker numbers, queen fecundity increased with the number of larvae but was unrelated to the number of workers over at least a 16-fold range of worker:larva ratio. The number of workers is therefore irrelevant, and subsequent experiments could simply use 3000 as a standard number.

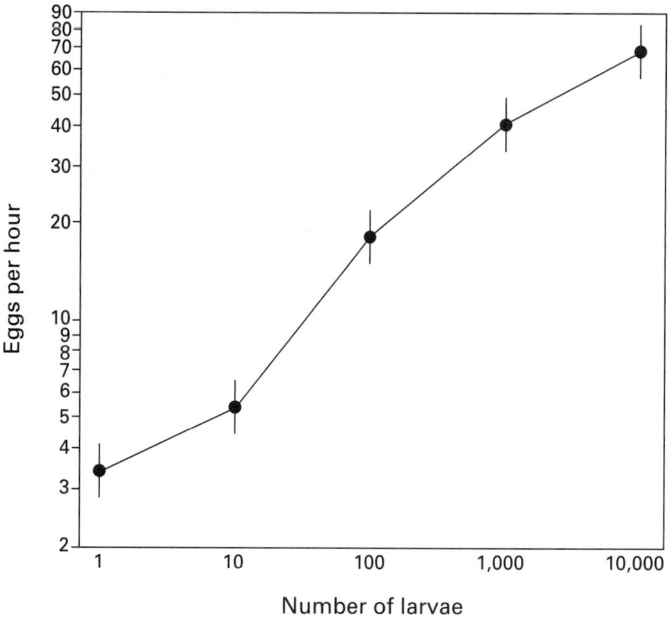

Figure 23.1. The relationship between the number of larvae and the queen's egg-laying rate, as determined in an experiment in which the number of larvae in nests was varied. Note the log-log scales. A 10-fold increase in number of larvae approximately doubles the egg-laying rate. Within wide limits, the number of workers has no effect. Data from Tschinkel (1988c).

When nests with 3000 workers were supplied with 1 to 10,000 larvae, the queen's egg-laying rate increased from about four eggs per hour to almost eighty (Figure 23.1). The relationship was allometric—every 10-fold increase in the number of larvae only approximately doubles the queen's egg-laying rate. Said another way, each *additional* larva had a smaller effect—adding one larva to a nest containing one larva increased the rate by 1.6 eggs per half hour, but adding that same larva to a nest already containing 10,000 larvae increased it by only 0.0074 eggs per hour, a 430-fold decline in effectiveness. The queen's fecundity is thus increasingly resistant to change as the larval population grows. Larval ability to stimulate fecundity approaches an upper limit.

Eggs, of course, do not appear out of nowhere. They are the product of a complex physical and metabolic machine consisting of the queen's ovaries, each composed of 80 to 100 ovarioles (Figure 23.2; Hermann and Blum, 1965), her fat body, and her endocrine glands. All these must be coordinated across an enormous range of activity. Whereas a queen with regressed ovaries weighs only about 9 mg, one at peak reproductive rate weighs 25 to 29 mg. Between 50 and 70% of this weight is ovaries and developing eggs. Such a queen may lay half her weight in eggs every day. The name "queen" suggests command and power, the peak of an immense hierarchy. The reality is far different—the queen is an egg-laying machine, the evolutionary victim of an extreme division of reproductive labor. Such are the hazards of anthropomorphic metaphors.

As reproductive demand increases, the egg-making machinery is gradually activated, probably by the secretion of juvenile hormone (Brent and Vargo, 2003). In a broodless nest, the queen's ovary contains only 70 to 90 yolk-sequestering (vitellogenic) follicles and many inactive ovarioles. As the number of larvae increases from zero to about 100, all the ovarioles are first brought into activity, and the number of vitellogenic follicles in each increases from less than one to 10 or more. The total number of vitellogenic follicles steadily increases until, in the presence of 10,000 larvae, they number 1500 to 2000. In parallel, the queen's retinue increases from about 35 workers to about 60, but most of this increase has occurred by the time the nest has 100 larvae. For reasons that are not clear, the rates of egg laying by queens in naturally growing colonies start lower than that in lab colonies and increase more rapidly as the number of larvae grows, but the final levels of fecundity in very large colonies in lab and field are similar.

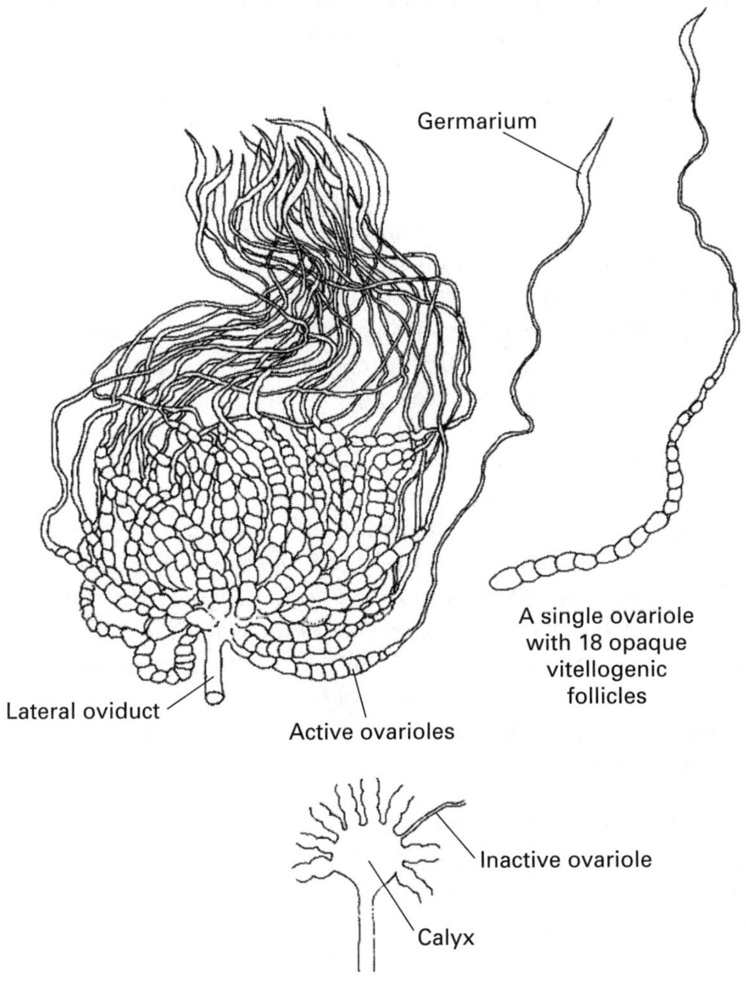

Figure 23.2. The ovary of an actively egg-laying fire-ant queen, dissected to show separate ovarioles and structure of the calyx (the base to which the ovarioles are attached). Meiosis takes place in the germarium. Fully developed eggs pass into the calyx in preparation for laying. Adapted from Tschinkel (1988c).

All these changes are, of course, interdependent, being regulated parts of an assembly line whose input is food and whose output is eggs. As a result, for every additional two eggs per hour that a queen lays in an experimental nest, she spends an additional 1.2 minutes feeding and has an additional 53 vitellogenic oocytes, or 0.25 oocytes per ovariole. The activity of the fat body, which synthesizes the yolk protein, probably undergoes correlated changes.

The size of the eggs decreases as larval number and egg-laying rate rise. Between 10 and 10,000 larvae, egg volume decreases about 22%. Therefore, for the same amount of material, the queen lays 22% more eggs in the presence of 10,000 larvae than in the presence of 10. The likely cause is a 26% decrease in the time it takes the follicles to move down the ovarioles. Given constant yolk uptake rate, faster movement means smaller eggs. Significantly, the largest eggs in the life cycle are laid by newly mated queens, who no doubt also have the lowest egg-laying rate. These giant eggs range from 17 to 20 nl in volume, as compared to the 3- to 4-nl eggs of highly fecund queens. Whether egg size itself has any effect on subsequent development is unknown, but interestingly, the smallest workers of the life cycle, the minims, develop from these largest eggs and vice versa.

Just as the addition of larvae to a broodless nest causes a surge in the queen's egg-laying rate, so removal of larvae causes a precipitous decline to near zero within 48 hours. In broodless nests, total vitellogenic follicles in field-collected queens dropped from about 4000 to almost none, and they lost 40% of their initial weight, mostly as eggs. Clearly, larvae initiate and maintain queen fecundity, but it is worth remembering that both rapid increase and rapid decrease are experimentally induced. In a natural colony, the larval population would grow gradually.

Most of the previous experiments simply used "large" larvae. Tested separately, large (late fourth-instar) larvae induced almost three times the egg-laying rate as did the same weight of small (second-, third-, or early fourth-instar) larvae. Pupae were equivalent to no brood at all, and a mixture of pupae and larvae did not differ from the larvae alone. On a weight basis, sexual larvae were only 5% as stimulatory as large worker larvae, but on an individual basis, they were about the same. Taken together, the fecundity-stimulating effect is a characteristic primarily of the older larvae (regardless of caste). It is either much weaker for or absent from younger larvae and ceases entirely upon pupation.

A Mechanism of Stimulation

How might late fourth-instar larvae stimulate the queen? Perhaps larvae produce a pheromone that activates the queen's reproductive machinery or produce a nutritive material transmitted to the queen by the workers. Of course, they might do both. Movement of fairly large

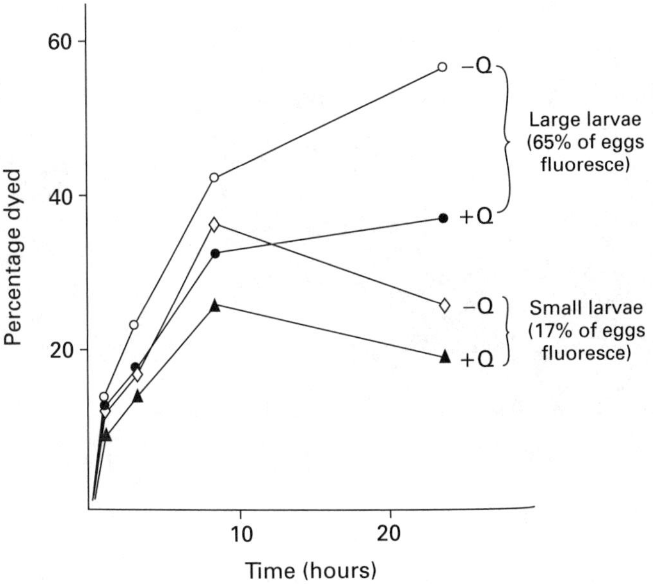

Figure 23.3. Fluorescent food moves from larvae to workers to the queen and eggs. More of it moves from large than from small larvae. Q+, colonies with queens; Q-, queenless colonies. Adapted from Tschinkel (1988c).

quantities of material should be detectable by means of some simple tricks. Let us feed larvae of diverse sizes in one set of colonies on food that has been dyed with the brilliantly fluorescent dye rhodamine B (Day-Glo orange). The next day, all the larvae are quite pink and shine like orange Japanese lanterns under ultraviolet light. These larvae are then separated into large and small size classes, equal weights of which are added to different, undyed experimental colonies, half of which are queenright and half queenless. At intervals, we crush a few workers and inspect for fluorescence under ultraviolet light. Fluorescence in workers could only have come from the larvae.

This experiment is quite illuminating (Figure 23.3). Fluorescence moves from the larvae into the workers, more rapidly so and to much higher levels from large larvae than from small. In both larval treatments, queenless colonies had more fluorescent workers than queenright colonies, suggesting that queens act as a sink for the fluorescent larval material. Sure enough, after 24 hours, the queens lay fluorescent eggs, and a much higher proportion of fluorescent eggs and brighter fluorescence appear in the large-larva colonies. Altogether, these findings show that a bulk material moves from large larvae to the workers to the queen and into the eggs. We do not yet know the nature of this material, nor do we know its role in stimulating queen fecundity.

The end of the larval period is signaled by the cessation of feeding, followed by apolysis (separation of the hypodermis from the cuticle) and the formation of the new pupal cuticle within the old larval cuticle (pupae thus doubly enclosed are called pharate pupae). One of the

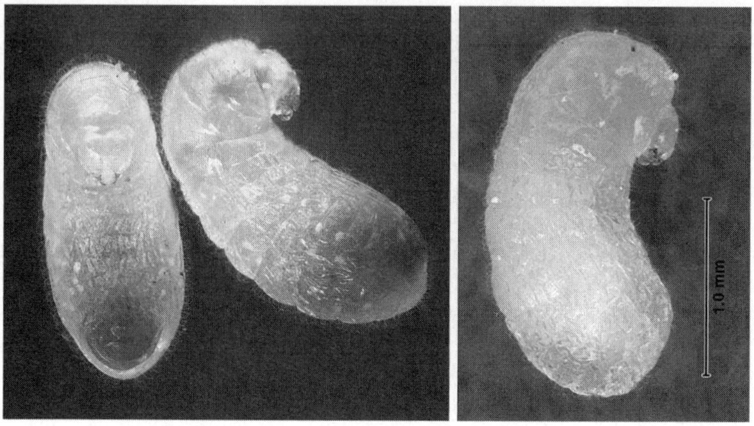

Figure 23.4. The meconial (left panel) and early post-meconial (right panel) stages of larval development. The meconium forms when the midgut and hindgut connect early in the pharate pupal stage. All food residues in the midgut then move into the hindgut to be voided as the meconium. The volume thus lost is so great that the post-meconial pharate pupa appears shriveled. The meconium is recognizable only for about two hours.

abiding peculiarities of the Hymenoptera is that, during the larval stage, the midgut does not connect to the hindgut. The larva therefore cannot defecate and retains all the undigested food it has ingested in the midgut. With the completion of the gut during the early pharate pupal stage, the entire contents of the gut are defecated as a gigantic, black mass called the meconium, leaving the pharate pupa considerably smaller, chalky white, and wrinkled (Figure 23.4). The brevity of the meconial stage (the interval between the formation and defecation of the meconium, one to three hours) allows us to separate four developmental stages rather precisely by reference to it (Tschinkel, 1995). The experimenter must scan thousands of larvae every few hours to collect sufficient numbers (100 brood of each timed stage) for assessment of their role in queen stimulation, and egg-laying rates in the resulting experiments are therefore relatively low.

Daily replacement of the treatment brood kept the developmental stage reasonably constant over the four days of the experiment. The meconial stage stimulated by far the greatest fecundity, 60% higher than larvae in the middle of the feeding stage and 170% higher than pupae. That is, the meconial larvae produce three times as much stimulatory material as feeding larvae and 13 times as much as pupae. Most probably, meconial larvae are the source of the material that regulates the queen's ovarian function.

Must the queen be in the same nest with the meconial larvae? Queens were transferred between two equivalent nests every 12 hours so that they were either present together with fresh meconial larvae or followed them just after their removal. This experiment showed that the meconial larvae and queen must be present in the nest at the same time for the larvae to produce the stimulating effect. Even a 0- to 12-hour lag eliminated the effect. Perhaps workers do not collect the magic material in the queen's absence, or the workers do not share it with the queen when larvae are absent, or the material is unstable, or the workers do not store it for any appreciable period. The correct explanation is currently unknown.

How does this stimulation make its way from the immobile meconial larvae to the queen? Not only are these larvae scattered through many nest chambers, but they are not visited by the queen, so we naturally suspect that workers are the agents of transport. Most logically, we ought to focus on workers that tend meconial larvae or that are members of the queen's retinue. For controls, we can use workers that simply feed larvae by trophallaxis, in the ordinary way. In each case, we are interested in what each worker does after the initial, identifying act. To track focal worker reliably, we use workers that have been lightly sprayed with blue paint. No two workers will have paint dots in exactly the same location, allowing recognition even after a brief lapse. Microscope fatigue is an occupational hazard.

Workers first observed during the act of imbibing material from the anal end of meconial larvae are highly likely to proceed immediately to the queen's retinue and to offer trophallactic feeding to the queen. Workers first observed during trophallactic feeding of larvae almost never join the retinue and are probably "ordinary" nurse workers. Those workers first observed in the retinue tend to stay there. Three behaviorally distinct types of workers therefore seem to exist—larva tenders, queen tenders, and pharate-pupa tenders. Both the queen tenders and pharate-pupa tenders can be found in the retinue, but the queen tenders do little pharate-pupa tending. Similarly, larva tenders rarely collected anal exudate from pharate pupae, and pharate–pupa tenders rarely fed larvae. Altogether, these results suggest a special group of workers who shuttle back and forth between meconial larvae and the queen, transferring the stimulatory material to her, where it regulates ovarian function. The shuttle workers continue in this role for at least an hour at a time and probably much longer. The proportion of the workers who play this role is unknown but is probably small. Such small groups of workers doing rare and specialized work are sometimes called elites.

What might the larvae be emitting from their anuses? The fluid that feeding larvae excrete from the anus contains a precipitate of uric acid and a clear fluid of water and salts, in other words, ordinary "urine" from the Malpighian tubules (Petralia et al., 1982). Perhaps a pheromone is added at the onset of metamorphosis. No glands opening near the anus have been described. In view of the circumanal origin of the stimulatory material, it seems unlikely that the bulk flow described earlier controls ovarian function. Of course, if it is simply nutrition, it may set an upper limit on a rate set by pheromonal stimulation. (Metamorphosing brood stimulate fecundity in the queens of at least two other species of ants, the pharoah's ant, *Monomorium pharaonis,* and the carpenter ant, *Camponotus pennsylvanica.* The phenomenon may be widespread but was first discovered in fire ants; Tschinkel 1988c.) Much remains mysterious.

If we could collect this magical material from postmeconial larvae, or from workers that had collected it from them, we could feed it to a broodless colony and stimulate egg-laying in its queen, using this response as a biological assay to isolate and identify the fecundity factor. Alas, in spite of much tedious dissection of worker crops, no increase in fecundity of the experimental queens could be generated. (This result would have been predictable if the "chasing" experiment had been done first. Oh, well.) Perhaps the precise focus on the queen by the shuttle workers is needed to give the system its specificity. In their absence, or in the absence of brood, the magical material probably simply enters the general food supply and is diluted into inactivity. It will be very difficult to develop a biological assay.

Coordination of Life Stages, Seasonality, and Upper Limits

The social control of queen function is obviously part of colony growth. Because each metamorphosing larva induces the queen to lay more eggs, which will in turn develop into more metamorphosing larvae, which will induce still more eggs, the queen and the larvae are linked in a positive feedback that drives egg laying and exponential growth of young fire-ant colonies. The duration of the stages is such that when the hatchlings from the eggs need care, the metamorphosing larvae that stimulated those eggs will be available as young workers to care for them. As the colony grows, each metamorphosing larva induces fewer eggs and cares for fewer larvae, causing important changes in the character of the worker force and colony growth rate.

Although it has not been directly tested, queen fecundity must vary seasonally with temperature, and winter curtailment must depend on latitude, but even when temperature is not limiting, one would expect the egg-laying rate during sexual production to be lower than that during worker production because colonies rear fewer sexual than worker larvae, but the two have the same egg-stimulating power per individual.

When the worker birth rate is equal to the worker death rate (averaged over the annual cycle), the colony can grow no more and is at its maximum size. The queen's egg laying rate is about three times the worker birth rate (calculated from sperm-use efficiency) and therefore ultimately limits colony size, but exactly what is the rate-limiting step? It may be the capacity of the ovaries to produce and nurture follicles or of the fat body to produce yolk. It may be the capacity of the queen to feed, although it probably is not because even very fecund queens do not feed continuously. Perhaps the digestion and absorption of food are limiting. It may be the declining ability of the larval population to stimulate more egg laying. Doubling the egg-laying rate of a queen already laying 80 eggs per hour would require increasing the larval population from 10,000 to 100,000, a level never observed in the field.

Worker behavior may also play a role in the declining efficiency. Pharate-pupa tenders coming from meconial larvae sometimes encountered and fed other workers before they joined the retinue. The larger the nest, the longer is the walk to the queen and the more likely that a shuttle worker will get sidetracked before reaching the queen. If this is the case, queen fecundity should depend on the worker-to-larva ratio, and we know that it does not. Many other questions remain open as well, about the chemical identity, glandular source, mode of action, and so on. What does seem clear is that, in natural populations, food limitation and competition among colonies assure that many queens are not at the limits of their fecundity. Here, I think we have arrived at the limit of our knowledge to date of how fecundity is controlled in fire ants.

Catching Queens

Even though each fire-ant colony contains only one queen, for a lot of studies, you need to get your hands on that single individual. Now, you wouldn't think that would be possible—finding one particular ant in a colony of 250,000—but it really is rather easy. Well, it's easy during part of the year. The rest of the year, it meets the original expectation—impossible (or at least very, very hard). From midwinter on, though, once the soil has cooled to sufficient depth, catching the queen is a piece of cake, because on sunny mornings, the mound warms rapidly to produce a nice cozy sunroom for the ants. Most of the workers and brood move into the warm mound after a night in a chilly nest, and the queen strolls up to warm her tarsi, too. So by 10:00 or 11:00 hours, the queen and a large fraction of her colony are huddled just under the surface of the mound, getting their metabolisms spinning to a high rate.

A small shovelful of the mound can easily be scooped off for close inspection. The chances are about 50–50 that it will contain the queen. The chances are 100-0 that it will contain masses of panicked workers and stacks of brood, like so much pale bird seed. But how to locate the queen? In the early days, we used the preference of fire ants colonies for roadsides to our advantage. The shovelful of mound dirt was tossed and spread thinly on the paved apron. We would then flatten some of the ever-handy Schlitz, Budweiser, or Coke cans that litter American roadsides and place them amid the ants. The workers would immediately begin moving brood under these shelters, and the queen, too, left the confusion to shelter under the nearest Bud can. All the colonies along a few hundred meters of road were treated this way, then patrolled repeatedly by researchers looking under the cans. Gloves or mighty quick hands were desirable. The queen is always conspicuous, not only because of her size (alates are the same size) but because the workers form an excited knot around her, and this ball of excitement readily catches the eye. You don't even have to bend down to spot it. A quick suck with the aspirator, and you've got a queen.

Of course, working along highways exposes one to the curious public, and rubberneckers are an occupational hazard of fire-ant research. Once, a Leon County sheriff's deputy who was just going off duty stopped to check us out. Hitching up his gun belt under his bulging belly, he said in a deep rumbling voice, "Wha' choo boys doin'?" We earnestly explained our procedure and its purpose as he kicked a couple of our flattened Bud cans with the tip of his mirror-shiny black boot (which matched the dark, shiny sunglasses that reminded us entomologists of a horsefly). The ants scattered, seeking new shelter. Pausing to think the situation through carefully, he finally asked, "You boys got the Highway Puhtrol's puh'mission to put dirt on their highway?"

Sometimes crimes must be committed so that science can progress.

We gradually refined the queen-capture method. The highway method had several drawbacks, including not only curious deputies but traffic and the linear distance that had to be patrolled. Tallahassee's first municipal airport was abandoned when the city built a larger airport farther out, but large expanses of its pavement still exist. Thus arose the Bag-and-Shingle Method. Each mound sample is shoveled into a heavy plastic bag, which is quickly taped shut. A supply of discarded asphalt shingles from redneck dump sites in the woods (another delightful feature of the southern landscape) is kept on hand at the old airport. As soon as we arrive at the airport with the ant bags, their contents are spread thinly, supplied with a few shingles, and patrolled. On a warm day, you have to work fast, because the ants spread quickly and climb your legs—then you see a lot of stomping, slapping, and rubbing. Tucking trousers into socks helps.

Within 10–15 minutes, most shingles shelter thousands of workers and brood, and with luck, the queen. On a good day, we can process 35 nests, capturing 15 to 20 queens. When you are trying to orphan 100 colonies, speed counts.

In cases when it is important to remove only the queen and to be able to return the workers and brood, we move the procedure into large photo trays, the sides treated with Fluon so the ants can't escape and the scientists don't get stung as often. A tray is placed next to the mound, dirt scooped and spread, shelter provided. Six or seven trays can be patrolled at once. A couple of minutes of hard staring at the chaotic spread of ants and dirt, cueing to any heightened excitement or clustering, soon reveal the queen. The rest of the ants and dirt are tossed back on the mound. If we fail to find the queen on one day, we can try again a couple of days later. We leave motherless colonies relentlessly in our wake.

By mid- to late May, none of these methods work anymore. The soil has warmed to considerable depth, and the early-morning mound is no longer attractive. In fact, by late morning, it is too hot for sensible ants. The fire-ant queens are then safe from us until the following January.

24

Necrophoric Behavior

Eventually, a worker lives out her life span and passes on to her heavenly reward. The majority of workers have the good grace to die outside the nest, while foraging. Their passing is unlikely to be noticed by their nestmates, although other species of ants, such as *Dorymyrmex* sp., may scavenge their corpses. In most ant species, however, should a worker kick the bucket within the nest, her nestmates soon remove her corpse and dispose of it. Some 19th-century myrmecologists even claimed the existence of "ant cemeteries." Dabbing the products of decomposition, such as oleic and other fatty acids, on live workers or inert objects will cause workers to carry these to the colony's trash pile and dump them (Blum, 1970; Wilson et al., 1958). Such "necrophoric" behavior has obvious importance to colony health because it reduces the spread of contagious disease within the colony. It is this very behavior that foils human attempts to create fungal epidemics by dousing fire ant colonies with fungal spores.

My second graduate student, Dennis Howard, carefully analyzed this important behavior (Howard and Tschinkel, 1976). In the laboratory, fire ants (and most other species, too) create "dead piles" in their foraging arenas, usually in the most distant corners. There they dump corpses, discarded food, feces, and any other trash they remove from the nest. When dead workers or worker-sized bits of wood, paper, or metal are placed in the arena, live workers carry most of the corpses toward the trash heap, but ignore all the other objects. They may not take corpses all the way immediately, but the latter invariably wind up on the heap within a day or so. This directional transport differentiates necrophoric from other kinds of transport, and provides a method for quantifying the cues associated with necrophory. In the bioassay, the transport of test objects placed near the nest entrance is compared to that of corpses, which serve as a control.

How might workers recognize that one of their nestmates has shuffled off her mortal coil, and how soon can they tell? Immobility is present immediately after death, but anaesthetized workers are not transported toward the trash pile. Moreover, when colonies are offered freeze-killed, then thawed corpses for five minutes at increasing elapsed times since death, they immediately remove almost none of those killed in the last five minutes, but they quickly remove those dead for 24 hours. By

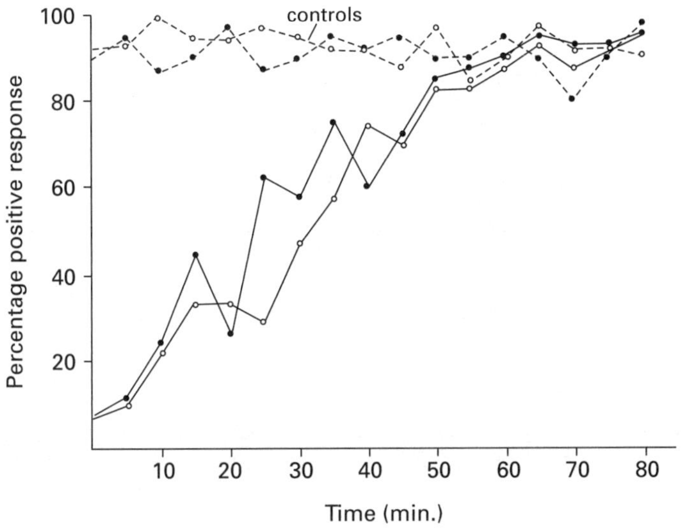

Figure 24.1. The cue that triggers necrophoric behavior in *Solenopsis invicta* reaches full strength within about one hour after death. This rate is the same for freeze-killed (open circles) and heat-killed ants (filled circles). Adapted from Howard (1974).

the end of an hour, they have removed nearly all of both categories (Figure 24.1). The "death cue" is therefore absent immediately after death but appears within an hour. If it is a chemical cue, it is certainly not a fatty acid, such as oleic acid, because these are produced more slowly through bacterial decomposition. Moreover, the cue appears at the same rate whether the ants are heat-killed, which would also kill bacteria, or freeze-killed which would not.

If the cue is chemical, it might be extractable with solvents. Indeed, extraction with a series of solvents of decreasing polarity showed that, whereas methanol did not reduce the necrophoric response to corpses, tetrahydrofuran or ether reduced it by 50% or more. Applying the methanol extract to bits of filter paper caused 90% of them to be transported to the trash heap. Methanol therefore extracted the cue, in spite of not reducing recognition of corpses. Further extractions produced ever-less necrophoric response when applied to filter paper bits, and reduced response to corpses by another 30%. The death cue was completely removed by treatment with potassium hydroxide solution (which hydrolyzes all soft tissue, fats, and proteins but leaves the cuticle intact) followed by a water rinse. Corpses so treated were ignored, although they were visually and morphologically similar to freshly killed workers. Adding methanol corpse-extract to these cueless corpses caused workers to treat them as corpses once again. Despite the lack of exact correspondence between the removal of the cue and its appearance in the filter-paper assay, the death cue is probably primarily a chemical one. Contact with the corpse is required for release of necrophoric behavior, so the cue is probably of very low volatility, but its nature is unknown.

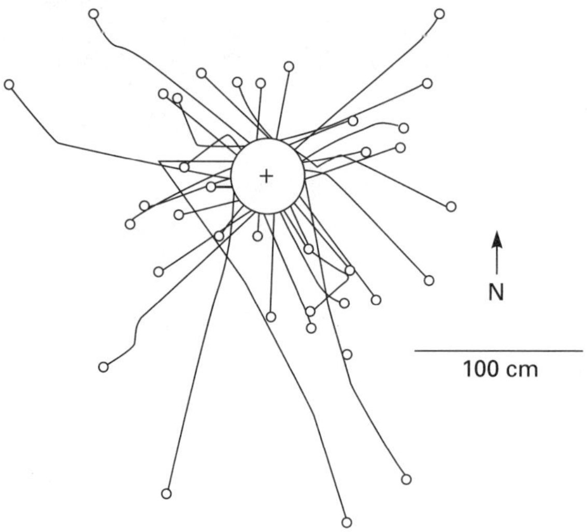

Figure 24.2. The paths of necrophoric workers relative to the mound (large circle) from which they came. Small circles mark the corpse deposition sites. Workers maintained radial paths by relying on sun-compass orientation. Adapted from Howard (1974) with permission from Dennis F. Howard.

What is the fate of corpses removed from the colony? Howard watched workers from colonies on a level lawn as they transported corpses, marking their paths with colored pins. Most workers headed radially outward from the colony in straight lines, thereby maximizing their distance from the colony with the least travel distance. Travel in such straight lines, in spite of the obstructions at ground level in a lawn, suggested that workers use some distant, parallax-free orientation cue, probably the sun, as a compass (Figure 24.2). In that case, shading the sun with a piece of cardboard and mirroring its reflection from a new angle (but at the same elevation) ought to cause the workers to change their direction by the same amount. Indeed, over sun-angle changes of up to ±150°, the necrophoric ants changed direction almost precisely as predicted by the sun-angle change, confirming their use of a sun compass. Howard did not test to determine whether fire ants compensate for the sun's movement during longer trips.

These observations suggest that corpses are scattered widely in the vicinity of the colony, and argues against the establishment of the "ant cemeteries" reported by 19th-century myrmecologists, but sometimes piles of dead ants are indeed formed. Howard killed 250 ants from each of 20 field colonies, marked them with fluorescent paint and returned them to their home colonies. The following evening, he searched the nest vicinity with an ultraviolet light. In this way, he discovered that 15 of the 20 colonies had refuse piles around their nests, though none was large. Colonies located on slopes were more likely to carry the marked workers downhill and were also more likely to have refuse piles. When the slope was less than 5°, 46% had refuse piles; between 6° and 10°, 69% had them; and between 11° and 15°, 86% had them.

To determine whether slope was the cause of these observations, Howard established lab colonies in the centers of foraging arenas, each 1 m in diameter; placed corpses in the nests; watched the travel direction of necrophoric workers (Figure 24.3). In level arenas, necrophoric workers showed no directional preference, but as he tilted the arenas up to 15°, they were ever more likely to head downhill with their burdens. Who wouldn't? It saves work and reduces the chances that anything once discarded will return to the mound, for example with runoff.

We can now picture the formation of refuse piles by following an imaginary worker carrying the corpse of a loved one, recently deceased. After she exits the colony, the greater the slope, the more likely she will walk downhill. Every worker's choice is similarly biased, so the proportion of workers heading downhill with their stiffs increases with slope. Eventually, necrophoric workers drop their corpses and are more likely to do so upon encountering an already discarded corpse. The likelihood that a necrophoric worker will come across corpses, will add her own burden alongside, and will thus help create dead piles increases with slope.

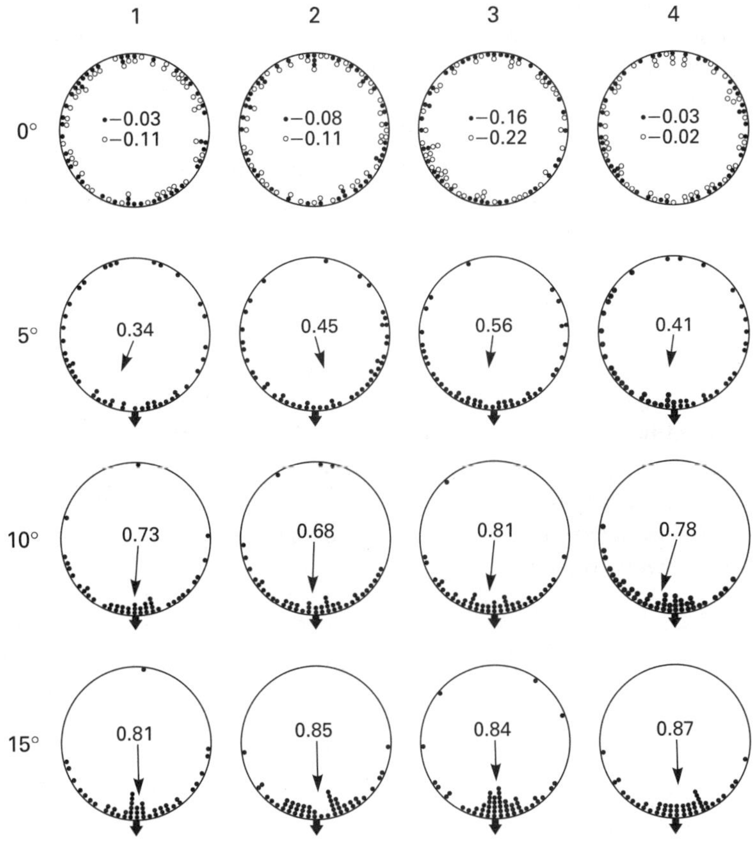

Figure 24.3. Necrophoric workers are ever more likely to travel downslope as the slope increases. The experiments were carried out in circular arenas whose tilt was varied. Arrows indicate the direction and length of the mean vector of travel (repeated in the number at the center of each circle), and the small circles the points of corpse deposition. Adapted from Howard (1974) with permission from Dennis F. Howard.

What cue emanating from trash piles tells necrophoric workers that their journey is at an end? Again, we suspect chemical cues. Again Howard set up experiments based on manipulated dead piles and their extracts, watching as necrophoric workers passed over these piles. Trash-pile recognition was signaled when a worker discarded a corpse. Ninety-six percent of necrophoric workers dropped their burdens on an open dead pile, and 76% did so on a screened dead pile that allowed contact with the dead ants. When the screen was raised to prevent contact, only 11% did so, no more than did so on the screen-only control. Clearly, recognition requires contact.

The same experiment was repeated with extracts of dead piles dried on filter paper. When contact with treated papers was allowed, 37% of workers dropped their corpses, when it was prevented, 11% did so. Finally, dead piles were treated with potassium hydroxide, rinsed in water, and dried. These corpses did not cause necrophoric workers to drop their burdens, but when an extract of dead pile was added and contact was allowed, they did so once again. Clearly, workers recognize dead piles by contact chemoreception.

Trash piles contain more than just corpses, and indeed, creation of a pile of 2-day-dead workers merely resulted in the dismantling of this human-created pile and its reassembly elsewhere by the ants. Howard noticed that the ant-created piles usually also contained fecal streaks. When he gave necrophoric workers the choice of pieces of tape with and without fecal streaks, 61% dropped their corpses on the streaked tape and 14% on the control tape. Once again, contact was required for the response. It thus appears that the cue that terminates necrophory emanates not from corpses but from fecal matter that is deposited with, or perhaps contaminates, the corpses. It makes sense that the cues initiating and terminating necrophory are different.

Necrophory should probably be regarded as a special case of sanitation behavior, the removal of anything unsavory, unhealthy, offensive, or toxic from the colony, which explains why a variety of chemicals produce removal behaviors that are not easily distinguished from necrophory. The bioassay we used probably would not distinguish between necrophory and other noxious cues, but the chemicals we tested were derived from dead ants and dead piles and are thus unlikely to produce the behavior artifactually.

IV

Polygyny

25

Discovery of Polygyny

For almost three decades after *Solenopsis invicta* forced itself upon southern consciousness, it was regarded as a nasty but otherwise ordinary ant, with an ordinary ant life cycle. Although Green (1952) had reported as many as 25 wingless queens in some fire-ant colonies (either *S. richteri* or *S. invicta*), this possible polygyny escaped serious notice. Then, in 1971, Glancey and his USDA colleagues in Gulfport, Mississippi, discovered several colonies of *S. invicta* with multiple dealate queens (Glancey et al., 1973a). Sperm in their spermathecae indicated they had mated. Isolated with a few workers, they laid eggs that developed into normal workers, as expected for a fully functional queen.

Although the significance of this discovery was not immediately apparent, the simple picture of fire-ant reproductive biology began, at that moment, to come apart. The find was initially viewed as a peculiar anomaly. After describing 36 polygyne mounds that yielded over 3000 inseminated queens and were scenically located next to a garbage dump in Hurley, Mississippi, the USDA group wrote, "We cannot explain this . . . extraordinary and intriguing case of polygyny." Soon it became less and less likely that polygyny was an anomaly as more and more polygyne populations were reported throughout the southeastern USA (Hung et al., 1974; Glancey et al., 1975; Fletcher et al., 1980; Mirenda and Vinson, 1982; Fletcher, 1983; Lofgren and Williams, 1984; Kintz-Early et al., 2003). Functional polygyny was also discovered in the native *S. geminata* (Banks et al., 1973b) and *S. xyloni* (Summerlin, 1976) and in the *S. invicta–S. richteri* hybrid (Glancey et al., 1989) in North America. It also turned up in several South American species of fire ants (*S. quinquecuspis, S. richteri, S. invicta*) (Jouvenaz et al., 1989; Trager, 1991).

Polygyny Is Functional

The possibility that these mated queens only commenced laying eggs when removed from inhibition by a single dominant queen was laid to rest by direct observations (Fletcher et al., 1980). About 90% of the queens in polygyne colonies laid from one to 75 eggs in a five-hour period (the oviposition test); colony means were between eight and 31 eggs. Polygyny was truly functional in *S. invicta*, and queen number ranged from low to extreme.

Most subfamilies of ants include some normally polygyne species. Some pairs of closely related species of ants also exist in which one species is monogyne and the other polygyne (Wilson, 1971; Hölldobler and Wilson, 1990), but *S. invicta* is different because, within what is without doubt a single species, strict monogyny exists alongside spectacular polygyny. Almost 700 queens from a single mound? That was enough to make myrmecologists sit up and take notice. The revelations that followed have been among the most interesting and rewarding chapters in fire-ant biology.

Even the earliest observations (Glancey et al., 1975) revealed radical biological differences from the monogyne form. Monogyne intolerance of workers from other colonies gave way to polygyne tolerance, and territoriality gave way to free exchange of workers. When foraging trails from different polygyne mounds crossed, the workers from each went peacefully on their ways, showing no hostility. Execution of extra queens added to monogyne laboratory colonies gave way to acceptance of multiple queens by polygyne colonies. These simple observations established some of the basic research questions that still occupy research on fire-ant polygyny.

Geographic Distribution of Polygyny

Surveys gradually made apparent that polygyne patches were distributed within a mostly monogyne population throughout the southeastern USA. The first actual distribution map confirmed polygyny in eight Florida counties (sometimes in the same fields with monogyne colonies) and possibly four more (Glancey et al., 1987). The geographic distribution of the two forms was still far from clear and demanded a much larger effort.

The Texas Study. Such an effort was made in Texas by Porter and his associates (Porter et al., 1991). Realizing that Texas was a large state, a political accident of which Texans seem inordinately proud, Porter and his coworkers organized a joint project including the Texas Department of Agriculture, the University of Texas, and Texas A&M University. During the spring of 1988, Texas Department of Agriculture inspectors sampled road rights-of-way at one site in each of the four corners of each of 168 Texas counties. At each site, data on mound density, size, and brood were collected. Brood is informative about the number of queens because in spring monogyne nests have much sexual brood and moderate worker brood, whereas polygyne nests have little sexual brood and copious worker brood. A search for queens in spread-out mound soil provided further evidence. Finally, researchers placed baits (hot-dog slices) along another transect to estimate the impact of fire ants on native ants.

A site was declared polygyne if at least one colony contained more than one inseminated queen. Certainty of polygyne designation was high, because multiple queens were found in over 80% of the inspected

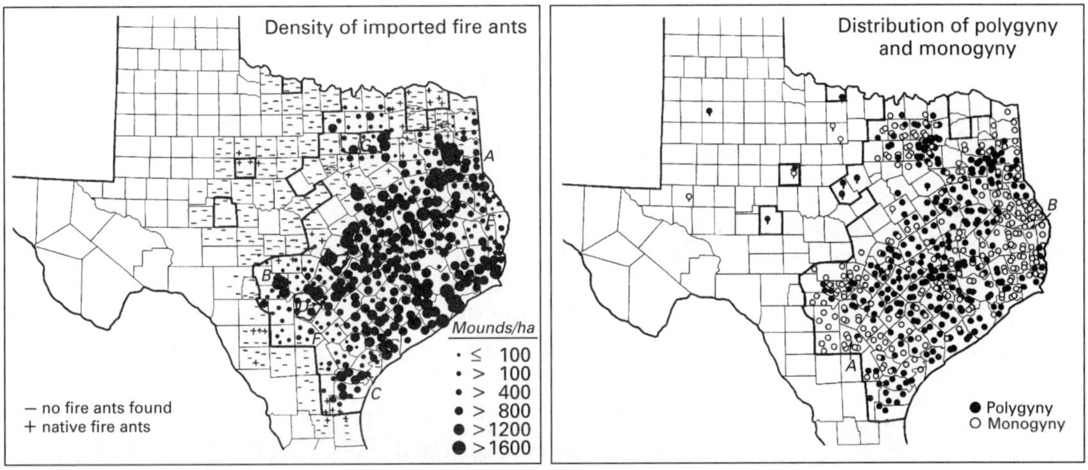

Figure 25.1. The distribution of fire-ant nest density in Texas in 1988. High nest density largely corresponds to areas of polygyny. Reprinted from Porter et al. (1991) with permission from the Entomological Society of America.

colonies. Monogyne status was less certain because the single queen could easily be missed. Sites were designated "possibly monogyne" if 15% of mounds yielded a single inseminated queen and the remainder yielded no queen.

All this effort resulted in two maps of Texas, one showing the density of fire ant mounds and the other showing whether those mounds were polygyne or monogyne (Figure 25.1). Areas with high mound densities are generally polygyne, and those with low densities, monogyne. Polygyne mound densities (680 mounds per hectare) averaged about twice as high as monogyne (about 300), but both were highly variable. Over 60% of polygyne sites sustained an amazing 500 mounds per hectare, as compared to only 10% of monogyne sites. The density-limiting effect of territoriality is absent in the polygyne form, as is true of other polygyne species of ants as well. This higher mound density was not the result of smaller mounds—the mounds of monogyne and polygyne colonies did not differ greatly in size. These polygyne densities were similar to those in other reports, but the monogyne densities were much higher. Sampling roadsides probably biased density upward, because edges of roads are a preferred location for colonies. Although the study is internally consistent, its density data should be compared only cautiously with those from other studies.

Polygyny was extremely widespread in Texas and was distributed in a mosaic fashion rather than contiguously (Figure 25.1). Sites were either polygyne or monogyne, not mixed on this scale. Overall, more than half (54%) of the roadside sites that had *S. invicta* were polygyne. This rate is more than three times as high as the rate later found in the remainder of the southeastern states. Two parallel broad belts of high density running southwest to northeast showed high frequencies of polygyny. Two pockets of monogyny, therefore low density, were conspicuous. One lay in the forested areas northeast of Houston (A and B in Figure 25.1), the other near the southwestern margin of the range (E and F in Figure 25.1). Density in the Dallas area was also mostly low, even though the ant had been there for over 30 years. This area, too, was monogyne.

Porter found no association of polygyny with historic quarantine zones, eradication efforts, vegetation type or soil type, and switching of social form seemed unlikely. This distribution most probably resulted from the history of the spread of *S. invicta* from monogyne and polygyne source populations. The polygyne form is especially easy to spread in soil-containing products because colony establishment requires only an inseminated queen and a few workers. Nevertheless, conditions in Texas may somehow favor polygyny, for it occurs even in the native fire ants there. Two of the 14 sites with *S. xyloni* were polygyne. Although monogyne in Florida, *S. geminata* is commonly polygyne in Texas.

When sites were resampled the following spring, the great majority (77 to 87%) retained their designation, a result that demonstrates both the reliability of the method and the stability of the population. Many of the switches resulted from small errors in relocating exact sample sites, but even when the resampled sites were several kilometers off, they tended to retain the same designation, indicating the regional scale of the mosaic.

The results of the bait study were depressing. A tremendous dominance by *S. invicta* reduced native ants on baits from about 11% to less than 1%. A decade after Hung and Vinson (1978) reported many occurrences of both species of native fire ants within the range of *S. invicta* in eastern Texas, Porter's study showed that these had all disappeared. This displacement is described in a later chapter.

Florida and the Southeast. Not wanting to give up on a good thing, Porter carried out a similar study from his new position with the USDA in Gainesville, Florida. He made two transects of 85 roadside sites throughout Florida and 52 between Florida and Texas. He sampled them in much the same way as in the Texas study but also recorded queen weights and worker size distributions (Porter, 1992; Porter et al., 1992). Queens from polygyne colonies average 14 mg, and those from monogyne ones 24 mg; the two weight distributions showed almost no overlap. Workers from monogyne colonies are both larger and 50% heavier. Addition of these measurements made the designation of a colony monogyne or polygyne more than 95% reliable.

Solenopsis invicta occurred at 74 of the 85 Florida sample sites and was polygyne at 15% of these (11), about one-third the rate of polygyny in Texas. Most of the polygyne sites were in the central Florida counties

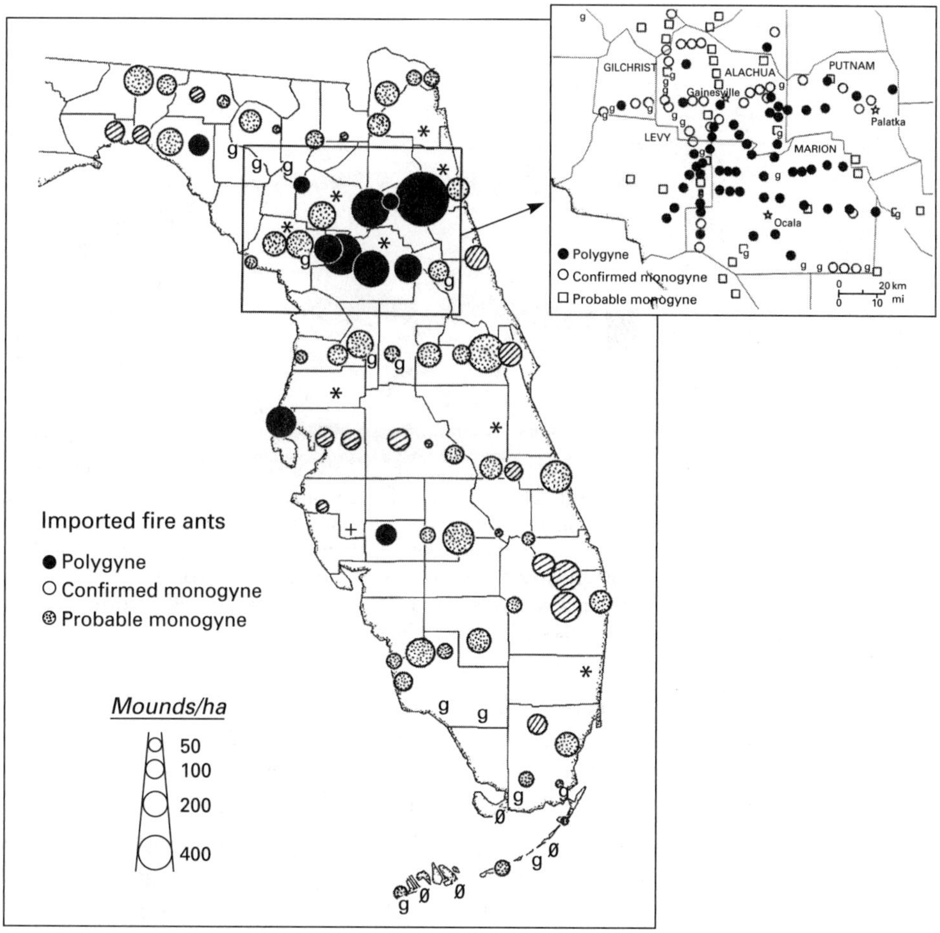

Figure 25.2. The distribution of polygyne and monogyne *Solenopsis invicta* colonies at roadside sites in Florida. "g" indicates sites with *S. geminata*; "Ø" indicates sites without either fire-ant species; asterisks indicate counties with confirmed polygyny. Reprinted from Porter (1992) with permission from Sanford D. Porter.

between Ocala and Gainesville (Figure 25.2), and another nine or ten were scattered among counties the length and breadth of Florida, but no polygyny was found in the Florida panhandle. On the Florida-to-Texas transects, nine sites were polygyne (17%; Figure 25.3). One of these was in eastern Alabama and one in eastern Mississippi, but the other seven were all located between central Mississippi and the eastern boundary of Texas. Of the sites in this segment, 40% were polygyne, far more than expected by chance.

As in Texas, polygyny in Florida occurred as a mosaic rather than contiguously. Adding another 113 sites in the central Florida area where polygyny was concentrated revealed that within an area of approximately 4600 km^2, most sites were occupied by polygyne colonies (Figure 25.2).

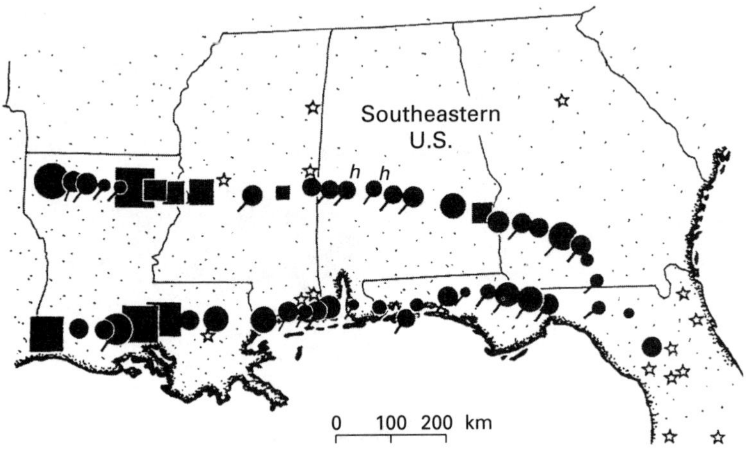

Figure 25.3. The distribution of polygyne *S. invicta* on transects between Florida and Texas. The areas of circles are proportional to mound density. Circles indicate monogyny, diagonal tails confirmed monogyny, squares polygyny. "h" indicates sites with the *S. invicta* × *S. richteri* hybrid. Stars indicate polygyne populations reported before Porter's study. Polygyny increases sharply west of Mississippi. Reprinted from Porter et al. (1992) with permission from Sanford D. Porter.

The greater detail of this map also revealed that six sites were occupied by both monogyne and polygyne colonies, so the two can exist side by side, but it casts no further light on the origin of the mosaic distribution. A resurvey 13 years later found that almost a third of the colonies collected from these central Florida polygyne sites were monogyne and that these were interspersed among polygyne colonies (Fritz and Vander Meer, 2003). The polygyne sites ranged from 4 to almost 60% monogyne.

The mound density of both the monogyne and polygyne forms increased from Florida to Texas (Figure 25.3). Monogyne densities averaged 115 mounds per hectare in Florida roadsides, 170 from Georgia to Louisiana, and 295 in Texas. For polygyne sites, these figures were 262, 544, and 680 mounds per hectare, a similar increase in density. Within each area, mound density was about twice as high at polygyne as at monogyne sites. Abundance of polygyne workers, as estimated from mound basal area, was about 30% higher than that of monogyne workers in Florida, but from Georgia to Texas, they were about twice as abundant. The reasons for these differences are not known.

Are Polygyne Populations Increasing?

The rather late recognition of scattered polygyne enclaves led early investigators to conclude that polygyne populations were increasing in frequency (Glancey et al., 1987). Such a trend would be cause for concern or even alarm, but is it really true? Being sure would require at least one resurvey on a later date, so Porter (1993) reinspected the sites of his two earlier studies to see whether they had switched from their earlier designation.

He found no support for the claim of a rapid increase of polygyne populations. One to three years after their original designation, 94% of the monogyne sites were still monogyne, and 97% of the polygyne sites were still polygyne. Switching was low and equally probable in the two directions. Of the 10 sites that switched (either way), seven had a mixture of monogyne and polygyne colonies. The switch could have been the result of slow replacement on the local level, of sampling error, or of colony movement. Overall, both social forms seemed to exist in very stable populations—even mound densities and the differences in mound densities between the two forms were essentially unchanged. Porter's methods could have detected only a net change of 3–5% per year, however. A lower net rate could be detected only with more intense sampling or greater elapsed times and therefore cannot be ruled out at this time.

Despite the lack of rapid change, population-genetic and historic work by Ross and Shoemaker (discussed below) suggests that new polygyne populations do occasionally arise from monogyne ones and grow outward from their points of origin. The rate of such expansion would depend in part on competition from existing monogyne populations and might well fall below the 3% annual increase detectable by Porter's methods. The situation is likely to be dynamic, and the subject is certainly not yet closed.

◂•• *I Want This to Be Accurate*

OK, here's a test. I am weighing a queen fire ant on a millibalance, and the readout says 25.68 mg. This doesn't seem just right to me, because the queen looks heavier, to judge from my experience with weighing hundreds of such queens. So I remove her from the balance, rezero the balance, and weigh her again, several times. The balance says 25.69 mg, then 25.66, 25.70, 25.68, and 25.66 mg. These average to 25.6783 mg, which, because of the significant-figure issue, must be rounded to 25.68 mg. The standard deviation is 0.016 mg, 0.062% of the mean value. That small figure tells me that the individual weights cluster pretty tightly around the mean.

But I'm still uneasy, so I open the drawer, and from the back, buried under instruction manuals for instruments I don't even own anymore, sheaves of misplaced memos, and dusty warranty statements, I pull out my set of standard weights, certified by the National Institute of Standards and Technology to weigh exactly what they say they weigh. I select the 10 mg weight, pick it up with forceps (no fingerprints—they have weight, too!) and put it on the millibalance. Aha! Just as I thought! The balance reads 8.87 mg. I put on the second 10 mg weight. Now the readout is 17.74 mg. I repeat this procedure six times. They produce a mean of 17.72 mg and vary around this mean by 0.058%.

So here are the questions that every student of science should be able to answer correctly, practically without reflection: (1) Was the first set of measurements (the queen) accurate? (2) Was the first set of measurements precise? If the student answers "no" and "yes," he or she can pass Go, collect $200, and go on to be a scientist. Unfortunately, many get it wrong and still go on to be scientists. Accuracy assumes that the true value is known and is an estimate of how close the measurement is to the true value. Precision, on the other hand, is an estimate of how closely a set of repeated measurements cluster. As in much of science, the true value is unknown—the set of measurements serves as an estimate of the true value. If you are a hunter, it is not enough that your rifle is precise in where it places the shots, it must be accurate, too. If measurements are made with an instrument that can be calibrated against a standard, as was my millibalance, we can be both accurate and precise, and go home smug and satisfied, having bagged our symbolic deer.

The general public, to the degree that this entity exists, seems to believe that scientists are always accurate and precise, wearing white lab coats and getting their measurements just right, including lots of numbers to the right of the decimal. The truth is that it isn't always equally important to be either accurate, or precise, or both, and a great deal of time and effort can be saved by knowing how much accuracy or precision is enough. Occasionally, you run into notions that are very, shall we say, individual? When I was still an undergraduate, the new graduate student in the lab, Bobbie, was asked by the senior grad student, Bill, to weigh out 100 g of agar to make a batch of culture medium. Twenty minutes later, Bill returned to find Bobbie still engrossed in the demanding process of weighing. With a tiny little spatula, she was drizzling a few flakes of agar at a time onto the growing pile on the balance. At the current rate, Bill would be finished with his master's degree, be married, and have two children in high school before Bobbie finished weighing the agar.

What are you doing? says Bill. Don't you know how to weigh stuff? You just put a 100 g weight on the left pan, dump agar on the right pan until the beam tips, and then you take away or add agar until it balances!

I know all that, says Bobbie, but I want this to be *accurate*.

Bobbie's notion of accuracy may be off the map, but many less spectacular errors have cost time, money, and effort. Developing judgment in these matters is an important, though little appreciated, skill of a working scientist. Sometimes accuracy is simply not very important, for example, when we are interested in the *difference* between two measurements. Accuracy is important if our measurements are to be compared with those made by others, because only if accuracy is assured can we conclude that differences or similarities in our measurements are real, not simply differences in our instruments. That is why we have standard units of weights and measures and their watchdog, the National Institute of Standards and Technology.

Sometimes even precision isn't too important. Let's say we want to determine the relationship between the length of a queen's gaster, swollen with ovaries, and the queen's body weight so that we can later estimate queen weight simply by measuring her gaster length through a microscope and thus avoid disturbing the queen. We can determine queen weight to four significant figures on our high-priced millibalance, but under an ocular micrometer we can measure her gaster length to only two significant figures. We take these measurements of 25 queens and plot our relationship.

To use this plot, under the microscope, we line up the ocular micrometer over an experimental queen's gaster and measure its length. Say it was 8.7 mm. We go to the x-axis of our graph and find 8.7 mm, run our finger up and over to the y-axis, and read off the answer, 24.25 mg. Is this precision justified? Are these last two figures meaningful, significant? We try to drum into the head of every student passing through our portals that they are not. The two significant figures in the length measurement limit the precision of the weight estimate to two significant figures as well. The last two figures are a meaningless pretense.

Computers have allowed us to carry this error to spectacular extremes. Column three is the product of column one, measured to two significant figures, and column two, weighing in at three significant figures. Yet the product in column three is reported to seven significant figures. Perhaps it seems a shame to throw away those five numbers. Perhaps it is the belief that computers, with their unfailing precision, can create something out of nothing, can rescue us from the limitations of physical measurements and the sad knowledge that all measurements are ultimately estimates. In physics, the estimate may have twelve or more significant figures. In biology, you are usually lucky to have three, but somehow, this limit hasn't slowed biology down much. We have muddled through on the backs of myriad sloppy measurements, best approximations, barn-door accuracy, shotgun stuff, eyeball estimates, measurements on the fly, our kid's school ruler, balances made from soda straws, pacing off in heavy boots, meat scales, shovel-and-bucket volumes, kitchen-cup measures. And in the end, it hasn't mattered much, because the things that interest us, and that are meaningful, happen at one, two, or (at most) three significant figures. Anything more is noise, pretend knowledge, painting a painting with a finer brush when the painting cannot be viewed from any closer than 20 m.

26

The Suppression of Independent Colony Founding in Polygyne Colonies

The polygyne and monogyne forms of the fire ant are very different creatures in fundamental biological ways. The difference is particularly conspicuous in the replacement of independent colony founding by dependent founding through colony fission as a consequence of several profound changes in colony life. Although alate females from polygyne colonies may occasionally succeed at independent colony founding, they are unlikely to do so for three reasons. (1) Multiple queens suppress the production of alates, so typical polygyne colonies produce far fewer alates than do monogyne colonies and have correspondingly puny mating flights. (2) The female alates that polygyne colonies do produce usually have insufficient body reserves for successful independent colony founding. (3) As a result of reduced genetic diversity in the USA, 80% of the males produced by polygyne colonies are diploid and therefore sterile, so polygyne sexual females have a reduced likelihood of successful mating.

Suppression of Sexual Reproduction in Polygyne Colonies

Suppression of Alate Production. Let us compare the reproductive performance of a representative polygyne queen to that of a monogyne queen (Vargo and Fletcher, 1986b, 1987, 1989). In north Georgia, our polygyne queen is probably one of 90 to 100 other egg-laying queens, about 30% of which are not inseminated. Larger colonies of both social forms produce more sexuals, but for any given colony size and during all seasons, the polygyne colony produces but a small fraction of the number of sexuals produced by its monogyne counterpart. In the summer and fall, she has a 30% chance of finding herself in a colony completely without sexuals. The more egg-laying queens are companions to our queen, the lower will be her egg-laying rate and theirs.

Are the multiple queens the cause of this suppression? Let us divide our queen's colony in half, placing her by herself in one half and all her companion queens, however numerous, in the other half. After a month, our (now single) queen is awash in sexuals, but none of those in the polygyne half have produced any to speak of (Figure 26.1). Moving all but one of the queens from the multiple-queen half to the single-queen half causes the sexual production to plummet to zero in the formerly single-, now multiple-queen half. Those sexual brood

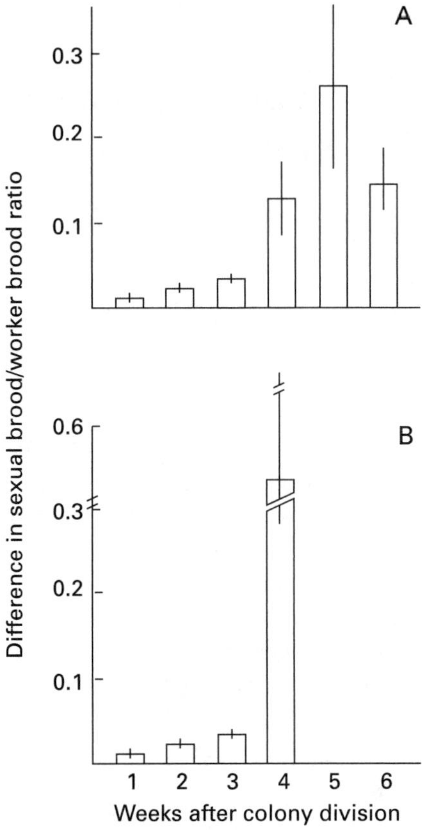

Figure 26.1. Sexual and worker brood production in the two halves of divided polygyne *Solenopsis invicta* colonies. A. The difference between polygyne and monogyne halves. B. The difference between the queenless and polygyne halves of formerly polygyne colonies. Adapted from Vargo and Fletcher (1986b).

beyond a critical stage continue to develop to adulthood, but no new ones appear. In the formerly multiple-, now single-queen half, sexual production rises. If we repeat the division, but this time place our single queen with her companions, leaving one half queenless, the queenless half produces even more sexuals than when our queen was present (but of course, without a queen, they cannot continue doing so for long).

The sexual larvae appear so rapidly, even in the absence of a queen, that they must develop from existing larvae, rather than from special sexual eggs laid by queens. Queens, it would appear, control the production of sexuals by inhibiting female larvae from sexualizing (becoming sexuals), probably by a nutritional mechanism. The rapid appearance of female sexual larvae suggests that sexualization occurs rather late, perhaps in the third larval instar. In contrast to females, males have no alternate developmental option because male "alateness" is genetically determined in the egg. Because male alates also fail to appear in polygyne nests, our queens must be inducing workers to execute male eggs or larvae. Because about 20% of the queens in polygyne colonies

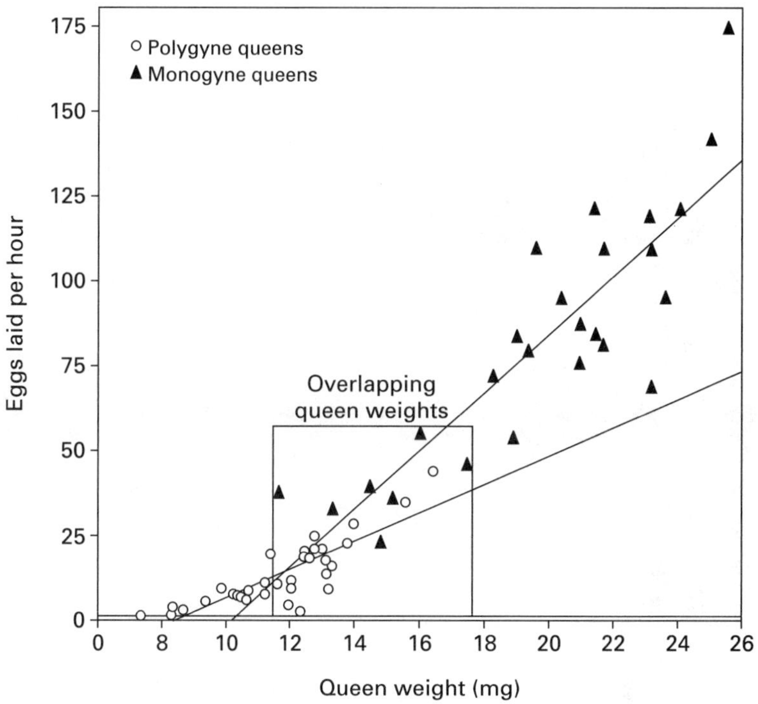

Figure 26.2. Relationships between egg-laying rate and queen weight in monogyne and polygyne colonies. The box shows the region in which polygyne and monogyne data overlap. Adapted from Vander Meer et al. (1992).

are match-mated and half of their eggs are therefore diploid males, the carnage must be quite considerable.

Suppression of Egg Laying. Let us compare our polygyne queen with a monogyne queen from a mature colony. In contrast with the glossy plumpness of the monogyne queen surrounded by her feverish retinue, our polygyne queen is pretty ho-hum, showing little sign of extreme ovarian development or the circle of worshippers that sets the monogyne queen apart. Our polygyne queen's low body weight implies low fecundity, if body weight parallels ovarian hypertrophy as it does in monogyne queens (it does, strongly so). In oviposition tests, she and her companions averaged only 3.0 eggs per hour, and even at their maximum laid far fewer than the average of 34 eggs per hour laid by queens in monogyne colonies. (These rates were determined after two weeks in the laboratory. Queens straight from the field are more fecund.) Polygyne colony means ranged from about 1.6 to 6.2 eggs per hour (Fletcher et al., 1980; Vargo and Fletcher, 1989; Vander Meer et al., 1992). Suppression seems to begin with fecundity.

We can use a queen's weight to estimate her fecundity in further investigations. Polygyne queens that weighed less than about 9 mg laid no eggs, but for every milligram over that weight, a queen laid about 3.6 to 4.2 eggs more every hour (Figure 26.2). This relationship was the same whether the queen was inseminated or not, representing, as it does,

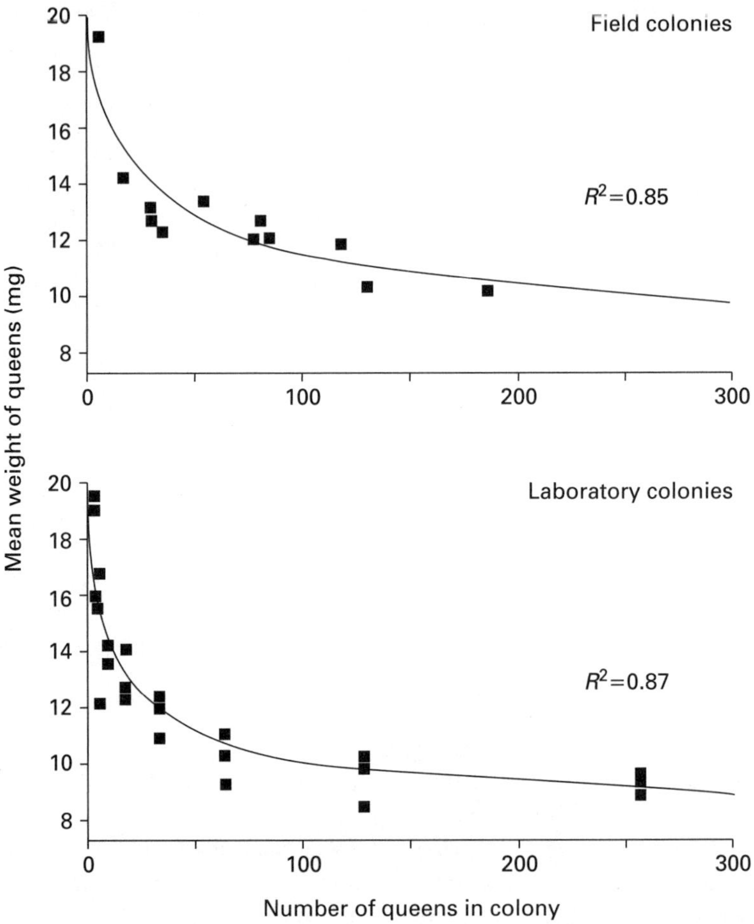

Figure 26.3. The relationship between average queen weight and the number of queens in nine-week-old polygyne colonies. Adapted from Vargo (1990).

ovarian hypertrophy. An egg is an egg, more or less. In monogyne queens, every additional milligram increased egg-laying rate by 8.6 per hour.

Monogyne queens lost about 8 μg for every egg laid, whereas polygyne queens lost about 48 μg. Monogyne queens therefore appear to convert their weight into eggs six times as "efficiently," but this difference is an illusion: assume that all queens lose the same amount of weight through non-egg processes, such as defecation, water loss, metabolism, etc. When a queen lays few eggs, these non-egg losses are divided among fewer eggs, and production of an egg therefore appears to require the loss of more weight. This phenomenon easily accounts for the sixfold difference, but a portion of this difference is nevertheless probably real, because in the region of weight overlap, polygyne queens lose more weight per egg and lay fewer eggs than monogyne queens of the same weight. The difference may have a genetic basis.

Knowing the relationship between queen weight and egg-laying rate, we can use queen weight as a tool to determine just how suppressive our queen and her companions are (Vargo and Fletcher, 1989; Vander Meer et al., 1992). Let us place our queen among ever-larger groups of queens, from 2 to 256, in standard homogeneous experimental colonies from pooled polygyne colonies. Pooling eliminates any possible intercolony differences in responsiveness to experimental conditions. Let us also sample field colonies to determine the natural distributions of queen number, colony size, and queen reproductive performance (body weight, egg-laying rate).

The polygyne field colonies from one of which our queen was plucked ranged in size from 12 to 267 g of workers and brood (mean 82 g) and contained 5 to 186 queens (mean 71). In both laboratory experiments and field studies, the mean weight of queens decreased as the number of queens increased (Figure 26.3). The more queens are already present in a colony, the less adding another queen reduced their mean weight further. The average polygyne queen weighed about 12 mg and laid 9.4 eggs every hour; the average monogyne queen weighed 23 mg and laid 48 eggs per hour, over five times as many. Of equal importance, the *variability* of queen weight also decreased with queen number. To the suppression of sexualization of larvae in polygyne colonies, we can now add mutual inhibition of queen fecundity. Similar suppression is common in other species of polygyne ants, termites, and even uninseminated replacement queens in orphaned colonies (Tschinkel and Howard, 1978).

So far, we have focused only on the average queen weights. Do some individual queens manage to lay a lot more eggs than the others? Body-weight frequency distributions do not expose distinct classes of egg layers and nonlayers, but they do often show a skew to the right (Figure 26.4). Some queens therefore achieve relatively high fecundity in spite of high queen numbers and thus contribute disproportionately to the egg pool. As estimated from their weights, the heaviest 10% of the queens lay about 25% of the eggs. They are relative winners.

Despite the decline in individual egg production, the summed egg production in polygyne colonies with more than nine queens exceeds that in monogyne ones, ultimately by up to 15-fold. Even in the average polygyne colony, eggs accumulate over twice as fast as in a monogyne one. In addition, whereas colony size had a positive effect on monogyne queen weight (Tschinkel and Howard, 1978), individual polygyne colonies show no such effect, even after adjustment for queen number. This result seems counterintuitive but might arise because the monogyne colonies spanned a 100-fold range of colony size, but the polygyne only a 30-fold range.

Did colonies with more queens grow faster? Does the number of eggs limit growth? The answer, based only on raw queen number, was no, but of course, the mutually inhibitory effect of queens makes queen

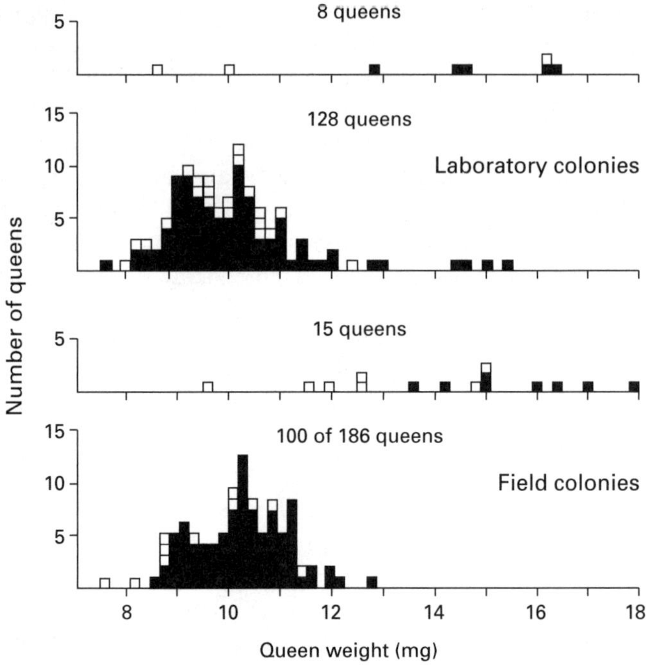

Figure 26.4. The frequency distributions of queen weights (and therefore fecundity) in representative field and laboratory colonies of polygyne *S. invicta* with high and low queen numbers. Solid bars indicate inseminated queens, open bars uninseminated queens. Adapted from Vargo and Fletcher (1989).

number a poor estimate of total reproductive output. Let us assume that only the weight in excess of 9.4 mg represents "fecundity," because queens weighing less lay no eggs. The sum of this "fecundity" over all queens in each nest explains 69% of the variation in growth. Eliminating unmated queens raises this figure to 74%. Colony growth rate is indeed related to the production of eggs. No such relationship occurs in the field colonies.

As far-fetched as it sounds, only the eggs of one or a few of these multiple queens might actually be reared. If so, then workers should be significantly more related to one another than to workers from the population at large, because they would predominately be the daughters of one or a few mothers and fathers. We know that queens mate with only one male, because the worker offspring of isolated polygyne (or monogyne) queens produce enzyme-banding patterns consistent only with single mating. (In each of 55 single-queen colonies, each allozyme occurred as either one or the other of the two alleles, or they bore both in a 1:1 ratio.) We also know that the queens within polygyne colonies are no more related to one another than they are to any queen in the population at large (on the basis of starch gel electrophoresis of the allozymes of 24 to 150 queens from each of 26 polygyne nests). Worker allozyme banding patterns for two independently segregating loci (*Agp-1* and *est-4*) tell us that none of the workers in 20 polygyne colonies were more related to one another than to any randomly chosen worker in the population (Ross and Fletcher, 1985a). We must conclude, therefore, that most or all queens contribute substantially to worker production. In fact, we will see later that they contribute fairly evenly. Polygyne colonies are truly polygynous.

Inhibition of Embryonation. Eggs laid by ants do not necessarily develop. Founding queens, workers, and sometimes virgin queens may lay trophic eggs that fail to form embryos and are eaten by larvae or the queen. In the trophic eggs of colony-founding queens, early development is arrested by an abnormal mitosis under juvenile hormone control (Voss, 1981). Might queens in polygyne colonies inhibit the development of each other's eggs in this manner?

To find out, let us rear single queens of three classes in small nests of a few thousand workers and brood each (Vargo and Ross, 1989): mated queens producing normal worker brood, unmated queens producing a few males or no brood at all, and match-mated queens, half of whose brood develops into diploid males. After about six weeks, all the pupae and larvae are the undoubted offspring of the respective resident queen. Now we take the eggs of each queen and give them to a few workers for 24 hours of foster care. Nuclear division produces many nuclei in embryo-forming eggs but not in nonviable, nonembryonated eggs. The Feulgen DNA stain speckles embryo-containing eggs with pretty pink nuclei, but nonembryonated eggs remain unstained.

Clearly, mating makes a tremendous difference—all mated queens lay more than 85% embryonated eggs; in fact, those of most are over 95% embryonated (Figure 26.5A). In contrast, no unmated queen lays more than 54% such eggs, and 38% produce no embryonated eggs at all.

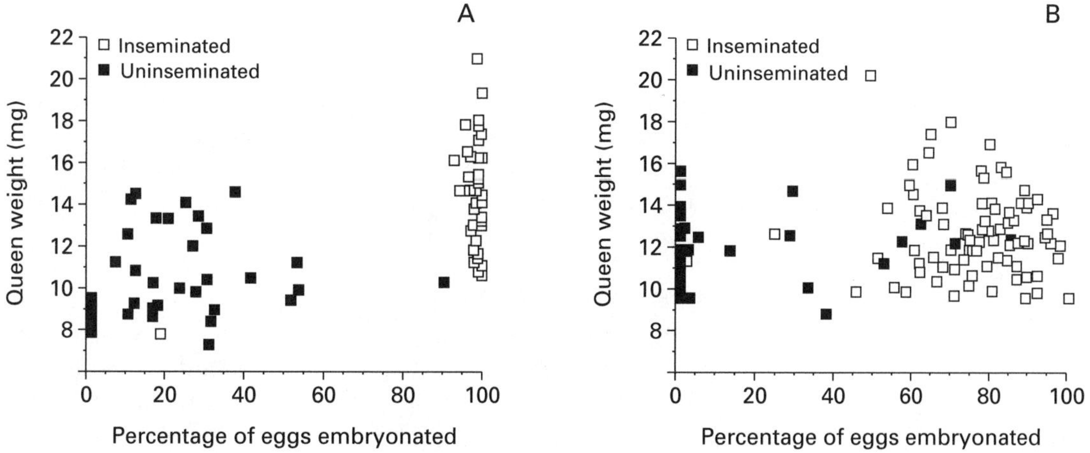

Figure 26.5. Relationship between proportion of viable eggs laid by polygyne queens and their insemination status and body weight. Queens were held individually in small nests. A. After several weeks in individual laboratory nests. B. Directly from the field. Adapted from Vargo and Ross (1989).

Match-mated queens do not differ significantly from normal mated queens. Mating, not the nature of the mate, makes the difference. A small problem remains: because unmated queens produce very little brood, little worker brood is present to stimulate their oviposition. Switching mated and unmated queens between nests stimulated the oviposition of the former and reduced that of the latter but had no effect on their respective rates of embryonation. Unswitched queens of both types served as controls. Embryonation and egg production are independent processes, and only egg production is controlled by larvae.

The rate of embryonation after an extended period in the laboratory might not be similar to that in natural field colonies, especially after the laboratory queens had been monogyne for six weeks. Staining the eggs of polygyne queens whisked directly from the field, we discover that the situation in natural polygyne colonies is quite different from that in the laboratory. In the field, both mated and unmated queens produce substantial fractions of nonembryonated eggs, although the average differences between the two are still very large (Figure 26.5B). Whereas almost all laboratory-held mated queens laid more than 85% embryonated eggs, only about a quarter of mated field queens do so. Only a tenth of field queens lay more than 95% embryonated eggs, but after six weeks in the laboratory this figure increases eightfold. In further contrast, embryonation rates of mated and unmated queens from the field overlap because about one-fifth of the unmated queens have embryonation rates between 50 and 80%. These figures do not imply that embryonation rate is high among unmated queens—almost 60% laid no embryonated eggs at all. Egg viability is again not related to fecundity, further evidence that these two phenomena are independent. The more queens, the more embryonation is suppressed. (Source colony significantly affected the rate of egg embryonation, another example of an important biological trait in which colonies are individualistic.)

In conclusion, polygyne queens suppress each other's reproduction by multiple means—they inhibit embryonation of eggs, reduce fecundity, and inhibit sexualization of larvae. These effects are not an artifact confined to an introduced species; polygyny in the native *S. geminata* is generally similar.

Suppression Is Pheromonal

How might queens be exerting all this negative influence on one another? Fire-ant queens show little aggression toward one another, so something more cryptic must be involved. The two most likely mechanisms are brood care and a primer pheromone, so we can formulate two hypotheses. Hypothesis 1: because removal of queens eliminates the production of new brood, the remaining brood receives more care and therefore sexualizes. Hypothesis 2: queens secrete a mutually

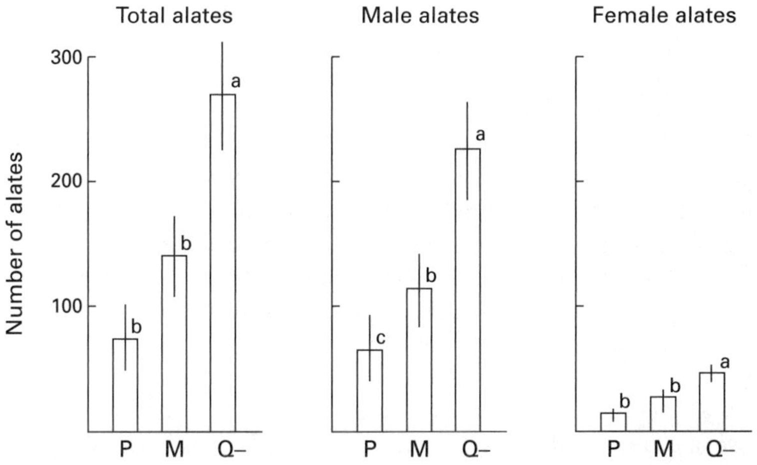

Figure 26.6. Number of adult alates produced in 10-g experimental colonies with 20 queens (P), 1 queen (M), or no queen (Q-). Bars with different letters are significantly different. Adapted from Vargo and Fletcher (1986a).

inhibitory primer pheromone. (Primer pheromones have a delayed physiological or developmental effect in the receiving animal, whereas "releaser" pheromones cause an immediate behavioral response.) Of course, both hypotheses could be true (and of course, other hypotheses are also possible but were not tested by Vargo). Experience with other ants pointed toward primer pheromones, but other queen cues are possible as well (e.g., body shape, tactile cues, brood production).

Full-sized colonies are impractical for such experimentation, so Vargo and Fletcher used experimental units each containing 10 g of workers (about 2500) plus some brood from pooled polygyne colonies. Each unit received a different number of polygyne queens (e.g., 20, 1, or 0). Sexual larvae were counted as they appeared (larvae were referred to as having been sexualized), or egg laying was estimated in five-hour oviposition tests. These experimental units responded strongly even to single queens in as little as two days. Queenless units produced two to four times as many sexuals as did monogyne units, which in turn produced two to 30 times as many as polygyne units (which often failed to produce any at all; Figure 26.6). Because unmated queens can only produce haploid, male-producing eggs, the data from any unit containing an unmated queen were discarded, as were those from diploid-male-producing units.

Fourth-instar larvae stimulate queen fecundity, but the reduced fecundity in polygyne colonies seems unlikely to result from sharing of a fixed amount of stimulation among more queens, because taking colony size (and therefore number of fourth-instar larvae) into account does not explain any additional variation in fertility of polygyne queens over and above that explained by queen number (Vargo and Fletcher, 1989).

Now we are ready to test the first hypothesis, brood care. Removal of queens would end new brood production. Perhaps sexualization results from improved care of the remaining brood? If so, then adding the daily egg production of a monogyne queen to queenless nests every day should prevent that sexualization. Control queenless nests would receive no eggs. From the first week until the end of the experiment, queenless units produce far more sexual larvae than monogyne or polygyne units, but the eggs-added treatment did not differ from the queenless control. This result shows that brood ratio (worker care) is *not* the mechanism by which sexualization is suppressed.

Let us turn then to the second hypothesis, that queens secrete an inhibitory pheromone. Ultimately, proof of a pheromone requires its isolation and identification, but the customary first step is simply to eliminate several other possible modes by testing dead queens for suppressive effect. Can queen corpses reverse the inevitable increase of a live queen's fecundity when she is made the sole queen of a colony? Of course, the test is not entirely straightforward. To assure equivalent stimulation of fecundity, we must maintain an equal number of fourth-instar larvae in each test unit (Tschinkel, 1988c). Vargo then added three fresh queen corpses to 15 of these experimental units every day and three virgin queen corpses (with wings removed) to another 15. The experiments on larval sexualization were generally similar, with the addition of a 10-queen polygyne treatment (suppressed control) and a queenless control. The other two units lacked live queens. One received 10 queen corpses from unrelated polygyne colonies every day (the previous day's corpses were removed), and the other received virgin queen corpses whose wings have been removed. Screens over the colony entrances prevent the workers from removing dead queens.

Adding corpses of reproductively active queens effectively inhibited the increase of fecundity (estimated from changes in body weight) that follows reduction to monogyny, but adding virgin corpses had no such effect. By the third day, monogyne queens in colonies receiving virgin corpses weighed 10% more than those in colonies receiving functional queen corpses, which changed little (Figure 26.7). Because the two types of corpses are similar in shape and tactile cues and differ primarily in their reproductive state, these results suggest that an inhibitory pheromone suppresses fecundity. Interestingly, uninseminated queens are more sensitive to inhibition (or perhaps less responsive to disinhibition), and therefore less fecund, than inseminated queens.

When the measure was sexualization of larvae, differences appeared by the second day. By the sixth day, the treatment receiving dead queens had produced significantly fewer sexual larvae than the queenless control and virgin-alate control but more than the polygyne control, and these differences increased for the remainder of the 10-day experiment (Figure 26.8A). Clearly, queens can suppress sexualization even after death, so they are unlikely to do so through sound or any other behavioral signal. Because body morphology and tactile cues are

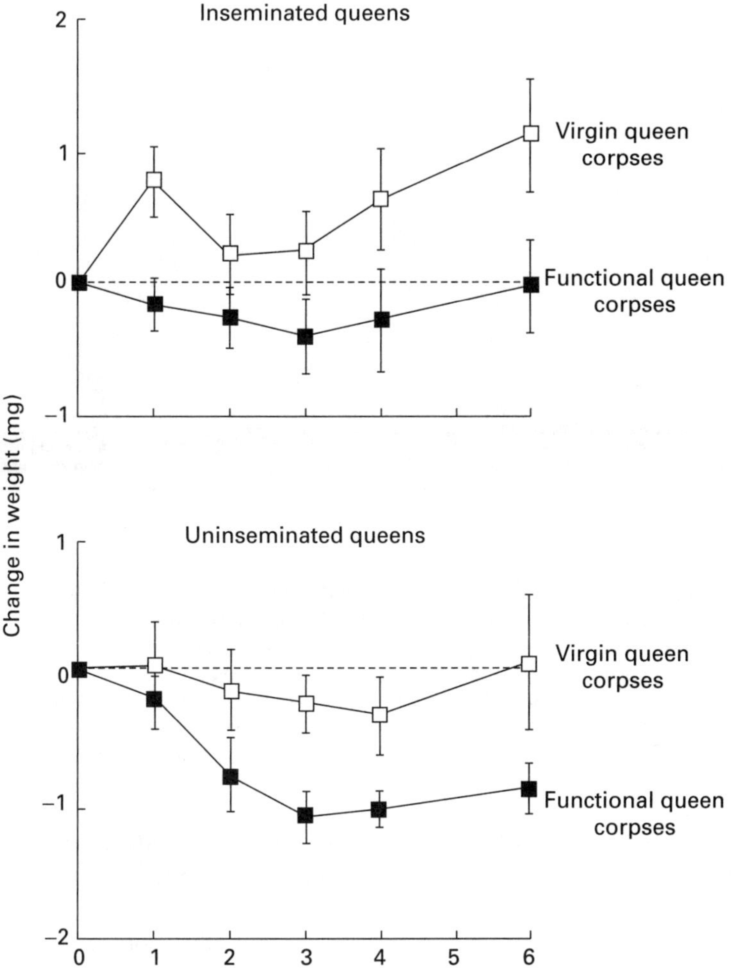

Figure 26.7. Inhibition of fecundity of inseminated and uninseminated queens by corpses of functional queens (open squares) or virgin queens (wings removed; filled squares). Change in fecundity was estimated as the change in weight from weight at setup. Adapted from Vargo (1992).

essentially the same for corpses of virgin and functioning queens, a pheromone secreted by egg-laying queens remains the most likely causative agent. However, because 10 dead queens were less effective at suppression than were the 10 live queens, other signals may still supplement the chemical mode, or the amount or dispersal of the material may be lower when the queens are dead. Fresh corpses were very attractive to workers, but within 24 hours, most had been dismembered and all but the gaster taken to the rubbish heap. The gasters remained attractive, and were kept in the middle of the nest, tended by workers like true believers around the corpse of Lenin.

The pheromone is rather nonvolatile, requiring contact for effectiveness—placing queen corpses in a mesh cage that prevents worker contact reduces inhibition to the level of that of the alate-corpse or queenless controls (Figure 26.8B). The pheromone is probably spread

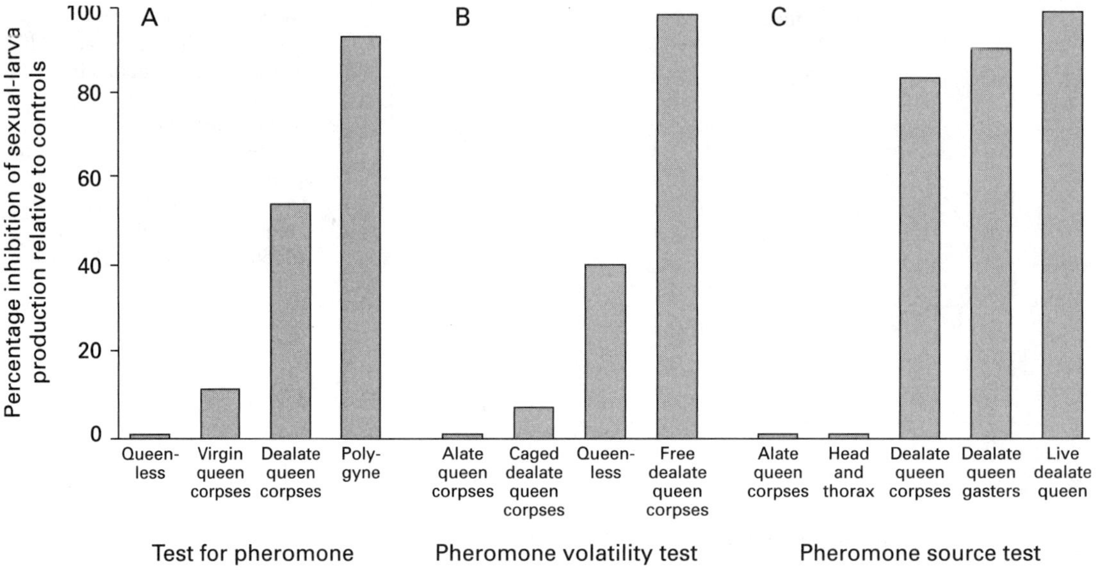

Figure 26.8. Results of experiments testing for the existence of a queen inhibitory pheromone, its volatility, and its source. Inhibition computed from the cumulative number of sexualized larvae produced after 10 days. A. Inhibition of sexual-larva production in experimental nests with 5 g of workers and 10 live queens, 10 dealate queen corpses daily, 10 virgin queen corpses daily, and no queens. B. Test for volatility of the queen pheromone. Corpses of functional queens were either caged (to prevent worker contact) or uncaged. Alate queen corpses and queenless nests served as controls. C. Test for pheromone source. The gaster alone of a dead functional queen has inhibitory power equal to that of the entire corpse. Data from Vargo and Fletcher (1986a) and Vargo (1988a).

by contact, trophallaxis, or both, as was the case with the primer pheromone that prevented dealation and ovarian development in virgin alates (Fletcher and Blum, 1981), as well as the colony odor. When queens are dipped briefly in solutions of radioactive isotope, the radioactivity spreads rapidly through the colony, primarily through contact (Sorensen et al., 1985c).

The effects of queenlessness are reversible. When 1 or 10 queens were added to queenless units after these had produced about 10 sexual larvae each, about half of the units made monogyne reduced the number of sexual larvae, and 80% of those made polygyne did so. In the extreme case, a queenless nest made polygyne, 19 of 20 sexual larvae disappeared within three days (Figure 26.9). As expected, units that continued queenless (controls) did not reduce the number of sexual larvae. Careful observations after queen addition dispelled the mystery of what happens to the sexual larvae—in many cases, workers surrounded the larvae and bit them repeatedly, the first step to execution. Interestingly, once the larvae began metamorphosis, they were no longer attacked.

Is this execution also under the command of a queen pheromone? Again, Vargo and colleagues allowed queenless units to produce sexual larvae, but this time, in addition to the 10-live-queens-per-day treatment, they added 10 dead queens to a second treatment, 10 virgin queens to the third, and left the fourth queenless. As expected, 80% of the units made polygyne killed some of their sexual larvae (control), whereas none of the queenless units did (control). Many units given dead queens showed decreases in the number of sexual larvae, but the effect was weaker than that of live queens (34% showed reductions), and the effect of dead virgin queens was very weak (7% reductions). Qualitatively similar results were obtained by Klobuchar and Deslippe (2002), but their claim that the active principle is a protein because it is extractable in buffered saline is unsatisfying because it is unsupported by tests for other protein properties, such as heat or salt denaturing and susceptibility to proteases. In any case, queens induce the execution of sexualized larvae at least partly through a pheromone that modifies worker behavior toward larvae. The queen herself remains innocent of actual murder.

A Role for Juvenile Hormone? Ovarian function, the source of fecundity, is under the positive control of juvenile hormone in many insects, providing a potential mechanism for mutual inhibition of fecundity. Queens topically treated with methoprene (a juvenile hormone analog that slowly diffuses through the cuticle) gain about 14% in weight, whereas acetone and untreated controls gain only 8.5 and 6%, significantly less (Figure 26.10). Because this juvenile hormone–like compound increases fecundity, mutual inhibition probably involves the

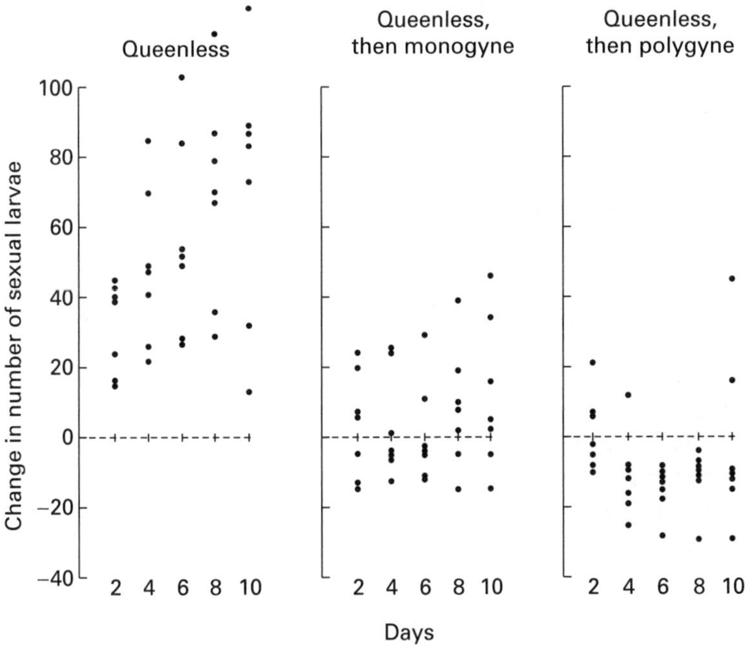

Figure 26.9. Changes in numbers of large sexual larvae in queenless experimental colonies in response to introduction of one or 10 live queens. Each dot on each day is a separate experimental nest. Adapted from Vargo (1988a).

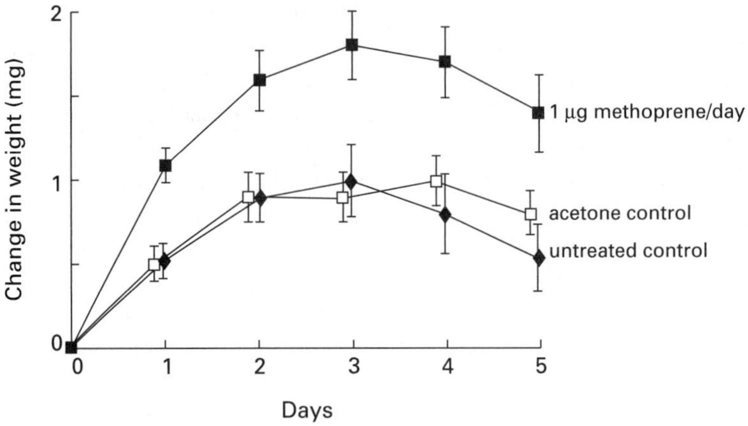

Figure 26.10. The effect of methoprene treatment on queen weight (an indicator of fecundity). Adapted from Vargo (1992).

suppression of juvenile-hormone secretion, possibly by direct action on the corpora allata, the glandular source of juvenile hormone. Caution is called for, however—methoprene is not a natural juvenile hormone, and although it has physiological effects generally similar to those of the natural hormone, it is not metabolized as rapidly. Nevertheless, involvement of juvenile hormone seems likely because control of ovaries by juvenile hormone predates the evolution of sociality. In the course of becoming social, ants have probably merely added a second, social level of control on top of an existing physiological level (Vargo, 1998). Juvenile hormone is also implicated in caste differentiation in social insects, possibly also playing a role in sexualization of larvae.

Source and Nature of the Pheromone. Evidence of primer pheromones is in the bag, so the next logical step is to locate the pheromone source. A simple first step is to test separated parts of queen corpses in the now-familiar standard experimental units (Vargo, 1988a). In the critical comparison, the effect of the gaster alone was indistinguishable from that of the whole queen corpse, whereas that of the head and mesosoma was indistinguishable from that of the alate control (Figure 26.8C). All the inhibitory power of the queen corpse therefore resides in the gaster.

Another queen pheromone, the one that prevents dealation of female sexuals, had also been localized to the gaster, specifically to the poison sac, in which all activity resided (Vargo, 1998). Might the pheromone suppressing queen fecundity also originate there? Might it be the same compound? If so, its removal should lead to the loss of a queen's capacity to suppress fecundity, to inhibit dealation of female alates, and (as will be discussed below) to attract workers. Vargo and Hulsey (2000) removed the poison sacs from functional, egg-laying queens, virgin replacement queens, and female alates. (Stretching the intersegmental membranes of the gaster reveals the poison sac, which can be seen through and removed through a small hole in the intersegmental membrane. Sham-operated queens controlled for the surgical trauma and handling.)

Surprisingly, removal of the poison sac from egg-laying queens did not affect their reproductive functioning or attractiveness. Both sac-less and control queens attracted 25 to 35 workers into their retinues, both weighed about 12.3 mg each, and neither changed much in these respects during the experiment. What is going on here? Attractant either also arises from another source or takes a long time to fade. When the poison sac was removed from virgin alates (which do not secrete the pheromone) that subsequently became virgin replacement queens, they developed normal attractiveness and laid eggs normally. A second source of queen attractant pheromone is therefore likely.

Whatever body region contains the source gland should be coated with more of the secretion and therefore be more attractive per unit surface area. The head of functional queen corpses attracted more workers than did the mesosomas or gasters, so the head, possibly the mandibular or postpharyngeal glands, might contain a second source of the queen pheromone. Attractiveness tests in which one queen-equivalent of extracts of these glands was spread on each of several glass "queen surrogates" were unequivocal—mandibular-gland extracts were no more attractive than hexane controls (attracting 12–15 workers), but postpharyngeal-gland extracts were just as attractive as the poison-sac extract (attracting 35–40 workers). Combining the extracts of the two head glands did not increase attraction over that of the postpharyngeal gland alone. As expected, the postpharyngeal glands of female alates were not attractive.

Poison sac–less, egg-laying, virgin replacement queens inhibited 90% of the dealation of female alates that occurred in queenless nests. Therefore, the pheromone that inhibits dealation also has multiple glandular sources. Location of the second source also in the postpharyngeal gland would seem to contradict the lack of inhibitory activity in whole heads and mesosomas. (These results seem to contradict the earlier experiments by Vargo in which only the gaster of queens could inhibit dealation. Perhaps, for anatomical reasons, the pheromone is continually and slowly released from the poison sacs of dead queens but not from their postpharyngeal glands. The issue is currently unresolved.)

The postpharyngeal gland is far more developed in the queen than in the workers. In several other species of ants, the queen's postpharyngeal gland is at the center of nestmate recognition through a distributional/sharing system, so it would be very suitable for the distribution of queen pheromone. In any case, queens without their postpharyngeal glands were reproductively normal, laying eggs and attracting workers (Vinson et al., 1980), as would be expected if the poison sac were a second source of queen pheromone(s), but these queens also gradually lost fat and weight, so fat metabolism may play a role as well. The gland may synthesize the queen pheromone directly, or it may be synthesized elsewhere (although not by the poison sac) and be absorbed by the gland. These patterns have yet to be clarified.

Secretion of a single pheromone by two different glands may seem odd, but in fact, the honeybee is known to have at least two sources of the queen pheromone that attracts workers and inhibits their reproduction. This pattern is remarkably parallel with that of *Solenopsis invicta* and may be general for social Hymenoptera (at least those with large colonies). From a practical point of view, it points out a pitfall of gland-removal experiments designed to locate pheromone sources. Biologically, it raises the question of why two glands are needed. Can a single gland not secrete enough material? Or are the secretions of the two glands somewhat different in important and functional ways? Or do they secrete different compounds with similar behavioral effects? We do not know the answers to these questions for either *S. invicta* or the honeybee.

The poison sac appears to be quite a little bag of marvels—in addition to primer pheromones, it contains a mixture (discussed in an earlier chapter) of antimicrobial alkaloids with which the queen coats eggs (Blum et al., 1958; Vander Meer and Morel, 1995), and a "queen recognition" releaser pheromone. Specifically, it attracts workers to the vicinity of the egg-laying queen and causes them to aggregate around her to form the retinue. Workers aggregate around objects experimentally treated with small amounts of queen extract or poison sac material and in places where queens have resided for a period of time (Jouvenaz et al., 1974; Glancey et al., 1983, 1984; Lofgren et al., 1983; Rocca et al., 1983a, b). Attraction of workers over a distance in *Solenopsis geminata* should be taken with a grain of salt, because tested concentrations were far in excess of the likely rate of emission by live queens (Cruz-Lopez et al., 2001). Disrupted *S. invicta* colonies display a dramatic sequence of queen-rescue behaviors in which they quickly clump around their queens, establish a wide trail back to the nest, and guide or coax the queen back to safety. Dummies treated with extracts of queens are "tended" in a queen-like manner. Attractive activity was localized to the queen's poison gland, although most of the tests did not distinguish clearly between attraction and aggregation, and many were poorly designed. The USDA group eventually isolated and identified three compounds from queens, a pyranone and two lactones, one of which was new to science and was named "invictolide" (Figure 26.11). The synthetic mixture of these is active in several attraction/aggregation assays, but the amounts needed to match the activity of the crude extract make it doubtful that they represent the full "queen-recognition" pheromone.

Both monogyne and polygyne egg-laying queens are attractive to workers, but they are unlikely to secrete identical pheromones and to

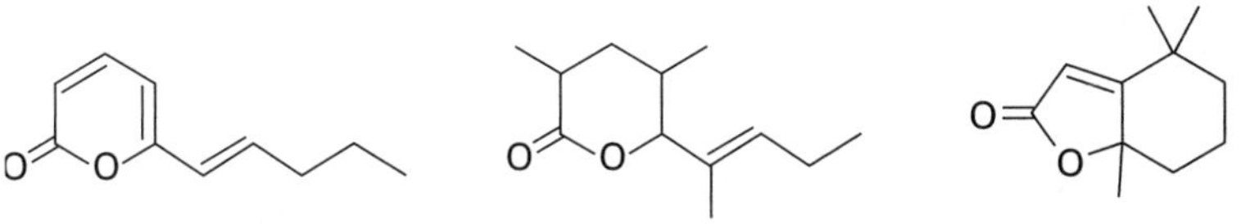

Figure 26.11. Chemical compounds isolated from the queen poison sac and shown to be active in the queen-recognition assay: a pyranone (left) and two lactones, one of which (center) was named "invictolide." Adapted from Rocca et al. (1983a).

differ only in the quantity secreted. A polygyne queen stimulated to the fecundity levels of a monogyne queen is still killed by monogyne workers, and a monogyne queen inhibited to the fecundity levels of a polygyne queen is still killed by polygyne workers. The role of genes in mono- and polygyny and in pheromone variants will be taken up in a later chapter.

These two little bags of chemicals, the poison sac and the postpharyngeal gland, have a lot of functions. Are all caused by the same pheromone or by several? Some properties are rather parallel, arguing for one or a small number of pheromones. For example, both the attraction and the dealation inhibition by queens increase with fecundity (Fletcher and Blum, 1983b; Willer and Fletcher, 1986). Up to a limit, more fecund queens have larger retinues, and their corpses inhibit dealation for much longer than do those of low-fecundity queens. Nonovipositing queens are neither attractive nor inhibitory. Many questions about how functions are assorted among chemicals and glands remain unanswered. Ultimately, the answers must wait until the pheromone(s) responsible for each effect have been isolated and chemically identified.

In an unpublished study, Vargo and Slessor took the first step toward isolating the dealation primer pheromone by developing a bioassay—as dose was increased from 0.05 to 2.0 queen equivalents, the time to dealation rose from about 50 hours (same as control) to over 100 hours. In the absence of knowledge of the pheromone release rate, we do not know whether this range accounts fully for the inhibitory effect of either living or dead queens. Separation and bioassay of poison-sac contents yielded a basic fraction containing mostly the alkaloids (mostly *cis*-2-methyl-6-undecyl-piperidine) and a neutral fraction containing a variety of minor compounds (Vargo, 1997). The basic fraction was inactive by itself, the neutral fraction inhibited dealation to a moderate degree, and the two together showed full activity, a synergy. The isolation is unlikely to proceed further for some time to come—Vargo has moved on to work on termites, and Slessor has retired. Present knowledge of queen pheromones is summarized in Table 26.1.

Queen Pheromone Ontogeny. Female alates produce none of these pheromones but begin doing so once they begin laying eggs. Whether they are mated queens or unmated replacement queens in orphaned

Table 26.1. ▶ Summary of knowledge about *Solenopsis invicta* queen pheromones.

Phenomenon caused	***Individuals affected***	***Properties of pheromone***
Attraction, aggregation	Workers	Originates in poison sac Also originates in postpharyngeal gland Consists of several attractants so far identified
Inhibition of dealation	Virgin queens	Is secreted by fecund queens Parallels queen fecundity in strength of inhibition Acts on queen behavior Does not affect virgin queens whose antennae have been removed Originates in poison sac Also originates in head
Repression of ovaries	Virgin queens	Is secreted by fecund queens Affects even virgin queens whose antennae have been removed
Suppression of fecundity	Polygyne queens	Is secreted by queens Parallels queen number in strength of inhibition Acts on queen endocrine system (probably) Originates in gaster (possibly in poison sac)
Suppression of egg embryonation	Polygyne queens (and possible eggs directly)	Inhibits in proportion to queen number May act on eggs through the queen endocrine system
Suppression of sexualization of larvae	Workers, larvae	Is secreted by queens Inhibits in proportion to queen number Originates in gaster (possibly in poison sac)

colonies, they become attractive to workers, develop their ovaries, inhibit the dealation and ovary development of other alates, and induce the execution of supernumerary queens in monogyne nests (Glancey et al., 1981a; Willer, 1984). Alate females in queenless colony fragments soon dealate and begin reproductive development. Vargo determined the time course of the appearance of these reproductive functions in such colonies.

The number of eggs laid increased sharply after day 3, so by day 7, test queens laid an average of six eggs per hour (Figure 26.12AD; Vargo, 1999). Simultaneously, the proportion of queens laying eggs increased

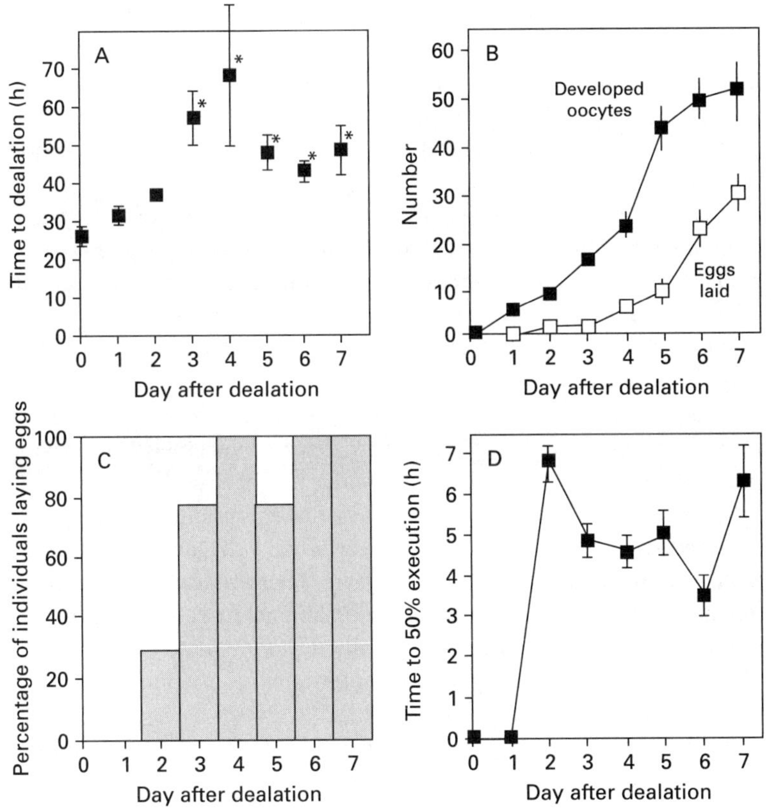

Figure 26.12. The ontogeny of queen-pheromone effects following dealation after removal of queen influence. A. Capacity of corpses of virgin dealated queens to inhibit dealation in other alates increased sharply between days 2 and 3. B. Ovarian development (number of developed oocytes; filled squares) and eggs laid (open squares) in five hours. C. Proportion of dealates laying eggs. D. Time to 50% execution when dealated queens were returned to queenright nests at increasing intervals after dealation. Bars are standard errors. Adapted from Vargo (1999).

sharply to 100% between day 2 and day 4. Queens became significantly attractive to workers on day 2 and generally increased in attractiveness until day 7. Execution of virgin replacement queens returned to their natal colonies also began on day 2 and continued at high rates thereafter. The ability of these replacements to inhibit dealation in winged females appeared on day 3 and remained significant thereafter.

Clearly, ovarian development and the appearance of several queen-pheromonal effects are tightly linked. A few insect examples of combined ovarian-pheromonal regulation have been reported. We must also distinguish between production of a pheromone and its release. For example, poison-sac pheromone, even though present continuously, is released only when eggs are coated during laying, perhaps explaining why several pheromonal effects appeared on the same day when virgin replacement queens laid their first eggs. Once a queen begins laying eggs and producing pheromones, a positive feedback loop quickly accelerates her fecundity and pheromone production. The first queens to dealate probably have the advantage because their pheromone begins to inhibit other queens from dealating and causes workers to execute dealates that lag in pheromone production. Alates that respond too readily to low queen pheromone may become reproductive in the presence of their queen mother, risking execution. Overwintering queens may play exactly this dangerous game, for many begin laying eggs while still in the natal nest, a process that primes them for their role in socially parasitic colony founding. Attrition of such overwintering alates has not been determined.

The detection of the dealation primer pheromone one day later than the recognition (execution, attraction) pheromone(s) could be the result either of real differences in secretion, suggesting two different chemicals, or of differences in the ability of the bioassays to detect the pheromones. Resolution must await isolation and identification of the compounds involved.

Fecundity in claustral-founding monogyne queens probably ramps up much more slowly than in virgin replacement monogyne queens because the strength of the positive feedback and the colony resources are much less. The egg-laying rate and pheromone production of queens generally tracks colony size, especially the larval population, although the relationship is nonlinear. Newly mated and adopted polygyne queens, if they survive, presumably increase their reproductive and pheromonal functions quickly, whether mated or not. The pheromonal/hormonal regulation of reproductive function is probably similar under several circumstances.

A recapitulation. In polygyne fire ants: Queens inhibit several reproductive processes in several different kinds of individuals, mutually in the case of other queens. At least four of these types of suppression are caused by a queen-secreted pheromone, acting in some cases as a primer and in some as a releaser. Specifically, queen pheromone (1) acts directly on virgin queens to inhibit the initiation of reproductive function through dealation and ovarian development; (2) suppresses the sexualization of larvae by acting on the behavior of workers, probably causing them to feed female larvae differently and to kill male larvae; (3) suppresses the ovarian development and egg laying of other queens, probably acting directly on the queens' endocrine system, and mated queens are less sensitive than unmated ones. (4) Finally, queens inhibit each other's ability to lay viable eggs, but the mode and target of this inhibition are not clear. It may suppress the queen's juvenile hormone, in turn affecting the early mitosis of the egg.

The Meaning of Suppression Is Competition

How should we interpret this rampant reproductive inhibition in polygyne colonies? Is it coordination or competition; chemical communication or chemical warfare? The answer depends on whether the polygyne fire-ant colony is a legitimate superorganism, defined as a collection of cohabiting organisms functionally organized to achieve a common goal. (Among ants, this common goal is the production of alates to which the workers are related, i.e., the reproduction of close relatives.) We already know that it is not. A monogyne fire-ant colony is a superorganism because it is a simple family in which workers share the genetic benefits of colony reproduction through relatedness. A polygyne fire-ant colony may not look very different superficially, but genetically and superorganismally, it is chalk to the monogyne colony's cheese. The multiple queens within it are unrelated to each other (Ross, 1989), so unrelated families cohabit in the same polygyne mound, and workers are more likely than not to rear ants to which they are not related. The benefit of this cohabitation must thus be mutualistic, but what it might be is unknown. The association of queens in polygyne colonies raises many of the same questions as their association in founding groups. What advantage counterbalances the risk to personal reproduction? Because the queens in polygyne colonies are not related, they must achieve their fitness by direct reproduction, by having their own offspring reared as alates. One queen's gain is therefore another's loss and tells us that we need to find out how sexual production is distributed among the queens within a polygyne nest. To the extent that a queen can bias the group's efforts to favor her own relatives, she is a social parasite exploiting the labor and resources of nonrelatives.

If reproduction in polygyne colonies is really a big competitive free-for-all, then it must include winners and losers. One indication of "victory" might be differences in the abilities of queens to attract worker attention. Almost all marked queens in undisturbed laboratory polygyne nests of several sizes (5 to 20 queens, 2000 to 10,000 workers) assorted themselves into a single aggregation (76 to 100%) within each nest over a 25-day period (Chen and Vinson, 1999, 2000). The queens showed no aggression within these clusters. Occasional small second aggregations occurred when queen numbers were high. All these aggregations moved very little, and when they did, the clusters of workers around them seem to retard, or possibly control, their movement (Kuriachan and Vinson, 2000).

When queens were separated, however, they differed greatly in the ability to attract workers to themselves. Every day for 10 days, workers and brood were dumped into the central arena of an apparatus from which five chambers opened, in each of which was a queen, confined within wire mesh. Every day, the workers (with brood) chose among the five queens, only to be removed and dumped back into the central arena to do it all over again. Ant workers, of course, can be relied upon not to balk at pointless, repetitive work. In one set of replicates, queens were returned to the same chambers, but the positions of the chambers around the arena were randomized. In the second set, queens were presented in fresh, uncontaminated chambers every day.

The results were unequivocal. One queen always attracted at least 70% of the workers and brood into her chamber, whether the chamber was lily-fresh or smelled of previous occupation. The remaining queens were about equally attractive to the few workers that remained. This dominance by one queen was remarkably stable from day to day; it changed in only nine of the 24 colonies. If the most attractive queen (the alpha queen) was removed (and clean chambers were being used), workers immediately chose a new alpha queen, and when she was removed, they chose still another. When the removed queens were returned (they had been maintained with workers, food, and brood during their absence), they reestablished the original hierarchy within a day or two, but when the removed alpha queen was starved during her absence and returned with a well-fed beta queen, it took her 10 days to regain her top status.

Upon removal of the alpha queen when queens kept the same chambers from day to day, her empty but contaminated chamber, rather than the remaining four queens, attracted most of the workers and brood for about seven to eight days. Attraction was probably due to the pheromone discussed above. Thereafter, workers chose a new alpha queen, but when the original alpha queen was returned (after being kept with workers and brood), she returned to her throne as the alpha queen.

Queen attractiveness and feeding rate (determined by feeding of radioactive protein) paralleled the egg-laying rate—highest for the alpha queen and lower for the lower-ranking queens. When the alpha queen's fecundity dropped for intrinsic or experimentally induced reasons, so did her attractiveness. Attractiveness, fecundity, and feeding rate go together, but which causes which? Increased feeding surely always precedes increased fecundity because production requires material, but do queens become more fecund because they are more attractive and are therefore fed more, or are they fed more because being more fecund makes them more attractive? The flow of this causation will color our interpretation. The typical American bias is to see competition, but workers might just as well simply be shunting food to the most productive queens, who are signaling their productivity through attractiveness. Queens might even compete while workers merely regulate.

The sharp distinction between the top of the hierarchy and all the rest decreased as the number of queens and chambers increased. The alpha queen attracted an average of 83% of the workers when competing with four other queens but only 45% when competing with 20. In experiments with more than 15 queens, a beta queen appeared, attracting about 30% of the workers. When queens are this numerous, individual fecundity and attractiveness are sufficiently suppressed and variability sufficiently lower that at least two of the queens are indistinguishable in attractiveness. Hierarchies are stable but not immutable, judging from the occasional shifts of workers from one queen to another.

Now for the problems. Whereas these experiments show that some queens are more attractive than others when separated, they do not show whether in their normal aggregations differences in attractiveness result in preferential treatment and higher fecundity. Testing separated queens may exaggerate initially small differences—workers bring pheromones from the brood, which stimulate fecundity of the initially most attractive queen, raising her fecundity and attractiveness and decreasing that of the remainder of the queens. In field colonies, the top 10% of queens on the fecundity scale laid about 25% of the eggs, fewer than the attractiveness of separated queens would imply. Because workers often push or pull queens back into aggregations, Chen suggests that workers control queen aggregations (Chen and Vinson, 1999), serving colony efficiency by aggregating care of the queen in space. A more appealing interpretation is that queens clump in order to compete more effectively with each other and especially with the top queen. Collecting crumbs from the alpha queen's table, so to speak, may leave a subordinate queen better off than not being at the table at all, as in Chen's experiment. A subordinate queen away from the locus of queen care may be able to attract less and less worker care and feeding and thus spiral down in fecundity and attractiveness. Many animals aggregate to compete, especially when the prospects of a loner are very poor (as in, e.g., mating aggregations).

Differential worker attraction to queens seems likely to be pheromonal, but the only test of this possibility was done in *S. geminata* from Chiapas, Mexico, not *S. invicta*. In a two-choice olfactometer, workers of polygyne *S. geminata* were more strongly attracted to heavier (more fecund) polygyne queens (Rojas et al., 2004). When queen weights were equal, they were more attracted to those from their own colony. An attractive chemical in the poison sac, discussed further below, increased with queen weight, but the relationship was not strong.

However they do it, some queens seem to attract more worker attention than others and therefore probably produce more than an even share of the eggs and workers, but workers are merely steps on the road to fitness. The real goal is alate production. To win the fitness contest, a queen must produce a disproportionate share of the colony's alates. Could the real meaning of the strong fecundity skew be that the most fecund queens also have more alate offspring? We cannot jump to such a conclusion, because suppression of embryonation is independent of egg-laying rate, reducing the link between high fecundity and offspring production. Moreover, disproportionate offspring production does not necessarily lead directly to disproportionate sexual production. To derive such estimates, Ken Ross applied genetic methods in two separate studies (Ross, 1988, 1993).

The question is simple enough. What proportion of the female sexuals (males were not considered) and what proportion of the workers are the daughters of each queen in a polygyne colony? Queens producing more than an even share of sexuals would then stand out as recognizable winners. The trick was to build a polygyne colony with queens whose daughters could be distinguished from one another. Ross combined four to six polygyne queens whose offspring had uniquely distinguishable genotypes (relative to one another) into experimental nests with their original workers. He collected and genotyped all the female alates at six-week intervals, along with a sample of worker pupae, until fewer than three queens remained alive (200 to 400 days). The result was the number of female sexual daughters and workers for every queen during every six-week period, their realized reproductive success.

Queens command widely different shares of the total reproduction, as we already know. That is not the issue here. Rather, we want to know whether queens produce similar shares of the worker and female alate pools. As a point of comparison, let us assume that all queens contribute equally to colony offspring. Then the fraction of each genotype in the colony would be 1/N, where N is the number of queens. In all but one of 18 samples, the proportion of sexuals differed more from this expectation than did the proportion of workers—queens contributed more evenly to the worker population than to that of alates.

Closer inspection revealed several interesting patterns. All but one of the 32 queens produced some daughters, but five produced no female alates at all, in spite of living long enough to do so. During the period when all queens were still alive, 26% of the queens produced 84% of the female sexuals. Three of the 32 queens managed to have 70 to 85% of their offspring reared as alates (Figure 26.13). In nine of the 14 samples from this period, a single queen produced 60% or more of the female sexuals, while one to five of her nestmate queens produced less than 5% (mostly none). In about half the samples, individual queens produced less than 5% of the female alates. In comparison, in only 16% of the samples did they produce less than 5% of the workers, and only once did a queen produce more than 60% of the workers. Over the course of the experiment, almost a quarter of the queens contributed substantially to worker production (16 to 39%) but produced less than a 5% share of the female alates.

A few queens dominated sexual production over considerable periods, often producing less than an even share of workers. More fecund queens were only moderately and variably more successful at sexual production. The contest is apparently not won by simply producing more eggs. High rates of egg laying may not even be important.

Ross also noticed that eight queens greatly dominated female sexual production (35 to 90%) during a relatively short period when they lost 40 to 60% of their weight and finally died. Other queens, however, managed to turn in high sexual production over long periods without weight loss and death. In any case, inseminated queens are clearly not

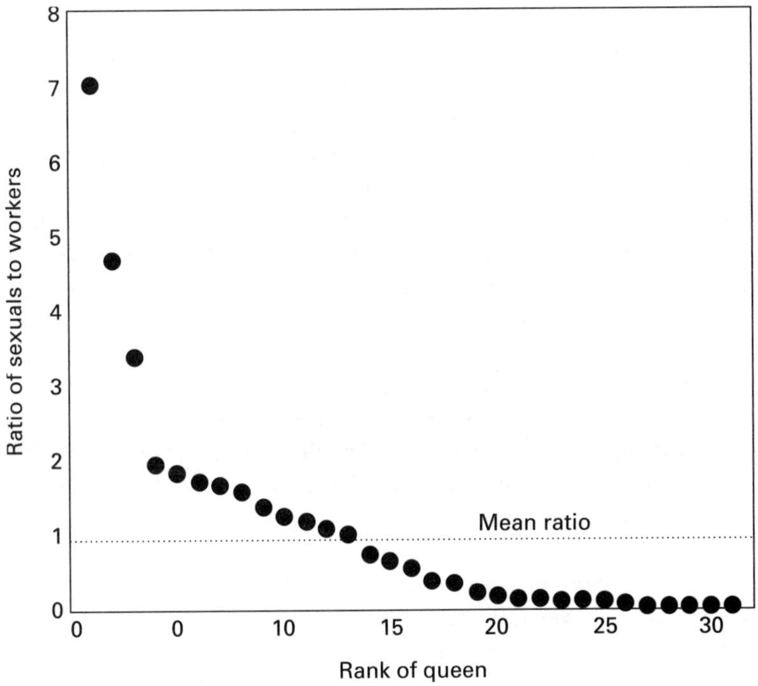

Figure 26.13. Three of 32 polygyne queens in experimental colonies managed to have 75 to 85% of their offspring reared as sexuals, thereby "winning" the competition for personal reproduction. Data from Ross (1988).

all equal, that a great competitive game is played out in polygyne colonies, and that some queens end up "winners" and others "losers."

Ross was reluctant to extend the findings from his "hand-built" laboratory colonies directly to natural colonies, because these have a lot more queens; live in large, complex earthen nests; and cope with a highly variable environment. In natural colonies, he had to rely on less-direct methods based on allozyme analysis of 31 polygyne nests. If some queens in natural colonies are more able than others to have their offspring reared as sexuals, then for a given number of queens, the relatedness among the sexuals should be higher than the relatedness among workers. Indeed, female alate pupae were consistently more related to one another than were worker pupae, suggesting that, sure enough, some queens were having more than an even share of their offspring reared as sexuals. For the nests with the highest numbers of queens, relatedness of worker pupae averaged about 0.08 and queen pupae 0.14.

Ross then reversed the equation and, from the relatedness values, calculated the "effective" numbers of queens contributing to worker and to female-alate production. In all comparisons, whether queen number was low or very high, the estimated mean number of queens contributing to worker production was 16 to 80% higher than that contributing to female-alate production, confirming that, on average, fewer queens contribute to sexual than to worker production.

Ever the thorough scientist, Ross then inspected the genotypes of sexuals and workers within 25 individual colonies and compared how frequently he observed genotype differences across worker life stages and between worker and queen pupae. If queen turnover explains the genotype differences between worker pupae and queen pupae, then genotype differences in the one should be strongly associated with those in the other, but this result was not observed. In only about half the nests did genotype differences among adult workers cooccur with genotype differences between queen and worker pupae. In one-third of the nests, queen and worker pupae differed in genotype even when adult workers did not. As in the laboratory experiments, this result suggests that some queens have a greater ability than others to have workers rear their eggs as alates or to bias their eggs toward sexual development.

What might be the basis of this ability? The short answer is that we do not know, except that simply laying more eggs does not make queens "winners." Perhaps nurse workers prefer rearing the larvae of "winners" as sexuals or are less likely to execute them. Perhaps nurse workers recognize their own siblings or mothers and bias their care toward these relatives (nepotism), but a specific test by Chris DeHeer found no such bias. The workers within any queen retinue were no more related to each other than to randomly chosen workers in their colony, nor were they related to the queen they were tending (DeHeer and Ross, 1997). A second test focused on young adult queens in the process of fattening up for their mating flight. Such queens were unrelated to the workers that fed them. If nepotism exists, then queens producing more workers should also produce more alates, but they do not, even over extended periods. Indeed, the existence of several queens with low worker but high sexual production also argues against kinship bias. Even diploid male–producing queens manage sometimes to become "winners." Winning is thus unlikely to involve kinship—workers in polygyne nests care indiscriminately for related and unrelated individuals. This pattern is somewhat puzzling, but under native conditions in South America, kinship within polygyne colonies is always high, so no mechanism biasing care toward kin was evolutionarily necessary. Selection at the colony level may have effectively overwhelmed selection for nepotism at the individual level.

Perhaps winning involves pheromones? Queen pheromones manipulate worker behavior both toward queens and toward larvae, as we saw earlier, but we do not know whether these pheromones have unique characteristics, so that queens can manipulate worker behavior to their personal benefit. Alternatively, "winners" may manage somehow to make their own offspring more sensitive to sexualization, perhaps by biasing the eggs or by making their larvae less sensitive to pheromonal suppression of juvenile hormone production or more sensitive to other caste-switching factors. The weight-loss syndrome might bias the eggs—as the queen's weight dwindles, her egg-laying rate would slow and her eggs would increase in size and perhaps other qualities that lead to sexual development. Early-spring sexual production in monogyne colonies is associated with relatively low egg-laying rates. Inhibition of egg embryonation seems to be a direct effect on the egg. Could a tendency toward sexualization also be affected by a direct effect on the egg?

The future of this research is certain to be extremely interesting and probably full of surprises. However it works, the polygyne colony is the scene of an intense competition among unrelated queens for individual reproductive success. The term "warfare" seems too strong because the object is not to conquer, kill, or eliminate but to use the colony's labor to one's own advantage. Queens struggle by means of chemical signals and physiological adjustments to bend the worker force to invest a disproportionate share of colony resources in their own alates to serve their personal fitness. Because the great majority of these workers are not their kin, this relationship can be regarded as social parasitism. Success comes unequally to queens. Interestingly, the tools they use are probably the same as those that coordinate the monogyne superorganism. The same (or very similar) pheromones that prevent daughter queens in a monogyne colony from commencing reproduction prematurely, and that regulate the timing of alate production, suppress the reproduction of competitor queens and their alate offspring in polygyne colonies. The difference is the context.

27

The Nature and Fate of Polygyne Alates

After the silent war of mutual suppression among queens has played itself out, and the pheromones have been emitted and done their work, polygyne colonies do succeed in rearing both male and female sexuals, but these differ in many ways from those produced by monogyne colonies, and the differences strongly affect both the likelihood and mode of colony reproduction.

Males: Diploidy and Sterility

Most of the male polygyne fire ants produced in the USA are sterile and reproductively useless. When Ake Hung (Hung et al., 1974) tried artificially inseminating female alates, he found that many of the males he used lacked sperm—both the accessory glands and the vasa deferentia were present, but the testis lobes never developed and the vasa deferentia remained empty. All nine of the polygyne colonies Hung sampled around Texas A&M University produced from 50 to 100% spermless males, whereas monogyne colonies produced normal males. Later, sterile males were also shown to be significantly larger (average 9 to 9.5 mg) than normal males (average about 6 mg), so much so that size-frequency distributions are bimodal. Chromosome squashes (in which tissue is squashed under a microscope-slide cover slip and stained for chromosomes) of male pharate pupae revealed that normal males had 16 chromosomes (as expected) but that the sterile males had 32; that is, they were diploid.

Hung's findings were confirmed by Glancey and colleagues, working with the now-eminent Australian geneticist Ross Crozier (Glancey et al., 1976a), who found that 96% of males from a polygyne colony were diploid and sterile, compared to only about 1% in normal, monogyne colonies. Crozier had previously published a paper (Crozier, 1971) proposing that sex-determining loci existed in Hymenoptera and that when these loci were heterozygous, a female resulted, and when hemizygous or homozygous, a male. The authors hypothesized that frequent male diploidy in the USA resulted from a loss of genetic diversity ("polymorphism") at the sex-determining locus during the population-founding event, which in turn caused many females to mate with males bearing the same sex allele. Half the offspring of such "match-mated" queens would be homozygous at the sex-determining locus and develop into sterile, diploid males.

Ken Ross, David Fletcher, and later several other colleagues at the University of Georgia (Ross and Fletcher, 1985b, 1986; Ross et al., 1993) tested this reduced-diversity hypothesis and determined the geographic extent of male diploidy. Of 57 male-producing polygyne colonies (of 264 sampled) from Georgia, Louisiana, Florida, and Mississippi, all but two produced mostly or all diploid males. Diploidy was confirmed by heterozygosity at one or both of two allozyme loci, which is possible only if the males are diploid. In all cases, the genotypes of the diploid males matched those of their worker sisters, showing that the males have a mother and a father and do not develop through parthenogenesis from diploid eggs.

Diploid males flew on nuptial flights like normal males did, making up about 70 to 75% of the males departing from polygyne colonies. Whether they succeed in mating up there in the sky no one can say, but one rather suspects that they often do and that their matings account for the 30% of polygyne queens that appear unmated. Queens that mate with diploid males receive no sperm for their trouble (recall that queens mate only once). Still, diploid males are probably less successful at mating than normal ones, for although 70 to 95% of males from polygyne colonies are sterile, 70% of polygyne females do manage to mate with normal males, most of whom fly in from surrounding monogyne populations (see below). More recently, Krieger (Krieger et al., 1999) looked more closely, and found that, of the 83% of diploid males departing from polygyne colonies, 2.4% produced sperm. When these diploid males mate, their female progeny are triploid, but because no triploid queens have ever been found, this mating is a biological dead end. Considered over two generations, all diploid males are sterile.

In contrast to the abundance of diploid males in polygyne colonies, none were found among 2500 males from monogyne colonies. Male diploidy was associated strictly with polygyne colonies, as Hung had reported. If matched mating was indeed the origin of diploid male production, then males should appear in monogyne colonies as well. What might account for this difference?

The answer lies in the large larvae that frequently occur in the first brood of monogyne founding queens. Allozyme studies proved these to be heterozygous, and hence diploid, males resulting from matched mating. Nests containing them always failed, because producing males uses up the queen's reserves, but they do no work. Several laboratory-rearing studies of founding queens from both monogyne and polygyne colonies pegged the proportion of diploid males at between 15 and 20%. Worker and male brood appeared in a one-to-one ratio, as expected if a single sex-determining locus exists (because the queen has two different alleles, one of which is identical to her mate's allele). Random combination during fertilization will yield half homozygous offspring. With a single sex-determining locus, the proportion of match-mated queens is simply the proportion of diploid male–producing queens. A single locus was assumed in all later calculations.

If the high rate of diploid-male production among polygyne *Solenopsis invicta* is the result of a founder event, then male diploidy should be much rarer in the native Argentine populations. Ken Ross and his three collaborators therefore compared two polygyne populations in Argentina (Corrientes and Formosa) to four USA polygyne populations. They isolated 95 Argentine queens and 250 North American queens from polygyne colonies with a few hundred workers each. After six weeks, the appearance of diploid males (confirmed by heterozygous allozymes) among their brood indicated those that had been match mated.

The differences were striking (Figure 27.1A). Whereas the frequency of diploid male–producing queens was 17% in North America, it was only about 2 to 3% in Argentina. Not surprisingly, then, the frequency of diploid males in polygyne colonies was between 10 and 20% in Argentina, but 75 to 100% in the USA (mean ~90%; Figure 27.1B). Correspondingly, less than 10% of Argentine queens are uninseminated, compared to about 30% in North America. This huge disparity in match-mated queens is strong support for a loss of genetic diversity through a founder event, as postulated. Incidentally, the similarity among the four USA polygyne populations failed to support the hypothesis of further reductions through secondary founder events associated with range expansion in the USA. The number of sex-determining alleles was calculated from the frequency of match-mated queens—86 occur in Argentina but only 15 in the USA, an 80% loss of genetic diversity.

What does a matched mating mean to the reproductive future of a queen? Ross and Fletcher reasoned that, for monogyne queens, the answer might depend on whether the queens founded alone (a process called haplometrosis) or in a cooperative group (called pleometrosis). They therefore set up 10 diploid male–producing queens (DMP queens, recognizable by their tell-tale large larvae) alone, with only their own brood, but gave another 10 the worker brood from four other worker-producing founding queens. This arrangement simulated the cooperative-founding situation, in which workers, in their ignorance, chose the DMP queen as the sole survivor during reduction to monogyny. Controls were normal founding queens, 10 solo and 10 cooperative.

The weight of first brood produced by the two types of solo queens was about the same, but each large, useless male weighs as much as 17 worker pupae, so the *number* of brood was much lower in the DMP colonies, and too few minims were available for effective brood care and colony growth. Most queen mortality before 10 weeks was of DMP queens. Colonies headed by normal queens grew three to seven times as rapidly as those headed by DMP queens, and normal queens increased

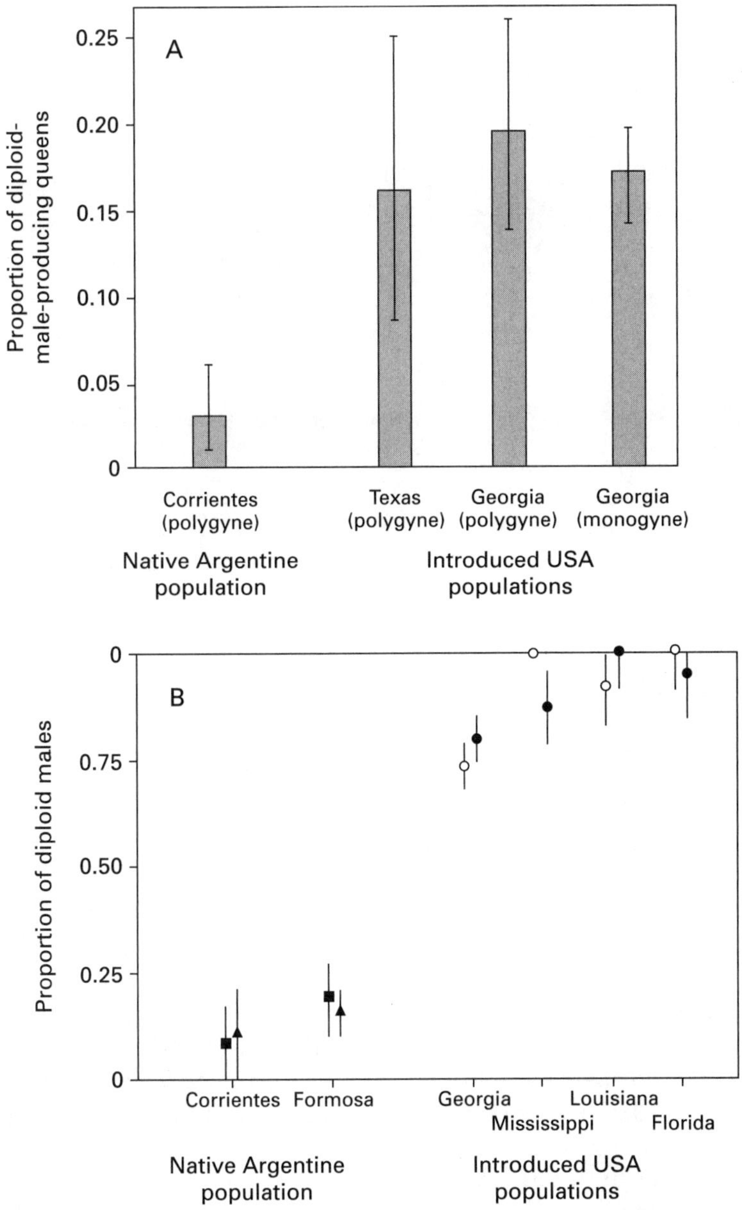

Figure 27.1. A. The frequency of diploid male–producing queens in one Argentine population and three USA populations. Monogyne queens were newly mated. Functional polygyne queens were collected from nests. B. The percentage of males that are diploid in polygyne colonies in Argentina and the USA. Bars are 95% confidence intervals. Adapted from Ross et al. (1993).

in weight and fecundity more rapidly to match their colony's productive capacity (Figure 27.2). Simulated "cooperative founding" merely postponed the inevitable—after three to six weeks, brood production lagged, and colony growth ceased. Whether they found alone or in groups, therefore, the piper ultimately has to be paid, and DMP queens are almost certainly doomed to failure in the intense competition that follows the claustral period.

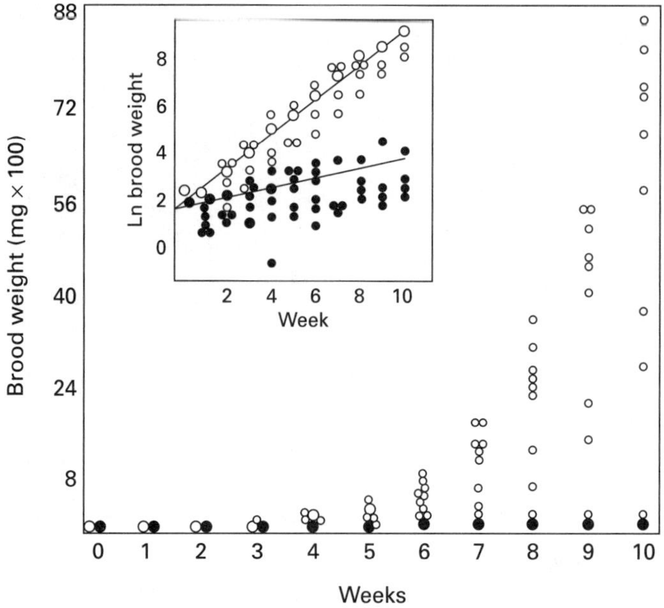

Figure 27.2. Brood production by solitary, monogyne, diploid male-producing queens (filled circles) and worker-producing queens (open circles). Inset: same data plotted on a semilog scale. Adapted from Ross and Fletcher (1986).

In a second experiment, Ross and Fletcher identified seven DMP queens from polygyne colonies by the heterozygous sons (recognizable by their allozyme patterns) they produced when isolated individually with small groups of workers. These queens were then adopted into queenless groups of monogyne workers (worker brood was added at intervals to keep the population up). Remarkably, workers not only accepted them but reared bunches of diploid males. Once again: the diploid sons of DMP queens are reared to adulthood both in mature monogyne colonies and in incipient ones, even though the colony takes a nosedive as a result. This pattern contrasts with the regulation of haploid male production through execution of male brood by queenright workers, as we saw above. Do workers not recognize diploid males as males? Perhaps the sex-recognition signals they produce are female as a result of their diploidy? We have no answer yet to these interesting questions.

Although match mating always spells doom for the monogyne queen, the fate of match-mated polygyne queens is not thus sealed. In fact, they are just as likely as nonmatched queens to be accepted by a polygyne host colony. The average polygyne colony contains 15 to 20% match-mated queens. The question has now been answered—polygyne and monogyne queens are equally likely to be match mated, but their chances of survival are quite different. Monogyne queens invariably perish during colony founding, not because of some personal shortcoming but only because of bad luck in mate choice. In contrast, match-mated polygyne queens are happily accepted by polygyne colonies, even though the males they produce represent a genetic and energetic burden on the host colony. The polygyne queen's bad luck is occasionally redeemed when some of her female offspring are reared as female sexuals.

Although male diploidy is a burden on the fire ants (though many think a well-deserved one), it has been a boon to fire-ant researchers. By capitalizing on the implications of male diploidy, researchers have gained two important pieces of information, discussed in earlier chapters: how many fire ants successfully made the trip from South America to Mobile and the success rate of parasitic colony founding among monogyne colonies.

Finally, its rapid spread and dominance in spite of the burdens of matched mating and male diploidy is testimony to both the congeniality of the North American environment and the vigor of *S. invicta*. Over the long term, this burden in the USA should gradually decrease as mutation produces new sex-determining alleles and frequency-dependent selection favors them.

Females: Body Weight, Social Environment, and Genetics

Let us now turn to the female sexuals produced by polygyne colonies. What are they like, how many are produced, and what are their reproductive options? As a result of the generous queen pheromone levels, polygyne colonies produce 30% fewer female alates than do monogyne ones (Porter et al., 1988). To this penalty are added sterile diploid males and 30% uninseminated females. Also added is that, among the 70% of females that do mate, most lack the body reserves for sufficient minim production. Mating flight–ready, they weigh about 30% less (11.8 mg) than their counterparts from monogyne colonies (15.5 mg), and only 20% of them produce sufficient minims (20 or more) to have a reasonable chance of founding a new colony. Many fail to produce workers at all, whereas about 75% of monogyne founding queens produce enough.

The body weight of mating-ready female alates is a key reproductive character, determining whether independent founding is possible or whether the newly mated queen must seek adoption. How is body weight determined, and why do female alates from the two social forms differ in this feature? These questions were addressed by Ken Ross and Laurent Keller, who was eventually to become the doyen of Swiss fire-ant research (Keller and Ross, 1993a, b, 1995, 1999). Female sexuals of ant species that engage in claustral, independent colony founding differ more in size from their own workers (Stille, 1996) and have higher body weight and fat content than do those that found dependently. In most ant species, including fire ants, the reserves needed for colony founding are accumulated in a week or two of heavy feeding after emergence from the pupa. Because fire ants practice both modes of founding,

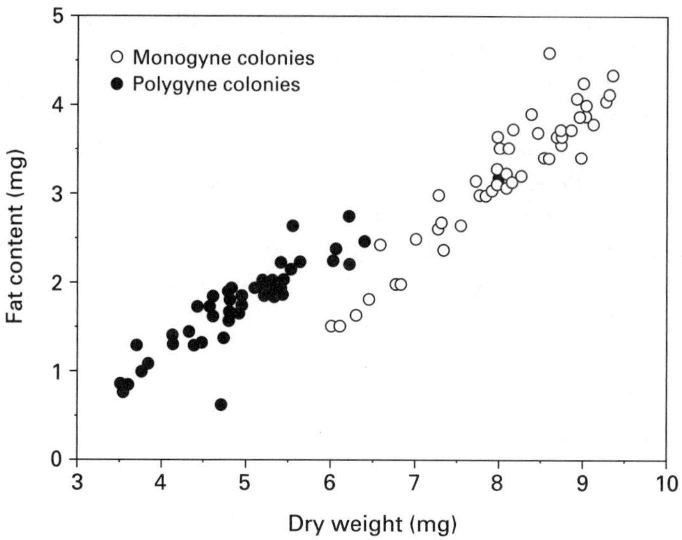

Figure 27.3. Dry weight and fat content of mature winged queens collected from the surfaces of monogyne and polygyne mounds just as they were departing on mating flights. Adapted from Keller and Ross (1993b).

alate body size, weight, and weight gain in the two social forms promised to be especially interesting.

Female alates from monogyne and polygyne colonies are similar in the dimensions of the mesosoma and head (Porter, 1992). Pupae and freshly emerged adults are only slightly heavier in monogyne than in polygyne colonies, a difference residing mostly in the gaster. By the time the female alates are ready for their nuptial flights, those from monogyne colonies are almost 50% heavier (15.8 mg) than those from polygyne colonies (10.6 mg), and the two groups overlap very little in weight (Figure 27.3). Their gasters are 70% heavier. Social form accounts for over 80% of the variance in live queen weight. Both fat and lean weights are approximately twice as high in monogyne alates, each accounting for about half the weight difference. Whereas mature monogyne alates are 40% fat, polygyne alates are only 32%. To put it still another way, dry weight increases about 240% in monogyne alates but only slightly more than 100% in polygyne alates. Most of this difference resides in the gaster, which increases almost 500% in monogyne but only 270% in polygyne alates. Fat increases almost eightfold in monogyne, but only fivefold in polygyne alates. These differences lead to successful claustral founding in one social form but failure in the other.

These differences might be caused by the social environment or by some genetic difference. If they are genetic, then rearing female pupae in the opposite social form (cross-fostering) should produce weights characteristic of their form of origin. If they are social, female weight should be characteristic of the colony of rearing. Cross-fostering experiments in which monogyne pupae were reared by polygyne workers and vice versa (plus same-form controls) showed that alate weight and fat content were mostly characteristic of the social form rearing the pupae—pupae were heavier when reared by monogyne workers and lighter if reared by polygyne workers. Although these studies suggest that social form and founding mode are passed on through the effect of rearing conditions, Ross and his colleagues later showed that this social trait, as well as many others, is largely under simple genetic control, and it is to this story that we now turn.

The tale is rather complicated and cannot easily be told in the manner in which it developed. It will help the reader if I divulge the punch line first, so here it is: a large number of the critical differences between the monogyne and polygyne social forms are associated with differences at a single gene locus, *Gp-9* (general protein-9, a non-enzymatic band on gel electrophoresis). (A brief digression on nomenclature: Initially, Ken Ross believed the allozyme *Pgm-3* (phosphoglucomutase-3) to be associated with social form. Eventually a second, closely linked gene that was almost perfectly associated with several important queen and social traits was identified on the same chromosome and named *Gp-9*. Practically all the effects originally attributed to *Pgm-3* genotype, including strong negative selection in polygyne nests, were explained by their *Gp-9* genotype (Keller and Ross, 1999). The reader should therefore bear in mind that, in the literature, the *Pgm-3* gene is really a surrogate for the *Gp-9* gene.)

The distribution of the two *Gp-9* alleles, B and b, is starkly simple: all reproductive queens in polygyne colonies are $Gp\text{-}9^{Bb}$, whereas all those in monogyne colonies are $Gp\text{-}9^{BB}$. This simple genetic difference controls many social and individual traits, indeed even social form itself. Such simple genetic control of complex social traits had never been proposed before. To begin, let us see what effects these genes have on body weight and colony founding of female alates (Ross and Keller, 1995a, 1998; Ross et al., 1996a, b, 1999; Ross, 1997; Keller and Ross, 1998).

Ross's students Chris DeHeer and Michael Goodisman captured polygyne and monogyne female alates departing from their nests on mating flights and others just after they had mated. Each hapless alate was checked for mated status, *Gp-9* genotype, and mitochondrial DNA haplotype (mtDNA). The mtDNA haplotype identified whether the queen came from a polygyne or a monogyne colony. This step was necessary because queens captured in polygyne areas could have flown in from surrounding monogyne ones and vice versa. Social form (mtDNA haplotype) and *Gp-9* genotype were thus determined independently.

Because the $Gp\text{-}9^{b}$ allele is absent from monogyne colonies, all monogyne alates are $Gp\text{-}9^{BB}$ and weigh 14 to 15 mg when mature. Alates from polygyne colonies come in three genotypes, as expected from mothers who are $Gp\text{-}9^{Bb}$. These alates tip the scale at about 13 mg if they are $Gp\text{-}9^{BB}$ and only 11.4 mg if they are $Gp\text{-}9^{Bb}$. The unfortunate $Gp\text{-}9^{bb}$

barely move the scales at 8.6 mg. Almost all of these differences reside in the gaster's metabolic reserves. The B and b alleles are codominant, that is, they affect the weight about equally. The approximately 10% lower weight of the polygyne than of the monogyne *Gp-9*^{BB} queens must be the result of maturing in a different social environment, whereas the differences among the polygyne queens are the result of their genetic differences interacting with their social rearing environment. These results were entirely due to the *Gp-9*, rather than the *Pgm-3*, gene (Keller and Ross, 1999).

What in the polygyne social environment causes these differences in the expression of the genes? Keller and Ross first tested high queen pheromone levels in cross-fostering experiments, but rearing alates in polygyne colonies reduced to a single queen (and therefore experiencing reduced pheromone level) did not produce alates any heavier than those reared with 15 to 20 queens (Table 27.1). Conclusion: queen pheromone does not mediate the weight differences among the genotypes.

Keller and Ross tested a second hypothesis: that the high brood-to-worker ratios of polygyne colonies result in less worker care and/or food for each maturing alate. This time, they divided polygyne colonies into three equal fragments. One received 60% of the worker brood (high-brood treatment), the other two 20% each. The high-brood treatment and one of the others (the control) were fed *ad libitum*, but the third received only about one-third as much food as it would normally consume (low-food treatment—the amount of brood declined as a result). Mature female alates from the high-brood treatment weighed significantly less than those from the control and low-food treatment, but once again, the genotype affected alate weight similarly within each treatment (oddly, low food had no effect on alate weight).

Table 27.1. ▸ Fresh weights (mg) of mature female alates of three genotypes at the *Pgm-3* locus (a surrogate for the *Gp-9* locus, which controls the difference between monogyny and polygyny in *Solenopsis invicta*) reared in polygyne-derived laboratory colonies with multiple queens or a single queen. AA, Aa, and aa correspond almost completely, respectively, to the BB, Bb, and bb genotypes at the *Gp-9* locus (all standard deviations are 1.0–1.2 mg). All these weights are less than those of mature alates from natural colonies.

Treatment	*Pgm-3* genotype		
	AA	Aa	aa
Monogyne environment	13.5	12.0	11.7
Polygyne environment	13.1	11.5	11.4

After the *Gp-9* gene was discovered, practically all of the differences in wild-caught alates could be attributed to differences at this locus (Keller and Ross, 1999). With this knowledge, revisiting the cross-fostering results is useful (Keller and Ross, 1993a). When *Gp-9*^{BB} female alate pupae from monogyne colonies are reared in polygyne colonies, they are a little lighter (12.4 mg) than when reared in monogyne ones (13.2 mg), indicating a difference in rearing environment. Pupae from polygyne colonies include genotypes BB, Bb, and bb. When reared in monogyne colonies, their genetic differences are not expressed, and their combined mean weight (13.82 mg) is indistinguishable from that of the monogyne BB alates (13.2 mg). In contrast, when they are reared by polygyne workers, their combined mean weight drops to 11.35 mg because the Bb and bb genotypes attain much lower mature weights. The effect of genotype is therefore expressed differently in the two social environments, although social form itself also makes a small independent contribution.

If not queen pheromones and brood ratios, what is the difference in these two social environments that leads to this outcome? The most likely culprit is the genetics of workers: those in monogyne colonies are all *Gp-9*^{BB}, whereas those in polygyne colonies always include at least a large fraction of *Gp-9*^{Bb} workers. As we shall see when we pick up the thread again later, these Bb workers recognize female alates of BB, Bb, and bb genotypes and probably provide them with different levels of care, producing different weights in the mature adults.

Capacity of *Gp-9* Genotypes for Independent Founding

When Porter showed that only 20% of newly mated polygyne queens have a reasonable chance of founding a new colony, compared to 75% of monogyne queens, he was unaware of genotypic differences. Chris DeHeer, and later Keller and Ross (1999), repeated these experiments but assessed genotype as well (Chris DeHeer, unpublished data, used with permission). The performance of the three genotypes of founding queens differed greatly on a number of counts. Three days after isolation, 66–82% of *Gp-9*^{BB} alates had shed their wings, but only 1–4% of the *Gp-9*^{Bb} had. No *Gp-9*^{bb} alates shed their wings. Monogyne queens (*Gp-9*^{BB}) laid their first eggs 2.4 days after mating, polygyne *Gp-9*^{BB} queens followed after 3.6 days, and polygyne *Gp-9*^{Bb} laid their first eggs only after 5.1 days. Whether originating in monogyne or polygyne colonies, *Gp-9*^{BB} alates seem poised and ready for rapid reproduction. Queen differences were not reflected in the developmental rate of their daughters: the first minims eclosed after about 22 days in all groups. In contrast, the number of workers produced varied widely: monogyne queens averaged 26, polygyne *Gp-9*^{BB} 19, and polygyne *Gp-9*^{Bb} a mere 12. Some, more commonly among polygyne than among monogyne queens, failed to produce any brood at all. All three types of queens weighed 7.5 mg at the end of the claustral period. Monogyne queens therefore lost 6.7 mg, polygyne BB 5.7, and polygyne Bb 3.8. The reserves accumulated during adult maturation are directly related to the number of minims produced and to weight lost. Among fire-ant alates, weight *really* counts.

Could a *Gp-9*^{Bb} queen found her own colony independently? An average of 12 workers might sometimes be sufficient, or perhaps groups of *Gp-9*^{Bb} queens might succeed in cooperative founding, leading directly to polygyny. The polygyne *Gp-9*^{BB} queens are clearly capable of founding their own colonies independently. But do they? Might the alate daughters of polygyne nests follow two different strategies, the one local mating and readoption (*Gp-9*^{Bb}) and the other high-altitude mating, dispersal, and independent founding (*Gp-9*^{BB})? These are the questions that DeHeer, Goodisman, and their professor Ross attacked next. But first, let us see whether *Gp-9* also affects the weight of male alates.

Male Alate Weight

Given the effect of *Gp-9* on polygyne female alates, it might well also affect male weight (Goodisman et al., 1999). In male alates, *Gp-9* produces no gene product detectable by gel electrophoresis, so *Gp-9* genotype cannot be determined, but in the North American *S. invicta* populations, *Pgm-3*^{A} is always linked to *Gp-9*^{B}, whereas *Pgm-3*^{a} is associated with both *Gp-9* alleles. The *Pgm-3* genotype therefore provides a good deal of information about the *Gp-9* genotype of an individual.

Males were collected from a polygyne laboratory colony that had been made monogyne for at least two worker generations. The worker genotypes showed that the queen *Pgm-3* and *Gp-9* genotypes were AB/ab and that the male she had mated with was AB. The sons of such queens were necessarily AB or ab. The *Gp-9* genotype is the one of interest here—the *Pgm-3* genotype merely informs us what that genotype is. If we exclude diploid males, polygyne haploid *Gp-9*^{B} males were heavier (7 mg) than *Gp-9*^{b} (5.9 mg), whether lab reared or field collected. For males from monogyne field colonies, the two *Pgm-3* genotypes (both *Gp-9*^{B}) both weighed 8.3 mg, significantly more than either type of polygyne male. Comparing the polygyne and monogyne *Pgm-3*^{A} males (both of whom are *Gp-9*^{B}), we see that monogyne social form results in significantly heavier *Gp-9*^{B} males, 8.3 rather than 7 mg. The lighter and smaller haploid males of polygyne colonies result from both an effect of the *Gp-9* genotype and an effect of the social form, independent of genotype. These weight differences are probably mostly the result of differences in adult dimensions, because, unlike females, males gain very little weight after adult eclosion. As with females, the smaller size of *Gp-9*^{b} males implies that polygyne workers recognize them as larvae and underfeed them, but this hypothesis has not been explicitly tested.

The meaning of these male weight differences is unclear. The heavier males may pursue a mating strategy different from that of the lighter ones. As we will see shortly, female alates do, but for males, this is currently merely an interesting conjecture.

◂•• *A Useful Tool*

When we receive unwelcome news, we don't usually throw the messenger out the window. At one time though, such rudeness seems to have been just part of doing business in central Europe and was common enough to gain its own name, "defenestration." The profession of messengery must have viewed this practice with a chary eye. Clear also is that the party who sent the messenger did not appreciate having his envoy treated in this manner. In Bohemia, back in 1618, the Defenestration of Prague started a war that raged for 30 years, killing one third of the population. A lot more windows were broken, and central Europe was reduced to smoking ruins and total exhaustion.

When I was last in Prague, I made my way up to the Prazsky Hrad, went inside, and found the very window through which the defenestration had taken place. It had been nicely repaired since 1618, but when I looked out through the window to see what was below, my sympathy for the Holy Roman Emperor's emissaries ballooned. It was a long way down. It was scary.

Being there in Prague, the capital of Bohemia, reminded me of the Defenestration of Middletown in 1960, in which I, a native-born Bohemian, participated. I can't claim that this event started another Thirty Years' War, or even a 10-minute one, but it gave me a certain claim to fame for the last two years of my college career.

Here is the scene. It is a brilliant spring day at Wesleyan University in Middletown, Connecticut, the kind of day on which it is a crime to do anything other than sprawl under an apple tree, letting the sun and soft breezes caress your winter-pallid skin.

That's outside.

Inside, where I am sitting in Dr. Odum's Chem 201 Analytical Chemistry class, we 11 students have our backs to the open windows, through which at intervals these same breezes softly brush the backs of our heads, reminding us of where we would rather be. At the blackboard, his back to the class, stands Dr. Odum (aka The Human Klein Bottle), fat chalk in hand, writing figure after figure on the board with painful care. A male cardinal's "weepa-weepa chee chee chee chee" drifts in languidly through the window, beckoning. Inside, the only sounds are the squeaking of the chalk and the rustle of fabric as Dr. Odum lifts and stretches to reach high on the board, then retracts to look at the paper in his hand, over and over. Not a word has been spoken by teacher or students for almost 10 minutes. Albert's head is on his desk. He is trying to make up for the sleep deficit acquired the previous night. Dave is twirling the eraser of his pencil in his right ear with a faraway dreamy look in his eye. Several guys exchange puzzled looks and shrug. The blackboard is already three-fourths covered with numbers, and not one of us has a clue what this is all about. Chalk dust drifts in thin wisps toward the floor. Odum's dark tweeds sport several chalk-dust zones, and on the left side of his nose is a white streak from the last time his nose itched.

I am perhaps suffering more than most of my classmates because I am in the last row, with the open window at my back, receiving wave after wave of reminders of more charming places on that very charming day. I am chafing to discover the results of my chromatography experiment over in the biology department, and its call is strong. Odum is the only professor at Wesleyan who requires attendance—and checks. I lean forward to ask Dave what all these numbers are about. Beats me, he whispers. The air of mystery deepens. Not that anyone cares. I fidget. Jason, three rows up, crosses his legs and starts to bounce his foot up and down absentmindedly.

I lean forward again and whisper to Dave. For two cents, I say, I'd jump out of that window. A pause, then he reaches into his pocket, turns around, and places two bright pennies on my desk. At nineteen, who has the sense to resist such a challenge? I take the pennies. Close up my books carefully. Put them under my arm. Stand up, watching Dr. Odum's back, still filling rows and columns with numbers. Step on my chair, the gently fizzing radiator, the windowsill. The change in my pocket clinks. The chance that Odum will hear and turn around sends a pang of fear shooting through my stomach. Too late to turn around to check. A quick step through the open window and a jump. It's eight feet to the ground below, just clear of the bushes. I land with a tremendous cascade of pocket-change noises, put my head down, and make a dash for cover.

What happens next in the classroom, of course, I know only from the later reports of my classmates. Exactly nothing happens, except that all 10 of my classmates know within seconds that I have jumped ship. Dr. Odum, android-in-tweed, still scratches digit after digit into his huge table of numbers and notices nothing. Finally, the table is complete, the double blackboard filled with rows and columns of numbers, like rank upon rank of soldiers on parade.

Dr. Odum steps back from the blackboard with a look of satisfaction, half faces the class, then views his mystery table. With the air of one about to bestow vision upon the congenitally blind, he says, "One can immediately see what a useful tool this is."

In this single sentence, these 10 little words, Dr. Odum has bequeathed to us, his students, a gift: an unforgettable line, a masterpiece of haiku, to be remembered, relished, and enjoyed for as long as our memories serve us. Perhaps he taught us some chemistry too. Who can remember where the first dim understanding of electromotive force began to glow or the ins and outs of gravimetric analysis first emerged from the fog? Was it Odum? None of us could be sure. But not one of us who sat in that warm, sleepy room that day will ever forget where we learned the line, *One can immediately see what a useful tool this is.*

28

Polygyne Mating, Adoption, Execution

Queen Genotype and Local Mating

Low body weight constrains an alate's reproductive options, but do alates actually act appropriately for their body weight? The dispersing mating flights of the monogyne form are described in earlier chapters. Polygyne ants are expected to mate near their natal nests and to seek readoption into them. Indeed polygyne *S. richteri* seem to do so in their native Argentina. There, males form low-flying aggregations at human head height over their home pasture, like clouds of smoke. They zoom around randomly in this cloud, but the cloud remains more or less in place. Females slowly fly straight into this cloud and hover in it until they are pounced on by males and fall to the ground *in copulo* (Wuellner, 2000). Large females are pounced on by large males and small ones by small males. Copulation only lasts about 10 seconds (life in the fast lane), and once the job is done, the female forcefully disengages herself and flies off, 3–5 m above the ground, landing from a few to 25 m away but sometimes out of sight. At least some females therefore land among colonies containing relatives and have a chance of being adopted by them. Males remain in the mating swarm until they die of exhaustion or are eaten by predators. The degree to which this mating syndrome is general to polygyne fire ants is unknown.

The clever and interesting studies by Ross and his students radically changed our view of how polygyne alates behave (Shoemaker and Ross, 1996; Ross et al., 1997; Goodisman and Ross, 1998, 1999; DeHeer et al., 1999; Goodisman et al., 2000; DeHeer, 2002). If female alates pursue reproductive strategies that are appropriate to their body reserves and genotype, their *Gp-9* genotype frequencies should depend on where they are collected and from which social form they originated. Over multiple mating flights and years, Ross and company collected and analyzed almost 4000 alates. *Gp-9*^{BB} alates with mtDNA haplotype B originated in monogyne colonies, those with haplotype C in polygyne colonies.

The study was carried out at 10 north Georgia sites, each with a bounty of polygyne colonies as well as areas of exposed soil. On mating-flight days, female alates nabbed from the aggregations waiting to take off from the tops of nests ("preflight alates") were unmated and represented a sample of the locally produced *Gp-9* genotypes. A second group of alates was netted during extended, level flights near the ground ("low-flight alates"), and a third was captured as they landed in open areas with sparse vegetation, but before they excavated a founding chamber ("postflight alates"). Strikingly, although polygyne *Gp-9*^{BB} alates made up only 2.3% of the preflight group, they made up a whopping 31% of the postflight group, probably because they were drawn to the disturbed habitat after a high-altitude mating and dispersal flight. If a site was strongly attractive to newly mated monogyne *Gp-9*^{BB} queens (mtDNA haplotype B), it was also attractive to newly mated polygyne *Gp-9*^{BB} queens (mtDNA haplotype C). These queens pursue the same dispersal strategy, and it is their genotype, not their social origin (or the devil), that makes them do it.

As a result of this preference, *Gp-9*^{BB} alates were significantly underrepresented in the low-flight group, 95% of which consisted of *Gp-9*^{Bb} queens from polygyne nests. In this hitherto undescribed phenomenon, tens to thousands of these female alates (no males), most of them mated, flew in more or less straight, horizontal lines 2–5 m above the ground, occasionally landing and even taking off again. The proportion mated was usually similar to that among functioning queens at the same site, suggesting local rather than regional mating swarms not more than a few hundred meters from home (the availability of fertile males varies widely across polygyne populations, causing the proportion mating to vary geographically). These female swarms therefore seem to be local, postmating dispersal or nest-selection flights.

The mtDNA haplotype patterns added interesting details. Recall that the mitochondria and their DNA are passed to offspring only by the mother (in contrast to the "nuclear" DNA, which each offspring receives from both mother and father) and therefore reflect her dispersal patterns. Low dispersal distances reduce population-wide mixing and allow haplotype frequencies to vary on a small geographic scale. This was the case for preflight alates—the 10 sample sites differed significantly in haplotype frequencies. Intersite differences were somewhat stronger among the low-flight alates (which are mostly *Gp-9*^{Bb}). Their haplotype frequencies were most similar to the preflight alates from the same sites and, most importantly, to those of functioning reproductive queens at the same site. This pattern suggests that these alates were produced by nearby colonies, mated locally, and were in the process of dispersing only modest distances. Because they never go far from home, local differences in haplotype frequencies can develop, as observed.

The patterns among postflight alates collected from open areas were quite different. Haplotype frequencies of the *Gp-9*^{BB} alates were similar at all sites, as would be expected if they had flown long distances and mixed with alates from many sites. Some of them were probably even newly mated monogyne queens from outside the polygyne areas.

Because monogyne alates disperse long distances, their populations show no local differentiation of haplotype frequencies, and their admixture would make the detection of local differences among polygyne alates more difficult.

A bigger surprise was that haplotype frequencies among postflight *Gp-9*^{Bb} also showed no intersite differences. Therefore, those polygyne *Gp-9*^{Bb} queens that landed in the open areas, presumably to attempt independent founding, did not come predominately from the same site but flew in from a distance, thereby following the same long-range dispersal strategy as *Gp-9*^{BB} alates and seeking out recently disturbed, open areas. If they did, then the more attractive an open area is to founding queens, the more the haplotype frequencies of postflight alates there should deviate from the preflight frequencies, because a higher proportion of the alates will have flown in from the outside rather than from local nests. This was indeed the case: sites that attracted more *Gp-9*^{BB} alates also drew more other dispersing alates, creating greater differences between the postflight alates on the one hand and pre- and low-flight on the other.

Gp-9^{bb} female alates were the most pathetic. They weighed less than 9 mg, compared to over 11 mg for the *Gp-9*^{Bb} and 13.5 mg for the *Gp-9*^{BB}. Although they made up almost a fifth of the alates preparing for take-off on top of nests, their frequency dropped by 80% in the low-flight crowd, and they were essentially absent from the pool of reproductive queens in polygyne nests. Perhaps, starved into punyness by workers, they simply do not have the reserves even to play the game.

Simply put then, two distinct dispersal syndromes exist, local and long-distance (Figure 28.1). Both are under the control of the *Gp-9* locus (or one closely linked to it), which also controls body weight. The heavy *Gp-9*^{BB} alates, whether born to monogyne or polygyne colonies, mate high and disperse long distances, and the light *Gp-9*^{bb} alates mate locally and seek adoption (and seem usually to fail). *Gp-9*^{Bb} alates may be of either syndrome: either they mate locally and disperse only small distances before seeking adoption or they join the *Gp-9*^{BB} alates to disperse long distances and (probably) found colonies independently. The basis on which they make this choice is unknown.

Even so, long-range-dispersing *Gp-9*^{Bb} alates rarely succeed in polygyne areas; otherwise little haplotype differentiation would be evident among polygyne sites. Nor do they often succeed in monogyne areas, for their most common mtDNA haplotype (haplotype C) is absent in the surrounding monogyne areas. Perhaps these *Gp-9*^{Bb} queens occasionally succeed in invading freshly disturbed habitats, establishing new polygyne populations through cooperative founding. The consistent differences among sites in haplotype frequencies suggest that long-distance dispersal is rarely successful for these alates, as it was also for the polygyne *Gp-9*^{BB} alates. Finally, monogyne *Gp-9*^{BB} alates

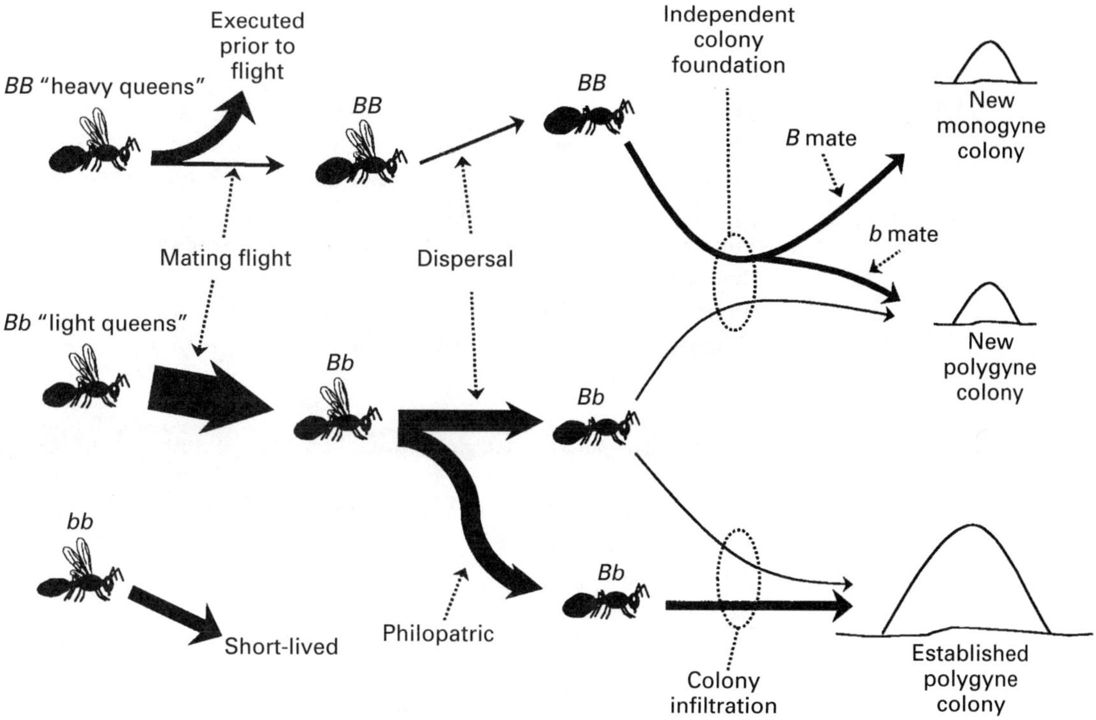

Figure 28.1. Schematic representation of the mating and dispersal syndromes associated with *Gp-9* genotypes of queens from polygyne colonies and, for *Gp-9*^{BB} queens, the genotypes of their mates as well. The arrow width is roughly proportional to the relative numbers succeeding in each pathway. "Philopatric" means "mating and settling locally." Adapted slightly from DeHeer et al. (1999).

also rarely succeed in polygyne areas, or their haplotype B would not be absent from polygyne populations. Gene flow by way of females is thus quite low within this polygyne population and between the neighboring polygyne and monogyne populations.

Queen Adoption and Colony Fission

Polygyne *Solenopsis invicta* colonies were expected, like other species of polygyne ants, to adopt newly mated queens and to reproduce by colony fission or budding. Ed Vargo and Sanford Porter tested for this type of reproduction at Brackenridge Field Laboratory near Austin, Texas. They first cleared two areas of fire ants by flooding nests with hot water (Vargo and Porter, 1989; Porter, 1991). In this area, they then planted five polygyne colonies whose queens they had marked with colored ink.

Four or five months later, queens from the five planted colonies were found to be distributed among 22 mounds, each of which contained 1–16 marked queens. No pattern (for example with distance) was apparent in the number of marked queens in each mound. Most interestingly, queens did not switch allegiance; that is, queens marked in one colony did not end up with queens marked in a different colony—queen colors did not mix. Colonies therefore reproduce by budding; a fraction of the worker and queen populations migrates through underground tunnels to construct a new mound a short distance (up to 50 m) away. Connections may persist, but the resulting nests are largely independent; exchange of workers is limited, exchange of queens rare. Budding produces populations of connected, closely spaced mounds that dominate the landscape in discrete patches, spreading "on foot," probably explaining why polygyne mounds often seem to be closely clumped together and separated from the monogyne mounds (Greenberg et al., 1985).

Colony reproduction was remarkably rapid—the original five mounds had split into at least 22 within four or five months (Figure 28.2). One nest yielded eight daughters. Total weight of ants increased between 140 and 640%, and the average mound more than quadrupled in size, an amazing growth rate indeed. Perhaps growth was so rapid partly because the area had been cleared of competing fire ants, but half-grown monogyne colonies grow by approximately 30 to 40% during a similar period, so other factors must also enter in. The final biomass of ants on the site was a remarkable 2 g, or 2200 workers, per square meter. As high as this density sounds, it is probably well below carrying capacity, because at the end of the study, only 30 mounds were present on the site from which Vargo and Porter had cleared 77 mounds.

The Source of Polygyne Queens. Colony reproduction by budding requires a source of additional queens; otherwise the colony would soon run out of queens to accompany the new buds. By now, it is no secret that the most likely source is the adoption of newly mated queens. Testing of this hypothesis got off to a bad start when Glancey and Lofgren (1988) released 800 paint-marked newly mated queens in a pasture but failed to recover any marked queens until nine months later, in the course of another study. Next, they dissected queens from polygyne mounds after mating flights, distinguishing recently mated from functional queens by the condition of the fat bodies, wing muscles, and ovaries. Colonies contained 1–52 queens that fit the criteria for recent mating. The evidence would have been stronger if they had sampled both before and after mating flights, to show that the number of these "adopted" queens had increased sharply.

As part of the budding experiment reported above, Sanford Porter marked all the alate and dealate queens from six weighed polygyne colonies (designated A–F) to identify both the colony of origin and initial queen status (alate or functioning queen; Porter, 1991). Once the planted colonies were well established, he scattered still more marked queens across the site. These included newly mated queens, queens

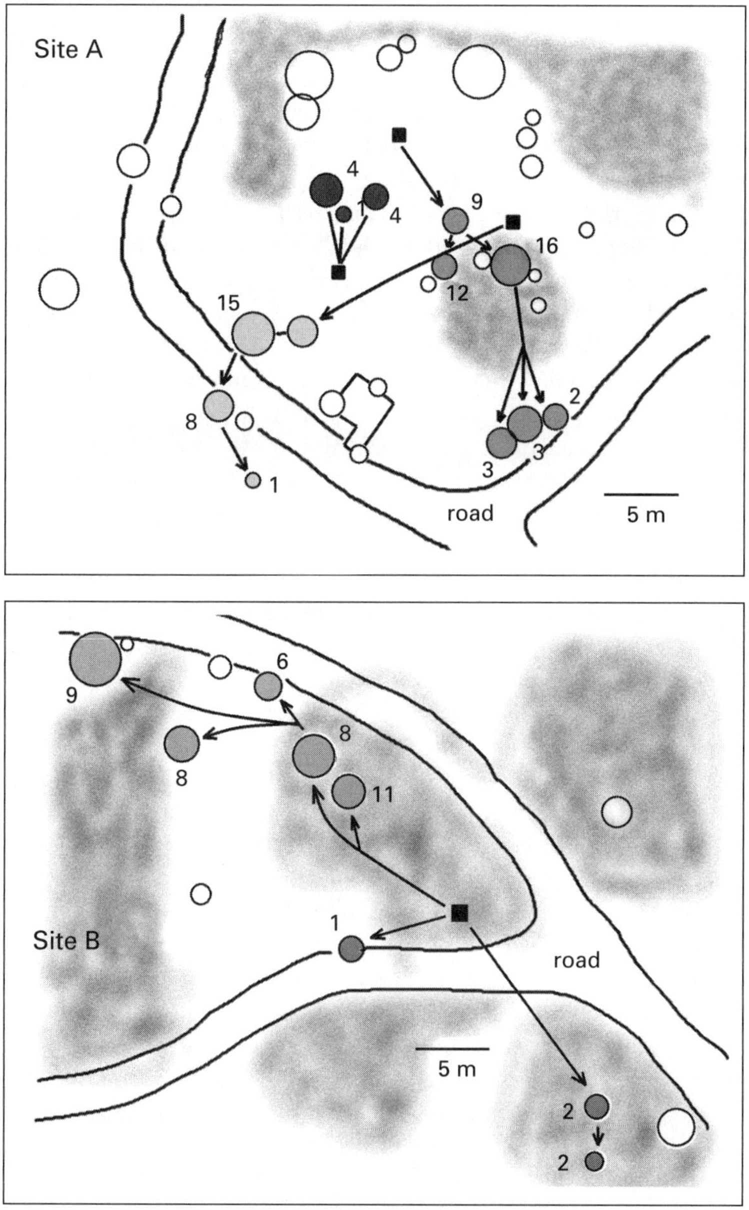

Figure 28.2. Colony budding in four polygyne colonies during a four-month period. Solid squares show locations of original colonies; shaded circles show locations of daughter colonies. Circles shaded similarly are progeny of the same initial colony. Numbers are the number of marked queens (i.e., queens from the original colony) found in each. Unshaded circles indicate unmarked colonies. Adapted, with permission, from a drawing by Sanford D. Porter.

mated two to five days previously, functioning colony queens from one of the six polygyne colonies, and queens that had completed colony founding and had minim workers (these incipient colonies were planted in the ground).

Of the 313 marked queens in the six transplanted colonies, 49 (16%) were recovered six months later in the 39 descendent colonies. In polygyne colonies, unlike monogyne ones, queens disappear, as did two of the six colonies. Almost all the queens stayed with their own descendent colonies, as in the earlier study. Only a single queen was found in a colony line different from the one she was marked in. Of the other types of marked queens, no recently mated or postclaustral queens were recovered, but a few of each of the other types were, including five (about 2%) of the colony queens from colony F scattered across the site and four (1.3%) of the newly mated queens. Even four (0.2%) of the unmated, winged queens turned up, three of them having been inseminated in the interval. All were in descendents of the colonies in which they had been marked; that is, they were with colony queens marked in the same original colony. They had probably mated on a local nuptial flight and returned to their nests, as described above, rather than mating in the nest. Finally, 198, 10, and 2 unmarked queens had joined colonies D, E, and F, an extremely uneven rate of queen acquisition.

Do Nests Regulate Queen Adoption? The low rates of queen adoption suggested that colonies might regulate queen number, i.e. that new queen acceptance might depend on the number already present. After several mating flights, Goodisman and Ross collected polygyne colonies, separating and dissecting all wingless queens (Goodisman and Ross, 1997, 1999). In addition, they differentiated between newly recruited queens and old queens by means of the enzyme glycerol-3 phosphate dehydrogenase-1 (G3pdh-1), whose function is to energize flight, and an alate-specific protein band of unknown function in the thorax that disappears during the week after mating. They could distinguish recently mated, newly recruited queens from the old queens for five to seven days after mating.

Of almost 2600 queens from 85 colonies, only 43 met the criteria for having been recently mated and recruited, an overall rate of 1.7%. (Note: the similarity of this number to Porter's is difficult to interpret, because Porter's estimates were made after a lapse of six months, confounding acceptance and survival, whereas Goodisman estimated numbers of queens adopted during the previous week.) For reasons unknown, adoption rates for individual mating flights varied from 0.5% to 5.5%. Newly mated queens seemed no more likely to get into colonies with low than high numbers of queens, or vice versa, and the numbers of old queens and new recruits were unrelated. Acceptance seems to be a random process, following the Poisson distribution characteristic of rare events. Because queen number is unregulated, it varies randomly among colonies, perhaps accounting in part for the wide range of queen number.

Random adoption is odd from the new recruit's point of view. Surely she can detect the queen pheromone level of the prospective host and thus judge her chances of successful reproduction? Perhaps the risk of testing several potential host colonies makes shopping around unfeasible. For that matter, the workers should also have an interest in keeping the number of queens low so that sexual production is not suppressed too much. Why they do not is a mystery.

We can roughly calculate the proportion of alates leaving on mating flights that actually end up safely back in a polygyne nest. With a mean queen number of about 30 and an acceptance rate of 1.7% per week for the duration of the flight season (four months), an average colony accepts about eight or nine new queens per season. The same average colony probably emitted between 200 and 500 female alates during that season, a success rate of 2–4%.

Of the older queens, about 50% were unmated in this survey, somewhat higher than the proportion at other sites, but remarkably, over 90% of the new recruits were unmated. The meaning of this discrepancy is not clear. Do unmated queens suffer higher mortality after adoption? Are they detectable as new recruits over a longer period? Are they better at getting themselves adopted? The new adoptees were also lighter than older queens, suggesting that it takes longer than a week to crank up to full reproductive output.

What Determines Success in Adoption? In most (but not all) polygyne ant species, queens within a colony are significantly related to one another, and the colony is a sort of extended family. Their willingness to accept a new queen being "interviewed" should increase with the degree to which the queen is related to them. In the USA, queen relatedness in polygyne colonies is essentially zero (Ross et al., 1996b). Still, particular subgroups of queens might be more related to other subgroups, but exhaustive testing of relatedness by Goodisman (Goodisman and Ross, 1997, 1998, 1999) showed that even after several refinements and much sophisticated statistics, older queens within polygyne nests were not related to each other, really, really (this time we mean it!). Nor were the newly recruited queens related to each other or to the older queens already in the nest. Furthermore, queen number and queen relatedness were uncorrelated and showed no pattern. Nor did colonies with very low queen numbers (< 9) have the highest relatedness (Figure 28.3). Overall, no evidence indicates that relatedness plays any role at all in getting queens adopted or in starting new colonies in groups. Adoption seems to be as random with respect to relatedness as it is with respect to queen number. Polygyne colonies simply do not behave as closed societies.

One more desperate attempt to indict relatedness—it could arise during budding if related queens moved into the new buds together, but the data do not support this hypothesis either. A significant relationship between queen number and relatedness could be obscured by free movement of queens and workers among interconnected mounds, as suggested by Bhatkar and Vinson (1987) because the queens counted in each mound would not be a meaningful measure of the number in the whole colony. In fact, queen number within a mound is quite a good

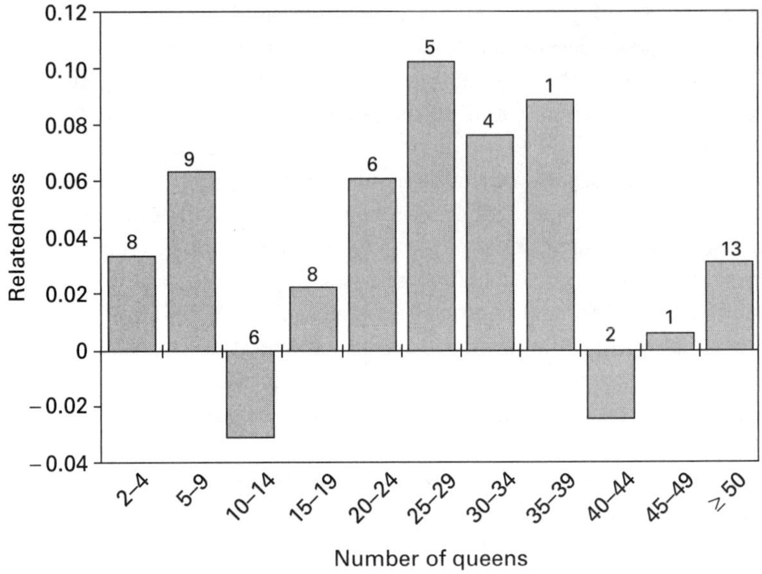

Figure 28.3. Relatedness of nestmate queens in polygyne *Solenopsis invicta* plotted against the numbers of queens (for nests containing similar numbers of queens, regardless of mating status). None of the values of relatedness is significantly greater than zero. Values above columns are the number of nests in each category. Adapted from Goodisman and Ross (1997).

predictor of worker relatedness and of worker size, so each polygyne mound must be a largely closed unit. Were it not, queen number could not leave this signature. In short, nothing can rescue us from the conclusion that queens within polygyne colonies, no matter how you slice it, are not related to one another.

Odd as it may seem, polygyne nests might still accept mostly nestmates as new recruits, nestmates to whom they are not related. Imagine that just a few related queens start colonies. They produce sexuals that mate with unrelated males and then reenter the natal nest. Relatedness will decline with every repetition of this cycle, until it is not distinguishable from zero. The patterns discussed in the next section rule out nestmate adoption.

The Scale of Mating Distance, Dispersal, and Adoption in Polygyne Populations. A comparison of nuclear and mitochondrial gene patterns can reveal both the scale and the sex through which gene flow occurs in a population (i.e., the distance from the natal site to the reproduction site), as described earlier. Local mating and adoption should allow mtDNA haplotype frequencies to accumulate on a small geographic scale, but wide dispersal should prevent such local differentiation through populationwide mixing. Similarly, local mating by males leads to local differentiation of nuclear genes, and long distance dispersal to populationwide homogenization. Male and female dispersal and gene flow can therefore be estimated separately. In keeping with the metaphor of flow and fluids, geographic differentiation of genes is referred to as population viscosity, or resistance to (gene) flow.

Significant population viscosity on a scale of kilometers is a common feature of polygyne ant populations, usually as a result of limited female alate dispersal, but how does viscosity in polygyne fire ants look on a still finer scale? At six sites, Goodisman collected 85 nests that were between 300 m and 2 km apart and determined genetic similarity (relatedness) on the basis of both nuclear and mitochondrial genes on increasing scales: within nests, within sites, and between sites. Repeated resampling (with return) from the data set allowed the relationship between genetic and geographic distance to be tested. The slope of a significant positive correlation between genetic distance and geographic distance is an estimate of population viscosity. It describes how fast genetic differences arise with physical distance and thus informs us about how far genes are carried by dispersing sexuals.

In this way, Goodisman compared the six sites, and as abstract as the results may seem, they paint a vivid picture of queen and male dispersal, mating and recruitment. Both nuclear and mitochondrial genes showed that queens within nests are entirely unrelated. Therefore, not only are the newly recruited queens not related (as we already knew) but they did not originate in the nest that adopted them (they are not nestmates). Nests must be adopting foreign queens regularly, for if they did not, because of chance loss of queens and therefore their mtDNA lines, all queens within a nest would eventually be the descendents of a single queen, and mtDNA relatedness would be 1.0. This is not the case. Once again, all indications are that mated alates that knock are let in at low frequencies without regard to their relatedness or nestmate identity, and of course this regular adoption of foreign queens is also in line with the high queen numbers often observed in polygyne nests.

Enlarging the scale from nests to sites, Goodisman found no correlation between the genetic and physical distances between nests within the sites. Queens therefore mix freely on the scale of site dimensions (less than 100 m). Budding might seem likely to lead to greater similarity between a nest and its daughters, but it can do so only if relatedness within nests is significant, and it is not.

When the scale was increased still further to all polygyne sites, genetic distance based on mtDNA increased significantly with geographic distance ($r=0.24$), whether computed from nests or the entire sites. Queens therefore do not often travel distances of 300 to 2000 m before seeking adoption. This localness is consistent with DeHeer's observations on mating flight syndromes and gives rise to the mtDNA haplotype differentiation we saw above. As a result, about one-half of the total genetic variation in polygyne mtDNA is found among sites and the other half among nests within sites. In both social forms together, about a quarter of mtDNA variation is between social forms.

In contrast to the mtDNA, the nuclear genes showed no geographic differentiation and no difference between social forms. Over 99% of the genetic variation occurred within sites (i.e., among nests), and less than

1% between sites or between forms. In other words, all sites and both social forms were homogeneous in their nuclear allele frequencies. Genetic distance based on nuclear genes remained uncorrelated with geographic distance ($r = -0.02$). These nuclear genes therefore flow and are mixed freely on scales that exceed 2 km, probably through long-distance dispersal of males during mating flights. The distance of nuclear gene flow is the average distance a male flies between his natal nest and the site of his aerial coupling, plus the distance between that coupling site and the place where the mated female becomes reproductive. In monogyne fire ants, both these distances are large. In the polygyne fire ants, the female contributes little to the nuclear gene flow created by male flight, and of course the male cannot affect mitochondrial gene flow at all. Polygyne populations therefore show little if any nuclear gene differentiation (actually, a hint is present), and monogyne populations show none on the scales discussed so far. When the geographic scale is increased to the Georgia–Texas axis, about a third of the monogyne nuclear allele frequencies differ significantly, because such distances are far beyond the flight distances of alates, making nuclear gene flow too low to homogenize allele frequencies. Most of these genetic patterns and deductions about gene flow were later confirmed by means of additional genetic markers (Ross et al., 1999).

Life Span of Polygyne Queens

Tenure among the functioning queens does not assure long life. If we assume that the average number of queens does not change over the year, then the recruitment rate equals the mortality rate. Of course, mating and recruitment occur mostly during the spring and summer, converting the weekly recruitment rate of 1.7% into a life-span estimate of almost three years. This figure agrees reasonably with Ross's estimate of 2.5 years in laboratory colonies. Estimates from other studies are lower. The six-month 84% queen turnover of Porter's study yields a life span of about seven months. Vargo estimated a life span of about 10 months (Vargo and Porter, 1989). Whether these represent real life-span differences among sites is unknown. In any case, the life span of a polygyne queen is at most about half the seven-year life expectancy of a monogyne queen. Reduced longevity is a common feature among polygyne ant species, but of course, whereas a monogyne queen must build up her colony for several years before investing heavily in alates, a polygyne queen can more or less jump right into producing them.

Genotype, Acceptance, and Execution of Queens

The generally low rate of acceptance of new queens suggests that 95–99% or more of newly mated queens either fail to be adopted or fail to survive long enough to reproduce. Of the factors that tilt a young queen's chances one way or the other, the best-understood are genetic

(Ross, 1992; Keller and Ross, 1993a, b, 1995; Ross and Shoemaker, 1993; Ross et al., 1996a, 1996b). This story began when Ross and Keller established that the *Pgm-3* genotypes occurred with the expected frequency (about 56% *Pgm-3*^{AA}, 40% *Pgm-3*^{Aa}, 6% *Pgm-3*^{aa}) among monogyne queens and prereproductive polygyne alates. This finding was not remarkable. What *was* remarkable was that in polygyne colonies, although *Pgm-3*^{AA} genotype was common among workers and alates awaiting their mating flights, no egg-laying queens *at all* were *Pgm-3*^{AA}. They simply vanished between emergence and the start of reproduction. About 75% of egg-laying queens were heterozygous (*Pgm-3*^{Aa}) and 25% were homozygous (*Pgm-3*^{aa}). As a result, allele A was only about half as abundant among polygyne laying queens (even unmated ones) as among monogyne ones. Practically all of the effects of *Pgm-3* genotype were later found to be caused by the closely linked *Gp-9* gene. In natural polygyne populations, sexual females with the genotype *Gp-9*^{BB} disappeared as they assumed reproduction, suggesting either that they were failing to be accepted by host colonies or that something nasty was going on as a new queen began to lay eggs.

Ken Ross teamed up with Laurent Keller to find out what was going on. Actually they worked initially with *Pgm-3*, which was linked to *Gp-9*. Ross estimated linkage from the recombination rate of double heterozygotic queens (*Pgm-3*^{Aa}, *Gp-9*^{Bb}) to be 0.16%, indicating that the genes are extremely close together on their chromosome and that AB was on one chromosome and ab on the other. Because *Pgm-3*^{A} is completely linked to *Gp-9*^{B}, and *Gp-9* is the gene associated with social form, I have substituted *Gp-9* for *Pgm-3* in this discussion. Although female polygyne fire ants begin from zygotes with six different *Pgm-3*/*Gp-9* genotypes, only two of these survive (AaBb, aaBb) in reproductive females. Selection against the BB and bb genotypes removes the AA, Aa, and aa genotypes unequally, making the B allele much more likely to be found on an A-chromosome than an a-chromosome and the A allele more likely to be found on a B- than a b-chromosome (linkage disequilibrium).

To learn the fate of the missing homozygous queens, Ross and Keller began by reasoning that a winged female alate might find herself assuming reproductive function under two circumstances. The first is a sudden reduction of queen number, which they mimicked by moving alates to queenless colony fragments. After 10 days, the onset of reproduction and culling by workers were well under way (12% of them were killed). If the *Gp-9* genotype had no effect on alate survival, about 36 queens should have carried the *Gp-9*^{BB} genotype, but only six queens did. Most of the 12% mortality must have been of *Gp-9*^{BB} queens.

In the second circumstance, an alate may leave the nest on a mating flight and be readopted in another colony, where she would initiate ovarian development. This situation was partially mimicked by isolation of individual unmated alates with small colony fragments for three days. The lack of queen pheromone leads to dealation and egg laying, and the *Gp-9^BB^* alates gained weight and function more rapidly than those of the other two genotypes. All these alates were then returned to their natal colonies (to avoid nestmate-recognition complications). At the end of one week, 68% of the 378 alates still survived. About 28 of these would be expected to be *Gp-9^BB^*, but none was.

Together, these experiments tell us that the shortage of the *Gp-9^BB^* homozygotes among functional queens results from their execution by workers, not their inability to enter the reproductive population. Possession of two *Gp-9^B^* alleles is the mark of death for queens once they become reproductively active. Oddly, the workers kill the more reproductive *Gp-9^BB^* polygyne queens, rather than the less, as unreasonable as this course seems. Among polygyne fire-ant queens, genotype is fate.

Selection on the Gp-9 *Gene.* In the spirit of Jerry Lee Lewis, there is thus a whole lot of selection going on, and it is time to look into the *Gp-9* gene in greater detail. The product of this gene is not one of the known enzymes, but the proteins it produces can be detected simply because they bind the general protein stain Coomassie blue on electrophoresis gels. *Gp-9^BB^* individuals have a single fast electrophoretic band, and *Gp-9^bb^* have a single, slower band. Heterozygotes (*Gp-9^Bb^*) have one of each.

Ross studied the inheritance and selection of *Gp-9* in wild populations and in laboratory-reared families. For the first, he determined both the *Gp-9* and *Pgm-3* genotype of a sample of queens, alates, and workers from both social forms from two sites in the USA and two sites in Argentina. Because genotypes determined on individuals from the same colony are not independent, Ross determined the population-wide allele and genotype frequencies by statistically resampling (by computer, with "return") a single ant at random 1000 times from each colony, then computing the populationwide frequencies.

The two social forms differed dramatically in *Gp-9* allele frequencies. In practically all monogyne *S. invicta*, the *Gp-9^B^* allele was the only allele present, and all females were *Gp-9^BB^*. In the polygyne form, the proportion of sexual females that were heterozygotes (Bb) increased as winged queens progressed from nonreproductive to reproductive status; practically all functioning queens were *Gp-9^Bb^*. We have seen this movie before. Because most polygyne queens carry the b allele, one would expect about 5% of females to be *Gp-9^bb^* at least among nonreproductive females. In reality, only 0.1 to 0.6% were *Gp-9^bb^*, indicating that 90 to 98% of this genotype disappears through selection. The *Gp-9^bb^* genotype was completely absent in the Argentine populations. The deviation from Hardy-Weinberg equilibrium showed that, among reproductive queens, heterozygotes were much more frequent than expected in the absence of selection. *Gp-9*, like *Pgm-3*, is under intense selection in polygyne *S. invicta*, both in the USA and in Argentina. The

$Gp\text{-}9^{bb}$ genotype was a clear loser, and the survival advantage of the $Gp\text{-}9^{Bb}$ genotype was essentially complete among reproductive polygyne queens. It was the only genotype present.

Ross then turned to family studies, to clarify the inheritance of the *Gp-9* alleles. He placed each of 60 queens from polygyne nests in her own nest with workers and brood, ending up with simple families after eight weeks. From the genotype of the queen and of her offspring, we can infer the genotype of the father. In all but five of 60 families, the genotypes occurred in the expected Mendelian ratios. In the five that deviated, the bb genotype showed a clear deficit, indicating that workers bearing this genotype survive poorly. We will return to this point below.

Queen Acceptance Experiments. Not a single fully reproductive $Gp\text{-}9^{BB}$ queen was found among over 2500 queens captured from polygyne field colonies. How do *Gp-9* genotype and worker behavior interact to produce this outcome? The first behavioral study determined alate genotype frequencies on a sample, then used alates isolated for three days to simulate queens that had mated, been readopted, and were now at the onset of egg laying. A second experiment simulated adoption of newly mated monogyne and polygyne queens by polygyne colonies (queenright monogyne colonies kill all foreign queens, so such tests were not run). The newly mated queens were collected after mating flights in areas of north Georgia known to be either monogyne or polygyne, and the social form of origin was confirmed on a sample by means of mtDNA haplotypes.

As before, the results were unambiguous. No matter what their origin or reproductive condition, all $Gp\text{-}9^{BB}$ queens were killed when returned to polygyne colonies (not their own). In contrast, a substantial fraction of polygyne $Gp\text{-}9^{Bb}$ queens, either newly mated or reproductive, survived for at least three days (Figure 28.4). Genotype, not social form of origin, made the difference, and selection was intense.

Ross and Keller next tested the different genotypes of *Gp-9* in older, fecund queens. They introduced the normal $Gp\text{-}9^{BB}$ queens from monogyne colonies and normal, polygyne $Gp\text{-}9^{Bb}$ queens into polygyne host colonies. Again, the polygyne workers killed all the monogyne $Gp\text{-}9^{BB}$ queens and accepted multiple polygyne $Gp\text{-}9^{Bb}$ queens, but were they responding directly to the higher weight and fecundity of the $Gp\text{-}9^{BB}$ queens or to some other characteristic? A strict diet caused a set of monogyne $Gp\text{-}9^{BB}$ queens to drop from their initial weight of over 17 mg to about 12 mg; their fecundity decreased accordingly. Conversely, polygyne $Gp\text{-}9^{Bb}$ queens overfed in individual nests increased in weight from less than 15 mg to about 19 mg. These treatments reversed the weight/fecundity characteristics of polygyne queens and monogyne queens. These "reversed" queens were then added to polygyne host colonies.

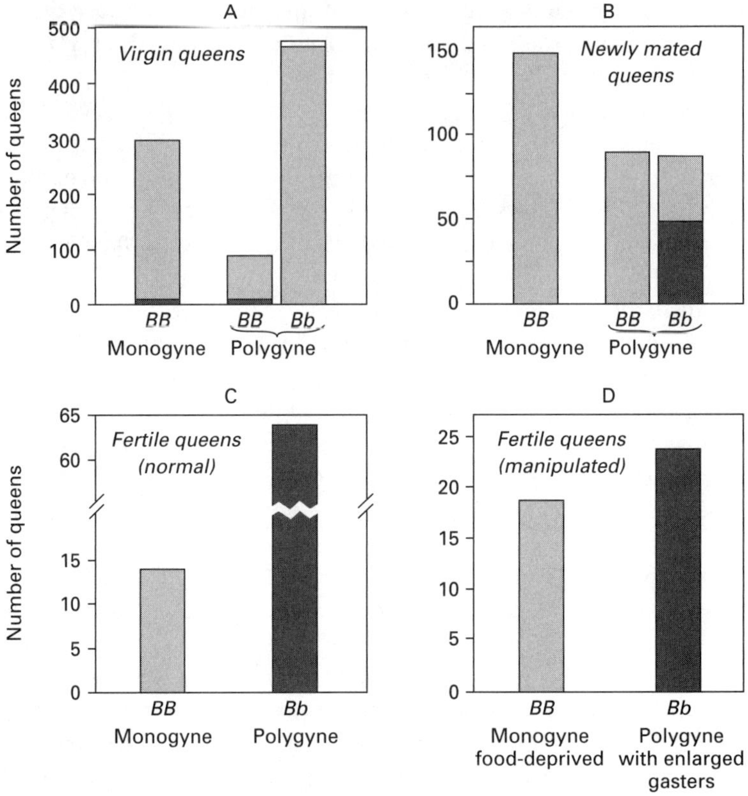

Figure 28.4. Acceptance or rejection of queens introduced into foreign polygyne test colonies. The *Gp-9* genotype and social form of origin of introduced queens are indicated under each bar. Light shading indicates queens rejected, darker shading queens accepted. Adapted from Ross and Keller (1998).

To no avail. Reversal of fecundity rates did not rescue queens from their fates—polygyne workers killed all *Gp-9*^{BB} queens and accepted multiple *Gp-9*^{Bb} queens (Figure 28.4D). Survival is based not on fecundity but perhaps on an odor or some other nonobvious queen attribute. This is a good point at which to reveal that general protein 9, the protein *Gp-9* codes for, appears only in females older than 8–14 days, i.e., mature adult females. This is also the age at which workers begin to thin the ranks of *Gp-9*^{BB} female alates. Just coincidence? Perhaps general protein 9 is involved in creating an odor signal in reproductive queens, and the two alleles produce different odors.

Worker Behavior and Gp-9. Although genes are fate for queens, the workers do the actual dirty work. Are the differences in their behavior in the two social forms also caused by genetic differences, or are they created by the social environment in which workers are reared? Is their murderous tendency related to their own genotypes or only to the queen's? The first test used artificial "monogyne" colonies: groups of polygyne workers, at least half of whom were *Gp-9*^{Bb}, and a single *Gp-9*^{Bb} queen. True monogyne colonies with *Gp-9*^{BB} queens and *Gp-9*^{BB} workers served as the control. Twice in five months all workers were removed from both types of colonies, so the final group of both monogyne and "monogyne" workers had experienced only a single queen during their entire lives. To make the test fair, the queen was removed three days before five *Gp-9*^{Bb} test queens were added to each queenless colony. Removal of the queen mother increases the likelihood that her orphaned workers will accept another queen.

The nests composed of half or more *Gp-9*^{Bb} workers always accepted multiple *Gp-9*^{Bb} queens, and the nests composed of *Gp-9*^{BB} workers always killed all of them. The social environments of the two groups of workers were identical. Only their *Gp-9* genotype differed. Interestingly, usually only half the workers are *Gp-9*^{Bb}.

Eighty percent of the workers attacking a *Gp-9*^{BB} queen were *Gp-9*^{Bb}, whereas only 62% of those surrounding a nonattacked *Gp-9*^{Bb} queen were, suggesting that the execution of *Gp-9*^{BB} queens is primarily the work of *Gp-9*^{Bb} workers. Somehow, the *Gp-9*^{Bb}-tolerant behavior of *Gp-9*^{Bb} workers becomes the characteristic of the entire worker population, even those of other *Gp-9* genotypes. Such a transfer of the genetically based behavioral characteristics from a fraction of the worker population to the whole population is also known from hygienic behavior of honeybees. How it occurs is unknown.

Ross and Keller next collected newly mated queens of both social forms from north Georgia and allowed them to found singly and rear colonies in the laboratory until they contained hundreds to thousands of workers, all the daughter of the single queen and all having experienced only monogyny. Some queens had mated with B males and others with b males, creating colonies with four different queen–worker combinations: (1) monogyne BB queens that had mated with B males and therefore had only BB workers; (2) polygyne BB queens that had mated with B males and produced BB workers (distinguished from the monogyne BB queens by their mitochondrial DNA haplotypes; Ross et al., 1999); (3) polygyne BB queens that had mated with b males and produced only Bb workers; (4) polygyne Bb queens that had mated with either B or b males and produced Bb (and other) workers. At the end of the experiment, the genotype of the father of each colony could be inferred from the combined genotypes of the queen and her daughters.

At test time, five fecund polygyne Bb queens were added to each nest. Again the results were unequivocal. Colonies including any Bb workers always accepted multiple Bb queens, whereas those lacking Bb workers killed them. That is, the BB workers of BB queens, whether originally from polygyne or from monogyne colonies, always killed the Bb queens. Again, it was genotype, not social form of origin or experience, that determined acceptance.

When single polygyne BB queens were added to orphaned colonies founded by polygyne Bb queens and having Bb workers, they were killed in half the instances. Apparently, a Bb queen must be present for the full effect, perhaps as some sort of reference for comparison.

When single BB queens were offered to orphaned colonies founded by BB queens and containing only BB workers, they were accepted, whether the founding queens came from polygyne or monogyne populations: both accepted the BB queen. More significantly, when two queens were added to each orphaned colony of BB workers founded by a polygyne BB queen, the workers killed one and let the other live, the hallmark of monogyny (but caution is advised because the experiment was replicated only twice).

Overview of Gp-9*'s Role.* In summary, colonies lacking workers with the $Gp\text{-}9^{Bb}$ genotype are very likely to accept only a single queen, but only if she lacks the b allele (i.e., is BB). Colonies with Bb workers are very likely to accept multiple queens as long as the queens have the b allele (i.e., are Bb, because the bb genotype acts like a lethal recessive, so bb queens are absent). Polygyny, the acceptance of multiple queens, therefore results only when both the candidate queens and the workers have the b allele, whereas monogyny, the tolerance of only a single queen, results only when both workers and queens lack the b allele.

Why Do These Alleles Still Exist in Polygyne Colonies? Gene Flow

In effect, in polygyne colonies the B allele of *Gp-9* acts like a lethal recessive—every time it appears as a homozygote, it is eliminated from the gene pool. Occurring generation after generation, this process should completely remove this allele from the polygyne population. For example, if we assume that the polygyne form arose from the monogyne form here in the USA, they would have been genetically identical, as in fact, they very nearly are. If the $Gp\text{-}9^B$ frequency were initially identical in the two social forms, the observed rate of selection would reduce its frequency to less than 1% in fewer than 10 generations. Yet, in the face of this intense selection for over 60 generations, the frequency of the $Gp\text{-}9^B$ allele remains substantial and shows no signs of decreasing. This is powerful evidence of continued flow of the $Gp\text{-}9^B$ allele from the monogyne to the polygyne form through interbreeding. But how does this flow occur?

Of the four possible routes of gene flow, three seem unlikely to operate effectively. Polygyne males seemed an unlikely agent of flow because only a small proportion of them carried the $Gp\text{-}9^B$ allele. On the other hand, all the males from monogyne colonies were $Gp\text{-}9^B$, suggesting that female alates from polygyne colonies mate primarily with males from monogyne colonies, thus counteracting the intense selection against the $Gp\text{-}9^B$ allele in polygyne colonies. To determine the genotype of queens' mates, Ross reared polygyne queens individually and genotyped the worker pupae, pinning down the genotypes of their fathers. Remarkably, 80–90% of mothers had mated with males from monogyne colonies (Ross and Shoemaker, 1993; Ross and Keller, 1995b).

The frequency of heterozygotes in polygyne colonies argued for similar rates. If polygyne females mated with monogyne males, the proportion of their offspring that was heterozygous would be higher than if they mated with polygyne males. Such "excess heterozygosity" arises whenever a gene occurs at different frequencies in the two sexes, and it indicated that polygyne females mated exclusively with monogyne males. Genes thus flow from the monogyne to the polygyne form by way of monogyne males. Flow in the opposite direction does not occur—monogyne queens mated only with males of their own kind.

Source of the Males. What was going on here? Were polygyne female alates preferentially mating with monogyne males, spurning their own kind? The simple answer is that most polygyne males are incompetent to mate, because they are sterile, diploid males. Six female alates pine to mate for every competent male in polygyne populations (Vargo and Fletcher, 1987). No wonder 30% of queens in polygyne colonies are uninseminated! In contrast, the nearby monogyne population produced only 0.7 females for every male, and mating success was never far short of 100%. The monogyne males with whom polygyne females mate must fly in from adjacent monogyne populations.

Geography of Gene Flow. If monogyne males migrate in, then frequencies of the *Gp-9*^{B} allele, mated females, and matings with monogyne (rather than polygyne) males should all decline with distance from the monogyne population, as the number of monogyne males dwindles from consummation, death, dispersal, and exhaustion (once again, this work was done with the surrogate *Pgm-3* gene, but I will again tell the story in terms of the *Gp-9* gene). The geographic details of the north Georgia polygyne population tell the story (Ross and Shoemaker, 1993; Ross and Keller, 1995b). This population is quite discrete, enveloping the town of Monroe and its vicinity and including as well several very small outlying polygyne pockets. Because male dispersal is probably aided by winds, the Ross group determined the average wind direction on mating-flight days during the previous four years; it was southwesterly 60% of the time, rarely coming from the northeast or southeast. One would therefore expect that the effects of monogyne male matings (and gene flow) would be greatest at the southwestern boundary between the polygyne and monogyne populations, decreasing in a northeasterly direction.

As before, the very large difference in the frequency of the *Gp-9*^{B} allele in the two social forms allowed the tracking of monogyne males because when these mate with polygyne females, they leave a quantitative genetic trace that "Kilroy was here" in the form of excess heterozygosity. These traces were quantified at six sites on a transect through the polygyne population in the same direction as the prevailing winds (SW to NE) and in two small, completely separated pockets of polygyne colonies, one windward and one leeward of the main population.

Collection of a single worker pupa from each of a large number of colonies at each site assured independence of samples. From most colonies, the investigators also collected one or more functional queens whose mated status and *Pgm* genotype were determined. The simple overall frequency of the A allele is too insensitive to reveal an effect of distance. Instead, Ross used the genotype frequencies at each site to estimate the proportion of monogyne mates from the excess of heterozygotes and deficit of homozygotes compared to the expected frequencies.

After some data manipulation, a clear geographic pattern appeared. The proportion of polygyne females mating with monogyne males was highest (about 100%) at the windward boundary, declined by 50% toward the downwind boundary (site E, Figure 28.5), and rose again for the separate downwind pocket (about 70%). Polygyne males capture a substantial fraction of the matings only in the leeward region beyond the dispersal range of most monogyne males. The average across all polygyne samples was similar to earlier estimates, about 91%.

These conclusions are strengthened by the geographic distribution of the proportion of mated queens, which roughly followed the previous estimates, declining from about 95% at the windward boundary to about 55% at the center, then increasing toward the downwind boundary (Figure 28.5, site E). Site E, the most leeward within the main population, was also the site at which polygyne males achieve a substantial fraction of the matings. Perhaps competent polygyne males displaced from further upwind are abundant enough at this site to equal monogyne males in mating success.

The proportion of mated females must be an index of competent male abundance of each social form at each site (Figure 28.5). The prevailing wind creates a striking geographic pattern in the abundance of the two types of males. Downwind drift creates "success zones" for each type of male, although the maximum for polygyne males is depressed by the prevalence of male diploidy.

Having established the direction, mode, and intensity of gene flow between the monogyne and polygyne forms, we can return to a closer inspection of how gene flow and selection operate together in the USA, based on *Gp-9*. Figure 28.5 summarizes what happens in polygyne populations during each episode of reproduction. Eighty percent of the matings are with monogyne males, who carry in only B gametes, because they come from a population in which the b allele is absent. Polygyne males cop only 20% of the matings and contribute gametes with B and b, the same as polygyne queens. The influx of monogyne males thus swells the population of gametes carrying the B allele. Assuming random fertilization, one can calculate the expected frequencies of zygote genotypes, as shown in Figure 28.6. Three phases of selection against certain genotypes then follow. First, all *Gp-9*^{bb} individuals die before they reach adulthood, then *Gp-9*^{BB} sexual females begin to disappear, and all are finally removed as they begin to lay eggs. When

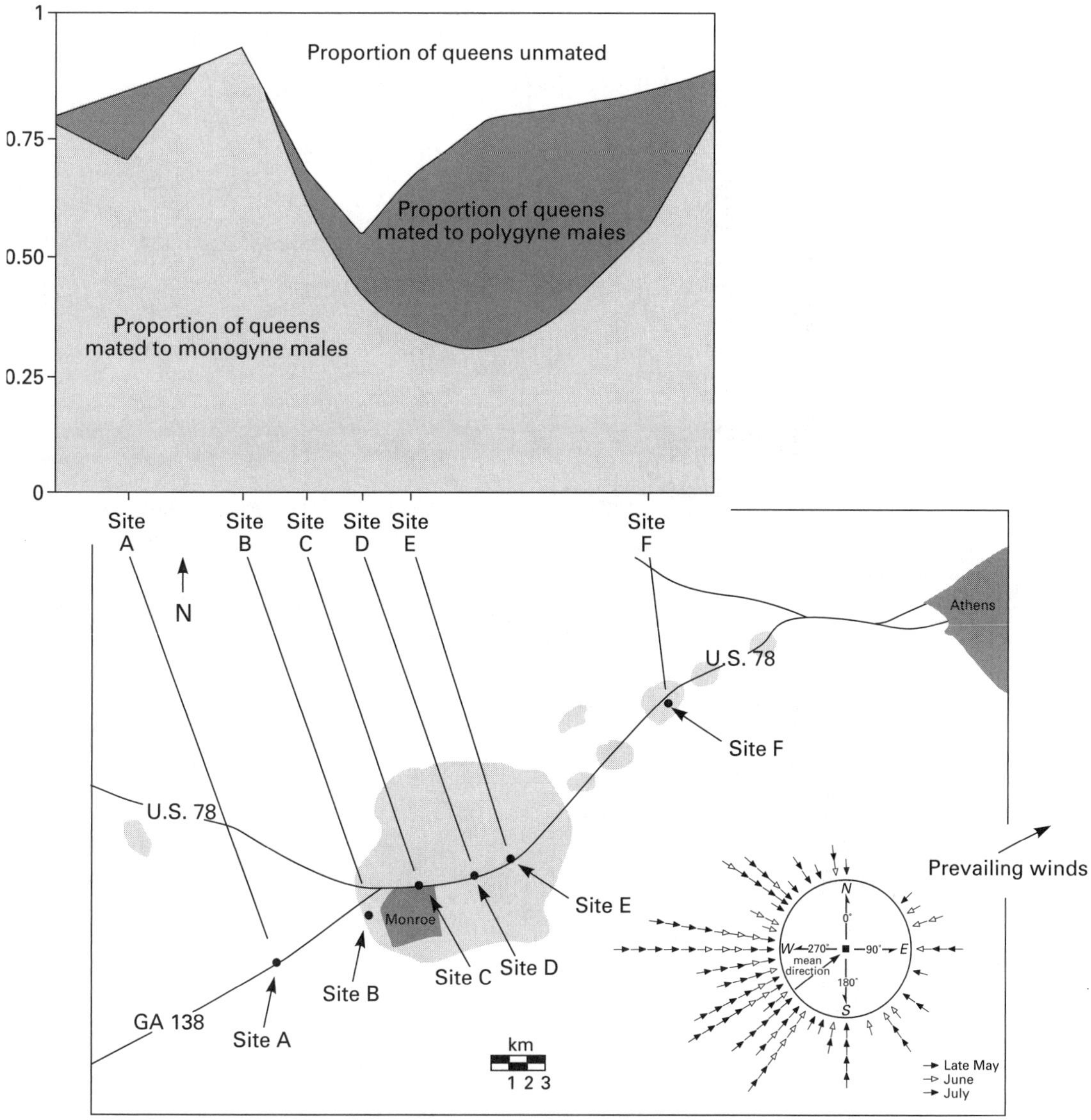

Figure 28.5. The proportion of polygyne queens mated to monogyne males, mated to polygyne males, and unmated along a transect parallel to the direction of prevailing winds during mating flights in a circumscribed north Georgia population. Stippling marks areas in which only polygyne nests were found. Polygyne queens mate primarily with monogyne males, whose dispersal is aided by winds. Adapted from Figures 1, 2, and 7 of Ross and Keller (1995b).

the process is finished, practically all functional polygyne queens are Bb. This powerful selection is repeated in each generation, and is counteracted by gene flow through males from the monogyne populations (where selection against B is absent).

Selection and fitness were somewhat different in the native Argentine polygyne populations. All $Gp\text{-}9^{Bb}$ queens had a relative fitness of

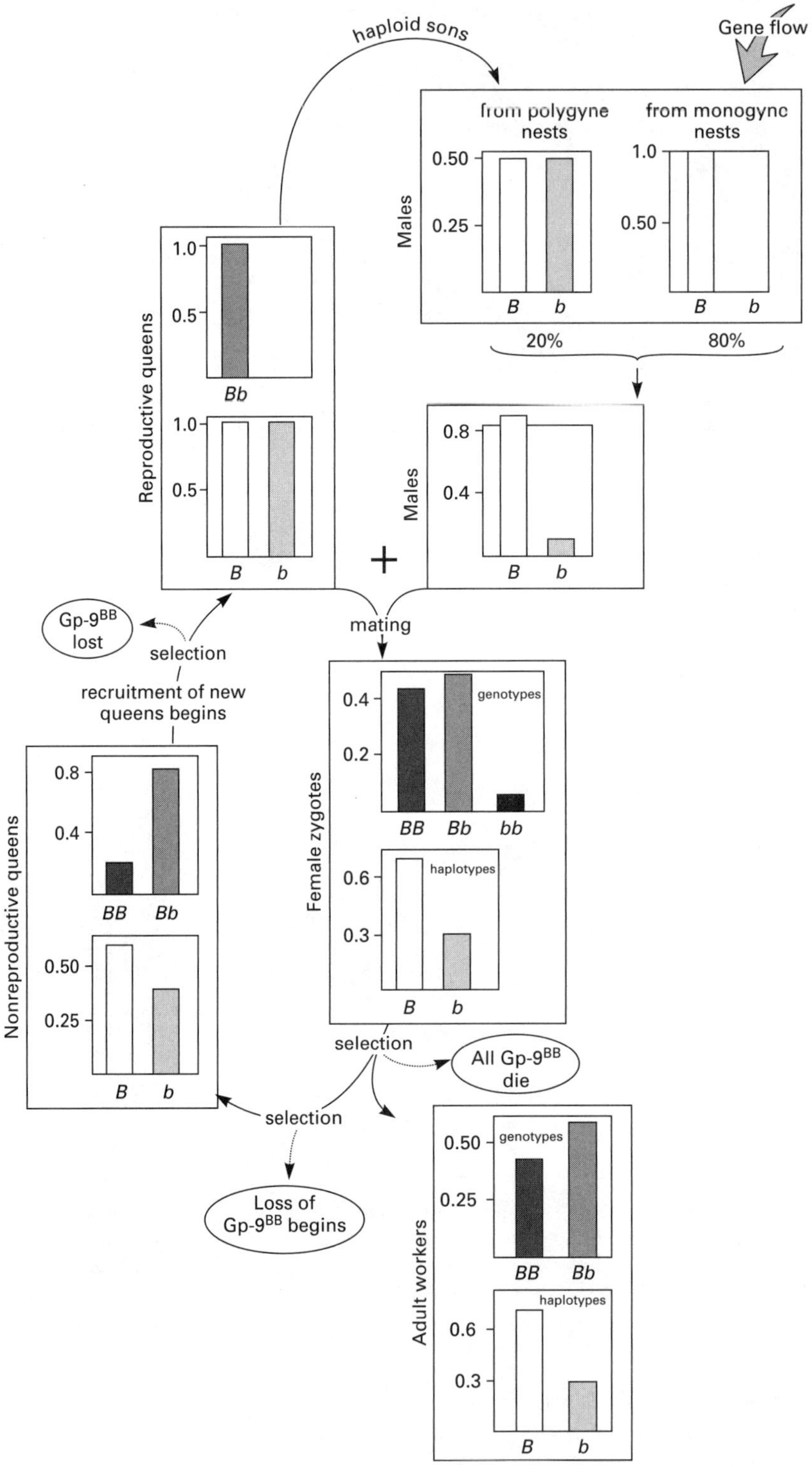

Figure 28.6. The combined effect of gene flow and postmating selection on the frequency of the *Gp-9* genes and genotypes in *Solenopsis invicta*. Adapted from Ross (1997).

1.0. Another three had a *Gp-9*^{BB} genotype but a relative fitness of 0.6. Although BB queens had a fitness of zero in the USA, they merely experienced a 40% reduction in survival in Argentina, probably because some of the B alleles are actually cryptic b-like alleles; that is, they act like b alleles but have the electrophoretic mobility of B alleles (see below). *Gp-9*^{bb} had a fitness of zero because the individuals with this genotype were eliminated or died.

Reviewing Gene Flow and Genetic Structure. We can now review the behaviors that create gene flow and geographic differences in mean allele frequencies (genetic structure). These patterns have been established and confirmed with several types of molecular markers (Ross et al., 1999), a thoroughness not duplicated with many other organisms. Mitochondrial DNA showed that most polygyne female alates (*Gp-9*^{Bb} and *Gp-9*^{bb}) do not disperse far and probably contribute little to nuclear gene flow. This pattern is mirrored in mtDNA haplotype-frequency differences on a very local scale. Polygyne *Gp-9*^{BB} and some *Gp-9*^{Bb} female alates disperse greater distances but seem rarely to succeed in becoming reproductive, so they also contribute little to nuclear gene flow. Monogyne *Gp-9*^{BB} females carry their genes long distances between their natal and founding colonies, assuring homogeneity of their gene frequencies over wide regions within monogyne populations, but their mtDNA signatures reveal that they rarely succeed in founding among polygyne populations, contributing little to gene flow between social forms. Therefore, although female alates are a major route of gene flow within monogyne populations, no females of any type are effective in dispersing genes over longer distances in polygyne populations.

In contrast, monogyne males, as we have seen, fly far into polygyne territory, carrying their nuclear genes long distances (they transmit no mitochondrial genes). They effectively homogenize the frequencies of all four types of nonselected nuclear alleles over large regions and both social forms. Whereas monogyne queens homogenize genes only within the monogyne populations, monogyne males do so for both social forms. The reciprocal is not true, at least in the USA, for most polygyne males are diploid and sterile and cannot contribute to gene flow of any kind. The existence of triploid workers indicates that diploid males sometimes do mate successfully, but the result is a reproductive dead end. Fertile polygyne males probably fly shorter average distances before mating. Because they are also relatively rare, they also contribute much less to gene flow than do monogyne males.

The patterns, rates, and routes determine whether differences in allele frequencies can accumulate and on what geographic scale (Figure 28.7; Ross et al., 1999). Variation in allele frequencies (genetic structure) is often described as the proportion of the total variation that occurs at

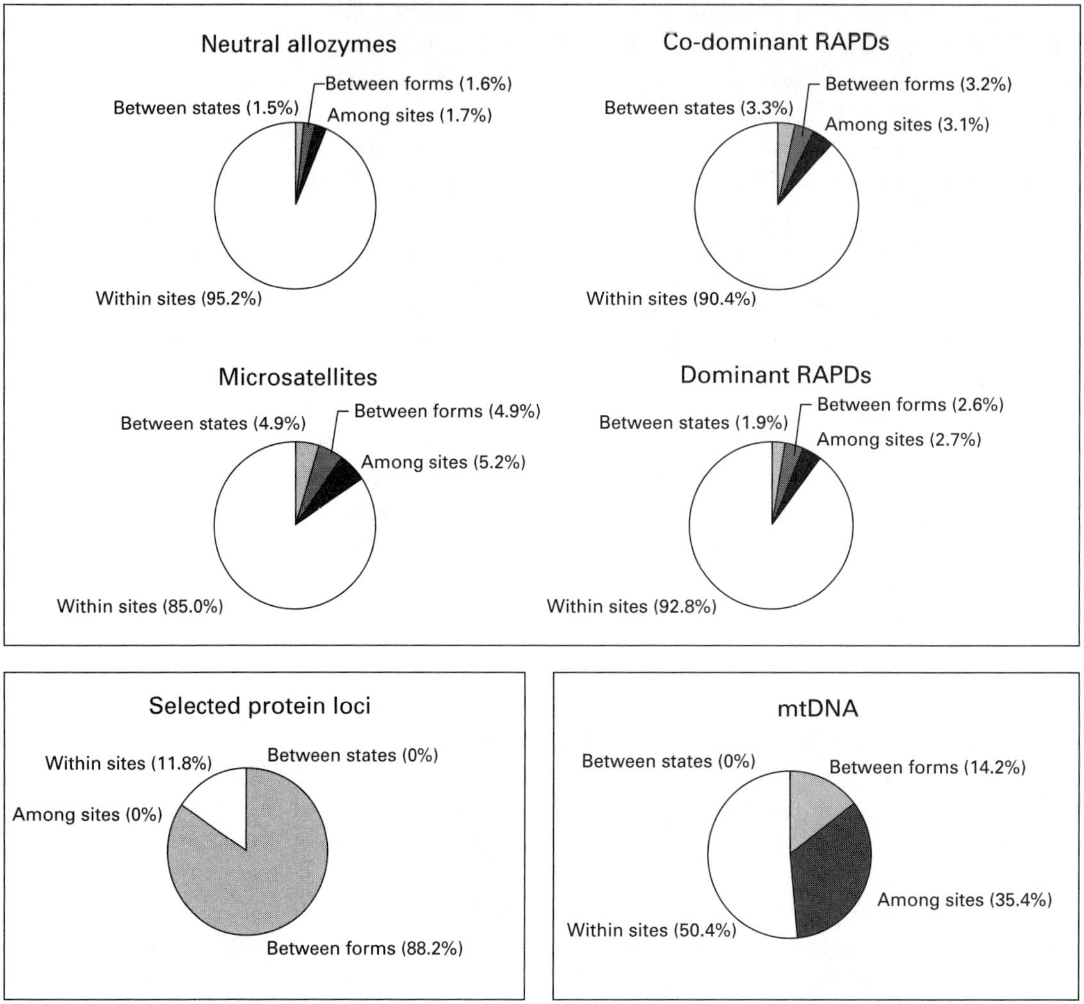

Figure 28.7. Partitioning of total genetic variance among different hierarchical levels of sampling in the study populations for each of six classes of molecular markers. The upper panel represents four classes of neutral (nonselected) nuclear markers. Adapted from Ross et al. (1999).

each scale. In fire ants, about 85–95% of total variation in nonselected nuclear allele frequencies occurs among colonies within sites. That is, only 5–15% of total variation occurs between sites, states, and social forms. In other words, social forms, states, and distant sites are much more like each other in their average allele frequencies than are colonies within sites.

On the other hand, 88% of the variation in the *selected* nuclear gene (*Gp-9*) occurs between social forms, and the remainder among colonies within sites. None occurs between sites or states because the continued elimination of *Gp-9*^{BB} and *Gp-9*^{bb} queens in polygyne colonies creates a sump that can never be quite filled by gene flow through monogyne males.

Finally, the very limited scale of movement and founding success by polygyne queens has allowed about 35% of the variation in mtDNA haplotype frequencies to occur among the averages of local sites and another 14% between social forms. Only about half of the mtDNA variation is among colonies within sites, contrasting with the 85–95% for nuclear alleles.

Patterns of gene flow must be rather different in native populations. There, diploid males are rare, and male-mediated gene flow between social forms may be similar in the two directions. Nevertheless, at least one Argentine monogyne/polygyne population pair showed significant differentiation of nuclear genes by social form, suggesting that larger barriers to gene flow exist in Argentina than in the USA. We can speculate that these barriers might involve more complete site-specific assortative mating, such that polygyne alates of both sexes mate locally and monogyne ones in swarms, with little crossing over. Mating near the polygyne nest is supported by high nestmate queen relatedness (about 0.45) in Argentina (Ross et al., 1996b). No doubt, much interesting research remains to be done in Argentina.

Is This a Case of Sympatric Speciation?

The restriction of gene flow between the monogyne and polygyne forms raises the question of whether the formation of two species from one in a single geographic location may be occurring. Currently, we do not know whether genetic differentiation between sympatric social forms is rapid enough to create separate species. High gene flow makes imminent speciation in the USA unlikely, but high differentiation in at least one Argentine site suggests that barriers exist there.

The Genetic Control of Social Form

The close association of two genes and their products with two distinct social forms raises the question of whether these genes themselves, or genes very closely linked to them, control social form. In the past, when biologists thought about the genetics of sociality at all, they assumed that queen number existed as a graded trait, from a single queen to a small number to a bunch, and that it was driven by a gradual increase in the number of "social" alleles through selection by different ecological conditions. Indeed some species fit such a pattern. Fire ants present a startling contrast to this largely hypothetical picture because this single species exists in two completely distinct social forms and does so in environments that are not perceptibly different. This binary state suggests that social form may be under the control of one or a very few genes. The logical candidate is the *Gp-9* gene (Keller and Ross, 1998; Ross and Keller, 1998).

Is Gp-9 *the Social-form Gene?* If *Gp-9* (or a gene very closely linked to it) is indeed the "social form gene," then genetic differences at the *Gp-9* locus should account for the main trait differences between the social forms. These are (1) differences in queen reproductive phenotype (high and low weight; rapid and slow development), (2) acceptance or rejection of queens of the two forms by workers of the other form; and (3) acceptance of one or of multiple queens by workers.

Reviewing the reproductive phenotype, we see that flight-ready $Gp\text{-}9^{BB}$ female alates from both polygyne and monogyne colonies have higher body weights, greater metabolic reserves, greater minim production, and more rapid reproductive development than $Gp\text{-}9^{Bb}$ and $Gp\text{-}9^{bb}$ polygyne females. The first condition is thus met—the nature of the queens, which is so closely tied to social form, is largely determined genetically. $Gp\text{-}9^{BB}$ females, whether from polygyne or monogyne societies, have the characteristics required for independent colony founding and pursue the same strategy of dispersal and independent founding. Conversely, the polygyne social form requires at least one $Gp\text{-}9^{b}$ allele and assures the lack of most of the characteristics necessary for founding independently, although a portion of its alates seem to pursue such a strategy.

The second condition, acceptance and survival of queens, also depends strictly on the *Gp-9* genotype. $Gp\text{-}9^{BB}$ queens, no matter what their reproductive condition or origin, invariably perish in polygyne colonies, whereas multiple $Gp\text{-}9^{Bb}$ queens are adopted. Masquerading as a low-fecundity polygyne queen does not prevent a $Gp\text{-}9^{BB}$ queen from being killed, nor does masquerading as a heavy, fecund monogyne queen prevent a $Gp\text{-}9^{Bb}$ queen from being accepted. Workers detect an attribute that is not obvious to human observers but that reveals the queen's genotype.

The third condition, worker behavior, is also controlled by the *Gp-9* genotype. When $Gp\text{-}9^{Bb}$ workers are abundant in the nest, they merrily accept multiple queens that have the b allele (i.e., are $Gp\text{-}9^{Bb}$), and they kill all $Gp\text{-}9^{BB}$ queens that come their way. They do so whether they have experienced only a single queen all their lives or come from a polygyne colony swimming in queens. Theirs is the mark of polygyny, and they impart this mark to their nestmate workers of other genotypes. Equally interesting, when all workers in an orphaned colony are $Gp\text{-}9^{BB}$, they accept only a single $Gp\text{-}9^{BB}$ queen, killing both $Gp\text{-}9^{Bb}$ and supernumerary $Gp\text{-}9^{BB}$ queens. This is the mark of monogyny.

All these results suggested a final experiment, one in which the *Gp-9* genotype of the workers was manipulated by switching of colony queens (Ross and Keller, 2002). Over time, the offspring of these adopted queens would gradually replace the original workers. The proportion of $Gp\text{-}9^{BB}$ workers in originally monogyne colonies would drop from 100% to low percentages and grow from low percentages to 100% in the originally polygyne. At what proportion of $Gp\text{-}9^{Bb}$ workers will colonies switch their collective behaviors?

The experiment is more difficult than it appears, because colonies cannot easily be induced to accept queens of the opposite form. Ross and Keller basically subjected them to torture. First, they kept the colonies queenless for five days, and then instead of just dropping in an adoptive queen, they added brood to these queens for several days, only later adding a few workers at a time. Whenever workers balled around the adopted queens, they blew on them to panic them, chilled them to reduce their activity, or misted them with water to break up the fight. They even carried the colonies home with them at night so as to continue their vigilance around the clock. In the end, they succeeded in getting 15 monogyne colonies to accept single polygyne queens and 11 polygyne colonies to accept monogyne queens.

From then on, it was a matter of testing worker behavior and *Gp-9* genotype at 30- to 40-day intervals, the approximate time it takes for an egg to grow into a young worker. The tests were somewhat different for polygyne queens in (originally) monogyne colonies and monogyne queens in (originally) polygyne colonies. For the former, a single polygyne ($Gp\text{-}9^{Bb}$) test queen was introduced in addition to the resident queen. If the test queen survived 24 hours, she was considered accepted. Whenever such a colony accepted a test queen, it also retained the adopted queen. For the latter, the adopted queen was removed for two days, then three polygyne ($Gp\text{-}9^{Bb}$) test queens were added, and the number surviving after 24 hours was noted. Any survivors were removed, and six hours later, the adopted queen was returned. These colonies never accepted just a single test queen, always two or three. Together, each colony unequivocally signaled its willingness or lack of willingness to tolerate multiple heterozygous (polygyne) queens.

At about 70 days, 10 of the 14 originally monogyne colonies switched from rejection of $Gp\text{-}9^{Bb}$ queens to acceptance (polygyne behavior), when an average of 11% of the worker population bore the $Gp\text{-}9^{Bb}$ genotype and 89% the $Gp\text{-}9^{BB}$ (Figure 28.8). By about 100 days, when they averaged 34% polygyne workers, practically all of them had switched to polygyne behavior. Conversely, by 74 days, all of the originally polygyne colonies were still showing polygyne behavior. At this time, almost 70% of their $Gp\text{-}9^{Bb}$ workers had been replaced by monogyne $Gp\text{-}9^{BB}$ ones. Only after 130 days, when the proportion of $Gp\text{-}9^{Bb}$ workers had dwindled to 5%, did these colonies switch from polygyne to monogyne behavior. Individual colonies with more than 10% $Gp\text{-}9^{Bb}$ workers behaved like polygyne colonies, and those with fewer than 5% like monogyne colonies. The *Gp-9* composition of the worker population determines social form. Not least remarkable, the collective action of the entire worker population, tolerance of multiple polygyne queens, depends on the genotype of a tiny minority of workers. It is easy to see how a small minority of workers that kills queens succeeds in forcing monogyny on the entire population, but how does a minority whose

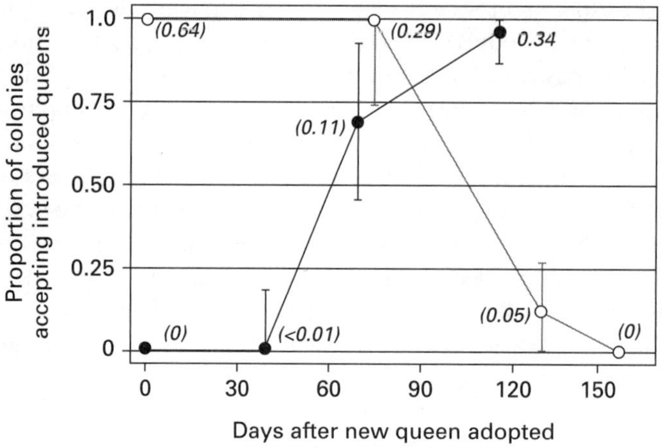

Figure 28.8. Progress of conversion of colonies from one social form to the other after they were induced to adopt a queen of the opposite social form. Open circles: Originally polygyne colonies with adopted monogyne queens. Filled circles: Originally monogyne colonies with adopted polygyne queens. Numbers in parentheses show the proportion of workers that were Gp-9^{Bb}. A small percentage of genetically polygyne workers can induce the entire colony (even an originally monogyne colony) to accept multiple queens, but the converse is not true. Bars represent 95% confidence intervals. Adapted from Ross and Keller (2002).

characteristic is the *failure* to kill imbue the entire population with this trait? How in the world does this tiny minority foil the apparent will of the vast majority and prescribe the social system for the overwhelming majority of monogyne workers?

By what sign do workers recognize each class of queens? Do these genes label the queens with some conspicuous mark, say a "green beard" (Dawkins, 1976), by which workers recognize queens of their own genotype and act nicely toward them (Keller and Ross, 1998)? What is the green beard? A simple experiment shows that it is probably a chemical odor. Keller rubbed randomly chosen workers against either *Gp-9^{BB}* queens that had been under attack or *Gp-9^{Bb}* queens that had not and then placed them with nestmates for observation. Forty percent of those rubbed against a *Gp-9^{BB}* queen were killed by their nestmates, but none of those rubbed against a *Gp-9^{Bb}* queen were killed. Of course using queens that had not first been attacked would have been better, because Keller's results confound prior worker attack of queens with their genotype—workers might mark attacked queens with an odor—but if the transferable odor label is indeed the queen's "green beard," it is probably a pheromone associated with reproductive activity.

Our review of the evidence is complete. The case for genetic control of social form is strong. The remarkable conclusion is that all these dramatically biological differences between monogyne and polygyne fire ants result from a simple genetic difference.

Characterizing the Gp-9 *Gene.* The application of modern molecular methods to the genetic analysis of polygyny has led to remarkable achievements, including the complete sequencing of the *Gp-9* gene (Krieger and Ross, 2002). By now, every school child knows that a typical gene codes the amino acid sequence of a particular protein. Of course the process that leads from a gene to a protein is far more complex than the school-child version, but the key to our story is that, with a protein in hand, one can also work backward from its sequence to that of the gene. And of course we have a protein that is tightly associated with polygyny in *S. invicta*—general protein 9—and it is precisely with this protein that Krieger and Ross started their quest. They separated general protein 9 from monogyne queens by electrophoresis and extracted it from the gel. It sounds simple to say that, if they could determine the amino-acid sequence of general protein 9, they could deduce the base-pair sequence of the *Gp-9* gene, but how is it actually done?

Here is the brief version: With general protein 9 in hand, one amino acid at a time is chemically cleaved from one end of the protein and identified, until about the first 20 have been identified. The *Gp-9* gene is also cleaved into large pieces at specific positions and some of the "internal" amino-acid sequences determined in the same way. The sequence of amino acids corresponds to the sequence of three-base "codons" in the messenger RNA (mRNA) template from which the protein was synthesized, and the sequence of the mRNA corresponds one-to-one with its nuclear DNA template. The amino-acid sequence doesn't uniquely specify the sequence of RNA codons, because 64 possible codons code for only 20 amino acids, so multiple codons correspond to each amino acid (the code is said to be "degenerate"). It does considerably narrow the search, however; it specifies multiple but limited sets of base sequences. These base sequences are then chemically synthesized as short polynucleotides called primers, which bind to the specific part of the mRNA and DNA that code for the corresponding part of the protein. RNA or DNA synthesis starts from these primers, and through a series of repetitions, eventually produces multiple copies of the full mRNA as well as the DNA complementary to it. We still don't have the full nuclear gene, however, because only certain sections of this gene, the coding regions, called exons, are transcribed into mRNA and translated into protein. The "silent" (untranscribed, noncoding) regions are called introns. Another cycle of very specific DNA primers is used in another series of steps to produce billions of copies of the full nuclear *Gp-9* gene. Because this gene is very long, it is cut into two pieces, and the base-pair sequence of each piece determined. The end result is the complete base-pair sequence of the nuclear gene for *Gp-9*, the mRNA transcribed from that gene, and the amino acid sequence of general protein 9.

So what do these look like? The full *Gp-9* gene plus a "flanking region" is 2200 base pairs long, but the *Gp-9* coding region is only 672 base pairs long, and only 459 of these are read into messenger RNA, which in turn is translated into a protein 153 amino acids long (each codon is three base pairs; 459/3 = 153). The transcribed portions of the gene are located in five exons located at intervals along the length of the gene, separated by four untranscribed introns and flanked at the ends by two more. Both the order of the exons along the gene and the base sequences within each exon are conserved in the transcribed mRNA; the introns are simply left out during transcription. The base sequence of the mRNA and the amino-acid sequence of general protein 9 are shown in Figure 28.9. A mere glance at this figure can dangerously numb the mind, but only a few significant features are important. First, exon 1 and part of exon 2 code for what is called the signal peptide of general protein 9, a tail that allows the newly synthesized general protein 9 precursor to be transported across the cell membrane, that is, to be secreted. In the process, the tail is cleaved off to yield the mature general protein 9. The sequence of the mature protein was stored in part of exon 2 and in exons 3–5.

Second, general protein 9 contains six well-spaced cysteine residues. Cysteine is a sulfur-containing amino acid that can link two amino-acid chains together through its side groups or link different regions of a single chain to each other. This linking activity is very important in determining the three-dimensional shape of the protein. Searching through libraries of sequences, Krieger and Ross found that in this feature their protein most resembled several odor-binding proteins, for example, the sex-pheromone-binding protein of the silkworm moth and others. The numbers of amino acids in all five intervals between cysteine residues were identical to those in the odor-binding proteins or nearly so.

Here is a plausible connection to polygyny. Perhaps the *Gp-9* gene produces an odor-binding protein specific for fire-ant queen pheromone, determining the response of workers to the odor of queens. They pursued this line of thought by computing the three-dimensional shape of general protein 9. The atomic and molecular forces that cause chains of amino acids (i.e., proteins) to kink, fold, and associate into their three-dimensional shapes are well enough understood that the shape can be computed from knowledge of the sequence of amino acids and a few environmental variables such as pH and ionic concentrations. This computation showed that general protein 9 was shaped like a vase with a cavity at the top and a "stem" below. Presumably, the cavity binds odorant molecules of a particular shape and chemistry (e.g., queen pheromone?) at the surface of the chemosensory organs and transports them to the chemosensory membrane, where the stem interacts with the receptors on the membranes of chemosensory cells to release an action potential (nerve impulse). The details have yet to be worked out, but *Gp-9* is at the heart of the difference between monogyne and polygyne colonies: workers either kill or tolerate queens in response to pheromones, and they do so on the basis of their *Gp-9* genotypes, and by extension, the nature of their odor-detection systems.

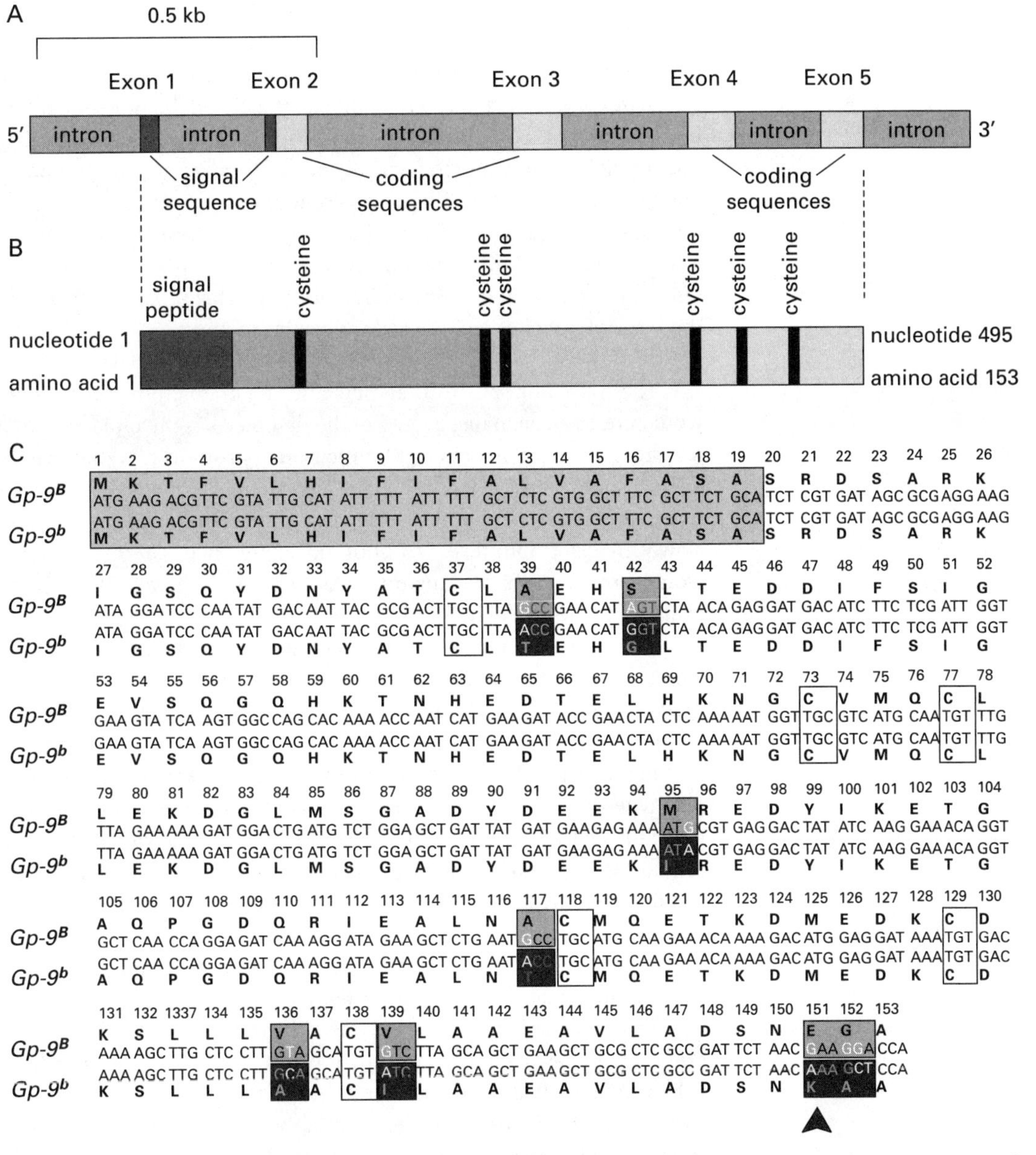

Figure 28.9. Structure and amino acid sequences of the *Gp-9* gene in fire ants. A. Intron and exon structure of *Gp-9*, showing signal and coding sequences. Intermediate shading indicates noncoding introns. B. Structure of *Gp-9* precursor protein, showing the signal peptide and six cysteine residues characteristic of pheromone-binding proteins. C. Coding region nucleotide and amino-acid sequences for the B and b alleles from *Solenopsis invicta* in the USA. The signal peptide is shaded in medium grey, and the six cysteine residues are enclosed in rectangles. Nucleotides in which alleles differ are shown in white. The charge-changing amino-acid substitution is marked with an arrowhead. Adapted from Krieger and Ross (2002).

If this gene is truly the one controlling polygyny, then the genes of polygyne and monogyne queens should differ in ways that produce the differences between the general protein 9 coded by *Gp-9*^{B} and that coded by *Gp-9*^{b}. Krieger and Ross therefore isolated the messenger RNA transcripts from five monogyne and five polygyne queens. Monogyne queens yielded only one transcript, corresponding to *Gp-9*^{B}, but polygyne queens yielded two slightly different ones, corresponding to *Gp-9*^{B} and *Gp-9*^{b}. Sequencing of these messenger RNAs revealed nine places at which bases differed. Eight of them resulted in substitution of a different amino acid in the transcribed protein. (This genetic difference is the basis of a recent, PCR-based molecular method for rapid determination of the social form of *S. invicta;* Valles and Porter, 2003.)

The most important difference, and the one without which *Gp-9* would not have been discovered in the first place, is the replacement of a basic lysine with an acidic glutamic acid at position 151, three amino acids from one end of the molecule. This change reduces the net negative charge on the protein by one, causing this variant to move more slowly in electrophoresis. Without this difference, *Gp-9*^{B} and *Gp-9*^{b} would not be visible as separate electrophoretic bands. What better example of the role of luck in science?

The degree of luck became even clearer when Krieger and Ross surveyed Argentine *S. invicta* queens. There, in addition to polygyne queens that were *Gp-9*^{Bb} (as in the USA), some polygyne queens appeared to be *Gp-9*^{BB} (i.e., their proteins showed a single band with the mobility of *Gp-9*^{B}). Sequencing the genes showed that the Argentine polygyne queens that appeared to be *Gp-9*^{BB} were actually heterozygous: they carried the B allele and a previously unknown allele whose amino-acid sequence was more like that of the b allele than like that of the B allele, but without the charge-changing substitution at position 151. This b' allele, as it was called, therefore had the same mobility as the B allele and was indistinguishable by electrophoresis. The functional differences obviously do not depend on a change of charge.

Further search among Argentine *S. invicta* queens turned up several more genes, which fell into two distinct families: those that were B-like in their base-pair and amino-acid sequences and those that were b-like. All the b-like alleles invariably conferred polygyny, and all the B-like alleles monogyny. *Gp-9* is indeed the gene for social form! It just so happened that the b allele that made the trip to the USA differed in charge from the B allele and showed up as a separate band when Ross ran his gels in the 1980s. *Double luck.*

Next, Krieger and Ross sequenced the entire *Gp-9* gene from nine other South American species of *Solenopsis* fire ants in order to determine how widespread this gene is and possibly to infer its evolutionary history. A similar gene was lacking in the related genus *Monomorium*, suggesting that the *Gp-9* gene arose rather recently in the evolution of

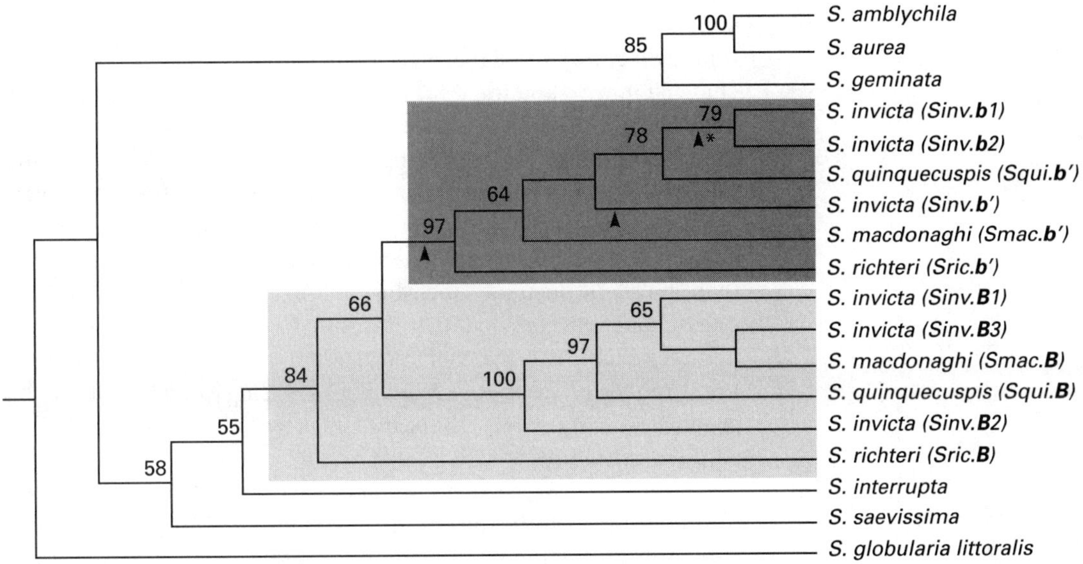

Figure 28.10. Evolutionary tree of the relationships among 18 *Gp-9* alleles found in 10 *Solenopsis* species. The top three species are North American; all others are native to South America. *Solenopsis globaria littoralis,* which is not a fire ant, serves as the outgroup. Alleles are shown in parentheses following the species bearing them. Alleles invariably associated with polygyny are shown in light shading; those associated with monogyny in darker shading. Arrowheads show branches for which episodes of positive selection were detected. The branch on which the charge-changing amino acid was acquired is marked by an asterisk. Numbers on branches show strength of the bootstrap analysis (a measure of the "strength" of the relationship). Adapted from Krieger and Ross (2002).

the *Solenopsis* line. By looking at the changes in the base-pairs and amino acids, they could arrange the genes on an evolutionary tree. The sequences of North American natives, *S. geminata, S. aurea,* and *S. amblychila,* fell into a separate group, distinct from the South American species (Figure 28.10). The South American *S. interrupta* and *S. saevissima* contained only B-like sequences and appeared at the base of the tree, consistent both with the absence of polygyny in these species and with other evidence of their basal position on the *Solenopsis* family tree. The ancestors of *Solenopsis,* as well as its earliest species, therefore appear to have been strictly monogyne. Polygyny arose only once, early in the evolution of the genus *Solenopsis,* when a B-like allele gave rise to a b-like one, changing the response of workers to queen odors. In *S. richteri, S. invicta, S. quinquecuspis,* and *S. macdonaghi,* polygyne colonies invariably yielded b-like sequences, whereas monogyne ones yielded only B-like ones. The charge-changing mutation occurs only in *S. invicta. Triple luck.* (An early reviewer of this volume added the note, "Quadruple luck: Ken Ross worked on the problem.")

Once the b-like allele arose, natural selection acted to increase its rate of divergence from the B-like alleles. We know this from comparison of the frequency of base substitution at positions that are "silent" (noncoding) with those that code for an amino acid. Silent positions include all introns and the third position in most codons. In these positions, substitution of one base for another does not matter. Natural selection is "blind" to such changes, and they accumulate at whatever rate they occur. In contrast, base substitutions at positions that matter are either eliminated or favored by natural selection, so they accumulate at lower or higher rates (antibody genes are examples of higher rates) than the silent, "neutral" ones. In the b-like alleles, but not the B-like ones, changes accumulated significantly faster than they would at the neutral rate. Selection seems to speed the divergence of the b-like alleles from the B-like ones, presumably increasing workers' ability to discriminate among queens, the basic means by which queen number and social form are regulated.

Once the b allele arose, workers carrying it began to eliminate queens not carrying it. The reproductive success of Bb queens therefore increased, and so did the number of copies of the b allele transmitted to the next generation through female alates. Under normal circumstances, this process should quickly eliminate the B allele from the population, but the inviability of the bb genotype in both workers and queens assures that this can never happen. All the fitness is associated with the heterozygous Bb queens. Viability and gene flow from monogyne males therefore assure that both alleles continue to exist in the polygyne fire ant. This is the fact that allowed the gene to be discovered.

Is *Gp-9* all there is to polygyny in fire ants? Probably not. *Gp-9* is part of a gene complex characterized by reduced recombination (remember *Pgm-3*?), that is, a group of genes that is kept together on the chromosome through positive selection. This selection creates a coadapted set of genes that function as a unit, in this case probably by determining both the queen odor and the worker response to it. They probably have other major effects on social form as well. The difference between the social forms is not merely a quantitative difference in queen pheromone or worker sensitivity to it. Bb (or better, $B_{\text{-like}}b_{\text{-like}}$) workers always discriminate against BB queens, even when these produce little pheromone, and BB workers never accept Bb queens, even when their pheromone production has been jacked up. The pheromones must be qualitatively different.

Most of the polygyne South American species belong to the *saevissima* species complex, relatively distantly related to the native North American *geminata* complex. Sequencing the *Gp-9* gene in our native *S. geminata* yielded another surprise—the *Gp-9* genes of monogyne and polygyne *S. geminata* were identical (Ross et al., 2003)! All were B-like and lacked sequence differences associated with social form.

Clearly, differences at *Gp-9* locus cannot explain polygyny in *S. geminata,* so we must conclude that acceptance of multiple queens has other causes in this species (indeed, in ants!). In searching for these causes, Ross and his colleagues found that, in contrast to the virtual genetic identity of the two social forms in native *S. invicta,* polygyne and monogyne *S. geminata* were genetically quite distinct in their allozymes, microsatellite DNA, and mtDNA. All of these polygyne colonies came from a single small patch of roadside in Gainesville, Florida, and were characterized by only a single (of 15) mtDNA haplotype (the monogyne form has 14) and only a single allele of each of six allozyme loci where the monogyne form had more than one. The forms also differed in satellite-DNA allele frequencies. Altogether, genetic-relatedness analysis indicated that all of these polygyne colonies probably represent an extended family, all descended from a few common ancestors through a single founding event that switched the social form from monogyne to polygyne, perhaps through a loss of genetic diversity that caused reduced capacity for nestmate and/or queen recognition. The existence of a historic population bottleneck was supported by the smaller number of nuclear alleles and especially mtDNA haplotypes in the polygyne population. A computer simulation showed that founding by 7 to 15 mated queens less than 100 years ago, followed by slow population growth and low gene flow from the parent monogyne population, could account for the observed genetic patterns in the polygyne population. Unlike the two social forms of *S. invicta* on both continents, which freely exchange genes, polygyne *S. geminata* are reproductively isolated even from their immediate monogyne neighbors.

Ross and his group have produced the most important and surprising findings in social biology in decades. Follow-up research is beginning to appear, for example, the determination that the genes for *Gp-9* and two forms of cytochrome P450 associated with odorant metabolism are expressed severalfold more strongly (i.e., are transcribed into more mRNA) in workers than in larvae, alates, and queens (Liu and Zhang, 2004). Although the meaning of such caste and stage differences in gene expression is not yet known, they are clearly an important lead to pursue. The future of this research will no doubt continue to be very exciting.

The Transition to and Distribution of Polygyny

In the USA. In *S. invicta,* the polygyne form probably arose from the monogyne one after entry into the USA, possibly several times. But how? From their study of the Monroe, Georgia, population, Ross and Shoemaker (1997) proposed a scheme of how polygyne populations began in the USA (Figure 28.11). The Monroe population arose in the late 1970s and spread gradually outward, and northeastward (downwind), producing a number of scattered, smaller populations. Within the core

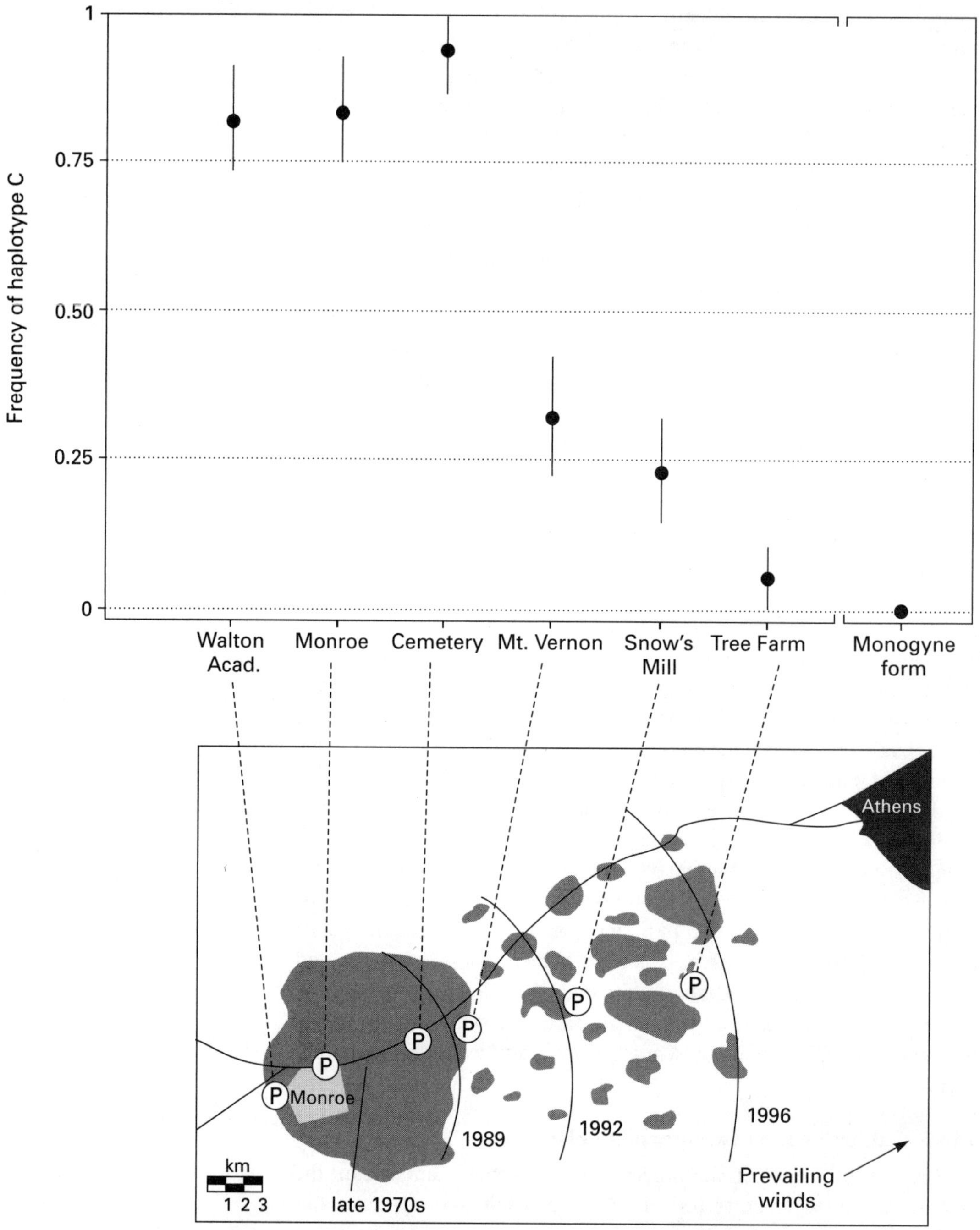

Figure 28.11. The eastward spread of the polygyne form of *S. invicta* from its original locus of establishment in Monroe, northern Georgia (largest shaded area). The upper panel shows frequency of the most common mtDNA haplotype, C, at six polygyne locations. Haplotype C has not been found in the monogyne form in Georgia. Adapted from Ross and Shoemaker (1997).

polygyne population, 80% of the mtDNA haplotypes are type C, but at the trailing edge, type C occurs in only 30% of queens. Haplotype A, which is characteristic of monogyne ants, makes up most of the rest. Haplotype C continues to decline with distance in the downwind polygyne populations until it is indistinguishable from zero. In other words, as one moves from the core population to the ever more recently founded downwind populations, the queens are ever less related to those of the polygyne population and ever more related to those of the monogyne one. This pattern suggests that new polygyne populations are founded by monogyne queens and that, as these populations grow, newly recruited queens are gradually ever more likely to come from the polygyne population.

The genetic factor that determines social form is the *Gp-9* locus. The conversion from monogyny to polygyny begins when a monogyne alate (*Gp-9*^{BB}) mates with a *Gp-9*^{b} male from a polygyne nest. This event is probably rare and requires presence of a nearby polygyne nest, probably dispersed to that location by human transport. The newly mated queen founds a monogyne nest, but all of her daughters will be *Gp-9*^{Bb}. Their body weight and behavior will be polygyne. Why these workers do not kill their mother is not clear, given that Bb workers execute BB queens. Perhaps they need a *Gp-9*^{Bb} queen as a cue-reference for full recognition of whom to kill, as in the experiments I discussed earlier in which orphaned *Gp-9*^{Bb} workers killed polygyne *Gp-9*^{BB} queens in only half the cases.

Let us assume, however, that this obstacle is overcome and that this monogyne colony with polygyne workers duly reaches maturity to produce female alates. All of these will be *Gp-9*^{Bb}. Their reproductive syndrome will be polygyne, not monogyne. If they succeed in nest founding, they will probably do so as a cooperative group, because they lack the reserves to found alone. They may be readopted by their natal colony, resulting in the demise of their mother and their own accession to queendom. Whichever route they take, the genotype *Gp-9*^{Bb} will be well represented among their worker offspring, who will thus be tolerant of multiple *Gp-9*^{Bb} queens but intolerant of *Gp-9*^{BB} queens. A polygyne colony has thus come into existence, with all the familiar reproductive features—low alate weight, local mating, adoption, and budding. Initially, the mtDNA of these nests will bear the monogyne imprint, dominated by haplotypes A and B (approximately 84–89% A, 5–16% B), but as the new polygyne population ages and grows, newly mated queens dispersing from the core polygyne population will find and enter these nests. They will bring with them haplotype C, and the newly founded population will gradually become 80–95% haplotype C. Looking back, the initial polygyne colony that provided the *Gp-9*^{b} males was like a seed or a condensation nucleus around which the polygyne population grew in collaboration with the monogyne population.

In Argentina. The alert reader will have spotted a problem. If the founding of a polygyne population requires a b-allele, and the b-allele is absent from monogyne populations, how did the first polygyne population arise in the USA? Diploid males are rare in Argentina, so mating between polygyne males and monogyne females (and therefore the transition between social forms) is more common there than in the USA. Indeed, in one of two monogyne Argentine populations, $Gp\text{-}9^{Bb}$ alates occurred with a frequency of about 4% (Ross, 1997)—their mothers had probably mated with polygyne males from a nearby polygyne population, and they would found polygyne colonies—but such cross-mated queens are too rare in Argentina to have contributed any of the original founders of the North American population. Perhaps at least one of the colonies that landed in Mobile was polygyne. From this Mobile beachhead, the b allele was probably scattered very widely and rapidly, judging from the occurrence of polygyne populations throughout the North American range of the introduced *S. invicta.*

Surveys in the early 2000s found polygyne populations to be much rarer and more geographically restricted in South America than in the USA (Mescher et al., 2003). The polygyne form was limited to the southern parts of *S. invicta's* native range, where it occurred in scattered patches within the monogyne population (Figure 28.12A). This pattern probably represents historical traces of chance events of population founding and extinction. The North American populations also consist of enclaves within mostly monogyne populations but contrast sharply in the much wider geographic distribution of these enclaves (Figure 28.12B). The much greater population densities in the USA, coupled with the greater density of roads and vehicular traffic, probably produced much greater fire-ant dispersal rates and distances, overcoming the naturally very low dispersal rates of the polygyne form and leading to similar geographic ranges for the two social forms in the USA. During range expansion, new polygyne populations were probably founded by ride-hitching mated queens or, more readily, by colony fragments containing one or more queens. This scenario is supported by the much higher local genetic (especially mtDNA) differentiation of polygyne populations in South America than in the USA.

South American polygyny is associated with two b-like alleles (b and b'), which differ in their geographic extent and range (Figure 28.12), whereas North American polygyny is associated with only one of these alleles (b). The presence of polygyny in the USA means that at least part of the initial *S. invicta* inoculum must have come from the southern half of *S. invicta's* native range. Whereas the absence of the b' allele from the USA suggests that populations having this allele are less likely sources, other genetic and morphological characters support northeastern Argentina, where both the b and b' alleles occur, as the most likely source.

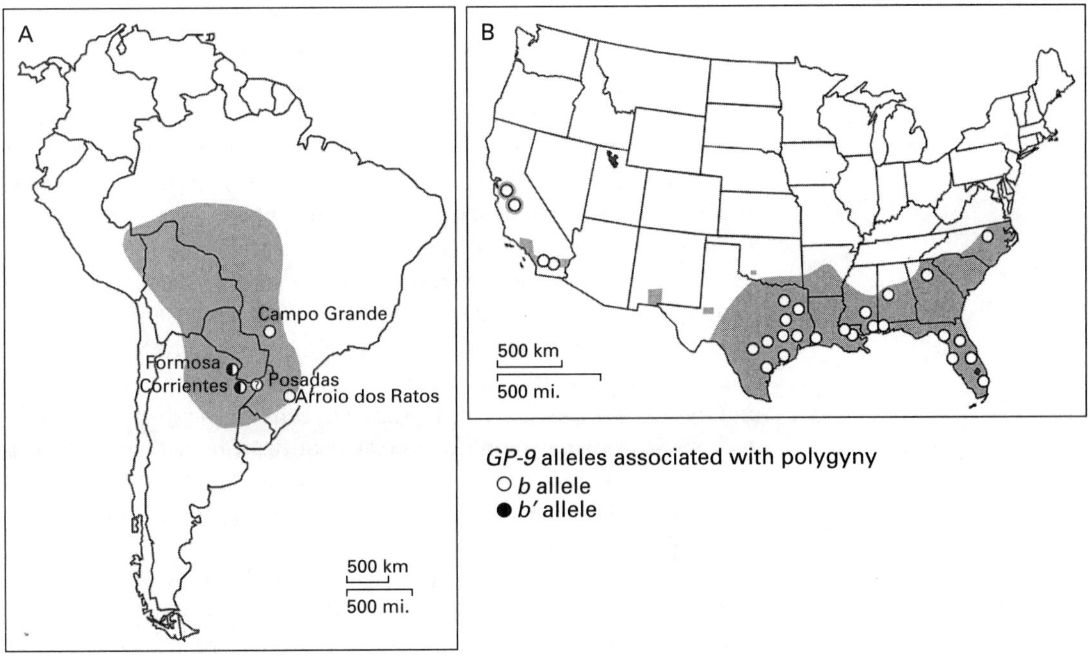

Figure 28.12. Distribution of polygyne *Solenopsis invicta* in South America (A) and North America (B). Open symbols indicate polygyny based on the b allele, filled the b′ allele, half-filled both alleles. Adapted from Mescher et al. (2003).

What Selects for the Maintenance of Polygyny? The force that holds a polygyne colony together in the short term is not obvious. Unrelated queens produce sexual brood of which only a few are relatives of a worker rearing them. The inclusive fitness that the worker gains for her trouble is very small. A number of hypotheses have been advanced to explain why polygyne colonies continue to exist. One option might be for workers to favor their own relatives within the colony (nepotism), but as already mentioned, DeHeer found no evidence that they do. Another is that, if queens that produced a lot of sexuals also produced a lot of workers, workers, on average, would be significantly related to the sexuals they were rearing, even if they could not recognize relatives. This hypothesis also goes down the drain, for the relatedness between workers and sexual brood is zero (Ross et al., 1996b). Another hypothesis is that polygyne colonies are strongly favored in certain kinds of habitats. Unfortunately, in both their native and their introduced range, polygyne and monogyne colonies occupy habitats that are indistinguishable, even intermingling on a small scale in Argentina. Still another hypothesis is that the excessive genetic diversity within polygyne colonies give them resistance to diseases and parasites in the USA, but as discussed elsewhere in this volume, the introduced population is mostly free of these banes already, so this selective factor is unlikely to be important in the USA. Moreover, the polygyne form is more susceptible to the pathogenic microsporidian *Thelohania* sp.

In the end, workers are probably doomed by their total sterility to rearing mostly unrelated sexuals in order to reap a small bit of inclusive fitness from the few that are relatives (Keller, 1995). They really have no other choice. Both the workers and the incumbent queens lose fitness by accepting more queens, so polygyny in the North American population is best seen as social parasitism on the labor of workers by unrelated queens. Unregulated queen numbers and the absence of related queens fit this interpretation. Interestingly, queens in the South American polygyne *S. invicta* are related (r = 0.46) and therefore probably admit mostly relatives as newly recruited queens. They are therefore able to benefit from the reproduction of relatives. Goodisman and Ross speculate that the extreme population densities and huge numbers of adoption-seeking newly mated queens in North America swamp the capacity of nests to recognize nestmates. In the absence of the fitness benefits of kinship, polygyne nests seem to exist solely as a venue for social parasitism, a communal location where queens struggle with one another for direct reproduction and workers have no choice but to let themselves be exploited for the occasional scrap of fitness that comes their way. They are the victims of their own success.

A postscript: the study of polygyny in *S. invicta* provides a clear demonstration of the power of genetic analysis. Much of what we know about polygyne biology and social structure, indeed the causes of polygyny itself, would not have been unraveled through ecological and behavioral methods. Genetic analysis will doubtless continue to enlighten us for years to come.

29

Biological Consequences of Polygyny

Lack of Territoriality

A widespread consequence of ant polygyny and nest budding is that territorial boundaries disappear from entire local (or even regional) populations (Hölldobler and Wilson, 1990). The polygyne workers' tolerance of nonnestmates eliminates the hostility between nonnestmate workers that gives rise to territorial defense in the monogyne form. Whereas the monogyne form occupies the landscape as a mosaic of distinct territories, like tiles on a floor, the polygyne form lacks distinct boundaries between foraging areas of neighbors. Underground tunnels continue to connect daughter nests to parent nests, forming a network of interconnected mounds. Some workers, or even queens, may move between mounds, but high rates of such movement are unlikely, because queen number predicts both mean worker size and worker relatedness within mounds quite well, a relationship that would be impossible if queens or workers were freely exchanged (Ross, 1988, 1993; Goodisman and Ross, 1997). In short, mounds are largely "closed," and free exchange of workers and queens is probably limited to the period when a new nest is budded.

Interestingly, interconnections among nests continue to exist despite the low exchange rate. Byron and Hays mapped these interconnections on four plots in South Carolina (see Figure 8.10; Byron and Hays, 1986). Although the social form was not specified, workers from different colonies lacking underground connections attacked one another and were probably monogyne, whereas workers from connected colonies tolerated one another and were probably polygyne.

The intense competition for space that occurs between neighboring monogyne colonies appears to have vanished from the polygyne form. In reality, the locus of competition has shifted. In the monogyne form, the worker forces of neighboring queens struggle with one another to limit their domains to relatives only. In polygyne colonies, instead, queens compete with one another chemically to exploit a homogenized worker force for their personal reproduction. These differences make the two social forms quite different in their ecological effects and their interactions with other ant species, topics that will be discussed in a later chapter.

Higher Population Densities

Lack of territoriality results in higher mound density, one of the more obvious ecological differences between the social forms. Does this change produce a higher density of ants as well, or are the nests of polygyne fire ants simply smaller than those of monogyne ones? Tom Macom and Sanford Porter found that, on 14 monogyne and 14 polygyne plots in Florida (Macom and Porter, 1996), monogyne mound volumes (14.7 liters) averaged about twice those of polygyne mounds (7.6 liters), but the threefold higher density of mounds on polygyne sites (470 as compared to 155 mounds per hectare) more than made up for this difference in mound size. Applying the relationship of mound volume to colony size (the same in both social forms), we calculate that each hectare there harbors almost twice as many polygyne workers (35,000,000) as monogyne ones (18,000,000); the densest polygyne site contained 100,000,000 workers per hectare. Correspondingly, a polygyne site supports close to twice as much ant biomass per unit area as does a monogyne one (13 versus 7 dry kg/ha; 1.30 versus 0.70 g/m^2), suggesting almost double the ecological impact, but trophic impact depends on the rate of energy use, i.e., metabolic rate, and this figure was 17% higher for polygyne than for monogyne ants (13.9 versus 11.5 cal/hour per gram), probably because of their smaller average body size. Therefore, polygyne fire ants extract 2.3 times more energy (18.3 versus 8.0 cal/hour) from each square meter than do monogyne ones. The polygyne ants that do this work contain about 1.9 times as much energy per square meter as do monogyne ones (8.4 versus 4.5 Kcal/m^2).

These comparisons paint a stark picture. In terms of energy use, polygyne fire-ant populations have about twice the ecological impact that monogyne ones do. One would expect a greater negative impact on other animals, a topic covered in a later chapter.

As to why the density of polygyne ants is higher, I can only offer speculation. As monogyne *Solenopsis invicta* foragers spread outward from their nest, they repel or avoid nonnestmates, creating an exclusive territory. The forager density responds to colony size and neighborhood interactions. When polygyne workers spread outward from their nests, they fail to discriminate and repel nonnestmates, so the foraging ranges of several colonies overlap, and their forager densities are summed. The lack of exclusiveness also removes a repellent force that regulates the distance between colonies and allows polygyne colonies to move closer to one another than monogyne colonies. One could readily test this mechanism by marking foragers in entire polygyne neighborhoods with colony-specific marks and determining the foraging patterns and distances for each colony. Lack of free worker exchange among polygyne nests means most foragers must return to their home nests. The few that enter other mounds are probably simply lost, as might be expected from overlapping foraging ranges.

Spread on Foot

Nest budding has a dramatic effect on the nature of a polygyne fire-ant invasion. A monogyne invasion progresses as scattered colonies gradually coalesce into a continuous population as colony density gradually approaches an upper limit. In contrast, from its initial nest, the polygyne form advances its mounds across the landscape as a coherent, sharply defined front, like a fungal colony on a petri plate. Porter chronicled this phenomenon at the Brackenridge Field Laboratory in Austin, Texas. The details of this invasion and its ecological consequences are discussed in a later chapter. The slow pace of polygyne invasions suggests that large, continuous populations of them probably arise through coalescence of multiple invasion foci, but the way polygyne spread proceeds when the area is already occupied by monogyne *S. invicta* is much less clear. Whether or not the polygyne can actively usurp the monogyne is unknown. If they do, it is a slower process still. Possibly, conversion of monogyne to polygyne populations by cross-mating, as proposed by Ross and Shoemaker, may also play a role.

Low Alate and High Worker Production, Low Seasonal and Lifetime Colony Size Variation

Because high levels of queen pheromone cause polygyne colonies to invest so little in alate production, they invest heavily in worker production, resulting in very high colony growth rates (fourfold in six months; Porter, 1991). Polygyne colonies might be said to have carried weediness one step further than monogyne colonies by emphasizing "vegetative" rather than sexual reproduction. For the same reason, their seasonal fluctuation in colony size is much less than that of monogyne colonies: the small allocation to alate production results in only minor seasonal reduction of worker production (and therefore in colony growth) (Vargo and Fletcher, 1987). Because colonies reproduce by budding, colony-size variation during the life cycle is usually less than two orders of magnitude; it is up to five orders of magnitude in monogyne colonies.

Small Workers

In nests of similar size, polygyne workers are smaller and less variable than monogyne ones—the modal size class of polygyne workers was 0.64 mm, that of monogyne workers 0.77 mm. Workers with head widths greater than about 1.2 mm are lacking in polygyne colonies (Figure 29.1; Greenberg et al., 1985). The mean head widths of the two social forms did not overlap (partly because ambiguous colonies were eliminated from consideration). The social form of most colonies could be correctly identified from mean worker size. Moreover, the mean sizes of monogyne workers from Texas were almost identical to those from Georgia (0.88 and 0.87 mm), and the same was true for polygyne workers (both 0.66 mm).

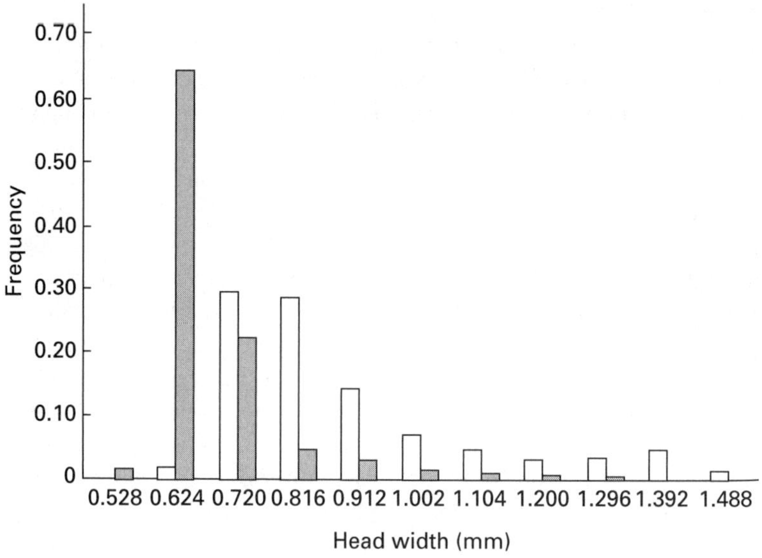

Figure 29.1. Worker size distribution of monogyne (open bars) and polygyne (shaded bars) colonies of *Solenopsis invicta*. Workers with head widths greater than 1.2 mm are more or less lacking in polygyne colonies. Adapted from Greenberg et al. (1985).

Of course worker size also varies greatly *within* social form, especially within the monogyne form, where it increases strongly with colony size. Within the polygyne form, colonies with more queens produce smaller workers. The number of workers in a nest probably plays a role too, but it is less well characterized. In both forms, nutrition and season probably also affect worker size. In light of all these effects on worker size, the degree of distinctness between the mean sizes of the two forms is almost surprising.

What causes these differences in worker size? Investigators long believed that most or all of these differences were environmentally induced through the high larva-to-worker ratios in polygyne colonies (Vander Meer et al., 1992), which assure that each adult worker cares for more larvae, producing smaller, less variable pupae. Lower larva-to-worker ratios in monogyne colonies produce larger and more variable workers (Porter and Tschinkel, 1985a; Vargo, 1998).

But not so fast! In light of the effects of the *Gp-9* gene on body weight of both sexes of alates, it seemed prudent to check out its effect on worker weight as well (Goodisman et al., 1999). After all, monogyne workers are all *Gp-9*^{BB}, whereas polygyne workers can be mixtures of all three *Gp-9* genotypes but always include a large number of *Gp-9*^{Bb} workers. Goodisman weighed and genotyped samples of workers from each of 13 laboratory colonies headed by single polygyne queens, as well as workers from over 100 field colonies. The laboratory-reared workers showed no significant effect of *Gp-9* genotype on weight, but among field-collected workers, BB workers averaged 1.73 mg, Bb 1.54 mg, and bb 1.20 mg (individual colony also had a significant effect, as noted in so many measures). BB workers from field colonies were heavier than their laboratory counterparts, Bb about the same, and bb quite a bit lighter. Conditions in field colonies seem to lead to a stronger expression of the *Gp-9* gene, in its effects on both worker body weight and worker viability.

Goodisman and Krieger and their coworkers (Goodisman et al., 1999; Krieger et al., 1999) also checked the effects of triploidy on worker body weight after discovering that about 12% of nonreproductive females in polygyne colonies were triploid (none in monogyne colonies), presumably fathered by one of the 2.4% of diploid males that produced sperm. Triploid workers heterozygous at a monomeric allozyme locus can be identified by the unequal intensity of the two monomeric bands on gel electrophoresis because they have twice as much of the product of one allele as of the other. In the USA, no allozyme loci are represented by more than two alleles, so Krieger also used microsatellite DNA loci that have multiple alleles. A triploid that has three different alleles will have three electrophoretic bands. Triploid workers were heavier, on average, than normal diploid workers, and the effect of triploidy overrode the effects of the *Gp-9* genotype; to turn the coin over, a higher proportion of large workers (14%) than smaller workers (10%) was triploid. Triploids occurred only in polygyne colonies, as expected if most polygyne males are diploid and mate only with polygyne alates, if at all. The biological importance of triploidy is uncertain.

V

Populations and Ecology

30

Hybridization between *Solenopsis invicta* and *S. richteri*

Most of the morphologically distinguishable species of North and South American fire ants seem, on the basis of genetic evidence, to be reproductively isolated from one another in their native ranges. An unknown number of cryptic species not distinguishable on morphological grounds also exist. As I have already noted, the situation is quite different for *Solenopsis invicta* and *S. richteri* in their nonnative range in the USA. Along the northern range limit in Alabama and Mississippi, these two species seem to hybridize freely. As mentioned above, hybrids were first detected when Bob Vander Meer found that the venom-alkaloid and cuticular-hydrocarbon patterns of specimens that looked like *S. richteri* were intermediate between those of *S. richteri* and *S. invicta* (Vander Meer et al., 1985; Vander Meer, 1986a; Vander Meer and Lofgren, 1988, 1989). Hybridization began with the very first contact between the two species but was not recognized because the hybrid simply looks like *S. richteri.* Analysis by gas-liquid chromatography of preserved specimens collected around Mobile in 1949 and around Starkville, Mississippi, in 1964 showed some of them to be hybrids.

Surveys for the hybrid relied initially on chemical analysis of venom and hydrocarbons. Later, more extensive surveys added four polymorphic enzymes (allozymes) to the list of diagnostic characters (Ross et al., 1987; Ross and Robertson, 1990). The two "pure" species differ greatly in allele frequency at two loci and have no alleles in common at the other two. Hybrids could thus be recognized if they had one allele of either or both from each parent species or if they had an allele of one species at one locus, and an allele of the other species at another locus. Chemical analysis confirmed these identifications. A later study also identified morphological characters that reliably distinguished the hybrids. Without a doubt, the species hybridize.

Surveys showed that pure *S. richteri* was limited to eight counties in north central Mississippi (Figure 30.1) and one county in north Alabama (Diffie et al., 1988). Hybrid populations or mixed populations of the hybrid and *S. invicta* occupied a broad band running from west central Mississippi across Alabama to northeastern Georgia. Interestingly, only in Mississippi does the hybrid population abut a pure *S. richteri* population to the north, suggesting that much of the range increase of the hybrid resulted from human transport.

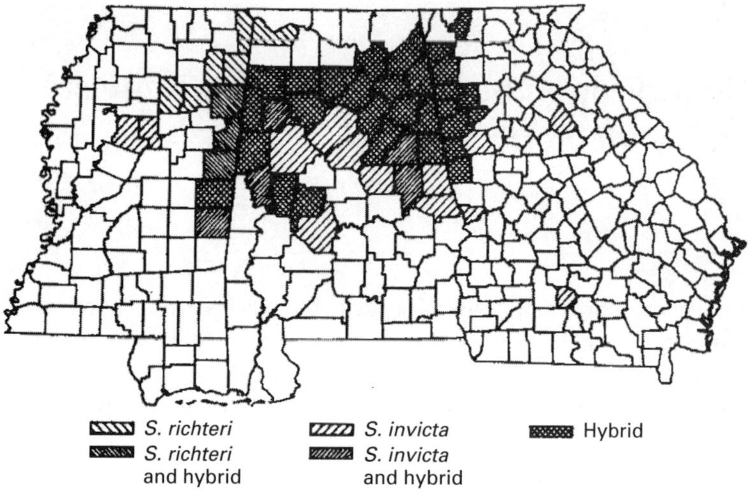

Figure 30.1. The zone of hybridization between *Solenopsis invicta* and *S. richteri* in November 1986. Reprinted from Diffie et al. (1988) with permission from the Georgia Entomological Society.

The allozymes and chemical analysis both showed that the two species were interbreeding freely along a broad contact zone running at least 120 km in eastern Mississippi. Colonies of pure *S. invicta* were found only at the southern end in Meridian, Mississippi, and pure *S. richteri* only at the northern end in Columbus. Even there, both occurred in mixed populations. Between these end points lay an abundance of hybrids or hybrids backcrossed to one of the species. No colonies were F1 hybrids, that is, founded by a pure queen of either species, mated to the male of the other. All seemed to be various hybrids backcrossed to other hybrids or parental species. This pattern is what is expected for a true hybrid zone in which the parental species and all hybrids readily interbreed. As a consequence of all this interbreeding, *S. invicta* genes are currently undergoing net flow northward and *S. richteri* genes net flow southward, such that *S. invicta* alleles become less frequent in a northerly traverse of the hybrid zone and *S. richteri* alleles less frequent in a southerly traverse (Figure 30.2).

Genetic tests sometimes failed to detect hybrids, because after a long hybridization period, hybrids sometimes, by pure chance, carried only the alleles of one parental species or the other (availability of more than four isozyme markers would make this problem less frequent). These would necessarily be assigned to that parental species, but their hybrid nature was nevertheless detectable through venom, hydrocarbon, and morphological characters. Tests of *S. invicta* and the more distantly related *S. geminata* did not detect any hybridization, at least not in the areas tested.

The hybrid population seemed normal in almost every way and included both social forms (Glancey et al., 1989). Monogyne colonies of the hybrid showed a greater ability to tell nestmates from nonnestmates

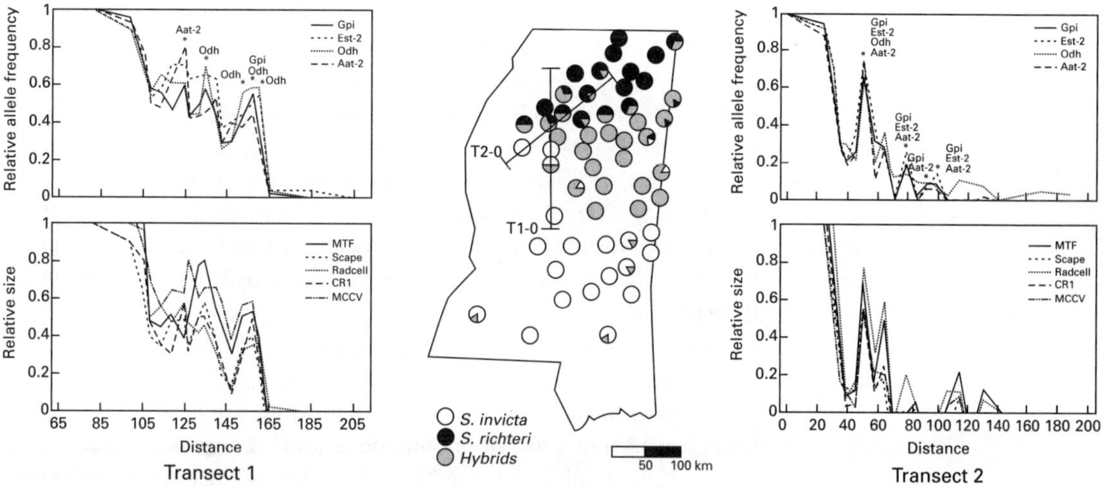

Figure 30.2. Transects of the hybrid zone between *Solenopsis invicta* and *S. richteri,* showing that the *S. invicta* alleles and morphological characters decrease in frequency in a generally northward direction. T1 and T2 are the two sampling transects. The "0" on each transect corresponds to the 0 on the graphs, the southernmost sample. The pie diagrams show the proportion of each species and their hybrids at each site. Adapted from Figures 1 and 4 of Shoemaker et al. (1996).

than did either of the parent species. The basis of this ability may have been the greater genetic diversity of the hybrid or the greater complexity of its nestmate-recognition cues (Obin and Vander Meer, 1989a). Workers of *S. invicta* and *S. richteri* were each attracted to spots of Dufour's gland secretion (part of the trail pheromone complex) of their own species but not the other. The hybrid was attracted to both secretions, and its Dufour's secretion attracted both of the parent species (Vander Meer and Lofgren, 1988, 1989). Gas chromatography of the secretions confirmed the intermediate nature of the hybrid secretion.

Initially, interbreeding seemed to entail no cost in viability or fertility, but more sensitive analyses showed this not to be the case. A subtle (but hotly contested) way to estimate the fitness cost (if any) of hybridization is based on body symmetry. As an animal develops from the egg to the adult, developmental genes regulate the expression of structural genes. When these genes are well adapted to one another, the system produces the same outcome on each side of the animal, resulting in an animal that is highly symmetrical. Degrees of genomic divergence and incompatibility show themselves as increased random differences between right and left, so-called "fluctuating asymmetry." In *S. invicta* and *S. richteri,* significantly greater fluctuating asymmetry among hybrids than among parents might mean that each species' regulatory genes are less effective at regulating the others' structural genes than its own (Ross and Robertson, 1990). High levels of fluctuating asymmetry are usually associated with reduced fitness.

Ross and Robertson collected alate females from the two ends of the hybrid zone (northwest Georgia and east central Mississippi) and from pure *S. richteri* and pure *S. invicta* populations. They measured seven bilateral characters from the wings, antennae, and legs and calculated the difference between the right and left measurements of each individual. They adjusted the differences to remove the effect of overall size and compared their magnitudes to measurement error. Significantly more variable measures with means of zero indicated significant fluctuating asymmetry (nonzero right or left bias was of no interest in this study).

Of the seven characters, six were significantly more variable than measurement error in the Mississippi hybrid population, five in the Georgia hybrid and pure *S. richteri*, but only three in the pure *S. invicta*. Comparison across the four populations showed only two characters to be significantly different; the Mississippi hybrid was at the high end, followed by the Georgia hybrid, pure *S. richteri*, and finally pure *S. invicta*. On the basis of allozyme analysis, the Mississippi hybrid contained 50% each of *S. invicta* and *S. richteri* genes, whereas the Georgia hybrid was over 60% *S. invicta*. A criticism of this approach is that other factors, including chemicals and environmental stress can also induce fluctuating asymmetry (indeed, fluctuating asymmetry has been used to detect pollution). One could counter that both species and the hybrid shared the same environment and that wing anomalies were more common in the hybrids.

All together, these results suggest that the genomes of *S. invicta* and *S. richteri* are modestly incompatible when they are mixed through hybridization (although obviously not enough to prevent hybridization), but all that matters is whether this incompatibility exacts a cost in reduced fitness. Reduced fitness would in turn mean that genes would flow more slowly through the hybrid region, creating a weak barrier to gene flow. Ross and Robertson calculated the strength of this selection against the hybrid from the average dispersal distance of fire ant queens and the width of the hybrid zone. Selection against the hybrid was only 0.1% to 0.01%. These low levels are also supported by the observation of hybrid genotypes in the proportions expected from random mating and the absence of selection.

The reasons for the stability of the hybrid zone are not immediately apparent. (Because the hybrid zone has not been revisited since 1990, even its stability is in some doubt.) Why does the superior species not simply swamp the inferior one, leading to its disappearance? Or if neither is superior, why do the populations not blend completely rather than forming a hybrid zone? Two types of explanations of stable hybrid zones have been proposed. In the first type, the hybrid is superior to the parents in the ecologically intermediate zone that it occupies, causing selection to favor its continued existence. Such hybrid zones are stable in space. In the second, the hybrid zone is the outcome of a balance between gene flow through dispersal of parental types (and alleles) and selection against the hybrid. If one of the parent populations is larger or fitter or disperses farther, such hybrid zones may slowly migrate in the direction of the weaker competitor. These questions, and the rarity of recently formed animal hybrid zones, led Ross's student DeWayne Shoemaker to map and analyze the hybrid zone in much greater detail (Shoemaker et al., 1996).

Shoemaker chose his two 200-km sample transects from prior knowledge of the hybrid zone's structure. One transect was oriented north-south across the width of the zone at its narrow end, the other was oriented northeast-southwest from the western end of the zone into the heart of *S. richteri* country. These transects presumably sampled the region of most recent contact.

Approximately every 10 km, Shoemaker stopped to collect a single female alate from each of 10 to 40 colonies, to assure independence of the data (i.e., to assure that the individuals he collected were not simply siblings and therefore very similar to one another). For morphology, the same measurements as in the above-mentioned study of fluctuating asymmetry were used. Genetic analyses were based on seven different allozymes known to distinguish *S. invicta* from *S. richteri* and a single diagnostic DNA gene. The morphological characters result from the action of multiple genes, but the enzyme and DNA characters are single-gene characters. These data allowed estimation of the proportion of each parent species' genes present in the genome, as well as detection of strong natural selection, high migration rates, or deviation from Hardy-Weinberg equilibrium (if two genes are in Hardy-Weinberg equilibrium, neither is being selected for or against, so deviations from it indicate the action of selection).

The picture appears when these genetic attributes are geographically mapped on the hybrid zone. Obviously, the fire ants south of the hybrid zone were all *S. invicta,* those north of it were *S. richteri,* and sandwiched in between lay the hybrids. The most dramatic patterns can be seen in Figure 30.2, which shows the relative proportion of each species' character in relation to distance along the two transects. On these graphs, zero represents pure *S. richteri* and 1.0 pure *S. invicta.* Each of the three classes of marker (morphology, allozyme, DNA) is shown separately. Clearly, the transition is anything but smooth, peaks and valleys are apparent in all three types of characters. In fact, the changeover showed four or five reversals along each transect, and pure populations were even flanked by hybrids.

Equally clearly, within each class of marker, the characters of the ants changed in parallel, at the same rate. We can demonstrate this pattern by plotting each allozyme against each other one, and each morphological character against each other one—all resulted in slopes equal to 1.0—but when each morphological character was plotted against each allozyme frequency, the slope was less than 1.0 in all 55 comparisons, showing that the changeover in morphological characters was more abrupt, less smooth, than that of the allozyme characters. In other words, allozyme genes seem to flow smoothly between the populations, but some barrier to the flow of morphological genes seemed to be present—the hybrid morphology must sometimes confer lower fitness than the parental morphologies. Given the nature of sexual reproduction, and its requirement for attraction and copulation, hybrid morphology might well put alates at a mating disadvantage, but in reality, we do not know which structures and which life phases are involved. Possibly, these classes of genes move through populations independently, each at rates dictated by specific resistance to its movement. Alternatively, perhaps interbreeding affects development in such a way that morphology and the genotype change at a different rate or that change is threshold-like.

From the proportion of parental species at each site, Shoemaker and Ross then calculated the expected proportion of hybrids. Of the 36 sites at which hybrids were present, about 55% had significantly fewer hybrids than expected. These populations did not meet the assumptions of the Hardy-Weinberg law, possibly because hybrids are less fit, parental types prefer to mate within their own species rather than randomly, immigration is significant, or the populations have been interacting over only a brief span of time. To determine whether certain classes of hybrids accounted for the deficit, Shoemaker classified hybrids by assigning one point for each *S. richteri* allele an alate possessed at three diagnostic allozyme markers. Pure *S. invicta* alates earned a score of zero and pure *S. richteri*, six. At 20 of 34 appropriate sites, although hybrids were fewer than expected, the "missing" hybrids were not of any particular class. On the other hand, distribution of the deficits along the transects was not uniform. Sites at which *S. invicta*-like hybrids were underrepresented had an excess of pure *S. invicta*, and the reverse was true for *S. richteri*-like hybrids. Each of these types of sites tended to be found close to the range of the pure parental species. Apparently, being a hybrid within a population of mostly pure species carries a disadvantage.

So the best interpretation, given current knowledge, goes something like this. The current geographic patterns were most probably initially laid down by the chance events of colonization combined with the chance availability of suitable habitat. When a patch of habitat in the current hybrid zone became available, neighboring source population densities combined with flight ranges, mating preferences, habitat preferences, and colony-founding ability to determine whether a patch was colonized by pure species or by various hybrids. As these chance-founded populations grew, they interbred to varying degrees with neighboring populations, creating the mosaic distribution of parental and hybrid types observed. The characteristics of each habitat patch might match up better with some parental or hybrid types than others, influencing which type dominated the patch. The selection against hybrids would reduce their frequency below that expected from random reproduction. Early on, these processes would create a strongly mosaic pattern, as in the northeast-to-southwest transect, but in time, hybridization and dispersal would blur the edges of the original patches to create a more gradual transition across the hybrid zone, as in the north-south transect, where the two fire ants have enjoyed longer contact.

More is probably at play here than historically established mosaics and their decay, however. The hybrid seems to be subject to some negative geographically specific selection; otherwise hybrid genotypes ought still to be present as far south as Mobile (where they once existed). This northward migration of the hybrid zone might be the result of the greater fitness of the *S. invicta* genotype (as indicated by the low level of fluctuating asymmetry) as well as its greater population densities and the ensuing greater northward gene flow. A series of warm winters probably also played a role. Possibly, *S. richteri*, which comes from temperate Argentina, is the superior competitor in the more northerly regions, whereas in the more southerly regions, the subtropical *S. invicta* is superior. In the transition zone between the two, the intermediate hybrid may be superior (Vander Meer and Lofgren, 1988). Such regional hybrid superiority could explain how the hybrid population can exist bounded on the south by *S. invicta*, but without pure *S. richteri*'s pumping genes southward from its northeastern boundary (presumably, the hybrid is bounded to the north by climate). Additional support comes from the observation that the hybrid seems slowly to be displacing the pure *S. richteri* population in northern Mississippi (but the data are thin). Understanding of these processes and patterns is far from complete, in part because the zone has not been monitored over a sufficiently long period.

Not far upstream from the Rio de la Plata, the southern extreme of *S. invicta*'s range overlaps the northern margin of *S. richteri's* (Figure 2.1). Unlike the populations in the Mississippi-Alabama contact zone, the two species do not hybridize in this contact zone. Although an early study of a sample of 39 colonies suggested that a small amount of gene flow might occur between the two species (Ross and Trager, 1990), a more thorough later study found these two species to be fully reproductively isolated in this contact zone (Ross and Shoemaker, 2005). Ross and Shoemaker's genetic toolbox contained seven polymorphic allozyme loci with 2–10 alleles each and the mtDNA gene for cytochrome oxidase I with 14 different sequence haplotypes. In the overlap zone, each species has a high-frequency allele at each of several loci that is either absent or rare in the other species. None of the mtDNA haplotypes are present in both species. This situation can exist only if barriers to gene flow are essentially complete and of long duration, and the two species are reproductively isolated.

Such barriers to interbreeding are very weak when *S. invicta* and *S. richteri* come into contact in Mississippi and Alabama, as we saw above. In South America, they must be "prezygotic," a fancy way of saying that they prevent the union of egg and sperm, perhaps as a result of differences in the timing or site of mating flights or other biological differences that would prevent cross-mating. Another possibility is that endosymbiotic bacteria, such as *Wolbachia* sp. (discussed in a later chapter), make cross-mated females infertile (*Wolbachia* is absent from both species in the USA). In the USA, "postzygotic" barriers are very weak: hybridization seems to exact a low price in viability or fertility. How interbreeding is prevented in South America but not North America remains to be determined.

Interestingly, small amounts of hybridization between other South American *Solenopsis* species were also detected through shared species-diagnostic allozymes, so low rates of gene flow among several species may be occurring, but Ross judged these rates to be too low to affect the population gene dynamics or to reduce genetic differentiation among the species. Allozyme patterns also suggested that several cryptic species of *Solenopsis* exist; for example, a sample of nine nests of what appeared to be *S. invicta* from Arroio dos Ratos was genetically completely distinct from the *S. invicta* around it. Cryptic species are species that cannot be distinguished morphologically but do not interbreed because of genetic, environmental, or life-cycle differences. They can be distinguished primarily on the basis of nonmorphological characters, such as genetics or behavior. Clearly, the population biology and systematics of South American *Solenopsis* are complex indeed.

The frequent interbreeding of *S. richteri* and *S. invicta* in the USA has caused some to ask whether these two can legitimately be regarded as separate species. A commonly used definition of animal species requires that gene flow not occur between them, a requirement clearly violated in this case. "Species" is defined in many ways, however, and the definition often does not include any requirement for reproductive isolation. For a considerable period of their evolutionary history, *S. richteri* and *S. invicta* have followed separate evolutionary paths in South America, qualifying them as pretty "normal" species, unable to interbreed in regions where they occur together. That they now interbreed in the USA is an artifact of human transport, not a reflection of their evolutionary history. Oddly, botanists define species very differently and have no problem with hybridization (even across generic lines). These difficulties show once again that nature abhors categories. In my opinion, understanding the biology is more important than crafting problem-free definitions.

Two species of polygyne fire ants native to the USA, *S. geminata* and *S. xyloni,* also have overlapping ranges in central and eastern Texas, and lo and behold, on morphological grounds, also seem to hybridize (Cahan and Vinson, 2003). In the hybrid zone, major workers either had *S. geminata* morphology or hybrid *geminata* × *xyloni* morphology but never *S. xyloni* morphology. When collected outside the range overlap zone, *S. geminata* had alleles at eight enzyme and two microsatellite loci that were not found in *S. xyloni* and vice versa. These loci were thus diagnostic for the species. Hybrids should have the alleles of both species at the diagnostic loci. Workers from the range overlap that appeared to be morphologically *S. geminata* were genetically *S. geminata* as well. Workers that appeared morphologically to be hybrids were indeed mostly hybrid at the diagnostic loci (a few appeared to be backcrosses to *S. geminata*). The surprise was that most male and female alates from these same hybrid colonies carried only the *S. xyloni* alleles. Was the expression of *S. geminata* traits being suppressed in sexuals during development? Apparently not, because the unexpressed DNA and the expressed protein results were exactly parallel; that is, most sexuals that had *S. xyloni* proteins also had only the *S. xyloni* DNA alleles.

The best explanation for these patterns is that, in the hybrid zone, all the offspring of *S. xyloni* females that mate with *S. geminata* males develop into workers (hybrids), whereas all the offspring of *S. xyloni* females mated with *S. xyloni* males develop into sexuals (nonhybrids). This pattern would explain why, within the hybrid zone, none of the 82 sampled colonies had pure *S. xyloni* workers, even where *S. geminata* was relatively rare. Within the hybrid zone, hybrid matings are obligate, and colonies must contain both *S. xyloni* queens mated to *S. geminata* males and *S. xyloni* queens mated to *S. xyloni* males in order to produce both workers and alates (for unknown reasons, colonies resulting from *S. geminata* females mating with *S. xyloni* males did not occur). Blocking the *S. geminata* genome from participation in colony reproduction has thus sidestepped the genetic cost of hybridization. When it comes to reproduction, colonies in the hybrid zone are essentially either *S. geminata* or *S. xyloni*. For any *S. geminata* male that mates with a *S. xyloni* female, it is the end of the genetic line, for his offspring will raise only *S. xyloni* sexuals. It doesn't seem fair, does it? Nevertheless, the situation is not without precedent; something similar occurs in two species of *Pogonomyrmex*.

31

Populations of Monogyne Fire Ants

This is the story of populations of colonies—how they get started, grow, age, and perhaps die. Fire ants exist simultaneously on several levels. Cells, organs, and organ systems are functionally organized into whole fire-ant individuals; individuals exist as the parts of functional superorganisms or colonies; and colonies aggregate into populations. Each of these levels has characteristics missing from the level below. In social insects, the colony is an additional functional, evolving unit, which is not found among nonsocial organisms. The collective properties of populations of social insects can thus often be expressed in units of both colonies and individuals. Many collective population properties are of great importance because they are related to ecological impacts. These include population size, biomass and biomass density, density and spatial distribution of colonies, total energy use, geographic extent, and numerous derivative properties.

For species that exploit disturbed habitats, it is especially obvious that populations also have a life cycle—they are "born" when disturbed habitat becomes available, grow to maturity, and eventually may also die as the habitat reverts to later stages of succession. Monogyne fire-ant populations come into existence when a number of newly mated queens, at the end of their dispersal flights, find and settle in unoccupied areas and successfully found new colonies there. From the survivors of this demanding process, the population takes shape and grows. What actions and interactions of colonies shape the population at each stage? What regulates population characteristics such as colony density, biomass density, and turnover rates? What determines the age and size distributions of colonies?

Many of the characteristics of a population can be understood through "life tables." These describe the probability that an average individual (or colony, for social insects) will still be alive in each of a succession of time intervals and the number of offspring (or new colonies) produced in that time interval. The values describe real populations and change as these develop. They may be very different for populations in different circumstances or habitats. The calculations are typically derived from large cohorts of newborn individuals or colonies (say 1000) whose survival and reproductive performance are followed until all have died. Such a table reveals how reproductive effort and the forces of mortality vary throughout life. When emigration and immigration are included, life tables allow complete numerical descriptions of population changes and represent the life of a typical individual. They can yield the net replacement rate, the number of offspring that replace an average individual or colony in one generation. When the replacement rate is equal to 1.0, the population is stable, because each individual or colony just replaces itself. Replacement rate therefore reveals whether the population is increasing, stable, or decreasing. Life tables also reveal the mean life expectancy and are the basis of the entire life-insurance industry. All this information and much, much more is available from detailed life tables.

No complete life tables have been published for fire ant colonies, but I have derived an approximate, composite life table from five studies, some already mentioned. Our Tharpe Street study (Tschinkel, 1992a) yielded survival of queens and colonies in newly founded populations and the growth of these colonies to maturity (Tschinkel, 1988a) but did not include alate production. Our Southwood Plantation studies produced size-specific survival and colony growth data for a mature population (Jerome et al., 1998; Adams and Tschinkel, 2001). Finally, our unpublished study at Southwood Plantation revealed the relationship between colony size, territory size, and alate production. Together, these data describe a prereproductive population and a mature, stable population but not an adolescent or dying population.

Establishment of Monogyne Populations and Population Properties

In the USA, populations of fire-ant colonies arise in two rather distinct situations. In the first, one or a few colonies or mated queens are transported far from the continuous population into countryside unoccupied by fire ants, as occurred when fire ants were transported throughout the southeastern USA in nursery and sod products. These "island" incipient populations then grew outwards, much like bacterial colonies on a petri dish. The details of this process are poorly understood, but it undoubtedly involved postmating dispersal flights, as well as human and water transport. The density of colonies was highest in the core of this spreading population, thinning toward zero at its expanding outer margins. At these outer margins, colony density probably increased logistically over a few years until existing colonies made the founding of new ones ever less likely. Finally, at a density of between about 100 and 150 colonies per hectare, the habitat was saturated with fire ant colonies of diverse ages, and the population was fairly stable.

In the second situation, the bulldozer or poisoned bait creates a new patch of vacant, ecologically disturbed land within the range of fire ants, and the surrounding fire-ant population expands to fill it. In contrast to isolated incipient population centers, where the supply of colony-founding queens and colony density gradually rise year by year, such patches or gaps are within flight range of huge numbers of alates, who probably saturate the new habitat with colonies in one or two seasons. This initial population is essentially an even-aged cohort. Over the next few years, the population develops by what is almost the reverse of a logistic function. The count of incipient nests is highest at its "birth" and declines to maturity as the incipient colonies eliminate one another through competition. Something similar is occurring at the same time among the thousands of plant seedlings that sprout alongside the fire ant colonies, a process botanists call "self-thinning." Even-aged cohorts are also apparent when pesticide-treated land is recolonized by *Solenopsis invicta.* For example, in a pesticide study near Albany, Georgia (Lofgren and Williams, 1985), 90% of all the smallest colonies ever found in the plot appeared in a single one of the six time samples.

A Newly Founded Monogyne Population. The best way to understand fire-ant populations is simply to follow one or several populations from birth through growth, maturation, and stasis. No studies of the gradual growth of populations at the range margins exist, but Tallahassee offers opportunities aplenty for studying colonization of newly disturbed patches, because it is one of those growing Sunbelt cities where clearing land for suburbs or trailer parks is something that happens between breakfast and lunch on most days. Our study site is located off Tharpe Street on the fringes of Tallahassee. It is a piece of land that had been forest a few months earlier but now looks like the surface of the moon, complete with a crater, except that the crater has water in it. Various weeds are beginning to colonize the site, so sparse green contrasts with the barren red to brown soil (Figure 31.1). Some areas have been scraped clean of topsoil, exposing the yellowish marl subsoil, and are only slowly being colonized by plants. The development company that created this charming landscape has gone belly-up, stopping the development and allowing us to follow events at this site for five years.

In a representative 30- by 40-m plot, after every mating flight, we mark the entrance of each newly founded nest. The site is so barren of plants hosting insects that very few ant nests survive the first summer; the rest probably starve to death (young nests persist in the more vegetated areas of the larger site). The process thus begins anew the next spring when a scant cover of vegetation has gained a foothold in the plot. Now we repeatedly map the locations and movements of all founding and incipient nests through the entire reproductive season and revisit occasionally for several years. This process produces a detailed history of a local population during the founding period, with a single visit at maturity (Tschinkel, 1992a), and teaches us much about how mating flights and brood raids generate population patterns.

Newly mated queens fly from their natal colonies in distinct waves

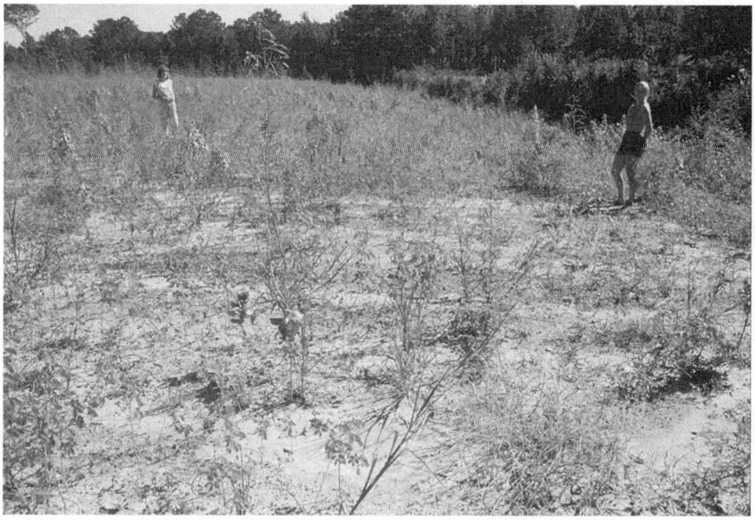

Figure 31.1. The Tharpe Street study site in 1985. The forest was bulldozed by a developer who subsequently went bankrupt, and plans for building were abandoned. Newly mated fire ant queens are attracted to such partially vegetated sites.

because, although their production during the season is probably continuous, mating flights are limited to a day or two after significant rains. Populations of nests are therefore founded in distinct, single-day waves. One might expect the surviving incipient nests to appear in equally distinct cohorts, but variable development rates, as well as variation in suitable conditions for nest opening, cause incipient nests founded the same day to emerge over periods of four to 15 days, smearing and overlapping the emergence of founding cohorts. In addition, cooler soils early and late in the season (on account of frequent rains) slow development and compress incipient nest emergence somewhat to a mid-season peak (Figure 31.2). The well-spaced rains of early season result in well-spaced, distinct flights, but during peak season, rains and flights come frequently, and incipient nest cohorts overlap in the timing of their appearance.

The season begins on the evening of 4 May. A weather front passes over Tallahassee, and a torrential rain soaks the ground. On 5 May the first wave of founding queens takes flight, mates, and then digs in (Figure 31.2). The nests they found become active on the surface after 10 June. Fifteen mating flights of significant size follow before the supply of alates is finally exhausted in early August. The last cohort emerges to the ground surface in early September. The number of incipient nests (and by implication newly mated queens) is about six to seven times greater in the earlier cohorts than in the last ones, declining gradually through the season. In the race for space, queens founding early in the season have greater value for the colony that produced them. Conversely, later queens have less value because the incipient nests already present have claimed an increasing share of the space. These earlier

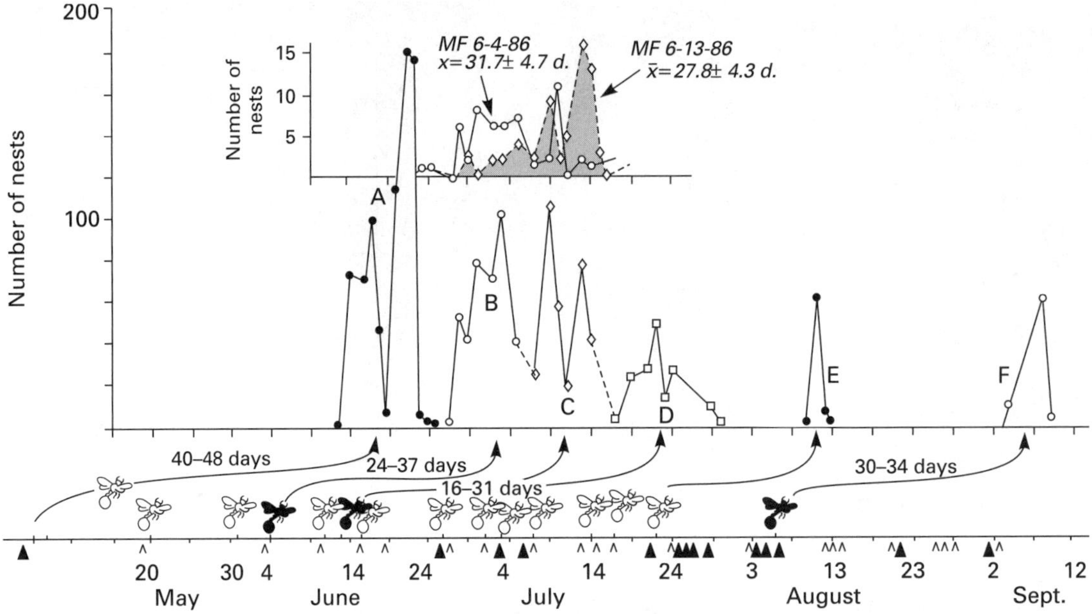

Figure 31.2. Each heavy summer rain triggers at least one mating flight, so newly mated queens settle in waves. The arrows and flying queens associate peaks in incipient nest emergence with earlier mating flights (although dates of emergence may overlap somewhat). Filled flying queens indicate flights for which founding chambers were tagged. Carets indicate rain and filled carets heavy rain. Inset shows emergence for chambers tagged for mating flights on 4 and 13 June 1986. Adapted from Tschinkel (1992a).

incipient nests are also likely to be large, and thus to have a competitive edge over come-latelies. The function of late cohorts is essentially to hedge possible failure of the early cohorts. The distinctly seasonal appearance of incipient colonies was apparent even to some of the earliest observers of fire ants (Green, 1962).

Incipient Monogyne Populations and Self-thinning. No matter when the queens fly, their chances of long-term survival are not great. Death comes in many guises and at this stage is one of the most important processes shaping the population. An unknown number of queens succumb to predators before even attempting the excavation of a founding nest. Of those queens that survive the mating flights to seal themselves in founding chambers, only 20 to 25% make it through the founding period. Survivors of the first cohort of about 2500 founding chambers establish a population of about 650 incipient nests on the 1200-m^2. This is the highest number of colonies that will ever exist on this plot during the entire life of the population. These incipient colonies immediately begin to "die" at the rate of 5–6% per day. Although true death must certainly be visited upon a substantial number of nests, most "deaths" are the result of loss of a brood raid. Each subsequent wave of newly mated queens replenishes the population of incipient colonies, so between about mid-June and mid-July, the population hovers between 500 and 650 incipient colonies, about 0.5 colonies per square meter. Each cohort also immediately begins dying at the rate of 5–6% per day. Only after mid-July, when mating flights become smaller and less frequent, does the population of incipient colonies begin to decline. By late August, 260 colonies are still alive, a density of about 0.2 nests per square meter, about half the peak density. Over the entire flight-founding-incipient period, a total of about 1700 incipient nests appear, a density of 1.4 nests per square meter (though the number alive at any time never exceeds a mean of about 0.5 nests per square meter). Each of these incipient nests is the survivor of about four to five attempts, and each attempt was made by one to several queens. From the relationship between nest density and queen number, we can deduce that perhaps 10,000 to 20,000 newly mated queens in repeated waves make their stands on this small plot in the course of one summer, establishing a population that never exceeds 650 live colonies. Life is not kind to fire-ant queens. By far the most important process shaping the early history of this fire-ant population is brood raiding, the first order of business for the minim workers after they open their founding chamber to the surface. Studying brood raiding in the field stretches both patience and thermoregulatory capacity. Every incipient nest on the plot must be checked for brood-raiding activity at half-hour intervals between early morning and midday, a total of five to seven checks per day. Raids are visible in the form of brood-carrying workers on odor trails between incipient nests. These raids peak in midmorning and decline as the sun heats the soil to lethal temperatures in the afternoon. As each cohort of incipient nests becomes active on the surface, they engage heavily in brood raiding, conducting over 100 raids on several days. More incipient nests are continuously making their appearance and also raiding. After a brief raidless period, synchronized with the emergence of the cohort founded in late June, a final synchronized burst of intense raiding follows in mid-July during which over 350 raids occur in a single day. Few further raids occur that summer, perhaps because the cohorts emerging in mid-August and early September are small.

The usual outcome of raiding among colonies is the demise of all but one of the colonies. Over the entire season, raids (1776) outnumbered incipient colonies (1690), and the same was true for most cohorts. In other words, on average, each colony participated in at least one bout of raiding, and many probably took part in multiple bouts. Raids result in the extinction of 42% of the raiding colonies. From this figure, we can calculate that about 750 "deaths" of incipient colonies resulted from raiding, about 60% of those that "died."

Observations of brood-raiding behavior leave little doubt that it is a form of competition. From their founding-chamber redoubts, queens send out their tiny minions to steal the very future from their neighbors, who do not take the theft lying down but return the thievery. The stakes are extreme, and the winner takes all—brood, workers, and future prospects—although the losing queen has another chance if she migrates with her workers. Cooperation certainly operates *within* nests, but it is hard to see events *between* nests as anything but a desperate competition among queens through the agency of their workers. The nature of brood raids is such that queens with more workers initially are more likely to win raiding bouts, enlarging their nests still more. Competition, in other words, is asymmetric—larger nests are likely to win more than a proportionate share of the resources (brood and workers). Such competition should increase both the size of some nests and the size differences among nests, but as the size of a few nests grows, the size differences may also decrease through the deaths of the small, losing nests. Theory is one thing, but what actually happens in the field?

One would expect that the closer the contesting neighbors are to one another, and the larger they are, the more likely they are to compete, and the more intense will be the competition. Among sessile social insects, competition has been invoked to explain regular spacing and correlations between colony size and density. Prior to the experiments I am about to describe, no social-insect biologist had ever thought of testing the importance of spatial patterns to competition and population development by simply planting ant colonies in desired patterns. Plant ecologists have been doing such experiments for years.

Experimental Studies of Self-thinning. Eldridge Adams and I carried out a series of experiments in which we farmed ants as though they were rutabaga seedlings to be set out in the field (Adams and Tschinkel, 1995a, b, c). The metaphor is more appropriate than it might first appear, because ant colonies do share some ecological properties with plants—they are founded by dispersed propagules, are more or less sessile, compete primarily with their neighbors, grow by adding modules, and derive their resources only from their neighborhood. We thus allowed thousands of ant "seeds" (newly mated queens) to found colonies in laboratory test-tube nests. When the first minim workers appeared, we transferred each colony to a small, hollow plaster block and planted these in a field cleared of other fire ants and most plants. Within a day, the workers opened these nests and began to scout the neighborhood, just as a naturally founded nest would have. Between 86 and 96% of them survived planting, considerably better than Farmer John's results with lettuce seedlings.

The first experiment simply tested the effect of density, that is, distance between incipient colonies, on brood raiding/survival. We planted groups of single-queen, incipient nests of random sizes in square arrays of 25 nests, pairing high and low density in eight replicates. Plots were checked regularly for raids and colony survival. After

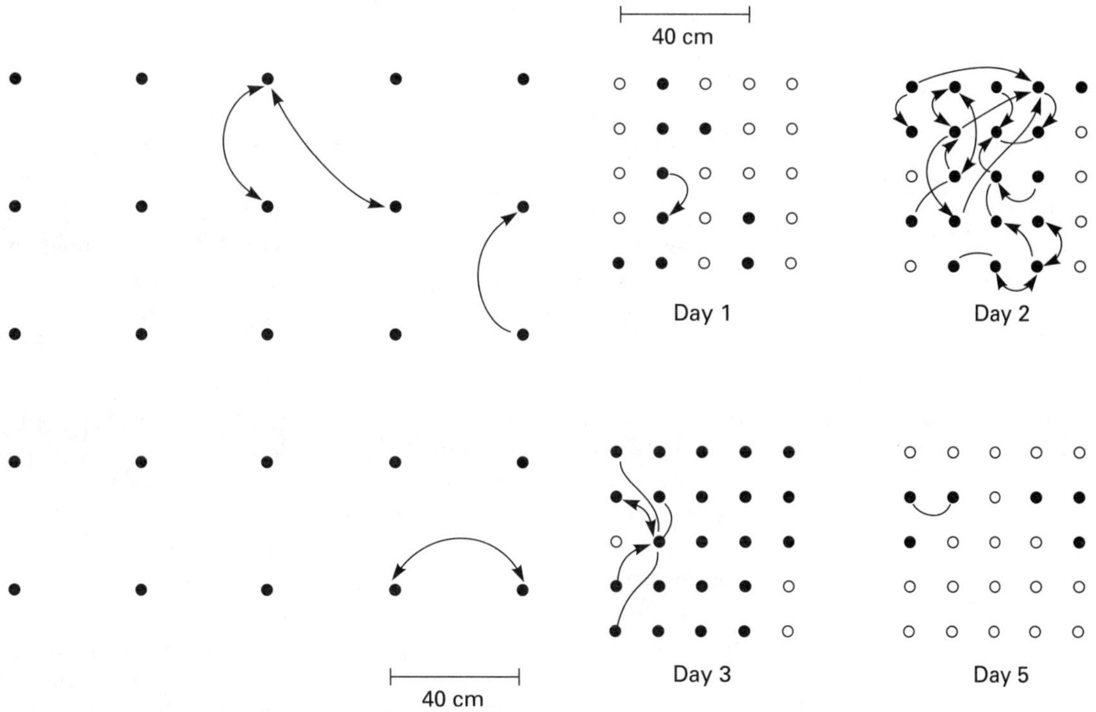

Figure 31.3. Brood raiding among planted incipient fire ant nests. Left panel: typical cumulative patterns of brood raiding on a low-density plot. Lines connect raid opponents; arrows show direction of brood-carrying workers. Solid circles show colonies that survived planting, open circles those that did not. Right panel: typical sequential maps of brood raids on a high-density plot. Lines without arrows indicate worker traffic without brood transport. Solid circles indicate colonies active on each illustrated day. No raids took place on the fourth day. Reprinted from Adams and Tschinkel (1995a) with permission from Blackwell Publishing Ltd..

the first replicate, we realized that raiding occurred primarily after rains, when the soil was damp, so we carried the farming analogy one step further and watered our delicate crop daily with sprinklers.

Simply put, the closer the colonies were to one another (the higher the density), the more likely they were to raid one another, verifying a causal connection between raiding and nest density. When the distance between nests was 1.2 meters, no raids occurred. At 60 cm, about 8% of the colonies raided; at 40 cm, about 22%; and at 13 to 15 cm, 68%. The mean number of raid opponents increased in parallel from one to 1.2 to about 2.1 at the closest spacing (Figure 31.3). As expected, the median life span of colonies declined from the lowest densities (17 to 22 days) to the highest (6.2 days). These numbers are in rough agreement with the theory that competition ought to be inversely proportional to the square of the distance between competitors. Certainly, the higher the density, the more intense the competition and the faster the population thins itself. At high densities, complex, multinest, simultaneous contests were common, whereas on low-density plots, most raids occurred between pairs of colonies. In all arrays, the nests within the core of each array raided over one-third more frequently and survived 25% fewer days than those on the edges, because edge nests have fewer neighbors.

Raiding carries a second expected outcome, namely that the variability of worker number increases, because some nests grow at the expense of others. Although uniformly variable initially, worker number is almost three times as variable on the high-density plots as on the low after raiding is complete. The consequence of this reassortment of workers is apparent—on high-density plots, more queens leave their nests and try to enter surviving nests. These are probably the losers of brood raiding contests. Their losses are not yet final, for they occasionally succeed not only in entering a surviving nest but in winning the contest for reproductive supremacy there. The population consequences of this struggle are difficult to judge.

The spacing of naturally founded colonies is not regular as it is in our arrays but varies from random at low queen densities to clumped at high densities. Regular arrays confound distance between nests with density, because the two vary exactly in parallel. Let us therefore plant incipient colonies in two kinds of arrays, both with the same overall density and outer dimensions. In the first, each of the 52 colonies is regularly spaced 50 cm from all of its nearest neighbors. In the second, the colonies are planted in eight more or less distinct clumps in which some neighbors are much less than 50 cm apart, some much more, and the distances between clumps mostly exceed distances within clumps (Figure 31.4). In these plots, density varies on a scale smaller than the plot, so this experiment will tell us the scale on which the minim workers perceive their neighborhoods during brood raiding.

An incipient nest in the clumped plots is more than twice as likely to brood raid as one in the regular plots (40% versus 19%), even though the overall density is the same. As a result, nests "died" faster in the clumped plots than in the regular ones (at ten days, 38% versus 22%). The minims' perception of neighborhood seems to be extremely local—in only 7% of raids were the opponent colonies more than 50 cm apart. Even when more than two colonies participated in raiding contests, only one-third of them were more than 50 cm apart and none more than 150 cm. The world of a minim fire-ant worker, not surprisingly perhaps, is tiny. Why should we expect otherwise of a creature whose body length is barely three millimeters?

What is much more surprising is that its world does not remain so tiny. At the conclusion of a brood raid, the victor's ranks are swelled when the losing minims defect to the victor's nest. Because numerical superiority is the chief determinant of success in brood raiding, each win makes the next victory more likely through the accretion of losing workers. When the size difference between raid opponents is extreme, raiding is not a contest but simply a one-time plundering of brood,

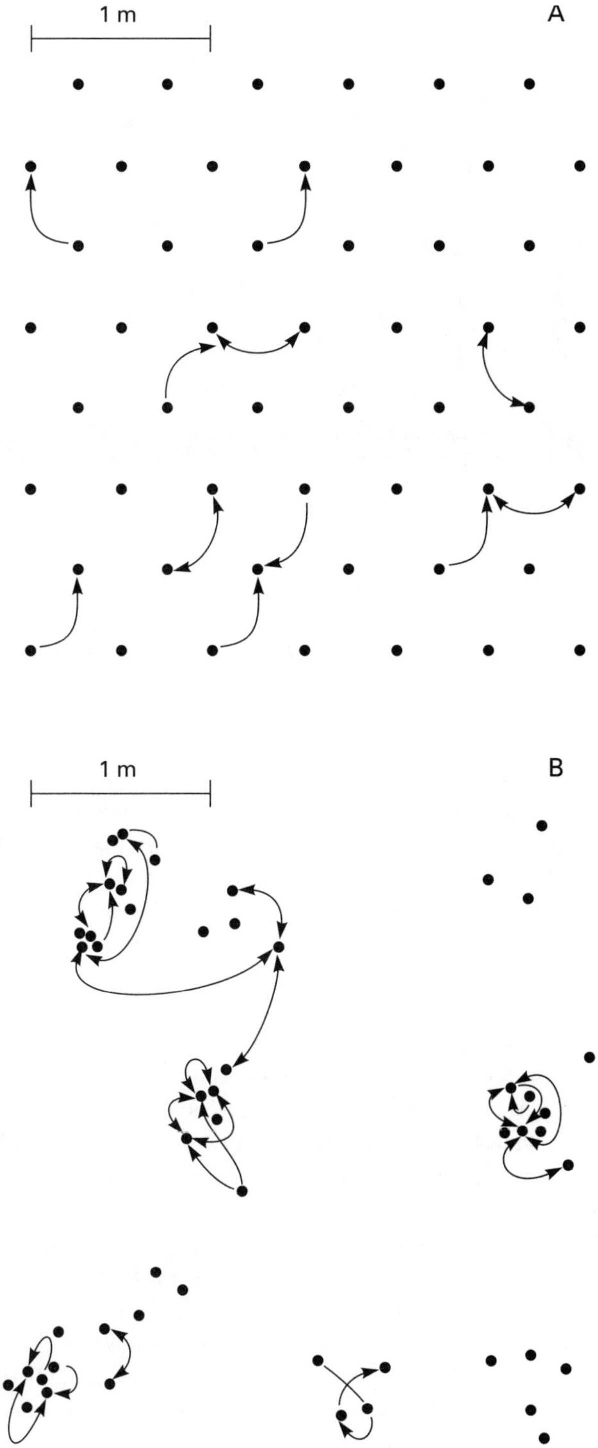

Figure 31.4. Brood raids among regularly spaced (A) and clumped (B) colonies of the same average density. Lines connect colonies between which workers moved, and arrows show direction of brood transport. Reprinted from Adams and Tschinkel (1995c) with permission from Springer.

Figure 31.5. As summer and raiding progress, raiding contests increased in size (involved more nests) and distance. The number of raiding contests on which each point is based is shown in parentheses. Histograms show the frequency of raiding contests with various numbers of raid opponents throughout the summer. Eight of nine raiding contests longer than 100 minutes occurred after July 17. Adapted from Tschinkel (1992a).

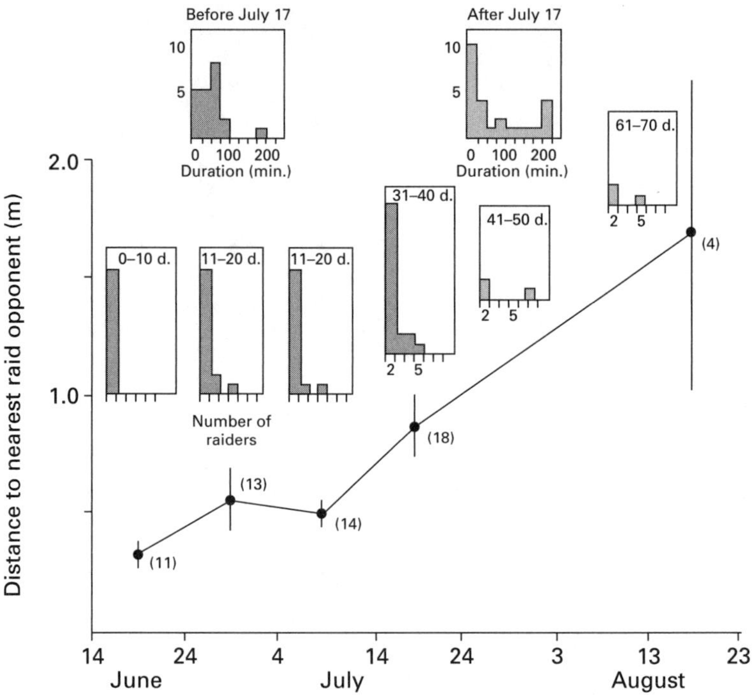

without return traffic. From mid-June until late August, raiding contests on our survey plot grow in size, in the number of participants, and in the raiding distance (Figure 31.5). Of the nine raiding contests lasting longer than 100 minutes, one occurs in the first half of the season and eight in the second half. Early in the season, raiding nests average only about one-third of a meter apart, but by the time raiding ends, this average increases to about 1.7 m. The final raid extends over 3.3 m. The complexity of raiding contests increases too—during the first 10 days, no raids involve more than two nests, but by the last 10-day period, five of 21 contests involve more than two raiding opponents. Raiding is a snowballing, self-catalytic process in which the size increase resulting from a win allows colonies to raid farther, longer, and more often. On rare occasions, this process may lead to truly enormous raiding contests on a mind-boggling scale (for the workers), in which thousands of workers raid dozens of colonies over distances of tens of meters in contests that last for weeks. But why should a nest with more workers raid farther? The workers are not larger; they are simply more numerous. Are larger multitudes simply more likely to include a few really adventuresome workers? Or do workers in larger nests feel emboldened to venture farther? Although such questions remain to be answered, raiding doubtless enlarges the world of the incipient nest.

If raiding is the chief cause of nest mortality and spacing is chiefly controlled by raiding, then logically clumped high-density incipient populations such as those in our 30- by 40-m plot should become ever more regular in spacing as raiding eliminates the closest colonies first. Indeed, a computer simulation confirms this expectation, but the real population did not behave in this way, a strong indication that ants either do not read the theoretical literature or refuse to abide by its assumptions. Although all four naturally founded cohorts of incipient nests were significantly clumped, mortality struck nests randomly and therefore did not change the nature of the spatial distribution, which remained clumped for most of the 60-day period. Even in the clumped planted arrays, the obviously neighborhood-focused brood raiding did not create regular spacing of the survivors.

How can we reconcile these observations? First, workers do not actually follow our rules very closely and often bypass their nearest neighbors to raid more distant ones (Figure 31.4). This situation might arise because workers have imperfect information about their neighborhoods, because they use criteria other than distance, because of some reason we have not yet thought of, or just by chance. Second, nonraiding mortality probably strikes without regard to the proximity of neighbors. Third, in naturally founded populations such as the 30- by 40-m plot, queens may be choosing better nest sites—those that improve their survival—causing founding nests to clump, as observed. This may be the purpose of the prolonged pedestrian search that queens often engage in before digging their founding chambers. This clumping also increases cooperative founding, which also contributes to nest survival and might counteract the negative effects of being close to neighbors. The effect of these factors is to loosen the bond between mortality and neighbor proximity enough that early nest mortality does not lead to regular spacing of colonies.

Yet mature colonies are significantly regularly spaced. The colonies in our Southwood Plantation populations are much less likely to be less than 3 m from neighboring colonies, and much more likely to be 5–10 m from them, than are randomly distributed points. When does this change in spatial distribution take place? It has not occurred during the first 60 days of the population's life, during which 90% of the incipient colonies died. The mutual repulsion of colonies is probably the outcome of territorial establishment, combined with colony movement, and is the next distinct phase in the population's life cycle.

Let us review the history of our young population so far. A rain of thousands of queens, 1.4 per square meter in the course of the season, arrived in waves to found a population of incipient colonies. Brood raiding and natural mortality immediately began thinning this population at a rate of 5–6% per day, so only 15% of the incipient colonies remained alive after two months, when brood raiding ended. At this time, density of surviving incipient colony was about one per five square meters, and the daily mortality rate dropped to about 1–2% per day. Roughly estimated, the biomass of fire ants per square meter of ground began at about 10 mg and increased to about 20–30 mg by the end of the raiding period.

No one has tracked a population from its infancy, where we now leave it, to its maturity, where we will rejoin it momentarily. Mature populations of fire ants support about 600 mg of ants per square meter and densities of one colony on every 60–70 m^2. Clearly, our infant population has a long way to develop. As the workers in the tiny colonies scour their neighborhoods for food and convert it into more ants who scour for still more food, the population's biomass will eventually increase by 25-fold or more, the number of live colonies will drop by more than 90%, and the area commanded by each colony will increase over 13-fold. The 650 incipient colonies once simultaneously alive on our 1200-m^2 plot dwindle to 250 at the end of raiding and continue to die until 18 or fewer remain at maturity, but how these changes occur has yet to be determined.

Instead of the history of a real population from infancy to maturity, we have a computer model (Korzukhin and Porter, 1994). By 1994, many of the important life-history parameters and spatial relationships could be assigned values, or at least bracketed. A model could be devised in which virtual colonies were allowed to found, raid, grow, and die on a virtual hectare, until the space was fully occupied by virtual territories that competed with one another. The resulting virtual population of colonies showed most of the known trends in young populations (described above) but also gave insights into the "adolescent" phase of population growth. The time it took to occupy the space fully was not much affected by whether all colonies were founded by a single cohort or they continued to be founded for extended periods, nor did incipient-nest mortality affect the outcome much. Under continual founding, the importance of being in the earliest possible cohort was very apparent (Figure 31.6). Founding one week earlier increased a colony's chances of survival by 40 to 80%. Virtual growth rate was important to success and survival. After very long times, two distinct episodes of reduced territorial coverage ensued, when the first- and second-generation queens died and were replaced with new colonies, but the third such episode was too spread out to be detectable. All in all, Korzukhin's spatial model behaved in a general way much as real fire-ant populations do, creating expected patterns on time courses that resembled those of real populations. Its predictions must still be compared to the behavior of real populations.

Survivorship and Age Structure in a Young Population. In the Tharpe Street study described above, the plots were almost completely saturated with young colonies in a single year. This population is essentially an even-aged cohort of colonies and very convenient for the

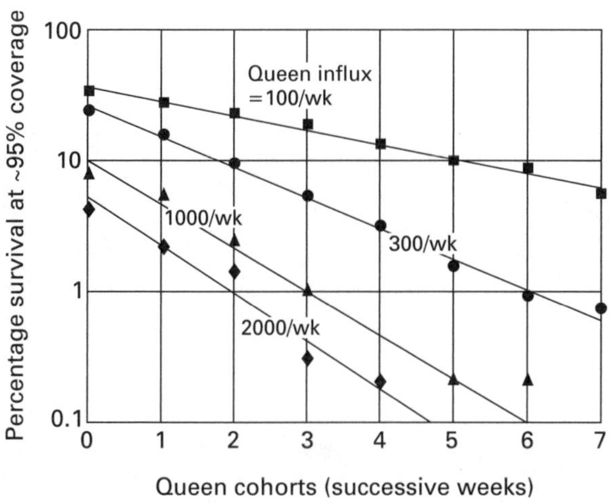

Figure 31.6. Results of a computer model of a hectare of fire-ant colonies. Founding success depended strongly on being in the earliest possible cohort, no matter what the founding queen density. Each line represents survival under different weekly rates of queen influx (per ha). Each point is the mean of 10 runs of the model. Adapted from Korzukhin and Porter (1994).

computation of colony survival, our current subject. Figure 31.7 begins with a wave of 1000 newly mated queens that eluded hungry birds and dragonflies, landed in the plot, and safely ensconced themselves in 450 claustral chambers (some of which they shared with other queens). The number of this original cohort still alive in each stage is shown on a logarithmic scale. Note that the x-axis is population stage and that the stages differ in length. Only 110 incipient nests result from the 450 founding chambers, a nest survival of 24%. Many of these may still contain more than one queen, but by the end of the raiding stage, all surviving colonies are monogyne, and only 17 queens (1.7%) of the original 1000 are alive, and these are found in 17 nests (3.7% of the original number). After another month, only three queens and colonies remain alive, and after three years, a single mature colony is the sole survivor of the original 1000 queens. During all but the last survey, colonies were too small to reproduce. Future reproduction requires survival of this prereproductive period.

If a mature colony "chose" to have all of its female alates settle on vacant land to found a new population, how many progeny colonies could an average mature colony expect to produce? In our alate-trapping studies at Southwood Plantation, the mean annual female alate production was about 1500, and in Morrill's (1974b) study, it was about 1700. If colonies reproduce for about six years before dying, the average colony produces about 9000 female alates in its lifetime. If all these female alates settle on land free of fire ants, 0.1% of them are likely to survive to produce nine progeny colonies, a ninefold increase in population.

We can make some educated guesses about the age structure of adolescent populations. After the founding year, colony survival in our new population is probably fairly high, new colonies rarely appear until

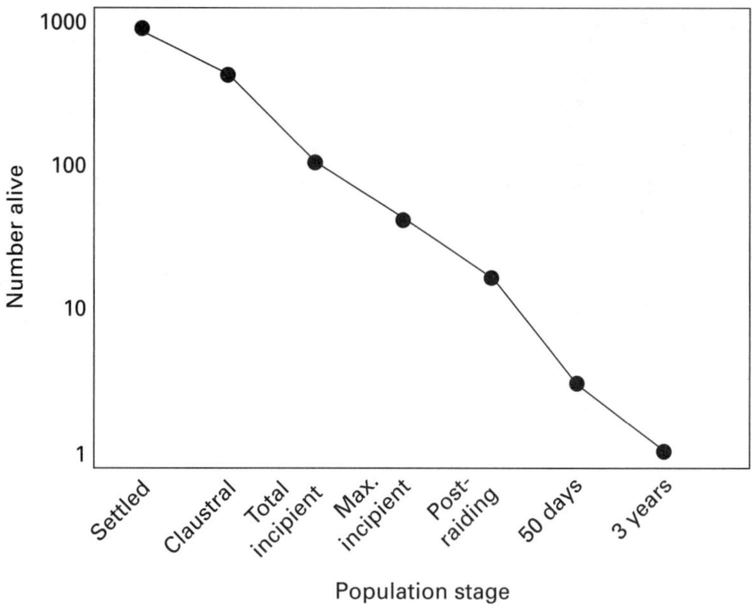

Figure 31.7. Survival of a cohort of 1000 newly mated queens through the stages of colony founding to maturity. "Max. incipient" indicates the time at which the greatest number of incipient nests was alive. Data from the unpublished Tharpe Street study.

the initial cohort begins to die. Then younger cohorts of colonies will appear, eventually producing an age pyramid characteristic of a mature and stable population. Such an age structure is presented in the next section.

Population Processes in a Mature, Stable Monogyne Population

Fortunately, we know a good deal more about old and stable populations, thanks to a study Eldridge Adams and I conducted at Southwood Plantation, near Tallahassee. There we established 12 circular plots, each 72 m in diameter, in an 18-ha cow pasture. This population had been undisturbed since fire ants first appeared in the late 1960s, so we assumed that its current condition was the outcome of natural processes. Three times a year, with the help of a small army of part-time helpers, we mapped the location of each colony within all 12 plots, using a compass and a meter tape pinned at the center of the plot. Each colony was tagged and its mound volume measured, as a surrogate measure of worker number and colony biomass. We added new colonies to the map as they appeared and noted colonies that had died or moved. The resulting series of 16 sequential maps allowed us to track individual colonies, following the births, early survival, growth, movement, and deaths of the 600 to 1200 colonies over five years. We can link colony fates to the neighborhoods they live in and to a number of population characteristics and changes. The analysis is not yet complete and has not been published as a research paper, but some of its early findings are revealing enough to be included here.

Annual Cohorts. Let us start with colony "birth." Newly founded colonies are detectable after 6–10 months, when they have grown to about 5 to 10 thousand workers and build the first visible mound, measuring a mere 0.02 liter, over their nest, popping up like mushrooms after a rain. Variation in founding date, survey date, mortality, and colony growth caused mound volume at first detection to vary greatly—80% of mounds fell between 3.5 liters (21 g of ants) and 0.13 liters (2.7 g of ants), and the median was about 0.7 liter (7.8 g of ants). For practical purposes, we defined new colonies in our May surveys as those smaller than 25 g.

New colonies that are too large to have been recently founded also appeared in every survey. Most of these were near the outer edges of the plots and probably migrated in from the unmapped surrounding area. Occasionally large colonies popped up in the plot's interior, where migration was unlikely. Spontaneous generation is a tempting explanation, but the origin of such colonies remains mysterious.

As expected, recently founded colonies appeared very seasonally (Figure 31.8). Colonies founded between May and July were detected as cohorts of 80 to 220 the following January through March. During other seasons, we never detected more than about 50 new colonies in all 12 plots together. This strong reproductive seasonality means that a mature population consists of overlapping annual cohorts, each produced mostly between May and July. Tracking such annual cohorts gives us a

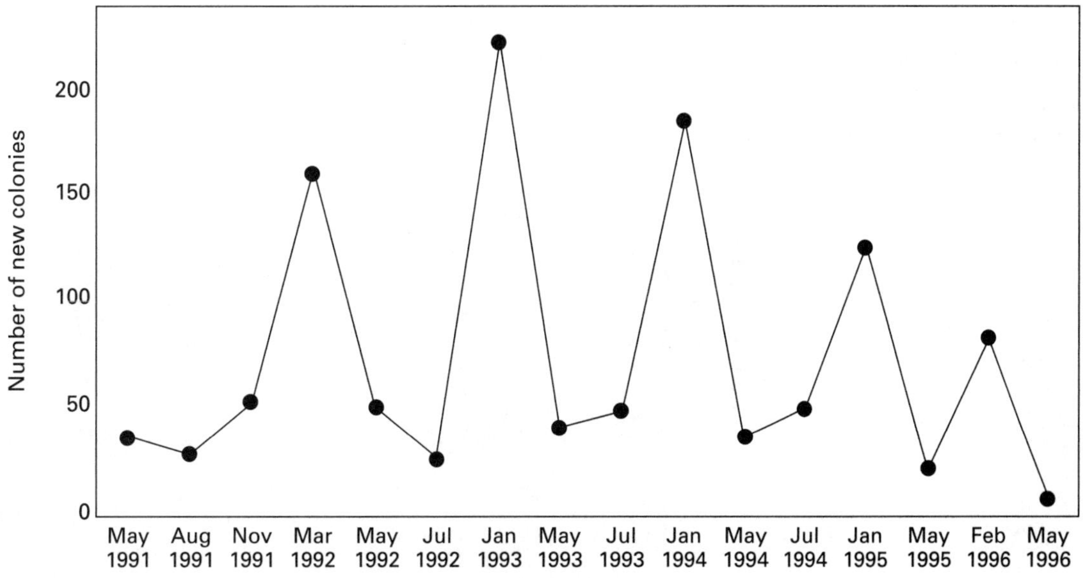

Figure 31.8. Colony founding was highly seasonal during the five years of the Southwood Plantation study of a mature fire-ant population. Colonies founded during the spring and summer that survived through January appeared as distinct peaks of new colonies.

meaningful and convenient way of understanding the fates of colonies and the nature of the population.

Some years are better for colony reproduction than others. New annual cohorts peaked in 1993 and declined thereafter (Figure 31.8). Either more newly mated queens were available in the spring of 1992, or the population of mature colonies was lower in that year, allowing more founding nests to survive. Generally, the number of new colonies was higher when the total ant biomass in the previous summer was lower, suggesting that more founding queens escaped destruction by mature colonies, but without a longer time series of samples, we can only speculate.

A Population of Cohorts. The colonies of each annual cohort grew, reproduced, moved, and died, describing the history of the cohort, and the sum of these patterns over all cohorts with live members described the entire population. In the best of all worlds, we would have tracked several annual cohorts until all their members had died. In the real world, leaseholders of cattle pastures can (and did) terminate long-term studies with a tractor and a set of discs when some of the colonies of even the first cohort were still alive. The best we could do was to build a composite picture of a typical annual cohort. Figure 31.9 shows the size distribution of all colonies in the population in 1991. Colonies smaller than 25 g were founded in the spring and summer of 1990, and those larger than 75 grams were mostly at least three years old. Their

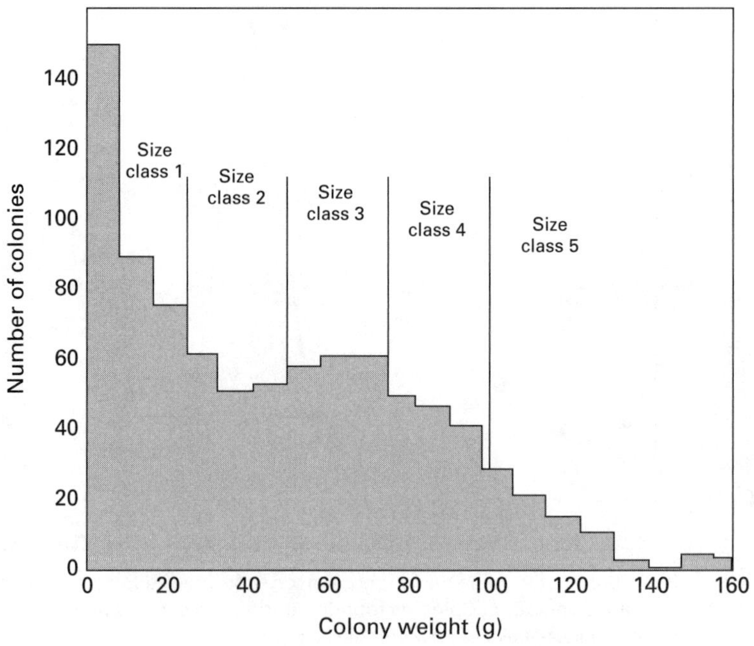

Figure 31.9. Size-frequency distribution of colonies, May 1991. Note the predominance of small colonies and the modal class of average-sized colonies.

exact age was unknown, for in a mature population, the fetters between colony size and age are loosened, much as they are in forest trees.

Survival and Growth of the Cohort Founded in 1990. When the spring/summer 1990 cohort was first detected in the June 1991 survey, the median mound volume was 0.7 liter and the median age 8 to 13 months. Almost a fifth of this cohort of 156 colonies was still alive in 1996. Whereas this cohort shows us the pattern of early growth, survival, and maturation, other cohorts must tell us about old age and extinction.

The new cohort in our May 1991 survey is itself the much winnowed remnant of a much larger cohort of founding queens from the previous year. Thanks to the work of Christopher Jerome, a student of Eldridge Adams, we can make a crude estimate of how many died to produce our cohort. Jerome placed one, two, or four newly mated queens in each of a number of perforated plastic tubes and planted these in our Southwood Plantation plots within the territories of mature colonies (Jerome et al., 1998). The holes in the tubes allowed worker ants access, but retained the queens, exposing these queens to whatever mature colonies doled out. At the end of the claustral period (three weeks), about 15% of the queens still survived, a daily survival rate of 91.6%. Survival thereafter was 93% per day, so after one more month, about 2% of the original queens were still alive. Here, though, we come to a long gap between the end of Jerome's study and our first-year cohort. During this gap, the daily mortality dropped from about 7% to about 0.2% per day (50% per year). Applying survival rates through midsummer, and assuming that the average founding chamber contains two queens, we conclude that the 156 colonies in our 1991 cohort were the survivors of 7800 founding nests containing 15,000 newly mated queens. This estimate is certainly a minimum, however, because of the noted "mortality gap." To estimate the other extreme, consider that, if all the queens produced by the approximately 500 mature colonies on our plots also tried to found there, about 1.0 million newly mated queens would have rained on these plots, of which a mere 0.03% survived (on the assumption of dual chamber occupancy). Even if this estimate is off by a factor of two or three, it is still a mighty low survival rate (though perhaps not unheard of for r-selected organisms).

Death continued to pursue the cohort, though at a declining rate (Figure 31.10). Only about half of these colonies (46%) were still alive in the following June (1992). Mortality threatens small colonies more than large ones, so colonies gradually escape high mortality as they grow. Although the causes of mortality are different, this trend in mortality began in the incipient stage—the larger the colony, the greater the survival advantage. Age was therefore correlated with mortality and size, but in ant competition, colony size is likely to be more important than age. Colonies of less than 15 g were about half as likely to survive as colonies

Figure 31.10. The numbers of the 1991 new-colony cohort that were still alive in successive years. The curve shows the annual mortality rate of this cohort. Mortality rate declines as colonies approach maturity.

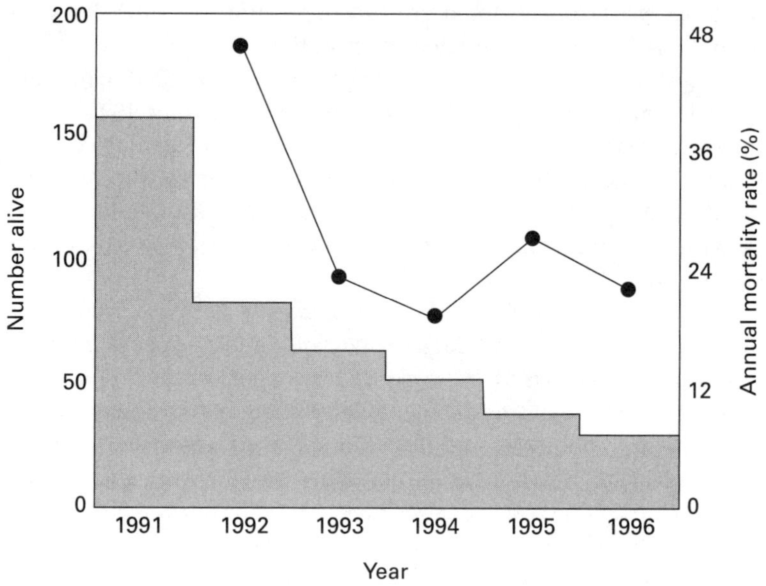

of more than 45 g. When colonies reached the average mature colony size in 1993, the annual mortality rate dropped to 20% and fluctuated around this value through the end of the study. The greatest increase in safety came from moving from the less-than-15-g size class to the 15-to-30-g size class (Adams and Tschinkel, 2001), which increased the chances that a colony lived to see another anniversary by 1.6-fold. The next increase of size class brought only a 1.1-fold improvement, and further increases brought none.

Although the cohort got smaller every year (Figure 31.11), the average size of its members increased rapidly for three years and then stabilized around 50 g, similar to the six annual population means of 42 to 55 g, where it remained until the end of the study. The amount of growth achieved from one year to the next decreased with colony size—the slope of initial colony size plotted against size one year later was always less than 1.0 (Adams and Tschinkel, 2001). The large standard deviations suggest that the annual means hide a great deal of individual variation, as we will see shortly.

Do colonies grow more slowly in a mature population than in a new one? Interestingly, colonies in mature and new populations both reach a size of about 65,000 workers in three years, but whereas those in mature populations grow no further, those in new populations continue to grow for two more years, stabilizing at about twice this size (in June). Perhaps the difference is the effect of territorial competition.

Decline and Extinction of Colonies Already Large in 1991. To learn about the decline and extinction of a cohort, we must follow the fate of colonies that were larger than 75 g in 1991. This group is almost certainly not a single annual cohort, but from the growth of the first five annual cohorts, we can reasonably assume that most of these colonies are more than three or four years old and thus well along in life. By 1996, even the youngest members of this group must be nine to 10 years old. Their size trends and survival contrast sharply with those of the cohort founded in 1990 (Figures 31.12, 31.13). Whereas colonies of new cohorts increase in size and survival, those of the "mature cohort" decrease in colony size and increase in mortality year by year. By 1991, mean colony size has decreased almost 50%, and annual mortality is pushing 35%. Projecting these trends suggests a final extinction of this group in 1998. The last colonies to die would thus have been about eleven or twelve years old, far older than the mean of about seven years.

A life-table calculation of these two cohorts yields the "expectation of life" for each year of age (Figure 31.14). Colonies that survive to one year of age can expect, on average, to live another 2.5 years; those that survive to age two, can expect to live another 3.5 years. Thereafter, life expectancy drops for every additional year of age. Beyond five years, the calculations are less certain because the cohort on which they are based (colonies already of mature size in 1991) was almost certainly of mixed ages. This uncertainty accepted, the expectation of life declines until it is 1.5 years in year 9. Thereafter, the calculation is pure guesswork, but 10-year-old colonies can almost certainly expect to live less than another year.

Why the mean size of older colonies should decrease is a mystery. Most probably, the trend results from the differences in the annual cycle

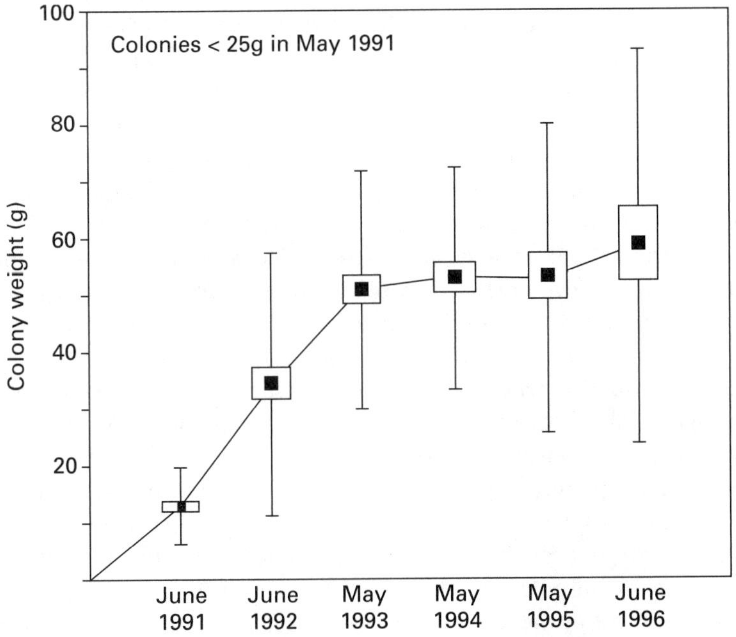

Figure 31.11. Growth of the 1991 new-colony cohort to maturity. Colony size was estimated as total colony weight, computed from mound volume. Boxes show standard errors, bars standard deviation.

Figure 31.12. Colonies that were of mature size in 1991 declined in size throughout the study.

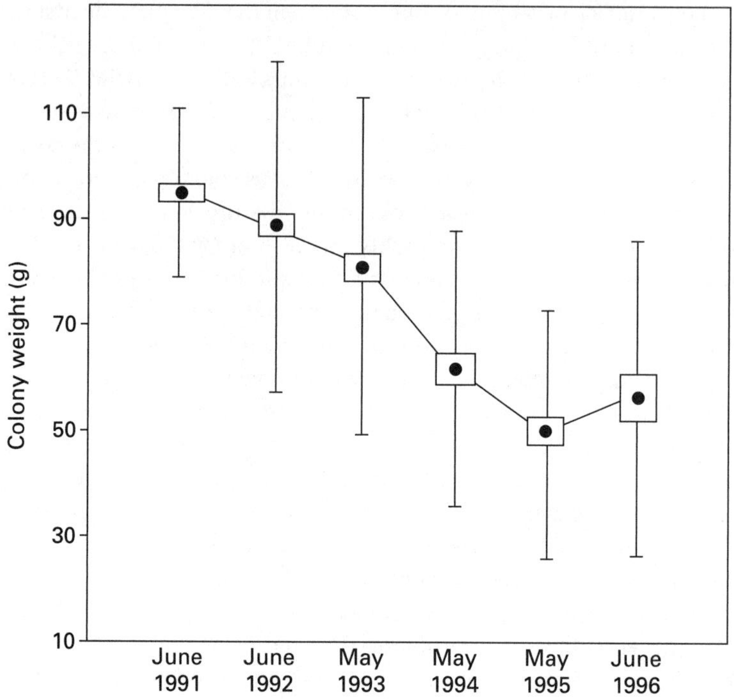

between immature and mature (sexual-producing) colonies. Large, sexual-producing colonies lose territory during the breeding season, whereas small ones gain it. If large colonies were unable to gain back all their losses, their size would ratchet down over several years.

Variation of Individual Life Histories. The size changes of many colonies are not simply variations on the average pattern. Of course, many newly founded colonies do grow to about 40 to 60 grams and see-saw around this size for several years before declining (Figure 31.15A), but others continue growing year by year and show no sign of slowing down even after six years and sizes of over 150 g, three times the mean size (Figure 31.15B). Some colonies in the five-year study exceeded 100 g by the time they were two years old and either stayed large or dropped rapidly to a much smaller size. Some changed size by two- to threefold (either way) in a single year. On the other hand, quite a few colonies grew to submean sizes and remained at these for the entire study. A few staggered along for several years without distinction, then suddenly grew to huge size within one year.

Similar patterns applied to colonies that were mature in 1991. Many colonies that were of average size during the first year remained in this category for the duration of the study, while others declined slowly or rapidly (Figure 31.16A), and a few showed spectacular rebounds. Most of the largest 1991 colonies generally shrank, but even this pattern was highly variable (Figure 31.16B). A few wrested victory from the jaws of extinction to rebound to huge sizes. Why do so many colonies (though not all by any means) not show an orderly tracking of the general pattern but deviate widely and irregularly from it? A number of explanations are possible, most of which will have to be tested in future research projects. (1) Neighborhood competition: a colony with many large neighbors may find itself on the short end of the competitive stick, with a smaller territory and fewer resources, and thus smaller colony size. The opposite is equally possible. (2) Winning the lottery: colonies may come into occasional, unpredictable food bonanzas that allow them to grow rapidly. (3) Greener pastures: colonies may change neighborhoods by moving or through the death of one or more large neighbors and thus be freed from particularly potent competitors and be able to grow rapidly. (4) Individual queens and colonies vary greatly in their capacities for worker production and therefore for producing large colonies. This variation is obvious to anyone who rears fire ants in the laboratory, where some grow like gangbusters, most shuffle along around the average, and a few poke along way behind. (5) Different investment strategy: perhaps colonies that invest a great deal in alates in one year are smaller the next year. No information is available on multiyear reproductive strategies in fire ants.

A Composite Full Life History. I noted the premature demise by tractor of our Southwood Plantation study, before the first-born annual cohort had all gone on to their higher rewards. For a full cohort life history, we must combine the histories of the various 1991 size groups.

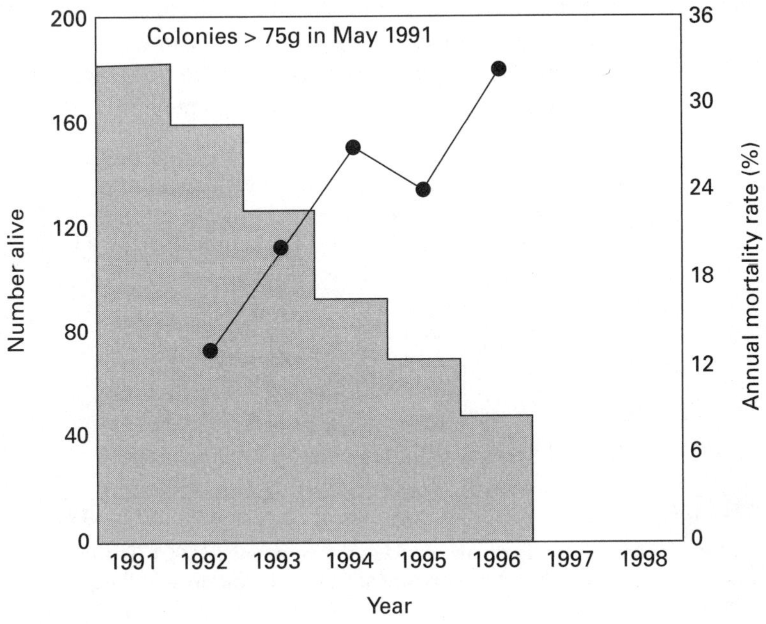

Figure 31.13. The number of colonies already mature in 1991 that survived in successive years. The curve shows the annual mortality rate, which increased with age.

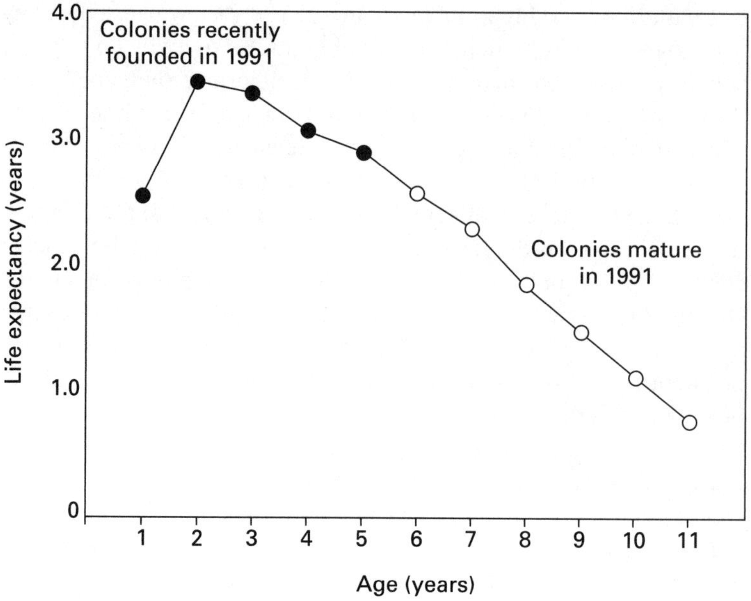

Figure 31.14. Life expectancy of colonies one year old and older. Filled circles represent the cohort that was recently founded in 1991; open circles represent colonies that were mature in 1991.

Assuming that the larger colonies are two to four years older than the smallest, we slide these data along the time axis until their "probable ages" coincide with the same ages in the 1991 annual cohort (Figure 31.17A). The result shows that, at about five years of age, the average colony begins a decline in size and eventually dies. Mortality rate is high (about 50%) in small, young colonies, drops to about 20% in midlife, and then climbs to 35% or more until all colonies in the cohort are dead (Figure 31.17B). Some colonies may live as long as 10 to 12 years.

Age and Size Structure of the Monogyne Population. The age structure of populations is usually presented as an "age pyramid," a graph

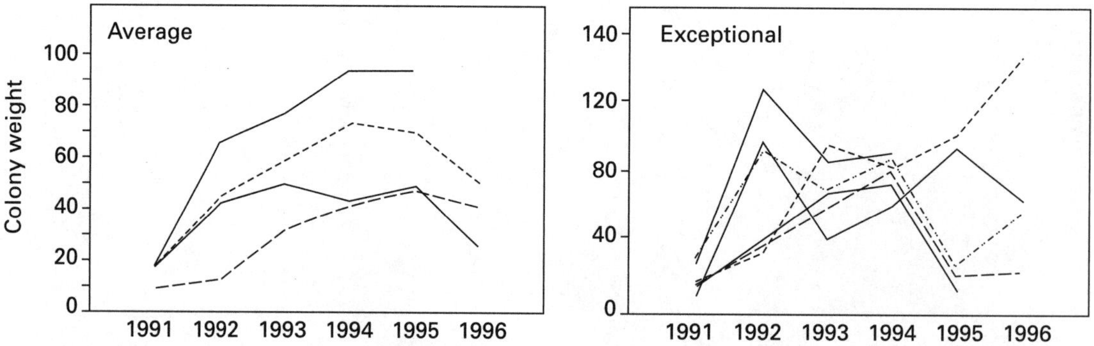

Figure 31.15. Examples of growth patterns of selected individual colonies that were new in 1991. The left panel shows colonies with average patterns, the left some with exceptional ones. Note the difference in the y-axis scales. Growth patterns varied greatly.

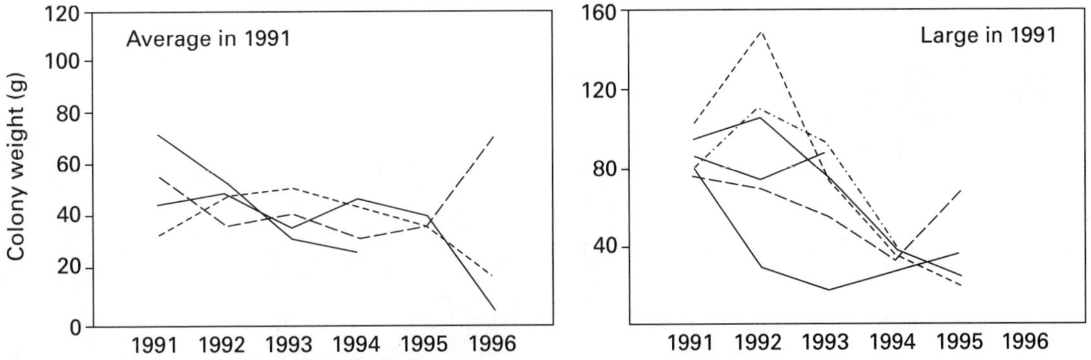

Figure 31.16. Growth patterns of selected individual colonies that were mature in 1991. Note the different y-axis scales. Patterns varied greatly.

showing the number or percentage of individuals in each age class. A population's age structure tells both the history of the population and its future. A pyramid greatly dominated by the youngest size classes indicates strong population growth; one with few members in the youngest classes indicates a population in decline.

The cohort structure of a fire-ant population simplifies the computation of the age structure. During the spring of year n, the colonies founded since the previous July are zero to one year old. In year n + 1, they are one to two years old, and a new zero-to-one cohort has appeared. The Southwood Plantation data reveal the survival rate of each cohort from year to year. The number of colonies of age j present in year n + j is simply the initial size of the cohort times the product of the survival rates of all j years. For example, if the initial cohort included 150 colonies, and survivals during the next three years were 0.4, 0.8, 0.7,

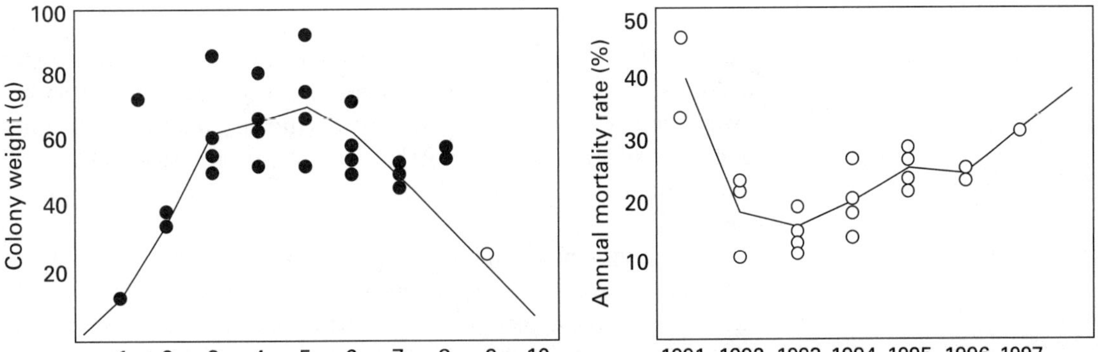

Figure 31.17. Lifetime colony size and mortality-rate patterns resulting from the combination of the 1991 new cohort and mature colonies. Extreme life span could reach 10 to 12 years, but the mean was around seven to eight years. Mortality rate declined for three years, then increased with age.

Figure 31.18. Applying founding, growth, and mortality rates allows the projection of the age structure of the 1991 cohort in year 6 (the last year of the study) to future years.

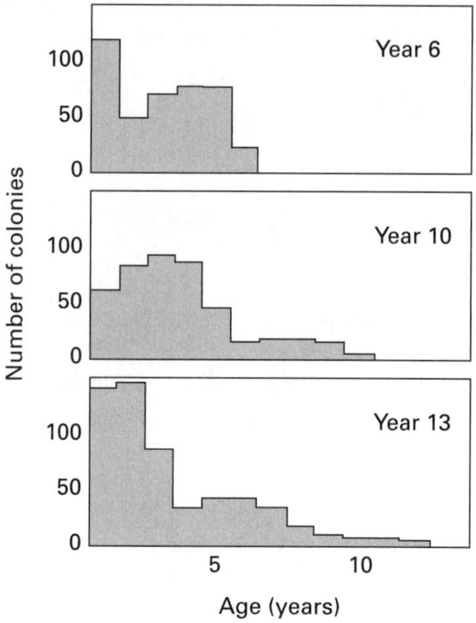

then the number of three- to four-year-old colonies present in year 4 is $150 \times 0.4 \times 0.8 \times 0.7 = 33.6$ colonies. I projected the calculations into the future by simply repeating the observed cohort sizes and using the mean survival rate over all cohorts.

The population age structure is shown for year 6, when about 20% of the first cohort is still alive (Figure 31.18). Colonies present before the beginning of the study are not shown. The next two panels show the age structure at years 10 and 13 when the entire population consists of colonies born after the beginning of the study. Some ages are more abundant and some less, causing the age structure to have a humped appearance. The humps are larger cohorts established during more favorable years. In addition, survival rates vary from year to year, even for colonies of the same age. Nevertheless, in the longer term, our Southwood population is neither in decline nor growing but simply fluctuating a little around a very stable mean. The age structure of the population is unlikely ever to be completely stable over even a few years.

The bonds between colony age and size are rather loose, as can be seen from the size distributions of a typical cohort year by year (Figure 31.19). The colonies within each panel are within a few months of the same age, but by the second year, they may span a fivefold difference in size. Colony size predicts alate production because workers are the machinery that makes the alates. As a consequence, fitness (alate production) is more tightly linked to colony size than it is to age. This pattern is analogous to that in many nonsocial organisms in which larger females are usually older and have more offspring, resulting in an age-size-reproduction correlation. In social insects, however, colony size can either increase or decrease, reducing both the age-size and the age-reproduction correlation. In a fire-ant population, the colony-size structure is therefore probably more important for alate production than is age structure. A representative size structure for summer and winter of 1994 is shown in Figure 31.20. Such size distributions permit certain predictions of population performance if the relationship between colony size and performance is known. For example, a dense population of small colonies is probably different from one of the same biomass distributed in fewer large colonies. We can expect differences in total energy use, forager density, and alate production because all of these are related to colony size in a nonlinear way.

Turnover and Mean Life Span. In nature, stability is a mere illusion on the surface of things. My cat Sniffles may look like the same cat I had

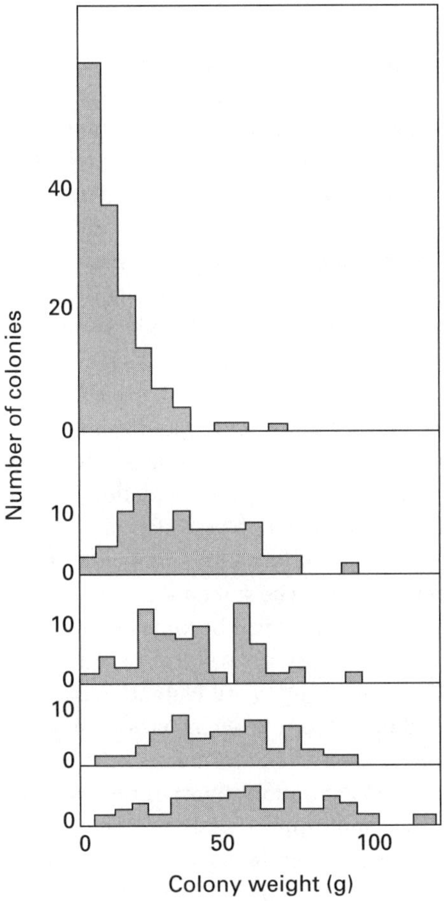

Figure 31.19. Colony-size distribution of the 1991 new-colony cohort during five successive years shows that colony size is not tightly linked to colony age.

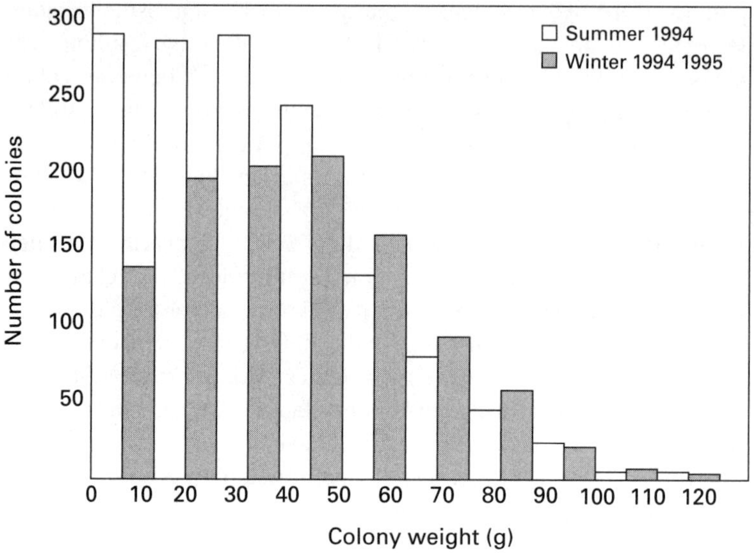

Figure 31.20. Comparing the colony sizes in summer and winter shows the effects of loss of colony size during the spring, when alates are produced.

a month before, but half her hair has been vacuumed off our furniture, and a substantial fraction of her cells and molecules have been replaced by new ones. Similarly, worker ants in a colony are constantly dying and being replaced, and the same is true of colonies within populations as well, as the discussion above makes obvious. When colonies die, they make spaces for new colonies to enter the population. The individuals making up a population thus change constantly, a process called turnover. In a stable population without significant immigration and emigration (as is the case for fire ants at small scales), the average rate at which colonies die is equal to the rate at which new ones appear. Under these conditions, the turnover rate is the reciprocal of the mean life span. We can calculate the turnover rate from the colony life span or vice versa. If one can estimate each independently, good agreement gives the estimates credence. Sperm depletion in queens predicted a colony longevity of about 6.8 years (standard deviation=two years) (Tschinkel, 1987). In a mature, stable population, this figure implies a turnover rate of about 15%. The standard deviation of two years implies that over 80% of the colonies will become "dear departeds" before 8.8 years have passed, and only 4% will live beyond 10.8 years. Tracking colonies directly on our Southwood Plantation plots reveals an annual turnover rate of about 12 to 13%, which yields a life span of about eight years (i.e., the inverse of 0.12), still close to the sperm-depletion estimate. The Southwood Plantation cohort-survivorship data yield a mean life span of 6.5 years and support the extreme longevity of 11 to 12 years (Figure 31.17). This relatively good agreement among these diverse methods indicates that our longevity estimates are pretty good.

Lifetime Alate Production (Fitness) of Monogyne Colonies. The object of the colony's lifetime of struggle is to produce as many daughter colonies as possible, either through female foundresses directly or through the males that mate with successful foundresses. (We speak of "daughter" colonies, but clearly colonies have no sex, or more precisely, they can be of either or both sexes.) Both provide routes for maximizing fitness, as biologists say. Alate production is a meaningful estimate of fitness, but direct measure of an entire lifetime of production has never been made. However, capture of alates during one reproductive season at Southwood Plantation, and my sociometric study (Tschinkel, 1993b), showed that larger colonies generally produce more sexuals. Presumably, that is the point of producing a larger colony. Therefore, as a final calculation from the Southwood Plantation study, let us sum a colony's total spring biomass over all of its reproductive seasons to give a weight that is proportional to the alates produced by that colony in its lifetime. Let us call this cumulated biomass "alate power" and compute it only for colonies that died during this study (to avoid the bias of unrealized future production).

The remarkable distribution in Figure 31.21 shows that a few colonies amass a disproportionate share of the alate power. The median lifetime value is 58 g, but the top-producing colonies have five to 10 times this capacity. If lifetime alate power of colonies is ranked from the lowest to the highest, we get the interesting relationship in Figure 31.22. When the lower half of all colonies has been summed, less than 10% of total alate power has been accounted for. When the lowest three-quarters

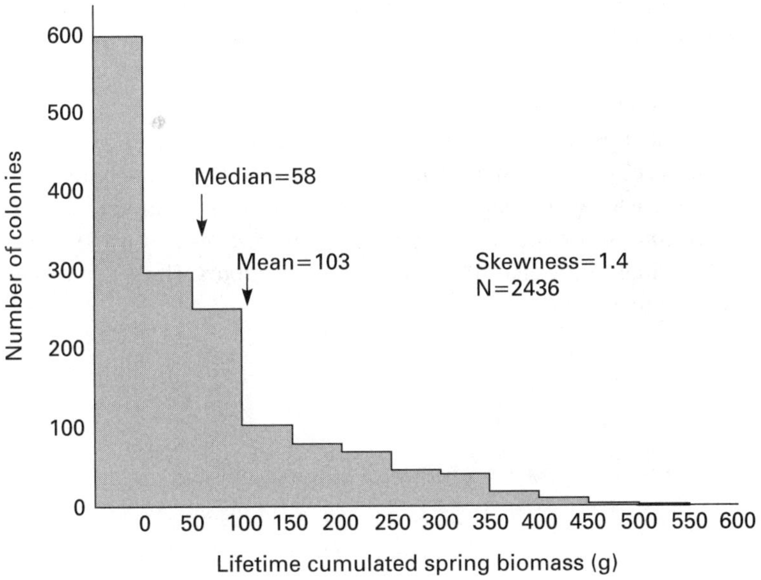

Figure 31.21. Distribution of lifetime alate-production capacity, calculated as the lifetime cumulation of spring biomass. A few very large colonies have a disproportionate capacity to produce alates, up to three to five times the mean and almost six to 10 times the median.

Figure 31.22. A small proportion of colonies produce a large proportion of a population's alates. Colonies were ranked by lifetime productive capacity (Figure 31.21) and their alates and numbers cumulated as percentage of total. The top 15% of colonies have more than half the alate-producing capacity of the population.

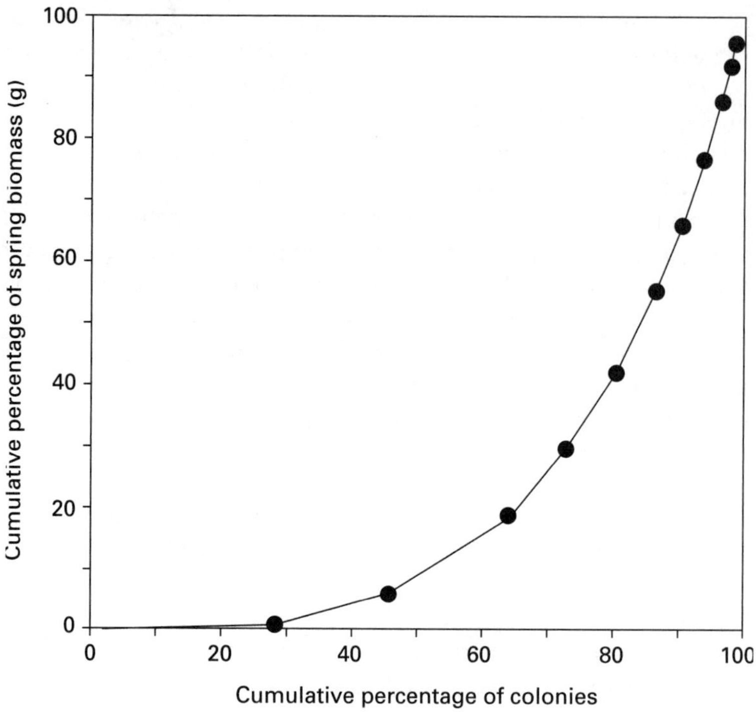

have been summed, we can still account for only 30% of the total alates. In fact, the top 15% of the colonies account for over 50% of the alate power, and the top 5% account for about 20% of the total. Most of the new colonies appearing as annual cohorts are probably the offspring of a small minority of the colonies that ever set foot in the population (so long as immigration is low). Success among fire ants is as unevenly distributed as it is among humans.

With numbers like these, who can doubt the importance of living as long as possible and fighting for the largest possible territory and colony size? Colonies that attain this goal probably leave behind many times the average number of daughter colonies. This extreme skewing means intense natural selection on those properties that lead to large size and high survival in competition with neighbors. The specific traits favored are unknown.

◂•• *Gang Wars*

Imagine yourself sitting at a roadside table in the Apalachicola National Forest in northern Florida, about to have your lunch. As you pick up the paper bag containing the deep fried chicken thigh and drumstick, the bottom breaks and the chicken plops onto the table, scattering crumbs of breading and bits of chicken. In accordance with the three-second rule you learned from your mother, you quickly grab the chicken and put it on top of the broken bag, and you sweep the archipelago of crumbs and chicken bits onto the ground. With this sweep of your hand, you have set the stage for one of life's tiny but raw dramas, and within 30 seconds, it begins to unfold on the ground below.

A tiny speck of a black ant (*Monomorium viride*) approaches one of the crumbs of breading, inspects it, and hurries back the way she came. Within two minutes, more specks arrive at the crumb, and within another minute, the crumb is peppered with several dozen. Meanwhile, a much larger orange ant (*Aphaenogaster floridanus*), elegant, long-legged, and thin like a greyhound, wanders through the crumb field at a steady, unhurried pace, selects a crumb of just the right size, lofts it over her head and walks off. Now both ends of the crumb field have been discovered by workers of a reddish, middle-sized ant (*Solenopsis geminata)*, who measure off several dimensions of the larger crumbs before setting off in opposite directions, dragging the tips of their gasters. At about the same time, a rather nervous, dark ant of about the same size (*Pheidole obscurithorax*) also discovers the bonanza and heads back to her nest, returning within minutes with a flock of recruits, some with very large heads. These swarm over a bit of breading far too large for any of them to carry, but after a minute of what looks like unorganized milling, the bit suddenly seems to float and is borne away over a crowd of workers.

When seven minutes have elapsed, the first recruits of *S. geminata* arrive at the far end of the crumb field, and after eight minutes others arrive at the near end, but from the opposite direction. Workers lay claim to some large bits, which they proceed to dismember and carry off. Some workers from each *S. geminata* group wander among the chicken dust, eventually coming into contact with one another. Momentarily, they freeze, antennate each other, and jerk back. One of the larger workers seizes another by the antennae, and the two grapple and roll, biting and struggling until the smaller worker breaks free and runs back toward her nestmates, dragging her gaster tip to lay an odor trail until she vanishes back into the hole from which she appeared. Soon, a line of workers, many very large, with huge heads, appear out of the hole and march down the odor trail to engage the enemies, who have also recruited reinforcements, but not as many. The battle zone among the crumbs is soon filled with grappling and biting workers. Antennae and legs are bitten off, gasters are severed, and decapitated bodies stand almost motionless, with tarsi aquiver. More recruits rush to battle, but after 15 minutes, the side with fewer recruits begins to lose ground, eventually abandoning the crumb field to the victors, who cut the hunks of chicken flesh into pieces and carry them back home. *Solenopsis geminata* is an extirpator that tolerates no members of other *S. geminata* colonies within its foraging area and fights to drive them out.

As the *S. geminata* consolidate their hold on the crumb field, they reach the zone initially occupied by the tiny black *M. viride*. As the two meet, the *S. geminata* seem occasionally to jerk back, then stop to clean their antennae or stumble about in a disoriented fashion, but as they advance, the *M. viride* invariably melt away, until *S. geminata's* conquest of the crumb field is complete. *Solenopsis geminata* clearly extirpates not only workers from other colonies of its own species but other species as well. *Momorium viride*, like *P. obscurithorax*, is an opportunist that relies on speed, not fighting power, to get its share of food. Rather than engaging the extirpator in a futile battle, it quits the field to be first at another patch of food.

Unnoticed by the combatants, tiny black ants, some with disproportionately huge heads (*Pheidole moerens*), have sneaked in to carry off tiny bits of chicken and breading. These are insinuators, who wander among the larger *S. geminata* without challenge. Occasional workers of the elegant, thin *A. floridanus* also continue to wander into the crumb field to pick up and cart off pieces. Several dull red ants (*Trachymyrmex septentrionalis*), their bodies studded with bumps and spines, also wander through, apparently oblivious to the murder and mayhem that surrounds them. They carry what appear to be caterpillar droppings. Climbing over a crescent-shaped mound of sand, they vanish into their nest hole a meter from one of the table legs.

The drama that has just been played out at your feet is easily overlooked because the actors are so small, but it has revealed several tactics ants use to compete, both within and between species, for the stuff they need: space, food, and nest sites. In comparison with that between the more familiar vertebrates, competition among ants can be particularly violent and more often lethal, as you have just learned through your observations of the Deep-fried Wars. Why is this the case? Although the colonies are locked in competition, only the sterile workers actually engage in the nitty-gritty of the contest. Their entire reproductive future is tied up in the queen back home safe in the nest. For the colony, the death of a worker is only a loss of a small investment in a laborer and does not threaten the colony's reproductive future. If the fitness of the colony is increased by the death of the worker, because, for example, it allows the colony to attain a large amount of food, then the death may have more than paid for itself. And in another demonstration of nature's parsimony, ants often eat the competitors they kill, blurring the lines between predation and competition (killing two ants with one stone?).

A second interesting feature of ant competition is that ant colonies resemble plants in several ecological features (Brian, 1965). Like plants, ant colonies remain more or less in one place and therefore compete only with their neighbors. Both harvest resources from a fixed space and bring them to a central place. Both grow by adding modules, and both can shrink by shedding modules. The sizes of both depend on the size and number of their neighbor-competitors and the area from which they draw resources, and both struggle with their neighbors for use of that space.

Although the form of competitive struggle is particular to ants, the struggle is universal. Each organism, be it plant, microbe, or animal, needs resources to live and reproduce, and one of those fundamental, if unpleasant, facts is that some of those resources are always in short supply, limiting the organism's growth, survival, or reproduction. A second fundamental, and again unpleasant, fact is that organisms must therefore struggle with one another over these key resources. One's loss is another's gain, and the degree to which an individual or colony succeeds in wresting critical resources away for its own use determines in large measure how successful it will be in leaving offspring behind. Plants therefore compete quietly, root-hair to root-hair for nutrients, leaf against leaf for light; microbes bathe each other in toxic substances, overgrow one another, or excrete stronger enzymes than their neighbors; animals threaten each other over a contested piece of food or have means of getting there first, or they battle over space or nest sites or mates. The keenest competitors are always other members of the same species, because their needs are exactly the same, but to the degree that species have overlapping needs, they compete across species lines as well. Even within social groups based on cooperation, such as our own, competition is ever present, if relatively subdued and ritualized.

32

Territorial Behavior and Monogyne Population Regulation

Brood-raiding behavior, which was so important in shaping populations of incipient nests, vanishes rather suddenly in late summer, both from naturally founded populations and from planted ones. From then on, when workers from different nests encounter one another, they often avoid or attack one another. Through mechanisms that are not yet well understood, this aggression and avoidance establishes an exclusive territory. Territorial behavior dominates the rest of the life of the population. Colonies convert their trunk raiding trails into underground foraging tunnels, establishing the first elements of a system of underground tunnels basic to territorial defense and exploitation.

The size and density of populations vary in response to external, physical factors such as climate or food, to internal interactions such as competition, or to an interaction of the two. Physical factors are usually independent of density, but many internal factors, such as competition, become more intense as the population density increases and are said to be density dependent. Interaction between these two types of factors can either increase or decrease population variation, and the two are difficult to distinguish without experimental tests. Ecologists generally assume that density-dependent processes regulate many populations, but supporting data are often merely correlative and thus open to debate. The degree to which density-independent factors are important is also debated.

Adams and I decided to weigh in on this issue by performing an experiment on the regulation of population density. No species is better suited for an experimental test of population regulation than *Solenopsis invicta* in the southeastern USA. Almost monospecific populations of *S. invicta* colonies interact mostly with each other, simplifying population processes. Populations are dense, so interactions among colonies are likely to be strong and readily detectable. Who could ask for more?

So after mapping and measuring all the colonies in six plots, we launched a scorched-earth, carpet-bombing, genocidal attack on the central quarter of each plot (ca. 1000 m^2; Adams and Tschinkel, 2001), killing each colony in this core area first with a dish of poisoned bait and then with hot water. Colonies in the surrounding annulus of approximately 3000 m^2 of each plot were not harmed. Six other plots were not manipulated in any way and served as controls. We then remapped all plots at least three times a year as described in the preceding chapter, tracking the cores and annuli separately. If regulated by density-dependent mechanisms, the "bombed" fire-ant populations should recover and converge toward the condition of the populations on the control plots. Because variation in environmental conditions is expected to affect the recovering and the untreated population similarly, the two will come to vary in parallel. Consequently, density-independent variation can be both recognized and separated from density-dependent variation.

We made year-to-year comparisons only on the May–June censuses to avoid the bias resulting from the annual cycle of colony size change. We tracked the populations in the core areas of both treated and control plots separately from the outer annuli of both (Figure 32.1). The recovery of the population was obvious. The death we wreaked upon the experimental plot cores in 1991 brought colony number and fire-ant biomass to zero. By the next census, in late 1991, ant biomass in the core areas had returned to about one-fifth of its original level and continued to rise for another year, gradually converging with the biomass in the cores of control plots. Thereafter the two fluctuated precisely together as though joined at the hip. In the outer annuli, the biomasses on the experimental and control plots fluctuated in parallel throughout the five-year period. The effects of colony removal on ant biomass seemed to be rather local and did not extend far into the outer annuli of the experimental plots.

The recovery of biomass in the experimental plots involved several simultaneous processes. Almost immediately, colonies just outside the treated cores expanded their territories into the now vacant territories of their former neighbors (see Figure 8.7 for examples). Within eight weeks, their territories increased by an average of 45 m^2, 122%, while the corresponding colonies on control plots lost 11 m^2, 7%. Over the following weeks, colonies on the experimental plots were more than twice as likely to move as were those on the control plots. Moreover, their moves were more frequently toward the vacant center during the first two years (but not thereafter). Colonies on control plots moved in random directions and less than half as frequently. This frequent oriented colony relocation moved 170 g of ant biomass into the vacated cores during the first year. The core areas of control plots lost about 8 g during the same period, and because there were no colonies to die in core areas of the treated plots, some of the difference in core biomass change might have been the result of reduced colony mortality in the treated plots.

Territorial expansion on experimental plots probably rippled outward from the core edge through several ranks of colonies. As the

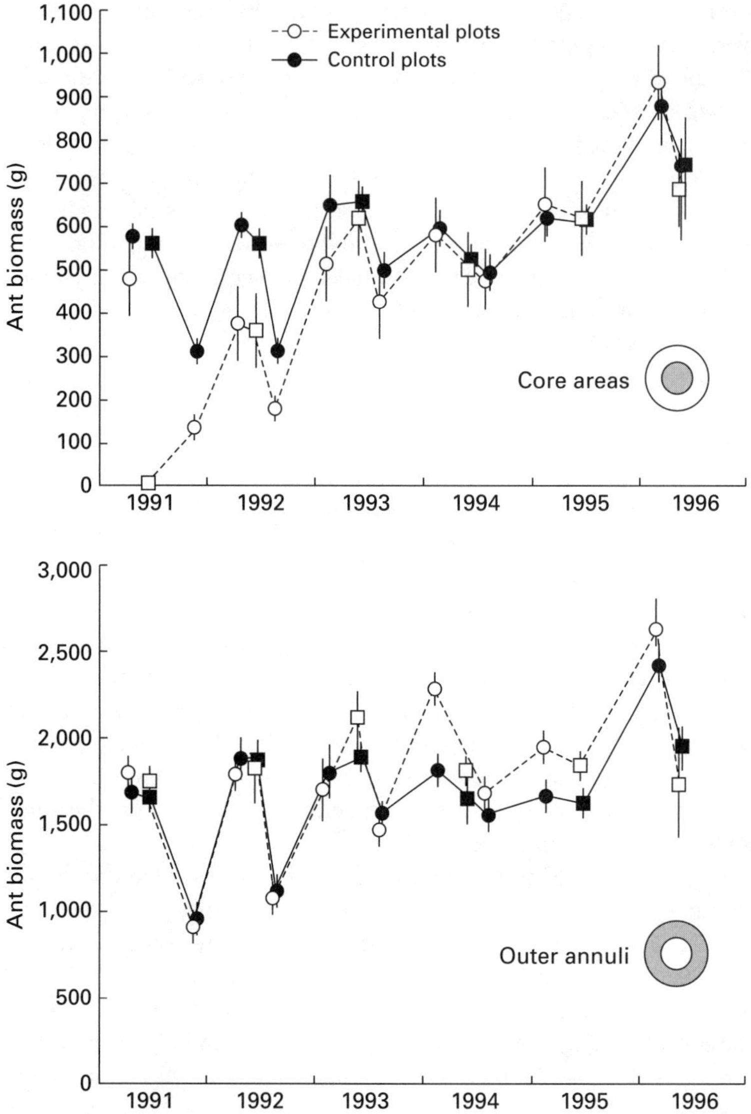

Figure 32.1. Recovery of biomass in experimental plots and control plots in core areas (from which fire ants had been eliminated; upper panel) and in outer annuli (where fire ants were undisturbed; lower panel). The area of the outer annulus of each plot was three times that of the core. The leftmost symbols indicate measurements taken prior to experimental manipulations. Squares indicate censuses taken in late spring. Bars show standard errors. Adapted from Adams and Tschinkel (2001).

innermost colonies expanded and moved into the removal area, their outer neighbors found the territorial pressure on their inner boundaries reduced and also expanded their territories in the direction of the core. Over time, and damped with distance, the entire neighborhood therefore shared the newly available space and resources, allowing colonies of all sizes to grow about 30% more than colonies on control plots (Figure 32.2) during the first and second years (but not thereafter). Colonies that moved into the vacated core areas also grew more rapidly, contributing further to the recovery of ant biomass in those areas.

Mature colonies are major killers of newly mated queens, so not surprisingly, the second major mode of biomass recovery was the survival of incipient colonies through the first year. Many more new colonies appeared in the core areas of experimental plots than in control ones, most probably because of increased survival rather than increased settlement—the outer annuli of control and experimental plots did not differ in the number of new colonies. Once again, the effects of removal were quite local, on the same scale as fire-ant territories. By the end of the first year, the *number* of colonies had more than recovered in the core areas of the experimental plots, but in all subsequent years experimental and control plots did not differ. The biomass required two years to recover fully because, at the end of the first year, the colonies in the core areas were a mixture of surviving, small young

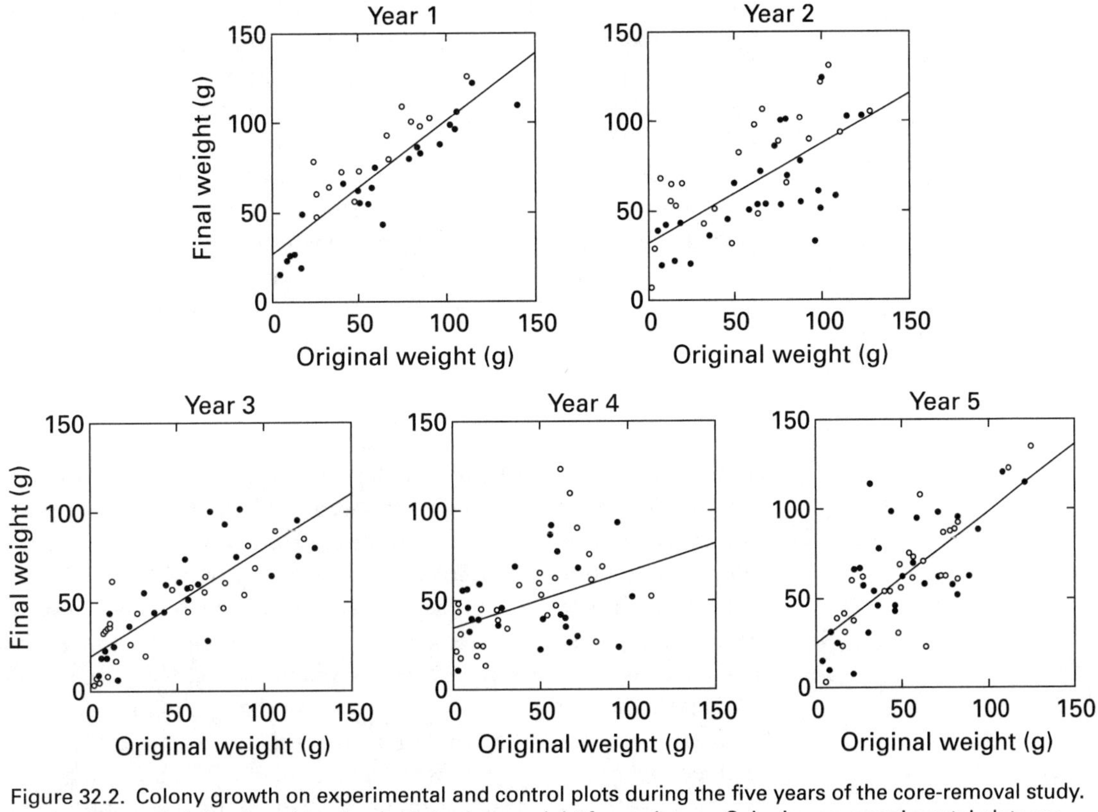

Figure 32.2. Colony growth on experimental and control plots during the five years of the core-removal study. Final colony weight is plotted against initial colony weight for each year. Colonies on experimental plots are shown as open circles, controls as filled circles. Growth on experimental plots was higher than that on control plots during the first two years. Adapted from Adams and Tschinkel (2001).

colonies and large colonies that had migrated in. Clearly, colony density is an insufficient measure of a population when colony size is highly variable. Biomass density, which is the most meaningful measure of the ants' ecological impact, depends on both the density and size of colonies.

All the processes leading to population recovery are density and/or size dependent. Their relative contributions tell us their relative importance. During the first year, the *number* of colonies in the core areas of experimental plots increased by about 14 colonies more than did control cores (because they started from zero). About 70% of this difference resulted from the increased survival of incipient colonies, 18% from the increased in-migration of mature colonies, and 12% from the decreased mortality of mature colonies (there were none to die in the experimental cores, but 12% died in the control cores). The fire-ant *biomass* in experimental core areas increased about 360 g more than did that in control cores in the first year. About 175 of these grams (49%) were the result of differences in colony movement, and about 135 g (38%) came from the establishment of new colonies. In control core areas, about 80 g were lost through colony mortality and 28 g were gained through growth of surviving colonies. During the second year, experimental cores gained about 170 g more than controls, completing the recovery, but most of this increase was due to the growth of colonies present at the end of the first year. Establishment of new colonies and net movement made only minor contributions. Getting there "the firstest with the mostest" is as important in mature fire ant populations as it is in incipient ones. In summary, early population recovery proceeds through oriented colony movement and the appearance of new colonies, ending when the space is fully occupied. The remaining recovery proceeds through colony growth, slowing as the biomass reaches previous levels, whereupon the perturbed area no longer differs from unperturbed areas.

The entire process is driven by competition. As the number and/or size of colonies increases, the space per colony decreases and the competition for foraging space becomes ever stiffer, with negative consequences for colony founding, growth, and movement. Territorial competition thus links the behavior of individual colonies with the nature of the population and regulates the population with remarkable precision. Our 0.4-hectare plots supported between 45 and 57 colonies weighing a total of about 2400 g. Over six years, biomass varied little from year to year. Although colony number increased significantly in 1993 (for unknown reasons) and then remained higher, this increase was accompanied not by an increase in the biomass but by a decrease in the average colony size. Clearly, territorial competition regulates primarily biomass density, not colony density.

Finally, variation occurs within years. It originates in the annual cycle of alternating alate and worker production and is very clear in the five-year record of biomass. Biomass peaks in midwinter and, after alates leave on spring mating flights, drops to the annual minimum in mid- to late summer.

33

Ecological Niche

When a new ant invades an ecosystem, it can interact with the ants, other animals, and plants already there in many ways, with a wide range of outcomes. The newcomer may fail to become established, it may become an addition to a less-than-fully exploited ecosystem, or it may replace an ecologically similar ant with little other effect. It may prey on some species, possibly suppressing their populations, but may in this way allow other competitors, and perhaps prey animals, to increase. The newcomer may compete to varying degrees with the ants already there, and the outcome can range from exclusion of the newcomer to varying degrees of suppression of some native species. Competition can take many forms, from aggressive territorial struggles to competition for nest sites to reduction of the food supply of some species through more effective foraging by others. When the newcomer has the advantage of leaving many of its natural enemies behind, the balance may be tipped in its favor, sometimes so far that the consequences for the ecosystem can be dramatic. Finally, the newcomer is also a resource for native creatures, as prey, host, and habitat modifier and in numerous other roles (Allen et al., 1974; Wojcik, 1983; Porter et al., 1990, 1991, 1992). All these considerations probably apply in various degrees to *Solenopsis invicta* and its invasion of North America.

The Ecological Niche of *Solenopsis invicta*

What is *S. invicta*'s particular place in its native ecosystem in South America, and what place has it assumed in its adopted home in North America? Characterizing *S. invicta*'s physical and biological niche can give us insight into how and why the ant has been such an enormously successful invader. Like every living creature, the boundaries of its "place" in the world are sketched first in physical needs and tolerances such as temperature, light, and water (which constitute its fundamental niche) and then restricted further through its biological relationships to all other creatures in its community (its realized niche). As a result, there are creatures of the desert or the rainforest, of water or land, the poles or the tropics, the mountains or the lowlands. At a finer scale, the creatures basking in the sun in the rainforest canopy do not thrive on the dim forest floor, nor are the soil dwellers found anywhere but there.

Niches can be narrow, as in the case of a highly specific parasitoid, or they may be as broad as a grizzly's backside.

The question of *S. invicta's* niche is further complicated because most of what we know of its biology has been learned in its adopted land, where it probably occupies a broader slice of the hosting ecosystem than it does in South America. The broad physical limits that the ant can tolerate are gradually being revealed as it spreads in North America. Already clear are that it is not a creature of honest cold, serious drought, or deep shade. Its absolute physiological limits have been revealed in the laboratory, but in nature, the ant lives in a subterranean microworld that often allows it to escape, for a time, the raw force of drought, cold, and heat. Even so, behaviors can postpone the inevitable for only so long. Physiological limits will stop its spread through Texas and along the northern boundary of its range before long (except where humans help out).

Association with Disturbance on Both Continents

The most prominent niche characteristic of *S. invicta* is that it is primarily a creature of disturbed habitats, whether in the USA, in the Galapagos Islands (Williams and Whelan, 1991), in Puerto Rico, or in its native homeland in South America. This is the reason why practically all investigators choose lawns, pastures, fields, and roadsides as study sites in both North and South America (Plate 1). In the Mato Grosso of South America, *S. invicta* is at home in an ecotype called *cerrado*, an open, grassy habitat with scrubby trees, but it is not common unless this habitat has been recently disturbed (Plate 16C; Wojcik, 1983, 1986; Banks et al., 1985). Then its colonies pop up in numbers in roadsides, recently cleared forest, pastures, cityscapes, parks, lawns, and the like. They persist there as long as the area is maintained by mowing, reclearing, and so on. It is also found in the Pantanal (Figure 33.1), a huge, seasonally inundated floodplain of the Paraguay River, where it occurs mainly on the perimeter of the wetlands and profits from the ecological disturbance provided by recurrent flooding. No one who has seen a fire-ant colony float can doubt that they are adapted to flooding. Such observations have been known since 1894 (Ihering, 1894; Reichensperger, 1927b).

The ecological disturbance with which the fire ant is associated may be physical or biotic. It may be as small as the gap caused by a single tree-fall in a forest or as great as the destruction of the entire forest or other natural vegetation type, especially if the destruction is accompanied by major soil disturbance and its many cascading effects. An experiment in South Carolina demonstrated the effect very clearly—*S. invicta* colonized clear-cut plots within months but remained absent from matched, forested control plots (Zettler et al., 2004). *Solenopsis invicta* was rare or absent in exhaustive surveys of five relatively

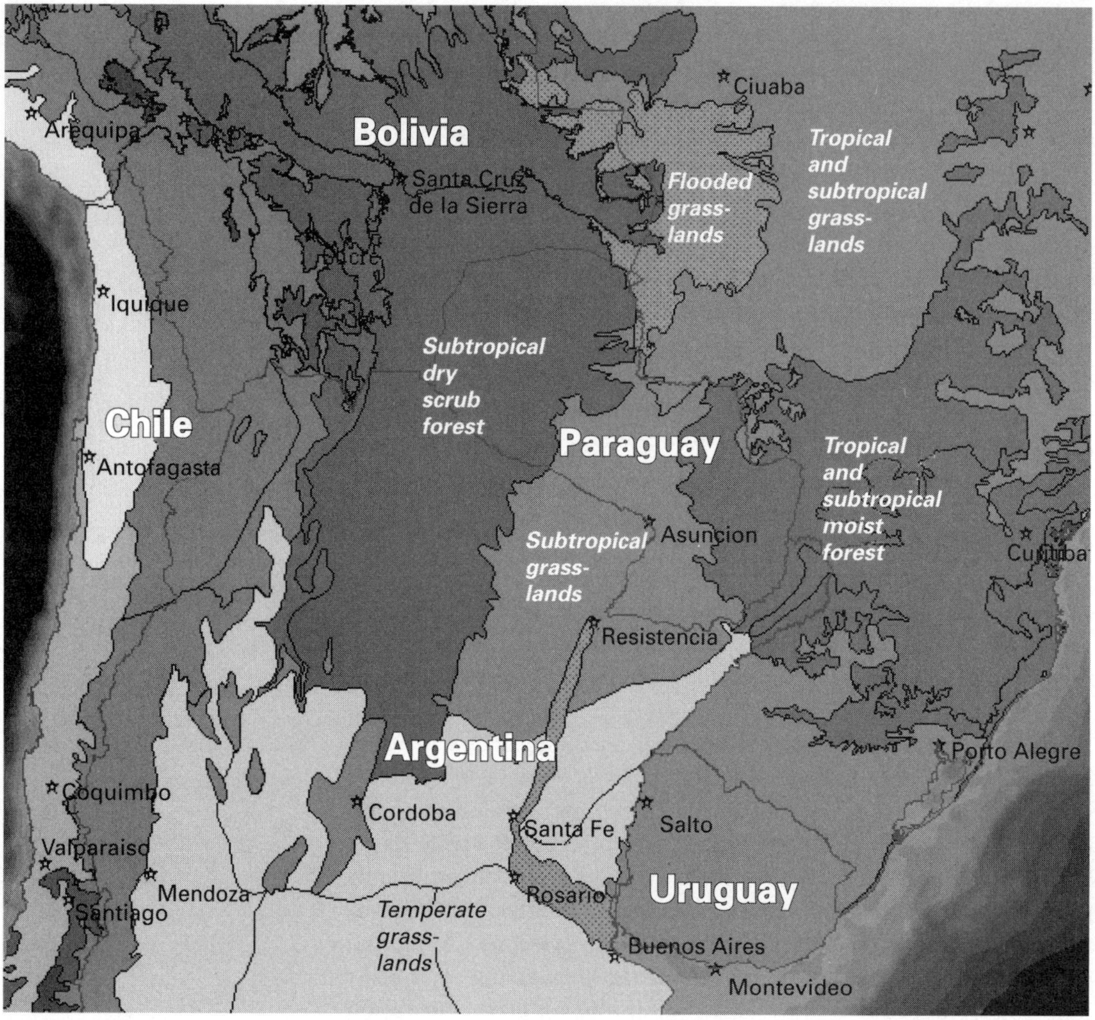

Figure 33.1. The Pantanal region of South America, showing ecological regions. Adapted from Hearn (2000).

undisturbed Florida ecosystems but was common and often dominant in disturbed areas (King, 2004). Disturbance also includes less drastic events such as flooding or, less certainly, fire. Disturbance is the bond that links the good fortunes of *S. invicta* in North America to human activity. These types of ecological disturbance were relatively limited and rare throughout most of evolutionary history. Once Europeans arrived in the Americas, they relentlessly created huge areas of ecological disturbance. By the time *S. invicta* set tarsus in Mobile, most of the USA was in some stage of recovery from disturbance, a paradise for weeds of every sort.

Studies in the Florida Keys (Forys et al., 2002) and the Savannah River Site (Stiles and Jones, 1998) illustrate the connection between human disturbance and *S. invicta.* Roads and development have often fragmented the natural pinelands, hardwood hammocks, and transitional salt marshes of the Florida Keys. In all three habitat types, the closer a bait transect was to a road, and the greater the area of development within 150 m of the transect, the more likely the baits were to draw *S. invicta.* For *S. invicta* to be absent, the transect had to be at least 1.3 km (median distance) from development and 2.2 km from a paved road. Transects with *S. invicta* were 300 m from development and 900 m from roads. Clearly, roads and development provide congenial habitat within undisturbed, uninhabitable natural habitats as well as corridors for further dispersal. From the edges of these disturbances, fire ants can forage into the natural habitat and possibly invade it to a limited extent. Roads and development menace natural ecosystems in direct and obvious ways but also in these more subtle indirect ways.

One of the few attempts to quantify disturbance is that of Judith Stiles and Robert Jones at the Savannah River Site, an 800-km^2 coastal-plain forest (Stiles and Jones, 1998). Stiles and Jones measured the correlations between the density of the monogyne *S. invicta* populations in road and power-line corridors and the degree and frequency of disturbance.

Disturbance frequency and corridor width affected the density of fire-ant mounds in a complex way. It was lowest for the least disturbed, narrow, closed canopy roads (9 mounds/ha). Once the canopy opening reached a critical width, density increased manyfold but gravel and paved roads did not differ clearly. For open-canopy habitats, mound density was highest in narrow habitat with the second-highest disturbance frequency, namely, open dirt roads (87 mounds/ha), and lowest in the widest with least frequent disturbance (power-line cuts, 28 mounds/ha). Average colony size decreased 40% as disturbance frequency increased. As a result of these opposing trends in mound density and colony size, the densest ant populations occurred in the second-most disturbed habitat, open dirt roads. Power-line cuts, the widest and second-least disturbed, had a much lower density than these open roads. Of the two factors, disturbance frequency was more obviously and simply related to fire ant density.

The distribution of mounds within each of these habitats hints at the conditions that fire ants seek (Figure 33.2). Colonies were much more likely to be located close to the forest edge or to the pavement or gravel edge in all corridors. This preference for edges is easy to see in any town in the Southeast. Colonies were also more likely to be located along the northern edges of these linear habitats, more strongly so in corridors running east–west. Presumably, the shading of the southern edge, especially during the winter, made it less suitable for *S. invicta.* The preference for open, disturbed areas clearly assured fire ants a

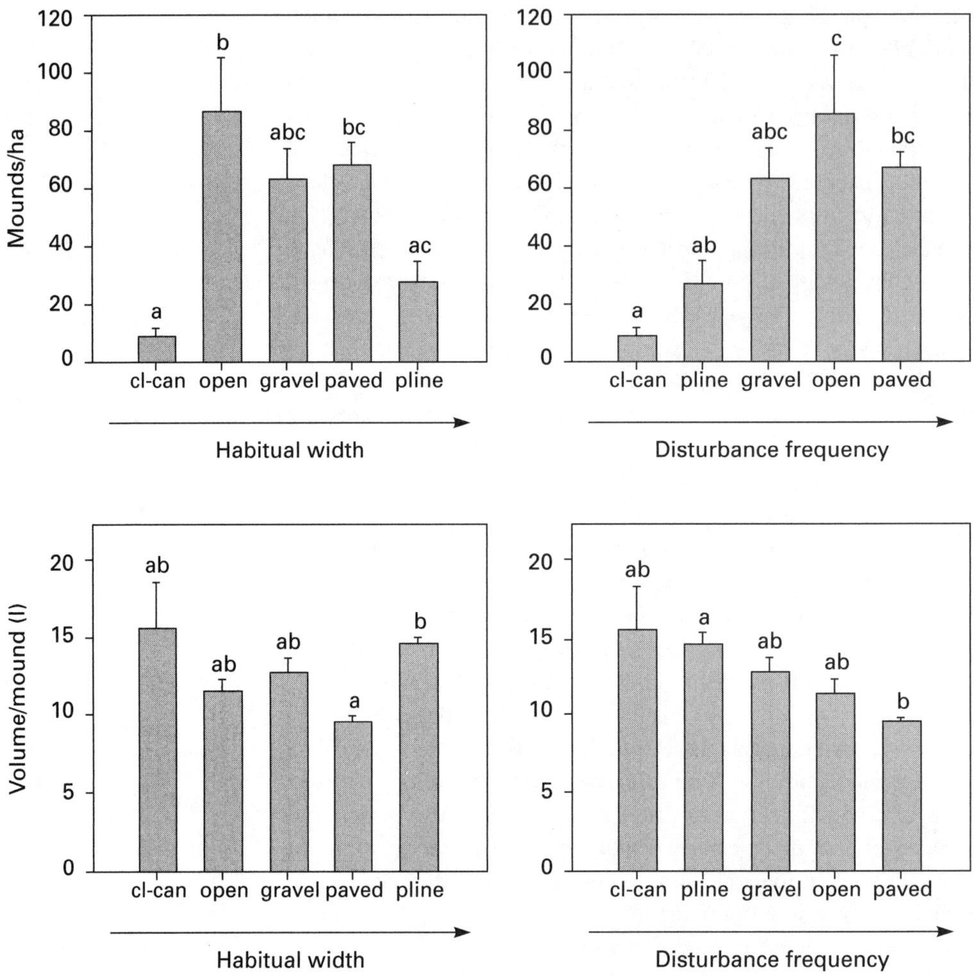

Figure 33.2. Density and size of fire-ant colonies in habitats of increasing width and frequency of disturbance. Habitats were closed canopy (cl-can), open dirt roads, gravel roads, paved roadsides, and power-line cuts (pline). Bars with different letters are significantly different; bars are standard errors. Adapted from Stiles and Jones (1998).

sunny exposure, but even within these habitats, they seemed to shun shade. The reason why fire ants prefer edges is unknown, but the preference was confirmed by a study along Texas highways (Russell et al., 2001).

These studies are exceptional. In most cases, the exact definition of "disturbance" is difficult to pinpoint. Without quantification, we cannot easily compare disturbance in different habitats, and we cannot say whether equal disturbance in a North American pasture and a Brazilian pasture leads to equal dominance by *S. invicta.* Even if we compare similar land-use types, or physical habitat types in North and South

America, many land-use practices differ and might contribute to the outcome. Therefore, a statement about disturbance is never precise.

If fire ants are indeed adapted for disturbed habitats, then newly mated queens should prefer to settle in such habitats and should home in on such suitable patches of habitat like heat-seeking missiles. The queen chooses the habitat while still airborne and travels no more than a few meters from her landing spot. Unfortunately, no studies have addressed aerial searching by newly mated queens or their settlement preferences. Notwithstanding, for decades my colleagues and I have collected thousands of newly mated queens in open sites with partly vegetated soil, along roadsides or pond margins, and in other highly disturbed areas. Additional support for selectivity is that male swarms are observed only over open land, not over forest or wetlands (Markin et al., 1971). Sometimes queens land in some particularly noxious, barren, sizzling place, take a few steps, and then take flight again. This behavior smacks of an active decision not to remain in a hellhole but to seek friendlier soils.

Colonies should also perform best in the preferred open, disturbed habitat, rather than in the unpreferred, wooded habitat—preference and performance should be linked. Otherwise, what would be the point of choice? (Of course, the choice may have evolved under conditions that no longer prevail and thus be a nonadaptive relict of the past.) Once again, no such experiments have been done, and the evidence is indirect. First, *S. invicta* obviously have the physiology to tolerate the harsher and more variable life in the open. Who, standing in an open Florida pasture amongst an acne of fire-ant mounds on a July day could doubt it? All evidence indicates that they not only tolerate life there but need openness to survive. Their mounds take advantage of solar heating but are useless in a shady wood. Their long underground foraging tunnels allow long trips even when it is too hot on the soil surface. Their temperature-limited capacity for foraging in open areas far exceeds that in woodland. Although they occasionally forage up in trees, this is not their regular habit. These and many other qualities suggest that habitat preference and performance are indeed linked. As forest succeeds field, *S. invicta* probably vanishes from the scene, to be replaced by ant species that tolerate cooler, shadier conditions. They disappear from young pine plantations when shade exceeds 40%. Other as yet unknown differences between forest and field, possibly including the ant fauna, may further reduce their capacity to inhabit forests. Unfortunately, no long-term successional studies exist.

Adaptations for a Weedy/Gap Existence

When a mature plant community is destroyed and then left to recover, the vacant site is first colonized by rapidly growing, mostly small and effectively dispersed plants collectively called weedy, pioneer, fugitive, ruderal, or gap species. Only later do the species in the original community gradually reappear in a process called succession. Weeds are adapted to taking advantage of early stages of such successional habitats, and the defining features of these habitats have shaped the life histories of these weeds. The same natural selective forces shape the animals of these disturbed habitats, justifying use of the same labels for the two. When it is viewed as a weedy/pioneer/gap species, many of the life history traits of *S. invicta* make sense, because they can be understood as responses to the challenges of disturbed habitats.

High Reproductive Failure Requires High Reproductive Effort. Under natural conditions (that is, in the absence of humans), early succession and gap habitats were relatively rare and short-lived patches, and the sum of their total areas was small. Weather, tree falls, hurricanes, sandbars, and landslides act capriciously, so the timing and location of patches of disturbed habitat were unpredictable. Any creature making its living in such patches or gaps must have a life history that meets the multiple challenges of rarity and unpredictability in time and space. These challenges impose a high failure (mortality) rate on colonization efforts.

To overcome this high failure rate, natural selection favors individuals or colonies that produce more "propagules," because doing so increases the probability that a few will find and colonize a suitable patch. Just as many plant weeds produce large numbers of seeds, fire-ant colonies produce large numbers of sexuals year after year. Both are a response to the low probability of either settling in a patch of new habitat or finding a gap in the carpet of mature colonies occupying the natal habitat. Comparative data for ants are scarce (Tschinkel, 1993b), but sexual production by *S. invicta* is apparently very high. Because sexual production is directly proportional to colony size across several species of *Solenopsis*, the need for more sexuals is met through the evolution of larger colony size; *S. invicta* is a case in point. A decrease in the size of sexuals may also contribute in some species.

Because suitable habitat or gaps can become available at any time, weeds typically have a long rather than a short reproductive season. In the Tallahassee area, mating flights of *S. invicta* occur in all months but December and January. *Solenopsis geminata* is a less weedy species, and its mating-flight season lasts only six months, rather than the ten of *S. invicta*. The earlier mating flights of *S. invicta* may in fact be one of the features that give it a jump over *S. geminata* in occupying vacant habitat. I know of many other ant species with much shorter mating-flight seasons, often only one to three summer months.

Unpredictability in Space Requires Effective Dispersal. Of course, seeds or queens must also be dispersed to who-knows-where, over unknown distances. Among plant pioneers, seeds are dispersed by wind, water, or animals, increasing the chances that at least a tiny fraction will fall on patches of suitable ground. Female alates ("ant seeds") come with their own power of dispersal, wings. Among ant species, either female alates mate near, on, or in the natal nest and found nests nearby, or they fly to male swarms and disperse longer distances. Monogyne *S. invicta*, with its high-altitude mating swarms and long postmating flights by queens, is an effective disperser, flying relatively great distances, as confirmed by the high rates of gene flow in monogyne populations. At least in part, this is an evolutionary response to the widely scattered, rare patches of suitable habitat and/or gaps in the continuous community in which *S. invicta* evolved.

Colonization of these scattered patches is essentially an Oklahoma Land Rush for space, a first-come, first-served scramble in which, at least initially, competition is very low. A characteristic of scramble competition is that, as the number of competitors increases, the resource is divided so many ways that it is no longer adequate for any of the competitors, and the great majority fail. In this respect, *S. invicta* departs from the pattern of many pioneer species, for although being early is highly advantageous, contest competition clearly develops as soon as the minims appear. These contests assure that the winners gain enough resources for their needs. In this regard, as in colony size and longevity, *S. invicta* is more like a "gap" species, those that depend on finding the gaps in a stable community where an individual has died. Gap species can be effective competitors and need not be small, but they play the same game of finding rare and unpredictable opportunities and therefore suffer high reproductive failure.

Ephemeral Habitat Requires a Short Life Cycle. Colonization and rapid succession make disturbed patches and gaps short-lived, as they begin on the road back to a climax community, a habitat unsuitable for the weed. Time is of the essence, and the prize goes to pioneers with speeded-up life cycles—early reproduction and early death. The generations of weeds thus endlessly chase disturbed patches, earning the alternate name of "fugitive" species.

Rapid growth assures that the newly arrived pioneer both meets the challenge of ephemerality and lays claim to enough habitat for successful reproduction. In *S. invicta*, this process takes the form of claiming as much territory as possible by growing as large as possible as fast as possible. The pleometrosis and brood raids of incipient colonies amplify colony size early and rapidly, leading to larger territory size. Multiplying colony size during this phase can mean multiplying territory size and much earlier reproduction down the line.

Most plant weeds are short-lived and small in comparison to the trees that succeed them. Small body size is usually understood as a response to the need for a fast life cycle. If colony size is the equivalent of body size, fire ants are not small like pioneer plant species, but the plants that exploit forest canopy gaps are often not small either. In this respect, fire ants are more like gap species than like pioneers. Colony size is more likely to be related to the need to produce large numbers of sexuals, as noted above.

Most weeds start reproduction long before they are full-grown and reproduce across an enormous range of body sizes. Again, few ant data are available for comparison, but *S. invicta* seems to reproduce unusually early in its life cycle, at only about 10% of maximum size. In favorable situations, colonies may reach this size in as little as a year. Once they begin reproduction, they invest about one-third of their annual production in sexuals but continue growing as they do.

The final element in a fast life cycle is low longevity. The short-lived nature of pioneer/gap plant species is obvious in comparison to the trees that eventually succeed them. Unfortunately, data on ant-colony longevity are woefully incomplete. In monogyne species, colony and queen longevity can usually be equated. The queen longevities compiled by Hölldobler and Wilson (1990) in *The Ants* showed that almost half of about 20 species lived from five to 10 years, only one less than five years, and only three more than 20 years. The six- to eight-year life span of fire-ant colonies places the species with the majority ants, not out of the ordinary and probably not specifically molded to the habitat characteristics (but only the *S. invicta* and *Pogonomyrmex owyheei* life spans were estimated from field populations; almost all others were from single laboratory colonies). Turning the question around, ant colonies in general, in contrast to plants, are not long-lived enough to require special adjustment for the ephemeral nature of early succession habitats. Colony longevity remains an unexplored subject.

In summary, several life-history attributes can be understood as adapting *S. invicta* to live as a pioneer in early successional communities: high reproductive effort, effective dispersal of queens, rapid colony growth, and early reproduction over a long season. Attributes not usually associated with pioneers, but perhaps compatible with gap species, include large colony size and "normal" longevity.

The Fire-ant Niche

The Displacement of Other Fire Ants. Describing *S. invicta* as a pioneer/gap species is a rather general statement. Can we refine our description? A possible handle on this challenge comes from a widely held belief about niches. To the extent that the niches of two similar species overlap, one will out-compete and eliminate the other. The USA is equipped with several species of similar fire ants, two native (*S. geminata, S. xyloni*) and two exotic (*S. invicta* and *S. richteri*). Although *S. invicta* has largely displaced the other three species, two continue to exist in parts of their original ranges.

As *S. invicta* spread, *S. xyloni* simply disappeared from the areas it took over, coexisting only briefly on the margins of *S. invicta*'s growing range. No one actually observed this process, but it is obvious from ant collections made before and after the advent of *S. invicta* and from the mutually exclusive distribution of the two species (Wilson, 1951; Wilson and Brown, 1958). Both species were most abundant in open, grassy land, cultivated fields, pastures, and roadsides and seemed to be ecological equivalents and thus incapable of coexistence.

The overlap was less with *S. geminata*, which, before *S. invicta* spread into Florida, was one of the most common ants there. It tolerated a wider range of habitats than either of the other two species, including some types of woodland not preferred by *S. invicta*. As *S. invicta* spread, the two species occurred side by side for a while, but within a few years, *S. geminata* could be found only as enclaves in more wooded locations. *Solenopsis xyloni* and *S. geminata*, although once common in the Mobile area (Creighton, 1930), were not among 16 species of ants taken on roadside bait transects in 1974 (Glancey et al., 1976c). Instead, almost 40% of the baits were dominated by *S. invicta*, followed at a distance by the Argentine ant. The other 14 species occupied only 2% of the baits. In nearby Baldwin County, Alabama, *S. geminata* occurred in 1973 but not in 1974.

As the range of *S. invicta* spread, this story was repeated over and over. In Arkansas, *S. xyloni* disappeared from the list of ants within *S. invicta's* range. By 1977, *S. geminata* and *S. xyloni* had disappeared from most of about 50 eastern Texas counties (where later work found many polygyne *S. invicta* populations; Hung et al., 1977; Hung and Vinson, 1978). When Porter surveyed Texas in 1988, he found these two native species only on the fringes of the expanding *S. invicta* population and not at its core. *Solenopsis xyloni* persists today only in the drier zones west of the 100th meridian, beyond the current *S. invicta* range (Moody et al., 1981; *S. geminata* is rare in this region). Furthermore, the number of native ant species Porter captured was inversely related to the density of *S. invicta* mounds in the surrounding area, suggesting sheer numerical dominance. In Florida, *S. geminata* was present at 83% of the roadside sites from which *S. invicta* was absent but only 7% at which it was present (Porter, 1992). Similarly, *S. geminata* largely disappeared from the Gainesville, Florida, vicinity (Whitcomb et al., 1972). Nevertheless, although eliminated from most of Florida's roadsides, *S. geminata* continues to persist in a few places, especially the drier, less-disturbed Florida sandhills habitat. There, *S. geminata* is the *only* fire ant species, giving way to *S. invicta* only upon gross disturbance or flooding (Tschinkel, 1988b). Whether these populations are stable is not known, but a decline cannot be very rapid because *S. invicta* has occupied most of Florida for about 20 to 30 years. At the very least, before *S. invicta* arrived on the scene, *S. geminata*'s niche was broader and included much of the habitat currently occupied by *S. invicta*. One would have to say that *S. invicta* occupies it more effectively.

The mechanism of these displacements is unknown. Does *S. invicta* prey on the other two species directly? Does it overwhelm and eliminate them in territorial competition, as suggested by Wuellner and Saunders (2003)? Does it monopolize food? The replacement of these other *Solenopsis* species in much less than a colony life span suggests that actual elimination of adult colonies is involved, rather than slow attrition through competition. Laboratory experiments found some differences in the abilities of fire-ant species to compete for different types of food (Morrison, 2000a), but these experiments probably tell us little about how and why *S. invicta* displaces the other two in the field. The observed differences may represent food preferences rather than food-collecting ability. More importantly, the opportunity for competition through food-collecting ability would simply not arise because all the forms are territorial, so their foraging areas do not overlap.

In tarsus-to-tarsus gladiator combat, body size was critically important—major workers of all forms killed seven to 10 times as many minors as the reverse. Pitted against the other species, *S. invicta* majors took greater losses than the majors of either native form, probably because they are smaller. Size differences among minors are, well, minor, and differences in their fighting ability and losses are minor as well. Nevertheless, fighting and territorial competition seem likely causes of displacement. Morrison set up multitray laboratory nests to simulate some aspects of the field situation. When colonies were matched for worker biomass, *S. invicta* enjoyed a 70% numerical superiority and controlled 89% of the foraging arenas against *S. geminata*. When worker numbers were equal, *S. invicta* controlled slightly less than half of the trays. The size of the army seems to be critical for gaining "territory." When armies are equal in size, then the size of the soldiers matters. The degree to which the more numerous majors of the native species contributed to the outcome is uncertain.

Extrapolating such laboratory results directly to the landscape-level displacement of native fire ants by *S. invicta* is risky. In the laboratory, the ants are forced to come into contact in the connected trays or because they are dumped into tiny dishes together. Nest relocation and individual avoidance are options in the field but not in the laboratory, so rates of fighting are probably unnaturally high. At best, these laboratory results are estimates of the intrinsic tendencies of the *Solenopsis* forms. Even transposed without reservation to the field, they do not clearly indicate why *S. invicta* should have displaced the native forms. After all, under some circumstances, the competitive advantage in the lab lies with the natives, under others with *S. invicta*. If numerical superiority is a key, then *S. invicta* is surely in the catbird seat. Its colonies are much larger than those of any native *Solenopsis*, and their vigor is enhanced by the relative absence of natural enemies, but currently we cannot be sure. Most of the important questions remain unanswered.

Refining the Niche. The displacement of native *Solenopsis* species by *S. invicta* can be seen as an opportunity, too, because by studying the habitats in which each continues to exist, we should be able to learn something about how the niches of these species differ. I will take a roundabout approach by contrasting the reproductive adaptations of *S. invicta* with those of its congener, *S. geminata.* The subtle differences between the two species will help define the niche of each more precisely. The reasoning goes like this. The monogyne forms of the two species have fundamentally similar life histories: both are capable of founding new colonies independently or dependently through social parasitism. The independent mode is much more successful when queens are invading suitable, fire ant–free areas, as would be the case for freshly disturbed land, whereas the dependent mode succeeds better when orphaned colonies are available, as would be the case in stable, aging populations. If the ant's habitat dictates that one of these modes pays more dividends than the other, natural selection will shift the colony's relative investment from the less successful to the more successful mode, until a unit of investment pays equal dividends in the two modes (the evolutionarily stable strategy). The balance between these modes therefore tells us something about the "normal" habitat of the ant, the ecological niche to which each species is adapted.

Seen this way, *S. invicta* "expects" to start most of its new colonies in land vacant of other *S. invicta* (and probably other ant competitors, too) and only much more rarely in occupied land. Less than 10% of its sexual investment is in dependently founding queens (the exact value is difficult to determine because overwintered queens may pursue either strategy). Such queens head only about 3% of colonies in a stable population. The fortunes of *S. invicta* depend on pioneering new land through independent queens, and its populations are typically young and expanding.

In contrast, *S. geminata* in the southern USA seems to "expect" its habitat to be much more saturated with other *S. geminata* and new, unoccupied land to be relatively rare. We form this conclusion because *S. geminata* expends a third of its queen investment on dependently founding (parasitic) queens. These dependently founding queens lack the large body size and high reserves necessary for independent founding, and about a third of the colonies in a stable population of *S. geminata* are headed by them. The fortunes of *S. geminata* depend on a steady supply of orphaned colonies, such as would be expected when unoccupied land is rare, stable populations are the norm, and queens die at a regular rate. The balance point has moved much further to the parasitic, dependent-founding side than in *S. invicta.* Even when *S. geminata* founds independently, the greater robustness of its young queens, weighing 25% more than those of *S. invicta,* suggests a more demanding, competitive founding environment, one in which the high queen numbers needed for locating scattered habitat patches have been traded off against improved founding success.

Sexual production is also in line with the more pioneering/gap nature of *S. invicta*, for it produces about four times as many sexuals as *S. geminata*, primarily because its colonies average four times as large. Furthermore, mating-flight season is much more restricted in *S. geminata*, as though the habitat its newly mated queens seek were more predictable.

All together, *S. invicta* seems to be a weedy beast pioneering gaps and recent ecological disturbance, persisting only so long as the habitat remains in an early stage of recovery from disturbance. In comparison, *S. geminata* is a less weedy creature of later, more stable communities, depending much less on recent ecological disturbance. The distribution of these two species of fire ants in the two major types of longleaf-pine forest in north Florida supports this characterization. The first type is an upland forest on rolling sandhills dotted with seasonal and permanent ponds that are highly sensitive to rainfall and the water table. Topographic relief is 5–10 m. The soil is excessively drained sand, and the water table may be up to 10 m below ground level. The other type, the flatwoods, occupies a gently undulating landscape that grades repeatedly from longleaf-pine forest into wetlands. Relief is less than 2 m, and the water table is close to the surface, as verified by the abundant turrets of the crayfish that burrow to the water table and forage on the surface at night.

The distribution of *S. invicta* and *S. geminata* is quite different in the two types. Only *S. geminata* is found within the undisturbed sandhills forests, at a density of 5–10 colonies per kilometer of transect. Moderate disturbance increases their density. For example, in clearcut and replanted areas with undisturbed soils, the abundance of *S. geminata* doubles, and along graded but unpaved forest roads, colony density increases to about 40 per kilometer of transect. Neither area supports significant numbers of *S. invicta* colonies. In contrast, when clearcut and replanted areas include heavy soil disturbance such as disking, a high density of *S. invicta* replaces *S. geminata*. Mowed roadsides of paved roads also represent heavy disturbance and are almost exclusively occupied by *S. invicta*, but the disturbance need not be human-caused—the margins of the seasonally fluctuating ponds (Figure 33.3) are exclusively occupied by high densities of *S. invicta* colonies. A few meters from these roadsides and pond margins, *S. invicta* is invariably replaced by *S. geminata*. In the sandhills region, *S. invicta* is almost completely limited to heavily disturbed areas.

In the flatwoods, *S. geminata* is almost absent. Although *S. invicta* is not frequent in undisturbed flatwoods, it appears abundantly after even mild disturbance. About 10 colonies per kilometer are found even along tracks worn in the forest ground cover by vehicles, certainly a

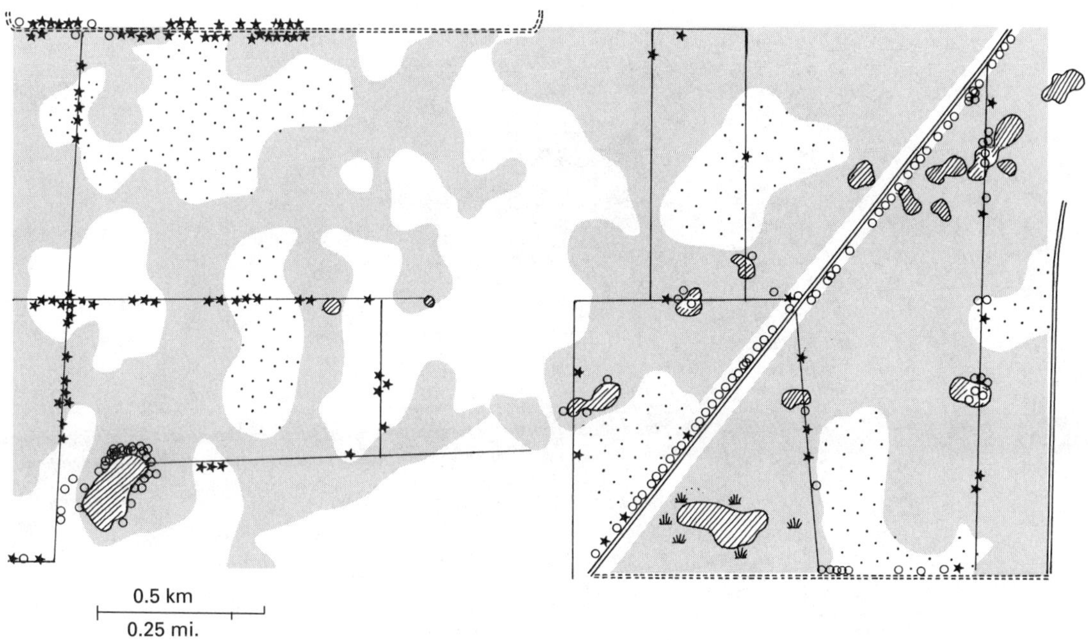

Figure 33.3. Distribution of *Solenopsis invicta* and *S. geminata* along transects in a portion of the sandhills region south of Tallahassee, Florida. Open circles indicate *S. invicta* nests, stars *S. geminata*. Shading indicates mature longleaf-pine forest, stippling thinned longleaf pine, and unshaded areas recently clearcut, site-prepared, and replanted with longleaf pine. Ponds are cross-hatched, paved roads are solid double lines, graded dirt roads are dashed double lines, and ungraded tracks are single dashed lines. Tuft symbols indicate wetland, mostly titi swamp. *Solenopsis invicta* is almost completely limited to heavily disturbed areas and to pond margins. Reprinted from Tschinkel (1988b) with permission from the Entomological Society of America.

small disturbance. Clearcuts in the flatwoods, even without heavy soil disturbance, are rapidly colonized by *S. invicta* at densities averaging over 20 colonies per kilometer of transect (Figure 33.4). Graded roadsides in the flatwoods supported over 50 colonies per kilometer.

What do these distributions tell us about fire-ant niches? First and foremost, they confirm that both ants are favored by ecological disturbance of the kind that humans perpetrate. Simply put, man is the fire-ant's best friend. Second, *S. invicta* is strongly favored by a high water table. The very deep (up to 8-m) nests of *S. geminata* may make it intolerant of the high water tables of the flatwoods and pond margins; the species probably never occurred in either historically. In contrast, *S. invicta* is clearly adapted to flooding. Colonies can float as mats of ants for weeks, reestablishing nests when they drift ashore or the flood recedes. *Solenopsis invicta* is one of only two species of ants found in the seasonally flooded savannas of the southeastern coastal plain (Figure 33.5; unpublished data); the other is *Tapinoma sessile*.

Solenopsis invicta displaced *S. geminata* from most of its historic range, implying that *S. geminata*'s niche has narrowed in the presence

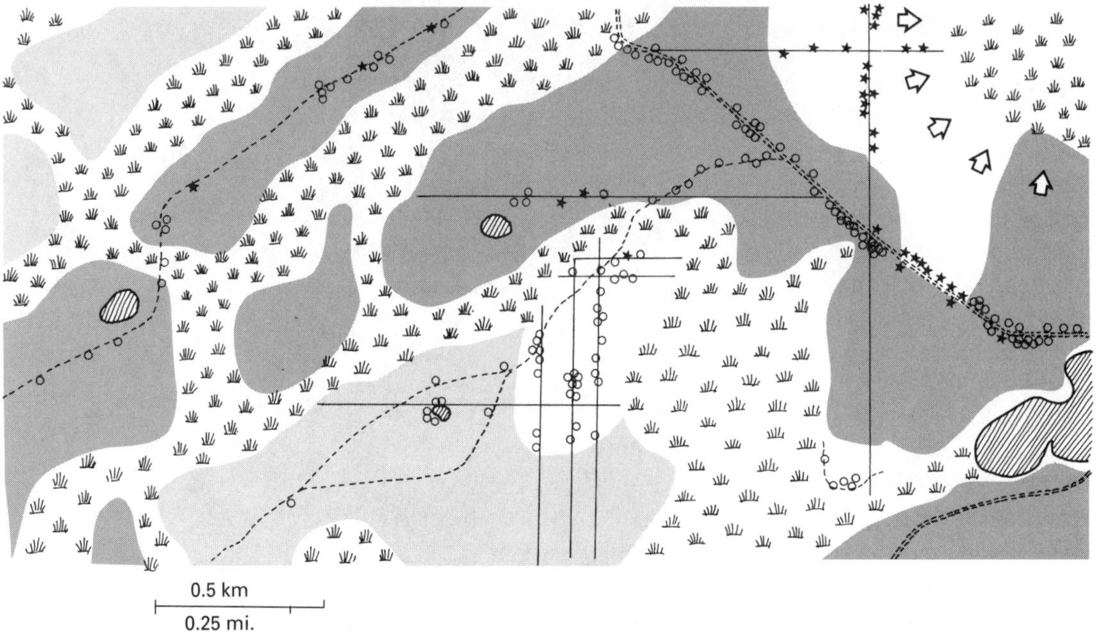

Figure 33.4. Distribution of *Solenopsis invicta* and *S. geminata* along transects in a portion of the flatwoods region southwest of Tallahassee, Florida. Symbols and shading are as in Figure 33.3. Arrows indicate brow of slope to wetlands. In the flatwoods, *S. geminata* is almost absent, except where the low-lying wetland depresses the water table. *Solenopsis invicta* becomes abundant after even mild disturbance. Reprinted from Tschinkel (1988b) with permission from the Entomological Society of America.

of *S. invicta* (some of *S. invicta*'s current niche, such as flatwoods, was never occupied by *S. geminata*). Persistence of *S. geminata* in some habitats indicates that the outcome is tempered by an interaction of biotic and physical factors. *Solenopsis geminata* is not adapted to whatever habitat condition is destroyed or altered by heavy disturbance or high water. When this condition is intact, *S. geminata* can resist the invasion by *S. invicta*. *Solenopsis invicta* is perfectly capable of thriving in the pine uplands region, provided that disturbance is severe enough. Conversely, *S. geminata* occupied the same disturbed habitats subsequently taken over by *S. invicta*. In Central America, *S. geminata* invades recently cleared land and is abundant in early successional fallow fields but absent from forest (Carroll and Risch, 1984).

Bill Buren speculated that, in its natural habitat in Brazil, subterranean ants that prey on *S. invicta*, especially species of *Diplorhoptrum* (a subgenus of *Solenopsis*) play a pivotal role. He suggested that flooding or heavy ecological stress eliminates *Diplorhoptrum* species and allows *S. invicta* to survive, accounting for the association of *S. invicta* with these factors in both North and South America. Both he and Creighton felt that *Diplorhoptrum* diversity was weak in North America

(Creighton remarked, "[t]he student of North American ants may count himself fortunate that so few species of this difficult genus occur in our latitudes"). Buren even proposed that control of *S. invicta* might be achieved by replacement of part of the North American ant fauna with complexes of ant species, especially underground predators, that would compete with or prey on *S. invicta* (Buren, 1983). Fortunately, no one has yet chosen to open this Pandora's box—subsequent work by Catherine Thompson (Thompson, 1980), a student of Buren's, and Joshua King (King, 2004) have discovered plenty of native *Diplorhoptrum* species. Bait grids, litter samples, and pitfall traps in several Florida habitats showed that these tiny subterranean predators, scavengers, and homopteran tenders were quite widely distributed and, contrary to Buren's belief, not conspicuously limited by a high water table or disturbance. Although some *Diplorhoptrum* species prey on newly mated *S. invicta* queens in their founding chambers, the extent to which they limit the distribution of *S. invicta* remains unknown.

Just what is it that keeps *S. invicta* and *S. geminata* to their particular, nonoverlapping habitats? My student Don McInnes (McInnes, 1994)

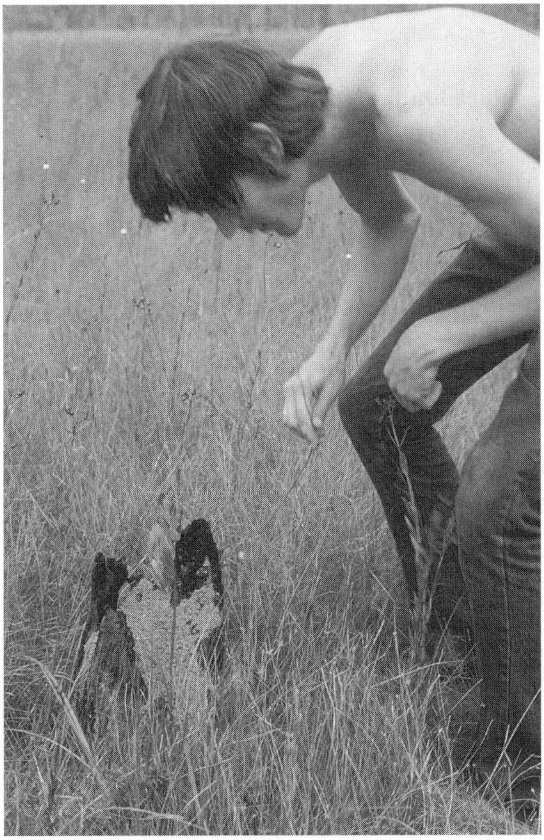

Figure 33.5. When fire frequency and high water table are extreme, as in these seasonally flooded savannas of the Apalachicola National Forest, *Solenopsis invicta* and *Tapinoma sessile* are the only ants able to exploit the habitat. They can nest only on small rises, such as this stump, or on fire lanes or wind throws, but once established, they create islands by moving soil upward during dry seasons.

expected that colony founding would be a particularly vulnerable phase, so he carried out reciprocal transplants of newly mated queens of each species into the habitat typical of the other (three pond-margin sites and three woodland sites). The controls were transplants within the habitat of each species. Newly mated queens of each species were imprisoned in small, buried plastic tubes, some with and some without tiny holes through which subterranean predators could enter.

When predators were excluded, habitat did not affect queen survival differently in the two species. Queens subject to predation died at a higher rate, no matter where they were planted, but *S. invicta* queens survived about equally well in the two habitats, whereas *S. geminata* generally outperformed *S. invicta* at the woodland sites but did worse than *S. invicta* at the pond-margin sites. Only two of 45 *S. geminata* queens at the pond margins lived long enough to produce minims, whereas 13 of the 45 *S. invicta* queens did. The predators were mostly fire ants, followed by thief ants (*Diplorhoptrum*) and occasionally other ant species. In the wooded plots, the fire ants were *S. geminata*, and in the pond plots, *S. invicta*. Interestingly, when predation by thief ants was high, fire ant predation was low to absent, suggesting mutual exclusion. Often, parts of dead worker ants were found in the nests of live queens, indicating that the queens had survived subterranean attacks by ants.

In sum, the advantage that *S. geminata* has when founding in the woods and *S. invicta* has when founding in pond margins appears to be the result of predation, not physical habitat differences.

Do these differences in survival fully account for the difference in distribution? Do mature colonies of each species survive differently in the habitat typical of the other? To find out, McInnes transplanted 64 mature colonies of each species into the habitat typical of the other, along with within-habitat control transplants. Because only the effects of the habitat were to be tested (at least initially), all plots were first cleared of any resident fire ants. This is a tougher experiment, because colonies move and must be repeatedly identified for up to two years. Confusion can also result when feral colonies move onto the plots, as they inevitably did.

Despite these weaknesses, this experiment established clearly that each species is capable of surviving in the habitat of the other—even after two years, some of the *S. invicta* colonies transplanted to the woods and *S. geminata* colonies transplanted to pond margins were still alive and holding their own against colonies of the other species. Colonies of *S. geminata* survived better in the woods than at pond margins, where *S. invicta* survived better than *S. geminata*, but differential survival alone cannot explain why *S. geminata* is essentially absent from pond margins and *S. invicta* from the woods. This exclusive distribution must arise either from differences in founding success, as noted above, or from long, drawn-out competitive interactions between mature colonies, in which each enjoys a small advantage in its typical habitat, but each is fully capable of existing in the other. This situation remains to be clarified.

Effect of Ecological Succession

Ant communities change as succession proceeds (Majer, 1983). *Solenopsis invicta* is generally absent from forests but appears rapidly after the forest is reduced to an early-succession habitat. Does the characteristic woodland ant fauna reappear in *S. invicta*'s place when plant succession is allowed to run its course? No one has actually followed the progress and rate of such successional changes, but in young pine plantations, fire-ant colony density began to decline at a crown cover of 15 to 20% (Pass 1960). By about 40% cover, *S. invicta* was essentially absent. What ant fauna replaced them was not determined. Why fire ant populations decline with shadiness is not known.

Humans act as a keystone species (a species that determines what other species can coexist with it), modifying ecosystems to *S. invicta*'s liking, but whether *S. invicta* in turn acts as a keystone species in the communities that receive it is not nearly so clear. That is the next question.

34

Solenopsis invicta and Ant Community Ecology

Soon after *Solenopsis invicta* took up residence in the USA, the claim arose that, when it was not decapitating full-grown bulls, it was displacing some native ant species and suppressing others. These conclusions are essentially correct with respect to its effect on other species of fire ants, as we saw above. With respect to other native ants species, these "facts" have been so consistently and uncritically repeated that today they are pretty much accepted as the Lord's Truth. A closer look, however, shows that much of the evidence is surprisingly inconclusive and weak and that some of these "facts" are not as factual as they seemed at first. The topic is an example of how difficult it can be to obtain clear answers in ecology. Studies ostensibly designed to answer questions about displacement or suppression were often not actually structured to do so. Instead of proper large-scale, long-range ecological experiments out in the sweaty, bramble-choked terrain, most studies have simply documented correlations between selected site differences and differences in the ant fauna, without experimental manipulation of putative causal factors. Even though they represent an enormous amount of work, such studies cannot identify what caused the observed differences. Conclusions about causality based on such studies are expressions of belief, not scientific facts. Community ecologists have traditionally been very cautious about claiming the existence of competitive suppression without experimental evidence. (My home institution, Florida State University, was the center of this debate in the 1970s and 1980s.) Fire ant ecologists seem to have forgotten this caution.

The South American Ant Community Is Richer

Before we take up the question of what sort of ant community *S. invicta* has joined in the USA, let us see what sort it came from in South America. Although it depends on disturbance on both continents, the host ant community is much more species-rich in South than in North America (Table 34.1; Allen et al., 1974; Wojcik, 1983; Porter et al., 1990, 1991, 1992). In Brazil, *S. invicta*'s native land, baits placed in a pasture attracted 48 species (Stimac and Alves, 1994), those in lawns 33 species, and those in cornfields 40, all far more than at comparable sites in the southeastern USA. In the Brazilian state of São Paulo, a six-month baiting and pitfall study in an old field that was reverting back to cerrado

Table 34.1. ► Comparable disturbed habitats are much richer in ant species in South America than in North America. Data from (Summerlin and Green, 1977; Claborn and Phillips, 1986; Fowler, 1990; Stimac, 1994).

Habitat	***South America***	***North America***
Pastures	48–56	4–13
Crop fields	30–40	4–9
Old fields	46–48	10–22

netted a total of 48 species (Fowler et al., 1990). Among them, *S. invicta* ranked 36th in its simple presence in pitfalls, 40th in number of individuals captured, 40th in density of activity, and 43rd in local dominance (Figure 34.1). In the exploitation of small food items (fruit flies) that could be carried away by a single individual, it ranked 38th. By these same measures, several *Pheidole* species dominated this community. Only at large baits did *S. invicta* make a relatively good showing, coming in sixth and showing that one of its strengths may be recruiting to and defending large food items (Figure 34.2). Such successes made little difference—the dominant species still made off with 25 times more baits than did *S. invicta.* Similarly, whether in a cornfield, a lawn, or a pasture in the Brazilian state of Mato Grosso (Fowler, 1988), *S. invicta* was never dominant in ant communities, even though the identity of the dominant ant species varied. The dominant *Pheidole* species in pasture and lawn were 10 to 300 times as likely to recover the large baits.

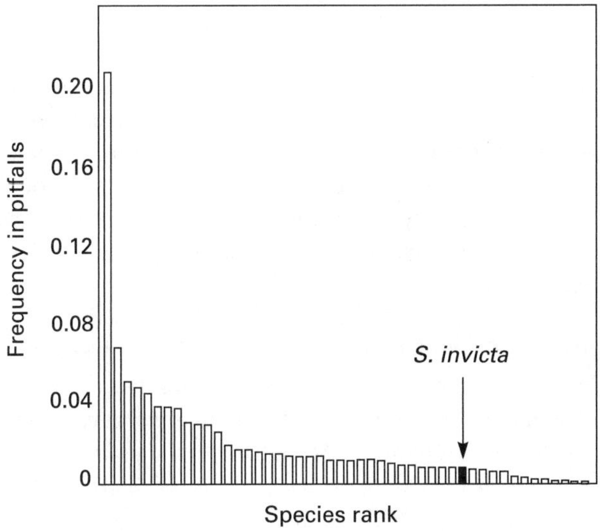

Figure 34.1. Abundance rank of ants in pitfall traps in Brazil. *Solenopsis invicta* is a relatively rare and subordinate ant in its native ecosystem. Data from Fowler et al. (1990).

Figure 34.2. Proportions of ant species at small and large baits in South America. Data from Fowler (1988).

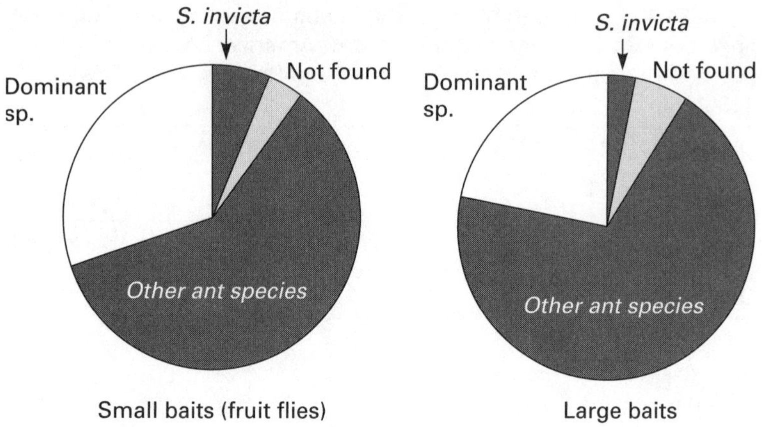

Only in the cornfield, competing against the dominant *Dorymyrmex*, did *S. invicta* have good success in dominating small and large baits. Generally, ant species in at least five genera were able to outcompete *S. invicta* through faster or more efficient recruitment and retrieval.

If we were searching for a likely virulent invader, would we have picked *S. invicta*? Probably not.

Similar pastures and old fields in North America (Florida, Louisiana, North Carolina, and Texas; Table 34.1) yielded only 4–13 species of ants (Howard and Oliver, 1979; Apperson and Powell, 1984; Phillips et al., 1986b; Fraelich, 1991), in contrast to the 50-odd species in Brazilian pastures. One might argue that the low diversity in the USA was the result of *S. invicta*'s presence, but even in the absence or near-absence of *S. invicta*, Texas pastures yielded only 7–12 species (Summerlin et al., 1977a; Claborn and Phillips, 1986). In a Texas Bermuda grass pasture at the western edge of *S. invicta*'s range, one year of trapping captured 16 species (for which at least eight individuals were trapped; Phillips et al., 1987). Agricultural land is even less diverse—Florida sugarcane fields yielded only six to nine species (Adams et al., 1981; Cherry and Nuessly, 1992). The diversity in cotton fields is so low that researchers do not even bother to report it. Under more natural conditions and with intense study, a piece of Texas savanna (Brackenridge Field Laboratory, near Austin) yielded 35 species of ants before *S. invicta* invaded (Porter and Savignano, 1990).

An Open Niche or Species Displacement in the USA?

Here then we have open habitats that appear superficially similar on two continents and lie at similar latitudes but differ greatly in the number of ant species they harbor. Why might this situation exist, and can it be a clue to the reasons for *S. invicta*'s success in the USA? Is there just simply more "room" in the USA; is the ant community not fully packed?

To develop this idea, we must turn to the study of historical biogeography and island biogeography. The latter showed that the number of species of a given taxon to which an island is home increases with the size of the island and decreases with its distance from the mainland. These principles also apply to continental areas. During the glacial periods of the Pleistocene epoch, the warm, moist habitats favored by ants contracted greatly into the southern USA and peninsular Florida. This reduced range and cooler climate probably caused many ant species to go extinct. With the postglacial warming, this reduced ant fauna was not readily replenished because migration from the warm, tropical species reservoirs of South and Central America would require species to island-hop to the Florida peninsula or cross the deserts of the southwestern USA. The southeastern USA may also always have been poor in the type of habitat that seems to favor the evolution of high ant diversity—hot, moist, open woodland.

The principles of island biogeography, extended to continental areas and habitat types, can also help us understand more recent events. If we could have flown over the southeastern USA before immigrants from Europe arrived in large numbers, it would have looked like the Amazon region of today—endless stretches of unbroken broadleaf or mixed forest, cut here and there by rivers and streams. Rarely would we have seen open patches of early successional habitat generated by various natural disasters. The same trip today reveals a patchwork landscape of forest remnants within a mostly open landscape crisscrossed by roads, dotted with towns, and peppered with houses. Today, the total amount of open, early successional area typically exceeds forested areas in most of the Southeast. This great reduction of forest area probably caused the extinction of a large number of species. Conversely, the greatly increased area of open habitat ecotype is probably "understocked,"—it has not had time to accumulate a rich fauna, and additional species could probably be accommodated there. Perhaps the open habitat in the southeastern USA was readily invaded in part because it contained too few species in relation to its total area. In South America, open, grassy habitats frequently disturbed by fire and flood have long occupied large areas of the cerrado and the Pantanal and have provided the opportunity for the evolution of many more open-habitat ant species. Such open habitat, tightly packed with ant species in South America, is a veritable cafeteria of opportunities in North America. This theoretical argument was once a favorite of Bill Buren, who trotted it out in a number of informal discussions at scientific meetings. He described our open-habitat native ant fauna as "weak." One could almost detect an upturning of the nose and an imperceptible sniff.

What Do the Data Say? If the "open-niche" line of reasoning is to bear fruit, one would expect the abundance of *S. invicta* in the USA to be added on top of the abundance and diversity of the ant fauna already present. To the extent that the natives were incapable of fully exploiting this sudden surplus of habitat, *S. invicta* should have found a space in the recently enlarged ecosystem with only minor effects on the abundance of the ants that were already there. On first perusal, this does not seem to be the case. After all, *S. invicta* eliminated its congeners, *S. geminata* and *S. xyloni*, as I described above, but these were more or less ecological equivalents whose place *S. invicta* usurped. If *S. invicta*'s relationship to the native ants is essentially similar to that previously "enjoyed" by *S. geminata* and *S. xyloni*, then it may have caused little if any reduction of the populations of native ants, even as the incumbent fire ants were usurped. The relevant question then is whether or not the invasion of *S. invicta* invariably suppressed native ant species other than native fire ants.

A preliminary reading of the literature offers little hope for this hypothesis. The results of one baiting or pitfall study after another reveals a terrific dominance by *S. invicta*. In roadsides, sugarcane fields, pastures, the story is always the same—*S. invicta* dominates the highest percentage of baits, is found in the most pitfalls, and makes up most of the specimens. Paper after paper concludes that *S. invicta* eliminates or reduces some native ants (Markin et al., 1974b; Lofgren et al., 1975; Phillips et al., 1987; Cherry and Nuessly, 1992; Stimac and Alves, 1994).

Proportion versus Number of Individuals. But not so fast! Such studies have two flaws. First, the question of suppression can only be answered by comparison of sites that have *S. invicta* with properly matched sites that lack *S. invicta* or the same site before and after invasion. Such comparisons have only rarely been made. Second, data are mostly given as percentage of the total, that is, as *relative* abundance, and this figure can be misleading. Let us say we caught 100 native ants in a pitfall sample collected in a fire ant–free zone. These make up 100% of the sample. Moving to an area where fire ants are present, we place similar traps and collect 900 fire ants along with 100 more native ants. The native ants' relative frequency has shrunk from 100% to 10%, but their actual number is still the same. To the extent that the number of trapped individuals bears some relationship to the abundance of the ants, the absolute abundance of native ants has not changed. Could the real percentages be fooling us in this way?

In a Texas study that included similar pasture within and outside the range of *S. invicta*, the number of non-*Solenopsis* ants captured in over 1400 trap-hours (8600) was actually higher in *S. invicta* areas than in *S. invicta*–free areas (5000; Claborn and Phillips, 1986). Consequently, the non-*Solenopsis* native ants composed 83% of the individuals in the absence of *S. invicta*, but only 38% in its presence, even though their actual numbers were higher. In another study, ants were collected in four different grassy habitat types at the western edge of *S. invicta*'s range for

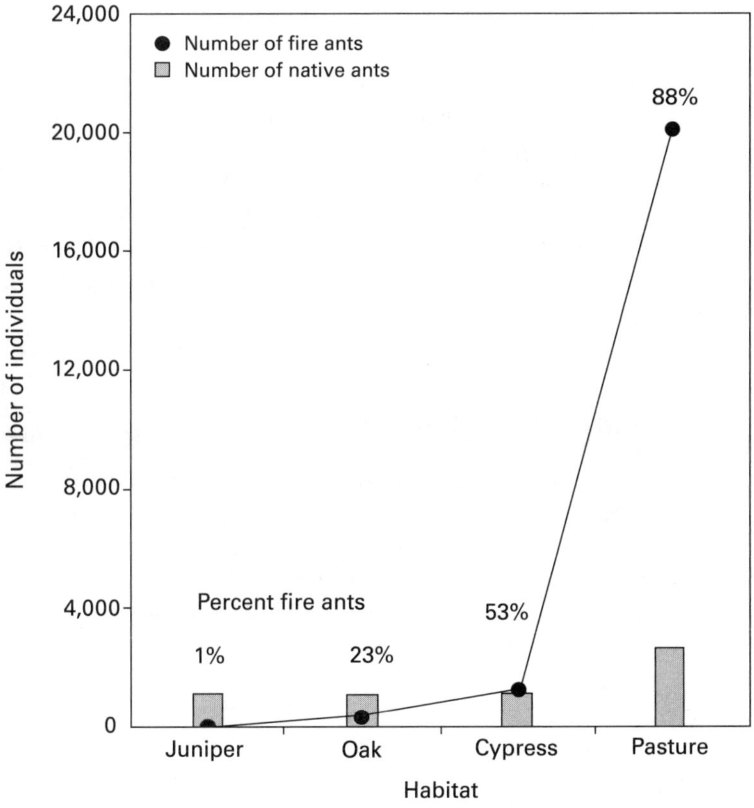

Figure 34.3. Abundance of fire ants and native ants in four Texas habitats. More native ants were found in pasture than in the other grasslands, even though they formed a smaller percentage of the total. Data from Claborn and Phillips (1986).

over 40,000 trap-hours spread over one year (Phillips et al., 1987). Three of these habitats were relatively natural and were characterized by scattered juniper, live oak, or cypress (Figure 34.3). The fourth was "improved" Bermuda grass pasture. Both species of *Solenopsis* were essentially absent from the juniper grassland, but either *S. invicta* or *S. geminata* was present in the other three at moderate to high levels (clearly, the displacement of *S. geminata* was not yet complete). Non-*Solenopsis* ants composed more than 99% percent of the individuals captured in the juniper grassland, 77% in the oak grassland, 47% in the cypress grassland, and only 12% in the improved pasture. This looks like a rout, but the numbers of individuals tell a different story—about 1100 individuals of non-*Solenopsis* ants in each of the three grassland types and 2700 in the Bermuda grass pasture. The corresponding numbers of species were 13, 13, 10, and 16. In other words, more individuals and more species of non-*Solenopsis* occurred in the Bermuda pasture than in the other habitats, but they formed only 12% of the total number of individuals captured there. No evidence here shows that fire-ant abundance is negatively related to the absolute abundance of native ants. Fire ants are an "add-on" in the highly disturbed pasture.

A similar flaw probably afflicts (although less severely) another study in central Texas near Abilene (Jusino Atresino and Phillips, 1994). In areas where it was present, *S. invicta* made up about 80% of the pitfall samples, but elsewhere, it made up less than 2%. The reported diversity index seems to show that native ants were routed, but traps in *S. invicta* areas captured about 60% as many native ants as traps in other areas. Four native species were more abundant in *S. invicta* areas, nine were less abundant, and six were too scarce to tell. Only seven of the 21 species yielded 100 or more individuals. The baiting results were even less decisive: six natives were more abundant in *S. invicta* areas, eight were less abundant, and 3 were too rare. The small sample size, low capture rate, lack of site replication, and absence of details about how the species were distributed among the pitfalls make this study hard to interpret.

If fire ants suppress native ants, then eliminating fire ants ought to increase or maintain native ants. The author of a recent experiment in Texas claimed that suppression of fire ants with poisoned bait kept ant diversity higher than in an untreated plot on which fire-ant mound density increased naturally during the course of the experiment (Cook, 2003), but the single replicate, low species richness, and lack of experimental manipulation of the high density of fire ants make this claim shaky.

The Role of Disturbance in the Solenopsis invicta *Invasion.* Everyone seems to agree that fire ants benefit from ecological disturbance. What if the same disturbance that increased the abundance of fire ants also directly decreased the abundance of native ants in North America? This pattern would create a negative correlation between fire-ant abundance and native-ant abundance, even though disturbance, not fire ants, was causing the decrease in native ants. The two factors are confounded and can only be separated by a proper experiment. A good deal of circumstantial evidence indicates that disturbance in its various forms may play this causative role.

In central Texas, Camilo and Phillips (1990) established 12 pitfall-trapped plots about 5 km apart along a transect that extended from within the range of *S. invicta* to areas beyond it. After 720 trap-days per plot, they had collected about 8000 ants belonging to 35 species. The interesting feature of this study was that the authors ranked these grassland habitats with respect to the frequency of disturbance (mowing or grazing) from one (least disturbed) to ten. The ant species fell into four distinct "assemblages" in relation to disturbance and *S. invicta.* Plots that were outside the range of *S. invicta* yielded 22 species of ants if they were undisturbed but only 13 if they were disturbed (Figure 34.4)—upon disturbance, 17 species disappeared and nine new ones appeared. Only four species were common to disturbed and undisturbed plots, illustrating that the ant fauna changed dramatically. The more

Species

Fire ants?	Disturbed?	1	2	3	4	5	6	7	8	9	10	11	12	13	14	15	16	17	18	19	20	21	22	23	24	25	26	27	28	29	30	31	32	33	34	35
No	No	■	■	■	■	■	■	■	■	■	■	■	■	■	■	■	■	■	■	■	■	■	■													
No	Yes																		■	■	■	■	■	■	■	■	■	■	■	■	■	■				
Yes	Low																		■	■	■	■									■	■	■	■	■	■
Yes	High																			■	■										■	■			■	■
Yes	Extreme																														■				■	

Figure 34.4. Occurrence of ant species in disturbed and undisturbed areas with and without fire ants. Disturbance changed the native ant fauna substantially with or without the presence of fire ants. Data from Camilo and Phillips (1990).

ecologically specialized ants were especially likely to be absent or rare in the more disturbed plots.

Within the range of *S. invicta*, ant diversity was even lower. Disturbed plots with low densities of *S. invicta* harbored 10 species of ants, of which six were the same as in the *S. invicta*–free disturbed plots. Disturbed plots with high *S. invicta* densities harbored only six species, all of which were also found in the low-density *S. invicta* plots. Finally, on two plots in heavily disturbed recreation areas, with lots of trash, trampling, and mowing, the daily pitfall catch was up to 1000 *S. invicta* per trap. Only one other ant coexisted with it on those plots.

The greatest turnover of ant species, 26 in all, was associated with disturbance. In disturbed plots, a turnover of 11 species was associated with the presence or absence of *S. invicta*. *Solenopsis invicta* might be causing this effect on the native ants, but we do not know whether the disturbance level was comparable in all of these plots because, although the authors ranked disturbance, they did not report the ranks. Finally, the loss of four of six species in the recreation areas is associated only with the highest level of disturbance. The richness of the ant community decreased as the disturbance and the density of *S. invicta* increased, but what was cause and what effect? Does disturbance reduce the native ants, allowing *S. invicta* to reach high densities? Or does *S. invicta* reduce native ant communities when it invades? Both? Can *S. invicta* even invade a native ant community without disturbance? Unfortunately, this study (and many others) cannot answer these questions because it confounds habitat disturbance with the presence or absence of *S. invicta*. It can only produce correlations. It also lacks undisturbed plots with *S. invicta*. Is that lack an accident, or does this combination not exist frequently in nature? This study leaves the door open to a primary role for disturbance and a secondary role, if any at all, for *S. invicta*.

The door is partly closed by one of the few experiments on the effect of drastic disturbance on ant communities. When forest plots in South Carolina were clearcut and converted to pine monocultures, native ant abundance and diversity decreased sharply from that on matched, but uncut, control plots (Zettler et al., 2004). *Solenopsis invicta* remained absent from forested control plots but appeared within three to 14 months on the clearcut plots, although even after two years its abundance was variable. Native ant abundance declined 40 to 80% even on plots with low *S. invicta* abundance, indicating that clearcutting, not *S. invicta*, caused most of this change in the ant community.

In this light, a study of Florida soybean fields is interesting. A survey of fields across the entire soybean-producing region uncovered 55 species of ants, but the great majority of these did not nest in the fields themselves. A few of the species that did seemed unable to produce alates. The authors (Whitcomb et al., 1972) reported that, in soybean fields lacking *S. invicta*, they often found 12 to 14 ant species but in those with *S. invicta* usually only five or six. They wrote, "[t]he most important single naturally occurring biotic factor, other than food, affecting species populations could be the presence or absence of *Solenopsis invicta*." Then in a masterpiece of understatement, they add, "Definite conclusions in this regard will have to await more precise data." Moreover, they opined that by far the greatest effect on the ant fauna resulted from cultivation practices, in other words, gross disturbance. The unanswered question is, does cultivation reduce the ant fauna and favor *S. invicta*, or does *S. invicta* reduce the ant fauna?

In 28 Florida sugarcane fields (Cherry and Nuessly, 1992), *S. invicta* occupied almost 40% of the baits, made up over 90% of the specimens, and was heavily overrepresented in pitfall traps as well. The authors concluded "since 1970 *S. invicta* has become the dominant ant species in Florida sugarcane, resulting in a large reduction in the *relative* [my emphasis] abundance of other ant species." The authors offer no data on the *actual* population sizes of native ants and their changes. Are native ants not very good at existing in the highly disturbed cane fields, whereas *S. invicta* is? None of the five native ant species occupied more than 6% of the baits, and more than half of all the baits were unoccupied. In more congenial habitats, 80 to 90% of baits are occupied, suggesting that cane fields are seriously underused by ants (or that oil is a poor bait).

Florida lawns are also highly disturbed ecosystems and are occupied by abundant *S. invicta*, along with only four or five other species of exotic ants that specialize in invading highly disrupted habitats (Cherry, 2001). In such lawns, as in sugarcane fields, no evidence indicates that in the absence of *S. invicta* the native ant fauna would be substantial or that any displacement has occurred. *Solenopsis invicta* has merely moved into an ecological vacuum and is actually an addition to the ant fauna there.

If you are still not convinced, consider another study, this time of *S. geminata*, the ecological equivalent of *S. invicta*, in milpas (fields in traditional tropical shifting agriculture) in Mexico (Risch and Carroll, 1982a, b, 1986). Milpas were created either within 40-year old forest or by plowing of a one-year-old existing milpa (a higher level of disturbance). The two types of fields were similarly planted, and their ant faunae tracked for about a year. After only five weeks, *S. geminata* dominated most of the baits in both milpas, even though it was absent from the forest surrounding the forest milpas. Although the more disturbed field milpa remained dominated by *S. geminata* through the end of the study, the ant fauna of the forest milpa gradually increased to 14 species, at which point *S. geminata* occupied only about a quarter of the baits. The more frequent the disturbance, the more abundant was *S. geminata* and the less abundant were other species of ants. Even within the two site types, *S. geminata* preferred the most open, most disturbed parts of them. As the forest began to regenerate in the forest milpa, the habitat became unsuitable for *S. geminata*, but the grassy and herbaceous field milpa continued to provide a suitable habitat. The habitat, as modified by disturbance, not competition, was the primary structuring agent.

The longest record of the progress of a *S. invicta* invasion is that produced by Wojcik et al. (2001), who baited the same 100 roadside transects in Gainesville, Florida, several times a year for 21 years, accumulating almost a million ants on over 13,000 baits. Because he used baits exclusively, the number of individuals of each species is a complex function of the abundance, recruitment style, food preference, and fighting ability of the ants present. With this caveat, the three most abundant species were *S. invicta*, *S. geminata*, and *Pheidole dentata*. Although 55 species of ants appeared on the baits, only 16 were abundant enough to test for trends. The patterns he observed show the value of patience. Many of the trends would not have been detected in the usual three- to four-year study.

Figure 34.5 shows that *S. invicta* was essentially absent at the beginning of the study in 1972, but by the early 1990s, it occurred at about half the sites and made up about 70% of the specimens. *Solenopsis geminata* showed the opposite trend. Initially, it was found at 35% of the sites and made up 55% of the specimens, but by 1992, it had declined to only 15% of the sites and 25% of the specimens. It continues to occur, but because of niche overlap, it has been usurped by *S. invicta*, possibly through territorial conflict (Wuellner and Saunders, 2003). *Pheidole dentata*, the third most abundant ant, also declined significantly, from about 20% of the sites to about 7%. The populations of 10 other native ants that prefer similar habitats decreased over these two decades, but all but one of these were rather rare even at the outset. A single native species, the trap-jawed ant *Odontomachus brunneus*, increased. In contrast, three species of introduced pest ants increased in parallel with *S. invicta*. Like *S. invicta*, these three are probably favored by disturbed

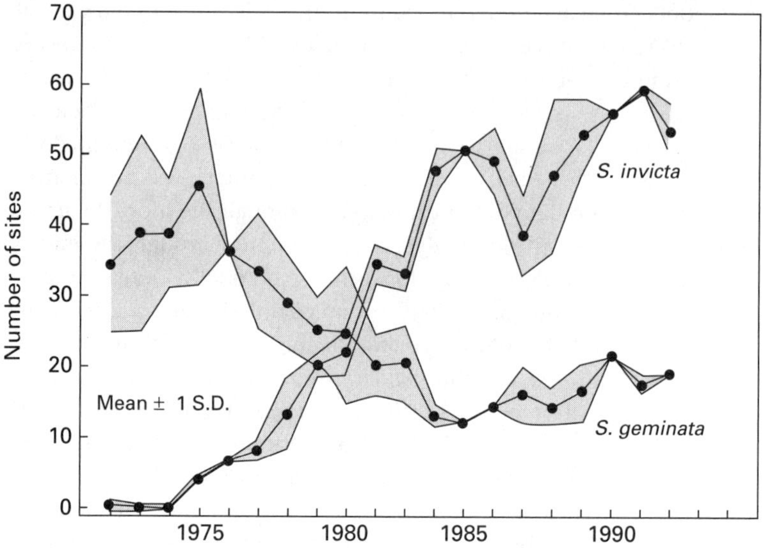

Figure 34.5. The replacement of *Solenopsis geminata* by *S. invicta* in Gainesville, Florida. Shaded areas show one standard deviation around the means. Data from Wojcik (1994).

habitat. Only one introduced species, *Cardiocondyla emeryi*, decreased as *S. invicta* increased.

Wojcik's study was not experimental and therefore cannot answer the question of what caused the observed trends. During the study, many of the sample sites became increasingly urbanized and disturbed. Roadside maintenance intensified, the property adjacent to the transects was converted from forest or field to houses and apartments, perhaps pesticide use increased, and many other changes associated with a rapidly growing town occurred. We do not know the exact nature of these changes, which transects were affected, or whether the changes were locally correlated with presence of native ants and *S. invicta*. Did urbanization make way for the invasion of *S. invicta*? Was the decline of the native ant community caused by *S. invicta* or by the habitat changes associated with urbanization or both? Was the high abundance of *S. geminata* already the product of pre-1970 urbanization and roadside disturbance? Probably so. Would habitat change have brought about the decline of the native species in the absence of *S. invicta*? The increase in other invasive ants of disturbed habitats alongside *S. invicta* points more strongly to urbanization and habitat change as the root cause. The competition between *S. invicta* and *S. geminata* is mediated by habitat disturbance; *S. geminata* wins in some habitats under low disturbance and *S. invicta* under high. So here we have another study, but we still do not really know whether *S. invicta* can invade an intact native ant community in an intact habitat. Its capacity to invade disturbed habitat is, however, beyond doubt.

My student David Lubertazzi carried out a detailed ant survey in the Florida coastal-plain pine flatwoods, a natural, though second-growth, longleaf-pine forest. The study's primary focus was not *S. invicta* but the association between recurrent ground fires (a disturbance), ground-cover vegetation differences, and the ant community. The more frequent the recurrent fires, the greater the percentage of grass (especially wire grass) in the ground-cover vegetation. Twelve plots covered the range from shrub-dominated ground cover to grass-dominated ground cover. The ant community in each plot was sampled in midsummer by means of pitfall traps, followed by ground baits and subterranean baits, as well as baits on tree trunks.

The species composition of the ground-foraging ant community was strongly correlated with the vegetation/fire frequency. By all sampling methods combined, the richest community—up to 55 species of ants—occurred in the shrubbiest, least grassy, least frequently burned plots. As fire frequency and grassiness increased, the number of ant species declined—the poorest community in the grassiest, most frequently burned sites contained only 31 species (Figure 34.6). Moreover, *S. invicta* was absent or nearly so in all except the grassiest, most frequently burned sites, and there it was quite dominant. These same sites also have a high water table and are frequently flooded for days to weeks at a time. Once again, reduced native ant diversity, disturbance, (fire, flooding) and *S. invicta* are all correlated, but which is cause and which effect? Clearly *S. invicta* cannot even be a candidate for the cause of most native ant decline, because *S. invicta* was absent or nearly so

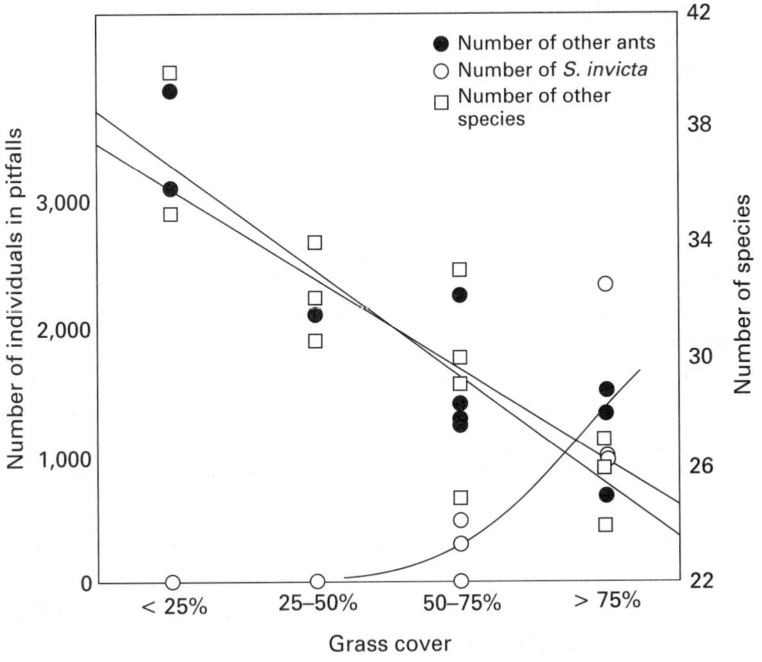

Figure 34.6. Abundance and species richness of ants at 12 Florida flatwoods sites. As disturbance by fire becomes more frequent, the ground cover becomes more grassy and the ant community poorer in species and individuals. Fire ants were abundant in only the grassiest, most frequently disturbed plots (open circles, curve). Number of species was adjusted to the same sample size in all plots by a method called rarefaction. Data from Lubertazzi (1999).

until grass exceeded 75%, yet species richness declined from about 38 to about 30 between 0 and 75%. Further decline is simply a continuation of this trend; we have no strong reason to attribute it to *S. invicta*. More probably, the changes in the ant community are caused by the same factors that sculpt the vegetation. When the fire/flooding disturbance in the pine flatwoods is intensified to extreme levels, the result is a treeless, grassy, seasonally flooded savanna. There, *S. invicta* and *Tapinoma sessile* are the only ants to be found, both frequently foraging in the grass over water (Figure 33.5). Extreme disturbance seems to have resulted in an extreme reduction of the ant community, leaving only two species standing. Interestingly, a low ant diversity and a dense population of *S. richteri* in South America were also associated with flooding (Folgarait et al., 2004).

My own study (discussed above) in the sandhills and flatwoods of northern Florida showed that *S. invicta* and *S. geminata* are favored by various kinds of habitat disturbance. An extreme case of attraction to a different kind of disturbance comes from a study in Nicaragua (Perfecto, 1991). There, the players are *S. geminata* and 11 other ant species associated with weedy fields. A small fallow field was plowed during the dry season and then repeatedly baited. Plowing is, of course, a major disturbance, and during the first 20 days, the percentage of baits occupied by *S. geminata* rose rapidly from 5% to 60% while those with other ants crashed from over 70% to about 25%. The *S. geminata* invasion occurred from the edges inward and probably emanated from a small number of mature colonies. This species is certainly not put off by gross disturbance. In the Philippines, *S. geminata* remained abundant in fallow, bare upland rice fields throughout the dry season even when fields were plowed (Way et al., 2002). I tentatively assume that *S. invicta* would behave similarly. After about a month of stability, the rainy season stimulated the growth of weeds, and the field slowly reverted to the initial ant community in which *S. geminata* was rare. A plowed field in Costa Rica was similarly invaded by *S. geminata*. Perhaps one should not make too much of such a small study, but it does demonstrate the attraction that disturbance has for a fire ant. Many other examples demonstrate the association of ecological disturbances with similar changes in the ant community. In places as widely scattered as Amazonia, Venezuela, Australia, Guadalcanal (Greenslade and Greenslade, 1977), the Solomon Islands, and Africa, heavy ecological disturbance usually caused the replacement of a diverse ant community of specialized species with a less diverse one containing many generalized, opportunistic, and often widespread species. In fact, such changes in ant communities have been used as bioindicators of ecological disturbance (King et al., 1998).

A final illustration from my own experience: once, on a field trip to view the management of mahogany in the Petén region of Guatemala, we bumped and bounced along logging tracks through an unbroken green tunnel of majestic, mature forest. Emergent giant trees hulked massively among the modest stems of the common soldiers, sunlight speckling the dark forest floor. At intervals, we stopped to heed nature's call or view something of significance, and I would take the opportunity to collect ants, finding no fire ants. After eight spine-punishing hours we reached a logging camp in a small clearing, the first clearing we had seen in eight hours and 60 km. In the clearing huddled three open-sided palm-thatched huts, and there, in a small pool of sunlight, against the foundation of one of the huts, was a single, large fire-ant mound (albeit *S. geminata*). How many dispersing queens died traversing this unbroken green ocean before one found this tiny island of disturbed habitat? Did they follow the road? Did they come by truck? Or does the harvest of scattered single mahogany trees create enough disturbance to maintain a sparse, highly dispersed population of fire ants in the region?

Ant Communities on a Geographic Scale. On a long road trip from Tallahassee, Florida, back to their home in Vermont, Nick Gotelli and Amy Arnett stopped every 50 to 60 km to trap ants by two days of pitfall trapping—25 pitfalls in a 5- by 5-m grid—in an open area and another, similar grid in a nearby wooded area (Gotelli and Arnett, 2000). Each location was thus sampled for 100 trap-days (a rather small sample, especially as the pitfalls were also small). The final total was over 14,000 ants of 82 species.

A plot of the number of native ant species in each sample location against the latitude of the site produced a hump-shaped distribution—low at the northern end (44.5° north longitude), maximum near the middle (35–38° north), low again at the southern end (about 31° north). The hump coincided broadly with the northern limit of *S. invicta's* range. A simulation showed that this result was unlikely to be a coincidence. In addition to the reduced number of species, particular native ant species were less likely to co-occur at sites outside *S. invicta*'s range than inside. This pattern suggested that *S. invicta* had disrupted the normal competitive relationships among native species, perhaps by reducing the density of native ants, by eliminating dominant native competitors from the community, or by other means. Conclusion: *S. invicta* was the "proximate cause" of the lower species richness of ant communities within its range.

Both of these findings are consistent with the widely held belief that *S. invicta* hammers native ant communities, but as a party-pooper scientist, I am obliged to point out other possible explanations as well as shortcomings, as the authors themselves did. The evidence is correlational, not experimental—after all, no similar fire ant–free control transect, historical or modern, is available for comparison to Gotelli and Arnett's. The model is geographic, not local, even though co-occurrence is a local phenomenon. The sample sizes were very small, and the spatial sampling scale was tiny—smaller than a single, mature fire-ant territory. Might the fire ant affect only the spatial distribution of the community, not its richness? Why were the forest samples not much more species rich than the field samples as expected from other studies? Which native ant species were affected, and what is their biology? The scant published data preclude a biological interpretation on a local scale.

The mid-latitude peak in species richness is fundamental to the case, but it too may have origins independent of *S. invicta*. The species richness of birds and trees in eastern North America also peak at middle latitudes, but surely these peaks are not related in a causal way to *S. invicta*? A recent publication by Colwell and Lees (2000) presents a much simpler explanation for the middle-latitude hump. When species ranges of random size are randomly cast (in a computer) into a fixed domain (imagine an island or a biome) such that each range lies entirely within the domain, transects through the domain show a maximum in species richness in the middle of the domain. No biology is involved here, only geometry. Several analyses of real biogeographic data have shown that the mid-domain effect can account for up to 85% of the variation in species richness. Gotelli's maximum ant species richness falls at the approximate middle of the eastern deciduous forest biome. The biologically neutral mid-domain effect offers a simpler explanation, especially in light of evidence above that casts doubt on the competitive displacement of native ants by *S. invicta*.

Poisoned Baits as a Disturbance of the Ant Community. A specific type of disturbance that can favor *S. invicta* is the disruption of ant communities with insecticidal baits. No such baits have a high degree of specificity; they reduce or eliminate most scavenging species of ants. If the toxicant is very persistent, as mirex was, it may poison ants that scavenge the first-killed ants as well. Evidence that such indiscriminate killing of ants can promote the invasion by *S. invicta* has been available since the mid-1970s (Summerlin et al., 1977a). In 1975, two years after *S. invicta* first appeared in Grimes and Brazos counties, Texas, its populations were still low. At that time, Summerlin and his colleagues surveyed an experimental plot for ants and then treated the plot with mirex. Within two months, the 70 colonies of 12 species had plummeted to four colonies of three species. One year later, 11 species once again occupied the plot, but the species composition had changed radically. The three original *S. invicta* colonies had been replaced by about 50 new ones, changing the status of *S. invicta* from a minor species to the dominant ant on the plot (Figure 34.7). Nests of *Dorymyrmex insana*, another ant of disturbed places, increased about sevenfold, but because its colonies are much smaller than those of *S. invicta*, they contain less than 10% as many ants as do mature *S. invicta* colonies. No other

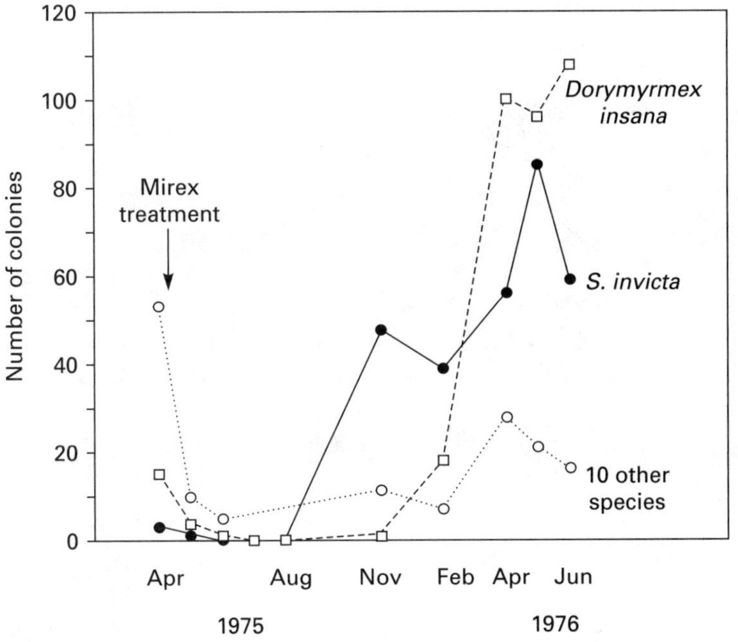

Figure 34.7. Changes in the ant community after treatment with the poisoned bait mirex. Fire ants (*Solenopsis invicta*) were a minor component before treatment but were dominant after recolonization. Data from Summerlin et al. (1977a).

previously abundant ant species rebounded beyond its previous levels. The disruption of a native ant community by mirex poisoning appears to have made it possible for *S. invicta* to invade and dominate a site rapidly. Unfortunately, the study has a serious flaw in that it lacked a control (minor detail, eh?). One could argue that the fire-ant invasion of Grimes and Brazos counties was just getting in gear and that the increase in the dominance of fire ants would have occurred in the absence of the mirex treatment as well.

However, a later experiment at the University of Florida strongly supported the role of the ant community in limiting *S. invicta* (Stimac and Alves, 1994). On circular, replicated plots in pastureland near Gainesville, Florida, Stimac and his assistants located nests of 17 species of ants, of which the six most abundant were *Pheidole morrisi, P. dentata, Dorymyrmex bureni, D. insana, S. geminata* and *S. invicta*. Next, Stimac treated some plots with Amdro, treated some with mirex, and left some untreated as controls. For the remaining 30 months of the study, populations on control plots remained fairly stable, and *S. invicta* made up less than 15% of all the colonies. The poisoned bait–treated plots contrasted sharply with the controls (Figure 34.8). Before treatment, these had harbored more than 50% *S. geminata* colonies and no *S. invicta*. Thirty months after treatment, *S. geminata* was almost absent, and the plots were more than 50% *S. invicta*, an almost exact replacement. The species of *Dorymyrmex*, previously so dominant, had been almost eliminated. Spreading indiscriminate death among the

Figure 34.8. Treatment with the poisoned bait Amdro brought about radical changes in the ant community. *Solenopsis invicta* almost completely replaced *S. geminata*. No such changes occurred in untreated controls. Adapted from Stimac and Alves (1994).

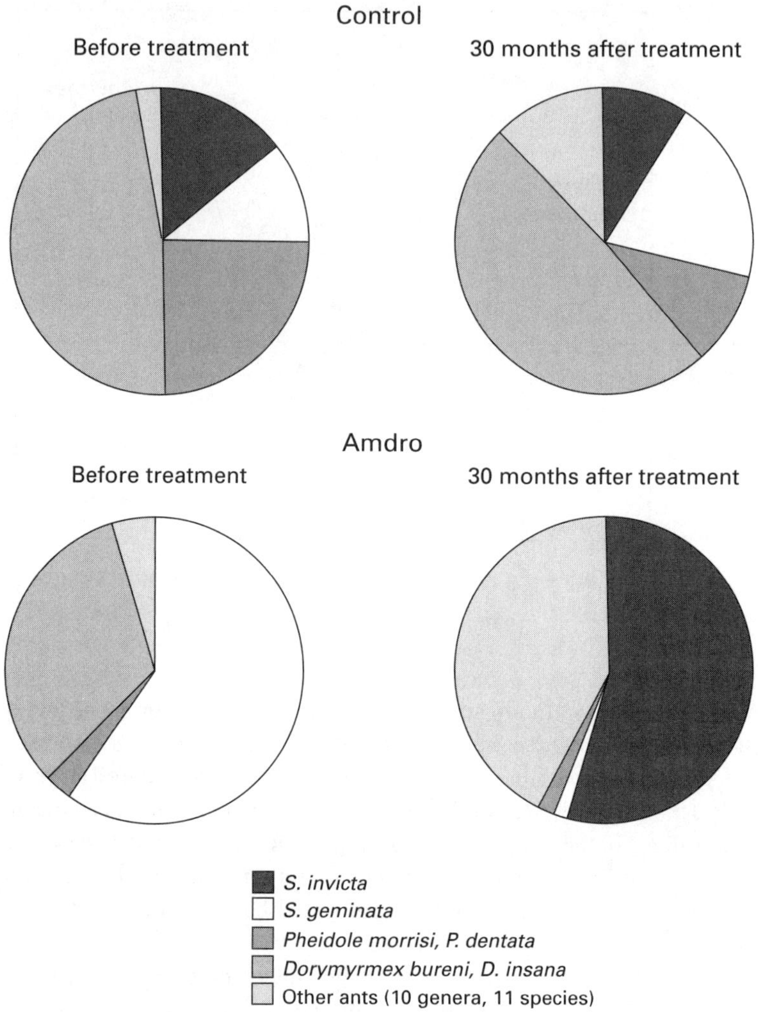

native ants had converted *S. invicta* from a minor player to the 800-pound gorilla. Although this result clearly has political implications for large-scale poisoned-bait programs such as the mirex program, the take-home message here is that the ecological disturbance favoring *S. invicta* need not be as gross as the bulldozer or plow—it can be quite specific to the ant community.

Many questions remain unanswered. The boost to *S. invicta* populations in both of these studies probably proceeded through colony founding in ant-free space; *S. invicta*, with its huge numbers of female alates, is particularly good at this process. Conversely, the stability of control plots suggests that *S. invicta* is not particularly good at invading the native community through the gaps left by the routine deaths of community members, i.e., by turnover. Or do they succeed but too slowly to be detected in three years? In our Southwood Plantation studies of *S. invicta* populations, replacement of dead *S. invicta* colonies with young ones was routine. How does this process differ when the community is mostly native ants? Are native ants simply efficient at extirpating colony-founding *S. invicta* queens? This process seems unlikely to account for *S. invicta*'s relative inability to invade intact ant communities. Mature colonies of *S. invicta* also kill such queens, yet new colonies appear every fall. Do the huge numbers of founding queens simply overwhelm the colonies' ability to eliminate them? Why would the same not be true in native ant communities?

A possible reading of the data in hand is that many different kinds of disturbance reduce native ant diversity/abundance and make room for *S. invicta*. Poisoned bait is merely the most effective reducer of diversity and density, driving both nearly to zero. This reduction would also make it the most effective enabler of *S. invicta* invasion. Flooding and soil disturbance may mediate *S. invicta*'s invasion by eliminating subterranean ant predators such as thief ants (Buren et al., 1978; Lammers, 1987). The dependence of these tiny ants on a system of widespread and diffuse tunnels (Thompson, 1980) ought to make them susceptible to soil disturbance, compaction and flooding, but work by Joshua King (2004) found these ants not to be particularly limited by these factors.

Turning the situation around, is an ant population greatly dominated by *S. invicta* stable, or does the native ant community gradually become reestablished, perhaps over years? No one has done such a study on monogyne fire ants (one on polygyne colonies is described below), but its results would be important.

A final comment about using poisoned baits to manipulate fire-ant densities in experiments: such has been common practice but is clearly a poor technique because it is so indiscriminate, killing not only many other ant species but other scavenging insects as well. Observed effects cannot be ascribed to the presence or absence of fire ants.

A Polygyne Invasion

So far, we have considered only the ecological effects of the monogyne form of *S. invicta* and have uncovered only inconclusive evidence that *S. invicta* reduces native ant diversity. When the invader is the polygyne form, the outcome looks radically different, but even this study has a surprise ending, as we shall presently see.

For drama, it is hard to beat the story of The Invasion of Brackenridge in Austin, Texas, the same state that gave us the Alamo. Unlike the monogyne *S. invicta*, the polygyne colonies at Brackenridge (a research station of the University of Texas) reproduced by budding like a gigantic

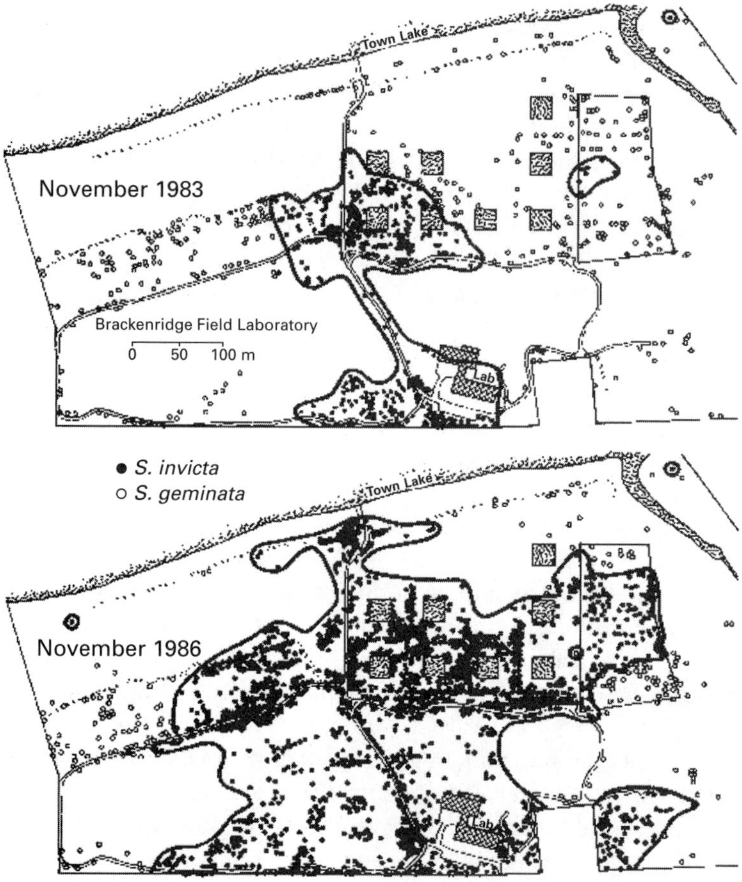

Figure 34.9. The invasion of Brackenridge, Texas, by polygyne *Solenopsis invicta*. Heavy solid lines separate the fire-ant population from the native ants. Reprinted from Porter et al. (1988) with permission from the Entomological Society of America.

bacterial culture. When Sanford Porter arrived at the laboratory in 1983, *S. invicta* was just beginning to appear in three localized areas of this former patch of prairie on the way to growing into scrub forest. Four times between October 1983 and November 1986, Porter, van Eimeren, and Gilbert mapped all the colonies of *S. geminata* and *S. invicta* and then compared the ant and arthropod fauna of areas that had and had not yet been invaded by *S. invicta*.

Porter's maps (Figure 34.9) convey the story much more powerfully than words (Porter et al., 1988; Hook and Porter, 1990; Porter and Savignano, 1990). The polygyne population advanced like an army along irregular but sharp fronts, about 35 m per year in the open, sunny areas along roads and at half that rate but no less inexorably in shady, wooded areas. In 1986, the isolated population fused with the main one spreading from the center. By this time, it occupied half of the Brackenridge Field Laboratory. The rate of spread implied that the population must first have appeared in 1980 in the center of the property and around the laboratory building and would complete its occupation of Brackenridge in 1989 or so (which it did). Certainly, *S. invicta* was absent in 1978, when Don Feener carried out detailed surveys of Brackenridge ants (Feener, 1978). The polygyne fire ant took about a decade to complete the invasion of about 0.33 km^2, a rate typical of other polygyne ant invasions. The invasion is not aerial but one that takes place on foot, through the construction of frontier outposts along a broad front, like the advance teams of road builders and their defenders in the invading Roman army. Although polygyne *S. invicta* produce fewer alates than monogyne *S. invicta*, both it and *S. geminata* do bombard each other's territories with thousands of newly mated queens, but these rarely succeed in establishing new colonies in enemy territory. Rather, budding is the principal mechanism of expansion.

The ecological consequences of this invasion were extreme. Pitfall traps, baits, litter samples, and hand searching all told the same story. As the invading front advanced, Brackenridge's moderately dense, diverse community of native ants, including the native fire ant *S. geminata* (Feener, 1978) was almost completely replaced, colony by colony, by a much higher density of *S. invicta*. The number of native ant species plummeted by 70%, and the numbers of individuals by 90%. Thirteen of 35 species of native ants vanished completely, and another 10 were significantly reduced. At least five colonies of harvester ants were seen to be attacked and destroyed by *S. invicta* (Hook and Porter, 1990). A later study in a native Texas prairie invaded by polygyne *S.invicta* (Morris and Steigman, 1993) generally confirmed these effects—ant species were reduced from 12 to four, whereas in an area in which *S. invicta* was suppressed with poisoned bait, nine species were found.

At the same time, the density of *S. invicta* was much higher than that of the ants they replaced. Over 1100 *S. invicta* mounds replaced 180 *S. geminata* colonies, a ratio of six to one. Pitfall traps captured three times as many *S. invicta* as *S. geminata*. The total number of *S. invicta* mounds rose from 500 in 1983 to over 2300 in 1986, and the density reached 400 mounds per hectare in sunny habitats. While the diversity of ants in pitfalls took a nosedive, the total number of ants increased almost 30-fold (from a mean of 165 to one of 4469). More than 99% of these were *S. invicta*. Baits were discovered 20 to 30 times more quickly after the invasion, and ants recruited to them four times as quickly, but practically all of this recruitment was by *S. invicta*. The original dominant, *S. geminata* (which is also typically polygyne at Brackenridge), as well as other native ants, had been replaced with many times the number and biomass of polygyne *S. invicta*. How can *S. invicta* manage this huge increase? Is it tapping an unused food source or exploiting food sources more efficiently? Perhaps the absence of natural enemies

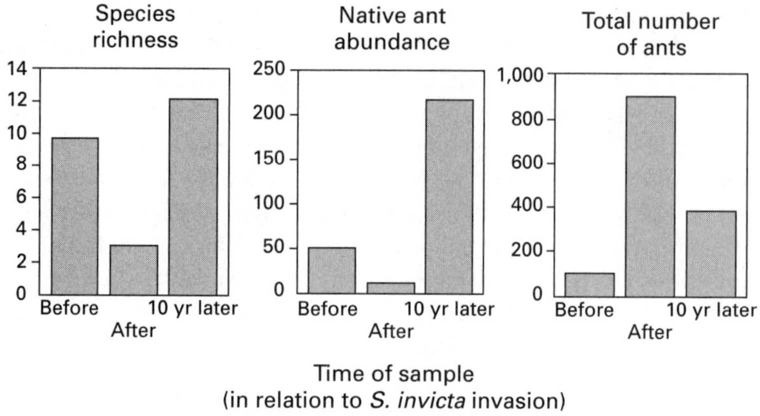

Figure 34.10. The ant community at Brackenridge Field Laboratory, near Austin, Texas, before invasion by the polygyne *S. invicta*, immediately after, and 10 years later. The community recovered or more than recovered from the invasion by polygyne *S. invicta*. Data from Morrison (2002).

contributes. At the very least, it has monopolized resources previously used by an entire feeding guild of ants and possibly other organisms as well. Perhaps it is feeding directly on plants, as has been proposed for a number of dominant ant species.

The impact of *S. invicta's* invasion extends beyond the ant community, affecting all animal communities—the biodiversity and abundance of arthropod species other than ants dropped by 30–60%, in parallel with the 70% decline in ant species. In Porter's words, ". . . monopolization rather than dominance may be a more appropriate description of how these ants affect their host communities." The exact mechanism of these revolutionary changes remains unknown, although interference and aggressive competition are the most likely. When Porter concluded his study in 1988, he noted that the consequences of the invasion seemed long lasting. He sought solace in taking an evolutionary point of view—over centuries, the species would adapt and reestablish a more balanced community, as had happened after the invasion of Caribbean islands by *S. geminata* four to five centuries ago (Hölldobler and Wilson, 1990).

This is a mighty gloomy picture, but wait! Now comes the surprise sequel! When Lloyd Morrison resampled Brackenridge 12 years after *S. invicta* completed its invasion, the picture had once again changed radically (Morrison, 2002). As exactly as possible, Morrison used the same sampling methods, sample locations, and sample sizes as Porter did in 1986. What he found was (perhaps) surprising—the native ant species' richness and abundance had rebounded to (at least) its preinvasion levels (Figure 34.10). Total ant abundance was higher than it was before the invasion but still much lower than it was in 1987, when *S. invicta* made up 99% of the captured ants. *Solenopsis invicta* remained a very abundant ant in 1999 but was only 10% as abundant as in invaded areas in 1987. Morrison collected 49 ant species at Brackenridge in 1999; Porter collected only 35 in 1987. Forty-two of these 49 had been collected at Brackenridge before 1989 and represent 86% of all the species ever found at Brackenridge. Seven were new—all were cryptic or rare species that are easily overlooked.

Although the dire prediction of permanent losses of ant species resulting from the invasion certainly did not come to pass, some species were clear losers. The most obvious was *S. geminata*, of which only about a dozen colonies remained in one isolated area in 1999. The harvester ant *Pogonomyrmex barbatus*, in spite of direct aggressive attack by *S. invicta*, was still present in 1999, although in greatly reduced numbers. Like the ants, all major arthropod groups rebounded to their preinvasion abundance and diversity. In 1999, arthropod diversity and abundance at all sites were similar to their 1987 preinvasion values (actually abundance and richness were higher, for unknown reasons).

Some of these reversions might have been the result of vegetation changes, but aerial photographs and other data showed that this was not the case. The final conclusion then is that the community changes and reversions were responses to the invasion by polygyne *S. invicta*. Clearly, this native ecosystem was resilient enough to recover, but the mechanism of the rebound is obscure (though of great interest). Among factors that might have been at work are (1) that *S. invicta* depleted or overexploited the food base it had commandeered; (2) that natural enemies adapted to using *S. invicta* as a host (*Thelohania solenopsae*, a microsporidian parasite of *S. invicta*, was found at Brackenridge in the fall of 2000, but its role in the reversion is not known); and (3) that native ants adapted behaviorally, becoming able to coexist with *S. invicta* and to reclaim some of the food base. Whether the decline of *S. invicta* is the cause of the return of native ants or their return has caused its decline is not known.

One should probably take care when extrapolating this good news to other biological communities. The outcome of an invasion is clearly tempered by habitat type and habitat changes, especially disturbance, and may differ greatly from ecotype to ecotype, as well as geographically. Finally, for invasions by the monogyne *S. invicta*, studies comparable to those of Porter, Morrison, and their coworkers are unavailable, and we are left hungry for answers.

So here we are again, this time with a more virulent polygyne invader, and we still cannot demonstrate a negative, long-term effect on the ant community, except for that on *S. geminata*. In fact, species richness has increased by one. Rosenzweig (2001) argues that the ultimate effect of species introductions will be to increase the species diversity of the world's biological provinces while at the same time decreasing the earth's total species diversity through extinctions. He points out that threats to species diversity from habitat loss far exceed those from exotic species introductions. Some of this same habitat loss is also what makes the invasion by *S. invicta* possible.

Overview of the Effect of *Solenopsis invicta* on Native Ants

We can accept as fact that *S. geminata, S. richteri,* and *S. xyloni,* all historically abundant ants, have been displaced by *S. invicta.* They are simply absent from vast areas where they were previously abundant. We cannot easily draw a similar conclusion with respect to other species of native ants, as I argued in the pages above. Yet, it seems so obvious that there *ought* to be an effect. Consider the situation: a new, vigorous competitor slips into the ecosystem and exploits the same resources as other ants—surely it must have an effect on them? The seductiveness of this belief is strong, and author after author has concluded that *S. invicta* suppresses native ants. Here is a typical statement: "[*S. invicta*] reduces native ant populations either by direct predation or indirectly by competition for food and nest sites. Over time without any additional ecological disturbance, the number of RIFA [red imported fire ant] colonies will increase to the carrying capacity of the habitat" (Wojcik, 1983). Here is another: "*Solenopsis invicta* is best known as an opportunistic predator that simplifies the ecosystem by displacing other ant species . . ." (Tedders et al., 1990). Such statements are not well substantiated. I say so for several reasons: (1) the evidence is always circumstantial rather than experimental; (2) it is usually based on relative abundance rather than absolute abundance; (3) it almost always confounds habitat disturbance, *S. invicta* density, and native ant abundance (in fact, evidence indicates that habitat disturbance directly reduces native ant diversity, even in the absence of *S. invicta;* poisoned bait certainly does); (4) ant communities were often sampled by baiting, a highly biased and inefficient method (King, 2004), and (5) many studies undersampled the study communities.

Because science is an endless argument, I feel obligated to acknowledge some possible counterarguments. First and foremost, most studies, although not strong by themselves, showed a negative rather than a neutral or a positive effect of *S. invicta* on native ants, so all together they are strongly suggestive. Perhaps this argument has merit, but it still leaves the effects of disturbance, *S. invicta,* and native ants badly tangled. Moreover, a paper by Morrison and Porter (2003) greatly weakens this argument: if *S. invicta* suppresses other ants, high densities of *S. invicta* within a habitat type should be associated with low densities and diversity of other ants, but in surveys of a large number of Florida pastures with naturally varying *S. invicta* densities, just the opposite was true—*S. invicta* and other ants (and other arthropods) were positively associated; perhaps both were regulated by some common factor(s).

Another counterargument might be that several of the studies I used to show the lack of an effect of *S. invicta* on native ants were in areas near the western edge of *S. invicta*'s range where the ant had been for 10 years or less. One could argue that suppression was in progress and, given a few more years, would become obvious. Alternatively, perhaps *S. invicta* was near the limits of its physiological tolerance and thus less able to suppress native ants, but *S. invicta*'s absolute abundance robs this argument of merit.

Only Porter's "natural experiment" at Brackenridge showed clearly that *S. invicta* is able to suppress native ants in the absence of any other habitat changes. Many authors carelessly cite this study as supporting their own claims of suppression, but few of them emphasize or even note that Porter's *S. invicta* were polygyne and their own were monogyne. In ecological terms, these are different creatures. In any case, the suppression was temporary, and a decade later, all the native ants are back to their preinvasion abundances. In the end, *S. invicta* has been added to the fauna without displacement of any native ants except its ecological equivalent. If this is the result with polygyne *S. invicta*, "Anttila the Hun" of the insect world, why should we believe that the monogyne form has any negative effect on native ants? Perhaps it does, but show us the data! At the very least, disturbance and competition probably interact.

Ecologists argue about the relative importance of biotic factors such as competition and the abiotic, physical characteristics of the habitat in structuring communities. Notions of competition have biased the interpretation of fire-ant community data for more than two decades. After all, all you needed was a piece of Spam to show that fire ants competed with native ants. Yet, ants fighting over Spam do not necessarily indicate an effect on coexistence in the ant community. Indeed, as I have attempted to show, the habitat probably has more explanatory power with respect to the native ant community than does competition. Only long-term, replicated experiments can finally settle this argument. (As this book is being completed, Joshua King and I have such a study under way in the Apalachicola National Forest and at Southwood Plantation.)

Mechanisms of Coexistence

Most evidence suggests that ant communities are structured first and foremost by the habitat. *Solenopsis invicta* is rare in or absent from undisturbed native habitats, and little evidence indicates that it has any impact on the native ants in them. It coexists only with native ants that also prefer disturbed habitats, as was shown by King's thorough inventory of five Florida ecosystems (King, 2004). Whether it suppresses these native ants there or not, much research has been stimulated by the widespread belief that it does. Differences in the details of ant species' biology were thought to allow each to exploit the resources in ways that determined its share of resources, its abundance, and the makeup of the community. Therefore, the details of the behavioral interactions between *S. invicta* and species of native ants, especially aggression and foraging, were subjected to repeated scrutiny (Baroni Urbani and Kannowski, 1974; Apperson and Powell, 1984; Claborn and Phillips, 1986; Phillips et al., 1986a; Claborn et al., 1988; Banks and Williams, 1989; Stein and Thorvilson, 1989; Jones and Phillips, 1990; Stein et al., 1990). Many authors suggested that several types of separation reduce direct competition between *S. invicta* and various native ants, allowing coexistence. Although no one has linked such observations of "microcompetition" to any changes in the ant community, reviewing their findings seems in order.

Separation in Time and Space. Two species might be able to coexist if they do not occur together in space or time. In a North Carolina pasture with abundant *S. invicta* and *Lasius neoniger* (a more northerly ant), pitfalls and baits that collected a lot of *S. invicta* tended to capture few *L. neoniger* and vice versa, suggesting spatial separation. The two never occurred simultaneously on the same baits. In an improved pasture in Bandera County, Texas, all three dominant ant species used the same areas, but both *S. invicta* and *Dorymyrmex* sp. were active primarily at night (more than 90% of all captures were nocturnal) and often ended up together in pitfalls, whereas *Forelius pruinosus* was active during the day (Claborn and Phillips, 1986; Phillips et al., 1986a; Claborn et al., 1988). Two-thirds of the pitfalls that captured *S. invicta* by night captured *F. pruinosus* by day, indicating that the separation was not in space but in time. This temporal separation probably resulted from differences in thermal tolerance, but whatever the origin, it reduced the interaction between potential competitors. Of course, it would not eliminate the indirect effects, if any, of resource depletion. Other claims of temporal adjustments in foraging by native ants in the presence of *S. invicta* have been made (Jusino Atresino and Phillips, 1994) but are unconvincing.

Living with Competitors. Each of four co-occurring Texas species (*Monomorium minimum, Forelius foetidus, S. invicta*, and *Pheidole dentata*) was tested alone for its intrinsic ability to find and retrieve food (Phillips et al., 1986a; Jones and Phillips, 1990). An overall composite index of time efficiency indicated that *S. invicta* outperformed only *M. minimum* and was outperformed by both *F. foetidus and P. dentata.* The importance of these findings to an explanation of coexistence is difficult to see, because the importance of foraging efficiency to competitive coexistence has not been demonstrated.

Ten individuals of the same four species were then placed in a dish to fight with 10 of another. When the workers were of different size, the larger ones usually killed more of the smaller ones than the reverse, and most confrontations among size equals were not significantly different. *Solenopsis invicta* beat both *F. foetidus* and *M. minimum*, and *P. dentata* beat *F. foetidus*. No other pairs differed significantly in kill rate. Only *S. invicta* and *P. dentata* attacked each other after a single contact—all others took 8–11 contacts before combat ensued. Again, despite these differences in aggressiveness, the value of these findings in explaining the relationship among these species in the field, where workers of two species tend to avoid one another after a single contact, is not clear. In a similar vein, several species of ants will readily invade small fire-ant colonies in the laboratory (Rao, 2002), but whether they do so in the field and the process is of any importance there are entirely unknown.

In a Louisiana pasture, *M. minimum* often bested *S. invicta* and several other ant species through the use of a powerful repellent. Adams had previously described such chemical interference competition (Adams and Traniello, 1981). In a truly strange twist, *S. invicta* exploited all the baits scattered generally in the plot, but 30% of the baits placed directly on the mounds of living *S. invicta* colonies were exploited by *M. minimum*.

Sometimes the speed of recruitment allows ants to exist in the presence of *S. invicta*. For example, in one Brazilian study, *Paratrechina longicornis* appeared on baits very quickly, only to be driven off eventually by the slower *S. invicta* workers by means of gaster flagging and aggressive lunging. In addition, *P. longicornis* was active during the heat of the day, but *S. invicta* became much more active after sundown, a difference that added another degree of separation.

Food Preferences. Coexistence might also be facilitated if species preferred different foods. In field studies near Houston, Texas, some ant species were found to "prefer" high-carbohydrate baits whereas other "preferred" high-protein baits (Stein and Thorvilson, 1989), but with respect to coexistence, several problems arise. First, the species are presumed to be limited by food. Second, appearance on baits may be as much a function of what the species succeed in competing for as what they "prefer." Third, food preferences almost certainly change with the condition of the colony—the relative absence of larvae probably explains why *S. invicta* colonies prefer sugars in the cool season, and their presence explains why the colonies prefer protein in the warm season (Stein et al., 1990).

An interesting aside with possible bearing on aggression and appeasement: Awinash Bhatkar found that, when an aggressor from another colony or species confronted a worker with a crop full of food (in the laboratory), the fed worker offered a droplet of food to the aggressor in a ritualized manner (Bhatkar and Kloft, 1977; Bhatkar, 1979a, b, 1983). This offering often deterred an attack by the aggressor, even postponing colony invasion in one experiment. In a more naturalistic setting, this behavior can result in a form of highway robbery when aggressive ants waylay foragers returning on an odor trail. Some ant guests may exploit the same tendency. The ecological significance of this behavior is unexplored, although roles in territorial behavior, intercolony competition, polygyny, and so on have been suggested.

◂•• *Membership in a Prestigious Organization*

Advancement through the ranks of a university faculty requires that you have an impressive curriculum vitae, with lots of publications, graduate students, committee work, honors, and professional society memberships. In the early 1970s, when I was just starting my research career (it's never just "research," always "research career"!), my vita looked pretty anemic. Sure, I'd been a teaching assistant at Berkeley, and sure, I had a few publications, and sure, I was a member of the AAAS and the ESA alongside about 50,000 other honorees. What I needed was something spicier, some special accomplishment or some prestigious organization that I and only a few others had been invited to join. The Nobel Prize seemed somewhat out of reach for the time being. The National Medal of Science was too political. What then? I shared my predicament with my students over Friday afternoon beers. As the beer dosage increased, imaginations expanded.

How about our own organization of fire ant researchers? said Bill. We could call it the Fire Ant Research Group or something, and the only people who could join are those we invite. I'd like Pancho to be a member.

Who's Pancho, asked John.

Bill sucked on his beer before answering, My dog.

I could see that this idea had potential, especially for discussions over beer. A charter was suggested, and more members, too. Eventually the pizza came and the ever-looser talk turned to other topics.

Monday morning, Bill and John came into my office. Bill said, We've figured out the name of our new organization—Fire Ant Research Team.

After a moment's reflection, during which I considered such things as acronyms, I knew we had hit pay dirt. Within half an hour, we had decided on a charter (which was to be kept strictly oral), a monthly publication whose first issue was to focus on a review of modern evolutionary biology as presented in the Uncle Milton's Ant Farm brochure, and a list of charter members. Pancho was fourth on the list. It was to be a democratically run society of which I was designated Founder and President For Life. Every member (including Pancho) had a vote, but I had as many votes as I wanted. New members were to be chosen by any member who wanted or could be self-chosen. The organization had no dues and no meetings, except unofficial ones that were required to feature beer.

We need a motto, said Bill. How about, "Today Florida, Tomorrow Dixie"?

Done!

Because these were the days of the Second Great Fire Ant Eradication and the USDA's mirex program was in full swing, we decided that a battle cry would also give us a certain panache lacking in stuffier scientific organizations. We were, after all, an Onward Formicine Soldiers sort of group, and indeed, inspired by one of the most popular bumper stickers of the times, I finally came up with the battle cry—"Christ, it's the ants, Sir!"

So now we had everything a statusy scientific organization required, and it remained only to design our own stationery. At the top, the letterhead was emblazoned with the organization's name, in which a large capital letter began each word—Fire Ant Research Team. To one side was our logo, a large fire ant standing menacingly over Florida's state capitol building. Around this logo was our motto, "Today Florida, Tomorrow Dixie," and at the bottom of the page, as a footer, was our inspiring battle cry. For years, the use of this stationery in my correspondence with colleagues brought me much prestige, admiration, and, yes, even envy.

The organization was and remains important in the field of scientific publishing. From the moment it came into existence, every paper on ants that we have published has borne the imprimatur, "This is paper number n of the Fire Ant Research Team." Some numbers may be missing, and others out of order, and a couple of papers unnumbered, but we keep adding to the list. A paper published just before this volume went to press bore the number 42.

Like the ants we study, we humans need meaningful organizations to bind us together in our work, to inspire us to greater productivity, and to mark our achievements. For me, the Fire Ant Research Team has made my name a household word among my colleagues, has brought me the prestige of having audiences remember my talks even years later, and (though harder to document) has assured the continuous funding of my research for over 35 years. When you are a member of a well-known and eclectic organization, funding agencies know that their money will be well spent. And it has been.

35

Solenopsis invicta and Other Communities

Solenopsis invicta is a scavenger and an opportunistic predator. One might expect it, in the process of being what it is, to suppress or increase the populations of other species of animals, either directly by killing them or indirectly by affecting their resources or natural enemies. Most research on this subject has been done in the highly simplified ecosystems of agriculture (agroecosystems, in the jargon). These may not respond like more complex, native ecosystems, on which very little research has been done. In either case, well-designed experimental studies (without the use of poisoned bait) are uncommon. Nevertheless, agricultural research can tell us some significant things about the effect of *S. invicta* on several animal communities.

Herbivore Communities

Some of the earliest attempts to determine the effects of fire ants on native ecosystems took place in the era of grand eradication programs. Rhoades compared three "ecologically similar," 500-ha areas of mixed cropland, pasture, and pine plantations in north Florida, one with *S. invicta* and treated with heptachlor, another with *S. invicta* but untreated, and a third naturally lacking *S. invicta* (the control; Rhoades, 1962, 1963; Rhoades and Davis, 1967; Rhoades and Murray, 1967). Heptachlor severely reduced a set of "indicator species," *S. invicta* among them, for several months. More to the point here, comparison of the untreated areas with and without *S. invicta* showed no detectable difference in the abundance of any of the indicator species. Rhoades did not subject his data to statistical analysis, living as he did in an era when a professor could tell a student (me) that, if you needed statistics to understand your results, you had better do the experiment over.

Later researchers detected significant impacts of *S. invicta* on selected species of agricultural importance. The two most thoroughly studied are sugarcane and cotton, in both of which the fire ant was deemed to be beneficial. Reagan and several coworkers (Reagan et al., 1972; Ali et al., 1984; Reagan, 1986; Showler and Reagan, 1991) found that Louisiana sugarcane fields treated once with mirex, showed a decrease in populations of arthropod predators, especially *S. invicta*, followed by a 53% increase in the pest, sugarcane borer, which did 69% more damage. Ant-free plots required more frequent treatment with insecticide to control the sugarcane borer. Louisiana cane farmers came to recognize that the fire ant provided them with a valuable ecosystem service in the form of reduced insecticide costs. As a result, *S. invicta* became a popular ant among Louisiana sugarcane farmers, who were reported to defend it vigorously against those who would try to eradicate it.

Claude Adams and several USDA coworkers claimed the situation was different in Florida sugarcane fields (Adams et al., 1981), where killing *S. invicta* and several other species of ants with mirex was initially followed by a doubling of damage by the sugarcane borer over that in untreated fields. By the end of the study, however, *S. invicta* had rebounded to 125% of its original level in the treated fields but dropped to about 25% of it in the untreated fields. This result, together with an obvious seasonal increase in the borer, makes the data complex to interpret. In the treated fields, *S. invicta* was positively correlated with the borer, probably because both were increasing, the first recovering from mirex and the second increasing seasonally. Two native species of ants, *Pheidole dentata* and *Camponotus floridanus*, were negatively correlated in untreated fields, probably a sign that they compete with each other over baits. The authors suggested that a complex of predaceous ants was probably more effective in controlling sugarcane borer than *S. invicta* alone, but their data are not clear enough to warrant such a conclusion. Overall, this study confirmed the benefits that fire ants bring to sugarcane growers, although the case is weaker than the one in Louisiana.

Meanwhile, in east Texas, *S. invicta* operates as a key predator of boll weevils and pink bollworm in cotton fields, while at the same time failing to reduce the populations of other arthropod predators on these pests (Sterling, 1978; Jones and Sterling, 1979; McDaniel and Sterling, 1979, 1982; Sterling et al., 1979, 1984; McDaniel et al., 1981; Reilly and Sterling, 1983a, b). Sterling and Jones staked weevil-infested cotton bolls in the shade of cotton plants throughout the production season, then retrieved them to determine weevil survival or cause of death. They also estimated the density of foraging ants by shaking cotton plants in a bucket and counting the ants dislodged. As *S. invicta* density increased during the growing season, so did ant-caused weevil mortality, which averaged 66% per week in one area but only 22% in another. Fire ants accounted for 84% of total weevil mortality; desiccation, egg infertility, and parasitoids accounted for the rest. Only 2% of weevils survived by the end of the summer, only just enough to replace the parent weevils. The weevil population therefore did not increase, and no insecticidal treatment was necessary—pretty good news in an insecticide-obsessed country.

These surveys were supplemented with experiments in which 100 fire-ant colonies were transplanted early in the cotton-growing season into one plot, and ants were eliminated with mirex bait from another. These experimental plots were deliberately infested with boll weevils—150 adult weevils per hectare were released during May (natural levels were too low for experimental purposes). In the plot with ants (mainly *S. invicta*), ants ate 66 to 83% of weevil larvae, and weevil-damaged squares (young bolls) never exceeded about 17%. During this same time, ants ate only 1–9% of the weevil larvae in the mirex-treated, ant-free plot, and the survivors damaged nearly 40% of the squares. No adult weevils emerged in the ant plot, but slightly more than replacement numbers emerged in the ant-free plot. Here then is experimental verification of the importance of ants, especially fire ants, as biological-control agents. Other studies by Sterling's group showed that fire ants were also effective predators on tobacco budworm (yes, a pest of cotton!) eggs and larvae (McDaniel and Sterling, 1982; Nuessly, 1986; Nuessly and Sterling, 1986) and on bollworm and beet armyworm eggs (Diaz et al., 2004). Farther afield, although fire ants tended pecan aphids, pecan weevil survival was 1.6 times higher when fire ants under trees were killed than when they were not (Dutcher and Sheppard, 1981). *Solenopsis invicta* reduces the survival of floodwater mosquito eggs by 50 to 100% in Texas rice paddies (Lee et al., 1994), but whether they reduce the population of adult mosquitoes is unknown. Leaf damage to that southern favorite collards was inversely related to fire-ant density, presumably because fire ants reduced the populations of several collard pests (Harvey and Eubanks, 2004). In upland rice cultivation in the Philippines, *S. geminata* was the most important predator on the brown planthopper, *Niliparvata lugens* (Way et al., 2002), quickly reducing their populations.

An objection that has often been leveled against *S. invicta* as a biological-control agent is that it exerts its effects indiscriminately, killing beneficial predators alongside harmful pests. The fire-ant transplantation experiment just described allowed Sterling and Jones to test this notion. In spite of the high densities of fire-ant foragers in the plots supplemented with fire ants, none of the 47 species of predaceous arthropods was significantly reduced, so at least in cotton, fire ants do not impoverish the predaceous arthropod fauna. Many predators are highly mobile or have defenses that make them less likely to fall prey to fire ants. The combination of effective predation on two major cotton pests with a lack of effect on other predators makes *S. invicta* a key biological-control agent in this agricultural system. Perhaps predatory fire ants exert a strongly density-dependent effect, preferring the more abundant pest prey well beyond what is expected on the basis of their abundance. The authors also go on to indict the use of insecticides, not *S. invicta*, as being the factor that has impoverished the native ant fauna.

The situation in peanut fields is similar to that in cotton—*S. invicta* are seven times as likely to prey on pest species such as the red-necked peanut worm, than on beneficial insects, and elimination of *S. invicta* from peanut fields did not result in an increase in predator populations (Vogt et al., 2001, 2002b). In some crops, however, including cotton and soybeans, high fire-ant densities were associated both with low pest abundance and low abundance of the pests' other natural enemies; suppression of fire ants with poisoned bait increased some beneficial insects but not others (Eubanks, 2001; Eubanks et al., 2002). Perhaps fire ants preyed directly on the natural enemies, or perhaps natural enemy populations were low as a result of low prey abundance. Or both. From the farmer's point of view, perhaps it does not matter—few samples containing abundant fire ants also contained pests above the action threshold (the density at which pesticide treatment is recommended), whereas the majority of samples that lacked fire ants were above the action threshold.

Can fire ants be encouraged in their beneficial role in agricultural fields? In newly planted sugarcane fields little food is available for fire ants until the cane is large enough to harbor an insect fauna. Insect prey was two to six times as abundant in weedy as in weed-free young cane fields, and fire-ant mounds were more than three times as plentiful (Showler et al., 1989). By allowing a level of weeds that did not compete with the growing cane, a farmer could increase the pest-control service that *S. invicta* renders after the cane canopy closes and the weeds die of shading. Cane and sugar yields (and therefore monetary returns) were about 20% higher from weedy fields than from weed-free ones. Some evidence even indicates that tolerating moderate levels of cane mealybugs might result in net benefit by encouraging fire ants to forage on the canes and by keeping them peppy and full of energy (Woolwine, 1998).

Strains of cotton that entice the ants with nectar might have a similar effect. Ancestral cotton plants bore nectar-secreting glands on the leaves (foliar nectaries) and on the bracts surrounding the blossom (extrafloral nectaries), but these have been bred out of many lines. *Solenopsis invicta* foragers prefer cotton strains with nectaries to those without, and they often fed on the nectar. Early in the season, ant abundance was similar on strains with and without nectaries, but once nectaries appeared, fire-ant abundance was 35 to 85% higher on the strains with nectaries (Agnew et al., 1982). Unfortunately, pest levels were so low in these experiments that no difference resulting from fire-ant visitation rate could be detected, but the accepted explanation for the evolution of nectaries outside flowers is this very attraction of predaceous insects to provide pest-control services in return for the nectar. We already know that predation rate increases with predator density.

Several less-well-documented studies of the effects of fire ants on biological-control agents in agricultural systems are summarized in Table 35.1.

Table 35.1. ▸ Effects of *Solenopsis invicta* on other biological-control agents.

System	*Results*	*Comments*	*Source*
Control of aphids and corn earworms by parasitoids	Emergence of adult parasitoids reduced	Did not determine whether the level of pest control was reduced	Lopez, 1982; Vinson and Scarborough, 1991
Control by the egg parasitoid *Trichogramma* sp.	Emergence of adult parasitoids reduced	Did not test for a population-level effect	Nordlund, 1988
Control of musk thistle by the weevil *Rhinocyllus conicus*	No reduction of weevil	Conducted a laboratory experiment followed by field experiment	Brinkman et al., 2001
Control of the citrus aphid *Toxoptera citricida* by the parasitoid wasp *Lipolexis scutellaris*	Sixfold higher adult emergence of the wasp when *S. invicta* was excluded	Did not estimate the effect on aphid control	Hill and Hoy, 2003
Control of cotton aphid by its natural enemies	Fewer aphids and more aphid predators upon suppression of fire ants	Did not estimate effect on cotton production	Kaplan and Eubanks, 2002
Control of pecan aphids by predators	Fire ants fed on honeydew and on two predators of aphids	Fire ant's role in aphid outbreaks not convincing	Tedders et al., 1990
Control of pecan aphids by predators	Reduction of fire ants had minor effects	Authors conclude control of fire ants unnecessary	Harris et al., 2003
Control of pests in soybean fields by ground arthropods	Suppression of fire ants with poison bait increased some species, reduced others, had no effect on some	Direct effects of bait and fire ants confounded	Seagraves et al., 2004

In the absence of *S. invicta, S. geminata* seems to play a very similar role, emphasizing the large degree of ecological equivalence between them. In the Mexican state of Tabasco, Risch and Carroll set up 12 small experimental plots within a larger area of early secondary succession, six with *S. geminata* colonies and six from which they had been removed (Risch and Carroll, 1982b). All plots were planted with the typical corn and squash crops of the region. Removal of *S. geminata* was followed by an eight- to 15-fold increase in arthropod abundance and a tripling of the number of species, followed by much greater crop damage. In contrast to the cotton case, *S. geminata* showed no preference for herbivores in this experiment, affecting most species about equally and acting as a keystone species by functioning as a generalist predator. Again, the importance of fire ants as general predators in early successional habitat is brought into sharp relief, reinforcing results from sugarcane, pastures, and cotton fields in the USA. Other *S. geminata* examples include the suppression of Mediterranean fruit fly in orange and coffee orchards in Guatemala (Eskafi and Kolbe, 1990); the 70% reduction in emergence of Mediterranean and oriental fruit flies beneath Hawaiian guava trees (both prey and predators are introduced pests; Wong et al., 1984); and predation on housefly and flesh-fly larvae in Puerto Rico (Pimentel, 1955).

Concern is particularly great about possible effects of *S. invicta* on endangered species in cases where the two come into contact. By definition, endangered species exist as small populations, vulnerable to extinction through chance events or continued stresses. Two focal areas have been mentioned: endangered cave invertebrates in Texas and several endangered species on Key Largo in the Florida Keys (Forys et al., 2001; Wojcik et al., 2001). Fire-ant foraging extends well into the caves in question, but no data are offered on the impact they have on the endangered animals there. In the Florida Keys, the Shaus swallowtail, an endangered butterfly that breeds at the edges of tropical hardwood hammocks, comes into contact with *S. invicta* along the roads and other developments that have fragmented much of this habitat. (The term "hammock" is applied in Florida to enclaves of hardwood forest surrounded by pine forest, saw-grass prairie, or other dissimilar habitat. Like most people, I always found it peculiar that a patch of woods should bear the same name as a sleeping arrangement. Then, on a visit to Hamburg, Germany, I learned that this city's name derives from an Old German word "hamma," meaning a forest—Hammaburg, the fortress in the forest. So this is probably the origin of "hammock" as well, although how the term made its way from early medieval Germany to Florida must be an interesting story in itself.) Although laboratory experiments show that *S. invicta* readily preys on a similar swallowtail species (a surrogate for the endangered species), the question of whether it actually affects the Shaus swallowtail in the wild remains open. A test in the hammock would have been more convincing. Still, we should clearly worry on behalf of the Shaus swallowtail. Similar data and concern exist for the Stock Island tree snail, another endangered animal of the Florida Keys. As an ounce of prevention, the authors recommended reducing *S. invicta*'s access to hammock animals by restoring abandoned roads and paths to native vegetation and reducing the level of disturbance on roadsides to make them less suitable for *S. invicta*. Makes sense to me.

Decomposer Communities

Plants and animals die, animals defecate, and both cast off a variety of exudates and structures. These represent opportunities for decomposers, and every ecosystem has its complement of bacteria, fungi, arthropods, and so on that exploit these resources. When the resource is abundant, widespread, and of low food value, like litter, the decomposers are also widespread and abundant and often have low powers of dispersal, forming a permanent community. When the resource is more localized, like dead trunks of trees, the decomposers are more localized and need greater powers of dispersion. Because decomposition is rather slow, the communities change rather gradually as decomposition proceeds. At the other extreme, resources with high food value, such as rotting fruit, feces, mushrooms, and dead animals, are ephemeral and highly localized but unpredictable in time and space. Creatures that exploit them must have great powers of detection and dispersal. Drawn by death and decay, by the odors of cadaverine, putrescine, skatole, or ethanol, the adults home in on their opportunity and reproduce rapidly before the resource becomes unsuitable, leaving their progeny to search the wind again.

Manipulating fire ants in a natural herbivore community is difficult, as we saw above. In contrast, fruit, carrion, and feces are like tiny islands that come and go. They can be readily manipulated to exclude or include fire ants, without effect on access by the decomposers, most of which can fly as adults. Rotting fruit, carrion, and feces are therefore very suitable for experimental studies on the effects of *S. invicta* on insect communities. A few such studies have been done.

Solenopsis invicta is especially fond of pastureland, so several investigators have reasonably asked whether *S. invicta* might not affect other arthropods in this habitat (Summerlin et al., 1977b, 1984a, b; Howard and Oliver, 1978, 1979; Hu and Frank, 1996). Pest flies that breed in cow dung, such as horn flies, cause annual losses of $870 million to the cattle industry in the USA. In the by now familiar experimental design, all these authors eliminated fire ants (and other ants) from one experimental pasture with mirex or Amdro bait but left another pasture untreated. Because dung and the community of insects exploiting it age rather quickly, the researchers hung around pastures to watch the cows defecate (need a job?). They noted the exact time when each new meadow muffin was deposited, then collected equal-aged pairs of them, one from the treated pasture and one from the untreated, at elapsed times of three to seven days and made pair-wise comparisons. After collection and before analysis, the dung was stored in a freezer (one assumes that food was kept in a different freezer).

When *S. invicta* abundance was reduced, horn-fly larvae were two to five times as abundant in dung as on untreated plots. Adult horn-fly emergence from laboratory-formed cowpats spiked with fly eggs and then placed in the field was 63–94% lower in the presence of fire ants than in their absence. Similar results were obtained with the ant bait Pro-Drone (Lemke and Kissam, 1988). A number of other species of dung-breeding flies were also reduced in untreated fields, and *S. invicta* was seen preying on fly larvae in the cowpats. All stages of the stable fly, *Stomoxys calcitrans*, especially pupae, were taken by fire ants (Summerlin and Kunz, 1978). The lower abundance of fly larvae in untreated pasture was probably the result of fire ant predation. A number of ground-dwelling and dung-inhabiting arthropods, most particularly scarab and staphylinid beetles, were more abundant in the treated plots lacking fire ants, but a few were less abundant. Whether these were direct or indirect effects of *S. invicta* was not determined.

Rapid removal of dung by dung-burying beetles has long been considered an important form of horn-fly control, but in areas inhabited by fire ants, the effectiveness of the dung beetles, such as the introduced Afro-Asian *Onthophagus gazella*, could potentially be reduced when fire ants prey on them. Summerlin and several coworkers investigated this possibility. First, they got themselves a steer and fed it a tasty mixture of alfalfa and Bermuda grass hay on a concrete platform so they could collect its manure, but because these fecal contributions to science were not standardized but were of size and shapes determined by the vagaries of the steer's digestive system and the physics of impact, the scientists used Mom's kitchen scale and a cake mold to form the manure into standard, 1-kg pats (Ah! Science!). These they placed in boxes on soil and added dung beetles, horn-fly eggs, and fire ants, alone and in combinations of two and all three. The results showed that fire ants did not affect the production of new dung beetles but reduced the emergence of horn flies by 95%. This was good news, for it suggested that fire ants would probably not interfere with the horn fly–reducing effect of the dung beetles in the field.

In a charming sequel to this study, some of the investigators marked "freshly deposited" cowpats (did they use deposit slips?) in a pasture and then collected them, along with surrounding and underlying soil, after four hours, one day, and three days. With a slow drip of water, they flooded out any ants from the pats and counted them. Invasion of pats during the cool months was low, but during the summer, the majority of pats contained fire ants within hours; ant numbers rose to about 600 in three days and higher in areas with more fire ants. Dung flies oviposit on fresh dung. The ants became more effective in preying on fly eggs and larvae as the dung dried and cracked over the first few days. Standard pats planted in the field showed that horn-fly emergence was significantly lower when ants had access to the pats than when they did not. In further field experiments, fire ants negatively affected colonization of dung by native scarabs and *O. gazella*, but these nevertheless managed to reproduce successfully in the presence of fire ants. Together, the disruptive effects of the dung beetles and the predation by fire ants suppressed horn-fly breeding success more than either alone.

The large reduction of emergence from experimental treatments makes it very likely that the population of adult flies was reduced, but adult fly populations were not directly measured, as the flies disperse widely and travel far. A second issue is that all experiments employing poisoned bait confound the reduction of fire ants with the reduction of other arthropods. A strict interpretation would therefore hedge the claim. As in many agricultural studies, the focus was on the effect of fire ants on the insects present in the dung. The question of how fire ants affect the decomposition of dung, either directly or indirectly, was not addressed.

Because *S. invicta* is so common along roadsides, a rich resource that it ought to have frequent access to is carrion, i.e., roadkill. Dead animals are decomposed by a predictable succession of insects, beginning with the maggots of several families of flies, which are soon followed by a variety of predaceous or scavenging beetles. As these waves of creatures finish their work, the desiccating bones and skin become the province of skin beetles and clothes moths, which reduce hair, skin, and cartilage to dust. The succession of insect communities is rather different in large and small carrion, as well as at different seasons. Because *S. invicta* is a potential predator on these carrion insects as well as a direct carrion consumer, its capacity to affect this community would seem considerable. Study of these effects is theoretically (community ecology) and practically (forensic entomology) interesting. Stoker and his coworkers at Texas A&M University placed dead mice or plucked chickens in sand-filled pans within mesh cages that excluded animals larger than insects and either prevented or allowed access by fire ants by placing them on Fluon-coated or uncoated supports. Every day, they removed a carcass to census the insects in it, then discarded it, and also sampled the flying insects over one of the remaining carcasses. Sounds like fun, eh?

Mouse carrion protected from *S. invicta* attracted small numbers of several fly and beetle families, but unprotected mice contained only *S. invicta* workers. Neither condition seems to have been appropriate for complete decomposition, however—all of the mice mummified. Under natural conditions, small carcasses are usually colonized by less specialized scavengers or are monopolized by burying beetles. In large carrion, insects specialize to varying degrees on different stages of decomposition. Blowflies (family Calliphoridae) arrived during the first week, muscid flies between three and 12 days, and flesh flies (family Sarcophagidae) between five and 15 days. The number of adult flies of all species captured over unprotected carrion was much lower than that over protected carrion for the entire 17 days of the study. Similar results were found for three families of beetles—rove beetles (Staphylinidae), burying beetles (Silphidae), and hister beetles (Histeridae).

The adult flies captured over the carcasses are on an egg-laying mission and do not themselves feed significantly on the carcass. The larvae that hatch from the eggs the adults leave behind actually consume the

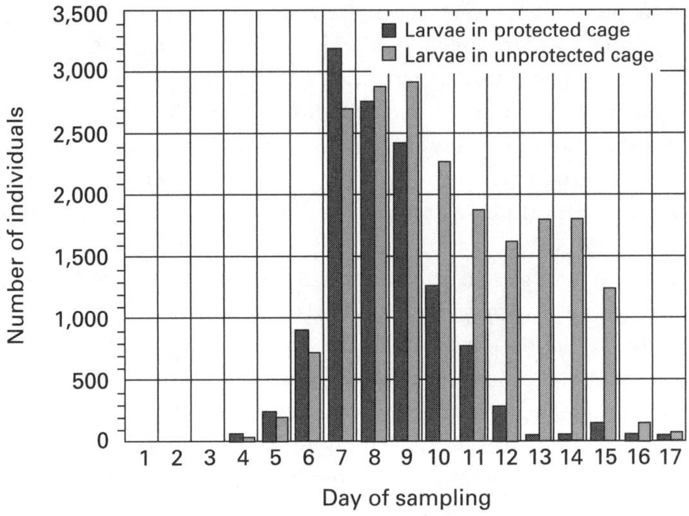

Figure 35.1. Numbers of fly larvae collected from carrion in cages protected or not protected from fire ants. The presence of fly larvae was greatly prolonged in the presence of fire ants. Adapted from Stoker et al. (1995).

flesh. The effect of *S. invicta* on the larval population is more complex than that on adults. Fly larvae (not identified to family) in both treatments peak at about 2500–3000 maggots per chicken between seven and nine days. In the protected cages, the larval population then dropped rapidly to nearly zero by day 13, whereas in the unprotected cages, larvae remained at 1500–2000 per chicken before finally dropping on day 16. The larval population pulse that reduces the carcasses' flesh is greatly elongated in the presence of fire ants (Figure 35.1). Whether or not these later larvae were of different species was not determined.

Pondering these results more deeply, we can ask what the decreased presence of adult flies over unprotected carcasses means. It probably simply means that the adult flies and beetles respond to the presence of fire ants by spending less time on the carcass (producing lower capture rates). Colonization and reproductive success of flies was not lower in the unprotected carcasses; in fact, summed over the entire experiment it was higher. The production of larvae was prolonged in the presence of fire ants. No information is available on beetle reproduction nor has the possible role of adult beetles (which do not simply lay eggs and run) been explored. So in the end, the effect of *S. invicta* on this decomposer community is not so large as it first appeared from adult fly captures, and the incompleteness of this study complicates interpretation of just how *S. invicta* alters the community. Even more obscure is any effect *S. invicta* might have on the rate and/or progress of the decomposition of the carcass. Can we expect our roadsides to become paved with undecomposed carcasses? Will sail-possums be as abundant as flattened Bud cans?

The community of insects that decompose rotting fruit is also greatly affected by fire ants (Vinson, 1991). As with carrion, fire ants were either allowed or not allowed access to ripe peaches or melon slices exposed for nine days on sand trays in areas inhabited by polygyne *S. invicta*. In the absence of ants, the rotting fruit attracted a fairly diverse community of decomposer insects and their parasites and predators. Fruit flies (*Drosophila* sp.) were most abundant, followed by several species of beetles (families Nitidulidae, Staphylinidae) and a few parasitic Hymenoptera. Large numbers of progeny emerged from these fruits. When the ants had access to the rotting fruit for the full nine days, almost no other decomposer insects successfully completed their life cycles. Reducing fruit exposure to three days allowed 10 to 20% as many flies and beetles to complete their life cycles as emerged from the ant-free controls. Clearly, fire ants are catastrophic to this decomposer community, but interestingly, little effect was seen on the decomposition of the fruit. The amount remaining after nine days was not different, whether it was decomposed by the ants or by other insects. In essence, the ants not only consumed the decomposers as prey but also commandeered the rotting fruit directly, replacing a community with a single species. Whether the effect is similarly severe in less dense monogyne areas has not been investigated, but it is probably not, because the study was carried out in a polygyne area with a mound density exceeding 1000 per hectare, very high indeed.

Other Arthropods and Communities

The lone star tick is a pest of cattle, people, and wildlife in the south central USA. The ticks are most abundant in brushy, open habitat and along forest–pasture margins and forest openings, where they wait on vegetation to drop on and attach to passing host animals (a process called "questing"). The habitat preference of the inactive, waiting ticks overlaps with that of fire ants, and it would appear that the ticks are vulnerable to foraging fire ants. Indeed, cattle ranchers noticed that a decline in tick populations seemed to coincide with the appearance of *S. invicta*. Wayne Harris, Ph.D. student at Louisiana State University, killed the fire ants on one set of pasture and woodland plots near Baton Rouge and Pine Grove, Louisiana, with mirex and used another, untreated set as controls (Harris and Burns, 1972; Burns and Melancon, 1977). He then exposed tick eggs, caged adult female ticks, and laboratory-hatched, engorged (and therefore quiescent) tick larvae in these plots for one, two, and three days. After three days, tick-egg survival was nearly 100% in treated woodland plots and 86% in treated pasture. In contrast, survival was only 46% on untreated woodland plots and 19% in untreated pasture, where the density of fire-ant mounds was about 200 per hectare. Over 60% of engorged larvae and 100% of engorged female ticks survived the three-day exposure in the treated, ant-free plots, but only 3% and 0% did so in the untreated plots. Many were attacked within hours of placement.

Driving the point home, Harris lovingly placed several thousand laboratory-reared, engorged tick larvae at the bases of grass clumps in his plots or mixed them with corncob grits and sowed them by hand, broadcast, like a medieval peasant. The beauty of ticks is that they can be recovered by attraction to stationary or dragged carbon-dioxide traps. Seven months after releasing the larvae in the Baton Rouge plots, Harris recovered 75% of them as adults in mirex-treated plots but none in untreated plots (he does not mention that the ticks were marked, so I am not certain that the recovered ticks were the same ones he released, although Harris writes as though they were). Recovery was lower in the Pine Grove plots (4%) but still significantly higher in the ant-free plots (0%).

Mirex is a big hammer, killing a number of other arthropods in addition to fire ants. Harris therefore offered additional observations—fire ants were 30 times more common in pitfall traps on the untreated plots than on treated plots, but spiders and beetles were about equally abundant. He also saw fire ants opening engorged female ticks like cans of pork and beans and feeding on the blood they contained. Several years later, the authors followed up this experimental study with a tick survey (using dry-ice traps) in northwestern Louisiana, before fire-ant invasion and soon after. In this natural experiment, tick populations were dramatically lower after invasion than before. Altogether, the evidence that fire ants reduce tick populations seems convincing.

The next proposed effect of *S. invicta* concerns a mutualistic, coevolved ant-plant relationship referred to as myrmecochory. Throughout the world, many species of plants produce seeds with an appendage called the eliaosome whose purpose seems to be to cause ants to transport the seed back to their nests. The ants eat the nutritious eliaosome and discard or abandon the seed, which has thus been dispersed to a favorable location for germination (which follows in the fullness of time). In the hardwood forests of the eastern USA, about 30% of the herbaceous plants depend upon myrmecochory for seed dispersal. The arrival of *S. invicta* raises the question of how this new ant might affect this old relationship, especially in light of *S. invicta's* known consumption of seeds—of 96 species of seeds offered to laboratory colonies of *S. invicta*, a substantial fraction were collected and damaged (Ready and Vinson, 1995). Zettler et al. (2001) found that *S. invicta* readily collected the seeds of five species of myrecochores but often damaged or destroyed the seeds before discarding them on the trash heap. The authors then pile "might" upon "may" to argue that *S. invicta* could "sour the relationship" between native ants and myrmecochorous plants—but consider: the seeds were collected from forest herbs in the Blue Ridge Mountains and were tested either in the confinement of a laboratory arena or on a lawn containing over 1300 mounds per hectare, an order of magnitude higher than ordinary populations. Neither seed nor ant set foot or eliaosome in the habitat in which this drama is supposed to occur. To conclude that *S. invicta* is a "potential threat to spring ephemerals" requires a lot of arm waving, because *S. invicta* typically does not live in the forests where the ephemerals live. Perhaps *S. invicta* is indeed a threat to myrmecochory, but in the face of this kind of evidence, that secret will remain quite safe. In contrast, the claim that selective seed feeding by *S. invicta* may affect the competition among early successional weeds with which they share a habitat is more credible (Seaman and Marino, 2003).

A study on the ant-dispersed seeds of the bloodroot in fragmented forests in northeast Georgia found that, when the forest edges were invaded by *S. invicta*, most of the seeds were collected by *S. invicta*, who dispersed them smaller distances than did the native ants in uninvaded edges or forest interiors (Ness, 2004). Whether this behavior affected the populations of bloodroot, independent of the disturbance associated with forest edges, was not tested.

A long-standing argument among ecologists is whether plant productivity is controlled from the top trophic level (predators) downward or from the bottom upward by physical factors affecting plant growth. Specifically, reduction of herbivores by *S. invicta* in early successional habitats might "cascade" down the trophic levels to reduce plant damage, allowing more vigorous growth and greater seed production. *Solenopsis invicta* seems particularly suited for answering this question because its high abundance and generally predaceous habits make detectable effects more likely. At the Savannah River Test Site in South Carolina, seedlings of several species of common weeds were planted in three-year-old gaps in slash-pine plantations, half of which served as untreated controls and the other half of which had been treated with Amdro bait and harbored only about one-fifth as many fire ants (Stiles and Jones, 2001). Although *S. invicta* significantly reduced leaf damage, this reduction did not result in greater plant biomass or seed production at the end of the growing season. A second experiment tested whether the various species of ants present in forest gaps could substitute for one another, but again plants exposed to fire ants did not differ from those exposed to other species in biomass or seed production. Neither experiment detected the expected top-down trophic control of production, but for unknown reasons, total leaf damage was very low. In the absence of fecundity-reducing levels of plant damage, the effects of ants may simply not be detectable, and bottom-up factors may influence plant productivity more strongly. Perhaps the experiment did not run long enough.

Solenopsis geminata is the most abundant ant in agricultural fields in most of tropical Mexico. The direct consumption of seeds by this

Figure 35.2. Seed collection by *Solenopsis geminata* in tropical Mexico. Ants collected proportionally more of an unpreferred seed when these were presented among an abundance of preferred seeds and took less of a preferred seed when it was "hidden" in an abundance of unpreferred. The mix always included 2 g of the seed shown on the y-axis. Adapted from Risch and Carroll (1986).

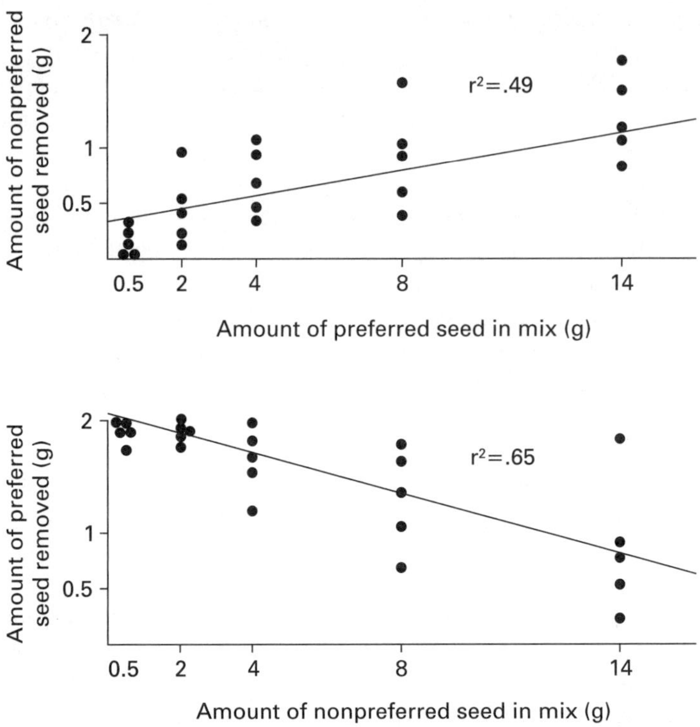

native fire ant suggested that it might have community-wide effects on the competition among weeds and between weeds and crop plants (Risch and Carroll, 1986). Seeds of eight species of common broadleaf and grassy weeds were mixed in all pair-wise combinations in the five different proportions by weight: 100%, 80%, 50%, 20%, and 0%. In some pairs, one weed species was capable of dominating the other, whereas in other pairs, the match was more even. These seed mixtures were sown in small plots, and ant access was either allowed or prevented. The biomass of each plant species was measured at several intervals during the growing season. *Solenopsis geminata* showed strong preferences for some species of seeds and avoided others. They took proportionally more of an unpreferred seed when these were presented among an abundance of preferred seeds and took less of a preferred seed when it was "hidden" in an abundance of unpreferred (Figure 35.2). Their preferences were thus density dependent as well as absolute. By exercising their preferences on the sown mixtures of seeds, *S. geminata* modified the outcome of competition between these weed species.

When the stronger weed competitor was the preferred seed, *S. geminata* reduced its biomass and allowed the weaker competitor to attain increased biomass. Preference for the seeds of the weaker competitor hastened its disappearance from the plots. When the preference for the stronger competitor was high enough, the outcome of the competition was actually reversed by *S. geminata*, and the stronger competitor disappeared from the plot. In other cases, the competitors reached a stable equilibrium in the ant's presence, although one of them disappeared in its absence. Total weed biomass was reduced only for the first 50 days. Thereafter, ants had no effect on weed biomass. Any crop benefits through weed reduction by *S. geminata* are thus likely to occur early in the season. In real life, many influences act simultaneously, so predicting specific effects on tropical agriculture is difficult, but the finding that *S. geminata* can modify the outcome of competition between plants is interesting. Such effects are actually commonplace; third-party diseases, parasites, predators, and competitors frequently modify the outcome of competition between focal pairs of species. Indeed, in the USA the outcome of the one-sided competition between *S. invicta* on the one hand and *S. geminata* and *S. xyloni* on the other might be partly the result of the absence of such third-party burdens on *S. invicta*.

Another possible third-party effect may be created by the preference of *S. invicta* to search more intensely on some species of plants than on others, irrespective of foraging success. Foragers visited aphid-free cowpea and indigo with equal frequency, but when both harbored cowpea aphids, the ants preferred to collect honeydew from the aphids on cowpeas. Similarly, they preferred aphids on indigo to those on sesbania (which they avoid completely; Kaakeh and Dutcher, 1992). To what extent this preference is for the plant or for the honeydew derived from the plant is currently unknown.

36

Fire Ants and Vertebrates

Observational Reports

Evidence strongly suggests that fire-ant venom contains a substance that affects the human mind. As little as a single dose of venom causes human persons to confuse correlation with causation and, on the flimsiest evidence, happily to credit *Solenopsis invicta* with practically unlimited capacity for mayhem. This equal-opportunity phenomenon affects backhoe operator and scientist alike. Among scientists, a minor industry has, for several decades, busily characterized *S. invicta* as The Ant From Hell, capable of sending almost any creature, large or small, to its heavenly reward and of bringing entire ecosystems to their knees. Although credible experimental work links *S. invicta* to effects on invertebrate populations, claims about vertebrate populations are built mostly on oft-repeated anecdotal, often unrefereed reports of fire ants devouring baby animals, a sort of "parade of horrors" (reviewed by Wojcik et al., 2001). No amount of hypothesizing about keystone species, cascading effects, indirect effects, and so on can make up for the lack of experimental work, modeling, and quality data. As a result, even 50 years after the first claims hit the press, we still cannot (with a couple of exceptions) say with any assurance whether *S. invicta* exerts any effects on populations of vertebrates. The literature remains short on data and long on conjecture and is highly problematic, speculative, and unsatisfying. A particularly annoying feature of this literature is the frequent citation of unrefereed papers such as conference proceedings or unpublished theses and dissertations. These have not been subjected to quality control by peer review, nor are they readily available for verification of details. Many are LPUs (least publishable units), the smallest amount of information that one can get into press, aka "wipes."

Many of the anecdotal reports that form this literature are listed in Table 36.1. A more complete listing can be found in a recent review by Allen et al. (2004). They confirm the opportunistic nature of *S. invicta*—most of the attacks are on relatively helpless, immobile nestlings or hatching eggs of ground-nesting birds or mammals or trapped animals. Undoubtedly, more such examples will surface in the future. Considering the widespread occurrence and abundance of *S. invicta*, reduction of these and other vertebrate populations is certainly possible, but

Table 36.1. ► Selected anecdotal reports of fire-ant attacks on vertebrates.

Vertebrate species	***Reported observations***	***Source***
Multiple species	Review articles compiling many reports	Allen et al., 1994; Wojcik et al., 2001
Houston toad	Fire ants killed several newly metamorphosed toadlets.	Freed and Neitman, 1988
Gopher tortoise hatchlings	Fire ants killed 10 hatchlings from 38 nests; most mortality was due to vertebrate predators.	Landers et al., 1980
Lizards eggs	Radioactivity from labeled lizard eggs found in nearby *S. invicta* colonies.	Mount et al., 1981
Ground-dwelling reptiles and birds	Fire ants attacked nests of several turtle and lizard species.	Mount, 1981
Wood ducklings	Fire ants were on pipped eggs and in three nest cavities.	Ridlehuber, 1982
Barn swallow nestlings	Solenopsis geminata killed nestlings in one of 25 nests.	Kroll et al., 1973
Cliff swallow nestlings	Nesting success in culverts was 34% lower in the presence of *S. invicta.*	Sikes and Arnold, 1986
Nestling quail	*S. geminata* kill pipping chicks.	Stoddard, 1931; Travis, 1938b
Nestling quail	Fire ants reduced nestling survival.	Wilson and Silvy, 1988
Purple martin	*Solenopsis* spp. may be a mortality factor.	Allen and Nice, 1952
California quail chicks	Nestlings in one of 32 nests were killed by *S. xyloni.*	Emlen, 1938
Nestling kestrels	Fire ants prey on nestlings.	Wojcik and Smallwood, 1993
Colonial water-bird nestlings	High nest failure was associated with *S. invicta;* mortality was reduced by fire-ant control.	Drees, 1992
Nesting terns	Article not seen	Lockley, 1995
Loggerhead shrike	Fledging success was positively related to fire-ant mound density, but reported densities are impossible.	Yosef and Lohrer, 1995
Loggerhead shrike	More shrikes and insects were present in areas where fire ants were reduced with poisoned bait; authors suggest shrikes choose wintering areas with more insect food.	Allen et al., 2001b
Nestling cottontail rabbits	Fire ants destroyed 68 of 231 nests of penned cottontails; evidence circumstantial.	Hill, 1969
Box turtles	Fire ants killed adult turtles.	Montgomery, 1996

Table 36.1. ▶ *(Continued)*

Vertebrate species	*Reported observations*	*Source*
Least tern	*Solenopsis xyloni* occasionally attacked pipped eggs and chicks.	Hooper-Bui et al., 2004
Egg-laying snakes	Author speculates on the decline of snakes at Payne's Prairie, Florida.	Bartlett, 1997
Loggerhead turtle hatchlings	Emergence was reduced 15% in nests invaded by *S. invicta;* about 8% of nests were invaded.	Moulis, 1996
Freshwater turtle hatchlings	In a laboratory experiment; hatchling survival was 67% lower in presence of *S. invicta*.	Allen et al., 2001a
Hatchling alligators	Hatchlings exposed to fire ants were 5% lighter than unexposed hatchlings three weeks after hatching.	Allen et al., 1997
Hatching alligators	Fire ants killed parts of two broods during hatching.	Watanabe, 1980
Trapped mammals	*Solenopsis invicta* was found on dead mammals in live traps.	Masser and Grant, 1986 Flickinger, 1988

until experimental tests are carried out, such reductions are purely hypothetical. In any case, juvenile mortality rates are not necessarily connected with size of a breeding population. For example, among territorial breeders, including many bird species, populations may be limited by the availability of breeding territories or nesting sites, and breeding population density may change little over very large ranges of juvenile mortality. The fire ant itself is another excellent example. The size of a fire ant population is limited by available space. Survival of newly mated queens (that is, juvenile colonies) is positively related to available territory, but the population density of mature colonies is practically independent of juvenile mortality rate. Reducing the supply of newly mated queens would have little effect on the size of this population.

A few of the cases listed in the table included experimental tests and deserve further mention. The tests consisted primarily of comparing mortality in the presence of *S. invicta* with that in its absence. During the fall of 1989 and 1990, Bart Drees treated half of a spoil island off the Texas coast with fire-ant bait, reducing the foraging population of *S. invicta* by 80 to 97% (Drees, 1994). The breeding success of several species of water birds that nested on the island was over 90% lower on the untreated half (i.e., in the presence of fire ants) than the treated half of the island. The effect of the ants was greatest after early June, when no additional young were produced on the island half with fire ants, although the birds continued to lay eggs. If such *S. invicta*–induced mortality were typical of all the breeding sites used by a species of water bird, one might rightfully expect their populations to decline, but a second breeding island harbored only the inconsequential ant *Monomorium minutum* and was not used in this experiment. The reasons are unknown why some but not other spoil islands support large populations of fire ants and why high *S. invicta* populations are frequent in water bird rookeries.

The presence of *S. invicta* on beaches has raised concern on behalf of the sea turtles that nest there. On nine of 19 Florida beaches, *S. invicta* made up 10 to almost 100% of the ants coming to baits. On the beaches of a Georgia wildlife refuge, about 8% of loggerhead-turtle nests were invaded by *S. invicta*. Hatchling emergence from invaded nests was about 15% lower than that from uninvaded ones (Moulis, 1996). Whether this 1.2% (0.08×0.15) reduction in total emergence is detectable at the population level and the extent of such nest invasions remain to be documented. For comparison, nest predation by vertebrates such as raccoons was 5–95%. Of course, with threatened species such as most marine turtles, one may not want to take chances.

Fire ants were shown in the laboratory to be capable of killing hatching turtles (Allen et al., 2001a); eggs of a common freshwater turtle were used as a surrogate for the marine species. The fire ants did not breach unhatched eggs but did invade pipped eggs; survival of hatchlings was reduced by two-thirds in the group exposed to fire ants. The conditions of this experiment were so unnatural that the results do not even justify a guess about the quantitative importance of *S. invicta* to marine turtles under natural conditions. Confinement of the experimental colonies in enclosures of only 0.2 m^2 on a diet of honey only (no protein) probably caused the ants to concentrate their attention on the turtle eggs. In addition, the eggs were placed in sphagnum moss rather than the sand marine turtle eggs would have been buried in.

A similar experiment exposed newly hatched alligators to fire ants in the laboratory. Eggs were placed in alligator nest material in five-gallon buckets containing fire-ant colonies (Allen et al., 1997), a kind of alligator version of *No Exit*. Once free of the eggshell, the hatchlings were removed. Three weeks after hatching, exposed hatchlings weighed 5% less than unexposed ones, a significant difference. Whether this small difference has any long-term impact remains unknown.

An attempt to explain population trends of the Texas horned toad, an iconic species in the region, deserves mention because *S. invicta* was one of the possible factors considered (Donaldson et al., 1994). Texans are awfully fond of their horned toad (actually not a toad at all, but a strange, camouflaged, flattened, spiny lizard). That is why they have, since time immemorial, been killing and stuffing them to sell as tourist curios. By the 1980s, the beloved lizards seemed to be becoming mighty scarce (except in curio shops, of course). In a series of surveys, interviews, and museum studies, Wendy Donaldson and two coauthors attempted to determine the status and future prospects of this lizard in Texas. Several authors have cited this as an example in which the fire ant has brought on a "population collapse." That ought to make it interesting to find out what the paper really said.

A survey of 78 selected Texas localities by professional herpetologists yielded only 145 lizards, 12 of which has been squashed even flatter on roads. The lizards were present at 48 of the localities; south Texas harbored the most. Significantly perhaps, no lizards or evidence of lizards (fecal pellets) were found in any of the survey sites in east Texas, although museum records indicated that the horned lizard once dwelt there. Because east Texas is also the area overrun with *S. invicta,* suggesting a connection seemed natural. And indeed, the lizard and *S. invicta* overlapped at only five of the 30 sites at which the horned lizards were seen or captured. (The most unusual finding was that "two lizards were seen driving between locations." Perhaps they were fleeing the invasion by *S. invicta.*) All but three of these 30 lizard-harboring sites had harvester ants, *Pogonomyrmex* spp. This coincidence is not surprising because the horned lizard eats mostly harvester ants, as can be readily seen from its fecal pellets. The hypothesis was that *S. invicta* wiped out *Pogonomyrmex,* in turn spelling curtains for the lizard. Attacks by *S. invicta* on *Pogonomyrmex* nests have been reported (Hook and Porter, 1990). Ironically, almost 40% of the surveyed Texan landowners used pesticides to kill harvester ants. Texans believe in tough love.

In the end, though, this correlational study, like most others, was inconclusive about why the Texas horned lizard is declining (which it seems to be). The strongest correlation was land use—the lizard was absent from most agricultural land, and urban land ran a close second. Plowing, vehicular traffic, and pesticides would surely be hard on a slow, ground-dwelling lizard. Furthermore, *S. invicta* is favored by many of the land uses associated with the absence of the lizard, possibly creating a noncausal correlation between lizard absence and fire-ant presence. Neither pesticide use nor *S. invicta* was clearly related to the lizards' absence. No one knows whether the lizard could or would subsist on a diet of fire ants, and no studies quantify the relationship between *S. invicta* and *Pogonomyrmex*. In the end, it is all pretty speculative, and the reality is that the decline of the horned lizards is correlated with a ragbag full of changes in human practices that have occurred in the last half-century. The authors, unlike many who cite their work, seem to lean toward land use as the primary explanatory factor, exacerbated by pesticides, domestic cats and dogs, commercial trade, and fire ants.

Another headscratcher is an early report of large numbers of bluegill sunfish dying in farm ponds in Mississippi (Green and

Hutchins, 1960). Dissections found their stomachs to contain large numbers of fire ants, and "it was assumed that the death of the fish was caused by the ants" (what ever happened to the constitutionally guaranteed presumption of innocence until proven guilty?). Laboratory tests in which fire ants were fed to fish were unreliable in creating toxic symptoms, although they did show that fire ant venom was toxic to fish. A follow-up study in which huge numbers of fire ants were fed to large numbers of fish—entire mounds were shoveled into several farm ponds—never found any adverse effects on the fish. The fish seemed happily to feed on and digest the ants. A closer reading of the initial report of the mysterious Mississippi deaths revealed a second correlated factor—heavy rains. Such are the dangers of drawing causal conclusions from correlated events. Perhaps if the researchers had been environmental engineers rather than agricultural entomologists, they would have concluded that the effects of storm-water runoff killed the fish.

Tests of Population-level Effects

Leaving this portfolio of inconclusive studies, let us turn to the three exceptional vertebrate species in which the effect of fire ants on populations have been tested. The first of these is the bobwhite quail, an intensely studied game species whose populations have been declining for decades throughout its range.

The only direct test of whether fire ants affect a focal species' population is to manipulate fire-ant populations (preferably in both directions) and measure the response of the focal species population. Removals were used to test the effects of *S. invicta* on bobwhite quail, white-tailed deer, and loggerhead shrike populations in Texas in the early 1990s (Allen et al., 1995, 2001b). Only the quail and deer stories will be covered in detail here. In the area near Corpus Christi, researchers matched five pairs of 200-ha plots by soil, vegetation, and other such characteristics. All but one were occupied by polygyne fire ants at high densities. They estimated the bobwhite populations by "whistle-counts" (listening for the calls of male bobwhites) and by counting flushed coveys along walked transect lines. In spring and fall of 1991 and again in the spring of 1992, they treated one member of each pair of plots with Amdro. Fire-ant populations were 70% lower on the treated than on the untreated plots by the first June and 96% lower by the second. One year after treatment ended, that is, in the third year, this difference disappeared. Note also that Amdro bait kills many ants other than fire ants. When fire-ant density is high, fire ants probably collect most of the bait, but after the fire ants have been reduced, the bait is available for other species. Strictly speaking, we cannot attribute effects entirely to *S. invicta*.

The first resurvey (fall 1991) revealed no difference in bobwhite population density (although whistle counts were significantly higher

Figure 36.1. The effect of fire-ant reduction on quail populations on five experimental areas in Texas. Data from Allen et al. (1995).

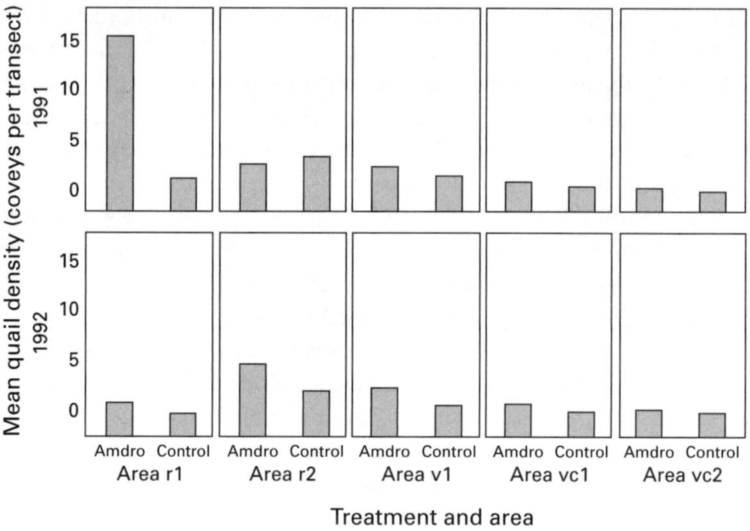

on treated plots), probably because of a single density estimate that was wildly out of line (Figure 36.1, area r1). In 1992, however, after three treatments with Amdro, the bobwhite populations on the treated areas were significantly higher (2.1 bobwhite/ha) than those on the untreated (0.9 bobwhite/ha). Over both years, in all but one of the 10 pair-wise comparisons, the treated areas had higher densities of quail than the controls (Figure 36.1). My own reanalysis of the authors' data found no significant effect of treatment on birds flushed per kilometer of transect ($p = 0.09$) but did with the authors' density estimates. These density estimates were calculated with a "Fourier series detection model for ungrouped perpendicular distances." How one should interpret differences in such highly derived density estimates is unclear; perhaps caution is advisable. Another problem is that these experiments cannot differentiate between direct effects of fire ants on quail and indirect effects on their food supply or some other necessity. Competition between quail and fire ants for insect prey could easily produce the same outcome.

Several subsequent studies attempted to determine how fire ants depress bobwhite populations (Giuliano et al., 1996; Pedersen et al., 1996; Mueller et al., 1999). The density response was quite rapid—four of five plots showed higher quail densities (although not significantly higher) within six months. Reports that fire ants prey on hatching quail eggs and newly hatched chicks have a long history, appearing as early as the 1930s (Stoddard, 1931; Travis, 1938a, b; Lehmann, 1946). The culprit in these early reports was the native fire ant, *S. geminata*. After *S. invicta* replaced *S. geminata*, reports (often simply observations) of fire ants attacking quail continued, but the subject has remained controversial to this day. Given these repeated observations, reduced chick survival might be the link between fire ants and lower bobwhite populations. (Perhaps the quail thrived by eating Amdro, a fat-rich diet of low toxicity to vertebrates, applied at the rate of 1.7 kg per hectare. Perhaps I speak with tongue in cheek, but the precisely correct control would still have been toxin-free Amdro.) A test of this hypothesis took place in Refugio County, Texas, an area inhabited by high densities (over 200 mounds per hectare) of polygyne *S. invicta*. Researchers trapped 77 female bobwhites in 1997 and 71 in 1998 (shouldn't females be called Roberta Whites?), provided each with a small radio collar, and released them. The investigators subsequently relocated these females and the nests they made by radio, taking care not to flush the females and risk nest abandonment. A 60- by 60-m area around each of half the (randomly chosen) nests was treated with Amdro, which suppressed fire ants in the vicinity of the nests. Controls received the same walk-around but no bait, so fire-ant density was left unchanged. The response variables were hatching success and survival to three weeks of age.

Of the nests found, 44% hatched in 1997 and 31% in 1998 (in rough agreement with a 46% nesting success in Texas; Lehmann, 1946). The rest fell prey to vertebrates, were abandoned by the hen, or were trampled by cattle. A total of 57 nests hatched in the two years, and hatching success within nests was 92 to 95%. More to the point, the hatch rate was the same whether fire ants had been reduced around the nest or not. Furthermore, in a number of cases, the chicks failed to hatch properly or died during pipping and were subsequently eaten by fire ants. Such hatch failure is especially common during times of drought, when water loss toughens egg membranes and makes hatching difficult (Johnson, 1961; Brennan, 1993). Hatch failure or death came before the fire ants. Apparently, observations of fire ants on hatching bobwhite chicks must be interpreted with caution; this result casts doubt on several reports based on such observations in which the chicks were already dead when first observed. In addition, chicks in nests abandoned by the hen often succumb to fire ants during or soon after hatching. Such chicks would have died in any case, and attributing their deaths to fire ants is not ecologically meaningful or scientifically defensible. Abandonment, not fire ants, sealed their fates. The implication is that hens protect their chicks from fire ants to a large degree.

Three weeks later, the two treatments differed greatly in survival of chicks. Survival was 60% where fire ants had been suppressed but only 20% in the control. Furthermore, all members of the broods died in over half the untreated nests; all members died in somewhat less than a quarter of treated nests. Almost half the difference in chick mortality could be attributed to fire ants, but the mortality rate contributed by the fire ants must be seen in light of total mortality. Mortality at hatching averaged 62% for both years. By the end of three weeks, total mortality from all causes, based on the total number of eggs laid, was 77% in the absence of fire ants and 92% in their presence. Total mortality was therefore about 15% higher in the presence of fire ants. To report only survival based on hatched eggs makes fire ants seem to double or triple mortality. Chick survival at three weeks decreased as the number of fire ants captured on baits in the nest shortly after the chicks' departure increased. Complete loss of broods was more frequently associated with very high captures of ants in the nest, and very high captures occurred only in the untreated areas. Complete brood loss in the untreated areas was more than twice that in the treated.

How did the fire ants kill the chicks? Those of us who know little of quail biology (like me) may not realize that, within a day of hatch, the parents and the chicks take up a wandering existence away from the nest. Therefore, one day after hatching, the quail from the treated and untreated nests shared the same environment. Whatever fire ants were doing to raise chick mortality, they must have been doing it in the 24 hours between hatch (which they do not affect) and departure from the nest.

Two possibilities were tested: delayed effects of stings acquired at the time of hatch and/or stress resulting from the need to avoid fire ants or to defend against them. In the first study, the authors bought two-day-old bobwhite chicks, weighed them, then plopped each into a beaker and added 0 to 200 fire ants for 60 or 15 seconds. Chicks that had been maximally exposed to fire ants (200 ants for 60 seconds) survived significantly less frequently (44% of the control). Such chicks averaged 34 ants on their bodies. Chicks subject to lower exposures did not survive significantly less than the controls (although the authors report that they did, the 90% confidence limits overlap the control). The maximally exposed chicks also gained weight significantly (15%) more slowly than did controls or chicks with lower exposure to fire ants.

The conclusion is tempting that the cause of decreased survival among wild chicks has been pinpointed, but these experiments are problematic. First, the exposure to fire ants was highly artificial, confined in a beaker with no escape. Second, the authors used two-day-old chicks, in spite of evidence that whatever reduces chick survival happens in the first 24 hours after hatching. Two-day-old chicks of this precocial species would probably not stand still long enough to allow dozens of fire ants to sting them. The authors state as much but go on to say that "stings by large numbers of [fire ants] may be possible." Not very confident, and for good reason. No one has quantified the number of stings sustained by bobwhite chicks in the wild or the age at which they sustain them. The relevance of this work is therefore in doubt. The question was and remains, what happens during the first 24 hours?

Might fire ants affect the ability of bobwhite chicks to feed and rest, thereby affecting their growth and health, in turn having possible indirect effects on survival? Randomly chosen bobwhite chicks less than six days old were placed individually in a penned area with a high density (580 polygyne mounds per hectare) of fire ants or in a similar area that had been freed of fire ants by poisoned bait. Each chick's behavior was observed for 15 minutes, and the process was repeated with many chicks. Chicks in the enclosure with fire ants spent less time sleeping and more time moving than those in the fire ant–free enclosure. They also spent more time "responding to ants" by preening, pecking, or jumping (although the authors do not say how they determined that fire ants prompted the responses). Other behaviors did not differ, and stinging by fire ants was not quantified. This experiment does not really seem to help explain the lower survival of chicks in the presence of fire ants. Could less sleep and more movement really have led to lower survival?

Although we still do not know how *S. invicta* decreases bobwhite populations, the story now goes back to a correlation tack. Allen and his colleagues reasoned that, if quail populations were indeed depressed by *S. invicta*, then a relationship should be apparent between the length of time since the ant appeared in a region and the quail population density. The USDA's quarantine surveys provided the fire-ant data, and the Audubon Society's annual Christmas Bird Count, standardized to quail per observer-hour, provided the quail population data for several decades that bracketed the appearance of *S. invicta* at each site. This method was used in two studies (Allen et al., 1995, 2000), the first including 15 counties in Texas and the second 19 sites across Florida, Georgia, and South Carolina.

Both among the Texas counties and across the three-state region, bobwhites were more abundant before invasion by *S. invicta* than after (quail per observer-hour in Texas, 0.044 before and 0.019 after; in the three-state region, 0.07 before and 0.02 after). Furthermore, no significant annual linear trend in the bobwhite populations was apparent before invasion. After invasion, bobwhite populations in Texas, Florida, and South Carolina, but not in Georgia, were negatively correlated with number of years elapsed since invasion. The authors claimed that time since invasion explained 75% of the variation in bobwhite populations and did so in spite of over 30 years of variation in invasion dates.

Had they found the smoking gun that convicts *S. invicta* of nationwide quailicide? Not so fast, Speedo. Things look bad at first blush, but closer inspection reveals serious doubts. To begin with, the question is not framed correctly. The bobwhite has long been declining throughout its range whether that range overlapped with *S. invicta* or not (Brennan, 1993). The question is not whether it is declining or not (it is) but whether *S. invicta* is speeding up this decline. Proof requires a significant difference in rates—the post–fire ant decline rate must be significantly greater than the decline rate by calendar year (which confounds fire ants with the general decline). Alternatively, the difference between the slopes before and after invasion should be significantly greater than zero. Another problem: the claim of 75% explained variation is misleading because the regression uses the annual mean data rather than the site counts. At best, 75% of the variance of the means (which are necessarily less variable) may be explained, but far less of the variation among sites is. In fact, because the repeated counts are not independent (regression assumes independent data), and the number of sites per year varies, no R^2 can be assigned to the regression. More yet: the use of mean values biases the slope of the regression because every mean value has equal weight in the regression, even though fewer data make up the means for longer elapsed times. And still more: even if the bobwhite count were unchanging from year to year, the index, bobwhites per observer-hour, would show a negative trend if the number of observers on the ever-more-popular Audubon bird counts increased with time. For these many reasons, wagering that fire ants are guilty of quailicide throughout the Southeast would be a bad bet, like investing in Enron stock.

But, you protest indignantly, the Texas Amdro experiment showed a population effect. True, but all but one of the ten experimental areas were occupied by polygyne fire ants, whose density is usually much higher than the monogyne form that dominates most of the southeastern USA. Effects of *S. invicta* on bobwhite are expected to be density dependent. Indeed, that is the central conclusion of the Amdro experiment. The effects of *S. invicta* on bobwhite should therefore be much lower in regions with monogyne populations, not higher as reported.

Now a postscript. Many reasons for the decline of bobwhite have been suggested. These include changes in farming practices, habitat degradation/loss, intensification of tree farming, increased use of pesticides, and changes in fire frequency, climate, predator abundance, and land-use trends. Changes that make the habitat less suitable for bobwhite may also make it more suitable for *S. invicta*, as I suggested in the case of native ants and *S. invicta*. Against this picture of decline, we should remember that bobwhite was probably not an abundant bird under precolonial conditions but was itself the beneficiary of the abandonment of farmland in much of the eastern USA, which created the brushy, open, fire-prone habitat the bird prefers. This habitat is not static and requires repeated disturbance of just the right type and amount if it is not to revert to a bobwhite-free late successional community. Although maintaining it is the intent of quail management, achieving this goal is no simple matter.

Effects on Mammal Populations

Alongside quail, deer and rodents have received considerable attention with respect to the effects of *S. invicta*. The same Texas Amdro experiment that tested the effect of *S. invicta* on quail populations simultaneously tested the effect on white-tailed deer recruitment (the number of new offspring entering the adult population). Remember that one 200-ha plot of each of five pairs of plots was treated with Amdro, which reduced fire-ant populations over a two-year period. White-tailed deer that spent their time on treated areas were thus exposed to greatly reduced fire-ant populations for over two years.

Before treatment began, and then in the fall of each year, deer were counted by spotlighting at night, which, as every backcountry redneck knows, means driving a vehicle equipped with a searchlight and scanning for the eye-shine of deer while trying to avoid the game warden. Some incidental sightings were also made during the pedestrian quail transects described above. The pairs of plots differed in the densities of deer (number per 100 ha), perhaps because of habitat differences. The treated and untreated plots did not differ in any of the three years, but about twice as many fawns per doe were observed on treated as on untreated plots in both years (year 1, 0.70 versus 0.34; year 2, 0.47 versus 0.23). These differences vanished in the year after fire-ant treatment ended and the two types of plots no longer differed significantly in fire-ant population. The number of fawns per doe was correlated with the fire-ant density in the second year but not in the first, reflecting the increased reduction of fire ants during the second year. Thus, reduction of fire ants was followed by greater reproductive success among white-tailed deer and/or migration into the treated areas. Better survival might have been a direct effect of reduced injury or death due to fire ants. The freeze response of fawn may protect them against large predators, but it is completely counterproductive against fire ants. Alternatively, if the fawns do respond to fire-ant irritation with movement, they may become more vulnerable to coyote predation, but this is speculation. The reason for the improved recruitment is unclear, and migration of does with fawns into the treated areas has not been ruled out. Generation of an effect on such large, mobile animals on relatively small plots may seem strange, but mother deer appear to be rather faithful to their fawning sites. When the fawns are yearlings, they tend to disperse from their natal area, perhaps explaining the lack of increase in deer density on the treated plots during the second year, even though recruitment was doubled during the first year. Unfortunately, the effect of fire-ant reduction on the deer population density therefore remains unclear. Higher recruitment says yes, but direct density estimates say no. A very large-scale experiment would probably be needed to settle the question. The claim that an increase in deer "harvest" of one per square mile would increase land value by $45 per hectare suggests that it might be worth the effort.

A final comment on this study: fire-ant densities on these plots were very high (170 to 270 mounds per hectare) because the plots were occupied by polygyne *S. invicta*. Given the lower densities characteristic of monogyne *S. invicta*, significant effects on deer recruitment would probably not be detectable in monogyne areas, that is, in most of the range of *S. invicta*. The threshold density for detecting such effects is currently unknown.

The more closely the scale of a creature's world matches that of fire ants, the less remarkable it seems that the animals collide. I reviewed the effects on insect populations in an earlier chapter; that fire ants also affect the populations of animals as large as deer is less intuitive. Animals living at intermediate scales, such as rodents and baby quail, are in the waffle range. Several studies of fire ants and rodents are thus of particular interest, especially because rodents and ants have been shown to compete in desert and forest ecosystems.

In east central Texas, researchers livetrapped, marked, released, and recaptured rodents for almost a year and compared their data with densities of vegetation and fire-ant mounds (Smith et al., 1990). High density of fire-ant mounds was present at 20 contiguous trap stations. The remaining 80 were low. (The reported mound densities were equivalent to one mound per 1600 m^2 at high density and only one per 5000 m^2 at low density. They must be in error, because the values typical of pastureland, fully stocked with monogyne colonies, are in the range of one mound per 100 m^2.) Of the five species of rodents captured, the pygmy mouse, *Baiomys taylori*, made up 80% of the captives and was the only species common enough for statistical analysis. Individuals captured during three or more monthly trappings were assigned to the "resident" category. The remainder was considered "transients" (homelessness may be a problem among rodents, too). Transients made up two-thirds of the samples in the summer, about three-quarters in the spring, and about half in the fall and winter.

The expectation was that the mice would avoid areas of high fire-ant density and that residents would be more likely to live in low-density areas than transients would. The results were rather more subtle and complex and much more difficult to interpret. Mound density, season, and mobility class had no simple effects on mouse capture. Even though the results are so complex as to make one's brain hurt, the authors attempted an explanation based on seasonal activity periods of the mice and ants. Perhaps I should leave it at this: during some seasons, both transient and resident mice are either positively or negatively associated with either high or low mound density. In the end, of course, the results are once again simply correlations, and the causes of mouse distribution may lie elsewhere.

Another, somewhat similar trap-grid study correlated density of fire ant mounds with small-mammal activity in two areas of east central Texas (Killion and Grant, 1993). (The authors' expectation that mound density and forager ant density should be correlated is a misapprehension. Their own data showed that they were not at any scale. Density of the forager population is more closely related to colony size than to colony density.) Rodents were trapped, marked, and released for four consecutive nights, in a search for pair-wise association between mound density, foraging, and trap captures. Mound density indicated that one site harbored polygyne *S. invicta,* the other monogyne colonies. Because rodent species differ greatly in their home-range size—*B. taylori* moved no more than 12 m between recaptures, whereas the two larger species moved about 40 m—the authors clumped grid units together and reanalyzed at these larger scales. Indeed, when the scale of the sample units was adjusted to the scale of the species' home ranges, the larger rodent species showed a significant negative correlation with mound density and fire-ant foraging activity. Mammals that range over many sample units are less likely to respond to local effects.

Fortunately, this correlational study was followed up with an experiment in which the density of a monogyne fire-ant population was manipulated and the response of *B. taylori* estimated (Killion et al., 1995). Fire-ant colonies were eliminated individually from half of each of two plots; periodic baiting followed, and care was taken to minimize the effect on other ant species. Small mammals were livetrapped monthly for 17 months, individually marked, and released. Because the analysis used only the first capture of animals and excluded animals captured only once (transients), it tells us where animals were choosing to settle rather than estimating abundance.

The numbers of new captures on the two halves of each plot were not significantly different before reduction of the fire-ant population, but after reduction, the number of new captures became increasingly greater in the areas of low ant density; after one year, about twice as many new captures were made in the low-density as in the original-density areas. Apparently, pygmy mice preferred to settle where the density of fire ants was lower, confirming the correlational study. Fire ants recruited to baits slightly but significantly more slowly near pygmy-mouse burrows than at random points in the plots. To the extent that recruitment time was a reliable estimate of fire ant presence, mice might be making settlement choices with respect to fire ants on a very fine scale. The reasons for the avoidance were not clear; the authors' observations indicate that the mice were not often stung. The generality of these findings is placed in some doubt by another study—use by the pygmy mouse of plots in which fire ants had been reduced with poisoned bait did not differ from that of untreated control plots (Pedersen et al., 2003), although that by the hispid cotton rat did.

Perhaps the psychology of mouse behavior can shed light on this question (Holtcamp et al., 1997). Foraging theory posits that an animal stops foraging in a patch of habitat when it perceives the cost to exceed the benefits. The benefits are obviously the rate of feeding. The costs include the risk of predation, food-handling time, missed opportunities elsewhere, metabolic costs, and so on. In theory, the animal divulges its perception of the costs and benefits by the time it chooses to stop foraging in patches of specified characteristics. Four different kinds of foraging patches were made in the laboratory: either 15 or six sunflower seeds were mixed with 250 ml of dry sand in a small box, and 100 fire ants were added to some. These combinations were tested on hungry deer mice (*Peromyscus maniculatus*).

Without fire ants, the mice foraged equally in rich and poor patches of sunflower seeds, but in the presence of ants, they biased their efforts toward the rich patches, visiting, leaving, and returning to them more often; spending more time there; and foraging faster while in the box. The number of times mice took a seed out of the box to process it nearly tripled when ants were present. Through these behavioral changes, the mice harvested the same total number of seeds, spent the same total amount of time foraging, and harvested at the same overall rate, in spite of the increased lost time and metabolic cost. A psychomurine interpretation runs as follows: in the presence of fire ants, it takes a higher rate of return to balance the increased cost (or risk) represented by the ants, and preferring the rich patch and working faster provides it. A similar line of thinking could explain why more seeds remained when the mice gave up harvesting rich patches in the presence of fire ants than in their absence. The point at which costs exceed benefits comes earlier in the presence of the ants. The ants cause the mice to change their behavior significantly. Nevertheless, extending these findings to a natural setting is difficult. First, the density of ants ($150/m^2$) is high compared to natural scout densities. Second, the geometry of patches and ant locations are not nearly so discrete in nature. In another study, *S. invicta* reduced foraging in trays by *Peromyscus polionotus* more than did other indications of risk, such as moonlight or predator urine (Orrock and Danielson, 2004). At the very least, mice take notice of and respond to fire ants.

Final Remarks

Solenopsis invicta kills a variety of baby animals. Essentially, if the babies are accessible, they are vulnerable. Nevertheless, convincing evidence of any population-level effects of *S. invicta* on vertebrate species is available only for three species: bobwhite quail, white-tailed deer, and pygmy mice. For pygmy mice, fire ants affected choice of settlement site, with unknown effects on breeding populations. All studies of bobwhite and deer were carried out in areas occupied by the polygyne *S. invicta*, whose densities are much higher than those of the monogyne form that occupies most of the Southeast. Monogyne *S. invicta* must have smaller effects. As a result, these findings cannot readily be extended to the entire range of *S. invicta*. Whether or not population effects on vertebrates can be demonstrated in monogyne regions remains to be seen.

Population size is sensitive to changes in any of several basic life-history parameters, each directly or indirectly affected by biotic and physical factors, the strength of whose effects may themselves depend on population density. Moreover, a positive effect at one level may well be compensated for by a negative one at another level. Disentangling a single factor, *S. invicta,* from this snarled ball of yarn is not a simple matter, nor is much necessarily to be gained by looking at one factor in isolation. In our ever-changing landscape, the effect of *S. invicta* is a moving target whose effects may differ from situation to situation or from time to time. In such complex ecological systems, a combination of experimentation and simulation modeling, shaped by educated guesses, can determine the direction and magnitude of effects and interactions and can detect density dependence of factors. These experimental findings can then be used to simulate multiple, simultaneous causation as it cascades through several levels and as factors interact. Complexity, interaction, and compensatory effects often make the predictions less than obvious. For example, increased mortality at one life stage may be partly balanced by decreased migration or increased reproduction at another—showing that fire ants kill some percentage of juveniles is not equivalent to showing that they depress breeding populations. Modeling must alternate with testing of predictions in further experiments. In this way, the trends of populations in response to various factors, including *S. invicta* density, can be predicted under a wide variety of conditions. This process would clearly require a long-term effort involving difficult experiments, which probably explains why it has not been done, but only such studies will permit us to explain why, even in the presence of high populations of *S. invicta,* one can manage land to increase bobwhite populations (Brennan, 1993) or why peaks in quail populations have occurred in areas well after fire ants became established there. Focusing only on *S. invicta* can be a distraction, a red herring in the words of Brennan (1993). Craig Allen said it well: "[a]dditional passive anecdotal evidence and experiments of small scope will not provide an understanding of the complex interactions between vertebrate populations and [*S. invicta*]. . . . [We] need long term comprehensive ecological studies . . . conducted with controls and adequate temporal and spatial replication" (Allen et al., 1994). Who could argue with that? But who will do it?

◂•• *A Microsafari in Antland*

Imagine yourself 5 mm tall, waiting quietly behind a small stone alongside a fire-ant foraging trail. You are on a microsafari to observe South American ants in their natural habitat and at their own scale of distance and time. Your camera is ready. Worker ants of diverse sizes hurry past, casting their antennae back and forth over their odor trail like bloodhounds on a convict's scent. Some of the workers on the way back to the nest are dragging their gasters on the ground, laying down more odor trail. Some carry unidentifiable pieces of a dead insect that is being cut to bits somewhere out of sight at the end of this trail, and some have bloated gasters full of prey juices. Gradually, you become aware of a very high pitched buzzing sound, and suddenly, from downwind appears what looks like a flying ball of gray fuzz. It is the female of the phorid fly *Pseudacteon tricuspis*, in search of worker ants in which to lay her eggs. Wings beating hundreds of times per second, she picks out a large worker on the trail and alternately hovers and darts over it. The worker stops, as if frozen. The fly feints toward the ant and the ant backs up. The fly follows. The worker tries to make a run for it, then bends into first one, then another defensive posture, but the fly keeps her position over the ant. As skillfully as a cowboy, the fly cuts the ant from the general mass and herds it to one side of the trail. A minute elapses. Suddenly, the fly darts in and strikes the ant in the side near the coxae, fast as a bullet (Figure 37.3). The impact knocks the ant onto her side, where she briefly curls her body before quickly righting herself. She walks a few steps then stops and rises onto the tips of her tarsi as if she has been stunned. A number of her nestmates come from some distance away and cluster around the stunned ant. After a few minutes, the ant recovers and once more joins the ants on the trail. The fly has picked another worker to harass a short distance away.

What has happened here? On our scale of time and distance, the crucial event is far too brief to see—we see the fly dart, the ant fall, and the fly zip away—but on the millimeter/millisecond scale of these tiny insects, we would have seen the female phorid fly leave her position over the ant and, in a sweeping arc, approach the ant low from the side and rear, her head facing the same direction as the ant's. As she neared the ant's coxa, her abdomen rotated forward, and the elaborately shaped, three-lobed ovipositor protruded as a grappling hook to attach to the ant. The moment the ant was in its grip, the powerful muscles surrounding the fly's genital tract (Figure 37.5) contracted with immense force, causing the hypodermic inner portion of the ovipositor to pierce the ant's intersegmental membrane and an egg to be fired into the ant like a tiny torpedo. The entire operation took only a tenth of a second or so. This is truly life in the fast lane. If your camera was ready, you would now have a sensational wildlife film, easily as exciting as anything on the Serengeti Plain.

37

Biological Control

All living things suffer at the hands of other living things, in part because all living things compete for resources with other organisms in their communities, thereby reducing the amount available to each competitor. Every organism also represents an opportunity for others, an opportunity that has led to the evolution of a vast array of predators, herbivores, pathogens, symbionts, and inquilines. To a dismayingly wide array of creatures, other organisms are little more than a nice lunch, a petri plate, a culture medium from which to extract the stuff of life and in which to reproduce. The effect on the host ranges from mildly debilitating to lethal. No animals, including social insects, are exempt (Schmid-Hempel, 1998). Wouldn't it be lovely, we can imagine all these sufferers saying, to be able to leave all these freeloaders behind? As in the case of the 19th century European peasant farmer who dreamed of escaping economic oppression by emigrating to America, lifting the burden of exploitative creatures ought to lead to a burst of population growth (and often does). This phenomenon exactly is thought to contribute to the sometimes-spectacular success of invasive exotic species, including the exotic fire ant *Solenopsis invicta* in the USA.

Why do the exploiters not come with the exploited? Obviously predators must make the trip separately from their prey. The case of parasites and diseases is more complex. In many cases, a diseased pioneer may be less capable of founding a new population. More generally, most diseases and parasites probably require a host population of a minimum size, a condition that may not exist in the early stages of an invasion. Small populations are more subject to extinction, and pathogen or parasite populations are always even smaller than those of their hosts. If a disease is uncommon, the chances a pioneer will be diseased may be quite small.

As a result of such processes, many invading exotic species arrive cleansed, to varying degrees, of their natural enemies. Energy expended on natural enemies can now be redirected to population growth. The new homeland is almost never completely free of competitors, but these will have a full complement of parasites and diseases that place them at a disadvantage with the invader. The success of invasive species ranges from legendary to downright frightening—kudzu strangles the southern forest, the zebra mussel alters the entire ecology of the Great Lakes, the gypsy moth defoliates thousands of square kilometers of forest, goats strip innumerable islands of greenery, and *Hydrilla* fills Florida's lakes so full that one can almost walk on the water. The fire ant, though not the worst of the invaders, is certainly in good company.

But is the fire ant really doing better in its adopted home in the USA than in South America? To answer this question, Porter and coworkers (Porter et al., 1992, 1997) compared the relative well-being of a dozen or more populations on road rights-of-way, grazed pasture, and lawns in North and South America, measuring colony size (mound volume) and density and collecting climatic and habitat variables as well. The results confirmed that the fire-ant populations were much more vigorous in North America than in their native home in South America. True, a diligent search discovered fire ants in almost all the sites on both continents, but the colonies were about 40% larger and four times as dense in North America. When colony size and density were combined into total ants per hectare, North America won over South America by a factor of five—North America averaged 1220 ants per square meter and South America 230. This result was also reflected in the bait samples. Fire ants of one species or another were found at almost all sites in both North and South America (93 to 98%). The picture is rather different at the level of baits. In South America, only about 13 to 20% of the baits were occupied by *Solenopsis* ants, whereas in North America, almost 80% to 97% were. Therefore, although *Solenopsis* ants, and *S. invicta* in particular, were present at most sites in South America, they were much less dominant. South American ant communities in disturbed habitats are much more complex and rich.

What might be the cause of this five- to sevenfold difference in fire-ant populations? Scientists often find themselves in the situation of having to plod along eliminating alternative explanations in the face of a clear favorite. Admittedly tempted by the "absence of natural enemies" explanation, Porter and his coauthors nevertheless carefully compared abundance with season of sampling, climate and climatic range, rainfall, temperature, seven other climatic variables, roadside and lawn maintenance patterns, type of adjacent habitat, and finally, polygyny. None of these explained much, if any, of the difference in ant abundance on the two continents. Of particular interest, about 20% of the sites on both continents had polygynous colonies, eliminating this factor as a source of the difference.

After winnowing these possible causes Porter was left with the final candidate, natural enemies. There is no doubt at all that North American *S. invicta* is living the easy life, relatively free of natural enemies—early intensive searches turned up only two or three of these. In contrast, *Solenopsis* in South America is host to at least 10 different microorganisms, three species of nematode worms, 18 species of parasitic phorid flies, a parasitic wasp, a parasitic ant, and dozens of other symbionts of undetermined importance. The species-rich ant community in South America also suggests that *S. invicta* has many competitors there, but this conclusion must be accepted only with some caution, for the presence of these other ants could as easily be the result of *S. invicta*'s lack of dominance as its cause.

What are the pathogens, parasites, and symbionts to which fire ants are host, and how does this burdensome fauna in the native habitat compare with that in introduced areas? This subject is of special interest, because the most likely agents of biological control are natural enemies in the fire ant's native homeland. Broadly speaking, these natural enemies fall into five broad categories: (1) pathogens, including bacteria, fungi, viruses, and protozoa; (2) parasitoids, mostly scuttle flies of the family Phoridae; (3) other ants, which act as competitors or predators; (4) socially parasitic ants; and (5) predators, including insects, vertebrates, and spiders. Much more information is available on some of these groups than on others.

Pathogens

A survey for pathogens in fire ants sounds like it ought to employ sophisticated and high-tech methods. The reality, at least in the early days, could more readily be described as "grind and find." Colonies of fire ants are shoveled into buckets and flooded from the soil. A sample of 1000 to 2000 mixed adult and immature ants from a single colony is then ground up with a little water in a tissue homogenizer (Jouvenaz et al., 1980), which breaks up the cells and frees their contents into the general soup. A drop of this soup is then examined under a phase-contrast microscope. This method is particularly good for detecting the spores of microsporidian (protozoan) parasites; investigators can easily detect one infected individual among 2000 uninfected ones. It is less good for bacteria and occluded viruses. The data reveal the percentage of infected colonies. Initial detection is followed by smears of individuals from infected colonies for determination of within-colony infection rates.

Such surveys of several species of fire ants in both North and South America found a number of pathogens (Jouvenaz et al., 1977, 1981; Jouvenaz and Anthony, 1979; Jouvenaz, 1983, 1984, 1986, 1990a, b; Jouvenaz and Ellis, 1986; Briano et al., 1995a). In the Mato Grosso of Brazil, up to 8% of colonies suffered from protist or nematode diseases (Wojcik, 1986). In Brazil, the fungus *Metarrhiium anasolpiaea,* which also causes "queen's disease" in *Atta sexdens,* is widespread among *S. invicta.* A fungus that may be specific to the fire ant, *Myrmecomyces annaellisae,* has been found in both South and North America (Briano et al., 1995a). A possible bacterial infection, several fungi, and virus-like particles were found in Brazilian *S. invicta,* but little is known of their pathogenicity or specificity. A nematode, *Tetradonema solenopsis*, infects less than 5% of fire-ant colonies in Brazil, but the life cycle and transmission mechanism are not known. A couple of unidentified mermithid nematodes from *S. richteri* were never reported again. A neogregarine protozoan, similar to *Mattesia geminata*, causes a disease in which pupae appear "sooty" and die. A similar disease occurs in a very low percentage of *S. geminata* in the USA. Early surveys in the southeastern USA detected no serious diseases of *S. invicta* but did find a very mildly pathogenic fungus (originally identified as a yeast). Even though most of these pathogens were discovered 20 to 25 years ago, we know little more about them today than at the time of their discovery. More recently, *Myrmicinosporidiaum durum*, a fungus pathogenic to several species of ants was also found in *S. invicta*, suggesting a possible host switch (Pereira, 2004). Another new disease of unknown origin is the yellow-head disease caused by a new species of *Mattesia* (Pereira et al., 2002). The cuticle of infected young adults fails to tan, so the spindle-shaped spores are visible within the head and thorax. Infected workers are likely to die sooner than uninfected ones. The disease occurred at about one-third of north Florida sites in 2002, infecting an average of 20% of colonies. A virus belonging to the picorna family was found to be widespread in Florida, infecting almost a quarter of the sampled nests in 2004 (Valles et al., 2004). Although no symptoms were discernible in the field, when infected colonies were brought into the lab, all the brood in them died within three months. Infection could be transferred to uninfected individuals, and all castes and stages were host to the virus. A mermithid nematode that parasitizes the alates of the native *S. geminata* and the thief ant *S. pergandei* also occasionally parasitizes the males of *S. invicta* (McInnes and Tschinkel, 1996). In the first two species, females are parasitized much more frequently than males. Worms up to 15 cm in length may emerge from ants less than 1 cm in length.

The first specific (and most common) pathogen of fire ants was discovered by accident when Bill Buren noticed cyst-like bodies in the Brazilian fire ants he was examining for a taxonomic study (Allen and Buren, 1974). Infected ant gasters contained four to six large cysts (up to 22) formed from tremendous enlargement of infected cells, each packed with the tiny spores of a new species of microsporidian, which was subsequently named *Thelohania solenopsae* (Knell et al., 1977) on the basis of the form, number, and staining of the intracellular spores it forms (Figure 37.1). The pathogen infects only the fat body and ovaries of its hosts. Vegetative stages are present in larvae and pupae, and sporulation occurs in adults, producing two types of spores in the same tissue. Formation of both free spores and membrane-enclosed octospores is a characteristic of the genus *Thelohania*.

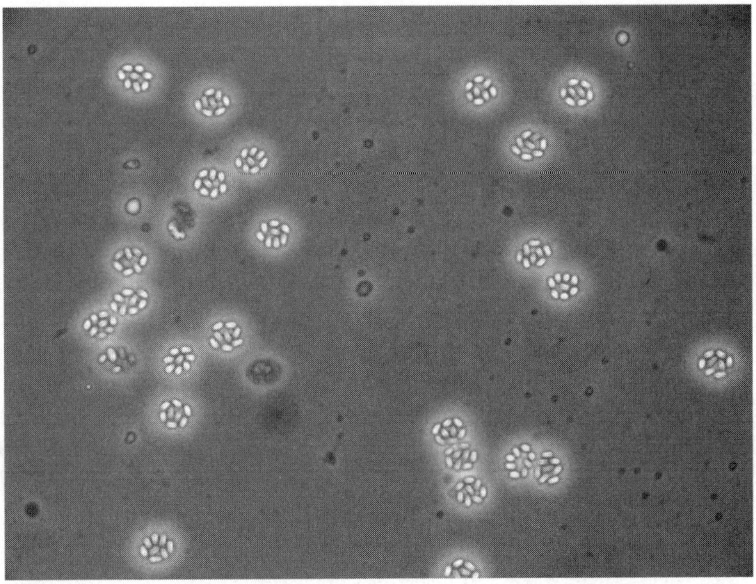

Figure 37.1. *Thelohanea solenopsae,* a microsporidian pathogen of fire ants. Photo courtesy of Roberto Pereira, U.S. Department of Agriculture, Fire Ant Unit.

Since this discovery, a good deal has been learned about microsporidian diseases of fire ants. *Thelohania solenopsae* and two other species of microsporidians have been found in almost a dozen species of *Solenopsis* in South America (Jouvenaz, 1983), often at substantial rates of infection, about 12% all together. In northern Argentina, *S. richteri* populations suffered 25 to 50% rates of infection with a microsporidian similar to *T. solenopsae* (Briano et al., 1995b). About 2% of Brazilian and Argentine *S. invicta* and *S. richteri* harbored another microsporidian, *Vairimorpha invictae,* producing both free and octospores. Both *T. solenopsae* and *V. invictae* were highly specific to *Solenopsis* species; infection rates varied widely (Briano et al., 2002b). Surveys in the southeastern USA uncovered four species of microsporidians in the native *S. geminata,* one of which (*Burenella dimorpha*) has been studied in some detail. The various microsporidians of fire ants are difficult to distinguish through microscopy or ultrastructure, but molecular systematics is beginning to sort out the phylogenetic and taxonomic relationships among the species of *Thelohania* and *Vairimorpha* (Moser et al., 1998, 2000). Ribosomal DNA sequences showed a 75% similarity between isolates of *Thelohania* and *Vairimorpha,* supporting their placement in different genera, but the 90% or greater similarity among the various *T. solenopsae*-like isolates from Argentina, Brazil, and the USA places them in a poorly resolved "species complex." An impediment to resolving their taxonomy has been the inability to cross-infect hosts with these strains in order to determine specificity. More recently, molecular probes against a specific piece of *T. solenopsae* DNA have been developed, allowing far more efficient and sensitive detection of infection in ground-up fire ants (Snowden et al., 2002; Valles et al., 2002).

Solenopsis invicta in the USA was thought to be free of *T. solenopsae*, but a very similar microsporidian pathogen was recently discovered at rather high rates of infection (23%) in polygyne *S. invicta* in Florida, Mississippi, and Texas, rates as high as those of *S. richteri* in Argentina (Williams et al., 1998). None of seven other species of ants had the pathogen. The ribosomal RNA of the newly discovered microsporidian was almost identical to that of *T. solenopsae* from South America in its base sequences, so they are very probably at least closely related species. How this microsporidian came to North America is unknown, but its life history is unlike that in South America; it seems to infect only the polygyne fire ant, passing from colony to colony through exchange of infected workers and brood. This limitation to the polygyne form does not have a genetic basis—$Gp\text{-}9^{BB}$ alates from polygyne nests can become infected and may even rear infected incipient nests after mating (Oi et al., 2004). In South America, the disease is common among both social forms of fire ants. Perhaps a vector is involved in the transmission of this disease there.

The disease caused by *T. solenopsae* seems not to be a ravaging one, producing no gross pathological signs or catastrophic mortality. Colonies can be infected by addition of infected brood, which they will readily accept and which develops into infected workers in whom the disease gradually destroys the fat body. Only about half the queens become infected, even after a long time. The reason is not apparent, but perhaps it is linked to the great variation in queen fecundity and the worker care queens receive.

Oddly, infection does not greatly reduce the longevity of workers, but infected laboratory colonies gradually dwindle away within a few months. This apparent contradiction is probably the result of reduced brood production by the infected queens, rather than early mortality of infected workers (Williams et al., 1999; Oi and Williams, 2002). From three until 10 months after infection, brood declined to zero, as did the egg-laying rates of queens, who lost weight and were more likely to die during that period. Even though some of the queens were uninfected, these broodless colonies never recovered. Although infected workers or larvae may be incapable of stimulating queens to lay eggs, a more likely explanation is that the ovaries and fat body of the queen become infected, debilitating her and reducing her fecundity.

Attempts to transmit the disease directly by feeding spores to ants have failed. Infected workers infect fourth-instar larvae but not younger ones, a pattern that suggests infection may occur when larvae are fed spore-contaminated solid food or oral secretion. Some infection can be transmitted from infected workers to pupae (which do not feed) by contact or grooming, but this route is relatively unimportant. In the absence of infected larvae, infected workers do not infect queens. When workers (infected or not) and infected larvae are present with queens, the queens quickly become infected (Oi et al., 2001). Most probably, infective spores are passed from fourth-instar larvae to the queen by workers, perhaps while they transmit the fecundity-stimulating factor from these larvae to the queen. Once she is infected, the microsporidian settles in the queen's ovary, and she passes the infection to her progeny in the egg, as is the case in the Argentine *S. richteri* (Briano et al., 1996). Incipient colonies founded by infected female alates may also provide a route for infecting new colonies when larger colonies raid the incipient ones and steal the infected brood (Oi and Williams, 2003). The microsporidian produces at least two and probably three different spore types (Shapiro et al., 2003; Sokolova et al., 2004), at least one of which is produced through meiosis and some of which occur in adult and larval feces and larval midguts. Some types probably transmit the disease horizontally, within a generation (Chen et al., 2004a), and others vertically between generations, but this possibility has not yet been tested experimentally.

Can the disease be experimentally caused and spread in nature? Williams and his coworkers inoculated widely separated polygyne colonies with infected brood. After five months, more than a quarter of the larvae in inoculated colonies were infected. After a year, the size of the inoculated colonies had declined by about 30% (size of controls had increased slightly) and their number had tripled; several neighbors of the inoculated colonies had become infected. After two years, more than 90% of the colonies in the plots were infected, and most of the originally inoculated colonies had disappeared, but new colonies often appeared, probably through fission, so the density did not decrease much. The lack of recovery in laboratory colonies suggests that these new colonies do not arise when uninfected queens assume the reproductive throne in infected colonies. Smaller colony size, rather than fewer colonies, caused the population to decrease. Clearly, *T. solenopsae* causes a sustainable and transmissible disease in polygyne fire ants in the USA, but equally clearly, this disease is not an epizootic that sweeps like wildfire over entire regions; it causes a creeping, slow debilitation. Its widespread distribution among North American polygyne fire ants is therefore puzzling. Perhaps it was transported in the ants themselves, through human commerce, or a highly mobile vector.

South American rates of infection with *T. solenopsae* are typically between two and 11% of *S. richteri* colonies but reach 40 to 80% in some areas of Argentina. Briano and his coworkers monitored over 1300 colonies in the polygyne population at Saladillo for over four years (Briano et al., 1995b, c, 1996). Infected colonies were only about a third the size of uninfected ones but did not have much less worker brood. They were also less than half as likely to contain sexuals, probably as a secondary effect of small colony size. The prevalence of infection within colonies was usually high, between 20 and 45% of the larvae and pupae. Most significantly, 20% of the eggs were infected, so the queen must pass the parasite to her progeny in the eggs (Briano et al., 1996). As in *S. invicta,* spore formation occurred in adults of all types at rates varying from 34 to 95%; 1 to 6 million spores were present per individual, and infected workers had life spans about 9 to 30% shorter than those of healthy ones. Infected colonies died at higher rates during the study (Briano and Williams, 1997). Interestingly, two species of myrmecophiles were also infected and may therefore play roles as vectors.

In the course of four years, the colony density at the infected Saladillo site dwindled irregularly from 162 per hectare to about 28. As the colony density declined, the number of infected colonies changed relatively little, so the infection rate approximately doubled. The authors suggest that the disease caused this decline, but as tempting as that conclusion may be, their data leave this question open. Although they give no data on colony size, the high initial density suggests a young population of small colonies, which would normally thin as the colonies grow. No comparable, disease-free control plots were monitored. Controls would have been nice.

The association of *T. solenopsae* with polygyny makes Texas, with its high rates of polygyny, prime territory for an investigation of the prevalence of this disease in the USA. Tamara Cook, intelligently staying out of the firing ranges, found infected colonies at three of four National Guard Training Camps in Texas (Cook, 2002). At one of these camps, infected colonies were smaller, and infection rate was higher when colony density was higher. Over a two-year period, the infection rate fluctuated between 9% and almost 50% of colonies, 20 to 60% of their workers were infected, and signs of debilitation were apparent, much as occurred in the Florida and Argentine sites. In contrast, she found no significant effect of infection on colony size at any of the other camps, and the infection even disappeared from one site. Perhaps the small sample size made small effects undetectable—colony size also varies greatly for reasons unrelated to infection. The high variation in infection rates suggests much coming, going, and who-knows-what between samples; contributions due to sampling error are unknown. Perhaps the best take-home lessons from this study are that natural *T. solenopsae* infections are widespread (Florida to Texas) in polygyne fire ants and that infection rates can be quite substantial as well as highly variable for unknown reasons on small geographic and time scales. In addition, the female alates produced by infected polygyne colonies are lighter than those from uninfected colonies, so their ability to found new colonies may be impaired (Cook et al., 2003). Whether this microsporidian is a good prospect for debilitating polygyne fire-ant populations across the South is still uncertain, despite its apparently large effects on *S. richteri* in Argentina.

A similar microsporidian disease of *S. invicta* occurs sporadically in Argentina (13% of 154 sites; Briano and Williams, 2002). The overall low infection rate of 2.3% hides occasional and local epidemics in which half or more of colonies are diseased. At some sites, these epidemics declined and disappeared over several years. Whether this disappearance stems from an inability of the pathogen to propagate and spread or from the rapid mortality of infected colonies or both is unknown. A large proportion of all stages and types of ants carry spores. Vegetative spores increased in proportion as larvae developed, reaching over 50% in pupae. Meiospores (sexual spores produced by meiosis) occurred only in adults—about half of workers and males contained up to half a million spores each (mean about 12,000 for minor workers and 64,000 for majors). Only 16% of sexual females carried meiospores. About 60% of worker pupae bore binucleate spores. As these numbers sum to more than 100%, individuals can clearly carry more than one type of spore. Even eggs carried some types of spores, evidence that the pathogen is transmitted through the ovaries of the queen to her progeny, but queens were not often infected (mean less than 5%).

Workers taken from the dead piles of infected laboratory nests were two to eight times as likely to contain spores, a result that suggests they died of the disease. Indeed, infection reduced the life span of workers by about 20 to 30%. Such virulence is consistent with the spotty but occasionally epizootic distribution of this disease.

These fire ants host two microsporidian pathogens, *T. solenopsae* and *V. invictae*. *T. solenopsae* is more abundant, affecting about 12% of colonies, whereas *V. invictae* affects only about 2.3%. The pathogens co-occurred at about 8% of sites, but only 0.24% of colonies were infected with both. This rate is precisely that expected from the frequency of occurrence of each, so the pathogens seem neither to encourage nor to discourage one another's infections.

By now, the reader should not be surprised to learn that, in the USA, the native fire ant *S. geminata* harbors at least four different species of microsporidians, at total infection rates of 10% or so. Little is known about three of these. The best studied is *Burenella dimorpha*, whose infection rates in Florida were usually less than 5% but were occasionally 40% locally (Jouvenaz, 1984, 1986; Jouvenaz and Lofgren, 1984a, 1984b; Jouvenaz et al., 1984). Unlike *T. solenopsae*, this pathogen can be transmitted experimentally to larvae if their food is wetted with spore suspensions, but only one of the two spore types is infective. The pathogen develops in the hypodermis and underlying fat body of pupae, causing clear areas in the vertex of the head and the petiole as a result of the destruction of adult cuticle and fat body. Eventually the cuticle ruptures, releasing two types of spores, one from the hypodermis and the other from the fat body. The adult *S. geminata* workers cannibalize these ruptured pupae, form them into pellets in the infrabuccal pocket, and feed these pellets to fourth-instar larvae, the only larval instar that eats solid food. Only the spores formed in the hypodermis are infective, afflicting about two-thirds of the fourth-instar larvae that eat them. Younger larvae are not infected because they are not fed solid food. Unlike infections with *Thelohania*, those with *B. dimorpha* are always fatal. Infection rates within colonies were usually less than 5% but can range up to 100% of pupae. In the laboratory, the infection can be transmitted by mouth to *S. invicta, S. richteri,* and *S. xyloni* but does not persist in these species. In the sense of indefinite persistence, *B. dimorpha* is specific to *S. geminata*.

The free spores that were infective in the laboratory experiments are probably formed without meiosis. The formation of meiotic spores, the octospores, depends strongly on temperature, the very sharp optimum is at 28°C. Below 25°C and above 30°C, octospore formation is essentially zero, but at 28°C, all pupae contain them, and they make up over one-third of all spores. In field colonies, octospores made up 25 to 40% of the total, probably because octospore formation has evolved to coincide with the preferred brood-rearing temperature of *S. geminata*. The function of these octospores is enigmatic—in laboratory experiments, they were not infective. Clearly, a piece of this puzzle is still missing.

Although both are microsporidians, the diseases caused by *T. solenopsae* and *B. dimorpha* are quite different. The former causes a slow debilitating disease whose causative organism insinuates itself into the very heart of the colony's reproductive machinery, the queen's ovaries. It has made the very cells of the host its homestead, where it cultivates its own progeny for long periods, gradually sapping its host's resources. In contrast, the latter is more akin to an ordinary disease, transmitted from individual to individual through feeding and growing explosively in its host's body, killing it quickly, and releasing millions of progeny. Of course, the list of unknowns in these two diseases is still much longer than the list of knowns, so perhaps such generalizations are premature.

Use of microsporidians as biological-control agents requires that they attack only the target species of fire ants. As a first step in determining the host ranges of *T. solenopsae* or *V. invictae,* Briano looked for microsporidian spores in non-*Solenopsis* ants that co-occurred with fire ants infected at moderate to high rates at several Argentine and Brazilian sites (Briano et al., 2002b). No non-*Solenopsis* species was infected. Specificity for *Solenopsis* species (including *S. invicta, S. richteri, S. macdonaghi, S. saevissima,* and *S. quinquecuspis*) appears to be high; infection rates of *S. invicta* and *S. richteri* by *T. solenopsae* were about the same but *V. invictae* was perhaps somewhat better adapted to *S. invicta*. *Thelohania solenopsae* occurs in *S. invicta* in the USA but has not been detected in the native *S. geminata* or *S. xyloni* so far. Whether this result stems from host preference or from the mutually exclusive distributions of these species with *S. invicta* is an unanswered question. Also unknown is whether *T. solenopsae* can be exploited as a biological-control agent beyond the effects it already has naturally. If a vector to move it among monogyne colonies in the USA cannot be found, its usefulness will be limited to polygyne areas.

Some fire-ant researchers have made much of the potential of a common soil fungus, *Beauveria bassiana*, for biological control of *S. invicta*. Strains of this fungus have been commercialized, but because large numbers of fungal spores must be applied to the target insects, it qualifies more properly as a biological insecticide (it kills quickly but does not persist) than a biological-control agent (which should persist in the host population and continue to control it). Use in *S. invicta* control requires application of a suspension of fungal spores to the fire-ant colony (Stimac et al., 1993a, b; Brinkman and Fuller, 1999; Brinkman and Gardner, 2000a, b). Ants directly touched by the spores, or that ingest them (Broome et al., 1976), are invaded by the fungus and killed, but the more than two-day lag between host death and fungal sporulation assures that no disease is sustained within the colony, for by the time new spores are produced, the dead workers have long been disposed of in the field by necrophoric workers. Moreover, pouring a fungal spore suspension on a nest mound is ineffective, because the soil filters out most of the spores before they get to the ants. Injecting powder containing fungal spores works fairly well, but use in the real world would require mound-to-mound treatment and would offer few advantages over chemical insecticides. Finally, when spores are mixed into liquid baits, the ants filter out most of them, form them into buccal pellets, and discard them (Siebeneicher et al., 1992), reducing the effectiveness of the fungus as a control agent.

The latest addition to the array of microorganisms to which fire ants are host is a most interesting one, *Wolbachia*. This parasitic rickettsial bacterium occurs in the cytoplasm of many insects, where it has a range of effects, sometimes symbiotic, sometimes pathogenic, and sometimes neutral. Transmission among individuals within a generation seems not to occur. Instead, because of its cytoplasmic location, *Wolbachia* is passed from one generation to the next in the host's egg. Sperm, lacking cytoplasm, cannot transmit this bacterium. This simple fact has enormous evolutionary implications; the bacterium should, and does, cause the host to produce more females than males, or even exclusively females, because its fitness is linked only to female hosts. At one extreme, it causes the host to produce females by parthenogenesis by simply duplicating the gametes to produce diploid clone females containing the *Wolbachia* in their cytoplasm. Other effects include killing males to produce highly skewed sex ratios (or all female offspring); producing cytoplasmic incompatibility between the gametes of infected males and uninfected females, causing reproductive failure by uninfected females and giving the reproductive advantage to infected females; and increasing host fitness, increasing its own.

Genetic work on other insects has produced several molecular tools for the detection and study of *Wolbachia*. A coalition of colleagues centered around Ken Ross and led by DeWayne Shoemaker used *Wolbachia*-specific probes for a gene that codes for the bacterium's surface protein coat (*wsp*) to hunt for *Wolbachia* in fire ants, both in its native South America and in North America (Shoemaker et al., 2000, 2003). These probes allowed the unambiguous detection of *Wolbachia* DNA from fire ants (a single individual from each colony to assure independence of samples), without any need for visual verification. *Wolbachia* was present in six of the nine species of the *S. saevissima* complex in South America. Typically, a high proportion of colonies, sometimes nearly all, in infected populations of both social forms was infected, as were all life stages, castes, and body regions. In the native *S. invicta*, infection rate varied greatly and was greater in the more southern populations (Figure 37.2). Infection was absent from the North American *S. invicta, S. richteri*, their hybrid, and the native *S. geminata*. The large sample size makes it unlikely that infection was present but undetected in North America. The ants that originally landed at Mobile either lacked the infection or lost it by chance while the population was small.

The gene sequences of the *wsp* gene revealed two major groups of *Wolbachia*, called simply A and B. Both occurred in *S. invicta*, but never in the same individual. *Solenopsis richteri* were infected by a strain that was almost identical to the A strain of *S. invicta*. A sequence analysis of the mitochondrial DNA of the hosts showed that the B strain of *S. invicta* was perfectly associated with one mitochondrial DNA haplotype, and the A strain with another, although this second haplotype included both *S. invicta* and *S. richteri*. This pattern raises the interesting possibility that the A strain entered the *S. invicta* population through hybridization with *S. richteri* followed by subsequent spread. In essence, the cytoplasmic genes of female *S. richteri* became associated with the nuclear genes of *S. invicta*. Because *Wolbachia* and cytoplasmic organelles (including mitochondria) are passed to the next generation only by the female in the egg, the descendents of the organelles and the bacterium never separate. The small difference between the A strains of *S. invicta* and *S. richteri* must be mutations accumulated after the divergence of two lines. To the extent that its various subterfuges spread *Wolbachia* in a population, diverse mitochondrial DNA haplotypes are displaced by the one associated with the parasite. In this light, perhaps the *S. invicta* carrying the A strain is actually a separate and cryptic species.

How *Wolbachia* affects its fire-ant hosts is unknown. In monogyne fire ants, we can expect strong selection against parthenogenesis by

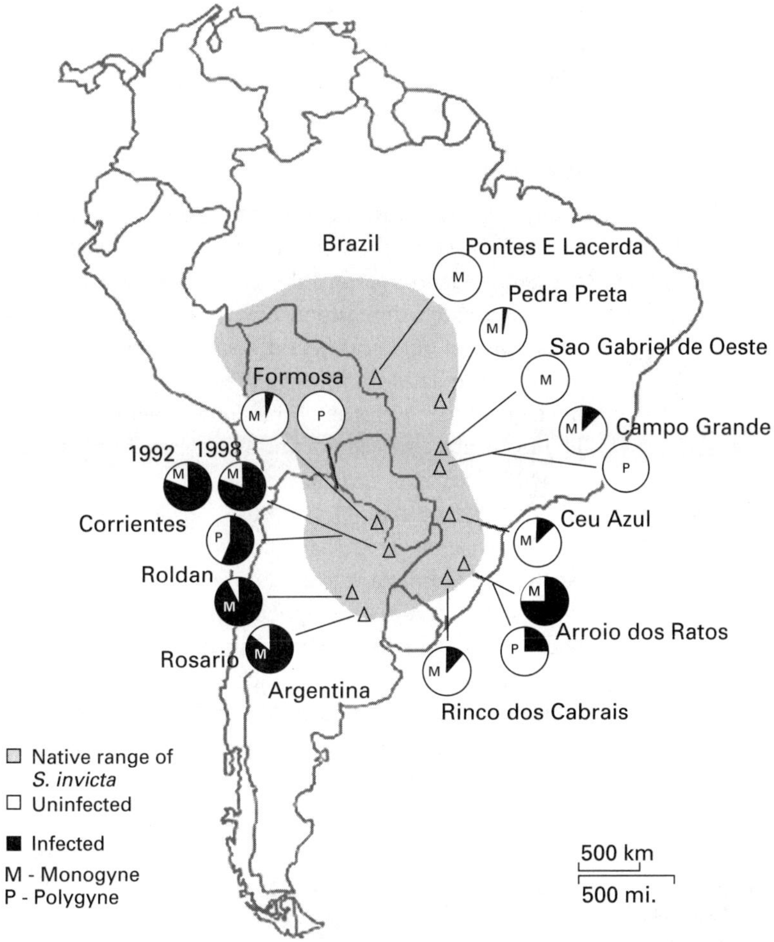

Figure 37.2. The distribution of the rickettsial bacterium *Wolbachia* in monogyne and polygyne *Solenopsis invicta* populations in South America. Pie diagrams show proportion of infected and uninfected individuals of each social form. Reprinted from Shoemaker et al. (2003) with permission from the Entomological Society of America.

gamete duplication. Parthenogenesis would produce homozygosity at the sex-determining locus, producing diploid males rather than females, and would cause failure during colony founding. Likewise, females carrying the cytoplasmic incompatibility type of *Wolbachia* would fail because the uninfected females their sons mate with would produce only haploid males and would likewise fail during colony founding. Moreover, infected females occur among the progeny of infected mothers, ruling out these two effects of the parasite. Although the parasite occurs in both social forms, it should spread more slowly and to lower equilibrium levels in polygyne populations because of lower female fecundity in polygyne colonies. In two of four South American populations in which both social forms were present, infection rate was lower in the polygyne form (Figure 37.2). Whether *Wolbachia* has any promise as a biological-control agent depends on whether infection entails any fitness costs. The answer is presently not known.

Pseudacteon Parasitoids

Many species of ants, including fire ants, are plagued by parasitoids. By far the best-studied parasitoids of *Solenopsis* species are phorid flies of the genus *Pseudacteon*. Unlike the ants' often secret, slow associations with pathogens, their union with insect parasitoids is full of drama. Because we live on a scale so different from that of ants, we do not easily perceive the action and excitement of their world. Behavioral studies have given us a glimpse of this tiny, insect equivalent of the Serengeti Plain—bizarre, gruesome, and heartless—but it includes no lions and no wildebeest. Mammalian predators are not analogs for insect parasitoids, which consume their hosts slowly, bite by bite, from the inside.

Many species of *Pseudacteon* attack *Solenopsis* species in both North and South America. The most-studied is *P. tricuspis*, which attacks *S. invicta*, but most species share the basic biology. Gravid female flies seek out worker fire ants in situations particular to each fly species. Each fly species prefers to attack workers of a particular size range, either attacking from the side (Figure 37.3) or herding the worker with a frontal assault. Attacks happen very fast—a female takes only a fraction of a second to grapple the ant with her specialized ovipositor (Figure 37.4)

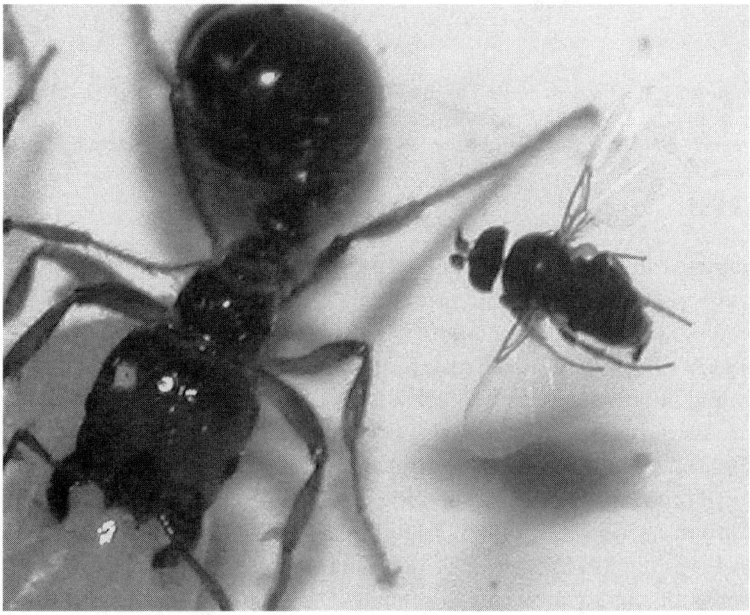

Figure 37.3. A female *Pseudacteon tricuspis* attacking a fire-ant worker. Note the fly's lowered ovipositor. Photo courtesy of Sanford D. Porter.

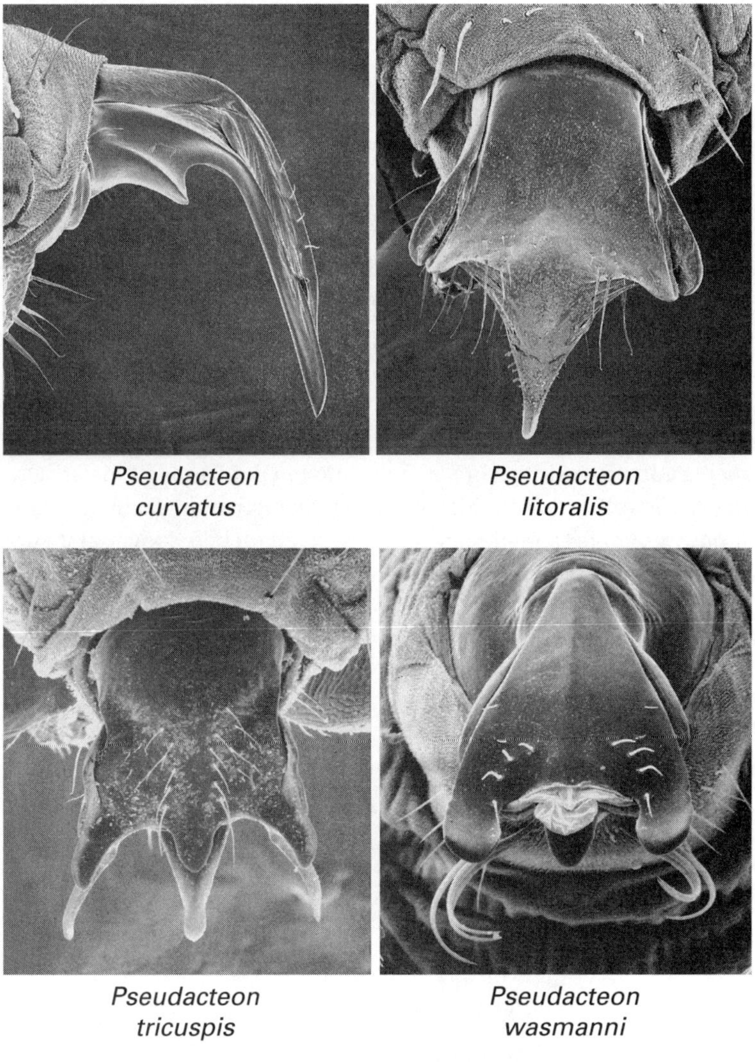

Figure 37.4. Scanning electron micrographs of *Pseudacteon* ovipositors. Note the radical differences in structure between species. Reprinted from Figures 5, 12, 26, and 19 of Porter and Pesquero (2001), with permission of Sanford D. Porter.

and to inject an egg through the ant's intersegmental membrane by means of a musculated injection apparatus (Figure 37.5). The attack stuns the worker briefly and causes nearby workers to tend her.

After recovering, the attacked ant goes about her business, but inside her body, the tiny egg, which the mother has endowed with minimal stores, increases over 100-fold in volume by taking up nutrients from the ant's hemolymph and synthesizing them into its own tissue. By the time it hatches into the first larval stage of *P. tricuspis*, it is a maggot perhaps a tenth of a millimeter long, taking up two-thirds of the thorax (Porter, 1998a; Consoli et al., 2001; Zacaro and Porter, 2003). By the fourth day, it has molted into the second instar and has slithered its way into the head of its host. Feeding on hemolymph, it has grown to be

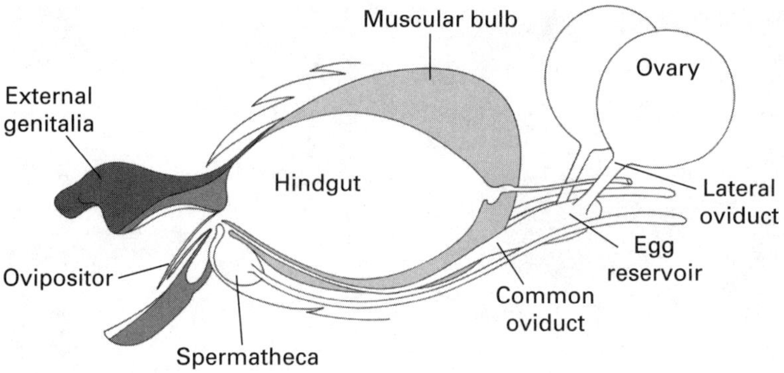

Figure 37.5. Schematic drawing of the genital tract of a *Pseudacteon* female. Piercing of the host's cuticle by the ovipositor, followed by the contraction of the muscular bulb, injects the egg into the host. Adapted from Zacaro and Porter (2003).

between 0.2 and 0.9 mm long and has a worm-shaped body of 12 segments. After a week or more, it molts into the third and last instar, leaving its cast-off cuticle to float in the ant's hemolymph, like space trash. It now resembles a planarian worm, with a flattened body that it curls into a horseshoe shape. Its fat bodies are arranged in a ladder-like pattern along its sides, so that "the live maggot . . . looks somewhat like a plastic bag of wet cotton balls twisting back and forth" (Porter et al., 1995b; Figure 37.6). Up to this point, the ant has shown no outward signs of being parasitized, but now it ceases feeding and moves about little, effects that appear to be caused by a chemical injected by the female fly. The maggot now secretes, or causes the ant to secrete, an enzyme that dissolves the intersegmental membrane between the head and thorax,

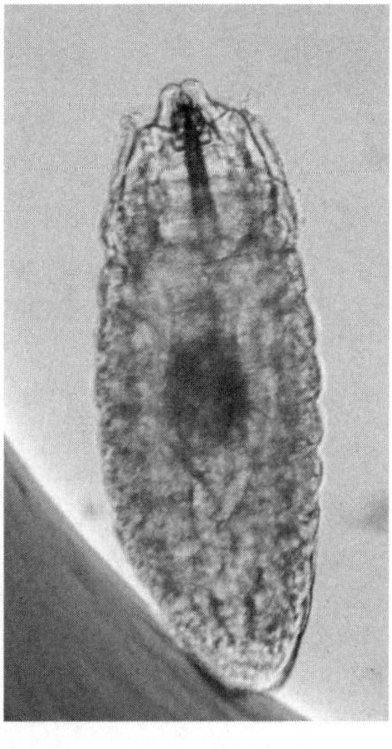

Figure 37.6. The larva of *Pseudacteon* (Consoli et al., 2001). Photo courtesy of Sanford D. Porter.

loosening the head, and often even the first pair of legs and the petiole, making them likely to fall off at the slightest touch. The maggot now consumes everything in the ant's head, down to the cuticle, causing the head to fall off with the maggot inside. The body of the ant is left standing motionless, legs and sting apparatus twitching gently.

Inside the head capsule, the maggot pushes out most of the ant's mouthparts so that its own anterior end is positioned in the resulting opening. It is preparing for pupariation inside the ant head. (In most higher flies, including the Phoridae, the last larval cuticle is not shed at the end of the larval period but is hardened into a protective case called the puparium. The process is therefore called pupariation, not pupation. The pupa develops into the adult within this puparium, and when time for adult emergence arrives, the puparium and the pupal cuticle within it are both shed at the same time.) Because the head serves as a protective case, the maggot parsimoniously sclerotizes (hardens) only the exposed first three segments that lie in the opening once occupied by the ant's mouthparts (Figure 37.7). The remainder of its cuticle remains pliable, thin, and soft. When the hardening is complete, the maggot, its exposed parts now tanned to the same color as the ant head, is barely noticeable. Three to four days later, the actual pupation takes place, and a pair of respiratory horns are pushed out of the corners of the host's mouth cavity. All of the 10 species of *Pseudacteon* that have been reared pupariate in this manner.

Very occasionally, workers recognize the interloper, cut it out, and kill it. Little is known about the later behavior of the parasitized workers, whether they behave normally, where they die, or what the fate of the ant-less heads is. Workers probably simply pick up any loose heads

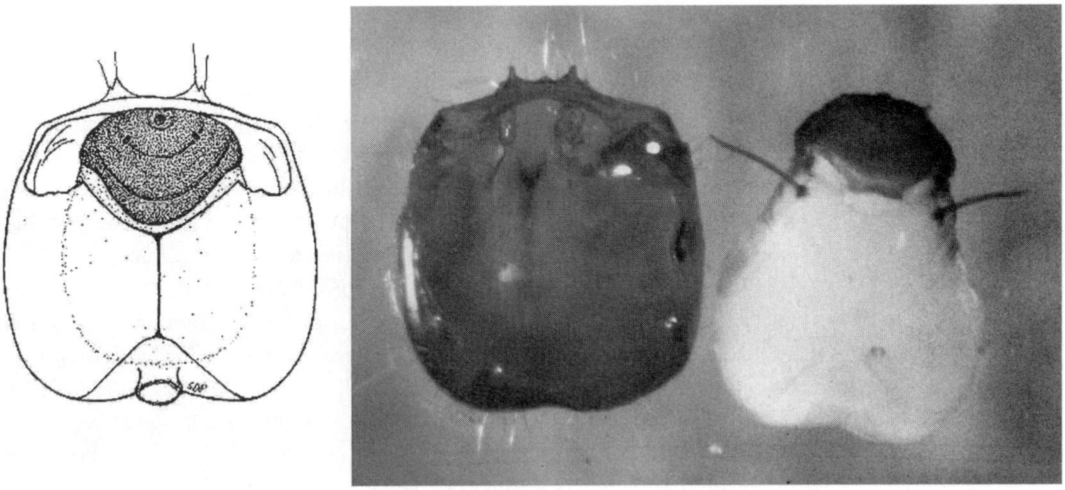

Figure 37.7. Pupa of *Pseudacteon* in the head of a fire-ant worker. Adapted from Porter et al. (1995b).

lying around in the nest and discard them as trash outside the nest. There, the flies would develop within their borrowed cases, emerging two to six weeks later. Adult emergence is unceremonious, taking only a few seconds in the early morning—the sclerotized cap pops off, and the adult slips out of its protective head capsule. Depending on the season, only five to 12 weeks have elapsed since this fly began its life as an egg shot into the body of a fire-ant worker.

Soon after emergence, the females fly off in search of fire-ant victims and the males in search of females, whom they find by searching for fire ants, too. Both sexes mate repeatedly (Wuellner et al., 2002). Their world, in keeping with their tiny bodies, is small, and they usually travel no more than a couple of hundred meters from where they were born. Still, a few adventurous *Pseudacteon* flies often appeared at trays full of *S. geminata* as far as 650 m from the nearest source populations of *S. geminata* (Morrison et al., 1999a). In spite of being only a millimeter or two in length, they seem to cover the huge stretches of hostless terrain largely under their own power rather than passively on the winds. Flies are not found preferentially downwind from source populations. The family Phoridae is full of tiny but intrepid travelers. Some species have been captured in airplane-mounted nets at altitudes of over 1000 m or on shipboard traps in midocean. Such gene flow between patches keeps the patches from drifting apart genetically and becoming separate species.

Their adult lives are short and are lived at an urgent pace symbolized by split-second matings and host attacks. In the laboratory, they are dead within three to seven days. Their life span in the field is unknown, but the females almost certainly have only a few days to dispense their 100 to 200 eggs into an equal number of fire-ant hosts. So compressed are their lives that complete egg development, followed by degeneration of the ovaries, occurred while they were still pupae inside their own personal ant heads. Despite this urgency, we know that they are active only during a particular, species-specific part of each day. When a female has located vulnerable fire ants, she attacks repeatedly in bouts that last about an hour, making perhaps 30 to 120 attacks. Of these, between 8 and 35% will result in the successful injection of an egg, but it is also dangerous to fool around with fire ants. Flies that fall among the ants, perhaps from exhaustion, perhaps by accident, perhaps while mating, are immediately killed by their intended victims.

Wherever they are native in North or South America, ants of the genus *Solenopsis* are harassed and parasitized by multiple such species of *Pseudacteon* parasitoids (Smith, 1928; Williams et al., 1973; Morrison et al., 1997b; Morrison and Gilbert, 1998, 1999; Porter, 1998a; Pesquero, 2000; Pitts and Pitts-Singer, 2001). At least 20 species parasitize South American *Solenopsis* (for a key, see Porter and Pesquero 2001). Another eight species attack the *Solenopsis* species native to North and Central

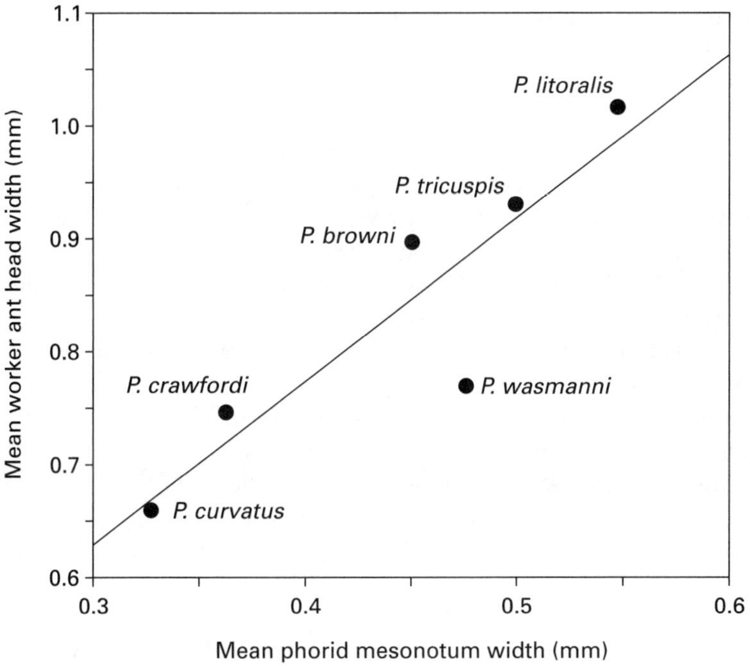

Figure 37.8. Parasitoid size is strongly correlated with host size across *Pseudacteon* and fire-ant species. Adapted from Morrison and Gilbert (1998).

America. The shape and configuration of the female ovipositor is an especially important taxonomic character, and that of each species is probably engineered to inject an egg into a particular part of a fire ant (Figure 37.4). Future exploration and taxonomic revision will undoubtedly increase the number of species. Each species of fly typically parasitizes several species within a group of related species, a species complex. South American fire ants belong to the *saevissima* complex and are attacked by one set of fly species, whereas North American native fire ants belong to the *geminata* complex and are host to another, nonoverlapping set of fly species.

Clearly then, because the same set of flies attacks several species of fire ants, each species of *Solenopsis* is host to multiple species of *Pseudacteon* flies—a conservative average seems to be 6.2 species. In contrast, of 22 other ant species parasitized by *Pseudacteon* species, 17 are host to a single species of fly. Of these, 14 are monomorphic and two are weakly polymorphic. All species of *Solenopsis* are polymorphic, so the large range of worker sizes has allowed the parasites to carve out a number of specialized niches based on host size. Although each fly species attacks multiple fire-ant species, within each species, it selects only a particular size range of workers to attack. Each fly species' niche is thus based largely on host size, not host species (Figure 37.8). Five to eight fly species may occur at the same site, attacking the same colony of fire ants and together covering the full range of worker sizes. For example, the four sympatric fly species tested on *S. invicta* and *S. saevissima* (Morrison et al., 1997b), arranged from the smallest to largest, are *P. curvatus, P. wasmanni, P. tricuspis,* and *P. litoralis*. The mean size of the workers in which females oviposit is in exactly the same order, ranging from 0.66 mm head width for *P. curvatus* to 1.03 mm for *P. litoralis*. A recent addition to this list is *P. cultellatus*, whose females prefer the very smallest *S. invicta* workers (Folgarait et al., 2002). Such size matching also characterizes the two species of *Pseudacteon* that attack *S. geminata* in North America—*P. crawfordi* attacks workers that are almost 20% smaller than those attacked by *P. browni*.

Still, the overlap of size preference among the fly species is very broad, suggesting additional factors separate them. For example, *P. tricuspis* is active during the middle of the day, whereas *P. litoralis* is active only in the early morning and late evening (Pesquero et al., 1996). The same is likely to be true of other species, so that fire ants get no rest from *Pseudacteon* flies during all daylight hours and during all seasons (Fowler et al., 1995b). Escape from this unrelenting harassment by *Pseudacteon* spp. may be one reason why fire ants evolved underground foraging tunnels (Markin et al., 1975c). Although less well documented, some evidence indicates that some of these species are more likely to seek out fire ants on trails and food sources, whereas others respond more strongly to disturbed mounds and mating flights in progress (Pesquero, 1993; Orr et al., 1995). Most of the *Pseudacteon* species attacking *S. saevissima* are not very seasonal and are found whenever the weather is suitable for their activity (Fowler et al., 1995b). The same is true for the four temperate species attacking *S. geminata*, which are active from April through December whenever the temperature exceeds 20°C (Morrison, 1999). Season is therefore an unlikely dimension for niche division. In addition, most *Pseudacteon* species are broadly distributed through a range of climates and habitat types, ranging from tropical rain forest to temperate grasslands, so geography and habitat are not niche dimensions either.

Putting out trays containing fire ants usually attracts *Pseudacteon* flies within minutes, possibly in response to an odor. Even to humans, fire ants have a very particular odor. If *Pseudacteon* flies are orienting to this odor, they are discriminating the smells of different species of fire ants at infinitesimally smaller concentrations than we can detect. A fly detecting essence of fire ant need only fly upwind to find the source, a ploy that is almost universal among insects orienting to smell. The speed with which the flies show up suggests that they are probably attracted from relatively close range, perhaps a few tens of meters.

Each fly species uses up to a four- to sevenfold *range* of head volumes. Because the maggot feeds primarily on tissue in the head, the head size determines the amount of larval food. When it is consumed, no more is available. Inevitably then, the bigger the ant head, the bigger

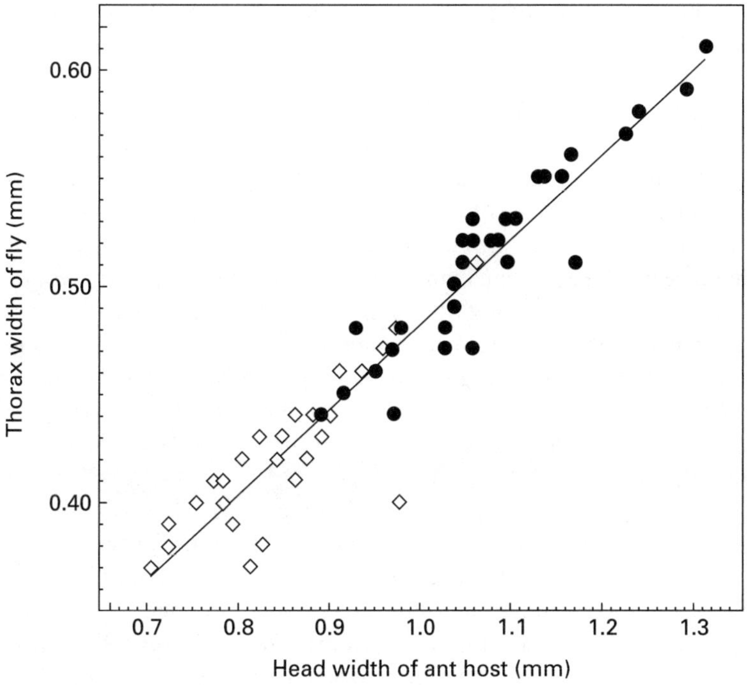

Figure 37.9. The size of the phorid parasitoid that emerges from the pupa is tightly correlated with the size of the worker-ant head in which it developed. Data are for *Pseudacteon tricuspis* on fire ants of the *S. saevissima* complex from Brazil. Males (diamonds) and females (filled circles) fall on the same regression line, but females develop in larger hosts. Patterns for *P. tricuspis* and *P. litoralis* in North American *S. invicta* workers are similar. Adapted from Morrison et al. (1999b).

the parasitoid that emerges from it. When the female parasitoid chooses a worker host of a particular size, she determines the future size of her offspring. For example, in *P. tricuspis,* females laid eggs in hosts whose head sizes ranged from 0.7 to about 1.2 mm. The flies that emerged from these varied in exact proportion (Morrison and Gilbert, 1998)—the largest ant heads, and emerging flies, were 170% the size of the smallest. In *P. litoralis,* a 170% increase in worker size resulted in only a 150% increase in fly size, suggesting that this species is less sensitive to host head size.

This resource limitation results in another remarkable phenomenon—female flies emerge from large worker hosts and males from the smaller ones (Morrison and Gilbert, 1998; Morrison et al., 1999b). *Pseudacteon litoralis* parasitized larger workers on average (1.13 mm) than *P. tricuspis* (0.95 mm), but in both fly species, females emerged from the upper end of the parasitized range of workers (Figure 37.9). In *P. tricuspis,* the switch point was 0.95 mm head width and included a narrow zone of overlap in which the sex was not easily predictable. For the larger *P. litoralis,* the sex-switch point was about 1.13 mm head width. The head sizes that yield each sex seem to be absolute rather than relative to their abundance—even when only a restricted range of worker sizes is available, the head sizes yielding females *P. tricuspis* always average about 1.05 mm and those yielding males 0.85 mm. For *P. litoralis,* these average sizes are about 1.17 mm and 1.02 mm.

When a female fly of either species has access to the full range of host sizes, she oviposits on about the same number of workers above and below the sex switch point, producing approximately the same number of male and female offspring, but when host size range is limited, the number of males and females may not be equal. Given only small workers, females produced predominately males; given only large workers, they produced predominately females. Such bias may even occur under natural circumstances: large workers are much less common in polygyne fire ants than in monogyne ones, so males greatly predominate among the offspring when *P. tricuspis* attacks polygyne *S. invicta* and when the native *Pseudacteon* attack polygyne *S. geminata* (Morrison, 1999). Sex ratio in the monogyne *S. geminata* is unknown. When *P. litoralis* attacks monogyne *S. invicta*, the sex ratio is biased toward females. Unfortunately, not much is known about the sex ratios in natural populations in South America.

What is the point of producing different sexes in hosts of different size? The reasoning goes like this: the size of the offspring depends largely on the size of the host's head, i.e., on the amount of available food. If one sex of offspring gains more fitness from being larger than does the other, then the mother will maximize her fitness if she causes the sex that gains more from body size to be reared in the larger host heads. Commonly, female fitness is more affected by body size than is male fitness, because larger females can mature more eggs. Males, on the other hand, increase their fitness primarily by mating with more females, an attribute less dependent on body size and therefore benefiting less from large size. These relationships probably hold for phorids—larger females probably enjoy greater fitness because they produce more eggs, enjoy greater longevity, and sustain longer flight periods or oviposition bouts, but larger males gain little fitness from their size.

This then is quite a neat trick, one shared with certain hymenopteran parasites, but whereas in Hymenoptera, the mother controls offspring sex simply by fertilizing an egg or not, in flies, both sexes develop from diploid fertilized eggs. The most likely mechanism of sex determination in *Pseudacteon* flies is therefore environmental, based on some size cue from the ant's head or the maggot's body after it has hollowed out the head. The mechanism is an intriguing subject for future research.

The Effects of Pseudacteon *on Fire-ant Behavior.* The appearance of *Pseudacteon* flies, or even a single fly, over a group of fire ants has an effect similar to that of an air-raid siren in a wartime city. Within a couple of minutes, hundreds of fire-ant workers stop foraging and seek shelter from airborne attack (Feener and Brown, 1992; Orr et al., 1995; Porter et al., 1995c; Porter, 1998a; Folgarait and Gilbert, 1999). They may cower under litter, stand motionless, return underground, disperse

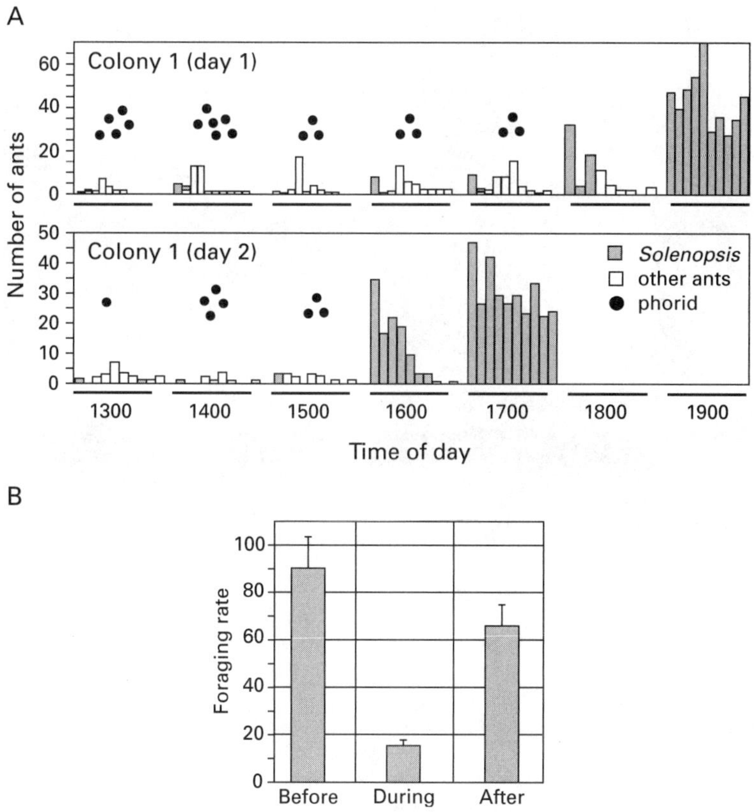

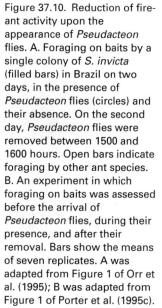

Figure 37.10. Reduction of fire-ant activity upon the appearance of *Pseudacteon* flies. A. Foraging on baits by a single colony of *S. invicta* (filled bars) in Brazil on two days, in the presence of *Pseudacteon* flies (circles) and their absence. On the second day, *Pseudacteon* flies were removed between 1500 and 1600 hours. Open bars indicate foraging by other ant species. B. An experiment in which foraging on baits was assessed before the arrival of *Pseudacteon* flies, during their presence, and after their removal. Bars show the means of seven replicates. A was adapted from Figure 1 of Orr et al. (1995); B was adapted from Figure 1 of Porter et al. (1995c).

from the foraging trail, or hide under the food. In field experiments with several fire-ant species, forager traffic and the number of workers on baits dropped precipitously by 80–85% as soon as *Pseudacteon* flies of any species or sex appeared (Figure 37.10). Foraging traffic remained low as long as the flies were present, up to several hours at a time. During this period, usually only a few workers in guarding postures (Figure 37.11) remained on the baits, assuming defensive postures under attack. When the flies were experimentally removed or left on their own, foraging returned to its initial level, but the buildup of workers on the baits was slower than it was before the flies appeared.

The behavioral adjustments that *S. richteri* makes to the presence of the six species of *Pseudacteon* that plague it in Argentina are particularly instructive (Folgarait and Gilbert, 1999). Each fly parasitoid has its special technique—some attack ants on foraging trails, some ants at baits, some isolated ants, some groups; some attack at the nest entrance, and some lie in wait to ambush passing ants. *Pseudacteon tricuspis* is strongly attracted to the sites of intercolony battles, whether these be over food or at the nest (Morrison and King, 2004). Perhaps the emission of venom or some alarm substance by the ants attracts the

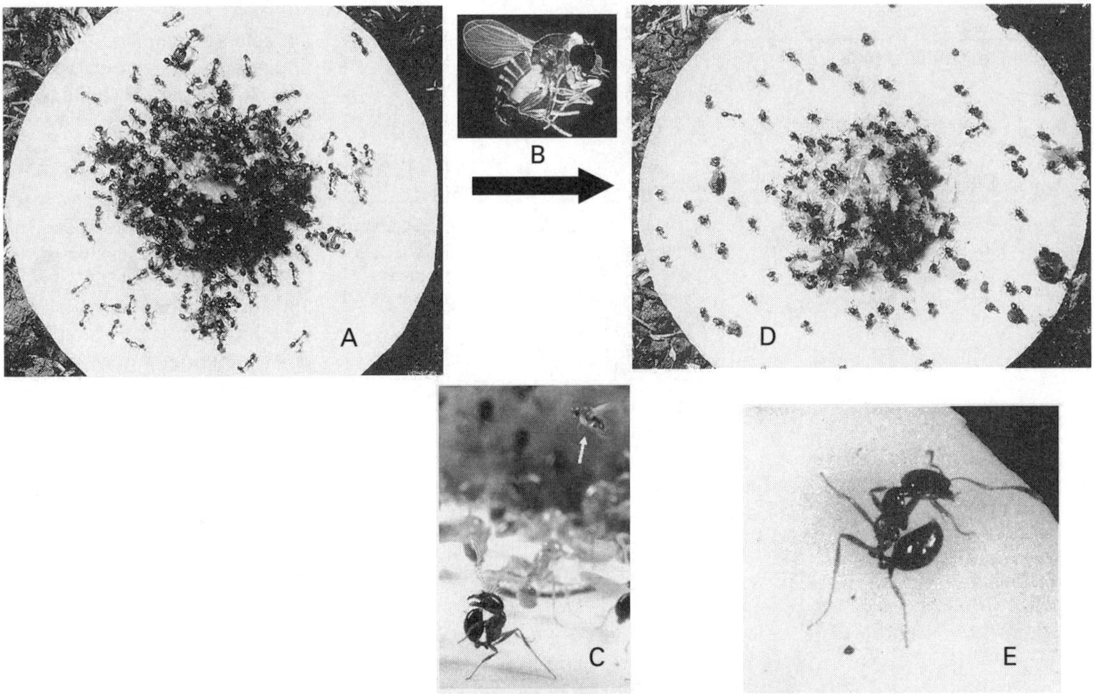

Figure 37.11. Defensive behavior of *S. geminata* workers in response to a *Pseudacteon* fly. A. Normal recruitment in the absence of flies. B. Female *Pseudacteon* fly. C. *Solenopsis geminata* worker in defensive posture. Note hovering fly (arrow). D. Response of workers to the presence of the flies. Note the prevalence of defensive postures. E. Worker in antiphorid defensive posture. A, D, and E adapted from Figures 1, 2, and 3 of Feener and Brown (1992); B and C from photographs by L. Gilbert at http://www.nysaes.cornell.edu/ent/biocontrol/parasitoids/pseudacteon.html by L. W. Morrison.

flies. When flies appeared while the ants were foraging, ant activity decreased by 75%, no matter how many ants were present. Larger ants, which are more suitable hosts for the flies and cost the colony more, became especially rare. The ants responded by freezing, running, or assuming various defensive postures, including the "hokey-pokey" posture, in which the worker fell on her side, lifted her superior legs, and waved them about (Wuellner and Saunders, 2003). At the group-response level, the ants neither ignore the risk presented by the flies nor do they avoid it completely by shutting down all activity. Rather, they reduce their foraging effort by a constant proportion, still allotting more foragers to a large food item than to a small one. They have thereby adjusted their potential losses in relation to the value of the food. Obviously, they must feed to exist, and ceasing foraging completely would cede the food to competing ants. The ants seem to be quite good accountants, balancing costs against benefits, although we cannot yet say how they read the bottom line.

Our native *S. geminata* is also plagued by its own species of

Pseudacteon parasitoids (Brown and Morrison, 1999; Morrison, 1999, 2000b; Morrison et al., 2000), especially *P. crawfordi* (which incidentally also parasitizes *S. aurea* in Arizona; Pitts and Pitts-Singer, 2001). Smith (1928) accurately described attacks by these flies on *S. geminata* as early as 1928 but failed to rear any progeny from parasitized workers because "every one of these specimens escaped from the vial in which they were confined, and an excellent opportunity to make further contribution on the biology of these flies was lost." In Texas, these parasitoids appeared at 60% of the baits occupied by *S. geminata*, reducing the rate of food retrieval but not strongly affecting the outcome of competition or the dominance order with other ant species. Similarly, laboratory experiments showed that *P. tricuspis* also reduced the foraging of its host *S. invicta* but did not affect the outcome of its competition with *S. geminata*.

Clearly, this interaction is a "war" between flies and ants, with defensive and attack strategies. On the scale of the ants' world, detecting a fly hovering a few millimeters overhead, its wings beating several hundred times per second, is probably no more difficult than would be detecting an Apache helicopter in our world. Neither one, in the words of Sanford Porter, is exactly a "subtle cue." Worker ants also seem to respond visually to the flies, tracking their movements. Whether or not the fire-ants' defenses are effective is unclear. Perhaps the low rates of parasitism are the result of the ants' defenses, both behaviors and underground foraging tunnels, as might be the low proportion of attacks that results in successful parasitism. Of course, the tunnels confer other benefits as well, but it is hard to imagine that fire ants could maintain such large territories if all traffic moved on the ground surface, vulnerable to strafing by phorids.

And as in many a war, while the combatants exchange volley upon volley on the field of battle, noncombatants quietly slip in and make off with the prize. When flies are present, other, less dominant species of ants can forage on the baits in relative peace. In South America, 10–20% of the baits with fire ants attract *Pseudacteon* flies, which in turn suppress foraging and make these baits available to other ants (Figure 37.10). With the help of these parasitic flies, therefore, species of ants that cannot compete directly with fire ants can nevertheless get a share of the food. Of course, only species that compete head to head with fire ants will be affected. Furthermore, even if *Pseudacteon* species suppress fire-ant feeding, no evidence so far indicates that this suppression actually affects abundance or distribution of ant species within these communities.

Biological Control and Species Specificity. On that happy day in the mid-1930s when *S. invicta* landed at Mobile Harbor in Alabama, it left all those pesky flies behind in South America. In its new home in the USA, therefore, nothing forces *S. invicta* to share food with less dominant ants, so it may have an "unfair" edge when competing with native ants over food. Bringing balance to this competition by reuniting these "pesky flies" with *S. invicta* in North America is the basis of a novel approach to biological control and is the source of the interest in *Pseudacteon* flies. In classical biological control, the agent kills the pest or prevents it from reproducing. The use of phorid flies for biological control of fire ants relies on a different effect. The specific defenses that fire ants have evolved suggest that parasitization by *Pseudacteon* flies may exact substantial costs in fire-ant fitness. These costs probably do not arise through direct mortality, which is hardly detectable and is limited to workers, but through the fly's substantial indirect effect on the foraging competition between *S. invicta* and other ants. Introducing *Pseudacteon* species into the southeastern USA might reduce the food intake of *S. invicta* and allow coexisting native species to increase theirs, promoting competing ant populations at the expense of *S. invicta*, whose dominance and abundance should decrease. Unfortunately, as discussed below, several key assumptions of this scheme are untested or incorrect, suggesting that *Pseudacteon* may not reduce *S. invicta* populations.

A good biological-control agent should attack only the pest species. Because the source of agents is commonly the native homeland of the pest, a great deal of travel, searching, and testing is invested in finding natural enemies. Before a natural enemy can be released, assurance must be ironclad that it will not create new problems by attacking nontarget organisms in its new home. Many behavioral and morphological adaptations of *Pseudacteon* flies, not least of which is their use of ant heads as pupation cases, make it obvious that they are specialized parasitoids of ants. None has ever been shown to be anything else. The question, therefore, is only whether the candidate *Pseudacteon* species attack ants other than *S. invicta.* Because the goal is the avoidance of "collateral damage," the parasite should therefore be tested not only under fairly natural conditions but also under the conditions most likely to cause it to accept a nontarget host. The most likely alternate hosts in the USA are those most closely related to *S. invicta,* namely *S. geminata* and *S. xyloni.*

Establishing host specificity calls for answers to three questions: is the *Pseudacteon* species attracted to each test ant species, does it attempt to oviposit, and if so, does the parasite develop successfully in the body of the host (Gilbert and Morrison, 1997; Morrison and Gilbert, 1999; Porter, 1998a, b, 2000; Porter and Alonso, 1999; Porter et al., 1995a)? In field studies, Porter and his coworkers (Porter et al., 1995a) tested 23 South American species of ants from 13 genera, exposed in trays, always in the presence of trays of several other species. The five species of *Pseudacteon* flies attracted were completely specific to *Solenopsis*—none was ever found over the trays of other species. Of the two *Solenopsis* species, *S. saevissima* attracted more flies than did *S. geminata* (an introduced species in Brazil). Even when non-*Solenopsis* species were presented alone in the laboratory, and the flies had no other choice, they were never parasitized (Porter and Alonso, 1999; Porter, 2000). An even stronger challenge followed (Porter, 1998b, 2000). *Pseudacteon* flies were attracted to *S. geminata* only when *S. saevissima* were nearby, and then at less than 1% as frequently, so after flies had been attracted, the trays of *S. saevissima* were removed. Would the flies, in full attack-and-oviposit mode, switch to *S. geminata*? The procedure resulted in 12 cases of parasitism, all by *P. wasmanni*, compared with the 588 cases of parasitism in *S. saevissima* by five species of *Pseudacteon* during the same period. Nevertheless, this result makes *P. wasmanni* somewhat less attractive for biological-control use. More important, *P. tricuspis* and *P. litoralis* were shown to be completely specific to *S. saevissima* under these conditions.

For the continuation of these studies, *Pseudacteon* flies (*P. tricuspis, P. litoralis, P. wasmanni*) were hand carried, lovingly packed in individual vials and protective wrapping, from South America to the USDA quarantine facility in Gainesville, Florida, to be tested in laboratory flight boxes against several native ant species, native *S. geminata,* and imported *S. invicta.* Whether offered a choice or not, each fly species overwhelmingly preferred *S. invicta,* but two of the species occasionally attacked *S. geminata,* although without a great deal of enthusiasm. Hundreds of flies emerged from attacked *S. invicta,* but only a single one from *S. geminata.* None of the other native ants stimulated any interest among the flies.

Similar studies in the quarantine facility at the Brackenridge Field Laboratory used a single female per test, thereby allowing estimation of attack and success rates and other individual characteristics (Gilbert and Morrison, 1997; Morrison and Gilbert, 1999). Appropriate controls assured that any lack of attack on *S. geminata* was not the result of a general lack of motivation to oviposit. Once again, females of *P. litoralis, P. wasmanni, P. obtusus,* and *P. tricuspis* were highly specific to *S. invicta,* attacking *S. geminata* at much lower rates or not at all, and even then, producing no signs of larval development. Two other species, *P. curvatus* and *P. borgmeieri,* turned out to be equal-opportunity parasitoids, attacking *S. geminata* 65% and 80% of the time, the latter parasitizing about 8% of the workers.

Overall, whereas *P. litoralis* and *P. tricuspis* greatly prefer *S. invicta* over *S. geminata,* the specificity is not absolute. Very rarely, when used for biological control, these flies may parasitize *S. geminata* or *S. xyloni.* Porter (Porter, 1998a, b) argued forcefully that this risk is not too great to be accepted. First, the two native fire ants already have three or four species of *Pseudacteon* parasitoids of their own. The introduced *Pseudacteon* species are highly unlikely to switch hosts and outcompete the native *Pseudacteon* species. The native *Pseudacteon* have had 60 years to switch to the introduced *S. invicta,* whose abundance ought to make them tempting hosts, but have not done so. Second, the ranges of fire ants of the *saevissima* and *geminata* complexes overlap broadly in northern South America, yet no switching of the *Pseudacteon* species from one complex to the other has occurred in millions of years. Third, *S. geminata* is not a rare or endangered species. In fact, in much of the world tropics, it is itself a pest species like *S. invicta* in the USA. Perhaps sometime in the future, the USA will contribute natural enemies for the biological control of *S. geminata* in distant lands. Finally, realistically, the threat to *S. geminata* in the USA is not some parasitoid that may (or may not) kill a few workers and disrupt its foraging. It is, in fact, the imported *S. invicta,* which has eliminated *S. geminata* from most of its previous range. The danger to *S. xyloni* is even greater, for this species has completely disappeared from all areas into which *S. invicta* has expanded. Porter argues that burdening *S. invicta* with its own, private set of natural enemies may give these native fire ants a fighting chance, but how the process would work in the mostly segregated populations is not immediately clear.

Adaptation of parasitoids to their hosts is sometimes quite local, population for population, biotype for biotype. A successful biological-control program must reestablish this exact host-parasite matching in the adopted country. Failure to do so may reduce the effectiveness of the control agent and has occasionally led to complete failure. For example, in South America *P. curvatus*'s range overlaps the ranges of both *S. invicta* and *S. richteri,* whose ranges are nearly mutually exclusive. *Pseudacteon curvatus* parasitizes the species available to it. Do flies from different parts of the range simply lack a preference, or is each specifically adapted to the available host? Given the choice, female flies from the *S. richteri* part of its range prefer *S. richteri* (whether Argentine or North American does not matter) or the *S. invicta* × *S. richteri* hybrid (Porter and Briano, 2000). When they are not given a choice, the development time of their offspring is 20% longer in *S. invicta* and 10% longer in the hybrid. Interestingly, the preference of these *P. curvatus* females for *S. richteri* remains even after two years of rearing only in *S. invicta* hosts.

Clearly, any biological control agent against *S. invicta* must be drawn from the *S. invicta* part of the range. This biotype preferred *S. invicta* three- to fourfold over *S. geminata* and *S. xyloni* in choice tests (Porter, 2000), parasitizing *S. geminata* and *S. xyloni* at much lower rates in no-choice laboratory tests, but the progeny reared from these alternate hosts still strongly preferred *S. invicta.* This biotype should thus be useful as a control agent against *S. invicta* and should present only a small threat to our native fire ants, so long as preferences do not evolve over time.

Release of Pseudacteon *Flies in the USA*. By the summer of 1997, all the hurdles for gaining approval for the release of a biological-control agent had been overcome, and over the next two years, Porter and his group released *P. tricuspis* at eight sites in northern Florida. In some cases they brought ants into the laboratory and allowed the flies to parasitize them before returning them to their home colonies with their little secrets. In other cases, they simply released adult flies over disturbed *S. invicta* nests.

Release was followed by periodic surveys for hovering flies over disturbed fire-ant nests. Assay is easy; if the flies are present in the area they soon appear. It soon became clear that populations had successfully been established at six of the eight sites and that these were expanding rapidly, soon fusing into a single population—125 km^2 in 1999, 3300 in 2000, and 8100 in 2001. By early 2003, the flies occupied an area that stretched from central Florida almost to the Georgia border. By late 2004, they appeared on our Southwood Plantation plots east of Tallahassee. The spread was so much faster than expected that the control sites that Porter planned to use for several years were overwhelmed by the third year. Five to 10 releases could probably cover a state the size of Florida in six to nine years.

Porter's group moved on to make *S. invicta* workers of all sizes miserable around the clock through the introduction of two more flies, *P. curvatus* and *P. litoralis*. The first was successfully established on *S. richteri* and the *S. richteri* × *S. invicta* hybrid in Alabama, but not in *S. invicta* populations in Tennessee and Florida (Graham et al., 2003). The release of other biotypes is planned. At this writing, eight years have passed since the first releases, and three species of flies have been released, but as yet no clear evidence indicates that the flies are depressing *S. invicta* populations.

Meanwhile, the laboratory of Larry Gilbert in Austin, Texas, was simultaneously developing *Pseudacteon* species for release in that state, where the challenges were greater than in the southeastern USA—the high rates of polygyny and short supply of large workers produce highly male-biased offspring in *P. tricuspis*, and the climate in much of Texas is far harsher, dryer, less predictable, and more extreme than the tropical/subtropical home of the *Pseudacteon* species. The tropical biotype of *P. tricuspis* was successfully established at only nine of 15 central and south Texas release sites. None has reached the density or expansion rates of the southeastern releases. Opportunities for the flies to parasitize the ants are rather rare in south Texas—disturbed mounds and mating flights are few and far between, and much foraging during the summer occurs at night. Large workers are uncommon, so the spring fly generation in the field was 80% males—hardly a good omen for rapid population expansion. Releasing genetically diverse flies drawn from several areas around Campinas, São Paulo, Brazil, as well as an admixture of the Gainesville stocks increased success somewhat. The group is also seeking flies in more Texas-like climatic areas of South America, expecting a better match for release in Texas. The various species of *Pseudacteon* are rather particular to different seasons and climatic conditions (Folgarait et al., 2003). Laboratory selection of lines of *P. tricuspis* that produce females in smaller host ants also offers promise. Particularly promising is the study of several other *Pseudacteon* species from varied South American climatic zones and of various sizes.

Overview and Critique of Biological Control Using Pseudacteon *Flies.* As I write this chapter, *Pseudacteon* flies are still far from certain to reduce the dominance of *S. invicta* by tipping the competition more in favor of native ants. I argued in an earlier chapter that native ants incompletely exploit the early successional niche that *S. invicta* entered in North America, i.e., that room is available for invaders and that severe displacement of native ants is unlikely. If this turns out to be substantially correct, harassing *S. invicta* with *Pseudacteon* flies may have no positive effect on native ants, who will therefore fail to reduce *S. invicta* densities. Of course, reduced feeding might directly decrease *S. invicta's* dominance, even without intervention by native ants, but in laboratory competition experiments, even though the presence of *P. tricuspis* allowed *Forelius mccooki* and *F. pruinosus* to forage more, and *S. invicta* to forage less, colonies of *F. mccooki* did not grow faster, nor did colonies of *S. invicta* grow more slowly (Mehdiabadi et al., 2004; Mottern et al., 2004). Laboratory experiments have tended to be fairly good predictors of field performance, so this is not good news for control efforts using *P. tricuspis* alone. Two of the corners of the *Pseudacteon* biological control concept are thus unsupported, at least in the laboratory. In the field as well, despite the presence of *Pseudacteon* flies for five or more years in many areas, no evidence indicates that fire-ant populations have declined (Morrison and Porter, 2005). After *P. tricuspis* had reached more or less stable populations, 3.5 years of monitoring discovered no discernible effect on fire-ant density, perhaps partly because of the naturally high variability of the fire-ant density estimate, but parasite and fire-ant densities were positively related—more hosts, more flies. Maybe fire ants will decrease in time, but so far, they have not.

The prospects for tilting competition would seem to be greater when *S. invicta* meets *S. geminata* or *S. xyloni*, because these are largely ecologically equivalent, but these species have largely been sorted into nonoverlapping habitats and are thus less likely to be fighting over food with *S. invicta*.

Of greater concern is the possibility that the release of *Pseudacteon* flies will not only fail to reduce *S. invicta* populations but will put additional stress on the native *S. geminata*. The existence of locally adapted biotypes with variable specificity within *Pseudacteon* species suggests that evolutionary host shifts come relatively easily to these flies. Species introduced against *S. invicta*, some of which already attack *S. geminata* at low rates, may evolve biotypes more adapted to *S. geminata*. Failure of *Pseudacteon* flies to reduce *S. invicta* populations would probably result in continuously high populations of *Pseudacteon* species. Even low rates of host-switching might be significant when these mobile flies spill over into *S. geminata* areas. High fly populations would also facilitate evolution of a transition to native *Solenopsis* species. Biological-control programs occasionally have unintended, negative consequences for native creatures. Let us hope that the *Pseudacteon* program does not join this list.

Other Parasitoids

All species of the chalcid wasp family, the Eucharitidae, parasitize ants. One species, *Orasema* sp., is a common parasite of *S. invicta* in Argentina, found in up to about 40% of colonies, each of which may harbor 1 to 600 wasps. Female wasps lay their eggs on plants, where they hatch into active larvae called planidia. These attach themselves to foraging ants for transport back to the nest, where they become ectoparasites (parasites that attach themselves to the external body surface) on ant brood. To do so successfully, the eucharitid larvae must avoid detection by the host ants, something they accomplish by passively acquiring the host colony's cuticular hydrocarbon profile, its putative nestmate recognition cue (Vander Meer et al., 1989). Wasp larvae can be transferred between colonies, much as *S. invicta* brood can be. The wasp larvae synthesize no hydrocarbons of their own until they leave the host nest as adults. There they can be captured as they swarm over the nest. Little else is known of the life history of this wasp.

Although probably not naturally associated with fire ants, the straw itch mite, *Pyomotes tritici*, has been proposed as a possible biological-control agent (Bruce and LeCato, 1980). This mite seems to attack a wide range of insects, and although its usefulness in biological control seems doubtful, its biology contains some pretty gruesome tidbits. Female mites make up 95% of the population and give birth to adult mites, having brooded the eggs within their bodies. Females attach themselves to the host and suck its fluids to nourish their developing embryos. These begin to distend the mother's body enormously within hours, and she produces about 250 new adults in a week. Seeding fire ant colonies with these mites led to colony death in most cases, but the lack of a sustained population suggests limited usefulness for practical control.

A similar situation also seems to pertain to several species of the insect-parasitizing nematode *Steinernema* spp. These nematodes are so nonspecific that they are used like a toxin against many different soil-dwelling insects. When fire-ant mounds were drenched with commercially available infective juvenile nematodes of several species, a modest level of control was achieved (Jouvenaz et al., 1990; Morris et al., 1990; Drees et al., 1992). Any advantage of such nonspecific nematodes over more ordinary control agents is not obvious. The observation that nematode-treated nests often moved to an uncontaminated area does not deepen confidence that this method is ever likely to be an effective control method, but it does suggest that the presence of pathogen may be one reason why fire ants sometimes move their mounds.

Finally, additional natural enemies may still be found through surveys. Indeed, reports of insects associated with *Solenopsis* nests in South America date back to the 1920s (Borgmeier, 1923 (1922)), although their biology remains mostly unknown.

Social Parasites

While some parasites leech the life out of the bodies of fire ants, others insinuate themselves into the very social fabric of the superorganism to bend its labor to their own purposes. Such social parasites are tremendously variable in their adaptation to and dependence on the parasitic way of life. Only one social parasite has been found in fire ants, but it is an extreme case (Figure 37.12; Bruch, 1930; Silveira-Guido et al., 1973; Briano et al., 1997; Calcaterra et al., 2000, 2001). Originally described as *Labauchena daguerri, Solenopsis daguerri* is an ant that is so completely dependent on social parasitism that it has lost its worker caste in the course of evolution. Its hosts are completely limited to *Solenopsis* fire ants and include several species and both social forms of the *saevissima* group, including *S. richteri, S. invicta, S. saevissima, S. quinquecuspis,* and *S. macdonaghi*. Parasitism rates seem usually to be low, and populations of the parasite are small and localized. Queens of the parasite somehow gain entrance into host colonies or are born there. They find the queen and grip her neck membranes or other body parts with their mandibles and legs in an unrelieved embrace. A single queen may have two or three such parasites attached to her neck and a half dozen to other parts of her body. The parasite queens are only one-tenth the size of fire-ant queens—about 2 mg—but they somehow seduce the fire-ant workers into caring preferentially for them and their progeny and neglecting the host queen they hold in their grip. As the parasites lay eggs, the brood comes to include large numbers of parasite progeny, all of which develop into sexuals. Any particular colony usually contains mostly males or mostly females, but the population's numerical sex ratio is one male to every three females. When the weather is warm and humid, these mate within or on the host nest before a dispersal flight, and females shed their wings upon landing. Males probably die in the nest, but the newly mated queens find new host colonies and clamp onto their host queens. They must do so quickly, because parasite queens that are not attached are quickly recognized and killed. Some

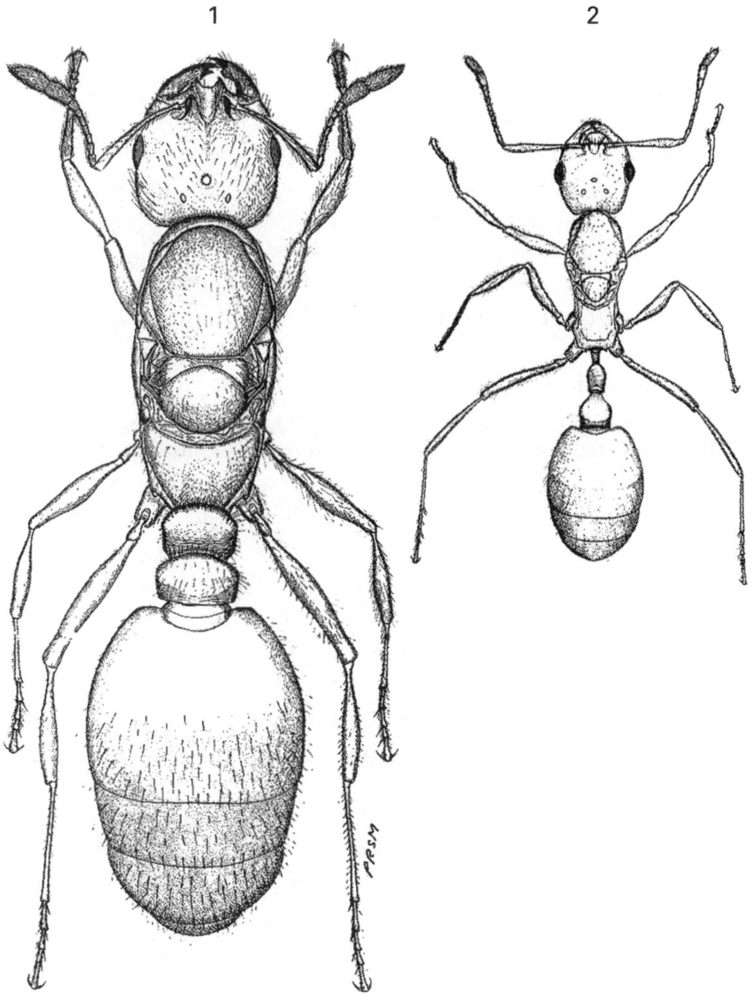

Figure 37.12. *Solenopsis richteri* (1) and its workerless social parasite *S. daguerri*. Reprinted from Silveira-Guido et al. (1973) with permission from Tall Timbers Research Station.

newly mated queens probably remain in their natal nests and attach to the host queen(s) there. Egg laying by the parasite queens, as by the host queens, is seasonal, at least in Argentina, beginning in the spring (October). *Solenopsis daguerri* sexuals appear two months later during summer (December) and are present into fall and winter. At the extreme, the fire-ant colony may "collapse" and die but less extreme outcomes seem to be more typical.

The parasite's effect on host populations was determined from a sample of almost 2600 colonies, mostly *S. richteri*, from 21 sites in the San Eladio region of Argentina (Calcaterra et al., 1999). This effect was not as dramatic as might be expected, in part because of the low and variable rate of parasitism, which ranged from 1.2% to 24% across sites (rates as high as 70% have been reported in other studies) and averaged about 5%. A much larger survey of over 12,000 fire-ant colonies from 726 collecting sites found infection rates 1.4–7.0% in three Argentine provinces, 1–6.2% in five Brazilian ones, less than 1% in Uruguay, and none in Paraguay and Bolivia. Why rates are not higher is unknown. Distribution of parasitized nests within sites was "contagious," that is, clustered, perhaps because the parasite queens moved easily among interconnected polygyne host nests or because parasites dispersed only locally. The negative effect of the parasite on fire-ant mound density should be confirmed because the unparasitized comparison site was distant and different from the parasitized sites.

At the colony level, the two social forms were equally likely to be parasitized, but parasitized colonies were not smaller, nor was their size distribution different from that of unparasitized colonies. Parasitized polygyne colonies had about half as many queens (2.9) per colony as unparasitized ones (5.5), suggesting that the parasite should drive the host toward monogyny. Parasite queens were more abundant in polygyne host nests than in monogyne ones (mean 6.2 versus 2.1 queens per nest). Their presence depressed worker production, but oddly, the parasitized and unparasitized colonies did not differ significantly in host sexual production or host queen weight. These are all rather modest effects—no exploding colonies, no dead bodies in the streets. Whether *S. daguerri* would make a good biological-control agent is uncertain. When Uruguayan scientists tried to transport it the rather moderate distance from Argentina to Uruguay to control *S. richteri*, the transplants failed. This extreme parasite may have extremely specific adaptations that limit its flexibility with respect to host, possibly even to geographic races. Its rarity suggests that it does not regulate its host's populations in their native range. Of course, the chance of detecting rapid colony collapse in surveys would be small. Experimental infection of a set of colonies with the parasite, along with proper controls, is needed to reveal the effects of infection and the potential of this ant as a control agent. Colonies have been transferred to the USDA quarantine facility in Gainesville, Florida, but attempts to establish the parasite in new laboratory or field colonies have been unsuccessful (Briano et al., 2002a), and its release as a biological-control agent seems uncertain and far in the future.

Other Predators and Myrmecophiles

Undoubtedly, large numbers of unspecific and opportunistic predators attack fire ants on both continents, for example those that prey on newly mated queens in North America. These predators eat fire ants because they are abundant and because the predators, be they birds, beetles, or dragonflies, eat insects. Occasionally, predators that specialize on native North American ants also prey on fire ants. For example,

Texas blind snakes normally follow army-ant trails and eat their brood, defending themselves from ant attack by expelling cloacal contents and having tough scales. Occasionally, these snakes invade fire-ant nests, where they similarly prey on brood and survive attack (Baldridge and Wivagg, 1992). Three-fourths of the snakes were able to follow *S. invicta* pheromone trails that were less than five minutes old, but only 20% could follow those one hour old and 7% those five hours old. Workers initially attack the snakes, doing them only minor injury, but eventually accept them, even piling brood next to them. Nevertheless. the rarity of these snakes in fire-ant nests suggests that they are not important in limiting *S. invicta* populations.

Although many species of arthropods have been found in fire ant nests, most have no specific association with them (Collins and Markin, 1971). The prolific myrmecologist Wasmann (Wasmann, 1918) identified inquilines from the nests of South American *S. geminata saevissima* (in the sense that Small used that name), but it is difficult to be sure just what currently recognized species of fire ant he was dealing with. From this species, he described 16 species of staphylinid beetles, four species of pselaphid beetles, two parasitic phorid flies (including *Pseudacteon wasmanni,* later named for him by Schmitz), a lygaeid bug, a cydnid bug, a millipede, two ant species, a bethylid wasp, and a silverfish. To this list, Reichensperger (Reichensperger, 1927a) added another lygaeid bug species, a species of mite, a flightless proctupid wasp, and several more species of staphylinid beetles. Wojcik's (1990) literature review is probably the best compendium of current knowledge of fire-ant inquilines and should be consulted by the interested reader. He listed an additional 45 species of inquilines, 25 of them beetles, none of whose biology is known, and reviewed what little is known of the biology of numerous other fire-ant inquilines. Mites are common in *Solenopsis* nests on alates, workers, or both. Those on alates may be phoretic, that is, hitching a ride, and may not be very host specific. The bodies of fire-ant queens are often covered with little, hemispheric mites, but what role these play in the ant's biology and vice versa is not clear. More recent surveys of fire ants in South America usually turn up large numbers of inquilines of unknown relationship to the host. Wojcik found up to 45% of colonies to host one or several of a variety of inquilines, including species of scarabs, hister beetles, silverfish, diplopods, and eucharitid wasps (Wojcik, 1986). Some of the scarabs and wasps are ant predators, and the millipedes are scavengers, but few details of their lives are known. Wasmann was convinced that only a fraction of all fire-ant inquilines had been discovered, and he is probably still correct.

The discovery of two species of South American myrmecophilous beetles, the scarab beetle *Myrmecophodius excavaticollis* (later changed to *Martinezia duterteri*) and the rove beetle *Myrmecosaurus*

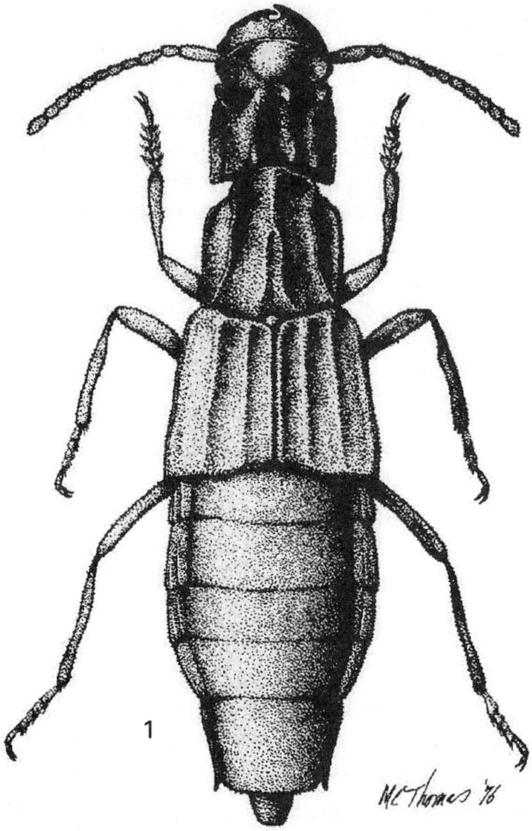

Figure 37.13. Drawing of a male *Myrmecosaurus ferrugineus*, an inquiline of *Solenopsis invicta*. Reprinted from Frank (1977) with permission from Howard Frank.

ferrugineus (Figure 37.13), in the USA argues that the original inoculum of *S. invicta* was a complete colony with brood, queen, and workers, rather than a newly mated queen (Wojcik, 1975, 1980; Frank, 1977; Wojcik and Habeck, 1977; Wojcik et al., 1977, 1978). The latter always occurs in *S. invicta* colonies in the USA, where it is widespread but not particularly common. The other six members of this South American genus are all associated with various species of *Solenopsis*. Observations on a single specimen revealed that *M. ferrugineus* does not prey on ant brood but solicits food from *S. invicta* workers by trophallaxis (Wojcik, 1980). I have seen *M. ferrugineus* follow the brood-raiding trails of incipient fire-ant colonies. What their purpose is in doing so, I do not know. Little more is known about the biology of these little myrmecophiles. Their usefulness for biological control is doubtful.

The scarab, *M. dutertrei*, was initially thought to be specific to *S. invicta* but was subsequently found in nests of all southeastern species of *Solenopsis*, both introduced and native, and even in Argentine ant nests. Its increasing abundance and range, together with the increasing range of its introduced host, seem to be reducing the abundance of the native myrmecophile *Euparia castanea,* which lives only in nests of the native fire ants *S. geminata* and *S. xyloni* and is not tolerated in the nests of *S. invicta. Martinezia dutertrei* is often found in abandoned mounds and has been found living and breeding in litter at a pond's edge, feeding on stranded aquatic organisms. It seems not even to be an obligate myrmecophile. These beetles would appear to be poor prospects for use in biological control.

The fire-ant workers attack the beetles when these enter their nests, but their hard cuticle, smooth shape, and passive behavior (playing dead) seem to make them impervious to attack. Within a few hours, they begin to acquire the colony-specific odor of their host, which makes them "invisible" in an olfactory sense (Vander Meer and Wojcik, 1982). Beetles transferred to three different species of fire ants acquired the cuticular hydrocarbon pattern of each of their hosts. Removal from fire ants caused the pattern to return to the innate beetle pattern. Such chemical mimicry is probably widespread among myrmecophiles, including the pselaphid beetle, *Fustiger elegans,* that lives with fire ants in Argentina.

Once accepted into the nest, the beetles solicit food from workers or feed on ant larvae, pupae, and dead workers (Wojcik and Habeck, 1977; Wojcik et al., 1977, 1978, 1991). They take dead flies if these are offered, so they may also be scavengers, perhaps stealing the booty foragers bring into the nest. The beetles return nothing to the ants. In contrast to *M. dutertrei, E. castanea* did not solicit food from live workers but ate brood and scavenged dead workers readily. On the other hand, it did obtain food from living larvae, perhaps by stimulating them to regurgitate.

Dissection of female *M. dutertrei* showed that they breed between May and September in northern Florida, so their larvae occurred in host nests in late summer to early winter. Both males and females fly to disperse during the summer, presumably searching for new opportunity. Most dispersing females had undeveloped ovaries or immature eggs. Egg maturation seems to occur only during the summer, probably in ant nests, and depletes fat bodies. Eggs mature continuously, rather than in batches, and most females contained 12 eggs or fewer (maximum 26). Lifetime egg production is probably much greater.

In the laboratory, the "sidewalk ant," *Tetramorium caespitum,* seems particularly efficient in killing small fire-ant colonies, attacking them in massed groups like Roman soldiers (King and Phillips, 1992). In one such episode, this battalion slowly advanced toward an *S. invicta* nest, clearing the arena of enemies. At the nest entrance, they yanked fire-ant workers from the nest one by one, pulled them into the massed group, and dismembered them. In this deliberate manner, they killed the entire nest. *Tetramorium caespitum* is known as a very territorial ant, famous for its "sidewalk wars" in urban areas. Its preference for disturbed habitats has brought it into contact with *S. invicta* in North Carolina, but whether or not it causes mortality of natural *S. invicta* colonies there is unknown.

◂•• *The Heartbreak of Parasitoids*

Most people, if they learn about parasitoids at all, don't do so until late in life. Ms. Hogarth is probably a pretty typical case. I picked up the phone in my office after the second toodle-oodle-oddle-oodle.

I got your name through the biology department, the voice said. They said you might know about this caterpillar I found yesterday.

I'll try, I said, knowing that over-the-phone insect IDs are often pretty hopeless. What does it look like?

Well, she said, it's really huge. I've never seen such a huge caterpillar, and it's got these big, prickly horns in front, and smaller horns all along its sides, and it's got sort of green with white diagonal lines on its side.

That, I said, is a *Citheronia regalis* (first you gotta snow 'em), also called the hickory horned devil (then you give 'em the plebe version). The adult is the regal moth. Where did you find it?

It was walking along the railing of our deck, she reported.

Well, it was probably looking for a place to pupate. It's in the wandering stage, just before it pupates, so you don't need to feed it. Just give it some leaf litter to hide in. As an afterthought, I added, but don't be surprised if you don't get an adult moth. You might end up with a bunch of flies. Let me know if that happens.

Three weeks later, Ms. Hogarth called again. Yes, she had put the caterpillar in a jar with leaf litter, it had pupated, and now she had several dozen bristly flies.

Tachinids, I told her, cryptically (the snow-job part). It's a kind of fly that lives within the bodies of other insects. Your *Citheronia* (more snow job) already had the developing fly larvae within its body when you found it. Sometimes the majority of caterpillars will be parasitized like that.

The fly family Tachinidae consists of nothing but parasitic species. Some tachinids have host ranges that include a hundred species from more than a dozen families and three different orders, but most parasitize a small number of related species. Collectively, tachinids attack the larvae (mostly) of every major order of insects; Lepidoptera are most "favored," Diptera the least. Lacking a piercing ovipositor, they deposit either fully developed eggs or freshly hatched larvae on the host. In either case, the larvae bore into the host and develop as internal parasitoids, slowly consuming the living host from the inside out. It is their answer to not having refrigeration for keeping their food fresh. They usually complete development only late in their host's larval stage, at which time they kill the host and emerge to pupariate.

A parasite is generally defined as a creature that lives on or in the body of a host, taking nourishment from the host without killing it—or at least death is not the immediate outcome of this one-way relationship. Whereas quite a few parasitic insect species fit this definition, the great majority, over 100,000 species, eventually kill their hosts. This habit can be regarded as a specialized form of predation, in which the prey is a single individual and provides sufficient nourishment to allow the parasitoid to complete its entire development. In most of these parasitoids, the adults are free-living, seeking out the host only to lay eggs in or on it. The two insect orders of which the most species have evolved into insect parasitoids are the Diptera (two-winged flies; about 11,000 species) and the Hymenoptera (wasps; over 100,000 species). Approximately 10 to 15% of all animal species are therefore insect parasitoids (Askew, 1971). The life histories of this vast array of species provide an almost endless *terra* poorly *cognita* for the curious naturalist to explore.

Hardly an insect exists that is not the victim of some parasitoid or other. Almost everyone who has tried to rear a few insects that they caught in the field has come face to frons with parasitoids. My daughter's second grade class learned about them early in their lives during a class project in which each child reared one caterpillar of the cloudless sulfur butterfly from egg to adult. This brilliant yellow butterfly migrates southeastward through Tallahassee by the thousands every September and October, depositing eggs singly on coffeeweed as they go. I collected enough of the pale yellow eggs to supply the whole class. Each student received a snap-lidded little plastic cup in which he or she placed the leaf-with-egg and waited the five days until it hatched. Thereafter, a fresh leaf was added daily, the caterpillar was measured, cuddled, talked to, and observed. The droppings were collected and their amount estimated, until finally, the plump, green, satiny caterpillar was ready to pupate. Trouble was, the eggs in six of the 28 cups failed to hatch, turning orange with a dark spot at the top end instead. Their owners were disappointed. Inside each cup, too small for the second graders to notice, were one or two moving specks of dust. When they learned that their eggs had fallen victim to egg parasitoids, the second graders were upset in that peculiar, contradictory way that humans, the most efficient and merciless of predators, often empathize with prey animals. These egg parasitoids were tiny, tiny chalcid wasps, so small that two of them found enough sustenance in a single butterfly egg to see them through their entire larval and pupal period. The microscope slide on which I mounted them looked blank, but under the 25x objective, their diaphanous wings and red eyes were perfectly clear.

Obviously a parasitoid can never be larger than its host. Therefore, some of the smallest insects are parasitoids of insect eggs, like the wasps that emerged from the butterfly eggs. Among the very smallest are the fairy "flies," of the wasp family Mymaridae. Some species measure less than 0.2 mm long, smaller than a large protozoan. Yet this is a complex, multicellular animal of long evolutionary pedigree. To an animal this small, air is as viscous a fluid as water is to us. As a result, the wings of these tiny wasps are oars with fringes of hairs, and the wasp literally rows its way through the atmosphere. No talk of airfoils, lift, and Bernoulli's principle, movement is as much swimming to the wasp as a lap in the pool is to us.

Our friend Terry also had an adventure that ended in tachinids. One day while harvesting his scuppernong grapes, he felt something on his neck and swatted it, only to experience an immediate and intense burning. A fuzzy, tortoise-shaped caterpillar dropped to the ground. I was the only entomologist known to Terry the lawyer, so the caterpillar ended up in my office for identification. It's a megalopygid, Terry, I phoned him. It belongs to a family that contains several urticating (snow job again, it means rash-causing) members. They have long, hollow, venom-filled hairs that break off in your skin when you whack 'em, like you did.

It's a bad idea to whack 'em, I added helpfully.

I kept the comatose caterpillar, and Terry kept the ulcerating, swollen neck and pulsating lymph nodes for several weeks. About the time that the swelling in his neck started receding, eleven tachinid flies emerged from the limp, shriveled body of the caterpillar. I was not surprised, for I had already figured out that the caterpillar was parasitized. Through its thin cuticle, from which most of the hair had fallen, I could see the fly larvae wrestling and jostling inside, making the caterpillar's body bulge and ripple, lethal babies kicking in a foreign womb. A few of them had made breathing holes in the caterpillar's cuticle. A couple of days later, they all cut holes, popped out, and formed little barrel-shaped, brown puparia, from which emerged bristly, adult flies three weeks later. I still have them in a vial on my desk.

The history of parasitoids and their hosts is an evolutionary war in slow motion, with defensive ploy, counter ploy, host shifts, novel tactics, feints, and frontal attacks. No wonder then that the relationships are intricate, subtle, and full of startling adaptations. Almost any insect community that you choose offers a panoply of host-parasite pairs with lots of amazing details—oak or willow galls, aphids, scales, dung insects, social wasps, and so on—but insects aren't the only hosts of insect parasitoids. Vertebrates, even humans, are tortured by a number of parasitic flies, including bot flies, warble flies, and heel flies. The larvae of these flies pass their lives under the skin, within the stomachs, or in the sinus cavities of their mammal hosts, and when abundant are capable of driving the host crazy or even killing it. I twice came home from Costa Rica with a bot fly under my skin. When I showed the wiggly hole in my calf to my insect biology class, none of the students would come closer to me than about two meters. They thought it was disgusting and possibly contagious. Sheltered urban kids have a hard time coming to grips with the realities of biology. I, on the other hand, felt privileged to have been chosen to host such an amazing animal. That's what a lifetime of studying biology will do to an originally perfectly normal brain.

Some Final Words

We have come to the end of our romp through fire-ant biology. I have tried to give a balanced account of what we know, think we know, and don't know about *Solenopsis invicta* and have repeatedly pointed out (perhaps *ad nauseam*) that the many interesting insights we have gained into *S. invicta's* alien world have more than repaid us for any aggravation that it has caused. Of course, you would expect a research biologist who has spent his life poking at *S. invicta* to say so, but those who would focus only on the fire ant's negatives are refusing to let their minds be enriched by what we have learned. One of the purposes of this book is to draw these insights together and to share them more widely, broadening the horizons of readers, be they professionals or not. Another purpose is to stimulate and guide future research into the unknown worlds that remain.

As for *S. invicta* itself, it will be with us for the indefinite future, as long as we make the habitat it needs. When we stop doing that, both the fire ant and we will vanish, for we need the same habitat it does. Nevertheless, the fire ant's fortunes will, without doubt, decline gradually—the only question is how fast it will do so. So abundant a creature is simply an opportunity waiting for exploitation by other creatures. Indeed, in the last few years, we have seen reports of new protozoan pathogens, microsporidians, viruses, and fungi, some probably recruited from North American natives, that have made *S. invicta* their host. Unknown numbers of predators probably now depend on fire ants for sustenance, and more will do so in the future. Moreover, new populations often increase well above their long-term stable levels before easing down to sustainable numbers. A common lament among my colleagues in fire-ant research is that *S. invicta* is not as abundant or dominant as it was 30 years ago—it takes so much more effort to collect sufficient colonies for experiments nowadays. Not hard data, perhaps, but a strong impression. My work is harder, though the general public may be pleased. The ultimate fate of *S. invicta* in its new home is to become just another ant among many. My fate as a fire ant researcher is to keep teasing secrets out of *S. invicta* for as long as I can—the attraction of its alien world is as strong now as it was when I started 35 years ago.

References

Adams, C. T. 1986. Agricultural and medical impact of the imported fire ants. Pages 48–57 in Fire Ants and Leafcutting Ants: Biology and Management. C. S. Lofgren and R. K. Vander Meer, editors. Westview Press, Boulder, Colorado.

Adams, C. T., and C. S. Lofgren. 1981. Red imported fire ants (Hymenoptera: Formicidae): frequency of sting attacks on residents of Sumter County, Georgia. Journal of Medical Entomology 18:378–382.

Adams, C. T., and C. S. Lofgren. 1982. Incidence of stings or bites of the red imported fire ants (Hymenoptera: Formicidae) and other arthropods among patients at Ft. Stewart, Georgia, USA. Journal of Medical Entomology 19:366–370.

Adams, C. T., T. E. Summers, C. S. Lofgren, F. A. Focks, and J. C. Prewitt. 1981. Interrelationship of ants and the sugarcane borer in Florida sugarcane fields. Environmental Entomology 10:415–418.

Adams, E. S. 1998. Territory size and shape in fire ants: A model based on neighborhood interactions. Ecology 79:1125–1134.

Adams, E. S. 2003. Experimental analysis of territory size in a population of the fire ant *Solenopsis invicta*. Behavioral Ecology 14:48–53.

Adams, E. S., and M. T. Balas. 1999. Worker discrimination among queens in newly founded colonies of the fire ant *Solenopsis invicta*. Behavioral Ecology and Sociobiology 45:330–339.

Adams, E. S., and J. F. A. Traniello. 1981. Chemical interference competition by *Monomorium minimum* (Hymenoptera: Formicidae). Oecologia 51:265–270.

Adams, E. S., and W. R. Tschinkel. 1991. Density-dependent self-thinning in populations of the fire ant, *Solenopsis invicta*. Bulletin of the Ecological Society of America 72(2, Supplement):54.

Adams, E. S., and W. R. Tschinkel. 1995a. Density-dependent competition in fire ants: effects on colony survivorship and size variation. Journal of Animal Ecology 64:315–324.

Adams, E. S., and W. R. Tschinkel. 1995b. Effects of foundress number on brood raids and queen survival in the fire ant *Solenopsis invicta*. Behavioral Ecology and Sociobiology 37:233–242.

Adams, E. S., and W. R. Tschinkel. 1995c. Spatial dynamics of colony interactions in young populations of the fire ant *Solenopsis invicta*. Oecologia 102:156–163.

Adams, E. S., and W. R. Tschinkel. 2001. Mechanisms of population regulation in the fire ant *Solenopsis invicta:* an experimental study. Journal of Animal Ecology 70:355–369.

Adkins, H. G. 1970. The imported fire ant in the southern United States. Annals of the Association of American Geographers 60:578–592.

Adrouny, G. A., V. J. Derbes, and R. C. Jung. 1959. Isolation of a hemolytic component of fire ant venom. Science 130:449.

Agnew, C. W., and W. L. Sterling. 1981. Predation of boll weevils in partially-open cotton bolls by the red imported fire ant. Southwestern Entomologist 6:215–219.

Agnew, C. W., W. L. Sterling, and D. A. Dean. 1982. Influence of cotton nectar on red imported fire ants and other predators. Environmental Entomology 11:629–634.

Ali, A. D., T. E. Reagan, and J. L. Flynn. 1984. Influence of selected weedy and weed-free sugarcane habitats on diet composition and foraging activity of the imported fire ant (Hymenoptera: Formicidae). Environmental Entomology 13:1037–1041.

Allen, C. R., S. Demaris, and R. S. Lutz. 1994. Red imported fire ant impact on wildlife: an overview. Texas Journal of Science 46:51–59.

Allen, C. R., R. S. Lutz, and S. Demaris. 1995. Red imported fire ant impacts on northern bobwhite populations. Ecological Applications 5:632–638.

Allen, C. R., K. G. Rice, D. P. Wojcik, and H. F. Percival. 1997. Effect of red imported fire ant envenomization on neonatal American alligators. Journal of Herpetology 31:318–321.

Allen, C. R., R. D. Willey, P. E. Myers, P. M. Horton, and J. Buffa. 2000. Impact of red imported fire ant infestation on northern bobwhite quail abundance trends in southeastern United States. Journal of Agricultural and Urban Entomology 17:43–51.

Allen, C. R., E. A. Forys, K. G. Rice, and D. P. Wojcik. 2001a. Effects of fire ants (Hymenoptera: Formicidae) on hatching turtles and prevalence of fire ants on sea turtle nesting beaches in Florida. Florida Entomologist 84:250–253.

Allen, C. R., R. S. Lutz, T. Lockley, S. A. Phillips, and S. Demaris. 2001b. The non-indigenous ant, *Solenopsis invicta,* reduces loggerhead shrike and native insect abundance. Journal of Agricultural and Urban Entomology 18:249–259.

Allen, C. R., D. M. Epperson, and A. S. Garmestani. 2004. Red imported fire ant impacts on wildlife: a decade of research. American Midland Naturalist 152:88–103.

Allen, G. E., and W. F. Buren. 1974. Microsporidan and fungal diseases of *Solenopsis invicta* Buren in Brazil. Journal of the New York Entomological Society 82:125–130.

Allen, G. E., W. F. Buren, R. N. Williams, M. de Menezes, and W. H. Whitcomb. 1974. The red imported fire ant, *Solenopsis invicta:* distribution and habitat in Mato Grosso, Brazil. Annals of the Entomological Society of America 67:43–46.

Allen, R. W., and M. M. Nice. 1952. A study of the breeding biology of the purple martin (*Progne subis*). American Midland Naturalist 47:606–665.

Alley, E. G. 1973. The use of mirex in control of the imported fire ant. Journal of Environmental Quality 2:52–61.

Alley, E. L. 1982. Ferriamicide: a toxicological summary (appendix E). Pages 144–153 in Proceedings of the Symposium on the Imported Fire Ant, June 7–10, 1982, Atlanta, Georgia. S. L. Battenfield, editor. EPA/USDA (APHIS) 0-389-890/70. U.S. Environmental Protection Agency and U.S. Department of Agriculture, Animal and Plant Health Inspection Service, Washington, D.C.

Alvarez, F. M., R. K. Vander Meer, and C. S. Lofgren. 1987. Synthesis of homofarnesenes: trail pheromone components of the fire ant, *Solenopsis invicta.* Tetrahedron 43:2897–2900.

Anderson, J. B., and R. K. Vander Meer. 1993. Magnetic orientation in the fire ant, *Solenopsis invicta.* Naturwissenschaften 80:568–570.

Anonymous. 1975a. Fire ant fiasco. Time Magazine, 12 May, p. 45.

Anonymous. 1975b. USDA suspends fire ant control program. Chemical and Engineering News, 14 April, p. 12.

Anonymous. 2001. *Solenopsis invicta* Buren, 1972 (Insecta, Hymenoptera): specific name conserved. Bulletin of Zoological Nomenclature 58:156–157.

Appel, A. G., M. K. Miller, and T. P. Mack. 1991. Cutaneous water loss of several stages of the red imported fire ant, *Solenopsis invicta* (Buren). Comparative Biochemistry and Physiology A Physiology 98:281–283.

Apperson, C. S., and E. E. Powell. 1984. Foraging activity of ants (Hymenoptera: Formicidae) in a pasture inhabited by the red imported fire ant. Florida Entomologist 67:383–393.

Arab, A., and F. H. Caetano. 2002. Segmental specializations in the Malpighian tubules of the fire ant *Solenopsis saevissima* Forel 1904 (Myrmicinae): an electron microscopical study. Arthropod Structure and Development 30:281–292.

Aron, S., E. L. Vargo, and L. Passera. 1995. Primary and secondary sex ratios in monogyne colonies of the fire ant, *Solenopsis invicta.* Animal Behaviour 49:749–757.

Askew, R. R. 1971. Parasitic Insects. American Elsevier, New York. 316 pp.

Ba, A. S., S. A. J. Phillips, and J. T. Anderson. 2000. Yeasts in mound soil of the red imported fire ant. Mycological Research 104:969–973.

Baer, B., E. D. Morgan, and P. Schmid-Hempel. 2001. A nonspecific fatty acid within the bumblebee mating plug prevents females from remating. Proceedings of the National Academy of Sciences of the USA 98:3926–3928.

Baer, H., D. T. Liu, M. Hooten, M. Blum, F. James, and W. H. Schmid. 1977. Fire ant allergy: isolation of three allergenic proteins from whole venom. Annals of Allergy 38:378.

Baer, H., T. Y. Liu, M. C. Anderson, M. Blum, W. H. Schmid, and F. J. James. 1979. Protein components of fire ant venom (*Solenopsis invicta*). Toxicon 17:397–405.

Baker, M. F. 1958. Observations of effects of an application of heptachlor or dieldrin on wildlife. Pages 12–21 in Twelfth Annual Conference of the Association of Game and Fish Commissioners. Association of Game and Fish Commissioners, Louisville, Kentucky.

Balas, M. T., and E. S. Adams. 1996a. The dissolution of cooperative groups: mechanisms of queen mortality in incipient fire ant colonies. Behavioral Ecology and Sociobiology 38:391–399.

Balas, M. T., and E. S. Adams. 1996b. Nestmate discrimination and competition in incipient colonies of fire ants. Animal Behaviour 51:49–59.

Balas, M. T., and E. S. Adams. 1997. Intraspecific usurpation of incipient fire ant colonies. Behavioral Ecology 8:99–103.

Baldridge, R. S., and D. E. Wivagg. 1992. Predation of imported fire ants by blind snakes. Texas Journal of Science 44:250–252.

Ball, D. E., and S. B. Vinson. 1984. Anatomy and histology of the male reproductive system of the fire ant, *Solenopsis invicta* Buren (Hymenoptera: Formicidae). International Journal of Insect Morphology and Embryology 13:283–294.

Ball, D. E., H. J. Williams, and S. B. Vinson. 1984. Chemical analysis of the male aedeagal bladder in the fire ant, *Solenopsis invicta* Buren. Journal of the New York Entomological Society 92:365–370.

Banks, W. A., and D. F. Williams. 1989. Competitive displacement of *Paratrechina longicornis* (Latreille) (Hymenoptera: Formicidae) from baits by fire ants in Mato Grosso, Brazil. Journal of Entomological Science 24:381–391.

Banks, W. A., B. M. Glancey, C. E. Stringer, D. P. Jouvenaz, C. S. Lofgren, and D. E. Weidhaas. 1973a. Imported fire ants: eradication trials with mirex bait. Journal of Economic Entomology 66:785–789.

Banks, W. A., J. K. Plumley, and D. M. Hicks. 1973b. Polygyny in a colony of the fire ant *Solenopsis geminata.* Annals of the Entomological Society of America 66:234–235.

Banks, W. A., C. S. Lofgren, D. P. Jouvenaz, C. E. Stringer, P. M. Bishop, D. F. Williams, D. P. Wojcik, and B. M. Glancey. 1981. Techniques for Collecting, Rearing, and Handling Imported Fire Ants. AATS-S-21. U.S. Department of Agriculture, Science and Education Administration, New Orleans, Louisiana. 9 pp.

Banks, W. A., D. P. Jouvenaz, D. P. Wojcik, and C. S. Lofgren. 1985. Observations on fire ants, *Solenopsis* spp., in Mato Grosso, Brazil. Sociobiology 11:143–152.

Barker, J. F. 1978. Neuroendocrine regulation of oocyte maturation in the imported fire ant *Solenopsis invicta*. General and Comparative Endocrinology 35:234–237.

Barker, J. F. 1979. Endocrine basis of wing casting and flight muscle histolysis in the fire ant *Solenopsis invicta*. Experientia 35:552–554.

Barlin, M. R., M. S. Blum, and J. M. Brand. 1976. Fire ant trail pheromones: analysis of species specificity after gas chromatographic fractionation. Journal of Insect Physiology 22:839–844.

Baroni Urbani, C., and P. B. Kannowski. 1974. Patterns in the red imported fire ant settlement of a Louisiana pasture: some demographic parameters, interspecific competition and food sharing. Environmental Entomology 3:755–760.

Bartlett, D. 1997. 40 years of thoughts on Paynes Prairie. Reptiles 5(7):68, 70–73.

Bartz, S. H., and B. Hölldobler. 1982. Colony founding in *Myrmecocystus mimicus* Wheeler (Hymenoptera: Formicidae) and the evolution of foundress associations. Behavioral Ecology and Sociobiology 10:137–147.

Bass, J. A., and S. B. Hays. 1979. Nuptial flights of the imported fire ant in South Carolina. Journal of the Georgia Entomological Society 14:158–161.

Battenfield, S. L. 1982. Proceedings of the symposium on the imported fire ant, June 7–10, 1982, Atlanta, Georgia. EPA/USDA (APHIS) 0-389-890/70. U.S. Environmental Protection Agency and U.S. Department of Agriculture, Animal and Plant Health Inspection Service, Washington, D.C. 256 pp.

Bernasconi, G., and L. Keller. 1996. Reproductive conflicts in cooperative associations of fire ant queens (*Solenopsis invicta*). Proceedings of the Royal Society of London B Biological Science 263:509–513.

Bernasconi, G., and L. Keller. 1998. Phenotype and individual investment in cooperative foundress associations of the fire ant, *Solenopsis invicta*. Behavioral Ecology 9:478–485.

Bernasconi, G., and L. Keller. 1999. Effect of queen phenotype and social environment on early queen mortality in incipient colonies of the fire ant, *Solenopsis invicta*. Animal Behaviour 57:371–377.

Bernasconi, G., M. J. B. Krieger, and L. Keller. 1997. Unequal partitioning of reproduction and investment between cooperating queens in the fire ant, *Solenopsis invicta*, as revealed by microsatellites. Proceedings of the Royal Society of London B Biological Science 264:1331–1336.

Bhatkar, A. P. 1979a. Evidence of intercolonial food exchange in fire ants and other Myrmicinae using radioactive phosphorus. Experientia 35:1172–1173.

Bhatkar, A. P. 1979b. Trophallactic appeasement in ants from distant colonies. Folia Entomologica Mexicana 41:135–143.

Bhatkar, A. P. 1983. Interspecific trophallaxis in ants, its ecological and evolutionary significance.

Bhatkar, A. P., and W. J. Kloft. 1977. Evidence, using radioactive phosphorus, of interspecific food exchange in ants. Nature 265:140–142.

Bhatkar, A. P., and S. B. Vinson. 1987. Colony limits in *Solenopsis invicta* Buren. Pages 599–600 in Chemistry and Biology of Social Insects. J. Eder and H. Rembold, editors. Verlag J. Peperny, München.

Bhatkar, A., W. H. Whitcomb, W. F. Buren, P. Callahan, and T. Carlysle. 1972. Confrontation behavior between *Lasius neoniger* (Hymenoptera: Formicidae) and the imported fire ant. Environmental Entomology 1:274–279.

Bigley, W. S., and S. B. Vinson. 1975. Characterization of a brood pheromone isolated from sexual brood of the imported fire ant, *Solenopsis invicta*. Annals of the Entomological Society of America 68:301–304.

Billen, J. P. J. 1990. Phylogenetic aspects of exocrine gland development in the Formicidae. Pages 317–318 in Social Insects and the Environment: Proceedings of the 11th International Congress of IUSSI. G. K. Veeresh, B. Mallik, and C. A. Viraktamath, editors. E. J. Brill, New York.

Blake, G. H., Jr., W. G. Eden, and K. L. Hays. 1959. Residual effectiveness of chlorinated hydrocarbons for control of the imported fire ant. Journal of Economic Entomology 52:1–.

Blomquist, G. J., and J. W. Dillworth. 1985. Cuticular lipids. Pages 117–154 in Comprehensive Insect Physiology, Biochemistry and Pharmacology, Volume 3, Integument, Respiration, and Circulation. G. A. Kerkut and L. I. Gilbert, editors. Pergamon Press, Oxford.

Blum, M. S. 1970. The chemical basis of insect sociality. Pages 61–94 in Chemicals Controlling Insect Behavior. M. Berozam, editor. Academic Press, New York.

Blum, M. S. 1984. Poisonous ants and their venoms. Pages 225–242 in Handbook of Natural Toxins, Volume 2, Insect Poisons, Allergens, and Other Invertebrate Venoms. A. T. Tu, editor. Marcel Dekker, New York.

Blum, M. S. 1985. Alkaloidal ant venoms: chemistry and biological activities. Pages 393–408 in Bioregulators for Pest Control. P. A. Hedin, editor. American Chemical Society Symposium Series 276. American Chemical Society, Washington, D.C.

Blum, M. S. 1988. Biocidal and deterrent activities of nitrogen heterocycles produced by venomous myrmicine ants. Pages 438–449 in Biologically Active Products. H. G. Cutler, editor. American Chemical Society Symposium Series 380. American Chemical Society, Washington, D.C.

Blum, M. S., and P. S. Callahan. 1960. Chemical and biological properties of the venom of the imported fire ant (*Solenopsis saevissima* var. *richteri* Forel) and the isolation of the insecticidal component. Proceedings of the 11th International Congress of Entomology 3:290–293.

Blum, M. S., J. R. Walker, P. S. Callahan, and A. F. Novak. 1958. Chemical, insecticidal, and antibiotic properties of fire venom. Science 128:306–307.

Blum, M. S., J. E. Roberts, Jr., and A. F. Novak. 1961. Chemical and biological characterization of venom of the ant *Solenopsis xyloni* McCook. Psyche 68:73–74.

Blum, M. S., J. M. Brand, R. M. Duffield, and R. R. Snelling. 1973. Chemistry of the venom of *Solenopsis aurea* (Hymenoptera: Formicidae). Annals of the Entomological Society of America 66:702.

Bolton, B. 1995. A New General Catalogue of the Ants of the World. Harvard University Press, Cambridge, Massachusetts. 504 pp.

Bonabeau, E., G. Theraulaz, and J. L. Deneubourg. 1996. Quantitative study of the fixed threshold model for the regulation of division of labour in insect societies. Proceedings of the Royal Society of London Series B Biological Sciences 263:1565–1569.

Bonner, C. 1980. Fire-ant insecticide to get initial OK. Tallahassee Democrat, 2 August, p. 6A.

Boomsma, J. J. 1989. Sex-investment ratios in ants: has female bias been systematically overestimated? American Naturalist 133:517–532.

Borgmeier, T. 1923 (1922). Beitrag zur Biologie der Feuerameise und iher Gäste (*Solenopsis geminata saevissima* Sm.). Zeitschrift Deutscher Verein für Wissenschaft und Kunst in São Paulo 3:1–9.

Boulay, R., L. M. Hooper-Bui, and J. Woodring. 2001. Oviposition and oogenesis in virgin fire ant females *Solenopsis invicta* are associated with a high level of dopamine in the brain. Physiological Entomology 26:294–299.

Bourke, A. F. G., and N. R. Franks. 1995. Social Evolution in Ants. Princeton University Press, Princeton, New Jersey.

Brand, J. M. 1978. Firc ant venom alkaloids: their contribution to chemosystematics and biochemical evolution. Biochemical Systematics and Ecology 6:337–340.

Brand, J. M., M. S. Blum, H. M. Fales, and J. G. MacConnell. 1972. Fire ant venoms: comparative analyses of alkaloidal components. Toxicon 10:259–271.

Brand, J. M., M. S. Blum, and M. R. Barlin. 1973a. Fire ant venoms: intraspecific and interspecific variation among castes and individuals. Toxicon 11:325–331.

Brand, J. M., M. S. Blum, and H. H. Ross. 1973b. Biochemical evolution in fire ant venoms. Insect Biochemistry 3:45–51.

Braulick, L. S. 1982. The Effect of Acute Exposure to Relative Humidity and Temperature on the Worker Caste of Four Species of Fire Ants. M.S. thesis, Texas Tech University, Lubbock, Texas. 59 pp.

Brennan, L. A. 1993. Fire ants and northern bobwhite quail: a real problem or a red herring? Wildlife Society Bulletin 21:351–355.

Brent, C. S., and E. L. Vargo. 2003. Changes in juvenile hormone biosynthetic rate and whole body content in maturing virgin queens of *Solenopsis invicta*. Journal of Insect Physiology 49:967–974.

Brian, M. V. 1956. Group form and causes of working inefficiency in the ant *Myrmica rubra*. Physiological Zoology 29:173–194.

Brian, M. V. 1965. Social Insect Populations. Academic Press, London. 135 pp.

Brian, M. V. 1978. Production Ecology of Ants and Termites. Cambridge University Press, Cambridge. 408 pp.

Brian, M. V., and A. Abbott. 1977. The control of food flow in a society of the ant *Myrmica rubra* L. Animal Behaviour 25:1047–1055.

Brian, M. V., R. T. Clarke, and R. M. Jones. 1981. A numerical model of an ant society. Journal of Animal Ecology 50:387–405.

Briano, J. A., and D. F. Williams. 1997. Effect of the microsporidium *Thelohania solenopsae* (Microsporida: Thelohaniidae) on the longevity and survival of *Solenopsis richteri* (Hymenoptera: Formicidae) in the laboratory. Florida Entomologist 80:366–376.

Briano, J. A., and D. F. Williams. 2002. Natural occurrence and laboratory studies of the fire ant pathogen *Vairimorpha invictae* (Microsporida : Burenellidae) in Argentina. Environmental Entomology 31:887–894.

Briano, J., D. Jouvenaz, D. Wojcik, R. Patterson, and H. Cordo. 1995a. Protozoan and fungal diseases in *Solenopsis richteri*, and *S. quinquecuspis* (Hymenoptera: Formicidae) in Buenos Aires Province, Argentina. Florida Entomologist 78:531–537.

Briano, J., R. Patterson, and H. Cordo. 1995b. Long term studies of the black imported fire ant (Hymenoptera: Formicidae) infected with a microsporidium. Environmental Entomology 24:1328–1332.

Briano, J., R. Patterson, and H. Cordo. 1995c. Relationship between colony size of *Solenopsis richteri* (Hymenoptera: Formicidae) and infection with *Thelohania solenopsae* (Microsporidia: Thelohaniidae) in Argentina. Journal of Economic Entomology 88:1233–1237.

Briano, J. A., R. S. Patterson, J. J. Becnel, and H. A. Cordo. 1996. The black imported fire ant, *Solenopsis richteri*, infected with *Thelohania solenopsae*: intracolonial prevalence of infection and evidence for transovarial transmission. Journal of Invertebrate Pathology 67:178–179.

Briano, J., L. Calcaterra, D. Wojcik, D. Williams, W. Banks, and B. Patterson. 1997. Abundance of the parasitic ant *Solenopsis daguerrei* (Hymenoptera: Formicidae) in South America, a potential candidate for the biological control of the red imported fire ants in the United States. Environmental Entomology 26:1143–1148.

Briano, J. A., L. A. Calcaterra, D. F. Williams, and D. H. Oi. 2002a. Attempts to artificially propagate the fire ant parasite *Solenopsis daguerrei* (Hymenoptera: Formicidae) in Argentina. Florida Entomologist 85:518–520.

Briano, J. A., D. F. Williams, D. H. Oi, and L. R. Davis. 2002b. Field host range of the fire ant pathogens *Thelohania solenopsae* (Microsporida: Thelohaniidae) and *Vairimorpha invictae* (Microsporida: Burenellidae) in South America. Biological Control 24:98–102.

Brinkman, M. A., and B. W. Fuller. 1999. Influence of *Beauveria bassiana* strain GHA on nontarget rangeland arthropod populations. Environmental Entomology 28:863–867.

Brinkman, M. A., and W. A. Gardner. 2000a. Enhanced activity of *Beauveria bassiana* to red imported fire ant workers (Hymenoptera: Formicidae) infected with *Thelohania solenopsae.* Journal of Agricultural and Urban Entomology 17:191–195.

Brinkman, M. A., and W. A. Gardner. 2000b. Possible antagonistic activity of two entomopathogens infecting workers of the red imported fire ant (Hymenoptera: Formicidae). Journal of Entomological Science 35:205–207.

Brinkman, M. A., W. A. Gardner, and G. D. Buntin. 2001. Effect of red imported fire ant (Hymenoptera: Formicidae) on *Rhinocyllus conicus* (Coleoptera: Curculionidae), a biological control agent of musk thistle. Environmental Entomology 30:612–616.

Brody, J. E. 1975. Agriculture department to abandon campaign against the fire ant. New York Times, 20 April, p. 46

Broome, J. R., P. R. Sikorowski, and B. R. Norment. 1976. A mechanism of pathogenicity of *Beauveria bassiana* on larvae of the imported fire ant, *Solenopsis richteri.* Journal of Invertebrate Pathology 28:87–91.

Brown, B. V., and L. W. Morrison. 1999. New *Pseudacteon* (Diptera: Phoridae) from North America that parasitizes the native fire ant *Solenopsis geminata* (Hymenoptera: Formicidae). Annals of the Entomological Society of America 92:308–311.

Brown, W. L., Jr. 1961. Mass insect control programs: four case histories. Psyche 68:76–109.

Bruce, W. A., and G. L. LeCato. 1980. *Pyemotes tritici:* a potential new agent for biological control of the red imported fire ant, *Solenopsis invicta* (Acari: Pyemotidae). International Journal of Acarology 4:271–274.

Bruce, W. G., J. M. Coarsey, Jr., M. R. Smith, and G. H. Culpepper. 1949. Survey of the imported fire ant, *Solenopsis saevissima* var. *richteri* Forel. U.S. Bureau of Entomology and Plant Quarantine Special Report S-15. U.S. Department of Agriculture, Savannah, Georgia. 25 pp.

Bruch, C. 1930. Notas preliminares acera de *Labauchena daguerrei* Santschi. Revista Sociedad Entomologica Argentina 3:73–80.

Buren, W. F. 1972. Revisionary studies on the taxonomy of the imported fire ants. Journal of the Georgia Entomological Society 7:1–26.

Buren, W. F. 1982. Red imported fire ant now in Puerto Rico. Florida Entomologist 65:188–189.

Buren, W. F. 1983. Artificial faunal replacement for imported fire ant control. Florida Entomologist 66:93–100.

Buren, W. F., G. E. Allen, W. H. Whitcomb, F. E. Lennartz, and R. N. Williams. 1974. Zoogeography of the imported fire ants. Journal of the New York Entomological Society 82:113–124.

Buren, W. F., G. E. Allen, and R. N. Williams. 1978. Approaches toward possible pest management of the imported fire ants. Bulletin of the Entomological Society of America 24:418–421.

Burns, E. C., and D. G. Melancon. 1977. Effect of imported fire ant (Hymenoptera: Formicidae) invasion on lone star tick (Acarina: Ixodidae) populations. Journal of Medical Entomology 14:247–249.

Burns, S. N., P. E. A. Teal, R. K. Vander Meer, J. L. Nation, and J. T. Vogt. 2002. Identification and action of juvenile hormone III from sexually mature alate females of the red imported fire ant, *Solenopsis invicta.* Journal of Insect Physiology 48:357–365.

Byrd, I. B. 1959. What are the side effects of the imported fire ant control program? Pages 46–50 in Biological Problems in Water Pollution: Transactions of the Second Seminar on Biological Problems in Water Pollution. C. M. Tarzwell, editor. U.S. Public Health Service, Cincinnati, Ohio.

Byron, D. W., and S. B. Hays. 1986. Occurrence and significance of multiple mound utilization by colonies of the red imported fire ant (Hymenoptera: Formicidae). Journal of Economic Entomology 79:637–640.

Cabrera, A., D. Williams, J. V. Hernandez, F. H. Caetano, and K. Jaffe. 2004. Metapleural- and postpharyngeal-gland secretions from workers of the ants *Solenopsis invicta* and *S. geminata.* Chemistry and Biodiversity 1:303–311.

Cahan, S. H., and S. B. Vinson. 2003. Reproductive division of labor between hybrid and nonhybrid offspring in a fire ant hybrid zone. Evolution 57:1562–1570.

Calabi, P., and S. D. Porter. 1989. Worker longevity in the fire ant *Solenopsis invicta:* ergonomic considerations of correlations between temperature, size and metabolic rates. Journal of Insect Physiology 35:643-649.

Calcaterra, L. A., J. A. Briano, and D. F. Williams. 1999. Field studies of the parasitic ant *Solenopsis daguerrei* (Hymenoptera: Formicidae) on fire ants in Argentina. Environmental Entomology 28:88–95.

Calcaterra, L. A., J. A. Briano, and D. F. Williams. 2000. New host for the parasitic ant, *Solenopsis daguerrei* (Hymenoptera: Formicidae), in Argentina. Florida Entomologist 83:363–365.

Calcaterra, L. A., J. A. Briano, D. F. Williams, and D. H. Oi. 2001. Observations on the sexual castes of the fire ant parasite *Solenopsis daguerrei* (Hymenoptera: Formicidae). Florida Entomologist 84:446–448.

Calder, W. A. I. 1984. Size, Function and Life History. Harvard University Press, Cambridge, Massachusetts. 431 pp.

Callahan, P. S., M. S. Blum, and J. R. Walker. 1959. Morphology and histology of the poison glands and sting of the imported fire ant (*Solenopsis saevissima* v. *richteri* Forel). Annals of the Entomological Society of America 52:573–590.

Callcott, A.-M. A., and H. L. Collins. 1996. Invasion and range expansion of red imported fire ant (Hymenoptera: Formicidae) in North America from 1918–1995. Florida Entomologist 79:240–251.

Callcott, A. M. A., D. H. Oi, H. L. Collins, D. F. Williams, and T. C. Lockley. 2000. Seasonal studies of an isolated red imported fire ant (Hymenoptera: Formicidae) population in eastern Tennessee. Environmental Entomology 29:788–794.

Camilo, G. R., and S. A. Phillips, Jr. 1990. Evolution of ant communities in response to invasion by the fire ant *Solenopsis invicta.* Pages 190–198 in Applied Myrmecology: A World Perspective. R. K. Vander Meer, K. Jaffe, and A. Cedeno, editors. Westview Press, Boulder, Colorado.

Canter, L. W. 1981. Final programmatic environmental impact statement for the cooperative imported fire ant program. APHIS-ADM-81-01-F. U.S. Department of Agriculture, Animal and Plant Health Inspection Service, Hyattsville, Maryland. 240 pp.

Carlin, N. F. 1989. Discrimination between and within colonies of social insects: two null hypotheses. Netherlands Journal of Zoology 39:86–100.

Carroll, C. R., and S. J. Risch. 1984. The dynamics of seed harvesting in early successional communities by a tropical ant, *Solenopsis geminata.* Oecologia 61:388–392.

Cassill, D. 2002. Brood care strategies by newly mated monogyne *Solenopsis invicta* (Hymenoptera: Formicidae) queens during colony founding. Annals of the Entomological Society of America 95:208–212.

Cassill, D. 2003. Rules of supply and demand regulate recruitment to food in an ant society. Behavioral Ecology and Sociobiology 54:441–450.

Cassill, D. L. 2000. Distributed intelligence: a mechanism for social response in an insect society. Available on the Internet at http://www.mbl.edu/CASSLS/deby_cassill.htm.

Cassill, D. L., and W. R. Tschinkel. 1995. Allocation of liquid food to larvae via trophallaxis in colonies of the fire ant, *Solenopsis invicta.* Animal Behaviour 50: 801–813.

Cassill, D. L., and W. R. Tschinkel. 1996. A duration constant for worker-to-larva trophallaxis in fire ants. Insectes Sociaux 43:149–166.

Cassill, D. L., and W. R. Tschinkel. 1999a. Regulation of diet in the fire ant, *Solenopsis invicta.* Journal of Insect Behavior 12:307–328.

Cassill, D. L., and W. R. Tschinkel. 1999b. Task selection by workers of the fire ant *Solenopsis invicta.* Behavioral Ecology and Sociobiology 45:301–310.

Cassill, D. L., W. R. Tschinkel, and S. B. Vinson. 2002. Nest complexity, group size and brood rearing in the fire ant, *Solenopsis invicta.* Insectes Sociaux 49:158–163.

Chen, J. S. C., K. Snowden, F. Mitchell, J. Sokolova, J. Fuxa, and S. B. Vinson. 2004a. Sources of spores for the possible horizontal transmission of *Thelohania solenopsae* (Microspora: Thelohaniidae) in the red imported fire ants, *Solenopsis invicta.* Journal of Invertebrate Pathology 85:139–145.

Chen, M. E., D. K. Lewis, L. L. Keeley, and P. V. Pietrantonio. 2004b. cDNA cloning and transcriptional regulation of the vitellogenin receptor from the imported fire ant, *Solenopsis invicta* Buren (Hymenoptera: Formicidae). Insect Molecular Biology 13:195–204.

Chen, Y. P., and S. B. Vinson. 1999. Queen attractiveness to workers in the polygynous form of the ant *Solenopsis invicta* (Hymenoptera: Formicidae). Annals of the Entomological Society of America 92:578–586.

Chen, Y. P., and S. B. Vinson. 2000. Effects of queen attractiveness to workers on the queen nutritional status and egg production in polygynous *Solenopsis invicta* (Hymenoptera: Formicidae). Annals of the Entomological Society of America 93:295–302.

Cherry, R. 2001. Interrelationship of ants (Hymenoptera: Formicidae) and southern chinch bugs (Hemiptera: Lygaeidae) in Florida lawns. Journal of Entomological Science 36:411–415.

Cherry, R. H., and G. S. Nuessly. 1992. Distribution and abundance of imported fire ants (Hymenoptera: Formicidae) in Florida sugarcane fields. Environmental Entomology 21:767–770.

Claborn, D. M., and S. A. Phillips, Jr. 1986. Temporal foraging activities of *Solenopsis invicta* (Hymenoptera: Formicidae) and other predominant ants of central Texas. Southwestern Naturalist 31:555–557.

Claborn, D. M., S. A. Phillips, Jr., and H. G. Thorvilson. 1988. Diel foraging activity of *Solenopsis invicta* and two native species of ants (Hymenoptera: Formicidae) in Texas. Texas Journal of Science 40:93–99.

Clemmer, D. I., and R. E. Serfling. 1975. The imported fire ant: dimensions of the urban problem. Southern Medical Journal 68:1133–1138.

Cokendolpher, J. C., and S. A. Phillips, Jr. 1989. Rate of spread of the red imported fire ant, *Solenopsis invicta* (Hymenoptera: Formicidae), in Texas. Southwestern Naturalist 34:443–449.

Cokendolpher, K. L., and S. A. Phillips, Jr. 1990. Critical thermal limits and locomotor activity of the red imported fire ant (Hymenoptera: Formicidae). Environmental Entomology 19:878–881.

Collins, H. 1992. Control of Imported Fire Ants: A Review of Current Knowledge. Technical Bulletin 1807. U.S. Dept. of Agriculture, Animal and Plant Health Inspection Service, Washington, D.C. 27 pp.

Collins, H. L., and G. P. Markin. 1971. Inquilines and other arthropods collected from nests of the imported fire ant, *Solenopsis saevissima richteri.* Annals of the Entomological Society of America 64:1376–1380.

Collins, H. L., T. C. Lockley, and D. J. Adams. 1993. Red imported fire ant (Hymenoptera: Formicidae) infestation of motorized vehicles. Florida Entomologist 76:515–516.

Colwell, R. K., and D. C. Lees. 2000. The mid-domain effect: geometric constraints on the geography of species richness. Trends in Ecology and Evolution 15:70–76.

Consoli, F. L., C. T. Wuellner, S. B. Vinson, and L. E. Gilbert. 2001. Immature development of *Pseudacteon tricuspis* (Diptera: Phoridae), an endoparasitoid of the red imported fire ant (Hymenoptera: Formicidae). Annals of the Entomological Society of America 94:97–109.

Cook, J. L. 2003. Conservation of biodiversity in an area impacted by the red imported fire ant, *Solenopsis invicta* (Hymenoptera: Formicidae). Biodiversity and Conservation 12:187–195.

Cook, T. J. 2002. Studies of naturally occurring *Thelohania solenopsae* (Microsporida: Thelohaniidae) infection in red imported fire ants, *Solenopsis invicta* (Hymenoptera: Formicidae). Environmental Entomology 31:1091–1096.

Cook, T. J., M. B. Lowery, T. N. Frey, K. E. Rowe, and L. R. Lynch. 2003. Effect of *Thelohania solenopsae* (Microsporida: Thelohaniidae) on weight and reproductive status of polygynous red imported fire ant, *Solenopsis invicta* (Hymenoptera: Formicidae), alates. Journal of Invertebrate Pathology 82:201–203.

Cottam, C. 1958. A commentary on the fire ant problem. Pages 31–34 in Twelfth Annual Conference of the Association of Game and Fish Commissioners. Association of Game and Fish Commissioners, Louisville, Kentucky.

Creighton, W. S. 1930. The New World species of the genus *Solenopsis* (Hymenoptera: Formicidae). Proceedings of the American Academy of Arts and Sciences 66:39–151.

Crider, B. 1977. War on fire ants resumes. Tallahassee Democrat, 27 February, p. 16.

Crozier, R. H. 1971. Heterozygosity and sex determination in haplo-diploidy. American Naturalist 105:399–412.

Cruz-Lopez, L., J. C. Rojas, R. De la Cruz-Cordero, and E. D. Morgan. 2001. Behavioral and chemical analysis of venom gland secretion of queens of the ant *Solenopsis geminata.* Journal of Chemical Ecology 27:2437–2445.

Culpepper, G. H. 1953. Status of the imported fire ant in the southern states in July 1953. U.S. Bureau of Entomology and Plant Quarantine Special Report E-867. U.S. Department of Agriculture, Savannah, Georgia. 8 pp.

Davidson, N. A., and N. D. Stone. 1989. Imported fire ants. Pages 196–217 in Eradication of Exotic Pests. D. L. Dahlsten, R. Garcia, and H. Lorraine, editors. Yale University Press, New Haven, Connecticut.

Davis, L. R. J., R. K. Vander Meer, and S. D. Porter. 2001. Red imported fire ants expand their range across the West Indies. Florida Entomologist 84:735–736.

Davis, W. L., R. G. Jones, and G. R. Farmer. 1989. Insect hemolymph factor promotes muscle histolysis in *Solenopsis.* Anatomical Record 224:473–478.

Davison, K. L., H. H. Mollenhauer, R. L. Yonger, and J. H. Cox. 1976. Mirex-induced hepatic changes in chickens, Japanese quail, and rats. Archives of Environmental Contamination and Toxicology 4:469–482.

Dawkins, R. 1976. The Selfish Gene. Oxford University Press, New York. 224 pp.

DeHeer, C. J. 2002. A comparison of the colony-founding potential of queens from single- and multiple-queen colonies of the fire ant *Solenopsis invicta.* Animal Behaviour 64:655–661.

DeHeer, C. J., and K. G. Ross. 1997. Lack of detectable nepotism in multiple-queen colonies of the fire ant *Solenopsis invicta* (Hymenoptera: Formicidae). Behavioral Ecology and Sociobiology 40:27–33.

DeHeer, C. J., and W. R. Tschinkel. 1998. The success of alternative reproductive tactics in monogyne populations of the ant *Solenopsis invicta:* significance for transitions in social organization. Behavioral Ecology 9:130–135.

DeHeer, C. J., M. A. D. Goodisman, and K. G. Ross. 1999. Queen dispersal strategies in the multiple-queen form of the fire ant *Solenopsis invicta.* American Naturalist 153:660–675.

deShazo, R. D., C. Griffing, T. H. Kwan, W. A. Banks, and H. F. Dvorak. 1984. Dermal hypersensitivity reactions to imported fire ants. Journal of Allergy and Clinical Immunology 74:841–847.

Deslippe, R. J., and Y. J. Guo. 2000. Venom alkaloids of fire ants in relation to worker size and age. Toxicon 38:223–232.

Detrain, C., and J. M. Pasteels. 1991. Caste differences in behavioral thresholds as a basis for polyethism during food recruitment in the ant, *Pheidole pallidula* (Nyl.) (Hymenoptera: Myrmicinae). Journal of Insect Behavior 4:157–176.

Diaz, D., A. Knutson, and J. S. Bernal. 2004. Effect of the red imported fire ant on cotton aphid population density and predation of bollworm and beet armyworm eggs. Journal of Economic Entomology 97:222–229.

Diffie, S., R. K. Vander Meer, and M. H. Bass. 1988. Discovery of hybrid fire ant populations in Georgia and Alabama. Journal of Entomological Science 23:187–191.

Diffie, S. K., and D. C. Sheppard. 1989. Supercooling studies on the imported fire ants: *Solenopsis invicta* and *Solenopsis richteri* (Hymenoptera: Formicidae) and their hybrid. Journal of Entomological Science 24:361–364.

Diffie, S. K., M. H. Bass, and K. Bondari. 1997. Winter survival of *Solenopsis invicta* and the *Solenopsis* hybrid (Hymenoptera: Formicidae) in Georgia. Journal of Agricultural Entomology 14:93–101.

Donaldson, W., A. H. Price, and J. Morse. 1994. The current status and future prospects of the Texas horned lizard (*Phrynosoma cornutum*) in Texas. Texas Journal of Science 46:97–113.

Drees, B. M. 1994. Red imported fire ant predation on nestlings of colonial waterbirds. Southwestern Entomologist 19:355–359.

Drees, B. M. 1995. Red imported fire ant multiple stinging incidents to humans indoors in Texas. Southwestern Entomologist 20:383–385.

Drees, B. M., L. A. Berger, R. Cavazos, and S. B. Vinson. 1991. Factors affecting sorghum and corn seed predation by foraging red imported fire ants (Hymenoptera: Formicidae). Journal of Economic Entomology 84:285–289.

Drees, B. M., R. W. Miller, S. B. Vinson, and R. Georgis. 1992. Susceptibility and behavioral response of red imported fire ant (Hymenoptera: Formicidae) to selected entomogenous nematodes (Rhabditida: Steinernematidae & Heterorhabditidae). Journal of Economic Entomology 85:365–370.

Dunning, W. C. 1957. U.S. plans aerial war on fire ants. Mobile Press Register, Mobile, Alabama, 7 April.

Dutcher, J. D., and D. C. Sheppard. 1981. Predation of pecan weevil larvae by red imported fire ants. Journal of the Georgia Entomological Society 16:210–213.

Eden, W. G., and F. S. Arant. 1949. Control of the imported fire ant in Alabama. Journal of Economic Entomology 42:976–979.

Eisner, T. 1957. A comparative morphological study of the proventriculus of ants (Hymenoptera: Formicidae). Bulletin of the Museum of Comparative Zoology of Harvard University 116:439–490.

Eisner, T., and E. O. Wilson. 1958. Radioactive tracer studies of food transmission in ants. Proceedings of the 10th International Congress of Entomology 2:509–513.

Emlen, J. T., Jr. 1938. Fire ants attacking California quail chicks. Condor 40:85–86.

Eskafi, F. M., and M. M. Kolbe. 1990. Predation on larval and pupal *Ceratitis capitata* (Diptera: Tephritidae) by the ant *Solenopsis geminata* (Hymenoptera: Formicidae) and other predators in Guatemala. Environmental Entomology 19:148–153.

Ettershank, G. 1966. A generic revision of the world Myrmicinae related to *Solenopsis* and *Pheidologeton* (Hymenoptera: Formicidae). Australian Journal of Zoology 14:73–171.

Eubanks, M. D. 2001. Estimates of the direct and indirect effects of red imported fire ants on biological control in field crops. Biological Control 21:35–43.

Eubanks, M. D., S. A. Blackwell, C. J. Parrish, Z. D. Delamar, and H. Hull-Sanders. 2002. Intraguild predation of beneficial arthropods by red imported fire ants in cotton. Environmental Entomology 31:1168–1174.

Fagen, R. M., and R. N. Goldman. 1977. Behavioural catalogue analysis methods. Animal Behaviour 25:261–274.

Feener, D. H., Jr. 1978. Structure and Organization of a Litter Foraging Ant Community: Roles of Interference Competition and Parasitism. Ph.D. dissertation, University of Texas at Austin. 168 pp.

Feener, D. H., Jr., and B. V. Brown. 1992. Reduced foraging of *Solenopsis geminata* (Hymenoptera: Formicidae) in the presence of parasitic *Pseudacteon* spp. (Diptera: Phoridae). Annals of the Entomological Society of America 85:80–84.

Fewell, J. H., J. F. Harrison, J. R. B. Lighton, and M. D. Breed. 1996. Foraging energetics of the ant, *Paraponera clavata*. Oecologia 105:419–427.

Fillman, D. A., and W. L. Sterling. 1983. Killing power of the red imported fire ant [Hym.: Formicidae]: a key predator of the boll weevil [Col.: Curculionidae]. Entomophaga 28:339–344.

Fletcher, D. J. C. 1983. Three newly-discovered polygynous populations of the fire ant, *Solenopsis invicta*, and their significance. Journal of the Georgia Entomological Society 18:538–543.

Fletcher, D. J. C., and M. S. Blum. 1981. Pheromonal control of dealation and oogenesis in virgin queen fire ants. Science 212:73–75.

Fletcher, D. J. C., and M. S. Blum. 1983a. The inhibitory pheromone of queen fire ants: effects of disinhibition on dealation and oviposition by virgin queens. Journal of Comparative Physiology A Sensory, Neural and Behavioral Physiology 153:467–475.

Fletcher, D. J. C., and M. S. Blum. 1983b. Regulation of queen number by workers in colonies of social insects. Science 219:312–314.

Fletcher, D. J. C., M. S. Blum, T. V. Whitt, and N. Temple. 1980. Monogyny and polygyny in the fire ant, *Solenopsis invicta.* Annals of the Entomological Society of America 73:658–661.

Flickinger, E. L. 1988. Observations of predation by red imported fire ants on live-trapped wild cotton rats. Pages 80–81 in Proceedings of the Governor's Conference on the Imported Fire Ant: Assessment and Recommendations. S. B. Vinson and J. Teer, editors. Sportsmen Conservationists of Texas, Austin, Texas.

Folgarait, P. J., and L. E. Gilbert. 1999. Phorid parasitoids affect foraging activity of *Solenopsis richteri* under different availability of food in Argentina. Ecological Entomology 24:163–173.

Folgarait, P. J., O. A. Bruzzone, and L. E. Gilbert. 2002. Development of *Pseudacteon cultellatus* (Diptera: Phoridae) on *Solenopsis invicta* and *Solenopsis richteri* fire ants (Hymenoptera: Formicidae). Environmental Entomology 31:403–410.

Folgarait, P. J., O. A. Bruzzone, and L. E. Gilbert. 2003. Seasonal patterns of activity among species of black fire ant parasitoid flies (*Pseudacteon*: Phoridae) in Argentina explained by analysis of climatic variables. Biological Control 28:368–378.

Folgarait, P. J., P. D'Adamo, and L. E. Gilbert. 2004. A grassland ant community in Argentina: the case of *Solenopsis richteri* and *Camponotus punctulatus* (Hymenoptera: Formicidae) attaining high densities in their native ranges. Annals of the Entomological Society of America 97:450–457.

Folkerts, G. W., M. A. Deyrup, and D. C. Sisson. 1993. Arthropods associated with xeric longleaf pine habitats in the southeastern United States: a brief overview. Proceedings of the Tall Timbers Fire Ecology Conference 18:159–192.

Forys, E. A., A. Quistorff, and C. R. Allen. 2001. Potential fire ant (Hymenoptera: Formicidae) impact on the endangered Schaus swallowtail (Lepidoptera: Papilionidae). Florida Entomologist 84:254–258.

Forys, E. A., C. R. Allen, and D. P. Wojcik. 2002. Influence of proximity and amount of human development and roads on the occurrence of the red imported fire ant in the lower Florida Keys. Biological Conservation 108:27–33.

Fowler, H. G. 1988. A organização das comunidades de formigas no Estado de Mato Grosso, Brasil. Anales del Museo de Historia Natural de Valparaíso 19:35–42.

Fowler, H. G., J. V. E. Bernardi, and L. F. T. di Romagnano. 1990. Community structure and *Solenopsis invicta* in São Paulo. Pages 199–207 in Applied Myrmecology: A World Perspective. R. K. Vander Meer, K. Jaffe and A. Cedeno, editors. Westview Press, Boulder, Colorado.

Fowler, H. G., S. Campiolo, M. A. Pesquero, and S. D. Porter. 1995a. Notes on a southern record for *Solenopsis geminata* (Hymenoptera: Formicidae). Iheringia Serie Zoologia 79:173.

Fowler, H. G., M. A. Pesquero, S. Campiolo, and S. D. Porter. 1995b. Seasonal activity of species of *Pseudacteon* (Diptera: Phoridae) parasitoids of fire ants (*Solenopsis saevissima*) (Hymenoptera: Formicidae) in Brazil. Cientifica (Jaboticabal) 23:367–371.

Fox, R. W., R. F. Lockey, and S. C. Bukantz. 1982. Neurologic sequelae following the imported fire ant sting. Journal of Allergy and Clinical Immunology 70:120–124.

Fraelich, B. A. 1991. Interspecific Competition for Food Between the Red Imported Fire Ant, *Solenopsis invicta,* and Native Floridian Ant Species in Pastures. M.S. thesis, University of Florida, Gainesville, Florida. 158 pp.

Francke, O. F., and J. C. Cokendolpher. 1986. Temperature tolerances of the red imported fire ant. Pages 104–113 in Fire Ants and Leafcutting Ants: Biology and Management. C. S. Lofgren and R. K. Vander Meer, editors. Westview Press, Boulder, Colorado.

Francke, O. F., J. C. Cokendolpher, and L. R. Potts. 1984. Supercooling studies on four species of fire ants. Page 72 in Proceedings of the 1984 Imported Fire Ant Conference, Gainesville, Florida. M. E. Mispagel, editor.

Francke, O. F., L. R. Potts, and J. C. Cokendolpher. 1985. Heat tolerances of four species of fire ants (Hymenoptera: Formicidae: *Solenopsis*). Southwestern Naturalist 30:59–68.

Francke, O. F., J. C. Cokendolpher, and L. R. Potts. 1986. Supercooling studies on North American fire ants (Hymenoptera: Formicidae). Southwestern Naturalist 31:87–94.

Frank, J. H. 1977. *Myrmecosaurus ferrugineus,* an Argentinian beetle from fire ant nests in the United States. Florida Entomologist 60:31–36.

Frank, W. A. 1988. Report of limited establishment of red imported fire ant, *Solenopsis invicta* Buren in Arizona. Southwestern Entomologist 13:307–308.

Franks, N. R. 1985. Reproduction, foraging efficiency and worker polymorphism in army ants. Pages 91–107 in Experimental Behavioral Ecology and Sociobiology: In Memoriam Karl von Frisch, 1886-1982, Volume 31. B. Hölldobler and M. Lindauer, editors. Sinauer Associates, Sunderland, Massachusetts.

Freed, P. S., and K. Neitman. 1988. Notes on predation on the endangered Houston toad, *Bufo houstonensis.* Texas Journal of Science 40:454–456.

Freeman, T. M. 1997. Hymenoptera hypersensitivity in an imported fire ant endemic area. Annals of Allergy, Asthma, and Immunology 78:369–372.

Fritz, G. N., and R. K. Vander Meer. 2003. Sympatry of polygyne and monogyne colonies of the fire ant *Solenopsis invicta* (Hymenoptera: Formicidae). Annals of the Entomological Society of America 96:86–92.

Gaines, T. B. 1969. Acute toxicity of pesticides. Toxicology and Applied Pharmacology 14:515–534.

Gaines, T. B., and R. D. Kimbrough. 1970. Oral toxicity of mirex in adult suckling rats. Archives of Environmental Health 21:7–14.

George, J. L. 1958. The Program to Eradicate the Imported Fire Ant. The Conservation Foundation, New York. 38 pp.

Gilbert, L., and L. Morrison. 1997. Patterns of host specificity in *Pseudacteon* parasitoid flies (Diptera: Phoridae) that attack *Solenopsis* fire ants (Hymenoptera: Formicidae). Environmental Entomology 26:1149–1154.

Giuliano, W. M., C. R. Allen, R. S. Lutz, and S. Demarais. 1996. Effects of red imported fire ants on northern bobwhite chicks. Journal of Wildlife Management 60:309–313.

Glancey, B. M., and J. C. Dickens. 1988. Behavioral and electrophysiological studies with live larvae and larval rinses of the red imported fire ant, *Solenopsis invicta* Buren (Hymenoptera: Formicidae). Journal of Chemical Ecology 14:463–473.

Glancey, B. M., and C. S. Lofgren. 1985. Spermatozoon counts in males and inseminated queens of the imported fire ants, *Solenopsis invicta* and *Solenopsis richteri* (Hymenoptera: Formicidae). Florida Entomologist 68:162–168.

Glancey, B. M., and C. S. Lofgren. 1988. Adoption of newly-mated queens: a mechanism for proliferation and perpetuation of polygynous red imported fire ants, *Solenopsis invicta* Buren. Florida Entomologist 71:581–587.

Glancey, B. M., C. E. Stringer, C. H. Craig, P. M. Bishop, and B. B. Martin. 1970. Pheromone may induce brood tending in the fire ant, *Solenopsis saevissima.* Nature 226:863–864.

Glancey, B. M., C. H. Craig, C. E. Stringer, and P. M. Bishop. 1973a. Multiple fertile queens in colonies of the imported fire ant, *Solenopsis invicta.* Journal of the Georgia Entomological Society 8:237–238.

Glancey, B. M., C. E. Stringer, and P. M. Bishop. 1973b. Trophic egg production in the imported fire ant, *Solenopsis invicta.* Journal of the Georgia Entomological Society 8:217–220.

Glancey, B. M., C. E. Stringer, Jr., C. H. Craig, P. M. Bishop, and B. B. Martin. 1973c. Evidence of a replete caste in the fire ant *Solenopsis invicta.* Annals of the Entomological Society of America 66:233–234.

Glancey, B. M., C. E. Stringer, C. H. Craig, and P. M. Bishop. 1975. An extraordinary case of polygyny in the red imported fire ant. Annals of the Entomological Society of America 68:922.

Glancey, B. M., M. K. St. Romain, and R. H. Crozier. 1976a. Chromosome numbers of the red and the black imported fire ants, *Solenopsis invicta* and *S. richteri.* Annals of the Entomological Society of America 69:469–470.

Glancey, B. M., M. K. Vandenburgh, and M. K. St. Romain. 1976b. Testes degeneration in the red imported fire ant, *Solenopsis invicta.* Journal of the Georgia Entomological Society 11:83–88.

Glancey, B. M., D. P. Wojcik, C. H. Craig, and J. A. Mitchell. 1976c. Ants of Mobile County, AL, as monitored by bait transects. Journal of the Georgia Entomological Society 11:191–197.

Glancey, B. M., A. Glover, and C. S. Lofgren. 1981a. Pheromone production by virgin queens of *Solenopsis invicta* Buren. Sociobiology 6:119–127.

Glancey, B. M., A. Glover, and C. S. Lofgren. 1981b. Thoracic crop formation following dealation by virgin females of two species of *Solenopsis.* Florida Entomologist 64:454.

Glancey, B. M., R. K. Vander Meer, A. Glover, C. S. Lofgren, and S. B. Vinson. 1981c. Filtration of microparticles from liquids ingested by the red imported fire ant, *Solenopsis invicta* Buren. Insectes Sociaux 28:395–401.

Glancey, B. M., C. S. Lofgren, J. R. Rocca, and J. H. Tumlinson. 1983. Behavior of disrupted colonies of *Solenopsis invicta* towards queens and pheromone-treated surrogate queens placed outside the nest. Sociobiology 7:283–288.

Glancey, B. M., J. Rocca, C. S. Lofgren, and J. Tumlinson. 1984. Field tests with synthetic components of the queen recognition pheromone of the red imported fire ant, *Solenopsis invicta.* Sociobiology 9:19–30.

Glancey, B. M., J. C. E. Nickerson, D. Wojcik, J. Trager, W. A. Banks, and C. T. Adams. 1987. The increasing incidence of the polygynous form of the red imported fire ant, *Solenopsis invicta* (Hymenoptera: Formicidae), in Florida. Florida Entomologist 70:400–402.

Glancey, B. M., R. K. Vander Meer, and D. P. Wojcik. 1989. Polygyny in hybrid imported fire ants. Florida Entomologist 72:632–636.

Glasgow, L. L. 1958. Studies of the effect of the imported fire ant program on wildlife in Louisiana. Pages 23–29 in Twelfth Annual Conference of the Association of Game and Fish Commissioners. Association of Game and Fish Commissioners, Louisville, Kentucky.

Glunn, F. J., D. F. Howard, and W. R. Tschinkel. 1981. Food preference in colonies of the fire ant *Solenopsis invicta.* Insectes Sociaux 28:217–222.

Goodisman, M. A. D., and K. G. Ross. 1997. Relationship of queen number and queen relatedness in multiple-queen colonies of the fire ant *Solenopsis invicta*. Ecological Entomology 22:150–157.

Goodisman, M. A. D., and K. G. Ross. 1998. A test of queen recruitment models using nuclear and mitochondrial markers in the fire ant *Solenopsis invicta*. Evolution 52:1416–1422.

Goodisman, M. A. D., and K. G. Ross. 1999. Queen recruitment in a multiple-queen population of the fire ant *Solenopsis invicta*. Behavioral Ecology 10:428–435.

Goodisman, M. A. D., P. D. Mack, D. E. Pearse, and K. G. Ross. 1999. Effects of a single gene on worker and male body mass in the fire ant *Solenopsis invicta* (Hymenoptera: Formicidae). Annals of the Entomological Society of America 92:563–570.

Goodisman, M. A. D., C. J. DeHeer, and K. G. Ross. 2000. Unusual behavior of polygyne fire ant queens on nuptial flights. Journal of Insect Behavior 13:455–468.

Gordon, D. M. 1988. Group-level exploration tactics in fire ants. Behaviour 104:162–175.

Gotelli, N. J., and A. E. Arnett. 2000. Biogeographic effects of red fire ant invasion. Ecology Letters 3:257–261.

Gotwald, W. H., Jr. 1969. Comparative Morphological Studies of the Ants, with Particular Reference to the Mouthparts (Hymenoptera: Formicidae). Memoir of the Cornell University Agricultural Experiment Station No. 408. 150 pp.

Graham, L. C., S. D. Porter, R. M. Pereira, H. D. Dorough, and A. T. Kelley. 2003. Field releases of the decapitating fly *Pseudacteon curvatus* (Diptera: Phoridae) for control of imported fire ants (Hymenoptera: Formicidae) in Alabama, Florida, and Tennessee. Florida Entomologist 86:334–339.

Green, H. B. 1952. Biology and control of the imported fire ant in Mississippi. Journal of Economic Entomology 45:593–597.

Green, H. B. 1959. Imported fire ant mortality due to cold. Journal of Economic Entomology 52:347.

Green, H. B. 1962. On the biology of the imported fire ant. Journal of Economic Entomology 55:1003–1004.

Green, H. B., and R. E. Hutchins. 1960. Laboratory study of toxicity of imported fire ants to bluegill fish. Journal of Economic Entomology 53:1137–1138.

Greenberg, L., D. J. C. Fletcher, and S. B. Vinson. 1985. Differences in worker size and mound distribution in monogynous and polygynous colonies of the fire ant *Solenopsis invicta* Buren. Journal of the Kansas Entomological Society 58:9–18.

Greenslade, P. J. M., and P. Greenslade. 1977. Some effects of vegetation cover and disturbance on a tropical ant fauna. Insectes Sociaux 24:163–182.

Gronenberg, W. 1995. The fast mandible strike in the trap-jaw ant *Odontomachus*. I. Temporal properties and morphological characteristics. Journal of Comparative Physiology. A Sensory, Neural and Behavioral Physiology 176:391–398.

Haight, K. L. 2002. Fire Ant Venom Economy: Patterns of Synthesis and Use. M.S. thesis, Florida State University, Tallahassee, Florida. 75 pp.

Haight, K., and W. R. Tschinkel. 2003. Patterns of venom synthesis and use in the fire ant, *Solenopsis invicta*. Toxicon 42:673–682.

Hangartner, W. 1967. Spezifität und Inaktivierung des Spurpheromons von *Lasius fuliginosus* Latr. und Orientierung der Arbeiterinnen im Duftfeld. Zeitschrift für Vergleichende Physiologie 57:103–136.

Hangartner, W. 1969a. Carbon dioxide, a releaser for digging behavior in *Solenopsis geminata* (Hymenoptera: Formicidae). Psyche 76:58–67.

Hangartner, W. 1969b. Structure and variability of the individual odor trail in *Solenopsis geminata* Fabr. (Hymenoptera, Formicidae). Zeitschrift für Vergleichende Physiologie 62:111–120.

Hannan, C. J., C. T. Stafford, R. B. Rhoades, and B. B. Wray. 1984. Seasonal variation in imported fire ant (IFA) antigens. Annals of Allergy 52:227.

Harris, M., A. Knutson, A. Calixto, A. Dean, L. Brooks, and B. Ree. 2003. Impact of red imported fire ant on foliar herbivores and natural enemies. Southwestern Entomologist Supplement 27:123–134.

Harris, W. G., and E. C. Burns. 1972. Predation on the lone star tick by the imported fire ant. Environmental Entomology 1:362–365.

Harvey, C. T., and M. D. Eubanks. 2004. Effect of habitat complexity on biological control by the red imported fire ant (Hymenoptera: Formicidae) in collards. Biological Control 29:348–358.

Hays, S. B., and K. L. Hays. 1959. Food habits of *Solenopsis saevissima richteri* Forel. Journal of Economic Entomology 52:455–457.

Hefetz, A., C. Errard, A. Chambris, and A. Le Negrate. 1996. Postpharyngeal gland secretion as a modifier of aggressive behavior in the myrmicine ant *Manica rubida*. Journal of Insect Behavior 9:709–717.

Helms, K. R., and S. B. Vinson. 2003. Apparent facilitation of an invasive mealybug by an invasive ant. Insectes Sociaux 50:403–404.

Hermann, H. R., Jr., and M. S. Blum. 1965. Morphology and histology of the reproductive system of the imported fire ant queen, *Solenopsis saevissima richteri*. Annals of the Entomological Society of America 58:81–89.

Hermann, H. R., and M. S. Blum. 1981. Defensive mechanisms in the social Hymenoptera. Pages 77–197 in Social Insects, Volume 2. H. Hermann, editor. Academic Press, New York.

Herzog, D. C., T. E. Reagan, D. C. Sheppard, K. M. Hyde, S. S. Nilakhe, M. Y. B. Hussein, M. L. McMahan, R. C. Thomas, and L. D. Newsom. 1976. *Solenopsis invicta* Buren: influence on Louisiana pasture soil chemistry. Environmental Entomology 5:160–162.

Hill, E. P. 1969. Observations of imported fire ant predation on nestling cottontails. Proceedings of the Annual Conference, Southeastern Association of Game and Fish Commissioners 23:171–181.

Hill, S., and M. A. Hoy. 2003. Interactions between the red imported fire ant *Solenopsis invicta* and the parasitoid *Lipolexis scutellaris* potentially affect classical biological control of the aphid *Toxoptera citricida.* Biological Control 27:11–19.

Hinkle, M. K. 1982. Impact of the imported fire ant control programs on wildlife and quality of the environment pesticide residues, United States. Pages 130–143 in Proceedings of the Symposium on the Imported Fire Ant, June 7–10, 1982, Atlanta, Georgia. S. L. Battenfield, editor. EPA/USDA (APHIS) 0-389-890/70. U.S. Environmental Protection Agency and U.S. Department of Agriculture, Animal and Plant Health Inspection Service, Washington, D.C.

Hoffman, D. R. 1987. Allergens in Hymenoptera venom XVII. Allergenic components of *Solenopsis invicta* (imported fire ant) venom. Journal of Allergy and Clinical Immunology 80:300–306.

Hoffman, D. R., D. E. Dove, and R. S. Jacobson. 1988a. Allergens in Hymenoptera venom XX. Isolation of four allergens from imported fire ant (*Solenopsis invicta*) venom. Journal of Allergy and Clinical Immunology 82:818–827.

Hoffman, D. R., D. E. Dove, and R. S. Jacobson. 1988b. Isolation of allergens from imported fire ant (*Solenopsis invicta*) venom. Journal of Allergy and Clinical Immunology 81:203.

Hoffman, D. R., D. E. Dove, J. E. Moffitt, and C. T. Stafford. 1988c. Allergens in Hymenoptera venom XXI. Cross-reactivity and multiple reactivity between fire ant venom and bee and wasp venoms. Journal of Allergy and Clinical Immunology 82:828–834.

Hölldobler, B., and E. O. Wilson. 1990. The Ants. Belknap Press of Harvard University Press, Cambridge, Massachusetts. 732 pp.

Holtcamp, W. N., W. E. Grant, and S. B. Vinson. 1997. Patch use under predation hazard: effect of the red imported fire ant on deer mouse foraging behavior. Ecology 78:308–317.

Hood, W. G., and W. R. Tschinkel. 1990. Desiccation resistance in arboreal and terrestrial ants. Physiological Entomology 15:23–35.

Hook, A. W., and S. D. Porter. 1990. Destruction of harvester ant colonies by invading fire ants in south-central Texas (Hymenoptera: Formicidae). Southwestern Naturalist 35:477–478.

Hooper-Bui, L. M., M. K. Rust, and D. A. Reierson. 2004. Predation of the endangered California least tern, *Sterna antillarum browni* by the southern fire ant, *Solenopsis xyloni* (Hymenoptera, Formicidae). Sociobiology 43:401–418.

Horton, P. M., S. B. Hays, and J. R. Holman. 1975. Food carrying ability and recruitment time of the red imported fire ant. Journal of the Georgia Entomological Society 10:207–213.

Howard, D. F. 1974. Aspects of Necrophoric Behavior in the Red Imported Fire Ant, *Solenopsis invicta.* M.S. thesis, Florida State University, Tallahassee, Florida. 94 pp.

Howard, D. F., and W. R. Tschinkel. 1976. Aspects of necrophoric behavior in the red imported fire ant, *Solenopsis invicta.* Behaviour 56:157–180.

Howard, D. F., and W. R. Tschinkel. 1980. The effect of colony size and starvation on food flow in the fire ant, *Solenopsis invicta* (Hymenoptera: Formicidae). Behavioral Ecology and Sociobiology 7:293–300.

Howard, D. F., and W. R. Tschinkel. 1981a. The flow of food in colonies of the fire ant, *Solenopsis invicta:* a multifactorial study. Physiological Entomology 6:297–306.

Howard, D. F., and W. R. Tschinkel. 1981b. Internal distribution of liquid foods in isolated workers of the fire ant, *Solenopsis invicta.* Journal of Insect Physiology 27:67–74.

Howard, F. W., and A. D. Oliver. 1978. Arthropod populations in permanent pastures treated and untreated with mirex for red imported fire ant control. Environmental Entomology 7:901–903.

Howard, F. W., and A. D. Oliver. 1979. Field observations of ants (Hymenoptera: Formicidae) associated with red imported fire ants, *Solenopsis invicta* Buren, in Louisiana pastures. Journal of the Georgia Entomological Society 14:259–263.

Hu, G. Y., and J. H. Frank. 1996. Effect of the red imported fire ant (Hymenoptera: Formicidae) on dung-inhabiting arthropods in Florida. Environmental Entomology 25:1290–1296.

Hubbard, M. D., and W. G. Cunningham. 1977. Orientation of mounds in the ant *Solenopsis invicta* (Hymenoptera: Formicidae: Myrmicinae). Insectes Sociaux 24:3–7.

Hung, A. C. F., and S. B. Vinson. 1978. Factors affecting the distribution of fire ants in Texas (Myrmicinae, Formicidae). Southwestern Naturalist 23:205–213.

Hung, A. C. F., S. B. Vinson, and J. W. Summerlin. 1974. Male sterility in the red imported fire ant, *Solenopsis invicta.* Annals of the Entomological Society of America 67:909–912.

Hung, A. C. F., M. R. Barlin, and S. B. Vinson. 1977. Identification, Distribution, and Biology of Fire Ants in Texas. Texas Agricultural Experiment Station Bulletin 1185. 24 pp.

Hylander, R. D., A. A. Ortiz, T. M. Freeman, and M. E. Martin. 1989. Imported fire ant immunotherapy:

effectiveness of whole body extracts. Journal of Allergy and Clinical Immunology 83:232.

Ihering, H. von. 1894. Die Ameisen von Rio Grande do Sul. Berliner Entomologische Zeitschrift 39:231–245.

Jaffe, K., and H. Puche. 1984. Colony-specific territorial marking with the metapleural gland secretion in the ant *Solenopsis geminata* (Fabr). Journal of Insect Physiology 30:265–270.

James, F. K., Jr., H. L. Pence, D. P. Driggers, R. L. Jacobs, and D. E. Horton. 1976a. Imported fire ant hypersensitivity: studies of human reactions to fire ant venom. Journal of Allergy and Clinical Immunology 58:110–120.

James, F. K., Jr., H. L. Pence, D. P. Driggers, R. L. Jacobs, and D. E. Horton. 1976b. In vivo and in vitro studies of human reactions to fire ant venom. Journal of Allergy and Clinical Immunology 57:209.

James, S. S., R. M. Pereira, K. M. Vail, and B. H. Ownley. 2002. Survival of imported fire ant (Hymenoptera: Formicidae) species subjected to freezing and near-freezing temperatures. Environmental Entomology 31:127–133.

Jensen, T. F., and I. Holm-Jensen. 1980. Energetic cost of running in workers of three ant species, *Formica fusca* L., *Formica rufa* L., and *Camponotus herculeanus* L. (Hymenoptera, Formicidae). Journal of Comparative Physiology 137:151–156.

Jerome, C. A., D. A. McInnes, and E. S. Adams. 1998. Group defense by colony-founding queens in the fire ant *Solenopsis invicta*. Behavioral Ecology 9:301–308.

Johansson, S. G. O., S. Nordvall, D. Ledford, and R. Lockey. 1985. Specific IgE and IgG responses to immunotherapy (IT) with imported fire ant whole body extract (IFA-WBE) in IFA hypersensitivity (H). Journal of Allergy and Clinical Immunology 75:208.

Johnson, A. S. 1961. Antagonistic relationships between ants and wildlife with special reference to imported fire ants and bobwhite quail in the southeast. Proceedings of the Annual Conference, Southeastern Association of Game and Fish Commissioners 15:88–107.

Johnson, E. L. 1976. Administrator's decision to accept plan of Mississippi Authority and order suspending hearing for the pesticide chemical mirex. Federal Register 41:56694–56703.

Jones, D., and W. L. Sterling. 1979. Manipulation of red imported fire ants in a trap crop for boll weevil suppression. Environmental Entomology 8:1073–1077.

Jones, R. G. 1977. Insemination-induced histolysis of the indirect flight muscle of the red imported fire ant *Solenopsis invicta:* an electron microscopic study. Anatomical Record 187:618A–619A.

Jones, R. G., and W. L. Davis. 1985. Muscle cell membranes from early degeneration muscle cell fibers in *Solenopsis* are leaky to lanthanum: electron microscopy and x-ray analysis. Anatomical Record 212:123–128.

Jones, R. G., W. L. Davis, A. C. F. Hung, and S. B. Vinson. 1978. Insemination-induced histolysis of the flight musculature in fire ants (*Solenopsis* spp.), an ultrastructural study. American Journal of Anatomy 151:603–610.

Jones, R. G., W. L. Davis, H. K. Hagler, and S. B. Vinson. 1981. Calcium and muscle degeneration in *Solenopsis:* histochemistry and electron microprobe analysis. Journal of Cell Biology 91:358A.

Jones, R. G., W. L. Davis, and S. B. Vinson. 1982. A histochemical and X-ray microanalysis study of calcium changes in insect flight muscle degeneration in *Solenopsis,* the queen fire ant. Journal of Histochemistry and Cytochemistry 30:293–304.

Jones, S. R., and S. A. Phillips, Jr. 1990. Resource collection abilities of *Solenopsis invicta* (Hymenoptera: Formicidae) compared to those of three sympatric Texas ants. Southwestern Naturalist 35:416–422.

Jouvenaz, D. P. 1983. Natural enemies of fire ants. Florida Entomologist 66:111–121.

Jouvenaz, D. P. 1984. Some Protozoa infecting fire ants, *Solenopsis* spp. Pages 195–203 in Comparative Pathobiology, Volume 7. T. C. Cheng, editor. Plenum Press, New York.

Jouvenaz, D. P. 1986. Diseases of fire ants: problems and opportunities. Pages 327–328 in Fire Ants and Leafcutting Ants: Biology and Management. C. S. Lofgren and R. K. Vander Meer, editors. Westview Press, Boulder, Colorado.

Jouvenaz, D. P. 1990a. Approaches to biological control of fire ants in the United States. Pages 620–627 in Applied Myrmecology: A World Perspective. R. K. Vander Meer, K. Jaffe, and A. Cedeno, editors. Westview Press, Boulder, Colorado.

Jouvenaz, D. P. 1990b. Biological control of fire ants: current research. Pages 621–622 in Social Insects and the Environment: Proceedings of the 11th International Congress of IUSSI. G. K. Veeresh, B. Mallik, and C. A. Viraktamath, editors. E. J. Brill, New York.

Jouvenaz, D. P., and D. W. Anthony. 1979. *Mattesia geminata* sp. n. (Neogregarinida: Ophrocystidae) a parasite of the tropical fire ant, *Solenopsis geminata* (Fabricius). Journal of Protozoology 26:354–356.

Jouvenaz, D. P., and E. A. Ellis. 1986. *Vairimorpha invictae* n. sp. (Microspora: Microsporida), a parasite of the red imported fire ant, *Solenopsis invicta* Buren (Hymenoptera: Formicidae). Journal of Protozoology 33:457–461.

Jouvenaz, D. P., and C. S. Lofgren. 1984a. Host specificity of *Burenella dimorpha* (Microspora: Microsporida). Journal of Invertebrate Pathology 43:441–442.

Jouvenaz, D. P., and C. S. Lofgren. 1984b. Temperature-dependent spore dimorphism in *Burenella dimorpha* (Microspora: Microsporida). Journal of Protozoology 31:175–177.

Jouvenaz, D. P., M. S. Blum, and J. G. MacConnell. 1972. Antibacterial activity of venom alkaloids from the imported fire ant, *Solenopsis invicta* Buren. Antimicrobial Agents and Chemotherapy 2:291–293.

Jouvenaz, D. P., W. A. Banks, and C. S. Lofgren. 1974. Fire ants: attraction of workers to queen secretions. Annals of the Entomological Society of America 67:442–444.

Jouvenaz, D. P., G. E. Allen, W. A. Banks, and D. P. Wojcik. 1977. A survey for pathogens of fire ants, *Solenopsis* spp., in the southeastern United States. Florida Entomologist 60:275–279.

Jouvenaz, D. P., C. S. Lofgren, D. A. Carlson, and W. A. Banks. 1978. Specificity of the trail pheromones of four species of fire ants, *Solenopsis* spp. Florida Entomologist 61:244.

Jouvenaz, D. P., W. A. Banks, and J. D. Atwood. 1980. Incidence of pathogens in fire ants, *Solenopsis* spp., in Brazil. Florida Entomologist 63:345–346.

Jouvenaz, D. P., C. S. Lofgren, and W. A. Banks. 1981. Biological control of imported fire ants: a review of current knowledge. Bulletin of the Entomological Society of America 27:203–208.

Jouvenaz, D. P., E. A. Ellis, and C. S. Lofgren. 1984. Histopathology of the tropical fire ant, *Solenopsis geminata,* infected with *Burenella dimorpha* (Microspora: Microsporida). Journal of Invertebrate Pathology 43:324–332.

Jouvenaz, D. P., D. P. Wojcik, and R. K. Vander Meer. 1989. First observation of polygyny in fire ants, *Solenopsis* spp., in South America. Psyche 96:161–165.

Jouvenaz, D. P., C. S. Lofgren, and R. W. Miller. 1990. Steinernematid nematode drenches for control of fire ants, *Solenopsis invicta,* in Florida. Florida Entomologist 73:190–193.

Jusino Atresino, R., and S. A. Phillips, Jr. 1994. Impact of red imported fire ants on the ant fauna of central Texas. Pages 259–268 in Exotic Ants: Biology, Impact, and Control of Introduced Species. D. F. Williams, editor. Westview Press, Boulder, Colorado.

Kaakeh, W., and J. D. Dutcher. 1992. Foraging preference of red imported fire ants (Hymenoptera: Formicidae) among three species of summer cover crops and their extracts. Journal of Economic Entomology 85:389–394.

Kaiser, K. L. E. 1974. Mirex: an unrecognized contaminant of fishes from Lake Ontario. Science 185:523–525.

Kaplan, I., and M. D. Eubanks. 2002. Disruption of cotton aphid (Homoptera: Aphididae)—natural enemy dynamics by red imported fire ants (Hymenoptera: Formicidae). Environmental Entomology 31:1175–1183.

Kaspari, M., and E. L. Vargo. 1994. Nest site selection by fire ant queens. Insectes Sociaux 41:331–333.

Keller, L. 1995. Social life: the paradox of multiple-queen colonies. Trends in Ecology and Evolution 10:355–360.

Keller, L., and K. G. Ross. 1993a. Phenotypic basis of reproductive success in a social insect: genetic and social determinants. Science 260:1107–1110.

Keller, L., and K. G. Ross. 1993b. Phenotypic plasticity and "cultural transmission" of alternative social organizations in the fire ant *Solenopsis invicta.* Behavioral Ecology and Sociobiology 33:121–129.

Keller, L., and K. G. Ross. 1995. Gene by environment interaction: effects of a single gene and social environment on reproductive phenotypes of fire ant queens. Functional Ecology 9:667–676.

Keller, L., and K. G. Ross. 1999. Major gene effects on phenotype and fitness: the relative roles of *Pgm-3* and *Gp-9* in introduced populations of the fire ant *Solenopsis invicta.* Journal of Evolutionary Biology 12:672–680.

Keller, L. A., and K. G. Ross. 1998. Selfish genes: a green beard in the red fire ant. Nature 394:573–575.

Killion, M. J., and W. E. Grant. 1993. Scale effects in assessing the impact of imported fire ants on small mammals. Southwestern Naturalist 38:393–396.

Killion, M. J., and W. E. Grant. 1995. A colony-growth model for the imported fire ant: potential geographic range of an invading species. Ecological Modelling 77:73–84.

Killion, M. J., W. E. Grant, and S. B. Vinson. 1995. Response of *Baiomys taylori* to changes in density of imported fire ants. Journal of Mammalogy 76:141–147.

King, J. R. 2004. Ant Communities of Florida's Upland Ecosystems: Ecology and Sampling. Ph.D. dissertation, University of Florida, Gainesville, Florida. 143 pp.

King, J. R., A. N. Andersen, and A. D. Cutter. 1998. Ants as bioindicators of habitat disturbance: validation of the functional group model for Australia's humid tropics. Biodiversity and Conservation 7:1627–1638.

King, T. G., and S. A. Phillips, Jr. 1992. Destruction of young colonies of the red imported fire ant by the pavement ant (Hymenoptera: Formicidae). Entomological News 103:72–77.

Kintz-Early, J., L. Parris, J. Zettler, and J. Bast. 2003. Evidence of polygynous red imported fire ants (Hymenoptera: Formicidae) in South Carolina. Florida Entomologist 86:381–382.

Klobuchar, E. A., and R. J. Deslippe. 2002. A queen pheromone induces workers to kill sexual larvae in colonies of the red imported fire ant (*Solenopsis invicta*). Naturwissenschaften 89:302–304.

Knell, J. D., G. E. Allen, and E. I. Hazard. 1977. Light and electron microscope study of *Thelohania solenopsae* n. sp. (Microsporida: Protozoa) in the red imported fire ant, *Solenopsis invicta.* Journal of Invertebrate Pathology 29:192–200.

Korzukhin, M. D., and S. D. Porter. 1994. Spatial model of territorial competition and population dynamics in the fire ant *Solenopsis invicta* (Hymenoptera: Formicidae). Environmental Entomology 23:912–922.

Korzukhin, M. D., S. D. Porter, L. C. Thompson, and S. Wiley. 2001. Modeling temperature-dependent range limits for the red imported fire ant (Hymenoptera: Formicidae: *Solenopsis invicta*) in the United States. Environmental Entomology 30:645–655.

Krieger, M. J. B., and K. G. Ross. 2002. Identification of a major gene regulating complex social behavior. Science 295:328–332.

Krieger, M. J. B., K. G. Ross, C. W. Y. Chang, and L. Keller. 1999. Frequency and origin of triploidy in the fire ant *Solenopsis invicta.* Heredity 82:142–150.

Kroll, J. C., K. A. Arnold, and R. F. Gotie. 1973. An observation of predation by native fire ants on nestling barn swallows. Wilson Bulletin 85:478–479.

Kuriachan, I., and S. B. Vinson. 2000. A queen's worker attractiveness influences her movement in polygynous colonies of the red imported fire ant (Hymenoptera: Formicidae) in response to adverse temperatures. Environmental Entomology 29:943–949.

Kutz, F. W., A. R. Yobs, W. G. Johnson, and G. B. Wiersma. 1974. Mirex residues in human adipose tissue. Environmental Entomology 3:882–884.

Lahav, S., V. Soroker, A. Hefetz, and R. K. Vander Meer. 1999. Direct behavioral evidence for hydrocarbons as ant recognition discriminators. Naturwissenschaften 86:246–249.

Lammers, J. M. 1987. Mortality Factors Associated with the Founding Queens of *Solenopsis invicta* Buren, the Red Imported Fire Ant: A Study of the Native Ant Community in Central Texas. M.S. thesis, Texas A&M University, College Station, Texas. 206 pp.

Lamon, B., and H. Topoff. 1985. Social facilitation of eclosion in the fire ant, *Solenopsis invicta.* Developmental Psychobiology 18:367–374.

Landers, J. L., J. A. Garner, and W. A. McRae. 1980. Reproduction of gopher tortoises (*Gopherus polyphemus*) in southwestern Georgia. Herpetologica 36:353–361.

Lanza, J. 1991. Response of fire ants (Formicidae: *Solenopsis invicta* and *S. geminata*) to artificial nectars with amino acids. Ecological Entomology 16:203–210.

Lanza, J., E. L. Vargo, S. Pulim, and Y. Z. Chang. 1993. Preferences of the fire ants *Solenopsis invicta* and *S. geminata* (Hymenoptera: Formicidae) for amino acid and sugar components of extrafloral nectars. Environmental Entomology 22:411–417.

Lavine, B. K., L. Morel, R. K. Vander Meer, R. W. Gunderson, J. H. Han, A. Bonanno, and A. Stine. 1990a. Pattern recognition studies in chemical communication: nestmate recognition in *Camponotus floridanus.* Chemometrics and Intelligent Laboratory Systems 9:107–114.

Lavine, B. K., R. K. Vander Meer, L. Morel, R. W. Gunderson, J. H. Han, and A. Stine. 1990b. False color data imaging: a new pattern recognition technique for analyzing chromatographic data. Journal of Microchemistry 41:288–295.

Lay, D. W. 1958. Fire ant eradication and wildlife. Pages 22–23 in Twelfth Annual Conference of the Association of Game and Fish Commissioners. Association of Game and Fish Commissioners, Louisville, Kentucky.

Lee, B. J. 1974. Effects of mirex on litter organisms and leaf decomposition in a mixed hardwood forest in Athens, Georgia. Journal of Environmental Quality 3:305–311.

Lee, D. K., A. P. Bhatkar, S. B. Vinson, and J. K. Olson. 1994. Impact of foraging red imported fire ants (*Solenopsis invicta*) (Hymenoptera: Formicidae) on *Psorophora columbiae* eggs. Journal of the American Mosquito Control Association 10:163–173.

Lehmann, V. W. 1946. Bobwhite quail reproduction in southwestern Texas. Journal of Wildlife Management 10:111–123.

Lemke, L. A., and J. B. Kissam. 1988. Impact of red imported fire ant (Hymenoptera: Formicidae) predation on horn flies (Diptera: Muscidae) in a cattle pasture treated with Pro-Drone. Journal of Economic Entomology 81:855–858.

Lennartz, F. E. 1973. Modes of Dispersal of *Solenopsis invicta* from Brazil into the Continental United States—A Study in Spatial Diffusion. M.S. thesis, University of Florida, Gainesville, Florida. 242 pp.

Levia, D. F., and E. E. Frost. 2004. Assessment of climatic suitability for the expansion of *Solenopsis invicta* Buren in Oklahoma using three general circulation models. Theoretical and Applied Climatology 79:23–30.

Lewis, D. K., J. Q. Campbell, S. M. Sowa, M. E. Chen, S. B. Vinson, and L. L. Keeley. 2001. Characterization of vitellogenin in the red imported fire ant, *Solenopsis invicta* (Hymenoptera: Apocrita: Formicidae). Journal of Insect Physiology 47:543–551.

Lewis, D. K., J. Q. Campbell, S. M. Sowa, M. E. Chen, S. B. Vinson, and L. L. Keeley. 2002. The biochemical characteristics of vitellogenin in the red imported fire ant, *Solenopsis invicta* (Hymenoptera: Formicidae). Southwestern Entomologist Supplement 25:71–79.

Li, J. B., and K. M. Heinz. 1998. Genetic variation in desiccation resistance and adaptability of the red imported fire ant (Hymenoptera: Formicidae) to arid regions. Annals of the Entomological Society of America 91:726–729.

Lighton, J. R. B. 1989. The energetics of locomotion in ants—current knowledge and future prospects. Society for Experimental Biology Meeting, Edinburgh. Abstract A47.

Lighton, J. R., G. A. Bartholomew, and D. H. Feener, Jr. 1987. Energetics of locomotion and load carriage and a model of the energy cost of foraging in the leaf-cutting ant *Atta colombica* Guer. Physiological Zoology 60:524–537.

Lind, N. K. 1982. Mechanism of action of fire ant (*Solenopsis*) venoms. I. Lytic release of histamine from mast cells. Toxicon 20:831–840.

Liu, N. N., and L. Zhang. 2004. CYP4AB1, CYP4AB2, and *Gp-9* gene overexpression associated with workers of the red imported fire ant, *Solenopsis invicta* Buren. Gene 327:81–87.

Livingstone, R. J., G. W. Cornwell, V. D. Cunningham, R. Harriss, M. W. Provost, and W. H. Whitcomb. 1970. Report of the Select Study Committee on Mirex. Tallahassee, Florida. 9 pp.

Lockaby, B. G., and J. C. Adams. 1985. Pedoturbation of a forest soil by fire ants. Soil Science Society of America Journal 49:220–223.

Lockey, R. F. 1974. Systemic reactions to stinging ants. Journal of Allergy and Clinical Immunology 54:132–146.

Lockey, R. F. 1980. Allergic and other adverse reactions caused by the imported fire ant. Pages 441–448 in Advances in Allergology and Clinical Immunology. A. Oehling, E. Mathov, I. Glazer, and C. Arbesman, editors. Pergamon Press, Oxford.

Lockley, T. C. 1995. Effect of imported fire ant predation on a population of the least tern-an endangered species. Southwestern Entomologist 20:517–519.

Lofgren, C. S. 1986a. The economic importance and control of imported fire ants in the United States. Pages 227– 256 in Economic Impact and Control of Social Insects. S. B. Vinson, editor. Praeger, New York.

Lofgren, C. S. 1986b. History of imported fire ants in the United States. Pages 36–47 in Fire Ants and Leafcutting Ants: Biology and Management. C. S. Lofgren and R. K. Vander Meer, editors. Westview Press, Boulder, Colorado.

Lofgren, C. S. 1988. Historical perspective: origin of the imported fire ant, its spread and taxonomic status. Pages 1–5, 121–125 in Proceedings of the Governor's Conference on the Imported Fire Ant: Assessment and Recommendations. S. B. Vinson and J. Teer, editors. Sportsmen Conservationists of Texas, Austin, Texas.

Lofgren, C. S., and C. T. Adams. 1982. Economic aspects of the imported fire ant in the United States. Pages 124–128 in The Biology of Social Insects. M. D. Breed, C. A. Michener, and H. E. Evans, editors. Westview Press, Boulder, Colorado.

Lofgren, C. S., and D. E. Weidhaas. 1972. On the eradication of imported fire ants: a theoretical appraisal. Bulletin of the Entomological Society of America 18:17–20.

Lofgren, C. S., and D. F. Williams. 1984. Polygynous colonies of the red imported fire ant, *Solenopsis invicta* (Hymenoptera: Formicidae) in Florida. Florida Entomologist 67:484–486.

Lofgren, C. S., and D. F. Williams. 1985. Red imported fire ants (Hymenoptera: Formicidae): population dynamics following treatment with insecticidal baits. Journal of Economic Entomology 78:863–867.

Lofgren, C. S., F. J. Bartlett, C. E. Stringer, Jr., and W. A. Banks. 1964. Imported fire ant toxic bait studies: further tests with granulated mirex-soybean oil bait. Journal of Economic Entomology 57:695–698.

Lofgren, C. S., W. A. Banks, and B. M. Glancey. 1975. Biology and control of imported fire ants. Annual Review of Entomology 20:1–30.

Lofgren, C. S., B. M. Glancey, A. Glover, J. Rocca, and J. Tumlinson. 1983. Behavior of workers of *Solenopsis invicta* (Hymenoptera: Formicidae) to the queen recognition pheromone: laboratory studies with an olfactometer and surrogate queens. Annals of the Entomological Society of America 76:44–50.

Lok, J. B., E. W. Cupp, and G. J. Blomquist. 1975. Cuticular lipids of the imported fire ants, *Solenopsis invicta* and *richteri*. Insect Biochemistry 5:821–829.

Long, W. H., E. A. Cancienne, E. J. Concienne, R. N. Dobson, and L. D. Newsom. 1958. Fire ant eradication program increases damage by the sugarcane borer. Sugar Bulletin 37:62–63.

Lopez, J. D. 1982. Emergence pattern of an overwintering population of *Cariochiles nigriceps* in central Texas. Environmental Entomology 11:838–842.

Lubertazzi, D. 1999. Ant (Formicidae) community change across a vegetational gradient in north Florida longleaf pine (*Pinus palustris*) flatwoods. Ph.D. dissertation, Florida State University, Tallahassee, Florida. 86 pp.

Lyle, C., and I. Fortune. 1948. Notes on an imported fire ant. Journal of Economic Entomology 41:833–834.

MacConnell, J. G., M. S. Blum, and H. M. Fales. 1970. Alkaloid from fire ant venom: identification and synthesis. Science 168:840–841.

MacConnell, J. G., M. S. Blum, and H. M. Fales. 1971. The chemistry of fire ant venom. Tetrahedron 26:1129–1139.

MacConnell, J. G., R. N. Williams, J. M. Brand, and M. S. Blum. 1974. New alkaloids in the venoms of fire ants. Annals of the Entomological Society of America 67:134–135.

MacConnell, J. G., M. S. Blum, W. F. Buren, R. N. Williams, and H. M. Fales. 1976. Fire ant venoms: chemotaxonomic correlations with alkaloidal compositions. Toxicon 14:69–78.

MacKay, W. P. 1988. The impact of the red imported fire ant on electrical equipment. Pages 57–63 in

Proceedings of the Governor's Conference the Imported Fire Ant: Assessment and Recommendations. S. B. Vinson and J. Teer, editors. Sportsmen Conservationists of Texas, Austin, Texas.

MacKay, W. P., and R. Fagerlund. 1997. Range expansion of the red imported fire ant, *Solenopsis invicta* Buren (Hymenoptera: Formicidae), into New Mexico and extreme western Texas. Proceedings of the Entomological Society of Washington 99:757–758.

MacKay, W. P., D. Sparks, and S. B. Vinson. 1990. Destruction of electrical equipment by *Solenopsis xyloni* McCook (Hymenoptera: Formicidae). Pan-Pacific Entomologist 66:174–175.

MacKay, W. P., L. Greenberg, and S. B. Vinson. 1991. Survivorship of founding queens of *Solenopsis invicta* (Hymenoptera: Formicidae) in areas with monogynous and polygynous nests. Sociobiology 19:293–304.

MacKay, W. P., S. Majdi, J. Irving, S. B. Vinson, and C. Messer. 1992a. Attraction of ants (Hymenoptera, Formicidae) to electric fields. Journal of the Kansas Entomological Society 65:39–43.

MacKay, W. P., S. B. Vinson, J. Irving, S. Majdi, and C. Messer. 1992b. Effect of electrical fields on the red imported fire ant (Hymenoptera: Formicidae). Environmental Entomology 21:866–870.

MacKay, W. P., S. Porter, H. G. Fowler, and S. B. Vinson. 1994. A distribução das formigas lavapés (*Solenopsis* spp.) no estado de Mato Grosso do Sul, Brasil (Hymenoptera: Formicidae). Sociobiology 24:307–312.

Macom, T. E., and S. D. Porter. 1995. Food and energy requirements of laboratory fire ant colonies (Hymenoptera: Formicidae). Environmental Entomology 24:387–391.

Macom, T. E., and S. D. Porter. 1996. Comparison of polygyne and monogyne red imported fire ants (Hymenoptera: Formicidae) population densities. Annals of the Entomological Society of America 89:535–543.

Majer, J. D. 1983. Ants: bio-indicators of minesite rehabilitation, land-use, and land conservation. Environmental Management 7:375–383.

Marak, G. E., Jr., and J. J. Wolken. 1965. An action spectrum for the fire ant (*Solenopsis saevissima*). Nature 205:1328–1329.

Markin, G. P. 1964. A lead-solder alloy casting technique for studying the structure of ants' nests. Annals of the Entomological Society of America 57:360–362.

Markin, G. P. 1968. Handling techniques for large quantities of ants. Journal of Economic Entomology 61:1744–1745.

Markin, G. P., and J. H. Dillier 1971. The seasonal life cycle of the imported fire ant, *Solenopsis saevissima richteri,* on the gulf coast of Mississippi. Annals of the Entomological Society of America 64:562–565.

Markin, G. P., and S. O. Hill. 1971. Microencapsulated oil bait for control of the imported fire ant. Journal of Economic Entomology 64:193–196.

Markin, G. P., J. H. Dillier, S. O. Hill, M. S. Blum, and H. R. Hermann. 1971. Nuptial flight and flight ranges of the imported fire ant, *Solenopsis saevissima richteri* (Hymenoptera: Formicidae). Journal of the Georgia Entomological Society 6:145–156.

Markin, G. P., H. L. Collins, and J. H. Dillier. 1972. Colony founding by queens of the red imported fire ant, *Solenopsis invicta.* Annals of the Entomological Society of America 65:1053–1058.

Markin, G. P., J. H. Dillier, and H. L. Collins. 1973. Growth and development of colonies of the red imported fire ant, *Solenopsis invicta.* Annals of the Entomological Society of America 66:803–808.

Markin, G. P., H. L. Collins, and J. H. Spence. 1974a. Residues of the insecticide mirex following aerial treatment of Cat Island. Bulletin of Environmental Contamination and Toxicology 12:233–240.

Markin, G. P., J. O'Neal, and H. L. Collins. 1974b. Effects of mirex on the general ant fauna of a treated area in Louisiana. Environmental Entomology 3:895–898.

Markin, G. P., J. O'Neal, J. H. Dillier, and H. L. Collins. 1974c. Regional variation in the seasonal activity of the imported fire ant, *Solenopsis saevissima richteri.* Environmental Entomology 3:446–452.

Markin, G. P., H. L. Collins, and J. O'Neal. 1975a. Control of the imported fire ants with winter applications of microencapsulated mirex bait. Journal of Economic Entomology 68:711–712.

Markin, G. P., J. O'Neal, and H. L. Collins. 1975b. Control of the red imported fire ant with microencapsulated baits containing reduced amounts of mirex. Journal of the Georgia Entomological Society 10:281–284.

Markin, G. P., J. O'Neal, and J. Dillier. 1975c. Foraging tunnels of the red imported fire ant, *Solenopsis invicta* (Hymenoptera: Formicidae). Journal of the Kansas Entomological Society 48:83–89.

Marshall, E. 1982. Mississippi Inc., pesticide manufacturer. Science 218:548–550.

Maschwitz, U. W. J., and W. Kloft. 1971. Morphology and function of the venom apparatus of insects—bees, wasps, ants, and caterpillars. Pages 1–60 in Venomous Animals and Their Venoms, Volume 3. W. Bücherl and E. E. Buckley, editors. Academic Press, New York.

Masser, M. P., and W. E. Grant. 1986. Fire ant–induced trap mortality of small mammals in east-central Texas. Southwestern Naturalist 31:540–542.

McDaniel, S. G., and W. L. Sterling. 1979. Predator determination and efficiency on *Heliothis virescens* eggs in cotton using 32P. Environmental Entomology 8:1083–1087.

McDaniel, S. G., and W. L. Sterling. 1982. Predation of *Heliothis virescens* (F.) eggs on cotton in east Texas. Environmental Entomology 11:60–66.

McDaniel, S. G., W. L. Sterling, and D. A. Dean. 1981. Predators of tobacco budworm larvae in Texas cotton. Southwestern Naturalist 6:102–108.

McInnes, D. A. 1994. Comparative Ecology and Factors Affecting the Distribution of North Florida Fire Ants. Ph.D. dissertation, Florida State University, Tallahassee, Florida. 137 pp.

McInnes, D. A., and W. R. Tschinkel. 1995. Queen dimorphism and reproductive strategies in the fire ant, *Solenopsis geminata* (Hymenoptera: Formicidae). Behavioral Ecology and Sociobiology 36:367–375.

McInnes, D.A., and W. R. Tschinkel. 1996. Mermithid nematode parasitism of *Solenopsis* ants (Hymenoptera: Formicidae) of northern Florida. Annals of the Entomological Society of America 89:231–237.

McIver, J. D., and C. Loomis. 1993. A size-distance relation in Homoptera-tending thatch ants (*Formica obscuripes, Formica planipilis*). Insectes Sociaux 40:207–218.

Mehdiabadi, N. J., E. A. Kawazoe, and L. E. Gilbert. 2004. Phorid fly parasitoids of invasive fire ants indirectly improve the competitive ability of a native ant. Ecological Entomology 29:621–627.

Mescher, M. C., K. G. Ross, D. D. Shoemaker, L. Keller, and M. J. B. Krieger. 2003. Distribution of the two social forms of the fire ant *Solenopsis invicta* (Hymenoptera: Formicidae) in the native South American range. Annals of the Entomological Society of America 96:810–817.

Metcalf, R. L. 1982. A brief history of chemical control of the imported fire ant. Pages 122–129 in Proceedings of the Symposium on the Imported Fire Ant, June 7–10, 1982, Atlanta, Georgia. S. L. Battenfield, editor. EPA/USDA (APHIS) 0-389-890/70. U.S. Environmental Protection Agency and U.S. Department of Agriculture, Animal and Plant Health Inspection Service, Washington, D.C.

Michener, C. D. 1964. Reproductive efficiency in relation to colony size in hymenopterous societies. Insectes Sociaux 11:317–341.

Mikheyev, A. S. 2003. Evidence for mating plugs in the fire ant *Solenopsis invicta*. Insectes Sociaux 50:401–402.

Milio, J., C. S. Lofgren, and D. F. Williams. 1988. Nuptial flight studies of field-collected colonies of *Solenopsis invicta* Buren. Pages 419–431 in Advances in Myrmecology. J. C. Trager, editor. E. J. Brill, New York.

Mills, H. B. 1967. Report of committee on imported fire ant to Administrator, ARS, USDA. National Academy of Science. Mimeographed. 15 pp.

Mirenda, J. T., and S. B. Vinson. 1979. A marking technique for adults of the red imported fire ant (Hymenoptera: Formicidae). Florida Entomologist 62:279–281.

Mirenda, J. T., and S. B. Vinson. 1981. Division of labor and specification of castes in the red imported fire ant *Solenopsis invicta* Buren. Animal Behaviour 29:410–420.

Mirenda, J. T., and S. B. Vinson. 1982. Single and multiple queen colonies of imported fire ants in Texas. Southwestern Entomologist 7:135–141.

Montgomery, W. B. 1996. Predation by the fire ant, *Solenopsis invicta,* on the three-toed box turtle, *Terrapene carolina triunguis.* Bulletin of the Chicago Herpetological Society 31:105–106.

Moody, J. V., O. F. Francke, and F. W. Merickel. 1981. The distribution of fire ants, *Solenopsis* (*Solenopsis*) in western Texas (Hymenoptera: Formicidae). Journal of the Kansas Entomological Society 54:469–480.

Morehead, S. A., and D. H. Feener. 1998. Foraging behavior and morphology: seed selection in the harvester ant genus, *Pogonomyrmex.* Oecologia 114:548–555.

Morel, L., and R. K. Vander Meer. 1988. Do ant brood pheromones exist? Annals of the Entomological Society of America 81:705–710.

Morel, L., R. K. Vander Meer, and C. S. Lofgren. 1990. Comparison of nestmate recognition between monogyne and polygyne populations of *Solenopsis invicta* (Hymenoptera: Formicidae). Annals of the Entomological Society of America 83:642–647.

Morrill, W. L. 1974a. Dispersal of red imported fire ants by water. Florida Entomologist 57:39–42.

Morrill, W. L. 1974b. Production and flight of alate red imported fire ants. Environmental Entomology 3:265–271.

Morrill, W. L. 1977a. Overwinter survival of the red imported fire ant in central Georgia. Environmental Entomology 6:50–52.

Morrill, W. L. 1977b. Red imported fire ant foraging in a greenhouse. Environmental Entomology 6:416–418.

Morrill, W. L. 1978. Red imported fire ant predation on the alfalfa weevil and pea aphid. Journal of Economic Entomology 71:867–868.

Morrill, W. L., and J. A. Bass. 1976. Flight and survival of alate red imported fire ants after mirex treatment. Journal of the Georgia Entomological Society 11:203–208.

Morrill, W. L., P. B. Martin, and D. C. Sheppard. 1978. Overwinter survival of the red imported fire ant: effects of various habitats and food supply. Environmental Entomology 7:262–264.

Morris, J. R., and K. L. Steigman. 1993. Effects of polygyne fire ant invasion on native ants of a blackland prairie in Texas. Southwestern Naturalist 38:136–140.

Morris, J. R., K. W. Stewart, and R. L. Hassage. 1990. Use of the nematode *Steinernema carpocapsae* for control of the red imported fire ant (Hymenoptera: Formicidae). Florida Entomologist 73:675–677.

Morrison, J. E., D. F. Williams, D. H. Oi, and K. N. Potter. 1997a. Damage to dry crop seed by red imported fire ant (Hymenoptera: Formicidae). Journal of Economic Entomology 90:218–222.

Morrison, L. W. 1999. Indirect effects of phorid fly parasitoids on the mechanisms of interspecific competition among ants. Oecologia 121:113–122.

Morrison, L. W. 2000a. Mechanisms of interspecific competition among an invasive and two native fire ants. Oikos 90:238–252.

Morrison, L. W. 2000b. Mechanisms of *Pseudacteon* parasitoid (Diptera: Phoridae) effects on exploitative and interference competition in host *Solenopsis* ants (Hymenoptera: Formicidae). Annals of the Entomological Society of America 93:841–849.

Morrison, L. W. 2002. Long-term impacts of the invasion of an arthropod community by the imported fire ant, *Solenopsis invicta*. Ecology 83:2337–2345.

Morrison, L. W., and L. E. Gilbert. 1998. Parasitoid-host relationships when host size varies: the case of *Pseudacteon* flies and *Solenopsis* fire ants. Ecological Entomology 23:409–416

Morrison, L. W., and L. W. Gilbert. 1999. Host specificity in two additional *Pseudacteon* spp. (Diptera: Phoridae), parasitoids of *Solenopsis* fire ants (Hymenoptera: Formicidae). Florida Entomologist 82:404–409.

Morrison, L. W., and J. R. King. 2004. Host location behavior in a parasitoid of imported fire ants. Journal of Insect Behavior 17:367–383.

Morrison, L. W., and S. D. Porter. 2003. Positive association between densities of the red imported fire ant, *Solenopsis invicta* (Hymenoptera: Formicidae), and generalized ant and arthropod diversity. Environmental Entomology 32:548–554.

Morrison, L. W., and S. D. Porter. 2005. Testing for population-level impacts of introduced *Pseudacteon tricuspis* flies, phorid parasitoids of *Solenopsis invicta* fire ants. Biological Control 33:9–19.

Morrison, L. W., C. G. Dall'Agilo-Holvorcem, and L. E. Gilbert. 1997b. Oviposition behavior and development of *Pseudacteon* flies (Diptera: Phoridae), parasitoids of *Solenopsis* fire ants (Hymenoptera: Formicidae). Environmental Entomology 26:716–724.

Morrison, L. W., E. A. Kawazoe, R. Guerra, and L. E. Gilbert. 1999a. Phenology and dispersal in *Pseudacteon* flies (Diptera: Phoridae), parasitoids of *Solenopsis* fire ants (Hymenoptera: Formicidae). Annals of the Entomological Society of America 92:198–207.

Morrison, L. W., S. D. Porter, and L. E. Gilbert. 1999b. Sex ratio variation as a function of host size in *Pseudacteon* flies (Diptera: Phoridae), parasitoids of *Solenopsis* fire ants (Hymenoptera: Formicidae). Biological Journal of the Linnean Society 66:257–267.

Morrison, L. W., E. A. Kawazoe, R. Guerra, and L. E. Gilbert. 2000. Ecological interactions of *Pseudacteon* parasitoids and *Solenopsis* ant hosts: environmental correlates of activity and effects on competitive hierarchies. Ecological Entomology 25:433–444.

Morrison, L. W., S. D. Porter, E. Daniels, and M. D. Korzukhin. 2004. Potential global range expansion of the invasive fire ant, *Solenopsis invicta*. Biological Invasions 6:183–191.

Moser, B., J. Becnel, J. Maruniak, and R. Patterson. 1998. Analysis of the ribosomal DNA sequences of the microsporidia *Thelohania* and *Vairimorpha* of fire ants. Journal of Invertebrate Pathology 72:154–159.

Moser, B. A., J. J. Becnel, and D. F. Williams. 2000. Morphological and molecular characterization of the *Thelohania solenopsae* complex (Microsporidia: *Thelohaniidae*). Journal of Invertebrate Pathology 75:174–177.

Mottern, J. L., K. M. Heinz, and P. J. Ode. 2004. Evaluating biological control of fire ants using phorid flies: effects on competitive interactions. Biological Control 30:566–583.

Moulis, R. A. 1996. Predation by the imported fire ant (*Solenopsis invicta*) on loggerhead sea turtle (*Caretta caretta*) nests on Wassaw National Wildlife Refuge, Georgia. Chelonian Conservation and Biology 2:433–436.

Mount, R. H. 1981. The red imported fire ant, *Solenopsis invicta* (Hymenoptera: Formicidae), as a possible serious predator of some native southeastern vertebrates: direct observations and subjective impressions. Journal of the Alabama Academy of Science 52:71–78.

Mount, R. H., S. E. Trauth, and W. H. Mason. 1981. Predation by the red imported fire ant, *Solenopsis invicta* (Hymenoptera: Formicidae), on eggs of the lizard *Cnemidophorus sexlineatus* (Squamata: Teiidae). Journal of the Alabama Academy of Science 52:66–70.

Mueller, J. M., C. B. Dabbert, S. Demarais, and A. R. Forbes. 1999. Northern bobwhite chick mortality caused by red imported fire ants. Journal of Wildlife Management 63:1291–1298.

Munroe, P. D., H. G. Thorvilson, and S. A. Phillips, Jr. 1996. Comparison of desiccation rates among three species of fire ants. Southwestern Entomologist 21:173–179.

Murphree, L. C. 1947. Alabama Ants: Description, Distribution and Biology, with Notes on the Control of the Most Important Household Species. M.S. thesis, Mississippi State College, Mississippi State, Mississippi. 144 pp.

Naves, M. A. 1985. A monograph of the genus *Pheidole* in Florida (Hymenoptera: Formicidae). Insecta Mundi 1:53–89.

Nelson, D. R., C. L. Fatland, R. W. Howard, C. A. McDaniel, and G. J. Blomquist. 1980. Re-analysis

of the cuticular methylalkanes of *Solenopsis invicta* and *S. richteri*. Insect Biochemistry 10:409–418.
Ness, J. H. 2004. Forest edges and fire ants alter the seed shadow of an ant-dispersed plant. Oecologia 138:448–454.
Newell, W. 1908. Notes on the habits of the Argentine or "New Orleans" ant, *Iridomyrmex humilis* Mayr. Journal of Economic Entomology 1:21–34.
Newsom, J. D. 1958. A preliminary progress report of fire ant eradication program Concordia Parish, Louisiana, June 1958. Pages 29–31 in Twelfth Annual Conference of the Association of Game and Fish Commissioners. Association of Game and Fish Commissioners, Louisville, Kentucky.
Nichols, B. J., and R. W. Sites. 1991. Ant predators of founder queens of *Solenopsis invicta* (Hymenoptera: Formicidae) in central Texas. Environmental Entomology 20:1024–1029.
Nickerson, J. C., W. H. Whitcomb, A. P. Bhatkar, and M. A. Naves. 1975. Predation on founding queens of *Solenopsis invicta* by workers of *Conomyrma insana*. Florida Entomologist 58:75–82.
Nielsen, M. G., T. F. Jensen, and I. Holm Jensen. 1982. Effect of load carriage on the respiratory metabolism of running worker ants of *Camponotus herculeanus* (Formicidae). Oikos 39:137–142.
Nijhout, H. F., and D. E. Wheeler. 1996. Growth models of complex allometries in holometabolous insects. American Naturalist 148:40–56.
Nordheimer, J. 1970. Environmentalists fight U.S. spraying plan on fire ants in South. New York Times, 13 December, p. 78.
Nordlund, D. A. 1988. The imported fire ant and naturally occurring and released beneficial insects. Pages 51–54 in Proceedings of the Governor's Conference on the Imported Fire Ant: Assessment and Recommendations. S. B. Vinson and J. Teer, editors. Sportsmen Conservationists of Texas, Austin, Texas.
Nowbahari, B., R. Feneron, and M. C. Malherbe. 2000. Polymorphism and polyethism in the formicinae ant *Cataglyphis niger* (Hymenoptera). Sociobiology 36:485–496.
Nuessly, G. S. 1986. Mortality of *Heliothis zea* Eggs: Affected by Predator Species, Oviposition Sites, and Rain and Wind Dislodgement. Ph.D. dissertation, Texas A&M University, College Station, Texas. 227 pp.
Nuessly, G. S., and W. L. Sterling. 1986. Distribution of 32P in laboratory colonies of *Solenopsis invicta* (Hymenoptera: Formicidae) after feeding on labeled *Heliothis zea* (Lepidoptera: Noctuidae) eggs: an explanation of discrepancies encountered in field predation experiments. Environmental Entomology 15:1279–1285.

Obin, M. S. 1986. Nestmate recognition cues in laboratory and field colonies of *Solenopsis invicta* Buren (Hymenoptera: Formicidae): effect of environment and the role of cuticular hydrocarbons. Journal of Chemical Ecology 12:1965–1975.
Obin, M. S., and R. K. Vander Meer. 1985. Gaster flagging by fire ants (*Solenopsis* spp.): functional significance of venom dispersal behavior. Journal of Chemical Ecology 11:1757–1768.
Obin, M. S., and R. K. Vander Meer. 1988. Sources of nestmate recognition in the imported fire ant *Solenopsis invicta* Buren (Hymenoptera: Formicidae). Animal Behaviour 36:1361–1370.
Obin, M. S., and R. K. Vander Meer. 1989a. Between- and within-species recognition among imported fire ants and their hybrids (Hymenoptera: Formicidae): application to hybrid zone dynamics. Annals of the Entomological Society of America 82:649–652.
Obin, M. S., and R. K. Vander Meer. 1989b. Mechanism of template-label matching in fire ant, *Solenopsis invicta* Buren, nestmate recognition. Animal Behaviour 38:430–435.
Obin, M. S., and R. K. Vander Meer. 1989c. Nestmate recognition in fire ants (*Solenopsis invicta* Buren). Do queens label workers? Ethology 80:255–264.
Obin, M. S., and R. K. Vander Meer. 1994. Alate semiochemicals release worker behavior during fire ant nuptial flights. Journal of Entomological Science 29:143–151.
Obin, M. S., L. Morel, and R. K. Vander Meer. 1993. Unexpected, well-developed nestmate recognition in laboratory colonies of polygyne imported fire ants (Hymenoptera: Formicidae). Journal of Insect Behavior 6:579–589.
Ofiara, D. D. 1983. Detrimental agricultural impacts of the imported fire ant: a review of the empirical studies. Faculty Service Paper FS83-4, Division of Agricultural Economics, University of Georgia, Athens, Georgia. 62 pp.
Oi, D. H., and D. F. Williams. 2002. Impact of *Thelohania solenopsae* (Microsporidia: Thelohaniidae) on polygyne colonies of red imported fire ants (Hymenoptera: Formicidae). Journal of Economic Entomology 95:558–562.
Oi, D. H., and D. F. Williams. 2003. *Thelohania solenopsae* (Microsporidia: Thelohaniidae) infection in reproductives of red imported fire ants (Hymenoptera: Formicidae) and its implication for intercolony transmission. Environmental Entomology 32:1171–1176.
Oi, D. H., J. J. Becnel, and D. F. Williams. 2001. Evidence of intracolony transmission of *Thelohania solenopsae* (Microsporidia: Thelohaniidae) in red imported fire ants (Hymenoptera: Formicidae) and the first report of spores from pupae. Journal of Invertebrate Pathology 78:128–134.
Oi, D. H., S. M. Valles, and R. M. Pereira. 2004. Prevalence of *Thelohania solenopsae* (Microsporidia:

Thelohaniidae) infection in monogyne and polygyne red imported fire ants (Hymenoptera: Formicidae). Environmental Entomology 33:340–345.

O'Neal, J., and G. P. Markin. 1973. Brood nutrition and parental relationships of the imported fire ant *Solenopsis invicta.* Journal of the Georgia Entomological Society 8:294–303.

O'Neal, J., and G. P. Markin. 1975a. Brood development of the various castes of the imported fire ant, *Solenopsis invicta* Buren (Hymenoptera: Formicidae). Journal of the Kansas Entomological Society 48:152–159.

O'Neal, J., and G. P. Markin. 1975b. The larval instars of the imported fire ant, *Solenopsis invicta* Buren (Hymenoptera: Formicidae). Journal of the Kansas Entomological Society 48:141–151.

Orr, M. R., S. H. Seike, W. W. Benson, and L. E. Gilbert. 1995. Flies suppress fire ants. Nature 373:292–293.

Orrock, J. L., and B. J. Danielson. 2004. Rodents balancing a variety of risks: invasive fire ants and indirect and direct indicators of predation risk. Oecologia 140:662–667.

Oster, G. F., and E. O. Wilson. 1978. Caste and Ecology in the Social Insects. Monographs in Population Biology Number 12. Princeton University Press, Princeton, New Jersey. 352 pp.

Pass, B. C. 1960. Bionomics of the Imported Fire Ant, *Solenopsis saevissima richteri* Forel. M.S. thesis, Auburn University, Auburn, Alabama. 54 pp.

Passera, L., S. Aron, E. L. Vargo, and L. Keller. 2001. Queen control of sex ratio in fire ants. Science 293:1308–1310.

Paull, B. R. 1984. Imported fire ant allergy: perspectives on diagnosis and treatment. Postgraduate Medicine 76:155–160.

Paull, B. R., T. H. Coghlan, and S. B. Vinson. 1983. Fire ant venom hypersensitivity. I. Comparison of fire ant venom and whole body extract in the diagnosis of fire ant allergy. Journal of Allergy and Clinical Immunology 71:448–453.

Paull, B. R., T. H. Coghlan, and S. B. Vinson. 1984. Fire ant venom hypersensitivity. II. Allergenic proteins of venom and whole body extract. Journal of Allergy and Clinical Immunology 73:142.

Pedersen, E. K., W. E. Grant, and M. T. Longnecker. 1996. Effects of red imported fire ants on newly-hatched northern bobwhite. Journal of Wildlife Management 60:164–169.

Pedersen, E. K., T. L. Bedford, W. E. Grant, S. B. Vinson, J. B. Martin, M. T. Longnecker, C. L. Barr, and B. M. Drees. 2003. Effect of red imported fire ants on habitat use by hispid cotton rats (*Sigmodon hispidus*) and northern pygmy mice (*Baiomys taylori*). Southwestern Naturalist 48:419–426.

Peloquin, J. J., and L. Greenberg. 2003. Identification of midgut bacteria from fourth instar red imported fire ant larvae, *Solenopsis invicta* Buren (Hymenoptera: Formicidae). Journal of Agricultural and Urban Entomology 20:157–164.

Pereira, R. M. 2004. Occurrence of *Myrmicinosporidium durum* in red imported fire ant, *Solenopsis invicta,* and other new host ants in eastern United States. Journal of Invertebrate Pathology 86:38–44.

Pereira, R. M., D. F. Williams, J. J. Becnel, and D. H. Oi. 2002. Yellow-head disease caused by a newly discovered *Mattesia* sp in populations of the red imported fire ant, *Solenopsis invicta.* Journal of Invertebrate Pathology 81:45–48.

Perfecto, I. 1991. Dynamics of *Solenopsis geminata* in a tropical fallow field after ploughing. Oikos 62:139–144.

Pesquero, M. A. 1993. A organização espacial de assembleias de formigas em areas de pastagem e duna. M.S. thesis, Universidade Estadual Paulista, Botucatu, Brazil.

Pesquero, M. A. 2000. Two new species of *Pseudacteon* (Diptera: Phoridae) parasitoids of fire ants (*Solenopsis* spp.) (Hymenoptera: Formicidae) from Brazil. Journal of the New York Entomological Society 108:243–247.

Pesquero, M. A., S. Campiolo, H. G. Fowler, and S. D. Porter. 1996. Diurnal patterns of ovipositional activity in two *Pseudacteon* fly parasitoids (Diptera: Phoridae) of *Solenopsis* fire ants (Hymenoptera: Formicidae). Florida Entomologist 79:455–457.

Petralia, R. S., and S. B. Vinson. 1978. Feeding in the larvae of the imported fire ant, *Solenopsis invicta:* behavior and morphological adaptations. Annals of the Entomological Society of America 71:643–648.

Petralia, R. S., and S. B. Vinson. 1979a. Comparative anatomy of the ventral region of ant larvae, and its relation to feeding behavior. Psyche 86:375–394.

Petralia, R. S., and S. B. Vinson. 1979b. Developmental morphology of larvae and eggs of the imported fire ant, *Solenopsis invicta.* Annals of the Entomological Society of America 72:iii, 472–484.

Petralia, R. S., and S. B. Vinson. 1980. Internal anatomy of the fourth instar larva of the imported fire ant, *Solenopsis invicta* Buren (Hymenoptera: Formicidae). International Journal of Insect Morphology and Embryology 9:89–106.

Petralia, R. S., A. A. Sorensen, and S. B. Vinson. 1980. The labial gland system of larvae of the imported fire ant, *Solenopsis invicta* Buren: ultrastructure and enzyme analysis. Cell and Tissue Research 206:145–156.

Petralia, R. S., H. J. Williams, and S. B. Vinson. 1982. The hindgut ultrastructure, and excretory products of larvae of the imported fire ant, *Solenopsis invicta* Buren. Insectes Sociaux 29:332–345.

Phillips, S. A., Jr., and J. C. Cokendolpher. 1988. Environmental limitation on the imported fire ant and the ants' response to environmental changes. Pages 11–17 in Proceedings of the Governor's Conference on the Imported Fire Ant: Assessment and Recommendations. S. B. Vinson and J. Teer, editors. Sportsmen Conservationists of Texas, Austin, Texas.

Phillips, S. A., Jr., and S. B. Vinson. 1980a. Comparative morphology of glands associated with the head among castes of the red imported fire ant, *Solenopsis invicta* Buren. Journal of the Georgia Entomological Society 15:215–226.

Phillips, S. A., Jr., and S. B. Vinson. 1980b. Source of the post-pharyngeal gland contents in the red imported fire ant, *Solenopsis invicta*. Annals of the Entomological Society of America 73:257–261.

Phillips, S. A., Jr., S. R. Jones, and D. M. Claborn. 1986a. Temporal foraging patterns of Solenopsis invicta and native ants of central Texas. Pages 114–122 in Fire Ants and Leafcutting Ants: Biology and Management. C. S. Lofgren and R. K. Vander Meer, editors. Westview Press, Boulder, Colorado.

Phillips, S. A., Jr., S. R. Jones, D. M. Claborn, and J. C. Cokendolpher. 1986b. Effect of Pro-Drone, an insect growth regulator, on *Solenopsis invicta* Buren and nontarget ants. Southwestern Entomologist 11:287–293.

Phillips, S. A., Jr., W. M. Rogers, D. B. Wester, and L. Chandler. 1987. Ordination analysis of ant faunae along the range expansion front of the red imported fire ant in south-central Texas. Texas Journal of Agricultural and Natural Resources 1:11–15.

Phillips, S. A., Jr., R. Jusino-Atresino, and H. G. Thorvilson. 1996. Desiccation resistance in populations of the red imported fire ant (Hymenoptera: Formicidae). Environmental Entomology 25:460–464.

Pimentel, D. 1955. Relationship of ants to fly control in Puerto Rico. Journal of Economic Entomology 48:28–30.

Pimm, S. L., and D. P. Bartell. 1980. Statistical model for predicting range expansion of the red imported fire ant, *Solenopsis invicta,* in Texas. Environmental Entomology 9:653–658.

Pitts, J. P. 2002. A Cladistic Analysis of the *Solenopsis saevissima* Species-group. Ph.D. dissertation, University of Georgia, Athens, Georgia 274 pp.

Pitts, J. P., and T. L. Pitts-Singer. 2001. A new host record for *Pseudacteon crawfordi* (Diptera: Phoridae). Florida Entomologist 84:310.

Porter, S. D. 1983. Fast, accurate method of measuring ant headwidths. Annals of the Entomological Society of America 72:472–484.

Porter, S. D. 1984. Fire Ant Polymorphism: Ergonomics of Brood Production. Ph.D. dissertation, Florida State University, Tallahassee, Florida. 83 pp.

Porter, S. D. 1988. Impact of temperature on colony growth and developmental rates of the ant, *Solenopsis invicta.* Journal of Insect Physiology 34:1127–1133.

Porter, S. D. 1989. Effects of diet on the growth of laboratory fire ant colonies (Hymenoptera: Formicidae). Journal of the Kansas Entomological Society 62:288–291.

Porter, S. D. 1990. Thermoregulation in the fire ant, *Solenopsis invicta.* Page 660 in Social Insects and the Environment: Proceedings of the 11th International Congress of IUSSI. G. K. Veeresh, B. Mallik, and C. A. Viraktamath, editors. E. J. Brill, New York.

Porter, S. D. 1991. Origins of new queens in polygyne red imported fire ant colonies (Hymenoptera: Formicidae). Journal of Entomological Science 26:474–478.

Porter, S. D. 1992. Frequency and distribution of polygyne fire ants (Hymenoptera: Formicidae) in Florida. Florida Entomologist 75:248–257.

Porter, S. D. 1993. Stability of polygyne and monogyne fire ant populations (Hymenoptera: Formicidae: *Solenopsis invicta*) in the United States. Journal of Economic Entomology 86:1344–1347.

Porter, S. D. 1994. Fire ant thermoregulation. Page 69 in Les Insectes Sociaux. A. Lenoir, G. Arnold, and M. Lepage, editors. Publications Université Paris Nord, Villetaneuse, France.

Porter, S. D. 1998a. Biology and behavior of *Pseudacteon* decapitating flies (Diptera: Phoridae) that parasitize *Solenopsis* fire ants (Hymenoptera: Formicidae). Florida Entomologist 81:292–309.

Porter, S. D. 1998b. Host-specific attraction of *Pseudacteon* flies (Diptera: Phoridae) to fire ant colonies in Brazil. Florida Entomologist 81:423–429.

Porter, S. D. 2000. Host specificity and risk assessment of releasing the decapitating fly *Pseudacteon* curvatus as a classical biocontrol agent for imported fire ants. Biological Control 19:35–47.

Porter, S. D., and L. E. Alonso. 1999. Host specificity of fire ant decapitating flies (Diptera: Phoridae) in laboratory oviposition tests. Journal of Economic Entomology 92:110–114.

Porter, S. D., and J. Briano. 2000. Parasitoid-host matching between the little decapitating fly *Pseudacteon curvatus* from Las Flores, Argentina and the black fire ant *Solenopsis richteri.* Florida Entomologist 83:422–427.

Porter, S. D., and C. D. Jorgensen. 1981. Foragers of the harvester ant, *Pogonomyrmex owyheei:* a disposable caste? Behavioral Ecology and Sociobiology 9:247–256.

Porter, S. D., and M. A. Pesquero. 2001. Illustrated key to *Pseudacteon* decapitating flies (Diptera: Phoridae) that attack *Solenopsis saevissima* complex fire ants in South America. Florida Entomologist 84:691–699.

Porter, S. D., and D. A. Savignano. 1990. Invasion of polygyne fire ants decimates native ants and disrupts arthropod community. Ecology 71:2095–2106.

Porter, S. D., and W. R. Tschinkel. 1985a. Fire ant polymorphism (Hymenoptera: Formicidae): factors affecting worker size. Annals of the Entomological Society of America 78:381–386.

Porter, S. D., and W. R. Tschinkel. 1985b. Fire ant polymorphism: the ergonomics of brood production. Behavioral Ecology and Sociobiology 16:323–336.

Porter, S. D., and W. R. Tschinkel. 1986. Adaptive value of nanitic workers in newly founded red imported fire ant colonies (Hymenoptera: Formicidae). Annals of the Entomological Society of America 79:723–726.

Porter, S. D., and W. R. Tschinkel. 1987. Foraging in *Solenopsis invicta* (Hymenoptera: Formicidae): effects of weather and season. Environmental Entomology 16:802–808.

Porter, S. D., and W. R. Tschinkel. 1993. Fire ant thermal preferences: behavioral control of growth and metabolism. Behavioral Ecology and Sociobiology 32:321–329.

Porter, S. D., B. Van Eimeren, and L. E. Gilbert. 1988. Invasion of red imported fire ants (Hymenoptera: Formicidae): microgeography of competitive replacement. Annals of the Entomological Society of America 81:913–918.

Porter, S. D., H. G. Fowler, and W. P. MacKay. 1990. A comparison of fire ant population densities in North and South America. Pages 623–624 in Social Insects and the Environment: Proceedings of the 11th International Congress of IUSSI. G. K. Veeresh, B. Mallik, and C. A. Viraktamath, editors. E. J. Brill, New York.

Porter, S. D., A. P. Bhatkar, R. Mulder, S. B. Vinson, and D. J. Clair. 1991. Distribution and density of polygyne fire ants (Hymenoptera: Formicidae) in Texas. Journal of Economic Entomology 84:866–874.

Porter, S. D., H. G. Fowler, and W. P. MacKay. 1992. Fire ant mound densities in the United States and Brazil (Hymenoptera: Formicidae). Journal of Economic Entomology 85:1154–1161.

Porter, S. D., H. G. Fowler, S. Campiolo, and M. A. Pesquero. 1995a. Host specificity of several *Pseudacteon* (Diptera: Phoridae) parasites of fire ants (Hymenoptera: Formicidae) in South America. Florida Entomologist 78:70–75.

Porter, S. D., M. A. Pesquero, S. Campiolo, and H. G. Fowler. 1995b. Growth and development of *Pseudacteon* phorid fly maggots (Diptera: Phoridae) in the heads of *Solenopsis* fire ant workers (Hymenoptera: Formicidae). Environmental Entomology 24:474–479.

Porter, S. D., R. K. Vander Meer, M. A. Pesquero, S. Campiolo, and H. G. Fowler. 1995c. *Solenopsis* (Hymenoptera: Formicidae) fire ant reactions to attacks of *Pseudacteon* flies (Diptera: Phoridae) in southeastern Brazil. Annals of the Entomological Society of America 88:570–575.

Porter, S. D., D. F. Williams, R. S. Patterson, and H. G. Fowler. 1997. Intercontinental differences in the abundance of *Solenopsis* fire ants (Hymenoptera: Formicidae): escape from natural enemies? Environmental Entomology 26:373–384.

Rao, A. 2002. Invasion of red imported fire ant nest by selected predatory ants: prospects of utilizing native ants in fire ant management. Southwestern Entomologist Supplement 25:61–70.

Ratnieks, F. L. W., and L. Keller. 1998. Queen control of egg fertilization in the honey bee. Behavioral Ecology and Sociobiology 44:57–61.

Read, G. W., N. K. Lind, and C. S. Oda. 1978. Histamine release by fire ant (*Solenopsis*) venom. Toxicon 16:361–367.

Ready, C. C., and S. B. Vinson. 1995. Seed selection by the red imported fire ant (Hymenoptera: Formicidae) in the laboratory. Environmental Entomology 24:1422–1431.

Reagan, T. E. 1982. Sugarcane borer pest management in Louisiana: leading to a more permanent system. Pages 100–110 (discussion pp. 152–158; Spanish version, pp. 320–330, 375–382) in Proceedings, Second Inter-American Sugar Cane Seminar. Inter-American Transport Equipment Company, Miami, Florida.

Reagan, T. E. 1986. Beneficial aspects of the imported fire ant: a field ecology approach. Pages 58–71 in Fire Ants and Leafcutting Ants: Biology and Management. C. S. Lofgren and R. K. Vander Meer, editors. Westview Press, Boulder, Colorado.

Reagan, T. E., G. Coburn, and S. D. Hensley. 1972. Effects of mirex on the arthropod fauna of a Louisiana sugarcane field. Environmental Entomology 1:588–591.

Reichensperger, A. 1927a. Eigenartiger Nestbefund und neue Gastarten neotropischer *Solenopsis*-Arten. Folia Myrmecologica et Termitologica 4:48–51.

Reichensperger, A. 1927b. Neue Myrmekophilen nebst einigen Bemerkungun zu Bekannten (Coleoptera, Paus., Clavig. Hist). Tijdschrift voor Entomologie 70:303–311.

Reilly, J. J., and W. L. Sterling. 1983a. Dispersion patterns of the red imported fire ant (Hymenoptera: Formicidae), aphids, and some predaceous insects in East Texas cotton fields. Environmental Entomology 12:380–385.

Reilly, J. J., and W. L. Sterling. 1983b. Interspecific association between red imported fire ant (Hymenoptera: Formicidae), aphids, and some predaceous insects in cotton agroecosystem. Environmental Entomology 12:541–545.

Rhoades, R. B. 1977. Medical Aspects of the Imported Fire Ant. University Presses of Florida, Gainesville, Florida. 75 pp.

Rhoades, R. B., W. L. Schafer, M. Newman, R. Lockey, R. M. Dozier, P. F. Wubbena, A. W. Townes, W. H. Schmid, G. Neder, T. Brill, and H. J. Wittig. 1977. Hypersensitivity to the imported fire ant in Florida:

report of 104 cases. Journal of the Florida Medical Association 64:247–254.

Rhoades, R. B., D. Kalof, F. Bloom, and H. J. Wittig. 1978. Cross-reacting antigens between imported fire ants and other Hymenoptera species. Annals of Allergy 40:100–104.

Rhoades, R. B., W. L. Schafer, W. H. Schmid, P. F. Wubbena, R. M. Dozier, A. W. Townes, and H. J. Wittig. 1975. Hypersensitivity to the imported fire ant: a report of 49 cases. Journal of Allergy and Clinical Immunology 56:84–93.

Rhoades, R. B., C. T. Stafford, and F. K. James, Jr. 1989. Survey of fatal anaphylactic reactions to imported fire ants stings. Journal of Allergy and Clinical Immunology 84:159–162.

Rhoades, W. C. 1962. A synecological study of the effects of the imported fire ant eradication program. I. Alcohol pitfall method of collecting. Florida Entomologist 45:161–173.

Rhoades, W. C. 1963. A synecological study of the effects of the imported fire ant eradication program. II. Light trap, soil sample, litter sample and sweep net methods of collecting. Florida Entomologist 46:301–310.

Rhoades, W. C., and D. R. Davis. 1967. Effects of meteorological factors on the biology and control of the imported fire ant. Journal of Economic Entomology 60:554–558.

Rhoades, W. C., and R. W. Murray. 1967. A Synecological Study of the Imported Fire Ant Eradication Program. Florida Agricultural Experiment Station Bulletin 720. 42 pp.

Ricks, B. L., and S. B. Vinson. 1972a. Changes in nutrient content during one year in workers of the imported fire ant. Annals of the Entomological Society of America 65:135–138.

Ricks, B. L., and S. B. Vinson. 1972b. Digestive enzymes of the imported fire ant, *Solenopsis richteri* (Hymenoptera: Formicidae). Entomologia Experimentalis et Applicata 15:329-334.

Ridlehuber, K. T. 1982. Fire ant predation on wood duck ducklings and pipped eggs. Southwestern Naturalist 27:222.

Risch, S. J., and C. R. Carroll. 1982a. The ecological role of ants in two Mexican agroecosystems. Oecologia 55:114–119.

Risch, S. J., and C. R. Carroll. 1982b. Effect of a keystone predaceous ant, *Solenopsis geminata,* on arthropods in a tropical agroecosystem. Ecology 63:1979–1983.

Risch, S. J., and C. R. Carroll. 1986. Effects of seed predation by a tropical ant on competition among weeds. Ecology 67:1319–1327.

Rocca, J. R., J. H. Tumlinson, B. M. Glancey, and C. S. Lofgren. 1983a. The queen recognition pheromone of *Solenopsis invicta,* preparation of (E)-6-(1-pentenyl)-2H-pyran-2-one. Tetrahedron Letters 24:1889–1892.

Rocca, J. R., J. H. Tumlinson, B. M. Glancey, and C. S. Lofgren. 1983b. Synthesis and stereochemistry of tetrahydro-3,5-dimethyl-6-(1-methylbutyl)-2H-pyran-2-one, a component of the queen recognition pheromone of *Solenopsis invicta.* Tetrahedron Letters 24:1893–1896.

Rojas, J. C., Y. Brindis, E. A. Malo, and L. Cruz-Lopez. 2004. Influence of queen weight and colony origin on worker response in *Solenopsis geminata.* Physiological Entomology 29:356–362.

Rosene, W. J. 1958. Whistling cock counts of bobwhite quail on areas treated with insecticide and untreated areas, Decatur County, Georgia. Pages 14–18 in Twelfth Annual Conference of the Association of Game and Fish Commissioners. Association of Game and Fish Commissioners, Louisville, Kentucky.

Rosenzweig, M. L. 2001. The four questions: what does the introduction of exotic species do to diversity? Evolutionary Ecology Research 3:361–367.

Ross, K. G. 1988. Differential reproduction in multiple-queen colonies of the fire ant *Solenopsis invicta* (Hymenoptera: Formicidae). Behavioral Ecology and Sociobiology 23:341–355.

Ross, K. G. 1989. Reproductive and social structure in polygynous fire ant colonies. Pages 149–162 in The Genetics of Social Evolution. M. D. Breed and R. E. Page, Jr., editors. Westview Press, Boulder, Colorado.

Ross, K. G. 1992. Strong selection on a gene that influences reproductive competition in a social insect. Nature 355:347–349.

Ross, K. G. 1993. The breeding system of the fire ant *Solenopsis invicta,* and its effects on colony genetic structure. American Naturalist 141:554–576.

Ross, K. G. 1997. Multilocus evolution in fire ants: effects of selection, gene flow and recombination. Genetics 145:961–974.

Ross, K. G., and D. J. C. Fletcher. 1985a. Comparative study of genetic and social structure in two forms of the fire ant *Solenopsis invicta* (Hymenoptera: Formicidae). Behavioral Ecology and Sociobiology 17:349–356.

Ross, K. G., and D. J. C. Fletcher. 1985b. Genetic origin of male diploidy in the fire ant, *Solenopsis invicta* (Hymenoptera: Formicidae), and its evolutionary significance. Evolution 39:888–903.

Ross, K. G., and D. J. C. Fletcher. 1986. Diploid male production—a significant colony mortality factor in the fire ant *Solenopsis invicta* (Hymenoptera: Formicidae). Behavioral Ecology and Sociobiology 19:283–291.

Ross, K. G., and L. Keller. 1995a. Ecology and evolution of social organization: insights from fire ants and other highly eusocial insects. Annual Review of Ecology and Systematics 26:631–656.

Ross, K. G., and L. Keller. 1995b. Joint influence of gene flow and selection on a reproductively important genetic polymorphism in the fire ant *Solenopsis invicta.* American Naturalist 146:325–348.
Ross, K. G., and L. Keller. 1998. Genetic control of social organization in an ant. Proceedings of the National Academy of Sciences of the USA 95:14232–14237.
Ross, K. G., and L. Keller. 2002. Experimental conversion of colony social organization by manipulation of worker genotype composition in fire ants *Solenopsis invicta.* Behavioral Ecology and Sociobiology 51:287–295.
Ross, K. G., and J. L. Robertson. 1990. Developmental stability, heterozygosity, and fitness in two introduced fire ants (*Solenopsis invicta* and *S. richteri*) and their hybrid. Heredity 64:93–103.
Ross, K. G., and D. D. Shoemaker. 1993. An unusual pattern of gene flow between the two social forms of the fire ant *Solenopsis invicta.* Evolution 47:1595–1605.
Ross, K. G., and D. D. Shoemaker. 1997. Nuclear and mitochondrial genetic structure in two social forms of the fire ant *Solenopsis invicta:* insights into transitions to an alternate social organization. Heredity 78:590–602.
Ross, K. G., and D. D. Shoemaker. 2005. Species Delimitation in native South American fire ants. Molecular Ecology, in press.
Ross, K. G., and J. C. Trager. 1990. Systematics and population genetics of fire ants (*Solenopsis saevissima* complex) from Argentina. Evolution 44:2113–2134.
Ross, K. G., R. K. Vander Meer, D. J. C. Fletcher, and E. L. Vargo. 1987. Biochemical phenotypic and genetic studies of two introduced fire ants and their hybrid (Hymenoptera: Formicidae). Evolution 41:280–293.
Ross, K. G., E. L. Vargo, L. Keller, and J. C. Trager. 1993. Effect of a founder event on variation in the genetic sex-determining system of the fire ant *Solenopsis invicta.* Genetics 135:843–854.
Ross, K. G., E. L. Vargo, and L. Keller. 1996a. Simple genetic basis for important social traits in the fire ant *Solenopsis invicta.* Evolution 50:2387–2399.
Ross, K. G., E. L. Vargo, and L. Keller. 1996b. Social evolution in a new environment: the case of introduced fire ants. Proceedings of the National Academy of Sciences of the USA 93:3021–3025.
Ross, K. G., M. J. B. Krieger, D. D. Shoemaker, E. L. Vargo, and L. Keller. 1997. Hierarchical analysis of genetic structure in native fire ant populations: results from three classes of molecular markers. Genetics 147:643–655.
Ross, K. G., D. D. Shoemaker, M. J. B. Krieger, C. J. DeHeer, and L. Keller. 1999. Assessing genetic structure with multiple classes of molecular markers: a case study involving the introduced fire ant *Solenopsis invicta.* Molecular Biology and Evolution 16:525–543.
Ross, K. G., M. J. B. Krieger, and D. D. Shoemaker. 2003. Alternative genetic foundations for a key social polymorphism in fire ants. Genetics 165:1853–1867.
Ruckelshaus, D. 1973. Mirex statement of issues. Federal Register 38:8616.
Russell, S. A., H. G. Thorvilson, and S. A. J. Phillips. 2001. Red imported fire ant (Hymenoptera: Formicidae) populations in Texas highway rights-of-way and adjacent pastures. Environmental Entomology 30:267–273.

Sauter, A., M. J. F. Brown, B. Baer, and P. Schmid-Hempel. 2001. Males of social insects can prevent queens from multiple mating. Proceedings of the Royal Society of London B Biological Science 268:1449–1454.
Schmid-Hempel, P. 1992. Worker castes and adaptative demography. Journal of Evolutionary Biology 5:1–12.
Schmid-Hempel, P. 1998. Parasites in Social Insects. Princeton University Press, Princeton, New Jersey. 409 pp.
Schmidt, J. O. 1990. Hymenopteran venoms: striving toward the ultimate defense against vertebrates. Pages 387–419 in Insect Defenses: Adaptive Mechanisms and Strategies of Prey and Predators. D. L. Evans and J. O. Schmidt, editors. State University of New York Press, Albany, New York.
Schoor, W. P. 1974. Accumulation of mirex-14C in the adult blue crab (*Callinectes sapidus*). Bulletin of Environmental Contamination and Toxicology 12:136–137.
Seagraves, M. P., R. M. McPherson, and J. R. Ruberson. 2004. Impact of *Solenopsis invicta* Buren suppression on arthropod ground predators and pest species in soybean. Journal of Entomological Science 39:433–443.
Seaman, R. E., and P. C. Marino. 2003. Influence of mound building and selective seed predation by the red imported fire ant (*Solenopsis invicta*) on an old-field plant assemblage. Journal of the Torrey Botanical Society 130:193–201.
Semenov, S. M., F. N. Semevski, and L. C. Thompson. 1997. Detecting imported fire ant (Hymenoptera: Formicidae) effects on crop production using a statistical model. Russian Entomological Journal 6:153–160.
Shapiro, A. M., J. J. Becnel, D. H. Oi, and D. F. Williams. 2003. Ultrastructural characterization and further transmission studies of *Thelohania solenopsae* from *Solenopsis invicta* pupae. Journal of Invertebrate Pathology 83:177–180.
Shapley, D. 1971a. Fire ant control under fire. Science 171:1131.
Shapley, D. 1971b. Mirex and the fire ant: decline in fortunes of "perfect" pesticide. Science 172:358–360.
Shatters, R. G. J., and R. K. Vander Meer. 2000. Characterizing the interaction between fire ants

(Hymenoptera: Formicidae) and developing soybean plants. Journal of Economic Entomology 93:1680–1687.

Shoemaker, D. D., and K. G. Ross. 1996. Effects of social organization on gene flow in the fire ant *Solenopsis invicta.* Nature 383:613–616.

Shoemaker, D. D., K. G. Ross, and M. L. Arnold. 1996. Genetic structure and evolution of a fire ant hybrid zone. Evolution 50:1958–1976.

Shoemaker, D. D., K. G. Ross, L. Keller, E. L. Vargo, and J. H. Werren. 2000. *Wolbachia* infections in native and introduced populations of fire ants (*Solenopsis* spp.). Insect Molecular Biology 9:661–673.

Shoemaker, D. D., M. Ahrens, L. Sheill, M. Mescher, L. Keller, and K. G. Ross. 2003. Distribution and prevalence of *Wolbachia* infections in native populations of the fire ant *Solenopsis invicta* (Hymenoptera: Formicidae). Environmental Entomology 32:1329–1336.

Showler, A. T., and T. E. Reagan. 1987. Ecological interactions of the red imported fire ant in the southeastern United States. Journal of Entomological Science Supplement 1:52–64.

Showler, A. T., and T. E. Reagan. 1991. Effects of sugarcane borer, weed, and nematode control strategies in Louisiana sugarcane. Environmental Entomology 20:358–370.

Showler, A. T., R. M. Knaus, and T. E. Reagan. 1989. Foraging territoriality of the imported fire ant, *Solenopsis invicta* Buren, in sugarcane as determined by neutron activation analysis. Insectes Sociaux 36:235–239.

Siebeneicher, S. R., S. B. Vinson, and C. M. Kenerley. 1992. Infection of the red imported fire ant by *Beauveria bassiana* through various routes of exposure. Journal of Invertebrate Pathology 59:280–285.

Sikes, P. J., and K. A. Arnold. 1986. Red imported fire ant (*Solenopsis invicta*) predation on cliff swallow (*Hirundo pyrrhonata*) nestlings in east-central Texas. Southwestern Naturalist 31:105–106.

Silveira-Guido, A., J. Carbonell, and C. Crisci. 1973. Animals associated with the *Solenopsis* (fire ants) complex, with special reference to *Labauchena daguerrei.* Proceedings Tall Timbers Conference on Ecology and Animal Control by Habitat Management 4: 41–52.

Skinner, G. J. 1980. The feeding habits of the wood-ant, *Formica rufa* (Hymenoptera: Formicidae), in limestone woodland in north-west England. Journal of Animal Ecology 49:417–433.

Smith, M. R. 1928. *Plastophora crawfordi* Coq. and *Plastophora spatulata* Malloch (Diptera: Phoridae), parasitic on *Solenopsis geminata* Fabr. Proceedings of the Entomological Society of Washington 30:105–108.

Smith, M. R. 1936. Consideration of the fire ant *Solenopsis xyloni* as an important southern pest. Journal of Economic Entomology 29:120–122.

Smith, T. S., S. A. Smith, and D. J. Smidly. 1990. Impact of fire ant (*Solenopsis invicta*) density on northern pygmy mice (*Baiomys taylori*). Southwestern Naturalist 35:158–162.

Smittle, B. J., C. T. Adams, and C. S. Lofgren. 1983. Red imported fire ants: detection of feeding on corn, okra and soybeans with radioisotopes. Journal of the Georgia Entomological Society 18:78–82.

Smittle, B. J., C. T. Adams, W. A. Banks, and C. S. Lofgren. 1988. Red imported fire ants: feeding on radiolabeled citrus trees. Journal of Economic Entomology 81:1019–1021.

Snelling, R. R. 1963. The United States species of fire ants of the genus *Solenopsis,* subgenus *Solenopsis* Westwood, with synonymy of *Solenopsis aurea* (Wheeler) (Hymenoptera: Formicidae). Bureau of Entomology Occasional Paper Number 3. California Department of Agriculture, Sacramento, California. 15 pp.

Snowden, K. F., K. S. Logan, and S. B. Vinson. 2002. Simple, filter-based PCR detection of *Thelohania solenopsae* (Microspora) in fire ants (*Solenopsis invicta*). Journal of Eukaryotic Microbiology 49:447–448.

Sokolova, Y. Y., L. R. McNally, J. R. Fuxa, and S. B. Vinson. 2004. Spore morphotypes of *Thelohania solenopsae* (Microsporidia) described microscopically and confirmed by PCR of individual spores microdissected from smears by position ablative laser microbeam microscopy. Microbiology-SGM 150:1261–1270.

Sonnet, P. E. 1967. Fire ant venom: synthesis of a reported component of solenamine. Science 156:1759–1760.

Sorensen, A. A., and S. B. Vinson. 1981. Quantitative food distribution studies within laboratory colonies of the imported fire ant, *Solenopsis invicta* Buren. Insectes Sociaux 28:129–160.

Sorensen, A. A., and S. B. Vinson. 1985. Behavior of temporal subcastes of the fire ant, *Solenopsis invicta,* in response to oil (Hymenoptera: Formicidae). Journal of the Kansas Entomological Society 58:586–596.

Sorensen, A. A., R. Kamas, and S. B. Vinson. 1980. The biological half life and distribution of ^{125}iodine and radioiodinated protein in imported fire ant, *Solenopsis invicta.* Entomologia Experimentalis et Applicata 28:247–258.

Sorensen, A. A., J. T. Mirenda, and S. B. Vinson. 1981. Food exchange and distribution by three functional worker groups of the imported fire ant *Solenopsis invicta* Buren. Insectes Sociaux 28:383–394.

Sorensen, A. A., T. M. Busch, and S. B. Vinson. 1983a. Behavior of worker subcastes in the fire ant, *Solenopsis invicta,* in response to proteinaceous food. Physiological Entomology 8:83–92.

Sorensen, A. A., T. M. Busch, and S. B. Vinson. 1983b. Factors affecting brood cannibalism in laboratory colonies of the imported fire ant, *Solenopsis invicta* Buren (Hymenoptera: Formicidae). Journal of the Kansas Entomological Society 56:140–150.

Sorensen, A. A., R. S. Kamas, and S. B. Vinson. 1983c. The influence of oral secretions from larvae on levels of proteinases in colony members of *Solenopsis invicta* Buren (Hymenoptera: Formicidae). Journal of Insect Physiology 29:163–168.

Sorensen, A. A., T. M. Busch, and S. B. Vinson. 1984. Behavioral flexibility of temporal subcastes in the fire ant, *Solenopsis invicta* in response to food. Psyche 91:319–331.

Sorensen, A. A., T. M. Busch, and S. B. Vinson. 1985a. Control of food influx by temporal subcastes in the fire ant, *Solenopsis invicta.* Behavioral Ecology and Sociobiology 17:191–198.

Sorensen, A. A., T. M. Busch, and S. B. Vinson. 1985b. Trophallaxis by temporal subcastes in the fire ant, *Solenopsis invicta,* in response to honey. Physiological Entomology 10:105–111.

Sorensen, A. A., D. J. C. Fletcher, and S. B. Vinson. 1985c. Distribution of inhibitory queen pheromone among virgin queens of an ant, *Solenopsis invicta.* Psyche 92:57–69.

Soroker, V., C. Vienne, A. Hefetz, and B. Nowbahari. 1994. The postpharyngeal gland as a "gestalt" organ for nestmate recognition in the ant *Cataglyphis niger.* Naturwissenschaften 81:510–513.

Stafford, C. T. 1996. Hypersensitivity to fire ant venom. Annals of Allergy, Asthma and Immunology 77:87–95.

Stein, M. B., and H. G. Thorvilson. 1989. Ant species sympatric with the red imported fire ant in southeastern Texas. Southwestern Entomologist 14:225–231.

Stein, M. B., H. G. Thorvilson, and J. W. Johnson. 1990. Seasonal changes in bait preference by red imported fire ant, *Solenopsis invicta* (Hymenoptera: Formicidae). Florida Entomologist 73:117–123.

Sterling, W. L. 1978. Fortuitous biological suppression of the boll weevil by the red imported fire ant. Environmental Entomology 7:564–568.

Sterling, W. L., D. Jones, and D. A. Dean. 1979. Failure of the red imported fire ant to reduce entomophagous insect and spider abundance in a cotton agroecosystem. Environmental Entomologist 8:976–981.

Sterling, W. L., D. A. Dean, D. A. Fillman, and D. Jones. 1984. Naturally-occurring biological control of the boll weevil [Col.: Curculionidae]. Entomophaga 29:1–9.

Stiles, J. H., and R. H. Jones. 1998. Distribution of the red imported fire ant, *Solenopsis invicta,* in road and powerline habitats. Landscape Ecology 13:335–346.

Stiles, J. H., and R. H. Jones. 2001. Top-down control by the red imported fire ant (*Solenopsis invicta*). American Midland Naturalist 146:171–185.

Stille, M. 1996. Queen/worker thorax volume ratios and nest-founding strategies in ants. Oecologia 105:87–93.

Stimac, J. L., and S. B. Alves. 1994. Ecology and biological control of fire ants. Pages 353–380 in Pest Management in the Subtropics, Biological Control—A Florida Perspective. D. Rosen, F. D. Bennett and J. C. Capineram, editors. Intercept Ltd., Andover, New Hampshire.

Stimac, J. L., R. M. Pereira, S. B. Alves, and L. A. Wood. 1993a. *Beauveria bassiana* (Balsamo) Vuillemin (Deuteromycetes) applied to laboratory colonies of *Solenopsis invicta* Buren (Hymenoptera: Formicidae) in soil. Journal of Economic Entomology 86:348–352.

Stimac, J. L., R. M. Pereira, S. B. Alves, and L. A. Wood. 1993b. Mortality in laboratory colonies of *Solenopsis invicta* (Hymenoptera: Formicidae) treated with *Beauveria bassiana* (Deuteromycetes). Journal of Economic Entomology 86:1083–1087.

Stoddard, H. L. 1931. The Bobwhite Quail, Its Habits, Preservation and Increase. Scribner, New York. 559 pp.

Stoker, R. L., D. K. Ferris, W. E. Grant, and J. Folse. 1994. Simulating colonization by exotic species: a model of the red imported fire ant (*Solenopsis invicta*) in North America. Ecological Modelling 73:281–292.

Stoker, R. L., W. E. Grant, and S. B. Vinson. 1995. *Solenopsis invicta* (Hymenoptera: Formicidae) effect on invertebrate decomposers of carrion in central Texas. Environmental Entomology 24: 817–822.

Storey, G. K. 1990. Chemical Defenses of the Fire Ant, *Solenopsis invicta* Buren, Against Infection by the Fungus, *Beauvaria bassiana* (Balsamo) Vuill. Ph.D. dissertation, University of Florida, Gainesville, Florida. 76 pp.

Storey, G. K., R. K. Vander Meer, D. G. Boucias, and C. W. McCoy. 1991. Effect of fire ant (*Solenopsis invicta*) venom alkaloids on the in vitro germination and development of selected entomogenous fungi. Journal of Invertebrate Pathology 58:88–95.

Stringer, C. E., Jr., B. M. Glancey, P. M. Bishop, C. H. Craig, and B. B. Martin. 1972. Air separation of different castes of the imported fire ant. Journal of Economic Entomology 65:872–873.

Stringer, C. E., Jr., B. M. Glancey, and B. B. Martin. 1973. A simple method for separating alate imported fire ants from workers and soil. Journal of Economic Entomology 66:295.

Stringer, C. E., Jr., W. A. Banks, B. M. Glancey, and C. S. Lofgren. 1976. Red imported fire ants: capability of queens from established colonies and of newly-mated queens to establish colonies in the laboratory. Annals of the Entomological Society of America 69:1004–1006.

Sudd, J. H., and N. R. Franks. 1987. The Behavioural Ecology of Ants. Blackie, Glasgow, U.K. 206 pp.

Summerlin, J. W. 1976. Polygyny in a colony of the southern fire ant. Annals of the Entomological Society of America 69:54.

Summerlin, J. W., and S. E. Kunz. 1978. Predation of the red imported fire ant on stable flies. Southwestern Entomologist 3:260-262.

Summerlin, J. W., A. C. F. Hung, and S. B. Vinson. 1977a. Residues in nontarget ants, species simplification and recovery of populations following aerial applications of mirex. Environmental Entomology 6:193–197.

Summerlin, J. W., J. K. Olson, R. R. Blume, A. Aga, and D. E. Bay. 1977b. Red imported fire ant: effects on *Onthophagus gazella* and the horn fly. Environmental Entomology 6:440–442.

Summerlin, J. W., R. L. Harris, and H. D. Petersen. 1984a. Red imported fire ant (Hymenoptera: Formicidae): frequency and intensity of invasion of fresh cattle droppings. Environmental Entomology 13:1161–1163.

Summerlin, J. W., H. D. Petersen, and R. L. Harris. 1984b. Red imported fire ant (Hymenoptera: Formicidae): effects on the horn fly (Diptera: Muscidae) and coprophagous scarabs. Environmental Entomology 13:1405–1410.

Taber, S. W., J. C. Cokendolpher, and O. F. Francke. 1987. Supercooling points of red imported fire ants, *Solenopsis invicta* (Hymenoptera: Formicidae) from Lubbock, Texas. Entomological News 98:153–158.

Tarzwell, C. M. 1958. The toxicity of some organic insecticides to fishes. Pages 7–13 in Twelfth Annual Conference of the Association of Game and Fish Commissioners. Association of Game and Fish Commissioners, Louisville, Kentucky.

Taylor, F. 1977. Foraging behavior of ants: experiments with two species of myrmecine ants. Behavioral Ecology and Sociobiology 2:147–167.

Tedders, W. L., C. C. Reilly, B. W. Wood, R. K. Morrison, and C. S. Lofgren. 1990. Behavior of *Solenopsis invicta* (Hymenoptera: Formicidae) in pecan orchards. Environmental Entomology 19:44–53.

Tennant, L. E., and S. D. Porter. 1991. Comparison of diets of two fire ant species (Hymenoptera: Formicidae): solid and liquid components. Journal of Entomological Science 26:450–465.

Thompson, C. R. 1980. *Solenopsis* (*Diplorhoptrum*) (Hymenoptera: Formicidae) of Florida. Ph.D. dissertation, University of Florida, Gainesville, Florida. 115 pp.

Thompson, C. R. 1989. The thief ants, *Solenopsis molesta* group, of Florida (Hymenoptera: Formicidae). Florida Entomologist 72:268–283.

Thompson, M. J., B. M. Glancey, W. E. Robbins, C. S. Lofgren, S. R. Dutky, J. Kochansky, R. K. Vander Meer, and A. R. Glover. 1981. Major hydrocarbons of the post-pharyngeal glands of mated queens of the red imported fire ant *Solenopsis invicta*. Lipids 16:485–495.

Thorvilson, H. G., J. C. Cokendolpher, and S. A. Phillips, Jr. 1992. Survival of the red imported fire ant (Hymenoptera: Formicidae) on the Texas high plains. Environmental Entomology 21:964–968.

Tian, H. S., S. B. Vinson, and C. J. Coates. 2004. Differential gene expression between alate and dealate queens in the red imported fire ant, *Solenopsis invicta* Buren (Hymenoptera: Formicidae). Insect Biochemistry and Molecular Biology 34:937–949.

Tofts, C. 1993. Algorithms for task allocation in ants. (A study of temporal polyethism: theory). Bulletin of Mathematical Biology 55:891–918.

Toom, P. M., E. Cupp, C. P. Johnson, and I. Griffin. 1976a. Utilization of body reserves for minim brood development by queens of the imported fire ant, *Solenopsis invicta*. Journal of Insect Physiology 22:217–220.

Toom, P. M., E. W. Cupp, and C. P. Johnson. 1976b. Amino acid changes in newly inseminated queens of *Solenopsis invicta*. Insect Biochemistry 6:327–331.

Toom, P. M., C. P. Johnson, and E. W. Cupp. 1976c. Utilization of body reserves during preoviposition activity by *Solenopsis invicta*. Annals of the Entomological Society of America 69:145–148.

Trager, J. C. 1991. A revision of the fire ants, *Solenopsis geminata* group (Hymenoptera: Formicidae: Myrmicinae). Journal of the New York Entomological Society 99:141–198.

Travis, B. V. 1938a. Fire ant problem in the Southeast with special reference to quail. Transactions of the North American Wildlife Conference 3:705–708.

Travis, B. V. 1938b. The fire ant (*Solenopsis* spp.) as a pest of quail. Journal of Economic Entomology 31:649–652.

Travis, B. V. 1939a. Poisoned-bait tests against the fire ant, with special reference to thallium sulfate and thallium acetate. Journal of Economic Entomology 32:706–713.

Travis, B. V. 1939b. Tests of soil treatments for the control of the fire ant, *Solenopsis geminata* (F.). Journal of Economic Entomology 32:645–650.

Travis, B. V. 1940. The Present Status of Methods of Control for the Fire Ant. Sherwood Plantation, Thomasville, Georgia. 9 pp.

Triplett, R. F. 1973. Sensitivity to the imported fire ant: successful treatment with immunotherapy. Southern Medical Journal 66:477–480.

Trivers, R. L., and H. Hare. 1976. Haplodiploidy and the evolution of the social insects. Science 191:249–263.

Tschinkel, W. R. 1987. Fire ant queen longevity and age: estimation by sperm depletion. Annals of the Entomological Society of America 80:263–266.

Tschinkel, W. R. 1988a. Colony growth and the ontogeny of worker polymorphism in the fire ant, *Solenopsis invicta.* Behavioral Ecology and Sociobiology 22:103–115.

Tschinkel, W. R. 1988b. Distribution of the fire ants *Solenopsis invicta* and *S. geminata* (Hymenoptera: Formicidae) in northern Florida in relation to habitat and disturbance. Annals of the Entomological Society of America 81:76–81.

Tschinkel, W. R. 1988c. Social control of egg-laying rate in queens of the fire ant, *Solenopsis invicta.* Physiological Entomology 13:327–350.

Tschinkel, W. R. 1992a. Brood raiding and the population dynamics of founding and incipient colonies of the fire ant, *Solenopsis invicta.* Ecological Entomology 17:179–188.

Tschinkel, W. R. 1992b. Brood raiding in the fire ant, *Solenopsis invicta* (Hymenoptera: Formicidae): laboratory and field observations. Annals of the Entomological Society of America 85:638–646.

Tschinkel, W. R. 1993a. Resource allocation, brood production and cannibalism during colony founding in the fire ant *Solenopsis invicta.* Behavioral Ecology and Sociobiology 33:209–223.

Tschinkel, W. R. 1993b. Sociometry and sociogenesis of colonies of the fire ant *Solenopsis invicta* during one annual cycle. Ecological Monographs 64:425–457.

Tschinkel, W. R. 1995. Stimulation of fire ant queen fecundity by a highly specific brood stage. Annals of the Entomological Society of America 88:876–882.

Tschinkel, W. R. 1996. A newly-discovered mode of colony founding among fire ants. Insectes Sociaux 43:267–276.

Tschinkel, W. R. 1998. An experimental study of pleometrotic colony founding in the fire ant, *Solenopsis invicta:* what is the basis for association? Behavioral Ecology and Sociobiology 43:247–257.

Tschinkel, W. R. 1999a. Sociometry and sociogenesis of colonies of the harvester ant, *Pogonomyrmex badius:* distribution of workers, brood and seeds within the nest in relation to colony size and season. Ecological Entomology 24:222–237.

Tschinkel, W. R. 1999b. Sociometry and sociogenesis of colony-level attributes of the Florida harvester ant (Hymenoptera: Formicidae). Annals of the Entomological Society of America 92:80–89.

Tschinkel, W. R. 2004. Nest architecture of the Florida harvester ant, *Pogonomyrmex badius.* Journal of Insect Science 4, Article number 21.

Tschinkel, W. R., and D. F. Howard. 1978. Queen replacement in orphaned colonies of the fire ant, *Solenopsis invicta.* Behavioral Ecology and Sociobiology 3:297–310.

Tschinkel, W. R., and D. F. Howard. 1980. Replacement of queens in orphaned colonies of the fire ant, *Solenopsis invicta.* Proceedings of the Tall Timbers Conference on Animal Control by Habitat Management 7:135–147.

Tschinkel, W. R., and D. F. Howard. 1983. Colony founding by pleometrosis in the fire ant, *Solenopsis invicta.* Behavioral Ecology and Sociobiology 12:103–113.

Tschinkel, W. R., and S. D. Porter. 1988. Efficiency of sperm use in queens of the fire ant, *Solenopsis invicta* (Hymenoptera: Formicidae). Annals of the Entomological Society of America 81:777–781.

Tschinkel, W. R., E. S. Adams, and T. Macom. 1995. Territory area and colony size in the fire ant, *Solenopsis invicta.* Journal of Animal Ecology 64:473–480.

Tschinkel, W. R., A. Mikheyev, and S. Storz. 2003. The allometry of worker polymorphism in the fire ant, *Solenopsis invicta.* Journal of Insect Science 2:12.

U.S. Department of Agriculture. 1958. Observations on the biology of the imported fire ant. U.S. Department of Agriculture Agricultural Research Service 33-49. 21 pp.

Valles, S. M., and S. D. Porter. 2003. Identification of polygyne and monogyne fire ant colonies (*Solenopsis invicta*) by multiplex PCR of *Gp-9* alleles. Insectes Sociaux 50:199–200.

Valles, S. M., D. H. Oi, O. P. Perera, and D. F. Williams. 2002. Detection of *Thelohania solenopsae* (Microsporidia: Thelohaniidae) in *Solenopsis invicta* (Hymenoptera: Formicidae) by multiplex PCR. Journal of Invertebrate Pathology 81:196–201.

Valles, S. M., C. A. Strong, P. M. Dang, W. B. Hunter, R. M. Pereira, D. H. Oi, A. M. Shapiro, and D. F. Williams. 2004. A picorna-like virus from the red imported fire ant, *Solenopsis invicta:* initial discovery, genome sequence, and characterization. Virology 328:151–157.

Van Pelt, A., Jr. 1958. The ecology of the ants of the Welaka Reserve, Florida (Hymenoptera: Formicidae). Part II. Annotated list. American Midland Naturalist 59:1–57.

Vander Meer, R. K. 1986a. Chemical taxonomy as a tool for separating *Solenopsis* spp. Pages 316–326 in Fire Ants and Leafcutting Ants: Biology and Management. C. S. Lofgren and R. K. Vander Meer, editors. Westview Press, Boulder, Colorado.

Vander Meer, R. K. 1986b. The trail pheromone complex of *Solenopsis invicta* and *Solenopsis richteri.* Pages 201–210 in Fire Ants and Leafcutting Ants: Biology and Management. C. S. Lofgren and R. K. Vander Meer, editors. Westview Press, Boulder, Colorado.

Vander Meer, R. K., and L. E. Alonso. 2002. Queen primer pheromone affects conspecific fire ant (*Solenopsis invicta*) aggression. Behavioral Ecology and Sociobiology 51:122–130.

Vander Meer, R. K., and C. S. Lofgren. 1988. Use of chemical characters in defining populations of fire ants, *Solenopsis saevissima* complex (Hymenoptera: Formicidae). Florida Entomologist 71:323–332.

Vander Meer, R. K., and C. S. Lofgren. 1989. Biochemical and behavioral evidence for hybridization between fire ants, *Solenopsis invicta* and *Solenopsis richteri* (Hymenoptera: Formicidae). Journal of Chemical Ecology 15:1757–1765.

Vander Meer, R. K., and C. S. Lofgren. 1990. Chemotaxonomy applied to fire ant systematics in the United States and South America. Pages 75–84 in Applied Myrmecology: A World Perspective. R. K. Vander Meer, K. Jaffe, and A. Cedeno, editors. Westview Press, Boulder, Colorado.

Vander Meer, R. K., and L. Morel. 1988. Brood pheromones in ants. Pages 491–513 in Advances in Myrmecology. J. C. Trager, editor. E. J. Brill, New York.

Vander Meer, R. K., and L. Morel. 1995. Ant queens deposit pheromones and antimicrobial agents on eggs. Naturwissenschaften 82:93–95.

Vander Meer, R. K., and L. Morel. 1998. Nestmate recognition in ants. Pages 79–103 in Pheromone Communication in Social Insects. R. K. Vander Meer, M. Breed, M. Winston and K. E. Espelie, editors. Westview Press, Boulder, Colorado.

Vander Meer, R. K., and S. D. Porter. 2001. Fate of newly mated queens introduced into monogyne and polygyne *Solenopsis invicta* (Hymenoptera: Formicidae) colonies. Annals of the Entomological Society of America 94:289–297.

Vander Meer, R. K., and D. P. Wojcik. 1982. Chemical mimicry in the myrmecophilous beetle *Myrmecaphodius excavaticollis*. Science 218:806–808.

Vander Meer, R. K., F. D. Williams, and C. S. Lofgren. 1981. Hydrocarbon components of the trail pheromone of the red imported fire ant, *Solenopsis invicta*. Tetrahedron Letters 22:1651–1654.

Vander Meer, R. K., B. M. Glancey, and C. S. Lofgren. 1982a. Biochemical changes in the crop, oesophagus and postpharyngeal gland of colony-founding red imported fire ant queens (*Solenopsis invicta*). Insect Biochemistry 12:123–127.

Vander Meer, R. K., C. S. Lofgren, B. M. Glancey, and D. F. Williams. 1982b. The trail pheromone of the red imported fire ant, *Solenopsis invicta*, chemistry, behavior and potential for control. Page 333 in The Biology of Social Insects. M. D. Breed, C. A. Michener, and H. E. Evans, editors. Westview Press, Boulder, Colorado.

Vander Meer, R. K., C. S. Lofgren, and F. M. Alvarez. 1985. Biochemical evidence for hybridization in fire ants. Florida Entomologist 68:501–506.

Vander Meer, R. K., F. Alvarez, and C. S. Lofgren. 1988. Isolation of the trail recruitment pheromone of *Solenopsis invicta*. Journal of Chemical Ecology 14:825–838.

Vander Meer, R. K., D. P. Jouvenaz, and D. P. Wojcik. 1989. Chemical mimicry in a parasitoid (Hymenoptera: Eucharitidae) of fire ants (Hymenoptera: Formicidae). Journal of Chemical Ecology 15:2247–2261.

Vander Meer, R. K., C. S. Lofgren, and F. M. Alvarez. 1990a. The orientation inducer pheromone of the fire ant *Solenopsis invicta*. Physiological Entomology 15:483–488.

Vander Meer, R. K., M. S. Obin, and L. Morel. 1990b. Nestmate recognition in fire ants: monogyne and polygyne populations. Pages 322–328 in Applied Myrmecology: A World Perspective. R. K. Vander Meer, K. Jaffe, and A. Cedeno, editors. Westview Press, Boulder, Colorado.

Vander Meer, R. K., M. S. Obin, and L. Morel. 1990c. Nestmate recognition in monogyne and polygyne populations of the fire ant *Solenopsis invicta* Buren. Pages 404–405 in Social Insects and the Environment: Proceedings of the 11th International Congress of IUSSI. G. K. Veeresh, B. Mallik, and C. A. Viraktamath, editors. E. J. Brill, New York.

Vander Meer, R. K., L. Morel, and C. S. Lofgren. 1992. A comparison of queen oviposition rates from monogyne and polygyne fire ant, *Solenopsis invicta*, colonies. Physiological Entomology 17:384–390.

Vander Meer, R. K., T. J. Slowik, and H. G. Thorvilson. 2002. Semiochemicals released by electrically stimulated red imported fire ants, *Solenopsis invicta*. Journal of Chemical Ecology 28:2585–2600.

Vargo, E. L. 1988a. A bioassay for a primer pheromone of queen fire ants (*Solenopsis invicta*) which inhibits the production of sexuals. Insectes Sociaux 35:382–392.

Vargo, E. L. 1988b. Effect of pleometrosis and colony size on the production of sexuals in monogyne colonies of the fire ant *Solenopsis invicta*. Pages 217–225 in Advances in Myrmecology. J. C. Trager, editor. E. J. Brill, New York.

Vargo, E. L. 1990. Social control of reproduction in fire ant colonies. Pages 158–172 in Applied Myrmecology: A World Perspective. R. K. Vander Meer, K. Jaffe, and A. Cedeno, editors. Westview Press, Boulder, Colorado.

Vargo, E. L. 1992. Mutual pheromonal inhibition among queens in polygyne colonies of the fire ant *Solenopsis invicta*. Behavioral Ecology and Sociobiology 31:205–210.

Vargo, E. L. 1996. Sex investment ratios in monogyne and polygyne populations of the fire ant *Solenopsis invicta*. Journal of Evolutionary Biology 9:783–802.

Vargo, E. L. 1997. Poison gland of queen fire ants (*Solenopsis invicta*) is the source of a primer pheromone. Naturwissenschaften 84:507–510.

Vargo, E. L. 1998. Primer pheromones in ants. Pages 293–313 in Pheromone Communication in Social Insects. R. K. Vander Meer, M. Breed, M. Winston and K. E. Espelie, editors. Westview Press, Boulder, Colorado.

Vargo, E. L. 1999. Reproductive development and ontogeny of queen pheromone production in the fire ant *Solenopsis invicta*. Physiological Entomology 24:370–376.

Vargo, E. L., and D. J. C. Fletcher. 1986a. Evidence of pheromonal queen control over the production of male and female sexuals in the fire ant, *Solenopsis invicta*. Journal of Comparative Physiology. A Sensory, Neural and Behavioral Physiology 159:741–749.

Vargo, E. L., and D. J. C. Fletcher. 1986b. Queen number and the production of sexuals in the fire ant, *Solenopsis invicta* (Hymenoptera: Formicidae). Behavioral Ecology and Sociobiology 19:41–47.

Vargo, E. L., and D. J. C. Fletcher. 1987. Effect of queen number on the production of sexuals in natural populations of the fire ant, *Solenopsis invicta*. Physiological Entomology 12:109–116.

Vargo, E. L., and D. J. C. Fletcher. 1989. On the relationship between queen number and fecundity in polygyne colonies of the fire ant *Solenopsis invicta*. Physiological Entomology 14:223–232.

Vargo, E. L., and C. D. Hulsey. 2000. Multiple glandular origins of queen pheromones in the fire ant *Solenopsis invicta*. Journal of Insect Physiology 46:1151–1159.

Vargo, E. L., and S. D. Porter. 1989. Colony reproduction by budding in the polygyne form of *Solenopsis invicta* (Hymenoptera: Formicidae). Annals of the Entomological Society of America 82:307–313.

Vargo, E. L., and K. G. Ross. 1989. Differential viability of eggs laid by queens in polygyne colonies of the fire ant, *Solenopsis invicta*. Journal of Insect Physiology 35:587–593.

Vinson, S. B. 1968. The distribution of an oil, carbohydrate, and protein food source to members of the imported fire ant colony. Journal of Economic Entomology 61:712–714.

Vinson, S. B. 1991. Effect of the red imported fire ant (Hymenoptera: Formicidae) on a small plant-decomposing arthropod community. Environmental Entomology 20:98–103.

Vinson, S. B., and T. A. Scarborough. 1991. Interactions between *Solenopsis invicta* (Hymenoptera: Formicidae), *Rhopalosiphum maidis* (Homoptera: Aphididae), and the parasitoid *Lysiphlebus testaceipes* Cresson (Hymenoptera: Aphidiidae). Annals of the Entomological Society of America 84:158–164.

Vinson, S. B., S. A. Phillips, Jr., and H. J. Williams. 1980. The function of the post-pharyngeal glands of the red imported fire ant, *Solenopsis invicta* Buren. Journal of Insect Physiology 26:645–650.

Vogt, J. T., and A. G. Appel. 2000. Discontinuous gas exchange in the fire ant, *Solenopsis invicta* Buren: Caste differences and temperature effects. Journal of Insect Physiology 46:403–416.

Vogt, J. T., A. G. Appel, and M. S. West. 2000. Flight energetics and dispersal capability of the fire ant, *Solenopsis invicta* Buren. Journal of Insect Physiology 46:697–707.

Vogt, J. T., R. A. Grantham, W. A. Smith, and D. C. Arnold. 2001. Prey of the red imported fire ant (Hymenoptera: Formicidae) in Oklahoma peanuts. Environmental Entomology 30:123–128.

Vogt, J. T., R. A. Grantham, E. Corbett, S. Rice, and R. E. Wright. 2002a. Dietary habits of *Solenopsis invicta* (Hymenoptera: Formicidae) in four Oklahoma habitats. Environmental Entomology 31:47–53.

Vogt, J. T., P. Mulder, A. Sheridan, E. M. Shoff, and R. E. Wright. 2002b. Red imported fire ants (Hymenoptera: Formicidae) fail to reduce predator abundance in peanuts. Journal of Entomological Science 37:200–202.

Vogt, J. T., W. A. Smith, R. A. Grantham, and R. E. Wright. 2003. Effects of temperature and season on foraging activity of red imported fire ants (Hymenoptera: Formicidae) in Oklahoma. Environmental Entomology 32:447–451.

Voss, S. H. 1981. Trophic egg production in virgin fire ant queens. Journal of the Georgia Entomological Society 16:437–440.

Voss, S. H., and M. S. Blum. 1987. Trophic and embryonated egg production in founding colonies of the fire ant *Solenopsis invicta* (Hymenoptera: Formicidae). Sociobiology 13:271–278.

Voss, S. H., J. H. D. Bryan, and C. H. Keith. 1986. Early developmental arrest in fire ant trophic eggs. Journal of Cell Biology 103:244A.

Walsh, C. T., J. H. Law, and E. O. Wilson. 1965. Purification of the fire ant trail substance. Nature 207:320–321.

Walsh, J. P., and W. R. Tschinkel. 1974. Brood recognition by contact pheromone in the red imported fire ant, *Solenopsis invicta*. Animal Behaviour 22:695–704.

Wasmann, E. 1918. Über *Solenopsis geminata saevissima* Sm. und ihre gäste. Entomologische Blätter 14:69–76.

Watanabe, M. E. 1980. An Ethological Study of the American Alligator (*Alligator mississippiensis* Daudin) with Emphasis on Vocalizations and Responses to Vocalizations. Ph.D. dissertation, New York University, New York. 182 pp.

Way, M. J., G. Javier, and K. L. Heong. 2002. The role of ants, especially the fire ant, *Solenopsis geminata* (Hymenoptera: Formicidae), in the biological control of tropical upland rice pests. Bulletin of Entomological Research 92:431–437.

Wehner, R., R. D. Harkness, and P. Schmid-Hempel. 1983. Foraging strategies in individually searching ants, *Cataglyphis bicolor* (Hymenoptera: Formicidae). G. Fischer, Stuttgart. 79 pp.

Wells, J. D., and G. Henderson. 1993. Fire ant predation on native and introduced subterranean termites in the laboratory: effect of high soldier number in

Coptotermes formosanus. Ecological Entomology 18:270–274.

West-Eberhard, M. J. 1981. Intragroup selection and the evolution of insect societies. Pages 3–17 in Natural Selection and Social Behavior. R. D. Alexander and D. W. Tinkle, editors. Chiron Press, Oxford.

Wheeler, D. E. 1990. The developmental basis of worker polymorphism in fire ants. Journal of Insect Physiology 36:315–322.

Wheeler, D. E., and N. A. Buck. 1992. Protein, lipid and carbohydrate use during metamorphosis in the fire ant, *Solenopsis xyloni.* Physiological Entomology 17:397–403.

Wheeler, D. E., and N. A. Buck. 1995. Storage proteins in ants during development and colony founding. Journal of Insect Physiology 41:885–894.

Wheeler, D. E., and T. Martinez. 1995. Storage proteins in ants (Hymenoptera: Formicidae). Comparative Biochemistry and Physiology B Comparative Biochemistry 112b:15–19.

Wheeler, W. M. 1910. Ants, Their Structure, Development and Behavior. Columbia University Press, New York. 663 pp.

Whitcomb, W. H., H. A. Denmark, A. P. Bhatkar, and G. L. Greene. 1972. Preliminary studies on the ants of Florida soybean fields. Florida Entomologist 55:129–142.

Whitcomb, W. H., A. Bhatkar, and J. C. Nickerson. 1973. Predators of *Solenopsis invicta* queens prior to successful colony establishment. Environmental Entomology 2:1101–1103.

Whitworth, S. T., M. S. Blum, and J. Travis. 1998. Proteolytic enzymes from larvae of the fire ant, *Solenopsis invicta:* isolation and characterization of four serine endopeptidases. Journal of Biological Chemistry 273:14430–14434.

Willer, D. E. 1984. Inhibitory Pheromone Production by Queens of *Solenopsis invicta.* M.S. thesis, University of Georgia, Athens, Georgia. 51 pp.

Willer, D. E., and D. J. C. Fletcher. 1986. Differences in inhibitory capability among queens of the ant *Solenopsis invicta.* Physiological Entomology 11:475–482.

Williams, D. F. 1983. The development of toxic baits for the control of the imported fire ant. Florida Entomologist 66:162–172.

Williams, D. F., and C. S. Logfren. 1988. Nest casting of some ground-dwelling Florida ant species using dental labstone. Pages 433–443 in Advances in Myrmecology. J. C. Trager, editor. E. J. Brill, New York.

Williams, D. F., and P. Whelan. 1991. Polygynous colonies of *Solenopsis geminata* (Hymenoptera: Formicidae) in the Galapagos Islands. Florida Entomologist 74:368–371.

Williams, D. F., C. S. Lofgren, and A. Lemire. 1980. A simple diet for rearing laboratory colonies of the red imported fire ant. Journal of Economic Entomology 73:176–177.

Williams, D. F., R. K. Vander Meer, and C. S. Lofgren. 1987. Diet-induced nonmelanized cuticle in workers of the imported fire ant *Solenopsis invicta* Buren. Archives of Insect Biochemistry and Physiology 4:251–259.

Williams, D. F., G. J. Knue, and J. J. Becnel. 1998. Discovery of *Thelohania solenopsae* from the red imported fire ant, *Solenopsis invicta,* in the United States. Journal of Invertebrate Pathology 71:175–176.

Williams, D. F., D. H. Oi, and G. J. Knue. 1999. Infection of red imported fire ant (Hymenoptera: Formicidae) colonies with the entomopathogen *Thelohania solenopsae* (Microsporidia: Thelohaniidae). Journal of Economic Entomology 92:830–836.

Williams, H. J., M. R. Strand, and S. B. Vinson. 1981a. Synthesis and purification of the allofarnesenes. Tetrahedron 37:2763–2767.

Williams, H. J., M. R. Strand, and S. B. Vinson. 1981b. Trail pheromone of the red imported fire ant *Solenopsis invicta* (Buren). Experientia 37:1159–1160.

Williams, R. N. 1980. Insect natural enemies of fire ants in South America with several new records. Proceedings Tall Timbers Conference on Ecology and Animal Control by Habitat Management 7:123–134.

Williams, R. N., J. R. Panaia, D. Gallo, and W. H. Whitcomb. 1973. Fire ants attacked by phorid flies. Florida Entomologist 56:259–262.

Wilson, D. E., and N. J. Silvy. 1988. Impact of the imported fire ants on birds. Pages 70–74 Proceedings of the Governor's Conference on the Imported Fire Ant: Assessment and Recommendations. S. B. Vinson and J. Teer, editors. Sportsmen Conservationists of Texas, Austin, Texas.

Wilson, E. O. 1951. Variation and adaptation in the imported fire ant. Evolution 5:68–79.

Wilson, E. O., Jr. 1952. A report on the imported fire ant *Solenopsis saevissima* var. *richteri* Forel in the gulf states. Journal of the Alabama Academy of Science 21-22:52.

Wilson, E. O. 1953a. The origin and evolution of polymorphism in ants. Quarterly Review of Biology 28:136–156.

Wilson, E. O. 1953b. Origin of the variation in the imported fire ant. Evolution 7:262–263.

Wilson, E. O. 1959. Source and possible nature of the odor trail of fire ants. Science 129:643–644.

Wilson, E. O. 1962a. Chemical communication among workers of the fire ant *Solenopsis saevissima* (Fr. Smith). 1. The organization of mass foraging. Animal Behaviour 10:134–147.

Wilson, E. O. 1962b. Chemical communication among workers of the fire ant *Solenopsis saevissima* (Fr. Smith). 2. An information analysis of the odour trail. Animal Behaviour 10:148–158.

Wilson, E. O. 1962c. Chemical communication among workers of the fire ant *Solenopsis saevissima* (Fr. Smith). 3. The experimental induction of social responses. Animal Behaviour 10:159–164.

Wilson, E. O. 1971. The Insect Societies. Belknap Press of Harvard University Press, Cambridge, Massachusetts. 548 pp.

Wilson, E. O. 1978. Division of labor in fire ants based on physical castes (Hymenoptera: Formicidae: *Solenopsis*). Journal of the Kansas Entomological Society 51:615–636.

Wilson, E. O. 1985. The sociogenesis of insect colonies. Science 228:1489–1495.

Wilson, E. O. 1994. Naturalist. Island Press, Washington, D.C. 380 pp.

Wilson, E. O., and W. L. Brown, Jr. 1958. Recent changes in the introduced population of the fire ant *Solenopsis saevissima* (Fr. Smith). Evolution 12:211–218.

Wilson, E. O., and J. H. Eads. 1949. A report on the imported fire ant *Solenopsis saevissima* var. *richteri* Forel in Alabama. Special report to Alabama Department of Conservation (mimeographed). 53 pp., 13 plates.

Wilson, E. O., N. I. Durlach, and L. M. Roth. 1958. Chemical releasers of necrophoric behavior in ants. Psyche 65:108–114.

Wilson, J. D. 1975. Mirex: Decision Making Problems in Pesticide Programs. M.S. thesis, University of Florida, Gainesville, Florida. 148 pp.

Wilson, N. L. 1969. Foraging Habits and Effects of Imported Fire Ant, *Solenopsis saevissima richteri* Forel, on Some Arthropod Populations in Southeastern Louisiana. Ph.D. dissertation, Louisiana State University, Baton Rouge, Louisiana. 80 pp.

Wilson, N. L., and A. D. Oliver. 1969. Food habits of the imported fire ant in pasture and pine forest areas in southeastern Louisiana. Journal of Economic Entomology 62:1268–1271.

Wilson, N. L., J. H. Dillier, and G. P. Markin. 1971. Foraging territories of imported fire ants. Annals of the Entomological Society of America 64:660–665.

Winston, M. L. 1987. The Biology of the Honey Bee. Harvard University Press, Cambridge, Massachusetts. 281 pp.

Wojcik, D. P. 1975. Biology of *Myrmecaphodius excavaticollis* (Blanchard) and *Euparia castanea* Serville (Coleoptera: Scarabaeidae) and Their Relationships to *Solenopsis* spp. (Hymenoptera: Formicidae). Ph.D. dissertation, University of Florida, Gainesville, Florida. 74 pp.

Wojcik, D. P. 1980. Fire ant myrmecophiles: behavior of *Myrmecosaurus ferrugineus* Bruch (Coleoptera: Staphylinidae) with comments on its abundance. Sociobiology 5:63–68.

Wojcik, D. P. 1983. Comparison of the ecology of red imported fire ants in North and South America. Florida Entomologist 66:101–111.

Wojcik, D. P. 1986. Observations on the biology and ecology of fire ants in Brazil. Pages 88–103 in Fire Ants and Leafcutting Ants: Biology and Management. C. S. Lofgren and R. K. Vander Meer, editors. Westview Press, Boulder, Colorado.

Wojcik, D. P. 1990. Behavioral interactions of fire ants and their parasites, predators and inquilines. Pages 329–344 in Applied Myrmecology: A World Perspective. R. K. Vander Meer, K. Jaffe, and A. Cedeno, editors. Westview Press, Boulder, Colorado.

Wojcik, D. P. 1994. Impact of the red imported fire ant on native ant species in Florida. Pages 269-281 in Exotic Ants. Biology, Impact, and Control of Introduced Species. D. F. Williams, editor. Westview Press, Boulder, Colorado.

Wojcik, D. P., and D. H. Habeck. 1977. Fire ant myrmecophiles: breeding period and ovariole number in *Myrmecaphodius excavaticollis* (Blanchard) and *Euparia castanea* Serville (Coleoptera: Scarabaeidae). Coleopterists Bulletin 31:335–338.

Wojcik, D. P., and J. A. Smallwood. 1993. Red imported fire ant predation on kestrels. Pages 120–121 in Proceedings of the 1993 Imported Fire Ant Research Conference. J. P. Ellis, editor. Charleston, South Carolina.

Wojcik, D. P., W. A. Banks, D. M. Hicks, and J. W. Summerlin. 1977. Fire ant myrmecophiles: new hosts and distribution of *Myrmecaphodius excavaticollis* (Blanchard) and *Euparia castanea* Serville (Coleoptera: Scarabaeidae). Coleopterists Bulletin 31:329–334.

Wojcik, D. P., W. A. Banks, and D. H. Habeck. 1978. Fire ant myrmecophiles: flight periods of *Myrmecaphodius excavaticollis* (Blanchard) and *Euparia castanea* Serville (Coleoptera: Scarabaeidae). Coleopterists Bulletin 32:59–64.

Wojcik, D. P., B. J. Smittle, and H. L. Cromroy. 1991. Fire ant myrmecophiles: feeding relationships of *Martinezia dutertrei* and *Euparia castanea* (Coleoptera: Scarabaeidae) with their ants hosts, *Solenopsis* spp. (Hymenoptera: Formicidae). Insectes Sociaux 38:273–281.

Wojcik, D. P., R. J. Burges, C. M. Blanton, and D. A. Focks. 2000. An improved and quantified technique for marking individual fire ants (Hymenoptera: Formicidae). Florida Entomologist 83:74–78.

Wojcik, D. P., C. R. Allen, R. J. Brenner, E. A. Forys, D. P. Jouvenaz, and R. S. Lutz. 2001. Red imported fire ants: impact on biodiversity. American Entomologist 47:16–23.

Wong, T. T. Y., D. O. McInnis, J. I. Nishimoto, A. K. Ota, and V. C. S. Chang. 1984. Predation of the Mediterranean fruit fly (Diptera: Tephritidae) by the

Argentine ant (Hymenoptera: Formicidae) in Hawaii. Journal of Economic Entomology 77:1454–1458.

Woolwine, A. E. 1998. Ecology and Loss Assessment of Selected Homopterans on Sugar Cane: Interactions with the Fire Ant *Solenopsis wagneri* Santschi. Ph.D. dissertation, Louisiana State University, Baton Rouge, Louisiana. 231 pp.

Wuellner, C. T. 2000. Male aggregation by *Solenopsis richteri* Forel (Hymenoptera: Formicidae) and associated mating behavior in Argentina. Journal of Insect Behavior 13:751–756.

Wuellner, C. T., and J. B. Saunders. 2003. Circadian and circannual patterns of activity and territory shifts: comparing a native ant (*Solenopsis geminata,* Hymenoptera: Formicidae) with its exotic, invasive congener (*S. invicta*) and its parasitoids (*Pseudacteon* spp., Diptera: Phoridae) at a central Texas site. Annals of the Entomological Society of America 96:54–60.

Wuellner, C. T., S. D. Porter, and L. E. Gilbert. 2002. Eclosion, mating, and grooming behavior of the parasitoid fly *Pseudacteon curvatus* (Diptera: Phoridae). Florida Entomologist 85:563–566.

Yosef, R., and F. E. Lohrer. 1995. Loggerhead shrikes, red fire ants and red herrings? Condor 97:1053–1056.

Zacaro, A. A., and S. D. Porter. 2003. Female reproductive system of the decapitating fly *Pseudacteon wasmanni* Schmitz (Diptera: Phoridae). Arthropod Structure and Development 31:329–337.

Zettler, J. A., T. P. Spira, and C. R. Allen. 2001. Ant-seed mutualisms: can red imported fire ants sour the relationship? Biological Conservation 101:249–253.

Zettler, J. A., M. D. Taylor, C. R. Allen, and T. P. Spira. 2004. Consequences of forest clear-cuts for native and nonindigenous ants (Hymenoptera: Formicidae). Annals of the Entomological Society of America 97:513–518.

Acknowledgments

E. O. Wilson is more responsible for this book than he probably realizes. In 1962, I was a young graduate student at the University of California, Berkeley, casting about for an interesting research area in which to apply my background in biochemistry and organic chemistry. One day I attended an Entomology Department seminar by Wilson, a young professor from Harvard University. I sat in the top balcony of the wonderful rotunda in Gilmer Hall, and heard the most brilliant, intriguing exposition on chemical communication in ants I could imagine. The world that Wilson opened up to me in that brief 50 minutes set the direction for my life of research. I will be forever grateful.

E. O. Wilson also provided the encouragement that finally pushed me over the threshold to write this book, and he put me in touch with Harvard University Press. The writing has not always been a happy process, but Wilson correctly perceived that it needed to be done and that I should be appointed to do it. Along the way, Michael Fisher of Harvard University Press provided gentle encouragement that kept me going, even when it seemed I would never finish.

Many other people contributed in important ways to this project. My wife, Vicki, was both encouraging and tolerant and at times provided necessary criticism. My daughter Erika's tolerance was also greatly appreciated. Too often, she came home from school to find me silhouetted against a computer screen.

No one can write a book without a lot of colleagues, and I am grateful to many of them for the kinds of interactions that keep a scientist on his toes, up to date, and inspired. Many surely got tired of asking, "So when's the book coming out?" I am particularly appreciative of those who read and commented on parts of this book—Joshua King, Kevin Haight, Erika Tschinkel, Edith Nierenberg, and Ed Wilson. Kenneth Ross, Sanford Porter, Joshua King, and Edward Vargo provided me with helpful discussions of their research in progress. It was great fun discussing the book's progress and possible titles with Marianne Tobias and Miriam Mulva; I apologize to them that I finally chose such an ordinary title in spite of their inspired suggestions.

No book is produced without technical help. For superb copy editing, I am grateful to Anne Thistle, Captain First Class of the Grammar Police. Kim Riddle helped with scanning electron microscopy, and Charles Badland with a number of the photographic images.

Two anonymous reviewers read the entire manuscript and provided insightful, detailed, and constructive criticism. Their comments greatly improved the book, and I am grateful for the care and effort they put into the task.

Index

Abdomen, 22
Aberdeen, Mississippi, 63
Abilene, Texas, 574
AC 217,300. *See* Amdro
Accessory gland 138–139, 313, 364, 442. *See also* Dufour's gland
Acclimation, 87, 253
Adams, Eldridge, 526
Adoption, queen, 447, 456, 458–460, 462–465, 469, 491. *See also* Usurpation
Aedeagal bladder, 138
Aedeagus, 138–139
Affirm, 67
Africa, 580
Age structure, 524–526, 534–537
Aggregation(s), of queens, 158, 175, 435, 437; to brood, 264; to pheromones, 321, 430, 432; mating, 437, 456, 456
Aggression, 422, 435, 544, 592–593
Agonistic behavior, 260
Agp-1, 420
Agp-141, 27
Agricultural Research Service, 50–51
Alabama, 41–43, 376, 409–410, 503–504, 509–510, 559; insecticide use in 46–48; reproductive cycle in 137, 148, 234; biological control in 657
Alabama Department of Conservation, 41–42
Alarm, 141, 252, 324, 651
Albany, Georgia, 514
Albumin, 331, 337
Albuquerque, New Mexico, 71
Alinotum, 275, 278
Alkaloids, 370–371, 367–369, 373, 378, 431, 503; alkylated piperidine, 368–369; effects of, 374, 376, 380–381, 430
Alkyl groups, 369
Allergens, 377
Allergic response, 370
Allergies, 375–376
Allied Chemical Corporation, 49–50, 55, 63, 65
Alligators, 614–615
Allocation: energy, 202–203, 206, 230, 383; food, 202–203, 330; labor, 202, 206, 326, 328–329; biomass, 205, 207, 210; sex, 239
Allometry, 275, 278–281
Allozymes, 27, 32, 240, 420, 439; for tracking parentage of hybrids, 259, 503, 506–510; for determination of di- and triploidy, 443, 499; amount of variation in 478, 489. *See also* Enzymes
Amazon River, 26
Amazonia, 15, 580
Amdro, 66, 73, 362, 583–584; research use of, 603, 609, 617–619, 622–623
American Cyanamid, 66
Anaphylaxis, 375, 376
Andes, 15
Antennae, 276–278, 336, 340, 506, uses of, 252, 304–305, 307, 312, 342, 432, 541–542
Antennation, 252, 265, 337
Antibiotic effects, 116
Antigens, 375, 377
Antigua, 71
Antihistamines, 374
Antiseptic, 380
Anus, 393
Apalachicola National Forest, 541, 591
Aphaenogaster floridanus, 541–542
Aphids, 121, 122, 601, 611, 669
Apolysis, 391
Architecture, nest, 99, 102, 292, 328
Argentina, 18–19, 456, 492, 552; origin of ant species in, 15–16, 23, 25–28, 492–493, 509; genetics in, 32, 444–445, 468, 477, 479; natural enemies in 633, 635–638, 641, 659, 661–662, 665
Argentine ant. *See Linepithema humile*
Aridity, 71, 75, 83
Arizona, 71, 76, 85, 653
Arkansas, 42–43, 65, 84, 148
Army ants, 278, 663
Armyworm, beet, 599
Arroio dos Ratos, Brazil, 493, 510, 641
Artesia, Mississippi, 39–40
Arthropods, 22, 367, 588–590, 598–599, 601, 603–604, 608, 663
Aspiration, 132–133, 165–167
Aspirator, 11, 292, 396
Assay(s), 251–252, 265, 320–322, 368, 394, 431, 657. *See also* Bioassay
Assimilation, 202–203, 218
Asymmetry, fluctuating, 505–507, 509
ATP, 374
Atta sexdens, 631
Auburn, Alabama, 234
Auckland, New Zealand, 71
Audubon Society, 65, 621–622
Aussendienst, 284
Austin, Texas, 240, 459, 497, 570, 585, 588, 657
Australia, 16, 71–72, 580

Backcrosses, 504, 511
Bacteria, 332, 374, 380–381, 399, 602, 631, 510; rickettsial, 639–640
Baiomys taylori, pygmy mouse, 624–626
Baldwin County, Alabama, 38, 559
Bandera County, Texas, 592
Baton Rouge, Louisiana, 607
Beauveria bassiana, 639
Beekeepers, 48, 377–378
Bees, 9, 54, 364, 384

Beetles, 120, 605–606, 608, 662, 665; ground, 57; tenebrionid, 90; tiger, 183; myrmecophile, 253; dung, 604; scarab, 30, 604, 663–665; staphylinid (rove, Staphylinidae), effects 30, 57, 604–605, 607, 663; burying (Silphidae), 605; hister (Histeridae), 605, 663; skin, 605; Nitidulidae, 607; pselaphid, 663, 665; tenebrionid, 265, 267
Benedryl, 375
Bioassay(s), 265, 320–322, 398–399, 402, 431–432. *See also* Assay
Bioconcentration, 52, 55
Biodiversity, 588. *See also* Diversity
Biogeography, 571
Biological control, 16–17, 631; of fire ants, 62, 638–639, 653, 655–657, 659, 662; by fire ants, 599
Biomagnification, 52, 56, 66
Birds, 46–47, 56, 120, 183, 525, 612–614, 662
Birth rate, 196, 198, 213, 217, 223, 225, 231–232, 235, 394. *See also* Natality
Blood cells, 368, 374
Blood pressure, 375
Bloodroot, 609
Blowflies, Calliphoridae, 605
Blue Ridge Mountains, 608
Bolivia, 15, 18–19, 552, 662
Boll weevil. *See* Weevil, boll
Bollworms, 598–599
Bootstrap analysis, 487
Bottleneck, population, 31–33
Boykin, Alabama Representative, 48
Brachymyrmex patagonicus, 29
Brackenridge Field Laboratory, 459, 497, 570, 585–589, 591, 655
Brain, 254
Brazil, 18–19, 26, 28, 131, 568, 570, 593; ants introduced from, 15–16, 23, 25, 27; ants introduced to, 16; other natural enemies from, 564, 631, 633, 638, 641, 661–662; phorid flies from, 649, 651, 655, 657
Brazos County, Texas, 582–583
Brisbane, Australia, 71
Bromine vapor, 267
Brood pile, 216, 348; as a position of dominance, 179, 181, 305; work on, 283, 289–290, 342, 345–346, 353–356
Brood raiding, 168–178, 184, 264, 270, 516, 523; factors affecting outcome of, 153, 155, 170–171, 175; queen competition and, 157, 176–178; timing and frequency of, 169–171, 517, 522; patterns of, 173–175, 515, 519–523; duration of, 175, 193, 524, 544
Brood, sexual, 415–416, 493
Brookley Air Force Base, 35
Brown, George, 65
Brownsville, Texas, 71
Budding, 242, 459–461, 463, 465, 491, 495, 497, 585, 587
Buenos Aires, Argentina, 16, 18–19, 29
Bugs, 663
Bullet ant, *Paraponera,* 384–385
Bull-horn acacia ant, 384
Bumblebees, 139, 206
Buren, Bill, 24–25, 27, 74, 375
Burenella dimorpha, 633, 637–638
Burning, 374
Bursa copulatrix, 138
Butterflies, 602, 668
Butz, Earl, 63

Calcium cyanide, 38
Calcium sulfate. *See* Plaster
California, 29, 71, 75–82
California Department of Food and Agriculture (CDFA), 75, 77–79, 81–82
Calliphoridae, blowflies, 605
Callows, 263, 284, 293
Calyx, 137–138, 389
Campinas, São Paulo, Brazil, 657
Camponotus floridanus, 598; *herculeanus,* 218; *pennsylvanica,* 394
Cannibalism, 161–164, 179, 200
Carbohydrates, 144, 152, 593. *See also* Sugar
Carbon dioxide, 144, 264, 299, 608
Cardiocondyla emeryi, 578
Caribbean, 15, 20, 588
Carpenter ant, *Camponotus pennsylvanica,* 394
Carrion, 122, 605
Carrying capacity, 460
Carson, Rachel, 46, 50, 361
Casamino acids, 334, 337–338, 346–348, 351–353. *See also* Protein
Cassill, Deby, 249–250
Cast(s), 97–98, 103–105, 116, 219, 292
Caste, 14, 273, 287–291, 297, 428, 489; worker, 12–14, 272, 285, 370, 660; larval, 332, 390; disease susceptibility and, 632, 640
Cataglyphis sp., 311; *niger,* 260, 280
Cattle, 5–6, 29, 48
Caves, 602
Cellulose, 203
Central America, 15–16, 21, 564, 571
Cerrado, 551, 568, 571
Chapada, Mato Grosso, Brazil, 25
Chaparral, 79
Charcoal, 292
Chemosensory membrane, 484
Chest pain, 375
Chiapas, Mexico, 437
Chiggers, 54
Chile, 15, 18–19, 552
China, 16, 72
Chitin, 203
Chlordane, 41, 45, 62
Chlorinated hydrocarbons, 41, 48, 50
Chlorocarbon, 50
Chorion, 137–138

Chromatography: gas-liquid, 130, 503; gas, 257–258, 263, 320, 322, 368, 382, 505; thin-layer, 265
Chromosomes, 243, 442
Cis-2–methyl-6–undecyl-piperidine, 431
Citheronia regalis, hickory horned devil, regal moth, 667
Claustral founding, 448
Claustral nests, 144–146, 157–158, 160–161, 525–526
Claustral period, 137, 150–164, 168, 183, 188, 445, 452, 528
Climate, 30, 630; range limits and, 70–71, 83, 85, 87, 89, 509; other effects of, 241, 544; Pleistocene, 292, 571; effects of on quail, 622; effects of on *Pseudacteon,* 648, 657
Clothes moths, 90
Codons, 483, 488
Coffee, 29, 602
Collards, 599
Collection methods, 130–132
Collembola, 122
Colobopsis, 35
Colon, 333
Colony cycle, 96, 110, 136–137
Colony odor. *See* Odor, colony
Color, 292–293
Coloration, 14
Columbus, Mississippi, 51, 504
Coma, cold, 84–85, 88–89
Committee on Agriculture (of U.S. Congress), 50
Competition, 280, 308, 379, 436, 542–543, 557, 594, 610–611, 618; modeling of, 111; during founding, 157–158, 160–161, 178; brood raiding and, 169, 172, 175, 517–518; among queens within a colony, 179, 395, 434, 441; mechanisms of, 182, 550, 593; and colony size, 200, 528, 530, 533, 540; territorial, 411, 495, 519, 530, 544, 548; matched mating and, 445; disturbance and, 557, 577–578, 591; displacement by, 560, 588, 590, 592; and natural enemies, 653–654, 658
Congress, U.S., 46, 50, 64
Conomyrma insana, 183. *See also Dorymyrmex*
Conservationists, 46
Convoluted gland, 314, 365
Convulsions, 375
Conway, South Carolina, 234
Coomassie blue, 468
Copulation, 456
Corn, 601
Corn earworm, 601
Corpora allata, 150–151, 428
Corpus Christi, Texas, 617
Corridors: dispersal, 553; power-line, 553–554
Corrientes, Argentina, 444–445, 493, 641
Costa Rica, 580
Cotton, 42, 58–60, 122, 570, 597–602
Cotton rat, hispid, 625
Council on Environmental Quality, 53
Cowpea, 611
Coxae, 628
Coyote, 623
Crickets, 122
Crop, 121, 131, 254, 311, 330–334, 353, 593; capacity of, 123, 279, 286, 335; contents of, 152, 254, 394; visualizing contents of, 254, 336–337, 348; trophallaxis from, 330–332, 340, 349, 355–356; residence time in, 335–338, 348
Cross-reactivity, 377
Crustaceans, 47–48
Cryptic species, 17, 503, 510, 589, 640
Cuiaba, Mato Grosso, Brazil 25–26
Cuticle, 125, 257, 265, 268–269, 365–366, 427, 644–645; desiccation resistance of, 89, 91; recognition cues on, 262, 266–267; apolysis from, 391; effects of disease on, 632, 637
Cyanide, 38
Cyanosis, 375
Cyphomyrmex rimosus, 29
Cytochrome oxidase, 509
Cytochrome P450, 489

Dade County, Florida, 359
Dallas, Texas, 408
Daphne, Alabama, 26
DDT, 38, 46, 55–56, 69
Dead piles, 398, 401–402. *See also* Trash piles
Dealation, 150–151, 426, 429, 432–434
Death rate, 196, 198, 213, 225, 231–232, 235, 394; *See also* Mortality
Decomposers, 602–607
Decomposition, 398–399
Deer, 617, 622, 626
Defecation, 418
Defense, 215, 367, 374, 383
Delaware, 71
Demography, 295–296. 303
Demopolis, Alabama, 148
Density, 460, 524, 562–563, 574, 587, 590, 607, 630; worker, 111, 115–116, 172, 221–222, 537, 598; queen, 158–159, 184; incipient, 172, 517–520, 523; factors affected by, 200, 494, 508, 512, 575, 581, 613, 617, 627; population, 492, 513, 621–625; factors affecting, 496–497, 544–545, 548, 553–554, 585, 635, 662; effects of prey, 599, 610; of vertebrates, 614, 617, 621, 623, 626; of *Pseudacteon,* 657
Desensitization, 376
Desiccation, 89–91, 120, 153, 160, 164, 183, 598. *See also* Aridity
Dieldrin, 45–49, 55, 68
Diet, 121–122, 124–126, 265, 346, 352, 615; and colony odor, 255–256, 259, 261
Digestion, 394
Digging, 79, 160, 177, 272, 292–293, 523. *See also* Excavation
Diploidy, male, 442–443, 447

Diplopods, 663
Diplorhoptrum, thief ants, 13, 120, 184, 564–566, 585
Diptera, 667–668. *See also* Flies
Disease(s), 381–382, 387, 398, 493, 611, 629; queen's, 631; yellow-head, 632; microsporidian, 638
Disequilibrium, linkage, 467
Dispersal, 457–459, 480, 506–507, 512–513, 556–557, 603; factors affecting, 142–144, 553; distance, 143–145, 464–465; factors affected by, 239–240; polygyne, 452, 464–465, 480
Displacement, 568
Dissection, 382
Disturbance, 5, 7, 78, 110, 252, efforts to quantify, 144–145, 361, 551–555, 574–582, 589–590; and competition, 562–564, 591; effects of on *Solenopsis geminata,* 562, 577, 581; and natural enemies, 565, 585; poisoned baits as, 582
Diversity: genetic, 33, 415, 442–444, 489, 493, 505; species, 564–565, 571, 576, 580, 589–590
Division of labor, 206, 295, 356; polymorphism and, 8, 207; experiments on, 134; in *Pogonomyrmex badius,* 293; in *Solenopsis invicta,* 301, 303, 305, 272–274, 281–283, 286–291; food traffic and, 330–331, 352; and the queen, 388
DNA haplotype. *See* mtDNA
Dorymyrmex, 398, 570, 592; *bureni,* 583–584; *insana,* 582–584
Dragonflies, 183, 525, 662
Drone cells, 239
Drosophila sp. *See* Flies, fruit
Ducks, 613
Duels, 304–307
Dufour's gland, 313–314, 319–324, 365, 387, 505. *See also* Accessory gland
Dwarf workers. *See* Minim workers
Dye(s), 107, 142, 143, 254, 335, 340; for tracing food distribution, 331, 349–350, 391; for marking larvae, 341, 343, 346, 348; for estimating crop volume, 353. *See also* Ink; Paint; Stains

E,E-a-farnesene, 321
E,E-homofarnesene, 321
Eads, J. H., 42
Earthworms, 120, 122, 349
Earwig(s), 59, 183
Ecological impact, 11, 87, 127, 496, 512, 548
Economic impact, 59–60
Economics, 198–199
Ecosystem(s), 67, 597–598, 602, 624; invasion of by *Solenopsis geminata,* 16; aquatic, 57; human effects on, 69, 73, 553, 567, 576; invasion of by *Solenopsis invicta,* 127, 550–552, 569, 572, 589–591, 597, 621
Efficiency, 203, 236, 309; of foragers, 217; of nurses, 217, 236, 301; determination of, 218, 296–297, 299, 301; division of labor and, 272–273, 298, 301; of sperm use, 394
Egg laying, 198, 387, 394, 441, 468; regulation of, 194, 386–388, 394; in polygyne nests, 417, 423, 434; by *Solenopsis daguerri,* 661. *See also* Oviposition
Ejaculatory duct, 138
El Paso, Texas, 71, 89
Electrical fields, 61
Electroantennogram, 264
Electrophoresis, 486; gel, 27, 449, 452, 468, 499; starch-gel, 420
Eliaosome, 608–609
Elites, 393
Embryonation, 421–422, 432, 437, 441
End plates, 374
Endangered species, 602
Endocrine system, 388, 432
Endoplasmic reticulum, 370
Energy flow, 202, 230, 232
Entomology, forensic, 605
Entre Rios Province, Argentina, 15
Environmental Defense Fund (EDF), 53, 65
Enzymes, 121, 332, 349, 351, 384, 644; in electrophoretic studies, 27, 420, 462, 468, 503, 511; unblocking by, 365, 370. *See also* Allozymes
Epidemics, 637
Epinephrine, 375
Epoxy, 269
Eradication, 48, 50–54, 61, 67, 77, 80, 408
Ergonomic phase, 206, 229
Escherichia coli, 381
Esophagus, 152, 332–334, 341, 351
est-4, 420
Esters, 139
Ether, 133, 270, 399
Eucharitidae, chalcid wasps, 659. *See also* Wasps
Euparia castanea, 665
Everglades, 148–149
Evolutionarily stable strategy, 191
Excavation, 103, 197, 216, 282, 292, 302. *See also* Digging
Excretion, 212
Execution, 182, 193; of excess queens, 160, 178, 180, 406, 432–433; factors leading to, 180–181, 183, 238, 434, 468; of sexual larvae, 426–427; of male brood, 446
Exocrine glands, 274, 365
Exons, 483–485
Exotic species, 23, 29, 34, 78, 123, 126, 629
Extinction, 571, 589
Exxant, 67
Eyes, 137–138, 207, 227, 274–275, 278, 668

Fairhope, Alabama, 38
Fairy flies, Mymaridae, 668
Farnesenes, 321
Fat(s), 144, 210–212, 279; in diet, 125, 254, 265; content of sexuals, 137–138; content of queen, 153, 273; content of workers, 299

Fat body, 460, 273, 388, 390, 394, 632, 637; of *Pseudacteon*, 644; of myrmecophile beetles, 665
Fatty acids, 116, 139, 254, 398–399
Fecundity, 225, 389, 395, 422, 434, 445, 634; and queen survival, 179, 431, 469–470; factors causing variation in, 223, 394, 420; of colonies, 225; stimulation of, 298, 387–388, 390–391, 424, 427–428; of polygyne queens, 417, 419, 641; inhibition of, 419, 425, 428–429, 432; and alpha status, 436–437; of plants, 609. *See also* Fertility
Federal Insecticide, Fungicide and Rodenticide Act (FIFRA), 54, 65–66
Femur, 278
Ferriamicide, 64–65, 67
Ferrous chloride, 64
Fertility, 180, 188, 423, 505, 510. *See also* Fecundity
Fertilization, 139
Feulgen DNA stain, 421
Fighting, 107, 112, 118, 155, 577; by queens, 179–181
Filamentous gland, 314
Fire, 552, 579–580, 622
Fire Ant Authority, Mississippi, 64
Fire Ant Research Team, 595–596
Fire ant, tropical. *See Solenopsis geminata*
Fish, 46–48, 53, 56
Fish and Wildlife Service, 53
Fission, 415, 459. *See also* Budding; Founding, dependent
Fitness, 191, 206, 300, 386, 542, 539–540; conflicts in, 181–182, 239, 493–494; embodiment of in sexuals, 182, 206, 236, 266, 536; factors increasing, 202, 251, 285, 295; contribution of male production to, 223; determination of, 296, 303, 505; of polygyne queens, 435, 441; of *Gp-9* genotypes, 475, 477; costs of hybridization, 505–506, 508; effects of parasites on, 642, 654; of phorid flies, 650
Flatwoods, 562–564, 578–580
Flies, 122, 605, 647, 650; bot, 669–670; dung, 604; face, 59; flesh (Sarcophagidae), 602, 605; fruit, 379, 569–570, 602, 607; heel, 669; horn, 59, 122, 603–605; horse, 54; house, 602; muscid, 605; phorid (Phoridae), 120, 628, 630–631, 645–646, 649–650, 653, 657, 663, 642; sand, 54; scuttle (Phoridae), 631; stable, *Stomoxys calcitrans*, 54, 59, 122, 604; tachinid (Tachinidae), 667, 669; warble, 669. *See also Pseudacteon.*
Floatation, 44, 131, 197
Flooding, 131, 551–552, 559, 564, 579–580, 585
Florida, 97, 101, 103, 147, 358–359, 361, 410, 571, 577–578; *Solenopsis geminata* in, 38, 489, 562–564; effects on vertebrates in, 38, 614–615, 621; ant surveys in, 41–43, 578–579; heptachlor and dieldrin in, 46, 48, 597; mirex in, 50–52, 53–54, 60, 62, 583, 598; effects of *Solenopsis invicta* on agriculture in, 59, 598; foraging rates in, 129; fire-ant seasonality in, 137, 139–140, 148–149, 189, 234–235; *Solenopsis invicta* habitat preferences in, 142, 358–359, 361, 562–564, 590; experiments in, 186, 260; sex allocation in, 240; *Pogonomyrmex badius* in, 292; hypersensitivity to *Solenopsis invicta* venom in, 376; polygyny in, 406, 408–410, 443; monogyny in, 410; male diploidy in, 445; ant diversity in, 570, 590; pathogens and parasites of ants in, 632, 634, 636–637, 655, 657, 662, 665
Florida Department of Agriculture, 52
Florida Keys, 553, 602
Florida State University, 55, 568
Florida, University of, 26, 583
Fluon, 7, 132, 396, 605
Follicles, 138, 298, 389–390, 394
Food basket, 349–350
Food exchange, 168
Food limitation, 395
Forager(s), 215, 217–218, 256, 285, 288; venom of, 367; density of, 222, 537, 624–625; number of, 216–217, 223; of *Pogonomyrmex badius*, 292; as food source, 284–285, 331, 337, 349, 356
Foraging, 272, 308–309, 330, 580, 592, 602, 628; by minim workers, 10; factors affecting success of, 111, 127–129, 212, 232–234, 555; along brood-raiding trails, 175; workers engaging in, 216, 282–287, 305, 330, 398; nestmate recognition and recruitment during, 252, 324; for work, 289–290; social form and, 406, 496; effects of on vertebrates, 625–626; parasitoids and, 650–654, 656
Forelius foetidus, 592, 593; *mccooki*, 658; *pruinosus*, 592, 658
Forest, 70
Formic acid, 380
Formica obscuripes, 309; *rufa*, 96, 121; *sanguinea*, 17
Formosa, Argentina, 444–445, 493, 641
Fort Stewart, Georgia, 58
Founder effects, 489
Founding, 136, 187, 189, 328, 548, 566; site selection, 142, 584; cooperative, 157–164, 175, 182–183, 239, 452, 458, 523; dependent, 186, 189–191, 237, 415, 447–448, 561; independent, 415, 447–448, 452, 458–459, 480, 561; parasitic, 434, 447; genetic effects on, 449, 451, 459; timing of, 524–525, 527. *See also* Budding; Fission; Pleometrosis
Freeze kill, 84–85, 87, 89
French Broad River, 44
Frogs, 47
Fruit flies. *See* Flies, fruit
Fugitive species, 555, 557
Fungi, 380–381, 398, 602, 631, 639, 671
Furcula, 366
Fustiger elegans, 665

Gainesville, Florida, 224, 408–409, 489, 577–578, 583; biological control at, 655, 658, 662
Galapagos Islands, 16, 551

Galls, 669
Gaster, 11, 22, 332–333, 432, 450, 542; size of, 211, 275, 278–279, 333, 353, 413, 448; queen's, 273, 386, 413, 425–426, 428–429; dragging, 314–315, 541, 628; flagging, 378–380, 593
geminata species complex, 14–15, 488, 647, 656
Gene flow, 472–474, 479, 488, 507, 557; and polygyny, 459, 464, 466, 472, 476–477, 479, 488; between species, 504, 510; in Phoridae, 646
General protein 9, 470, 483–484, 486. *See also Gp-9*
Genitalia, external, 139
Georgia, 44, 87, 242, 410, 497, 609; effects on vertebrates in, 38, 615, 621; surveys and experiments in, 42–43, 469, 471, 473, 514; pesticides in, 46, 51–53, 65, 514; hypersensitivity to *Solenopsis invicta* venom in, 58, 376; fire-ant seasonality in, 149; sex ratio in, 240; monogyny in, 415; polygyny in, 415, 443, 456, 489–491; male diploidy in, 445; hybridization and gene flow in, 466, 503–504, 506; pathogens and parasites of ants in, 657
Georgia, University of, 26, 443
Germarium, 138, 389
Germination, 381
Gilbert, Larry, 657
Glacial period, 571
Glands. *See* individual types
Global warming, 85
Glossae, 336–337, 340, 342
Glucose, 370
Glucosidase, 370
Glunn, John, 317
Glutamic acid, 127
Glutamine, 127
Glycerol-3 phosphate dehydrogenase-1 (G3pdh-1), 462
Glycogen, 127, 137–138, 144, 152
Gonopod, 314
Gonostylus, 366
Government Accountability Office (GAO), 50–51
Gp-9, 449–452, 456–459, 467–473, 475–489, 491, 498–499, 634
Grass, 382
Grasslands, 122
Great Pee Dee River, 44
Greenhouses, 122
Grimes County, Texas, 582–583
Grits, corncob, 50, 54, 66, 264, 608
Grooming, 168, 253, 272, 283–284, 380, 634; stimulation of, 237, 264, 267–268
Guadalcanal, 580
Guaporé River, Brazil, 26
Guatemala, 580–581, 602
Guava trees, 602
Guianas, 15
Gulf of Mexico, 143–144
Gulfport, Mississippi, 133–134, 234
Habituation, 262
Hammocks, hardwood, 358, 553, 602
Haplodiploid sex determination, 9–10
Haplometrosis, 444
Haplotype. *See* mtDNA
Hardy-Weinberg equilibrium, 468, 507–508
Harris, Judge David, 55
Harvard University, 26
Harvester ants. *See Pogonomyrmex*
Hawaii, 16, 602
Head width, 207,–208, 227, 274–276, 278, 281, 287, 300, 497–498, 647–649
Health effects, 58
Hemocoel, 365
Hemocytometer, 223
Hemolymph, 123, 125, 152–153, 254, 365, 643
Heptachlor, 45–50, 57–58, 68–69, 71, 597
Herbivory, 123
Heritability, 255, 259
Hexamerins, 127
Hexane, 257, 264, 266, 378, 429
Hibernation, 230
Hickel, Walter, 53
Hickory horned devil, regal moth, *Citheronia regalis*, 667
Histamine, 374
Holland, Spessard L., 50–51
Holotype, 25
Homeostasis, 212
Homopterans, 121, 123, 126. *See also* Bugs
Honey, 255, 284, 338–339, 364, 377, 615
Honeybee(s), 239, 430, 471
Honeydew, 123, 601–611
Hong Kong, 16
Hooker Chemical Co., 50, 64
Hopewell, Virginia, 56
Hormone, juvenile, 150, 181, 428, 434, 440; functions of, 127, 150–151, 207, 209, 389, 427–428
Hornet, bald-faced, 384
Horseflies, 54
House Appropriations Committee, U.S., 45
Housefly, 602
Houston, Texas, 408, 593
Humidity, 140
Hurley, Mississippi, 405
Hyaluronidase, 370
Hybrid(s), 88, 259, 405, 503–509, 511, 656, 640
Hybridization, 16, 40, 371, 503–506, 509–511, 640
Hydrocarbons, 17, 253–254, 257–260, 320–321, 503–504, 659, 665
Hydrogen bromide, 267
Hymenoptera, 125, 273, 392, 430; sex determination in, 31, 239, 442, 650; stinging by, 313, 364–365, 377, 384; parasitic, 607, 668. *See also* individual taxa
Hypersensitivity, 58–59, 375–377

Hypertrophy, 417
Hypodermis, 391

Ice age, 292
Identification, chemical, 424, 434
Ileum, 333
Imaginal disc, 279–280
Immokalee, Florida, 148–149
India, 16
Indigo, 611
Infection, 360, 374–375
Inflammation, 374, 384
Infrabuccal pocket, 332, 334, 637
Ink, 269–270, 309, 459. *See also* Dye; Paint; Stains
Innendienst, 284
Inquilines, 663
Insect(s), 22, 433, prey, 106, 120, 212, 215, 301, 349, 379, 514; in diet of *Solenopsis invicta*, 122, 125–126. *See also* individual taxa
Insecticide(s), 41, 598–600, 639. *See also* individual types
Insects Affecting Man and Animals Research Laboratory, 51
Integrated pest management, 38
Interbreeding, 504–505, 510. *See also* Hybridization
Introns, 483–485, 488
Invertebrates, 106. *See also* individual taxa
Invictolide, 430–431
Iodine, 331
Iodoplatinate, 378
Ipecac, 29
Isolation, chemical, 424, 431, 434, 451
Isometry, 275, 278–279, 281
Itch, itching, 374–375

Jacarepaguá, Rio de Janeiro, Brazil, 16
Jacksonville, Florida, 376
Jaw(s), 276, 283, 304
Jerome, Christopher, 528
Johnson, Lyndon, 51
Juvenile hormone. *See* Hormone, juvenile

Kansas, 84
Kempf, Father, 25
Kepone, 49–50, 55–57, 64–65
Kestrels, 613
Key Largo, 602
Keystone species, 567, 601, 612
Kin recognition, 162, 177, 251, 256, 263. *See also* Nestmate recognition
King, Joshua, 591
Kinship, 179, 440

La Pampa Province, Argentina, 15
Labauchena, 13, 660
Label, colony. *See* Odor, colony
Labium, 332, 340
Labor sectors, 215–217, 355
Lactones, 430–431
Lasius fuliginosus, 312; *neoniger*, 380, 592
lawn(s), 23, 96, 148, 358, 551; preference of *Solenopsis invicta* for, 80, 167, 551, 576; experiments on, 400, 551, 609, 630; species diversity on, 568–569
Lead, 116, 219
Leaf-cutter ants, 290
Leon County, Florida, 1
Leon Scrap Metal and Recycling, 270
Lepidoptera, 667. *See also* individual taxa
Leptothorax unifasciatus, 290
Lesser Antilles, 16
Life span(s), 3, 299, 398; effects of worker, 213, 223, 284–285, 287; effects on worker, 201, 335; effects of infection on, 636–637; of queens, 223–224, 274, 466; of colony, 189, 224, 383, 519, 535, 537–538, 558, 560; of *Pseudacteon* parasites, 654
Life table(s), 512–513, 531
Linepithema humile, Argentine ant, 29–30, 32, 35, 78, 559, 664
Linkage disequilibrium, 467
Linoleic acid, 139
Lipid(s), 127, 253–254, 257, 260, 265, 254. *See also* Fats
Lipolexis scutellaris, 601
Lipoproteins, 127
Live oak, 5–6
Lizards, 613, 615–616
Löding, H. P., 24–25
Logic, 67
Logistic function, 197–198
Longevity, 206, 215, 225, 299–300, 558; of queens, 466; of colonies, 538; effects of infection on, 634; of phorid flies, 650
Los Angeles, 75, 77
Louisiana, 410, 443, 445, 570, 593; surveys and experiments in, 42–43, 607; pesticides in, 46–48, 56, 597–598; seasonality in, 148, 234
Louisiana State University, 607
Lubbock, Texas, 71, 87–88
Lubertazzi, David, 578–579
Lyme disease, 59
Lymph nodes, 384
Lysis, 374

Macrogynes, 190
Magnetic fields, 312
Major worker(s), 14, 16, 134, 194, 287, 367; size and growth of, 207, 209–210, 280; trends in production of, 209–211, 215, 223; behaviors and work performed by, 282, 284–285, 287–288, 335, 560
Malpighian tubules, 333, 351
Mammals, 106, 120, 370, 378, 381, 612, 614, 624–625
Mandibles, 275–276, 284, 324, 336–337, 340

Mandibular glands, 141, 429
Marking, 269–271
Mark-recapture studies, 216, 309–310
Martin, purple, 613
Martinezia dutertrei, 30, 253, 663–665
Maryland, 71
Mast cells, 374
Matched mating, 31–32, 190, 443–444, 446
Mating flight(s), 142, 159, 166, 243–244, 459, 514, 516; social form and, 12, 443, 448–449, 456, 459; timing and seasonality of, 139–143, 147–149, 510, 517, 556; conditions triggering, 147–149, 165; by overwintered sexuals, 186–187; preparation for, 324, 440, 467; *Pseudacteon* and, 648, 657
Mating plug, 139
Mato Grosso, Brazil, 15, 25, 27, 551, 569, 631
Mato Grosso do sul, Brazil, 15–16, 27
Mattesia geminata, 632
McGee, Senator Gale, 51
Mealybugs, 126, 600
Meconium, 392
Media, 207
Medical complications, 358–360
Megalopygid, 669
Meiosis, 137, 151, 389, 635, 637–638
Meiospores, 637
Memphis, Tennessee, 148
Meridian, Mississippi, 39–40, 504
Mesonotum, 647
Mesosoma, structure and definition of, 7–10, 22, 279–280, 428–429, 448. *See also* Thorax
Messenger RNA (mRNA), 483–484, 486
Metabolic rate, 212, 299–300, 335, 496
Metamorphosis, 9, 207, 280, 330, 393, 426
Metapleural glands, 116
Metarrhiium anasolpiaea, 631
Metasoma, 22. *See also* Gaster
Methanol, 399
Methods Development Laboratory, Gulfport, Mississippi, 49, 264
Methoprene, 427–428
Methyl groups, 369
Mexico, 15, 83, 437, 601, 609–610
Miami, Florida, 358–359, 361
Mice, 625–626
Microbes, 387. *See also* microorganisms
Microgynes, 190–191
Micrometer, wedge, 227–228
Microorganisms, 380, 630. *See also* microbes
Micropil, 138
Microsatellites, 478, 489, 499, 511
Microscope, 227, 270–271, 348, 387, 393, 413
Microsporidans, 494, 631, 633, 589, 632–635, 637, 671; *See also Thelohania; Vairimorpha*
Microvilli, 365
Midges, 54
Migration, 302, 324, 507–508, 623, 627; of colonies, 117, 527, 548; of queens, 172, 178
Mikheyev, Sasha, 139
Millipede, 663
Milpas, 576–577
Mimicry, 665
Minim workers, importance of in colony founding, 153–156, 183, 188, 207, 390, 451–452, 461; during pleometrosis, 163–164; rate of production of, 155–157, 161, 194; number produced, 155, 193, 237, 444, 447; in brood raiding, 168–171, 517–518, 520; effects of, 168, 178–180, 557; marking of, 270; venom of, 369; of *Solenopsis geminata,* 566
Minor workers, 14, 24, 134, 210, 287, 335, 367, 370; size and growth of, 194, 207, 209–210, 280; behaviors and work performed by, 282, 284–285, 287–288, 560
Mirenda, John, 270
Mirex, 49–58, 60–66, 69, 71–74, 142, 361, 582–584, 595; effects of on nontarget organisms, 56–58; effects of on agriculture, 60, 597–598; scientific uses of, 143, 197, 599, 603, 607–608
Misiones, Argentina, 15
Mississippi, 38, 445, 503–506, 509–510, 616–617, 634; surveys and studies in, 41–43, 133–134, 264; mirex and ferriamicide in, 49–53, 63–66; fire-ant reproduction in, 137, 148–149, 234–235; hypersensitivity to *Solenopsis invicta* venom in, 376; polygyny in, 405, 409–410, 443
Mississippi State Chemical Laboratory, 64
Mississippi State University, 41
Mite(s), 659, 663
Mitochondria, 370
Mitochondrial DNA. *See* mtDNA
Mitosis, 151, 434
Mobile, Alabama, 26, 29–31, 37–38, 447, 559, 640; *Solenopsis invicta* in, 23, 25, 28, 32, 69, 492; *Solenopsis richteri* in, 24, 28, 503; E. O. Wilson in, 29, 34–36; hybridization near, 503, 509
Mobile Press Register, 45
Model(s), modeling, 101, 612, 627; of spread and effects of *Solenopsis invicta,* 83–86, 581; of population development, 111–113, 115, 524–525, 618; of brood dynamics, 240; of imaginal disc growth, 280; of task distribution, 291. *See also* Simulation, computer
Mold, 387
Molting, 9
Monogyny, 32, 95, 117–118, 237, 406, 411, 415; effects of on colony traits, 124, 142, 144, 146, 184, 286; reduction of colonies to, 178, 181, 424, 444; sex ratio and, 240, 244; nestmate recognition in, 251, 256–257, 260; distribution of, 407–409; genetic control of, 450, 480; conversion to polygyny from, 491
Monomorium, 367, 486; *minimum,* 592–593; *minutum,* 615; *pharaonis,* 393; *viride,* 541–542

Monroe, Georgia, 473, 475, 489–491
Monroe, Louisiana, 234
Monroeville, Alabama, 48
Montevideo, Argentina, 29
Morphometry, 205
Mortality, 203, 213, 225, 230, 235–236, 556, 598, 627; of colonies, 214, 225, 522–524, 527, 531, 533–534, 545, 548, 637; of queens, 528–530; of vertebrates, 613–614, 619–620; due to natural enemies, 634, 654, 666. *See also* Death rate
Mosquito(es), 54, 122, 599
Moth(s), 59, 484, 605
Mound(s), 96, 97, 232, 302, 319, 359, 361, 396–397; temperature regulation by, 87, 100–102, 129, 206, 234, 382, 555; size, 96–97, 99, 219–220, 234, 526–528, 630, 406–407, 496; polygyne, 261, 406, 460; density, 406–407, 409–411; disturbance of, 648, 657
Mouse, pygmy. *See Baiomys taylori*
Mouse, deer. *See Peromyscus maniculatus*
Mouthparts, 194
mtDNA, 449, 457–458, 479, 492, 640; for diagnosis of social form, 456–457, 464–465, 469, 477–478, 489, 491; for diagnosis of hybrids, 509
Muscles, 366–367, 460
Musk thistle, 601
Mutation(s), 244, 447
Mymaridae, fairy flies, 668
Myrmecochory, 608
Myrmecocystus, 168
Myrmecomyces annaellisae, 631
Myrmecophiles, 30, 44, 253, 636, 662, 664. *See also* individual taxa
Myrmecophodius excavaticollis. See Martinezia dutertrei
Myrmecosaurus ferrugineus, 30, 663–664
Myrmicinae, 13
Myrmicinosporidiaum durum, 632

N-acetyl-β-glucosaminidase, 370
Nanitic workers. *See* Minim workers
Natality, 214, 235. *See also* birth rate
National Academy of Sciences, 51–52, 54
National Cancer Institute, 64
National Cash Register Corporation, 54
National Environmental Policy Act, effect 53
National Wildlife Federation, 65
Natural enemies, 589, 597, 600–601; for biological control, 26, 631, 654, 656, 660; of exotic species, 550, 560, 587, 629–630
Natural selection, 206, 230, 377, 488, 507, 540, 556; on queens, 184; ;on sex ratio, 245; on worker roles, 273–274, 302. *See also* Selection
Nausea, 375
Necrophory, 398–402
Nectar, 123, 600
Nematode(s), 630, 632, 659–660
Neogregarine, 632
Nepotism, 440, 493
Nerve cord, 351
Nestmate recognition, 169, 178, 255, 259–260, 262–263, 429; in minim workers, 168–169; and kin recognition, 251; studies of, 251–252; and brood recognition, 267; in hybrids, 505; mimicry of, 659
Nests, laboratory, 132
Nevada, 76
New Mexico, 71, 75, 85
New Orleans, Louisiana, 30, 32, 376
New Orleans ant. *See Linepithema humile*
New Zealand, 16, 71–72
Nicaragua, 580
Niche, 550–551, 558, 561, 564, 570–571, 577, 658; of *Solenopsis geminata,* 559, 561, 653, 577; partitioning of, 647–648
Nicotine, 379
Niliparvata lugens, 599
Nitidulidae, 607
Nixon, Richard, 53
Normal-score analysis, 207, 209
North Carolina, 42–44, 570, 592, 666
Nurse(s), nursing, 215, 287–288, 342, 354, 440; numbers of, 216–217; role fidelity of, 285, 393; trophallaxis by, 344–346, 349, 355
Nurseries, plant, 42–43, 45, 75–76, 78, 81, 513
Nutrition, 112, 211–212, 335, 393, 498

Ocala, Florida, 409
Ocelli, 274
Octospores, 632–633, 638
Odontomachus, 290, 305–307; *brunneus,* 121, 304, 577; *insularis,* 35–36
Odor, 648; colony, 252–257, 260, 262, 264, 426, 665; of *Gp-9* genotype, 470, 482, 484
Odor trail(s), 178, 312–314, 319, 663; during brood raiding, 169, 172, 517; strength and durability of, 312, 315; to food, 310–311, 541, 594, 628
Odum, Dr., 454–455
Oil, 254, 317, 331, 335, 339, 346; storage of, 152, 279; preferences for, 317–319, 338
Oklahoma, 71, 76, 83–84, 122, 129
Oklahoma City, Oklahoma, 71, 84
Old World, 260
Oleic acid, 398–399
Olfactometer, 264, 266, 437
Olfactory cues, 252. *See also* Odor, colony
Onthophagus gazella, 604
Ontogeny, 431–432
Oocytes, 138, 390, 433
Orange orchards, 602
Orasema sp., 253, 659
Oregon, 85
Orinoco drainage, 16

Ortho, 362
Ortstreue, 310
Ovary, ovaries, 137–138, 273–274, 364, 388–389, 394, 413; development of, 137–138, 152, 187, 305–306, 460; suppression of, 150, 432; microsporidians in, 632, 635, 637; of *Pseudacteon,* 644, 646; in myrmecophiles, 665
Ovarioles, 137–138, 273, 389–390
Oviduct, 138, 389, 644
Oviposition, 387, 422, 650. *See also* Egg laying
Ovipositor, 364, 366, 642–644, 647
Ovulation, 137
Oxygen, 144, 299–300
Ozone, 267

Pain, 384–385
Paint, 269, 393, 400, 460. *See also* Dye; Ink; Stains
Palatka, Florida, 148
Palpation, 265
Pantanal, 26, 131, 551–552, 571
Paraffin, 266
Paraguay, 15, 18–19, 26–27, 131, 552, 662
Paraguay River, 25–26, 29, 32, 551
Paralysis, 379
Paraná, Rio, Brazil, 16, 27
Paraponera spp., 384
Parasite(s), 42, 607, 611, 629, 631, 641, 667; social, 13, 631, 660–662; resistance to, 493; transmission of, 640; phorid, 647; of humans, 669–670
Parasitism, 120, 653, 655; social, 187–189, 191, 226, 272, 441, 494, 561, 661
Parasitoid(s), 551, 647, 649, 653, 656, 667–670; of boll weevils, 598; as biological-control candidates, 601, 631, 642, 654–655
Paratrechina longicornis, 593
Paratypes, 26
Parthenogenesis, 9, 639–641
Pasture(s), 122, 129, 551, 555, 573, 624, 630; experiments in, 197, 526, 528, 597; interactions with *Solenopsis invicta* in, 559, 592–593, 603–604, 607; ant diversity in, 568–569, 590; predators in, 602
Pathogen(s), 380, 629, 631, 637, 671
Paynes Prairie, Florida, 614
Peanut worm, red-necked, 599–600
Peanuts, 42, 599–600
Penis, 138–139
Pensacola, Florida, 30
Pepsis wasp, 385
Peromyscus maniculatus, deer mice, 626; *polionotus,* 626
Peru, 15, 18–19
Pesticide(s), 514, 578, 600, 616, 622. *See also* individual types
Petén, Guatemala, 580–581
Petiole(s), 22, 275, 278, 299, 304, 637
Pets, 46, 48
Pgm-3, 449–450, 452, 467–468, 473–474, 488. *See also* *Gp-9*
Pharoah's ant, *Monomorium pharaonis,* 393
Pharynx, 254, 334, 351
Pheidole spp., 569; *dentata,* 145–146, 577, 583, 592–593, 598; *floridanus,* 35; *moerens,* 542; *morrisi,* 583–584; *obscurithorax,* 29, 541–542
Pheromone(s), 11–12, 223, 257, 367, 390, 393, 433; effects of queen on sexuals, 12, 237–239, 426–428, 434, 440–442, 450–451, 462, 468; territorial, 116; alarm, 141; effects of queen on workers, 150, 237–239, 429, 434, 436; brood, 264–265, 267–268, 346; sources and distribution of queen, 274, 238, 254, 386, 429, 430, 434, 447; trail, 311–317, 319–324, 327, 330, 346, 365; primer, 422–426, 431, 434, 387, 428; releaser, 423, 430; queen-recognition, 430, 434; genetics of, 431; current knowledge of queen, 431–432; sex, 484; proteins binding, 484–485
Philippines, 580, 599
Phoridae, scuttle flies, 120, 631, 645–647. *See also* *Pseudacteon*
Phosphate, 96
Phosphoglucomutase-3. *See Pgm-3*
Phospholipase, 370
Phosphorus (P^{32}), 331
Photoperiod, 239
Phylogenetic relationships, 13–14, 260, 633
Piedmont, Atlantic, 44
Pine: longleaf, 562–564, 579; slash, 609
Pine Grove, Louisiana, 607–608
Pinelands, 553
Pine plantations, 122, 567
Planidia, 253, 659
Plant Commissioner of Florida, 48
Planthopper, brown, *Niliparvata lugens,* 599
Plants, 106, 121–123. *See also* individual taxa
Plasma, 384
Plaster, 103–105, 292
Plata, Rio de, 26
Plates, cuticular, 366
Pleistocene epoch, 571
Pleometrosis, 157–164, 444. *See also* Founding, cooperative
Pogonomyrmex spp., 278, 511, 587, 616; *badius,* 292–294, 384; *barbatus,* 589; *owyheei,* 558
Poison duct, 138
Poison sac, 313–314, 365–366, 370, 431–432; of queen, 238, 373, 386–387, 428–431, 437
Poisson distribution, 462
Polyergus breviceps, 175
Polygyny, 32, 95, 116–118, 405, 487, 493–494; genetics of, 9, 33, 450, 472, 480, 483, 486–489, 492–493; biological consequences of, 124, 184, 283, 286, 297, 495–499, 630; colony founding in, 142, 144–146, 415; sex ratio in, 241–242, 244; nestmate recognition and, 251, 256–257, 260–261; geographic distribution of, 406–411, 489–494; food robbery in, 594; and natural enemies, 636, 657

Polymorphism, 8, 13, 207–208, 274, 281–282, 296, 301–303, 442
Ponerinae, 121, 304. *See also Odontomachus brunneus*
Porter, Sanford, 227, 270–271, 296, 653, 657
Postpharyngeal gland, 238, 253–254, 260, 335, 429–432
Posture, defensive, 652
Potassium, 96; hydroxide, 399
Powell, Scott, 304–307
Precocene, 151
Predation, 121–122, 124, 374, 379, 600, 668; by fire ants on agricultural pests, 59, 599, 604; on fire ants, 183–184, 566, 543; by vertebrates, 364, 623; on vertebrates, 625
Predator(s), 120, 590, 597, 599–600, 611; geographical escape from, 42, 629; biomagnification in, 56; during mating and founding, 141, 160, 183, 456, 516; subterranean, 564–566; fire ants as, 601; for biological control, 631, 662–663
Preference(s): habitat, 42, 109, 168, 396, 508, 553–555, 607, 611, 666; food, 61, 121, 317–319, 348, 352–353, 560, 577, 593, 601, 610; settlement, 159, 457, 555; task, 299; directional, 401; mating, 508; foraging, 611; host, 639, 648, 656
Prey, 122, 313, 330, 349, 351, 379, 550, 600
ProDrone, 67, 603
Proteases, 152, 349–350, 427
Protein(s), 139, 427, 462; dietary, 121, 125, 152, 206, 255, 615; storage of, 127, 137, 211, 273; preferences for, 317–318, 593; distribution of, 330, 337–338, 346, 352; venom, 367, 370, 376; yolk, 390; genetic coding for, 468, 478, 483. *See also* Reserves, protein
Protozoans, 631–632
Proventricular valve, 335
Proventriculus, 332
Pseudacteon, 120, 642–644, 646–647, 650–656, 659; *borgmeieri,* 655; *browni,* 647, 648; *crawfordi,* 647–648, 653; *cultellatus,* 648; *curvatus,* 643, 647–648, 655–657; *litoralis,* 643, 647–650, 655, 657; *obtusus,* 655; *tricuspis,* 628, 642–645, 647–651, 655, 657–658; *wasmanni,* 643, 647–648, 655, 663
Pseudomutant technique, 295–296
Pseudotrophallaxis, 344. *See also* Trophallaxis
Puerto Rico, 16, 71, 551, 602
Pupariation, 645
Puparium, 645, 669
Pustule, 374–375
Pyomotes tritici, straw itch mite, 659
Pyranone, 430–431

Quail, 38, 47, 613, 617–624, 626–627
Quarantine, 38, 45, 64, 70, 72–73, 408, 621, 655
Quarles, John, 63
Quebracho, 29
Queen(s), 210, 229, 254, 256–257, 274, 296, 298, 462, 524, 561, 566; genetics of, 31, 420, 452, 463–464, 467, 471–472, 475, 489, 494; flight range of, 43, 51, 75; transport of, 43, 131; recognizing and capturing, 101, 145, 396–397; settlement by, 108, 140, 143, 165–166, 184–185, 513, 515–517, 523, 525, 547, 555, 562, 565, 585, 614; nutrition of, 127, 391; during claustral period, 136, 150–151, 155, 161–164, 179, 390, 434, 445–447, 527–528; egg laying by, 138, 388–389, 394, 418–420, 634; mating by, 139, 473; competition among, 168, 241–242, 435–437, 439, 495; brood raiding and, 170, 175–178, 520; adoption, 186, 226, 262, 434; sperm supply and longevity of, 223–224; size and weight of, 237, 386, 408, 418–420; bee, 239; control of sex ratio by, 239, 243–244; polygyne, 241–242, 405–406, 415–422, 434–437, 439–441, 460–463, 475, 491, 493, 495; pheromone(s), 262, 429–434, 440, 462, 488, 497; colony care and control of, 263, 273, 283–284, 392–394, 481; in *Odontomachus brunneus,* 305–306; venom, 365, 371, 373, 387; embryonation of eggs by, 421–422; mutual inhibition by, 423–427, 431, 434, 442; alpha, 436–437; match-mated, 445–446; diseases and parasites of, 634–635

Rabbits, 613
Raccoons, 615
Radio-collar, for tracking bobwhite quail, 619
Radioactive label (radiolabel), 254, 332, 334–335, 340
Radioactivity, 338, 426
Radioisotopes, 331, 340, 370
Rain, 128, 319, 519, 617; triggering of mating flights by, 140–141, 165, 187, 189, 515–516
Rainfall, 85, 562, 564, 630
Ramus, 366
Range expansion, 1918–1995, 69–73
Range limits, 83–91
RAPDs, 478
Rarefaction, 579
Rash, 375
Rattlesnake, 384–385
Recruitment, to food, organization and mechanisms of, 127, 215–216, 308–313, 316–317, 319, 321–322, 324–327, 330, 346, 593; to food, by *Solenopsis invicta* in South America, 570; to food, age at first, 287; to food, measurement of, 128, 317, 577, 587, 625; to food, by *Solenopsis richteri,* 319; to food, by *Solenopsis geminata,* 652; of queens, 466, 476; of deer populations, 623–624
Rectal pad, 351
Rectum, 333, 351
Red hills country, 1
Refugio County, Texas, 619
Regal moth, hickory horned devil, *Citheronia regalis,* 667
Repellency, 378
Repellent, 593
Reproductive isolation, 17
Reptiles, 613. *See also* individual taxa
Rescue behaviors, 430
Research methods, low-tech, 130–135
Reserve workers, 349, 355. *See also* Reserve(s), labor

Reserve(s), fat, 127, 152, 187, 206, 223, 231–233, 239; metabolic, of founding queens, 150, 152–153, 187, 225, 447, 450, 480; metabolic, of the colony, 199, 202, 206, 211, 458; metabolic, annual cycle of, 204, 230, 232–233; storage, 203; labor, 215–217, 285, 287–288; protein, 223, 231–233. *See also* Fat; Protein
Respiration, 126, 144, 201, 212, 299
Retinue, 386, 389, 393, 417–418, 431, 440
Retrieval: of brood, 264–267; of food, 319, 570. *See also* Recruitment
Rhinocyllus conicus, 601
Ribosomes, 370
Rice, 599
Rickettsia, 639
Rio de Janeiro, 16, 18–19
Rio de la Plata, 509
Roadside(s), 122, 409–410, 489, 559, 577; *Solenopsis invicta*'s preference for, 109, 396, 406–407, 551, 572, 605; disturbed nature of, 578, 630
Rocky Mountain spotted fever, 59
Rodents, 622, 624. *See also* individual taxa
Rolândia, Paraná, Brazil, 16
Ross, Jim Buck, 64
Ross, Ken, 27–28
Rotap machine, 134
Ruckelshaus, William, 54
Running, 218

saevissima complex, 14, 18, 488, 640, 647, 649, 656, 660
St. Croix, U.S. Virgin Islands, 71
St. Elmo, Alabama, 38
St. Petersburg, Florida, 51
Saladillo, Argentina, 635–636
Salivary gland(s), 152, 350–351
Salt marshes, 553
San Eladio, Argentina, 661
Sandhills, Florida, 559, 562–563
Sanitation behavior, 402. *See also* Necrophory
Santa Barbara, California, 71, 76
Santa Catarina, Brazil, 15
Santa Fe Province, Brazil, 15
São Paulo, Brazil, 568, 657
Sarcophagidae, flesh flies, 602, 605
Savanna, 26, 563, 565, 570, 580
Savannah, Georgia, 51
Savannah River, 44
Savannah River Test Site, 552–553, 609
Scales, 669
Scavenger(s), 597, 605, 663, 665
Scavenging, 122, 124
Schmidt, Justin O., 377, 384
Scout(s), scouting, recruitment of nestmates by, 8, 107, 323; territorial marking by, 116; 128, 175, 272, 309, 626; discovery of food by, 310–311, 317, 324–326
Season, 206, 213, 224, 237, 498; effects of on diet and foraging, 121, 123–124, 127–129, 221; effects of on colony size, 198, 223, 231; effects of on energy allocation, 203, 206, 212
Secretions, defensive, 367
Seeds, 122–124, 282–283, 292–293
Select Study Committee on Mirex, 53
Selection, 447, 487, 506–507, 640; against *Gp-9* genotypes, 449, 467–469, 472, 474–475. *See also* Natural selection
Self-thinning, 514, 516, 518
Selma, Alabama, 39–40
Seminal vesicle(s), 138–139
Sensillae, 367
Sesbania, 611
Sesquiterpenoids, 387
Sex allocation. *See* Allocation, sex
Sex determination, 31, 239
Sex ratio, 239–245, 650
Sexualization, 419, 422–424, 428, 432–434, 440–441
Shaus swallowtail, 602
Shrike, loggerhead, 613, 617
Shuttle workers, 393–395
Sidewalk ant, *Tetramorium caespitum,* 665–666
Sierra club, 65
Silent Spring (1962), 46, 50
Silicone cement, 367
Silkworm moth, 484
Silphidae, burying beetles, 605
Silverfish, 663
Simulation, computer, 489, 523, 581, 627. *See also* Model(s), modeling
Site selection, 142–144
Snakes, 614, 663
Social form, 451, 470–471, 478–479, 492, 497–498; sexuals and, 240, 415, 448–449; stability of, 408, 411; genetics of, 449–450, 465–469, 474, 477, 479–483, 486, 488; in hybrids, 504; in *Solenopsis geminata,* 489; and natural enemies, 634, 640–642, 662. *See also* Monogyny; Polygyny
Sociality, 428
Sociogenesis, 193, 205–206, 222–223
Sod, 44–45, 78–79, 513
Sodium iodide, 331
Sodium phosphate, 331
Soil, 96, 129, 380–382, 387, 408; transport of fire ants in, 29, 31, 44–45
Solenopsins, 369–370
Solenopsis, 5, 13–18, 22–30, 33–74, 647, 654; phylogeny of, 14–16, 487; *altipunctata,* 14; *amblychila,* 15, 21, 281, 487; *aurea,* 15, 20, 87, 90, 281, 371–373, 487, 653; *blumi,* 281; *bruesi,* 15, 281; *daguerrei,* 14, 16, 18, 660–61; *electra,* 14–15, 19; *gayi,* 15, 281; *geminata,* 38, 90, 320, 430, 577–578, 611, 663; *geminata,* origin and range of, 16, 20, 40, 83, 87, 559–564, 572–573, 576–577, 581, 583–584, 586–590; *geminata,* foraging by, 116, 309, 313, 378, 541–542; *geminata,* diet of, 122–124; *geminata,* colony founding in, 190–192; *geminata,*

polymorphism and division of labor in, 281–283; *geminata,* venom of, 370–373, 377; *geminata,* polygyny in, 405, 408, 422, 437, 488–489; *geminata,* hybridization of, 504, 511; *geminata,* niche of, 556, 558, 561–563, 565–566, 580; *geminata,* biological control by, 599, 601, 609–611; *geminata,* effects of on vertebrates, 618; *geminata,* natural enemies of, 632, 637, 640, 646, 648, 650, 652–655, 658–659; *globularia littoralis,* 487; *hostilis,* 16; *interrupta,* 14–15, 18, 281, 487; *macdonaghi,* 14–15, 19, 487, 638, 660; *megergates,* 14–15, 19; *pergandei,* 632; *pusillignis,* 14–15, 18; *pythia,* 14–15; *quinquecuspis,* 14–15, 17–18, 281, 405, 487, 638, 660; *richteri,* 14–16, 24–25, 51, 258–261, 281, 320, 487; *richteri,* range and spread of, 17–18, 37–42, 44–45, 69, 71, 503, 590; *richteri,* origin of, 18, 24–25, 27–30, 32; *richteri,* natural enemies of, 30, 633–636, 638, 640, 651, 656, 660–662; *richteri,* niche of, 88; *richteri,* venom of, 371–373, 377; *richteri,* polygyny in, 405, 456, 558, 580; *richteri,* hybridization of, 503–511, 640; *saevissima,* 14–15, 19, 23–24, 281, 487, 638, 648, 655, 660; *substituta,* 14–15, 281; *tridens,* 14–15, 281; *virulens,* 15, 281; *wagneri,* 24; *weyrauchi,* 14–15, 18; *xyloni,* 38, 127, 281, 320, 611; *xyloni,* natural enemies of, 15, 638, 654–656, 658; *xyloni,* origin and range of, 20–21, 40, 83, 572, 590, 559; *xyloni,* niche of, 87, 90, 558; *xyloni,* venom of, 371–373, 377; *xyloni,* polygyny in, 405, 408; *xyloni,* hybridization of, 511
Solomon Islands, 580
Sound, 263, 266, 378, 424
South America, 35, 371, 510, 554–555, 568–569, 571; fire ants originating in, 5, 15–16, 20, 25, 447, 487, 571; natural enemies of fire ants in, 106, 120, 564, 631, 633, 640, 642, 646, 650, 653, 655, 658; *Solenopsis invicta* in, 184–185, 550, 570, 630; polygyny in, 405, 440, 492–493
South Carolina, 42–44, 139, 234, 551, 575–576, 609; mirex in, 51; sex ratio in, 240; polygyny in, 495; effects on vertebrates in, 621
Southeastern Association of Game and Fish Commissioners, 48
Southern Association of Commissioners of Agriculture, 45
Southern Plant Board, 45, 51
Southwood Plantation, 5, 140, 533, 539, 585, 591; territory size at, 112–113; population characteristics at, 224, 513, 523, 526–528, 536, 538; sex ratios at, 240, 242–243; parasitoids at, 657
Sows, 48
Soybean(s), 58–60, 123, 576, 600–601
Spanish Fort, Alabama, 35
Sparkman, Senator, 48
Speciation, 479
Species richness, 574–575, 579–582, 588–590
Spectracide, 67
Sperm, 9, 31, 138, 141, 499, 639; production and storage of, 10, 138–139, 166, 196, 330; depletion of, 11, 223, 226, 244, 538; number of in male, 138; number of in female, 223–224; efficiency of use of, 224–225, 239, 394
Spermatheca(e), 31, 138–139, 141, 223, 273–274, 405, 644
Spermathecal duct, 225
Spiders, 120, 183, 608, 631
Spiracular plate, 366
Spiracular valves, 89
Spring Hill, 41–42
Squash, 601
Stains, 269. *See also* Dye; Ink; Paint
Staphylinidae, rove beetles, 30, 57, 604–605
Staphylococcus aureus, 381
Starkville, Mississippi, 51, 61, 503
Starvation, 88, 123, 183, 200, 313, 335; effects of on trophallaxis, 337–339, 343, 353–354
State Plant Board of Mississippi, 41
Steinernema spp., 659
Stereomicroscope, 227
Sterility, 442
Sting, 364–366, 377, 386, 620–621; structure and function of, 138, 314, 366–367; use of to lay odor trails, 311, 313–315, 365; symptoms of, 367, 376, 384–385
Stinging, 168, 367
Stinkbug, southern green, 59
Stomoxys calcitrans, stable flies, 54, 59, 122, 604
Succession, 6, 512, 555–558, 567, 622
Sucrose, 314–315, 338, 347. *See also* Sugar
Sugar, 121, 123, 125–126, 255, 311; preference for, 317–319, 324–326, 593; allocation of, 330, 344–335, 337–348; trophallaxis of, 335, 337–339, 346, 351–353. *See also* Sucrose
Sugarcane, 57–59, 122, 570, 572, 576, 597–598, 600, 602
Sugarcane borer, 48, 59, 122, 597–598
Sun compass, 311–312, 400
Sunfish, bluegill, 616–617
Supercooling, 88
Superorganism, 96, 98, 106, 251, 295, 435, 441, 512; sociogenesis of, 193, 205; structure of, 210–211, 292; death of, 223; *Odontomachus brunneus* as, 306; and polygyny, 434; social parasitism of, 660
Swallows, 613
Swallowtail, Shaus, 602
Swarms, male, 557
Sweat bee, 384
Sweating, 375
Swelling, 374–375
Swine. *See* Sows
Symbionts, 631
Synergy, pheromonal, 431

Tabasco, Mexico, 601
Tachinidae (parasitic flies), 667, 669
Taiwan, 72
Talcum powder, 131, 132
Tallahassee, Florida, 1, 5, 11, 514–515, 564; sperm depletion in, 224; temperature limitation in, 232; sex allocation in, 240; parasitoids in, 657
Tampa, Florida, 51, 61, 137, 234

Tapinoma sessile, 563, 565, 580
Tarsus, 278
Teflon, 7, 132. *See also* Fluon
Temperature, 213, 232–233, 299, 550, 630; effects of on species ranges, 83, 85–89; effects of on desiccation, 90, 91; within the nest, 100, 120, 200–202, 212; effects of on foraging, 120, 127–129; role of in triggering mating flights, 140, 187; effects during the claustral period, 151; effects of on brood raiding, 172; effects of on the queen, 181, 238; effect of on brood development, 194–195, 230, 234–235, 239; effects of on colony growth, 195–196, 229; effects of on allocation rates, 203, 206; effects of on natural enemies, 638
Template, odor, 253, 255–256, 260. *See also* Odor, colony
Tennessee, 42–43, 84, 87–88, 148, 657
Termites, 122, 141, 379, 419, 431
Terns, 613–614
Territoriality, 175, 406–407, 495, 577
Territory, territories, 193, 220, 308–309, 544, 581; defense of, 8, 10, 106, 252, 257, 272, 301; boundaries and spacing of 107, 109, 524, 532, 540, 545–547; size of, 108–120, 182, 220–222, 533, 557; in vertebrates, 614
Testis, testes, 138–139, 442
Tetradonema solenopsis, 632
Tetrafluoroethylene, 7, 132. *See also* Fluon
Tetrahydrofuran, 399
Tetramorium caespitum, sidewalk ant, 665–666
Texas, 42–43, 61, 89, 240, 445, 466, 497; polygyny in, 11, 406–410, 459, 497, 585–589; spread of *Solenopsis invicta* in, 44, 71, 83, 85, 87, 551; pesticides in, 46–48, 65, 73, 582–583; effects on vertebrates in, 47, 614–619, 621–624; diet and foraging in, 122, 129, 593, 598; mating flights in, 148; natural enemies in, 184, 634, 636, 653, 657–658, 663; density and distribution in, 410, 554; hybridization in, 511; ant diversity in, 559, 570, 572–574, 592
Texas, University of, 406
Texas A&M University, 406, 442, 605
Texas Department of Agriculture, 406
Texas Tech University, 90
Thallium salts, 38
Tharpe Street, 171, 241, 513–515, 524, 526
Thelohania, 494; *solenopsae,* 632–639, 589
Thief ants, *Diplorhoptrum,* 13, 120, 184, 564–566, 585
Thorax, 22, 138, 643–644. *See also* Mesosoma
Threatened species, 615
Threshold, odor detection, 319
Threshold model, 291
Tibia, 278
Tick fever, 59
Tick, lone star, 59, 122, 607–608
Toad: Houston, 613; Texas horned, 615–616
Tobacco budworm, 59, 599
Tortoise, Gopher, 613
Toxins, 364. *See also* Venom
Toxoptera citricida, 601. *See also* Aphids
Tracer, radioactive, 339. *See also* Radioactivity
Tracheal system, 89
Trachymyrmex septentrionalis, 542
Trails, 113, 115, 264, 324, 356, 664. *See also* Pheromone(s), trail; Tunnels, foraging
Tramp species, 16, 20
Transport, 264, 272, 293, 308, 398, 513
Trap nest, trap nesting, 121–132, 197
Trap-jaw ant. *See Odontomachus brunneus*
Trap(s), 128, 572, 574; for alates, 132; aerial, 143; carbon-dioxide, 608; pitfall, 565, 569, 574–576, 579, 581, 587, 592, 608
Trash piles, 398–399, 401–402. *See also* Dead piles
Tree snail, Stock Island, 602
Trees, 555
Tricogramma sp., 601
Triglycerides, 265
Trinidad, 71
Triolein, 265
Triploidy, 499
Trophallaxis, 336–337, 349, 664; methods for study of, 331, 340–341, 350, 352; feeding of the queen by, 386, 393; feeding of larvae by, 355–356, 393; spread of pheromones by, 426. *See also* Pseudotrophallaxis
Trophic eggs, 151–152, 161, 306
Tularemia, 59
Tunnels, 8–9, 116–119, 121, 312; interaction of with territory size and shape, 113, 115–119, 222; mapping of, 116–117; construction of, 117–119, 174, 544; advantages of, 119–120, 128, 287, 308–311, 555, 648, 653; use of by thief ants, 120, 585; in polygyny, 460, 495
Tupelo, Mississippi, 148
Turks and Caicos, 71
Turnover: soil, 96; colony, 184, 212, 224, 440, 512, 537–538, 575, 585; worker, 196, 206, 213–214, 223, 225, 235; queen, 466
Turtles, 613–615
2-methyl-6–alkylpiperidines, 369

U.S. Budget Bureau, 50
U.S. Department of Agriculture (USDA), 37, 67–68, 335, 405, 621; research by, 41–42, 133–134, 265, 430; pesticides and, 45–46, 48–55, 61–63, 66, 74, 595; biological control and, 62, 655, 662
U.S. Department of Fish and Wildlife, 46
U.S. Department of Health, Education and Welfare, 52
U.S. Department of the Interior, 53
U.S. Environmental Protection Agency (EPA), 53–55, 57–59, 63–66, 68
U.S. Food and Drug Administration, 48
U.S. National Museum, Washington, D.C., 26
U.S. Office of Management and Budget, 64
U.S. Virgin Islands, 71
Ultraviolet light, 270
Urethane, 269
Uric acid, 393

Urine, 393
Uruguay, 15, 17–19, 25, 662
Usurpation, 169, 176–178, 183–184. *See also* Adoption
Uterine pouch, 138

Vairimorpha invictae, 633, 637. *See also* Microsporidians
Valvulae, 364, 366
Vasa deferentia, 442
Velvetbean caterpillar, 59
Venezuela, 580
Venom, 367, 386, 612, 617, 651; composition of, 367–373; synthesis and storage of, 314, 365, 370, 382–383; of *Pogonomyrmex badius,* 384; use of, 17, 364, 367, 374–382; of hybrids, 503–504
Venom gland, 313
Venom sac, 367
Vertebrates, 122, 364, 374, 612–613, 615, 617, 619, 626–627, 631
Video recording, 339–340, 346
Videotape, 250, 282, 324, 340, 354
Virginia, 71, 85
Viruses, 631, 671, 632
Viscosity, population, 464–465
Vision, 263
Visual spectrum, 312
Vitellogenin, 150–151, 187
Vomiting, 375
Vulva, 138, 386

Waggle-walking, 324–325
Walsh, John, 265
Washington (state), 85
Wasp(s), 9, 54, 364, 367, 384, 669; myrmecophile, 253; parasitic, 253, 601, 631, 659, 668; bethylid, 663; chalcid, 659, 668; eucharitid, 253, 659, 663; paper, 206, 384; *Pepsis,* 385; proctupid, 663
Wax(es), 257, 266, 269
Weather, 127–129, 200
Wedge micrometer, 227–228
Weeds, 123, 173, 199, 514, 580, 600, 609; adaptations of, 552, 556
Weevil(s): boll, 36, 42, 59, 122, 598–599; pecan, 59, 122, 599
Wesleyan University, 454–455
West Indies, 71, 16. *See also* individual localities
Westpoint, Mississippi, 234
Wetlands, 26, 551, 555, 562–563
Whistler, Alabama, 38
Whiteflies, 122
Whitten, Jamie, 50–51
Wichita, Kansas, 84
Wilson, E. O., 23–25, 29, 34–36, 41–42, 63, 68
Wind, 141, 473–475, 489–491, 646
Wire, 186, 270–271, 283, 299, 304
Wire grass, 579
Wolbachia sp., 510, 639–642
Wood ant, *Formica rufa,* 96, 121
Wood, Lois, 134
Workers, nanitic. *See* Minim workers
Workers, dwarf. *See* Minim workers
Workers, major. *See* Major workers
Workers, minor. *See* Minor workers
Worms, nematode, 630, 632, 659–660
wsp, 640

Yeast(s), 381, 632
Yellow jacket, 384
Yellow-head disease, 632
Yolk, 137, 150, 187, 273

Z,E-homofarnesene, 321
Z,E-a-farnesene, 321–323
Z,Z,Z-allofarnesene, isolation of from trail pheromones, 321